UNITED STATES
DEPARTMENT OF AGRICULTURE

NATIONAL AGRICULTURAL STATISTICS SERVICE

AGRICULTURAL

STATISTICS

2010

UNITED STATES
GOVERNMENT PRINTING OFFICE
WASHINGTON: 2010

For sale by the Superintendent of Documents, U.S. Government Printing Office
Internet: bookstore.gpo.gov   Phone: toll free (866) 512-1800;   DC area (202)512-1800
Fax: (202) 512-2104 Mail: Stop IDCC, Washington, DC 20402-0001
ISBN 978-0-16-088287-6

Agricultural Statistics 2010

Agricultural Statistics, 2010 was prepared under the direction of Rich Holcomb, Agricultural Statistics Board, National Agricultural Statistics Service.

The USDA and NASS invite you to explore their information on the Internet. The USDA Home Page address is **http://www.usda.gov/** and the NASS Home Page address is: **http://www.usda.gov/nass/.**

For information on NASS products you may call the **Agricultural Statistics Hotline, 1–800–727–9540 or send e-mail to nass@nass.usda.gov.**

The cooperation of the many contributors to this publication is gratefully acknowledged. Source notes below each table credit the various Government agencies which collaborated in furnishing information.

CONTENTS

	Page
Introduction	iii
Weights, measures, and conversion factors	iv
I—Grain and feed:	
Total grain supply	I-1
Food grains:	
Wheat	I-1
Rye	I-13
Rice	I-16
Feed grains:	
Corn	I-24
Oats	I-32
Barley	I-36
Sorghum	I-41
Grain consumption	I-47
Animal units fed	I-48
Feedstuffs	I-49
II—Cotton, tobacco, sugar crops, and honey:	
Cotton	II-1
Sugarbeets	II-13
Sugar	II-16
Honey	II-25
Beeswax	II-24
Tobacco	II-26
III—Oilseeds, fats, and oils:	
Cottonseed	III-1
Flaxseed	III-5
Peanuts	III-8
Soybeans	III-13
Sunflower	III-20
Peppermint and spearmint	III-23
Olive oil	III-24
Margarine	III-24
Shortening	III-25
Fats and oils	III-26
IV—Vegetables and melons:	
Vegetables and melons	IV-1
Vegetable and shipments	IV-28
Vegetable utilization	IV-29
Commercial pack	IV-30
V—Fruits, tree nuts, and horticultural specialties:	
Fruits	V-1
Tree nuts	V-40
Cocoa beans, coffee, and tea	V-45
Mushrooms	V-52
Flowers	V-54
VI—Hay, seeds, and minor field crops:	
Hay	VI-1
Seeds	VI-9
Beans, dry edible	VI-10
Peas, dry	VI-13
Hops	VI-14

	Page
VII—Cattle, hogs, and sheep:	
Cattle and calves	VII-1
Hogs	VII-18
Sheep and lambs	VII-27
Wool	VII-35
Goats and mohair	VII-40
Meats	VII-41
Hides	VII-47
Livestock numbers	VII-50
VIII—Dairy and poultry statistics:	
Cows, milk	VIII-1
Chickens	VIII-26
Dairy products	VIII-23
Turkeys	VIII-34
Eggs	VIII-37
IX—Farm resources, income, and expenses:	
Economic trends	IX-1
Farm property	IX-2
Farm labor	IX-13
Farm production and distribution	IX-16
Prices and income	IX-26
Costs and expenses	IX-39
X—Taxes, insurance, credit, and cooperatives:	
Taxes and insurance	X-1
Credit and loan programs	X-9
Farmers' cooperatives	X-11
Rural utilities	X-14
XI—Stabilization and price-support programs:	
Price support	XI-1
Payments to producers	XI-8
Marketing agreements and orders	XI-14
XII—Agricultural conservation and forestry statistics:	
Conservation Reserve Programs	XII-1
Soil conservation programs	XII-15
Forestry	XII-17
XIII—Consumption and family living:	
Population	XIII-1
Food consumption and nutrition	XIII-1
Prices at retail levels	XIII-10
XIV—Fertilizers and pesticides:	
Field crops	XIV-1
Fruits	XIV-19
Vegetables	XIV-20
XV—Miscellaneous agricultural statistics:	
Agricultural imports and exports	XV-2
Fishery statistics	XV-14
Refrigeration statistics	XV-31
Alaska statistics	XV-33
Crop rankings	XV-34
Crop progress	XV-35
Appendix I:	
Telephone contact list	Appendix-1
Index	Index-1

Introduction

Agricultural Statistics is published each year to meet the diverse need for a reliable reference book on agricultural production, supplies, consumption, facilities, costs, and returns. Its tables of annual data cover a wide variety of facts in forms suited to most common use.

Inquiries concerning more current or more detailed data, past and prospective revisions, or the statistical methodology used should be addressed directly to the agency credited with preparing the table. Most of the data were prepared or compiled in the U.S. Department of Agriculture.

The historical series in this volume have been generally limited to data beginning with 2000 or later.

Foreign agricultural trade statistics include Government as well as non-Government shipments of merchandise from the United States and Territories to foreign countries. They do not include U.S. shipments to the U.S. Armed Forces abroad for their own use or shipments between the States and U.S. Territories. The world summaries of production and trade of major farm products are prepared by the U.S. Department of Agriculture from reports of the U.S. Department of Commerce, official statistics of foreign governments, other foreign source materials, reports of U.S. Agricultural Attaches and Foreign Service Officers, and the result of office research.

Statistics presented in many of the tables represent actual counts of the items covered. Most of the statistics relating to foreign trade and to Government programs, such as numbers and amounts of loans made to farmers, and amounts of loans made by the Commodity Credit Corporation, etc., are data of this type. A large number of other tables, however, contain data that are estimates made by the Department of Agriculture.

The estimates for crops, livestock, and poultry made by the U.S. Department of Agriculture are prepared mainly to give timely current State and national totals and averages. They are based on data obtained by sample surveys of farmers and of people who do business with farmers. The survey data are supplemented by information from the Censuses of Agriculture taken every five years and check data from various sources. Being estimates, they are subject to revision as more data become available from commerical or Government sources. Unless otherwise indicated, the totals for the United States shown in the various tables on area, production, numbers, price, value, supplies, and disposition are based on official Department estimates. They exclude States for which no official estimates are compiled.

DEFINITIONS

"Value of production" as applied to crops in the various tables, is derived by multiplying production by the estimated season average price received by farmers for that portion of the commodity actually sold. In the case of fruits and vegetables, quantities not harvested because of low prices or other economic factors are not included in value of production. The word "Value" is used in the inventory tables on livestock and poultry to mean value of the number of head on the inventory date. It is derived by multiplying the number of head by an estimated value per head as of the date.

The word "Year" (alone) in a column heading means calendar year unless otherwise indicated. "Ton" when used in this book without qualifications means a short ton of 2,000 pounds.

WEIGHTS, MEASURES, AND CONVERSION FACTORS

The following table on weights, measures, and conversion factors covers the most important agricultural products, or the products for which such information is most frequently asked of the U.S. Department of Agriculture. It does not cover all farm products nor all containers for any one product.

The information has been assembled from State schedules of legal weights, various sources within the U.S. Department of Agriculture, and other Government agencies. For most products, particularly fruits and vegetables, there is a considerable variation in weight per unit of volume due to differences in variety or size of commodity, condition and tightness of pack, degree to which the container is heaped, etc. Effort has been made to select the most representative and fairest average for each product. For those commodities which develop considerable shrinkage, the point of origin weight or weight at harvest has been used.

The approximate or average weights as given in this table do not necessarily have official standing as a basis for packing or as grounds for settling disputes. Not all of them are recognized as legal weight. The table was prepared chiefly for use of workers in the U.S. Department of Agriculture who have need of conversion factors in statistical computations.

WEIGHTS, MEASURES, AND CONVERSION FACTORS
(See explanatory text just preceding this table)

WEIGHTS AND MEASURES

Commodity	Unit[1]	Approximate net weight U.S. Pounds	Metric Kilograms
Alfalfa seed	Bushel	60	27.2
Apples	....do	48	21.8
Do	Loose pack	38–42	17.2–19.1
Do	Tray pack	40–45	18.1–20.4
Do	Cell pack	37–41	16.8–18.6
Apricots	Lug (brent-wood)[2]	24	10.9
Western	4–basket crate[3]	26	11.8
Artichokes:			
Globe	Ctn, by count and loose pack	20–25	9.1–11.3
Jerusalem	Bushel	50	22.7
Asparagus	Crate (NJ)	30	13.6
Avocados	Lug[4]	12–15	5.4–6.8
Bananas	Fiber folding box[5]	40	18.1
Barley	Bushel	48	21.8
Beans:			
Lima, dry	....do	56	25.4
Other, dry	....do	60	27.2
	Sack	100	45.4
Lima unshelled	Bushel	28–32	12.7–14.5
Snap	....do	28–32	12.7–14.5
Beets:			
Topped	Sack	25	11.3
Bunched	½ crate 2 dz-bchs	36–40	16.3–18.1
Berries frozen pack:			
Without sugar	50–gal. barrel	380	172
3 + 1 pack	....do	425	193
2 + 1 pack	....do	450	204
Blackberries	12, ½-pint basket	6	2.7
Bluegrass seed	Bushel	14–30	6.4–13.6
Broccoli	Wirebound crate	20–25	9.1–11.3
Broomcorn (6 bales per ton)	Bale	333	151
Broomcorn seed	Bushel	44–50	20.0–22.7
Brussels sprouts	Ctn, loose pack	25	11.3
Buckwheat	Bushel	48	21.8
Butter	Block	55,68	25,30.9
Cabbage	Open mesh bag	50	22.7
Do	Flat crate (1¾ bu)	50–60	22.7–27.2
Do	Ctn, place pack	53	24.0
Cantaloups	Crate[6]	40	18.1
Carrots	Film plastic Bags, mesh sacks & cartons holding 48 1 lb. film bags	55	24.9
Without tops	Burlap sack	74–80	33.6–36.3
Castor beans	Bushel	41	18.6
Castor oil	Gallon	[7]8	3.6
Cauliflower	W.G.A. crate	50–60	22.7–27.2
Do	Fiberboard box wrapper leaves removed film-wrapped, 2 layers	23–35	10.4–15.9

Commodity	Unit[1]	Approximate net weight U.S. Pounds	Metric Kilograms
Celery	Crate[8]	60	27.2
Cherries	Lug (Campbell)[9]	16	7.3
Do	Lug	20	9.1
Clover seed	Bushel	60	27.2
Coffee	Bag	132.3	60
Corn:			
Ear, husked	Bushel	[10]70	31.8
Shelled	....do	56	25.4
Meal	....do	50	22.7
Oil	Gallon	[7]7.7	3.5
Syrup	....do	11.72	5.3
Sweet	Wirebound crate	50	22.7
Do	Ctn, packed 5 oz. ears	50	22.7
Do	WDB crate, 4½–5 oz. (from FL & NJ)	42	19.1
Cotton	Bale, gross	[11]500	227
Do	Bale, net	[11]480	218
Cottonseed	Bushel	[12]32	14.5
Cottonseed oil	Gallon	[7]7.7	3.5
Cowpeas	Bushel	60	27.2
Cranberries	Barrel	100	45.4
Do	¼–bbl. box[13]	25	11.3
Cream, 40-percent butterfat	Gallon	8.38	3.80
Cucumbers	Bushel	48	21.8
Dewberries	24–qt. crate	36	16.3
Eggplant	Bushel	33	15.0
Eggs, average size	Case, 30 dozen	47.0	21.3
Escarole	Bushel	25	11.3
Figs, fresh	Box single layer[14]	6	2.7
Flaxseed	Bushel	56	25.4
Flour, various	Bag	100	45.4
Do	Ctn or Crate, Bulk	30	13.6
Garlic	Ctn of 12 tubes or 12 film bag pkgs 12 cloves each	10	4.5
Grapefruit:			
Florida and Texas	½–box mesh bag	40	18.1
Florida	1⅗ bu. box	85	38.6
Texas	1⅖ bu. box	80	36.3
California and Arizona	Box[15]	[16]67	30.4
Grapes:			
Eastern	12–qt. basket	20	9.1
Western	Lug	28	12.7
Do	4–basket crate[17]	20	9.1
Hempseed	Bushel	44	20.0
Hickory nuts	....do	50	22.7
Honey	Gallon	11.84	5.4
Honeydew melons	⅔ Ctn	28–32	12.7–14.5
Hops	Bale, gross	200	90.7

See footnotes on page ix.

WEIGHTS AND MEASURES—Continued

Commodity	Unit [1]	Approximate net weight		Commodity	Unit [1]	Approximate net weight	
		U.S.	Metric			U.S.	Metric
		Pounds	*Kilograms*			*Pounds*	*Kilograms*
Horseradish				Do	Std box, 4/5 bu	45–48	20.4–21.8
roots	Bushel	35	15.9	Do	Ctn, Tight-fill		
Do	Sack	50	22.7	pack		36–37	16.3–16.7
Hungarian millet				Peas:			
seed	Bushel	48–50	21.8–22.7	Green,			
Kale	Ctn or crate	25	11.3	unshelled	Bushel	28–30	12.7–13.6
Kapok seed	......do	35–40	15.9–18.1	Dry	......do	60	27.2
Lard	Tierce	375	170	Peppers, green	......do	25–30	11.3–13.6
Lemons:				Do	1½ bu carton	28	12.7
California and				Perilla seed	Bushel	37–40	16.8–18.1
Arizona	Box [18]	76	34.5	Pineapples	Carton	40	18.1
Do	Carton	38	17.2	Plums and			
Lentils	Bushel	60	27.2	prunes:	Ctn & lugs	28	12.7
Lettuce, iceberg	Iceberg, carton			Do	½-bu. basket	30	13.6
	packed 24	43–52	19.5–23.6	Popcorn:			
Lettuce, hot-				On ear	Bushel	[10]70	31.8
house	24-qt. basket	10	4.5	Shelled	......do	56	25.4
Limes (Florida)	Box	88	39.9	Poppy seed	......do	46	20.9
Linseed oil	Gallon	[7]7.7	3.5	Potatoes	Bushel	60	27.2
Malt	Bushel	34	15.4	Do	Barrel	165	74.8
Maple syrup	Gallon	11.02	5.0	Do	Box	50	22.7
Meadow fescue				Do	......do	100	45.4
seed	Bushel	24	10.9	Quinces	Bushel	48	21.8
Milk	Gallon	8.6	3.9	Rapeseed	......do	50–60	22.7–27.2
Millet	Bushel	48–60	21.8–27.2	Raspberries	½-pint baskets	6	2.7
Molasses:				Redtop seed	Bushel	50–60	22.7–27.2
edible	Gallon	11.74	5.3	Refiners' syrup	Gallon	11.45	5.2
inedible	......do	11.74	5.3	Rice:			
Mustard seed	Bushel	58–60	26.3–27.2	Rough	Bushel	45	20.4
Oats	......do	32	14.5	Do	Bag	100	45.4
Olives	Lug	25–30	11.3–13.6	Do	Barrel	162	73.5
Olive oil	Gallon	[7]7.6	3.4	Milled	Pocket or bag	100	45.4
Onions, dry	Sack	50	22.7	Rosin	Drum, net	520	236
Onions, green				Rutabagas	Bushel	56	25.4
bunched	Ctn, 24-dz bchs	10–16	4.5–7.3	Rye	......do	56	25.4
Oranges:				Sesame seed	......do	46	20.9
Florida	Box	90	40.8	Shallots	Crate (4–7 doz.		
Texas	Box	85	38.5		bunches)	20–35	9.1–15.9
California and				Sorgo:			
Arizona	Box [15]	75	34.0	Seed	Bushel	50	22.7
Do	Carton	38	17.2	Syrup	Gallon	11.55	5.2
Orchardgrass				Sorghum			
seed	Bushel	14	6.4	grain [19]	Bushel	56	25.4
Palm oil	Gallon	[7]7.7	3.5	Soybeans	......do	60	27.2
Parsnips	Bushel	50	22.7	Soybean oil	Gallon	[7]7.7	3.5
Peaches	......do	48	21.8	Spelt	Bushel	40	18.1
Do	2 layer ctn or			Spinach	......do	18–20	8.2–9.1
	lug	22	10.0	Strawberries	24-qt. crate	36	16.3
Do	¾-Bu, Ctn/crate	38	17.2	Do	12-pt. crate	9–11	4.1–5.0
Peanut oil	Gallon	[7]7.7	3.5	Sudangrass			
Peanuts,				seed	Bushel	40	18.1
unshelled:				Sugarcane:			
Virginia type	Bushel	17	7.7	Syrup			
Runners,				(sulfured or			
South-east-				un-sulfured)	Gallon	11.45	5.2
ern	......do	21	9.5	Sunflower seed	Bushel	24–32	10.9–14.5
Spanish:				Sweetpotatoes	......do	[20]55	24.9
South-				Do	Crate	50	22.7
eastern	......do	25	11.3	Tangerines:			
South-				Florida	Box	95	43.1
western	......do	25	11.3	Arizona	Box	75	34.0
Pears:				California	Box	75	34.0
California	Bushel	48	21.8				
Other	......do	50	22.7				

See footnotes on page ix.

WEIGHTS AND MEASURES—Continued

Commodity	Unit [1]	Approximate net weight		Commodity	Unit [1]	Approximate net weight	
		U.S.	Metric			U.S.	Metric
		Pounds	*Kilograms*			*Pounds*	*Kilograms*
Timothy seed	Bushel	45	20.4	Turnips:			
Tobacco:				Without tops ..	Mesh sack	50	22.7
Maryland	Hogshead	775	352	Bunched	Crate [6]	70–80	31.8–36.3
Flue-cured	.....do	950	431	Turpentine	Gallon	7.23	3.3
Burley	....do	975	442	Velvetbeans			
Dark air-cured	....do	1,150	522	(hulled)	Bushel	60	27.2
Virginia fire-				Vetch seed	.....do	60	27.2
cured	....do	1,350	612	Walnuts	Sacks	50	22.7
Kentucky and				Water 60° F	Gallon	8.33	3.8
Tennessee				Watermelons	Melons of aver-		
fire-cured	....do	1,500	680		age or me-		
Cigar-leaf	Case	250–365	113–166		dium size	25	11.3
Do	Bale	150–175	68.0–79.4	Wheat	Bushel	60	27.2
Tomatoes	Crate	60	27.2	Various com-			
Do	Lug box	32	14.5	modities	Short ton	2,000	907
Do	2-layer flat	21	9.5	Do	Long ton	2,240	1,016
Tomatoes, hot-				Do	Metric ton	2,204.6	1,000
house	12-qt. basket	20	9.1				
Tung oil	Gallon	[7]7.8	3.5				

See footnotes on page ix.

To Convert From Avoirdupois Pounds

To	Multiply by
Kilograms	0.45359237
Metric tons	0.00045359237

Conversion Factors

1 Metric ton=2,204.622 pounds
1 Kilogram=2.2046 pounds
1 Acre=0.4047 hectares
1 Hectare=2.47 acres
1 Square mile=640 acres=259 hectares
1 Gallon=3.7853 liters

CONVERSION FACTORS

Commodity	Unit	Approximate equivalent
Apples	1 pound dried	7 pounds fresh; beginning 1943, 8 pounds fresh
Do	1 pound chops	5 pounds fresh
Do	1 case canned [21]	1.4 bushels fresh
Applesauce	do [21]	1.2 bushels fresh
Apricots	1 pound dried	6 pounds fresh
Barley flour	100 pounds	4.59 bushels barley
Beans, lima	1 pound shelled	2 pounds unshelled
Beans, snap or wax	1 case canned [22]	0.008 ton fresh
Buckwheat flour	100 pounds	3.47 bushels buckwheat
Calves	1 pound live weight	0.611 pound dressed weight (1999 average)
Cattle	do	0.607 pound dressed weight (1999 average)
Cane syrup	1 gallon	5 pounds sugar
Cherries, tart	1 case canned [21]	0.023 ton fresh
Chickens	1 pound live weight	0.72 pound ready-to-cook weight
Corn, shelled	1 bushel (56 lbs.)	2 bushels (70 pounds) of husked ear corn
Corn, sweet	1 case canned [22]	0.030 ton fresh
Cornmeal:		
Degermed	100 pounds	3.16 bushels corn, beginning 1946
Nondegermed	do	2 bushels corn, beginning 1946
Cotton	1 pound ginned	3.26 pounds seed cotton, including trash [23]
Cottonseed meal	1 pound	2.10 pounds cottonseed
Cottonseed oil	do	5.88 pounds cottonseed
Dairy products:		
Butter	do	21.1 pounds milk
Cheese	do	10 pounds milk
Condensed milk, whole	do	2.3 pounds milk
Dry cream	do	19 pounds milk
Dry milk, whole	do	7.6 pounds milk
Evaporated milk, whole	do	2.14 pounds milk
Malted milk	do	2.6 pounds milk
Nonfat dry milk	do	11 pounds liquid skim milk
Ice cream [24]	1 gallon	15 pounds milk
Ice cream [24] (eliminating fat from butter and concentrated milk).	do	12 pounds milk
Eggs	1 case	47 pounds
Eggs, shell	do	41.2 pounds frozen or liquid whole eggs
Do	do	10.3 pounds dried whole eggs
Figs	1 pound dried	3 pounds fresh in California; 4 pounds fresh elsewhere
Flaxseed	1 bushel	About 2½ gallons oil
Grapefruit, Florida	1 case canned juice [22]	0.64 box fresh fruit
Hogs	1 pound live weight	0.737 pound dressed weight, excluding lard (1999 average)
Linseed meal	1 pound	1.51 pounds flaxseed
Linseed oil	do	2.77 pounds flaxseed
Malt	1 bushel (34 lbs.)	1 bushel barley (48 lbs.)
Maple syrup	1 gallon	8 pounds maple sugar
Nuts:		
Almonds, imported	1 pound shelled	3½ pounds unshelled
Almonds, California	do	2.22 pounds unshelled through 1949; 2 pounds thereafter
Brazil	do	2 pounds unshelled
Cashews	do	4.55 pounds unshelled
Chestnuts	do	1.19 pounds unshelled
Filberts	do	2.22 pounds unshelled through 1949; 2.5 pounds thereafter
Pecans:		
Seedling	do	2.78 pounds unshelled
Improved	do	2.50 pounds unshelled
Pignolias	do	1.3 pounds unshelled
Pistachios	do	2 pounds unshelled
Walnuts:		
Black	do	5.88 pounds unshelled
Persian (English)	do	2.67 pounds unshelled
Oatmeal	100 pounds	7.6 bushels oats, beginning 1943
Oranges, Florida	1 case canned juice [22]	0.53 box fresh
Peaches, California, freestone	1 pound dried	5⅓ pounds fresh through 1918; 6 pounds fresh for 1919–28; and 6½ pounds fresh from 1929 to date
Peaches, California, clingstone	do	7½ pounds fresh
Peaches, clingstone	1 case canned [21]	1 bushel fresh
Do	do	0.0230 ton fresh
Peanuts	1 pound shelled	1½ pounds unshelled
Pears	1 pound dried	6½ pounds fresh
Pears, Bartlett	1 case canned [22]	1.1 bushels fresh
Do	do	0.026 ton fresh

See footnotes on page ix.

CONVERSION FACTORS—Continued

Commodity	Unit	Approximate equivalent
Peas, green	1 pound shelled	2½ pounds unshelled
Do	1 case canned [22]	0.009 ton fresh (shelled)
Prunes	1 pound dried	2.7 pounds fresh in California; 3 to 4 pounds fresh elsewhere
Raisins	1 pound	4.3 pounds fresh grapes
Rice, milled (excluding brewers)	100 pounds	152 pounds rough or unhulled rice
Rye flour	do	2.23 bushels rye, beginning 1947
Sheep and lambs	1 pound live weight	0.504 pound dressed weight (1999 average)
Soybean meal	1 pound	1.27 pounds soybeans
Soybean oil	do	5.49 pounds soybeans
Sugar	1 ton raw	0.9346 ton refined
Tobacco	1 pound farm-sales weight	Various weights of stemmed and unstemmed, according to aging and the type of tobacco (See circular 435, U.S. Dept. of Agr.)
Tomatoes	1 case canned [22]	0.018 ton fresh
Turkeys	1 pound live weight	0.80 pound ready-to-cook weight
Wheat flour	100 pounds	2.30 bushels wheat [25]
Wool, domestic apparel shorn	1 pound greasy	0.48 pounds scoured
Wool, domestic apparel pulled	do	0.73 pound scoured

[1] Standard bushel used in the United States contains 2,150.42 cubic inches; the gallon, 231 cubic inches; the cranberry barrel, 5,826 cubic inches; and the standard fruit and vegetable barrel, 7,056 cubic inches. Such large-sized products as apples and potatoes sometimes are sold on the basis of a heaped bushel, which would exceed somewhat the 2,150.42 cubic inches of a bushel basket level full. This also applies to such products as sweetpotatoes, peaches, green beans, green peas, spinach, etc.

[2] Approximate inside dimensions, 4⅝ by 12½ by 16⅛ inches.

[3] Approximate inside dimensions, 4½ by 16 by 16⅛ inches.

[4] Approximate dimensions, 4½ by 13½ by 16⅛ inches.

[5] Approximate inside dimensions, 13 by 12 by 32 inches.

[6] Approximate inside dimensions, 13 by 18 by 21⅝ inches.

[7] This is the weight commonly used in trade practices, the actual weight varying according to temperature conditions.

[8] Approximate inside dimensions, 9¾ by 16 by 20 inches.

[9] Approximate inside dimensions, 4⅛ by 11½ by 14 inches.

[10] The standard weight of 70 pounds is usually recognized as being about 2 measured bushels of corn, husked, on the ear, because it required 70 pounds to yield 1 bushel, or 56 pounds, of shelled corn.

[11] For statistical purposes the bale of cotton is 500 pounds or 480 pounds net weight. Prior to Aug. 1, 1946, the net weight was estimated at 478 pounds. Actual bale weights vary considerably, and the customary average weights of bales of foreign cotton differ from that of the American square bale.

[12] This is the average weight of cottonseed, although the legal weight in some States varies from this figure of 32 pounds.

[13] Approximate inside dimensions, 9¼ by 10½ by 15 inches.

[14] Approximate inside dimensions, 1¾ by 11 by 16⅛ inches.

[15] Approximate inside dimensions, 11½ by 11½ by 24 inches.

[16] Beginning with the 1993-94 season, net weights for California Desert Valley and Arizona grapefruit were increased from 64 to 67 pounds, equal to the California other area net weight, making a 67 pound net weight apply to all of California.

[17] Approximate inside dimensions, 4¾ by 16 by 16⅛ inches.

[18] Approximate inside dimensions, 9⅞ by 13 by 25 inches.6 by 16 by 16⅛ inches.

[19] Includes both sorghum grain (kafir, milo, hegari, etc.) and sweet sorghum varieties.

[20] This average of 55 pounds indicates the usual weight of sweetpotatoes when harvested. Much weight is lost in curing or drying and the net weight when sold in terminal markets may be below 55 pounds.

[21] Case of 24 No. 2½ cans.

[22] Case of 24 No. 303 cans.

[23] Varies widely by method of harvesting.

[24] The milk equivalent of ice cream per gallon is 15 pounds. Reports from plants indicate about 81 percent of the butterfat in ice cream is from milk and cream, the remainder being from butter and concentrated milk. Thus the milk equivalent of the milk and cream in a gallon of ice cream is about 12 pounds.

[25] This is equivalent to 4.51 bushels of wheat per barrel (196 pounds) of flour and has been used in conversions, beginning July 1, 1957. Because of changes in milling processes, the following factors per barrel of flour have been used for earlier periods: 1790–1879, 5 bushels; 1880–1908, 4.75 bushels, 1909–17, 4.7 bushels; 1918 and 1919, 4.5 bushels; 1920, 4.6 bushels; 1921–44, 4.7 bushels; July 1944–Feb. 1946, 4.57 bushels; March 1946–Oct. 1946, average was about 4.31 bushels; and Nov. 1946–June 1957, 4.57 bushels.

CHAPTER I

STATISTICS OF GRAIN AND FEED

This chapter contains tables for wheat, rye, rice, corn, oats, barley, sorghum grain, and feedstuffs. Estimates are given of area, production, disposition, supply and disappearance, prices, value of production, stocks, foreign production and trade, price-support operations, animal units fed, and feed consumed by livestock and poultry.

Table 1-1.—Total grain: Supply and disappearance, United States, 2001–2010 [1]

Year [2]	Supply				Disappearance			Ending stocks
	Beginning stocks	Production	Imports	Total	Domestic use	Exports	Total disappearance	
	Million metric tons	Million metric tons	Million metric tons	Million metric tons	Million metric tons	Million metric tons	Million metric tons	Million metric tons
2001	77.4	321.4	5.8	404.7	253.1	83.5	336.5	67.4
2002	67.4	294.0	4.9	366.6	248.7	72.8	321.5	45.2
2003	45.2	345.1	4.5	394.9	262.0	89.2	351.2	44.4
2004	44.4	385.4	4.6	434.3	275.8	83.0	358.7	74.7
2005	74.7	362.9	5.2	442.4	280.0	91.8	371.8	71.7
2006	71.7	335.3	6.7	413.5	277.7	87.0	364.7	49.8
2007	49.8	411.8	7.2	468.9	307.1	105.8	412.9	54.3
2008	54.3	400.3	6.9	461.7	314.2	81.6	395.9	65.9
2009 [3]	65.9	416.4	6.2	488.5	339.0	82.1	421.1	67.3
2010 [4]	67.3	418.9	5.4	491.6	337.6	95.3	432.9	58.6

[1] Aggregate data on corn, sorghum, barley, oats, wheat, rye, and rice. [2] The marketing year for corn and sorghum begins September 1; for oats, barley, wheat, and rye, June 1; and for rice, August 1. [3] Preliminary. [4] Projected as of January 11, 2010; World Agricultural Supply and Demand Estimates. Totals may not add due to independent rounding.
ERS, Market and Trade Economics Division, (202) 694–5296.

Table 1-2.—Wheat: Area, yield, production, and value, United States, 2000–2009

Year	Area		Yield per harvested acre	Production	Marketing year average price per bushel received by farmers [2]	Value of production [2]
	Planted [1]	Harvested				
	1,000 acres	1,000 acres	Bushels	1,000 bushels	Dollars	1,000 dollars
2000	62,549	53,063	42.0	2,228,160	2.62	5,771,786
2001	59,432	48,473	40.2	1,947,453	2.78	5,412,834
2002	60,318	45,824	35.0	1,605,878	3.56	5,637,416
2003	62,141	53,063	44.2	2,344,415	3.40	7,927,981
2004	59,644	49,969	43.2	2,156,790	3.40	7,277,932
2005	57,214	50,104	42.0	2,103,325	3.42	7,167,166
2006	57,334	46,800	38.6	1,808,416	4.26	7,694,734
2007	60,460	50,999	40.2	2,051,088	6.48	13,289,326
2008	63,193	55,699	44.9	2,499,164	6.78	16,625,759
2009	59,133	49,868	44.4	2,216,171	4.85	10,626,176

[1] Includes area seeded in preceding fall for winter wheat. [2] Includes allowance for loans outstanding and purchases by the Government valued at the average loan and purchase rate, by States, where applicable.
NASS, Crops Branch, (202) 720–2127.

GRAIN AND FEED

Table 1-3.—Wheat, by type: Area, yield, production, and value, United States, 2000–2009

Year	Area		Yield per harvested acre	Production	Marketing year average price per bushel received by farmers [2]	Value of production [2]
	Planted [1]	Harvested				
			Winter wheat			
	1,000 acres	1,000 acres	Bushels	1,000 bushels	Dollars	1,000 dollars
2000	43,313	35,002	44.6	1,561,723	2.51	3,883,640
2001	40,943	31,165	43.4	1,353,119	2.72	3,661,591
2002	41,766	29,742	38.2	1,137,001	3.41	3,810,235
2003	45,384	36,753	46.7	1,716,376	3.27	5,596,916
2004	43,320	34,432	43.5	1,497,979	3.32	4,943,118
2005	40,418	33,779	44.3	1,497,764	3.32	4,950,001
2006	40,565	31,107	41.6	1,294,461	4.17	5,367,806
2007	45,012	35,938	41.7	1,499,241	6.13	9,077,574
2008	46,307	39,608	47.1	1,867,333	6.57	11,936,139
2009	43,311	34,485	44.2	1,522,718	4.70	7,060,386
			Durum wheat			
	1,000 acres	1,000 acres	Bushels	1,000 bushels	Dollars	1,000 dollars
2000	3,937	3,572	30.7	109,805	2.66	301,356
2001	2,910	2,789	30.0	83,556	3.08	269,391
2002	2,913	2,709	29.5	79,960	4.05	329,936
2003	2,915	2,869	33.7	96,637	3.97	396,905
2004	2,561	2,363	38.0	89,893	3.85	347,336
2005	2,760	2,716	37.2	101,105	3.46	353,223
2006	1,870	1,815	29.5	53,475	4.43	243,992
2007	2,156	2,119	34.1	72,224	9.92	692,512
2008	2,721	2,574	32.6	83,827	9.26	731,445
2009	2,554	2,428	44.9	109,042	5.80	613,103
			Other spring wheat [3]			
	1,000 acres	1,000 acres	Bushels	1,000 bushels	Dollars	1,000 dollars
2000	15,299	14,489	38.4	556,632	2.85	1,586,790
2001	15,579	14,519	35.2	510,778	2.90	1,481,852
2002	15,639	13,373	29.1	388,917	3.82	1,497,245
2003	13,842	13,441	39.5	531,402	3.62	1,934,160
2004	13,763	13,174	43.2	568,918	3.51	1,987,478
2005	14,036	13,609	37.1	504,456	3.66	1,863,942
2006	14,899	13,878	33.2	460,480	4.46	2,082,936
2007	13,292	12,942	37.1	479,623	7.16	3,519.240
2008	14,165	13,517	40.5	548,004	7.31	3,958,175
2009	13,268	12,955	45.1	584,411	5.25	2,952,687

[1] Seeded in preceding fall for winter wheat. [2] Obtained by weighting State prices by quantity sold. [3] Includes small quantities of Durum wheat grown in other States.
NASS, Crops Branch, (202) 720–2127.

Table 1-4.—Wheat: Stocks on and off farms, United States, 2000–2009

Year beginning September	All wheat							
	On farms				Off farms [1]			
	Sept. 1	Dec. 1	Mar. 1	Jun. 1	Sept. 1	Dec. 1	Mar. 1	Jun. 1
	1,000 bushels	1,000 bushels	1,000 bushels	1,000 bushels	1,000 bushels	1,000 bushels	1,000 bushels	1,000 bushels
2000	808,390	623,420	384,750	197,270	1,544,280	1,182,705	953,648	678,912
2001	696,850	517,890	338,500	216,830	1,458,964	1,105,565	871,268	560,282
2002	578,200	384,800	236,300	132,110	1,170,787	935,069	670,333	359,306
2003	687,320	491,925	257,890	131,880	1,351,652	1,028,359	762,727	414,559
2004	790,600	531,020	304,710	161,275	1,147,807	899,306	679,681	378,825
2005	721,360	513,010	256,000	111,010	1,201,931	916,414	716,215	460,180
2006	572,020	403,250	192,450	73,190	1,178,525	911,408	664,278	382,963
2007	495,000	289,540	91,990	25,635	1,221,927	842,398	617,280	280,183
2008	635,700	454,400	280,400	140,745	1,222,186	968,089	759,664	515,760
2009	836,000	558,800	NA	NA	1,373,338	1,222,891	NA	NA

Year beginning September	Durum wheat [2]							
	On farms				Off farms [1]			
	Sept. 1	Dec. 1	Mar. 1	Jun. 1	Sept. 1	Dec. 1	Mar. 1	Jun. 1
	1,000 bushels	1,000 bushels	1,000 bushels	1,000 bushels	1,000 bushels	1,000 bushels	1,000 bushels	1,000 bushels
2000	85,700	72,000	44,200	29,100	37,573	32,306	28,616	16,073
2001	63,300	49,600	30,200	20,600	33,779	26,997	21,690	12,390
2002	66,000	50,800	31,700	15,100	26,854	25,917	25,149	13,008
2003	58,000	41,400	24,800	13,600	29,241	25,569	19,447	12,712
2004	65,600	51,800	35,200	24,100	25,508	26,805	20,496	13,494
2005	70,200	57,700	39,700	23,100	31,135	24,384	25,795	17,251
2006	31,500	25,900	17,100	8,950	31,524	25,447	21,736	12,430
2007	34,700	17,600	8,100	2,350	35,764	22,170	17,058	5,938
2008	36,200	26,100	18,700	13,300	22,599	18,405	13,571	11,774
2009	74,100	50,600	NA	NA	27,686	25,181	NA	NA

[1] Includes stocks at mills, elevators, warehouses, terminals, and processors. [2] Included in all wheat. NA-not available.
NASS, Crops Branch, (202) 720–2127.

Table 1-5.—Wheat: Supply and disappearance, by class, United States, 2005–2009 [1]

Item	Year beginning June				
	2005	2006	2007	2008	2009
	Million bushels	Million bushels	Million bushels	Million bushels	Million bushels
All wheat:					
Stocks, June 1	540	571	456	306	657
Production	2,103	1,808	2,051	2,499	2,216
Supply [2]	2,725	2,501	2,620	2,932	2,991
Exports [3]	1,003	908	1,263	1,015	881
Domestic disappearance	1,151	1,137	1,051	1,260	1,137
Stocks, May 31	571	456	306	657	973
Hard red winter:					
Stocks, June 1	193	215	165	138	254
Production	930	682	956	1,035	919
Supply [2]	1,123	898	1,121	1,174	1,175
Exports [3]	428	280	536	447	370
Domestic disappearance	481	453	448	472	420
Stocks, May 31	215	165	138	254	....
Soft red winter:					
Stocks, June 1	88	106	109	55	171
Production	308	390	352	614	404
Supply [2]	422	515	475	702	607
Exports [3]	76	145	208	199	109
Domestic disappearance	241	261	212	332	256
Stocks, May 31	106	109	55	171	241
Hard red spring:					
Stocks, June 1	159	132	117	68	142
Production	467	432	450	512	548
Supply [2]	638	614	615	625	731
Exports [3]	280	248	304	210	214
Domestic disappearance	226	249	243	273	282
Stocks, May 31	132	117	68	142	234
Durum:					
Stocks, June 1	38	40	21	8	25
Production	101	53	72	84	109
Supply [2]	171	135	134	130	169
Exports [3]	45	40	45	24	44
Domestic disappearance	82	74	81	81	90
Stocks, May 31	40	21	8	25	35
White:					
Stocks, June 1	63	78	44	37	64
Production	297	251	221	255	237
Supply [2]	370	339	275	300	310
Exports [3]	174	195	169	136	143
Domestic disappearance	118	100	68	100	88
Stocks, May 31	78	44	37	64	79

[1] Data except production are approximations. [2] Total supply includes imports. [3] Imports and exports include flour and products in wheat equivalent.
ERS, Market and Trade Economics Division, (202) 694–5285.

Table 1-6.—Wheat: Area, yield, and production, by State and United States, 2007–2009

State	Area planted [1]			Area harvested			Yield per harvested acre			Production		
	2007	2008	2009	2007	2008	2009	2007	2008	2009	2007	2008	2009
	1,000 acres	1,000 acres	1,000 acres	1,000 acres	1,000 acres	1,000 acres	Bushels	Bushels	Bushels	1,000 bushels	1,000 bushels	1,000 bushels
AL	120	240	220	76	200	180	42.0	71.0	55.0	3,192	14,200	9,900
AZ	89	159	132	86	155	129	101.4	97.9	99.4	8,724	15,172	12,825
AR	820	1,070	430	700	980	390	41.0	57.0	44.0	28,700	55,860	17,160
CA	640	840	770	345	545	485	85.4	90.3	87.0	29,465	49,225	42,200
CO	2,520	2,190	2,630	2,369	1,936	2,479	39.2	30.8	40.6	92,980	59,700	100,610
DE	57	80	70	55	79	67	68.0	77.0	62.0	3,740	6.083	4,154
FL	13	25	17	9	23	14	55.0	55.0	43.0	495	1,265	602
GA	360	480	340	230	400	250	40.0	56.0	42.0	9,200	22,400	10,500
ID	1,235	1,400	1,310	1,175	1,330	1,250	71.2	73.8	79.3	83,645	98,170	99,130
IL	1,000	1,200	850	890	1,150	820	55.0	64.0	56.0	48,950	73,600	45,920
IN	420	580	470	370	560	450	56.0	69.0	67.0	20,720	38,640	30,150
IA	35	40	28	28	35	22	48.0	48.0	45.0	1,344	1,680	990
KS	10,400	9,600	9,300	8,600	8,900	8,800	33.0	40.0	42.0	283,800	356,000	369,600
KY	440	580	510	250	460	390	48.0	71.0	57.0	12,000	32,660	22,230
LA	235	400	185	220	385	175	54.0	57.0	56.0	11,880	21,945	9,800
MD	220	255	230	160	180	195	66.0	73.0	60.0	10,560	13,140	11,700
MI	550	730	620	530	710	560	65.0	69.0	69.0	34,450	48,990	38,640
MN	1,765	1,925	1,655	1,710	1,870	1,595	47.9	55.9	52.8	81,900	104,440	84,175
MS	370	520	180	330	485	165	56.0	62.0	50.0	18,480	30,070	8,250
MO	1,050	1,250	780	880	1,160	730	43.0	48.0	47.0	37,840	55,680	34,310
MT	5,170	5,740	5,520	5,065	5,470	5,305	29.6	30.1	33.3	149,820	164,730	176,625
NE	2,050	1,750	1,700	1,960	1,670	1,600	43.0	44.0	48.0	84,280	73,480	76,800
NV	23	21	20	13	11	13	99.2	100.1	97.8	1,290	1,101	1,272
NJ	31	35	34	28	33	29	51.0	61.0	51.0	1,428	2,013	1,479
NM	490	430	450	300	140	140	28.0	30.0	25.0	8,400	4,200	3,500
NY	100	130	115	85	122	105	53.0	63.0	65.0	4,505	7,686	6,825
NC	630	820	700	500	720	600	40.0	60.0	49.0	20,000	43,200	29,400
ND	8,595	9,230	8,680	8,405	8,640	8,415	35.6	36.0	44.8	298,875	311,200	377,190
OH	820	1,120	1,010	730	1,090	980	61.0	68.0	72.0	44,530	74.120	70,560
OK	5,900	5,600	5,700	3,500	4,500	3,500	28.0	37.0	22.0	98,000	166,500	77,000
OR	855	960	890	835	945	877	52.3	55.7	55.7	43,680	52,600	48,858
PA	170	195	190	155	185	175	58.0	64.0	56.0	8,990	11,840	9,800
SC	160	220	165	135	205	150	30.0	54.0	47.0	4,050	11,070	7,050
SD	3,508	3,661	3,209	3,327	3,420	3,009	43.1	50.5	42.9	143,515	172,540	129,147
TN	420	620	430	260	520	340	41.0	63.0	51.0	10,660	32,760	17,340
TX	6,200	5,800	6,400	3,800	3,300	2,450	37.0	30.0	25.0	140,600	99,000	61,250
UT	146	150	154	132	139	147	42.8	41.4	49.5	5,656	5,756	7,278
VA	230	310	250	205	280	210	64.0	71.0	58.0	13,120	19,880	12,180
WA	2,170	2,290	2,290	2,137	2,255	2,225	58.7	52.7	55.3	125,342	118,790	123,085
WV	8	11	9	6	8	5	57.0	60.0	50.0	342	480	250
WI	299	373	335	278	357	315	67.1	64.5	68.0	18,640	23,012	21,420
WY	146	163	155	130	146	132	25.4	29.4	38.0	3,300	4,286	5,016
US	60,460	63,193	59,133	50,999	55,699	49,868	40.2	44.9	44.4	2,051,088	2,499,164	2,216171

[1] Includes area planted preceding fall.
NASS, Crops Branch, (202) 720–2127.

Table 1-7.—Wheat: Supply and disappearance, United States, 2000–2009

Year beginning June	Supply				Disappearance						Ending stocks May 31
	Beginning stocks	Production	Imports [1]	Total	Domestic use				Exports [1]	Total disappearance	
					Food	Seed	Feed [2]	Total			
	Million bushels	Million bushels	Million bushels	Million bushels	Million bushels	Million bushels	Million bushels	Million bushels	Million bushels	Million bushels	Million bushels
2000	950	2,228	90	3,268	950	79	300	1,330	1,062	2,392	876
2001	876	1,947	108	2,931	926	83	182	1,192	962	2,154	777
2002	777	1,606	77	2,460	919	84	116	1,119	850	1,969	491
2003	491	2,344	63	2,899	912	80	203	1,194	1,158	2,353	546
2004	546	2,157	71	2,774	910	78	181	1,168	1,066	2,234	540
2005	540	2,103	81	2,725	917	77	157	1,151	1,003	2,154	571
2006	571	1,808	122	2,501	938	82	117	1,137	908	2,045	456
2007	456	2,051	113	2,620	948	88	16	1,051	1,263	2,314	306
2008	306	2,499	127	2,932	927	78	255	1,260	1,015	2,275	657
2009 [3]	657	2,216	119	2,991	917	70	149	1,137	881	2,018	973

[1] Imports and exports include flour and other products expressed in wheat equivalent. [2] Approximates feed and residual use and includes negligible quantities used for distilled spirits. [3] Preliminary. Totals may not add due to independent rounding.

ERS, Market and Trade Economics Division, (202) 694–5296.

Table 1-8.—Wheat, by type: Area, yield, and production, by State amd United States, 2007–2009

State	Area planted [1]			Area harvested			Yield per harvested acre			Production		
	2007	2008	2009	2007	2008	2009	2007	2008	2009	2007	2008	2009
	1,000 acres	*1,000 acres*	*1,000 acres*	*1,000 acres*	*1,000 acres*	*1,000 acres*	*Bushels*	*Bushels*	*Bushels*	*1,000 bushels*	*1,000 bushels*	*1,000 bushels*
Winter wheat												
AL	120	240	220	76	200	180	42.0	71.0	55.0	3,192	14,200	9,900
AZ	6	9	7	4	6	5	90.0	95.0	85.0	360	570	425
AR	820	1,070	430	700	980	390	41.0	57.0	44.0	28,700	55,860	17,160
CA	550	680	590	265	400	315	81.0	85.0	80.0	21,465	34,000	25,200
CO	2,500	2,150	2,600	2,350	1,900	2,450	39.0	30.0	40.0	91,650	57,000	98,000
DE	57	80	70	55	79	67	68.0	77.0	62.0	3,740	6,083	4,154
FL	13	25	17	9	23	14	55.0	55.0	43.0	495	1,265	602
GA	360	480	340	230	400	250	40.0	56.0	42.0	9,200	22,400	10,500
ID	750	850	740	710	800	700	73.0	75.0	81.0	51,830	60,000	56,700
IL	1,000	1,200	850	890	1,150	820	55.0	64.0	56.0	48,950	73,600	45,920
IN	420	580	470	370	560	450	56.0	69.0	67.0	20,720	38,640	30,150
IA	35	40	28	28	35	22	48.0	48.0	45.0	1,344	1,680	990
KS	10,400	9,600	9,300	8,600	8,900	8,800	33.0	40.0	42.0	283,800	356,000	369,600
KY	440	580	510	250	460	390	48.0	71.0	57.0	12,000	32,660	22,230
LA	235	400	185	220	385	175	54.0	57.0	56.0	11,880	21,945	9,800
MD	220	255	230	160	180	195	66.0	73.0	60.0	10,560	13,140	11,700
MI	550	730	620	530	710	560	65.0	69.0	69.0	34,450	48,990	38,640
MN	65	75	55	60	70	45	45.0	52.0	45.0	2,700	3,640	2,025
MS	370	520	180	330	485	165	56.0	62.0	50.0	18,480	30,070	8,250
MO	1,050	1,250	780	880	1,160	730	43.0	48.0	47.0	37,840	55,680	34,310
MT	2,240	2,600	2,550	2,190	2,420	2,420	38.0	39.0	37.0	83,220	94,380	89,540
NE	2,050	1,750	1,700	1,960	1,670	1,600	43.0	44.0	48.0	84,280	73,480	76,800
NV	17	12	16	12	7	11	100.0	103.0	102.0	1,200	721	1,122
NJ	31	35	34	28	33	29	51.0	61.0	51.0	1,428	2,013	1,479
NM	490	430	450	300	140	140	28.0	30.0	25.0	8,400	4,200	3,500
NY	100	130	115	85	122	105	53.0	63.0	65.0	4,505	7,686	6,825
NC	630	820	700	500	720	600	40.0	60.0	49.0	20,000	43,200	29,400
ND	465	630	580	445	550	545	49.0	41.0	48.0	21,805	22,550	26,160
OH	820	1,120	1,010	730	1,090	980	61.0	68.0	72.0	44,530	74,120	70,560
OK	5,900	5,600	5,700	3,500	4,500	3,500	28.0	37.0	22.0	98,000	166,500	77,000
OR	735	780	760	720	775	750	53.0	58.0	56.0	38,160	44,950	42,000
PA	170	195	190	155	185	175	58.0	64.0	56.0	8,990	11,840	9,800
SC	160	220	165	135	205	150	30.0	54.0	47.0	4,050	11,070	7,050
SD	2,100	2,050	1,700	1,980	1,890	1,530	46.0	55.0	42.0	91,080	103,950	64,260
TN	420	620	430	260	520	340	41.0	63.0	51.0	10,660	32,760	17,340
TX	6,200	5,800	6,400	3,800	3,300	2,450	37.0	30.0	25.0	140,600	99,000	61,250
UT	135	130	140	125	120	135	42.0	41.0	50.0	5,250	4,920	6,750
VA	230	310	250	205	280	210	64.0	71.0	58.0	13,120	19,880	12,180
WA	1,720	1,750	1,700	1,690	1,720	1,640	62.0	56.0	59.0	104,780	96,320	96,760
WV	8	11	9	6	8	5	57.0	60.0	50.0	342	480	250
WI	290	350	335	270	335	315	68.0	66.0	68.0	18,360	22,110	21,420
WY	140	150	155	125	135	132	25.0	28.0	38.0	3,125	3,780	5,016
US	45,012	46,307	43,311	35,938	39,608	34,485	41.7	47.1	44.2	1,499,241	1,867,333	1,522,718
Durum wheat												
AZ	83	150	125	82	149	124	102.0	98.0	100.0	8,364	14,602	12,400
CA	90	160	180	80	145	170	100.0	105.0	100.0	8,000	15,225	17,000
ID	15	10	20	15	10	20	81.0	73.0	81.0	1,215	730	1,620
MT	480	590	570	475	570	535	24.0	19.0	31.0	11,400	10,830	16,585
ND	1,480	1,800	1,650	1,460	1,690	1,570	29.5	25.0	39.0	43,070	42,250	61,230
SD	8	11	9	7	10	9	25.0	19.0	23.0	175	190	207
US	2,156	2,721	2,554	2,119	2,574	2,428	34.1	32.6	44.9	72,224	83,827	109,042
Other spring wheat												
CO	20	40	30	19	36	29	70.0	75.0	90.0	1,330	2,700	2,610
ID	470	540	550	450	520	530	68.0	72.0	77.0	30,600	37,440	40,810
MN	1,700	1,850	1,600	1,650	1,800	1,550	48.0	56.0	53.0	79,200	100,800	82,150
MT	2,450	2,550	2,400	2,400	2,480	2,350	23.0	24.0	30.0	55,200	59,520	70,500
NV	6	9	4	1	4	2	90.0	95.0	75.0	90	380	150
ND	6,650	6,800	6,450	6,500	6,400	6,300	36.0	38.5	46.0	234,000	246,400	289,800
OR	120	180	130	115	170	127	48.0	45.0	54.0	5,520	7,650	6,858
SD	1,400	1,600	1,500	1,340	1,520	1,470	39.0	45.0	44.0	52,260	68,400	64,680
UT	11	20	14	7	19	12	58.0	44.0	44.0	406	836	528
WA	450	540	590	447	535	585	46.0	42.0	45.0	20,562	22,470	26,325
WI [2]	9	23	....	8	22	....	35.0	41.0	....	280	902	....
WY [2]	6	13	....	5	11	....	35.0	46.0	....	175	506	....
US	13,292	14,165	13,268	12,942	13,517	12,955	37.1	40.5	45.1	479,623	548,004	584,411

[1] Includes area planted preceding fall. [2] Estimates discontinued in 2009.
NASS, Crops Branch, (202) 720–2127.

Table 1-9.—Wheat: Support operations, United States, 2000–2009

Marketing year beginning June 1	Income support payment rates per bushel [1]	Program price levels per bushel		Put under loan		Acquired by CCC under loan program	Owned by CCC at end of marketing year [5]
		Loan [2]	Target [3]	Quantity	Percentage of production [4]		
				Million bushels	*Percent*	*Million bushels*	*Million bushels*
	Dollars	*Dollars*	*Dollars*				
2000/2001 ...	1.23	2.58	NA	181	8.1	27	97
2001/2002 ...	1.01	2.58	NA	197	10.1	17	99
2002/2003 ...	0.52/0.00	2.80	3.86	120	7.5	2	66
2003/2004 ...	0.52/0.00	2.80	3.86	186	7.9	3	61
2004/2005 ...	0.52/0.00	2.75	3.92	178	8.3	10	54
2005/2006 ...	0.52/0.00	2.75	3.92	170	8.1	1	43
2006/2007 ...	0.52/0.00	2.75	3.92	94	5.2	0	41
2007/2008 ...	0.52/0.00	2.75	3.92	36	1.8	0	0
2008/2009 ...	0.52/0.00	2.75	3.92	84	3.4	0	0
2009/2010 ...	0.52/0.00	2.75	3.92	....	....	....	....

[1] Payment rates for the 1998/1999 through 2001/2002 crops were calculated according to the Production Flexibility Contract (PFC) program provisions of the Federal Agriculture Improvement and Reform Act of 1996 (1996 Act) and include supplemental PFC payment rates for 1998 through 2001. Payment rates for the 2002/2003 and subsequent crops are calculated according to the Direct and Counter-cyclical program provisions, following enactment of the Farm Security and Rural Investment Act of 2002 (2002 Act). Beginning with 2002/2003, the first entry is the direct payment rate and the second entry is the counter-cyclical payment rate. [2] Starting in 2009, producers who participate in the Average Crop Revenue Election (ACRE) program get a 30 percent reduction in their loan rate, not calculated in this table. [3] Target prices were reestablished under the 2002 Act. [4] Percentage of production is on a grain basis. [5] CCC ownership includes 93 million in Food Security Reserve for 1998/1999 through 2001/2002, 66 million in 2002/2003, 59 million in 2003/2004, 52 million in 2004/2005, and 33.6 million in 2005/2006 through 2006/2007. The Food Security Reserve became the Food Security Commodity Trust in July of 1999 and the Bill Emerson Humanitarian Trust in July of 2002. NA-not applicable.

FSA, Food Grains, (202) 720–3134.

Table 1-10.—Wheat: Marketing year average price and value, by State amd United States, 2007–2009

State	Marketing year average price per bushel			Value of production		
	2007	2008	2009 [1]	2007	2008	2009 [1]
	Dollars	*Dollars*	*Dollars*	*1,000 dollars*	*1,000 dollars*	*1,000 dollars*
AL	5.30	5.95	4.60	16,918	84,490	45,540
AZ	7.03	8.27	8.85	61,329	125,993	112,970
AR	4.72	5.88	4.85	135,464	328,457	83,226
CA	5.41	7.08	5.70	159,583	352,644	240,600
CO	6.01	6.62	4.50	561,326	397,140	451,962
DE	5.56	5.96	3.50	20,794	36,255	14,539
FL	4.00	5.50	4.30	1,980	6,958	2,589
GA	6.50	5.95	4.30	59,800	133,280	45,150
ID	6.56	6.38	4.75	549,000	626,694	469,179
IL	5.37	5.89	3.85	262,862	433,504	176,792
IN	5.20	5.91	4.20	107,744	228,362	126,630
IA	5.25	5.90	3.95	7,056	9,912	3,911
KS	5.93	6.94	4.85	1,682,934	2,470,640	1,792,560
KY	5.28	5.60	4.60	63,360	182,896	102,258
LA	5.20	5.50	4.70	61,776	120,698	46,060
MD	5.97	5.89	3.60	63,043	77,395	42,120
MI	5.01	5.63	4.25	172,595	275,814	164,220
MN	7.28	7.06	4.80	595,467	739,133	402,825
MS	4.30	5.36	4.50	79,464	161,175	37,125
MO	5.17	5.35	4.30	195,633	297,888	147,533
MT	7.14	6.84	5.15	1,075,754	1,138,548	906,149
NE	5.82	6.68	4.90	490,510	490,846	376,320
NV	6.50	6.79	4.65	8,363	7,478	5,934
NJ	5.80	6.15	3.75	8,282	12,380	5,546
NM	5.50	7.70	4.70	46,200	32,340	16,450
NY	6.92	6.16	4.70	31,175	47,346	32,078
NC	4.90	5.80	4.35	98,000	251,424	127,890
ND	7.74	7.31	4.85	2,339,614	2,296,523	1,822,071
OH	5.37	5.82	4.35	239,126	431,378	306,936
OK	6.22	6.93	4.80	609,560	1,153,845	369,600
OR	8.23	6.56	4.60	358,968	343,104	223,633
PA	6.60	5.42	4.10	59,334	64,173	40,180
SC	4.55	5.95	4.85	18,428	65,867	34,193
SD	6.42	6.92	5.10	899,263	1,199,255	661,874
TN	5.05	5.71	4.65	53,833	187,060	80,631
TX	6.40	7.58	5.25	899,840	750,420	321,563
UT	8.30	7.97	6.30	46,822	45,855	40,090
VA	5.78	5.88	4.05	75,834	116,894	49,329
WA	7.58	6.26	4.80	949,132	745,163	585,473
WV	6.17	5.85	4.20	2,110	2,808	1,050
WI	5.30	5.47	4.10	99,002	125,803	87,822
WY	6.68	6.51	4.70	22,048	27,921	23,575
US	6.48	6.78	4.85	13,289,326	16,625,759	10,626,176

[1] Preliminary.

NASS, Crops Branch, (202) 720–2127.

Table 1-11.—Wheat: Area, yield, and production in specified countries, 2007/2008–2009/2010

Country	Area			Yield per hectare			Production		
	2007/ 2008	2008/ 2009	2009/ 2010	2007/ 2008	2008/ 2009	2009/ 2010	2007/ 2008	2008/ 2009	2009/ 2010
	1,000 hec- tares	*1,000 hec- tares*	*1,000 hec- tares*	*Metric tons*	*Metric tons*	*Metric tons*	*1,000 metric tons*	*1,000 metric tons*	*1,000 metric tons*
Australia	12,578	13,530	13,788	1.08	1.58	1.63	13,569	21,420	22,500
Canada	8,640	10,032	9,500	2.32	2.85	2.79	20,054	28,611	26,500
China, Peoples Rep.	23,721	23,617	24,200	4.61	4.76	4.75	109,298	112,464	115,000
EU-27	24,712	26,983	25,722	4.86	5.60	5.37	120,133	151,114	138,195
India	28,000	28,150	27,900	2.71	2.79	2.89	75,810	78,570	80,680
Kazakhstan	12,900	13,500	14,700	1.28	.93	1.16	16,450	12,550	17,000
Pakistan	8,578	8,550	9,046	2.72	2.45	2.66	23,295	20,959	24,033
Russian Federa- tion	24,400	26,650	28,700	2.02	2.39	2.15	49,400	63,700	61,700
Turkey	7,700	7,700	7,800	2.01	2.18	2.31	15,500	16,800	18,000
Ukraine	5,950	7,050	6,750	2.34	3.67	3.10	13,900	25,900	20,900
Others	40,090	37,252	37,302	2.44	2.23	2.55	98,001	83,159	95,029
Total foreign	197,269	203,014	205,408	2.82	3.03	3.02	555,410	615,247	619,537
United States ...	20,639	22,541	20,181	2.70	3.02	2.99	55,821	68,016	60,314
Total	217,908	225,555	225,589	2.81	3.03	3.01	611,231	683,263	679,851

FAS, Office of Global Analysis, (202) 720-6301.

Table 1-12.—Wheat and flour: United States imports,1999–2008

Year beginning June	Wheat grain	Flour (wheat equivalent)	Other products (wheat equivalent) [1]	Total wheat, flour, and other products
	1,000 bushels	*1,000 bushels*	*1,000 bushels*	
1999	72,408	7,116	14,986	94,511
2000	66,313	8,863	14,649	89,825
2001	82,615	9,907	15,029	107,551
2002	49,741	11,946	15,687	77,374
2003	37,156	11,363	14,508	63,026
2004	44,499	11,146	14,925	70,570
2005	54,073	11,258	16,023	81,354
2006	92,928	11,853	17,089	121,870
2007	85,806	10,710	16,115	112,631
2008	101,964	9,785	15,221	126.970

[1] Includes macaroni, semolina, and similar products.
ERS, Market and Trade Economics Division, (202) 694–5285.

Table 1-13.—Wheat, flour, and products: International trade, 2006/2007–2008/2009

Country	2006/2007	2007/2008	2008/2009
	1,000 metric tons	*1,000 metric tons*	*1,000 metric tons*
Principle exporting countries:			
Argentina	11,209	6,767	4,500
Australia	7,487	14,747	14,500
Brazil	770	400	1,200
Canada	16,116	18,812	18,500
China	2,835	723	892
EU-27	12,271	25,318	21,500
Kazakhstan, Republic of	8,181	5,701	7,800
Russian Federation	12,552	18,393	18,500
Turkey	1,722	2,238	4,374
Ukraine	1,236	13,037	9,300
Others	8,674	9,643	7,185
Total Foreign	83,053	115,779	108,251
United States	34,363	27,635	23,977
Total	117,416	143,414	132,228
Principle importing countries:			
Algeria	5,904	6,359	5,000
Brazil	6,772	6,403	6,500
Egypt	7,700	9,900	10,200
EU-27	6,942	7,740	5,500
Indonesia	5,224	5,423	5,800
Iran	200	6,700	4,500
Iraq	3,414	3,869	3,700
Japan	5,701	5,156	5,502
Korea, South	3,092	3,371	4,470
Nigeria	2,677	3,550	3,900
Others	62,975	74,930	72,640
Total Foreign	110,601	133,401	127,712
United States	3,065	3,456	3,228
Total	113,666	136,857	130,940

FAS, Office of Global Analysis, (202) 720-6301. Prepared or estimated on the basis of official USDA production, supply, and distribution statistics from foreign governments.

GRAIN AND FEED

Table 1-14.—Wheat and flour: United States exports by country of destination, 2007–2009

Country of destination	Year		
	2007	2008	2009
	Metric tons	*Metric tons*	*Metric tons*
Wheat:			
Japan	3,374,041	3,629,462	3,035,944
Nigeria	2,621,174	2,607,341	2,935,188
Mexico	2,516,499	2,804,365	1,921,255
Philippines	1,482,138	1,775,074	1,261,834
Korea, Republic of	1,304,811	1,321,414	1,108,795
Taiwan	1,245,788	752,714	861,826
Yemen	1,087,055	384,700	733,407
Egypt	3,119,708	2,160,589	681,728
Indonesia	1,041,985	927,340	669,699
Rest of World	15,197,426	13,657,863	8,717,848
World Total	32,990,625	30,020,862	21,927,524
Wheat flour:			
Canada	112,915	144,207	87,535
Mexico	47,309	31,627	31,137
Pakistan	2,771	0	29,120
United Arab Emirates	19	114	21,463
Belgium-Luxembourg(*)	17	1	15,850
Kenya	19,314	13,548	12,340
Nicaragua	0	0	9,328
Israel(*)	51,919	27	7,901
Sri Lanka	3,589	0	6,852
Rest of World	96,968	74,962	56,177
World Total	334,819	264,487	287,701

FAS, Office of Global Analysis, (202) 720-6301. Prepared or estimated on the basis of official USDA production, supply, and distribution, supply, and distribution statistics from foreign governments. Note: (*) Denotes a country that is a summarization of its component countries.

Table 1-15.—Rye: Area, yield, production, disposition, and value, United States, 2000–2009

Year	Area		Yield per harvested acre	Production	Marketing year average price per bushel received by farmers	Value of production
	Planted [1]	Harvested				
	1,000 acres	*1,000 acres*	*Bushels*	*1,000 bushels*	*Dollars*	*1,000 dollars*
2000	1,329	296	28.3	8,386	2.60	21,830
2001	1,328	250	27.6	6,896	2.86	19,752
2002	1,355	263	24.7	6,488	3.32	21,549
2003	1,348	319	27.1	8,634	2.93	25,336
2004	1,380	300	27.5	8,255	3.22	26,551
2005	1,433	279	27.0	7,537	3.30	24,890
2006	1,396	274	26.3	7,193	3.32	23,895
2007	1,334	252	25.0	6,311	5.01	31,604
2008	1,260	269	29.7	7,979	6.32	50,452
2009	1,241	252	27.8	6,993	4.78	33,427

[1] Area planted in preceding fall.
NASS, Crops Branch, (202) 720–2127.

Table 1-16.—Rye: Supply and disappearance, United States, 2000–2009

Year begin-ning June	Supply				Disappearance							Ending stocks May 31
	Begin-ning stocks	Produc-tion	Imports	Total	Domestic use					Exports	Total dis-appear-ance	
					Food	Seed	Industry	Feed [1]	Total			
	1,000 bushels	*1,000 bushels*	*1,000 bushels*	*1,000 bushels*	*1,000 bushels*	*1,000 bushels*	*1,000 bushels*	*1,000 bushels*	*1,000 bushels*	*1,000 bushels*	*1,000 bushels*	*1,000 bushels*
2000 ..	1,589	8,386	3,230	13,205	3,300	3,000	3,000	2,325	11,625	390	12,015	1,190
2001 ..	1,190	6,896	4,945	13,031	3,300	3,000	3,000	2,970	12,270	193	12,463	568
2002 ..	568	6,488	6,140	13,196	3,300	3,000	3,000	3,329	12,629	122	12,751	445
2003 ..	445	8,634	3,286	12,365	3,300	3,000	3,000	2,415	11,715	56	11,771	594
2004 ..	584	8,255	5,626	14,475	3,300	3,000	3,000	4,237	13,537	145	13,682	793
2005 ..	793	7,537	5,481	13,811	3,300	3,000	3,000	3,791	13 091	14	13,105	706
2006 ..	706	7,193	5,899	13,798	3,300	3,000	3,000	3,947	13,247	70	13,317	481
2007 ..	481	6,311	7,064	13,856	3,300	3,000	3,000	3,909	13,209	251	13,460	396
2008 ..	396	7,979	3,953	12,328	3,300	3,000	3,000	2,203	11,503	316	11,819	509
2009 [2]	509	6,993	4,251	11,753	3,300	3,000	3,000	1,448	10,748	73	10,821	932

[1] Residual, approximates total feed use. [2] Preliminary. Totals may not add due to independent rounding.
ERS, Market and Trade Economics Division, (202) 694–5302.

Table 1-17.—Rye: Area, yield, and production, by State and United States, 2007–2009

State	Area planted [1]			Area harvested			Yield per harvested acre			Production		
	2007	2008	2009	2007	2008	2009	2007	2008	2009	2007	2008	2009
	1,000 acres	*1,000 acres*	*1,000 acres*	*1,000 acres*	*1,000 acres*	*1,000 acres*	*Bush-e ls*	*Bush-els*	*Bush-els*	*1,000 bush-els*	*1,000 bush-els*	*1,000 bush-els*
GA	230	200	200	40	40	25	20.0	30.0	21.0	800	1,200	525
OK	300	280	270	60	55	40	18.0	19.0	14.0	1,080	1,045	560
Oth Sts [2]	804	780	771	152	174	187	29.2	33.0	31.6	4,431	5,734	5,908
US	1,334	1,260	1,241	252	269	252	25.0	29.7	27.8	6,311	7,979	6,993

[1] Includes area planted preceding fall. [2] Other States include IL, KS, MI, MN, NE, NY, NC, ND, PA, SC, SD, TX, and WI.
NASS, Crops Branch, (202) 720–2127.

Table 1-18.—Rye: Marketing year average price and value, by State and United States, 2007–2009

State	Marketing year average price per bushel			Value of production		
	2007	2008	2009 [1]	2007	2008	2009 [1]
	Dollars	*Dollars*	*Dollars*	*1,000 dollars*	*1,000 dollars*	*1,000 dollars*
GA	6.00	7.00	6.50	4,800	8,400	3,413
OK	6.10	7.00	8.20	6,588	7,315	4,592
Oth Sts [2]	4.56	6.06	4.30	20,216	34,737	25,422
US	5.01	6.32	4.78	31,604	50,452	33,427

[1] Preliminary. [2] Other States include IL, KS, MI, MN, NE, NY, NC, ND, PA, SC, SD, TX, and WI.
NASS, Crops Branch, (202) 720-2127.

Table 1-19.—Rye: Area, yield, and production in specified countries, 2006/2007–2008/2009

Country	Area			Yield per hectare			Production		
	2006/ 2007	2007/ 2008	2008/ 2009	2006/ 2007	2007/ 2008	2008/ 2009	2006/ 2007	2007/ 2008	2008/ 2009
	1,000 hec- tares	1,000 hec- tares	1,000 hec- tares	Metric tons	Metric tons	Metric tons	1,000 metric tons	1,000 metric tons	1,000 metric tons
Argentina	39	48	48	1.97	1.15	1.15	77	55	55
Australia	35	35	35	0.57	0.57	0.57	20	20	20
Belarus	600	540	600	2.17	2.78	2.83	1,300	1,500	1,700
Canada	115	132	1.20	2.19	2.39	2.33	252	316	280
EU-27	2,580	2,746	2,695	2.98	3.36	3.45	7,679	9,236	9,306
Kazakhstan, Re- public	70	70	70	0.71	0.71	0.71	50	50	50
Norway	8	8	7	5.00	6.00	3.86	40	48	27
Russian Federa- tion	2,100	2,200	2,150	1.86	2.05	2.00	3,900	4,500	4,300
Turkey	130	130	130	2.04	1.92	2.08	265	250	270
Ukraine	350	450	460	1.57	2.33	2.07	550	1,050	950
Others	59	59	59	1.36	1.36	1.36	80	80	80
Total foreign	6,086	6,418	6,374	2.34	2.67	2.67	14,213	17,105	17,038
United States ...	102	109	102	1.57	1.86	1.75	160	203	178
Total	6,188	6527	6,476	2.32	2.65	2.66	14,373	17,308	17,216

FAS, Office of Global Analysis, (202) 720-6301. Prepared or estimated on the basis of official USDA production, supply, and distribution, supply, and and distribution statistics from foreign governments.

Table 1-20.—Rye:[1] International trade, 2007/2008–2009/2010[2]

Country	2007/2008	2008/2009	2009/2010[3]
	1,000 metric tons	1,000 metric tons	1,000 metric tons
Principle exporting countries:			
Belarus	75	50	50
Canada	191	76	125
EU-27	76	114	100
Russia	119	16	10
Ukraine		6	20
Others	3		
Total Foreign	464	262	305
United States	6	8	2
Total	470	270	307
Principle importing countries:			
Croatia	2	3	1
EU-27	94	9	10
Israel	7	11	10
Japan	83	57	75
Korea, South	6	7	5
Norway	21	11	15
Switzerland	6	3	5
Turkey	13	8	5
Others		3	
Total Foreign	232	112	126
United States	179	100	108
Total	411	212	234

[1] Flour and products reported in terms of grain equivalent.   [2] Year beginning July 1.   [3] Preliminary.
FAS, Office of Global Analysis, (202) 720-6301: Prepared or estimated on the basis of official USDA production, supply, and distribution statics from foreign governments.

Table 1-21.—Rice, rough: Area, yield, production, and value, United States, 2000–2009 [1]

Year	Area planted	Area harvested	Yield per acre	Production	Marketing year average price per cwt. received by farmers	Value of production
	1,000 acres	*1,000 acres*	*Pounds*	*1,000 cwt.*	*Dollars*	*1,000 dollars*
2000	3,060.0	3,039.0	6,281	190,872	5.61	1,049,961
2001	3,334.0	3,314.0	6,496	215,270	4.25	925,055
2002	3,240.0	3,207.0	6,578	210,960	4.49	979,628
2003	3,022.0	2,997.0	6,670	199,897	8.08	1,628,948
2004	3,347.0	3,325.0	6,988	232,362	7.33	1,701,822
2005	3,384.0	3,364.0	6,624	222,833	7.65	1,738,598
2006	2,838.0	2,821.0	6,898	194,585	9.96	1,990,783
2007	2,761.0	2,748.0	7,219	198,388	12.80	2,600,871
2008	2,995.0	2,976.0	6,846	203,733	16.80	3,603,460
2009	3,135.0	3,103.0	7,085	219,850	14.30	3,145,521

[1] Sweet rice yield and production included in 2003 as short grain but not in previous years.
NASS, Crops Branch, (202) 720–2127.

Table 1-22.—Rice, rough: Stocks on and off farms, United States, 2001–2010

Year beginning previous December	On farms			Off farms [1]		
	Dec. 1	Mar. 1	Aug. 1	Dec. 1	Mar. 1	Aug. 1
	1,000 cwt.	*1,000 cwt.*	*1,000 cwt.*	*1,000 cwt.*	*1,000 cwt.*	*1,000 cwt.*
2001	38,085	18,715	921	95,842	67,305	21,097
2002	52,680	31,725	5,180	101,881	81,783	26,629
2003	53,220	27,505	1,225	103,850	75,073	18,846
2004	43,165	18,325	571	92,154	69,515	18,944
2005	57,545	37,590	2,815	109,151	81,193	28,822
2006	58,630	30,865	1,553	101,518	80,416	35,825
2007	52,420	28,015	1,220	97,706	76,145	33,713
2008	48,250	22,923	395	102,815	81,623	23,981
2009	47,530	21,286	876	91,071	70,042	23,787
2010 [2]	51,880	NA	NA	104,726	NA	NA

[1] Stocks at mills and in attached warehouses, in warehouses not attached to mills, and in ports or in transit.   [2] Preliminary.   NA-not available.
NASS, Crops Branch, (202) 720–2127.

Table 1-23.—Rice, by length of grain: Area, yield, and production, United States, 2000–2009

Year	Area harvested			Yield per acre			Production		
	Long grain	Medium grain	Short grain	Long grain	Medium grain	Short grain	Long grain	Medium grain	Short grain
	1,000 acres	*1,000 acres*	*1,000 acres*	*Pounds*	*Pounds*	*Pounds*	*1,000 cwt.*	*1,000 cwt.*	*1,000 cwt.*
2000	2,189.0	814.0	36.0	5,882	7,311	7,228	128,756	59,514	2,602
2001	2,697.0	591.0	26.0	6,213	7,801	6,192	167,555	46,105	1,610
2002	2,512.0	668.0	27.0	6,260	7,815	5,615	157,243	52,201	1,516
2003	2,310.0	644.0	43.0	6,451	7,481	6,293	149,011	48,180	2,706
2004	2,571.0	705.0	49.0	6,630	8,325	6,588	170,445	58,689	3,228
2005	2,734.0	575.0	55.0	6,479	7,375	6,000	177,125	42,408	3,300
2006	2,186.0	574.0	61.0	6,727	7,631	6,098	147,063	43,802	3,720
2007	2,052.0	630.0	66.0	6,980	8,105	6,197	143,235	51,063	4,090
2008	2,350.0	575.0	51.0	6,522	8,203	6,490	153,257	47,166	3,310
2009	2,265.0	786.0	52.0	6,743	8,052	7,373	152,725	63,291	3,834

NASS, Crops Branch, (202) 720–2127.

Table 1-24.—Rice, rough, by length of grain: Stocks in all positions, United States, 2001–2010

Year beginning previous December	Dec. 1	Mar. 1	Jun. 1	Aug. 1	Oct. 1[1]
	Long grain				
	1,000 cwt.	1,000 cwt.	1,000 cwt.	1,000 cwt.	1,000 cwt.
2001	82,718	51,428		8,305	116
2002	109,953	83,723		22,743	434
2003	113,897	75,733		11,673	59
2004	93,881	59,671		8,035	169
2005	112,799	79,994		19,026	172
2006	124,485	86,108		28,571	(3)
2007	109,301	76,127		25,738	77
2008	103,620	69,207	35,580	16,101	(3)
2009	96,994	64,226	34,293	17,698	372
2010[2]	103,430	NA	NA	NA	NA
	Medium grain				
	1,000 cwt.	1,000 cwt.	1,000 cwt.	1,000 cwt.	1,000 cwt.
2001	48,438	32,504		12,841	5,066
2002	42,525	28,515		8,477	2,691
2003	40,918	25,529		7,760	2,688
2004	38,736	26,562		10,887	4,261
2005	51,005	36,761		11,791	4,413
2006	32,802	23,299		8,012	2,921
2007	37,225	25,857		8,372	2,506
2008	43,520	32,584	16,284	7,196	1,084
2009	37,989	24,755	12,722	6,093	938
2010[2]	49,264	NA	NA	NA	NA
	Short grain				
	1,000 cwt.	1,000 cwt.	1,000 cwt.	1,000 cwt.	1,000 cwt.
2001	2,771	2,088		872	732
2002	2,083	1,270		589	363
2003	2,255	1,316		638	407
2004	2,702	1,607		593	370
2005	2,892	2,028		820	470
2006	2,861	1,874		795	(3)
2007	3,600	2,176		823	412
2008	3,925	2,755	1,554	1,079	(3)
2009	3,618	2,347	1,301	872	522
2010[2]	3,912	NA	NA	NA	NA

[1] California only. [2] Preliminary. [3] Not published to avoid disclosing individual reports. NA-not available.
NASS, Crops Branch, (202) 720–2127.

Table 1-25.—Rough and milled rice (rough equivalent): Supply and disappearance, United States, 2000–2009 [1]

Year beginning August	Supply				Disappearance					Ending stocks July 31
	Beginning stocks	Production	Imports[2]	Total	Food, industrial, & residual[3]	Seed	Total	Exports[2]	Total disappearance	
	Million cwt.	Million cwt.	Million cwt.	Million cwt.	Million cwt.	Million cwt.	Million cwt.	Million cwt.	Million cwt.	Million cwt.
2000	27.5	190.9	10.9	229.2	113.0	4.1	117.2	83.5	200.7	28.5
2001	28.5	215.3	13.2	257.0	119.0	4.0	123.0	95.0	218.0	39.0
2002	39.0	211.0	14.8	264.8	110.4	3.7	114.2	123.9	238.1	26.7
2003	26.7	199.9	15.0	241.6	109.1	4.1	113.2	104.7	217.8	23.8
2004	23.8	232.4	13.2	269.4	116.9	4.2	121.1	110.4	231.5	37.9
2005	37.9	222.8	17.1	277.8	114.4	3.5	117.9	116.8	234.7	43.1
2006	43.1	194.6	20.6	258.3	123.2	3.4	126.6	92.3	218.9	39.4
2007	39.4	198.4	23.9	261.7	121.8	3.7	125.5	106.6	232.1	29.6
2008	29.6	203.7	19.2	252.6	122.5	3.9	1264	95.9	222.2	30.6
2009[4]	30.6	219.9	19.0	269.4	118.1	4.5	122.6	105.0	232.7	36.7

Totals may not add due to independent rounding.
[1] Consolidated supply and disappearance of rough and milled rice. Milled rice data converted to a rough basis using annually derived extraction rates as factors. [2] Trade data from Bureau of the Census. [3] The residual includes unaccounted losses in transporting, processing, and marketing. [4] Forecasted September 2010.
ERS, Market and Trade Economics Division, (202) 694–5292.

Table 1-26.—Rice, by length of grain: Area, yield, and production, by State and United States, 2007–2009

State	Area harvested			Yield per acre			Production		
	2007	2008	2009[1]	2007	2008	2009[1]	2007	2008	2009[1]
	Long grain								
	1,000 acres	*1,000 acres*	*1,000 acres*	*Pounds*	*Pounds*	*Pounds*	*1,000 cwt.*	*1,000 cwt.*	*1,000 cwt.*
AR	1,180.0	1,295.0	1,245.0	7,230	6,640	6,760	85,314	85,988	84,162
CA	9.0	9.0	5.0	7,100	6,900	6,600	639	621	330
LA	355.0	450.0	410.0	6,150	5,820	6,320	21,833	26,190	25,912
MS	189.0	229.0	243.0	7,350	6,850	6,700	13,892	15,687	16,281
MO	177.0	197.0	197.0	6,900	6,620	6,710	12,213	13,041	13,219
TX	142.0	170.0	165.0	6,580	6,900	7,770	9,344	11,730	12,821
US	2,052.0	2,350.0	2,265.0	6,980	6,522	6,743	143,235	153,257	152,725
	Medium grain								
	1,000 acres	*1,000 acres*	*1,000 acres*	*Pounds*	*Pounds*	*Pounds*	*1,000 cwt.*	*1,000 cwt.*	*1,000 cwt.*
AR	144.0	99.0	224.0	7,250	6,960	7,010	10,440	6,890	15,702
CA	459.0	458.0	500.0	8,500	8,550	8,740	39,015	39,159	43,700
LA	23.0	14.0	54.0	6,040	6,050	6,120	1,389	847	3,305
MO	1.0	2.0	3.0	6,600	6,600	6,800	66	132	204
TX	3.0	2.0	5.0	5,100	6,900	7,600	153	138	380
US	630.0	575.0	786.0	8,105	8,203	8,052	51,063	47,166	63,291
	Short grain								
	1,000 acres	*1,000 acres*	*1,000 acres*	*Pounds*	*Pounds*	*Pounds*	*1,000 cwt.*	*1,000 cwt.*	*1,000 cwt.*
AR	1.0	1.0	1.0	6,000	6,000	6,000	60	60	60
CA	65.0	50.0	51.0	6,200	6,500	7,400	4,030	3,250	3,774
US	66.0	51.0	52.0	6,197	6,490	7,373	4,090	3,310	3,834

[1] Preliminary.
NASS, Crops Branch, (202) 720–2127.

Table 1-27.—Rice: Area, yield, and production, by State and United States, 2007–2009 [1]

State	Area planted			Area harvested			Yield per harvested acre			Production		
	2007	2008	2009[2]	2007	2008	2009[2]	2007	2008	2009[2]	2007	2008	2009[2]
	1,000 acres	*1,000 acres*	*1,000 acres*	*1,000 acres*	*1,000 acres*	*1,000 acres*	*Pounds*	*Pounds*	*Pounds*	*1,000 cwt.*	*1,000 cwt.*	*1,000 cwt.*
AR	1,331.0	1,401.0	1,486.0	1,325.0	1,395.0	1,470.0	7,230	6,660	6,800	95,814	92,938	99,924
CA	534.0	519.0	561.0	533.0	517.0	556.0	8,200	8,320	8,600	43,684	43,030	47,804
LA	380.0	470.0	470.0	378.0	464.0	464.0	6,140	5,830	6,300	23,222	27,037	29,217
MS	190.0	230.0	245.0	189.0	229.0	243.0	7,350	6,850	6,700	13,892	15,687	16,281
MO	180.0	200.0	202.0	178.0	199.0	200.0	6,900	6,620	6,710	12,279	13,173	13,423
TX	146.0	175.0	171.0	145.0	172.0	170.0	6,550	6,900	7,770	9,497	11,868	13,201
US ...	2,761.0	2,995.0	3,135.0	2,748.0	2,976.0	3,103.0	7,219	6,846	7,085	198,388	203,733	219,850

[1] Sweet rice acreage included with short grain.   [2] Preliminary.
NASS, Crops Branch, (202) 720–2127.

Table 1-28.—Rice: Marketing year average price and value, by State and United States, 2007–2009

State	Marketing year average price per cwt.			Value of production		
	2007	2008	2009 [1]	2007	2008	2009 [1]
	Dollars	Dollars	Dollars	1,000 dollars	1,000 dollars	1,000 dollars
AR	12.10	15.00	13.40	1,159,349	1,394,070	1,338,982
CA	16.20	27.50	18.60	707,681	1,183,325	889,154
LA	12.70	15.40	12.60	294,919	416,370	368,134
MS	12.60	15.40	12.80	175,039	241,580	208,397
MO	11.90	13.80	13.10	146,120	181,787	175,841
TX	12.40	15.70	12.50	117,763	186,328	165,013
US	12.80	16.80	14.30	2,600,871	3,603,460	3,145,521

[1] Preliminary.
NASS, Crops Branch, (202) 720–2127.

Table 1-29.—Rice, milled, by length of grain: Stocks in all positions, United States, 2001–2010

Year beginning previous December	Whole kernels (head rice)				
	Dec. 1	Mar. 1	Jun. 1 [1]	Aug. 1	Oct. 1 [2]
	Long grain				
	1,000 cwt.	1,000 cwt.	1,000 cwt.	1,000 cwt.	1,000 cwt.
2001	3,624	2,470	....	2,287	26
2002	3,796	3,222	....	2,788	23
2003	4,390	3,656	....	2,739	9
2004	3,338	2,682	....	1,622	8
2005	3,089	2,796	....	2,629	*
2006	3,305	2,552	....	2,880	7
2007	2,803	2,454	....	1,989	*
2008	2,638	2,546	2,015	2,065	*
2009	2,504	2,300	3,251	1,658	*
2010 [3]	2,022	NA	NA	NA	NA
	Medium grain				
	1,000 cwt.	1,000 cwt.	1,000 cwt.	1,000 cwt.	1,000 cwt.
2001	1,348	1,164	....	1,207	342
2002	986	622	....	1,032	388
2003	1,674	1,351	....	543	277
2004	2,000	2,194	....	547	322
2005	917	1,925	....	804	363
2006	1,247	1,136	....	395	422
2007	653	792	....	536	*
2008	958	1,735	850	508	*
2009	1,531	978	823	689	*
2010 [3]	1,496	NA	NA	NA	NA
	Short grain				
	1,000 cwt.	1,000 cwt.	1,000 cwt.	1,000 cwt.	1,000 cwt.
2001	67	84	....	87	57
2002	62	110	....	72	53
2003	58	59	....	60	30
2004	114	122	....	77	31
2005	31	69	....	56	*
2006	75	53	....	53	36
2007	55	98	....	48	*
2008	92	69	78	59	*
2009	80	69	57	51	36
2010 [3]	73	NA	NA	NA	NA

See footnotes at end of table.

Table 1-29.—Rice, milled, by length of grain: Stocks in all positions, United States, 2001–2010—Continued

Year beginning previous December	Broken kernels [4]				
	Dec. 1	Mar. 1	Jun. 1 [1]	Aug. 1	Oct. 1 [2]
	Second heads				
	1,000 cwt.	1,000 cwt.	1,000 cwt.	1,000 cwt.	1,000 cwt.
2001	1,006	1,035	....	667	403
2002	825	648	....	696	246
2003	1,026	1,190	....	1,066	587
2004	968	1,199	....	515	167
2005	460	512	....	619	*
2006	795	370	....	235	128
2007	240	562	....	307	*
2008	853	852	906	488	*
2009	661	794	828	1,465	*
2010 [3]	1,374	NA	NA	NA	NA
	Screenings				
	1,000 cwt.	1,000 cwt.	1,000 cwt.	1,000 cwt.	1,000 cwt.
2001	66	3	....	72	....
2002	123	139	....	133	....
2003	91	146	....	62	....
2004	71	22	....	77	....
2005	21	28	....	40	....
2006	198	162	....	84	....
2007	90	*	....	81	....
2008	195	163	145	206	....
2009	42	64	61	3	....
2010 [3]	52	NA	NA	NA	....
	Brewers				
	1,000 cwt.	1,000 cwt.	1,000 cwt.	1,000 cwt.	1,000 cwt.
2001	251	228	....	117	31
2002	115	72	....	209	16
2003	242	225	....	104	12
2004	125	114	....	113	13
2005	123	89	....	152	*
2006	320	299	....	297	175
2007	163	*	....	150	*
2008	533	239	379	249	*
2009	437	527	704	21	*
2010 [3]	662	NA	NA	NA	NA

[1] Estimates began in 2008. [2] California only. [3] Preliminary. [4] Screenings included in second heads in California. * Not published to avoid disclosing individual operations. NA-not available.
NASS, Crops Branch, (202) 720–2127.

Table 1-30.—Rice, rough: Support operations, United States, 2000–2010

Marketing year beginning August 1	Income support payment rates per cwt [1]	Program price levels per cwt		Put under loan		Acquired by CCC under loan program [4]	Owned by CCC at end of marketing year
		Loan [2]	Target [3]	Quantity	Percentage of production		
	Dollars	*Dollars*	*Dollars*	*Million cwt*	*Percent*	*Million cwt*	*Million cwt*
2000/2001	5.42	6.50	NA	97.4	51.0	0.0	0.0
2001/2002	4.49	6.50	NA	128.0	59.5	0.3	0.0
2002/2003	2.35/1.65	6.50	10.50	132.8	62.5	0.0	0.0
2003/2004	2.35/0.07	6.50	10.50	91.2	45.6	0.0	0.0
2004/2005	2.35/0.82	6.50	10.50	147.3	63.4	0.8	0.0
2005/2006	2.35/0.50	6.50	10.50	138.3	61.9	0.1	0.0
2006/2007	2.35/0.00	6.50	10.50	92.6	47.8	0.0	0.0
2007/2008	2.35/0.00	6.50	10.50	84.7	42.9	0.0	0.0
2008/2009	2.35/0.00	6.50	10.50	72.4	35.5	0.0	0.0
2009/2010	2.35/0.00	6.50	10.50	....	....	....	....

[1] Payment rates for the 1998/1999 through 2001/2002 crops were calculated according to the Production Flexibility Contract (PFC) program provisions of the Federal Agriculture Improvement and Reform Act of 1996 (1996 Act) and include supplemental PFC payment rates for 1998 through 2001. Payment rates for the 2002/2003 and subsequent crops are calculated according to the Direct and Counter-cyclical program provisions, following enactment of the Farm Security and Rural Investment Act of 2002 (2002 Act). Beginning with 2002/2003, the first entry is the direct payment rate and the second entry is the counter-cyclical payment rate.   [2] Starting in 2009, producers who participate in the Average Crop Revenue Election (ACRE) program get a 30 percent reduction in their loan rate, not calculated in this table.   [3] Target prices were reestablished under the 2002 Act.   [4] Acquisitions for 2006/2007 as of September 30, 2007.   NA-not applicable.
FSA, Food Grains, (202) 720-5653.

Table 1-31.—Rice: United States exports (milled basis), by country of destination, 2007–2009 [1]

Country of destination	Year		
	2007	2008	2009
	1,000 metric tons	*1,000 metric tons*	*1,000 metric tons*
Mexico	829	779	834
Japan	303	275	402
Haiti	277	289	277
Canada	254	248	219
Honduras	105	139	135
Iraq	223	64	121
Saudi Aeabia	127	131	114
Costa Rica	183	113	114
El Salvador	97	93	110
Rest of World	1,088	1,677	1,125
World Total [2]	3,486	3,810	3,449

[1] Year beginning Jan 1.   [2] Includes countries not shown.
FAS, Grain and Feed Division, (202) 720–6219. www.fas.usda.gov/grain/default.html.

GRAIN AND FEED

Table 1-32.—Rice, milled: Area, yield, and production in specified countries, 2007/2008–2009/2010

Country	Area			Yield per hectare			Production		
	2007/ 2008	2008/ 2009	2009/ 2010	2007/ 2008	2008/ 2009	2009/ 2010	2007/ 2008	2008/ 2009	2009/ 2010
	1,000 hec- tares	*1,000 hec- tares*	*1,000 hec- tares*	*Metric tons*	*Metric tons*	*Metric tons*	*1,000 metric tons*	*1,000 metric tons*	*1,000 metric tons*
Bangladesh	11,100	11,100	11,600	3.89	4.19	3.94	28,800	31,000	30,500
Brazil	2,874	2,909	2,800	4.20	4.33	4.11	8,199	8,569	7,820
Burma, Union of	7,085	6,700	7,000	2.61	2.61	2.61	10,730	10,150	10,597
China, Peoples	28,919	29,240	29,680	6.43	6.56	6.59	130,224	134,330	137,000
India	43,770	45,400	41,000	3.31	3.28	3.20	96,690	99,180	87,500
Indonesia	11,900	12,170	12,000	4.82	4.88	5.01	37,000	38,300	38,800
Japan	1,673	1,627	1,624	6.51	6.78	6.52	7,930	8,029	7,711
Philippines	4,346	4,528	4,404	3.83	3.77	3.52	10,479	10,753	9,757
Thailand	10,830	10,800	10,940	2.77	2.78	2.81	19,800	19,850	20,300
Vietnam	7,412	7,334	7,370	4.98	5.30	5.28	24,375	23,393	24,312
Others	24,023	24,816	25,437	2.21	2.30	34	53,197	57,049	59,454
Total foreign	153,932	156,624	153,855	2.78	2.82	2.82	427,424	441,603	433,751
United States	1,112	1,204	1,256	8.09	7.68	7.94	6,149	6,400	6,917
Total	155,044	157,828	155,111	4.17	4.25	4.25	433,573	448,003	440,668

FAS, Office of Global Analysis, (202) 720-6301. Prepared or estimated on the basis of official USDA production, supply, and distribution, supply, and and distribution statistics from foreign governments.

Table 1-33.—Rice, milled equivalent: [1] International trade, 2007/2008–2009/2010 [2]

Country	2007/2008	2008/2009	2009/2010 [3]
	1,000 metric tons	*1,000 metric tons*	*1,000 metric tons*
Principle exporting countries:			
Argentina	443	554	625
Burma	541	1.052	300
Cambodia	500	800	850
China	969	783	850
Egypt	750	550	600
India	4,654	2,090	2,200
Pakistan	2,982	2,910	3,750
Thailand	10,011	8,570	9,500
Uruguay	778	987	730
Vietnam	4,649	5,950	5,750
Others	1,611	1,717	1,539
Total Foreign	27,888	25,963	26,694
United States	3,305	3,004	3,429
Total	31,193	28,967	30,123
Principle importing countries:			
Brazil	422	675	950
EU-27	845	800	860
Cote d'Ivoire	1,568	1,339	1,350
Iran	1,550	1,470	1,300
Iraq	975	1,089	1,100
Malaysia	799	1,039	1,070
Nigeria	1,800	1,750	1,800
Phillippines	2,570	2,600	2,200
Saudi Arabia	961	1,166	1,095
South Africa	1,030	580	800
Others	16,105	13,973	14,281
Total Foreign	28,625	24,481	26,806
United States	759	610	619
Total	29,384	27,091	27,425

[1] Includes milled, semi-milled, broken, and rough rice in terms of milled equivalent. [2] Year beginning Jan 1. [3] Preliminary.
FAS, Office of Global Analysis, (202) 720-6301. Prepared or estimated on the basis of official USDA production, supply, and distribution statistics from foreign governments.

Table 1-34.—Food grains: Average price, selected markets and grades, 2002–2009 [1]

| Calendar year [2] | Kansas City | | | Minneapolis (rail) | | | Portland Wheat No. 1 Soft White | St. Louis Wheat, No. 2 Soft Red Winter (truck) |
	Wheat, No. 1 Hard Winter, Ordinary Protein (rail)	Wheat, No. 1 Hard Winter, 13% protein (rail)	Wheat, No. 2 Soft Red Winter (rail)	Wheat, No. 1 Hard Amber Durum (milling) (rail)	Wheat, No. 1 Dark Northern Spring (rail), 14% protein	Rye, No. 2, 20 day delivery (truck)		
	Dollars per bushel	Dollars per bushel	Dollars per bushel	Dollars per bushel	Dollars per bushel	Dollars per bushel	Dollars per bushel	Dollars per bushel
2002	3.94	3.99	3.50	4.97	4.15	3.57	3.89	3.28
2003	3.86	3.97	3.60	5.30	4.26	3.09	3.69	3.47
2004	4.14	4.28	3.97	5.32	4.63	3.49	4.07	3.66
2005	4.10	4.17	3.92	....	4.89	3.63	3.72	3.06
2006	5.11	5.21	4.27	NA	5.19	3.25	4.07	3.47
2007	6.85	7.06	6.27	16.33	7.01	6.24	7.29	5.96
2008	8.92	9.82	7.72	23.25	11.16	7.12	7.93	6.32
2009	5.80	6.29	5.06	....	7.21	4.35	5.28	5.55

| Calendar year [2] | Chicago Wheat, No. 2 Soft Red Winter (rail) | Denver Wheat, No. 1 Hard Winter (truck red) | S.W. Louisiana Milled Rice | | Arkansas Milled Rice | | Texas Milled Rice |
			Medium	Long	Medium	Long	Long
	Dollars per bushel	Dollars per bushel	Dollars per cwt.	Dollars per cwt.	Dollars per cwt.	Dollars per cwt.	Dollars per cwt.
2002	2.30	3.53	....	....	....	....	....
2003	3.40	3.35	....	20.82	....	21.51	22.91
2004	3.36	3.53	19.36	16.47	19.22	17.22	18.65
2005	3.01	3.37	16.55	14.22	15.94	14.32	16.53
2006	3.58	4.47	22.50	71.46	21.56	17.82	19.38
2007	5.85	6.05	23.44	19.28	22.81	19.50	21.58
2008	6.75	7.85	36.49	34.97	38.85	35.93	36.41
2009	4.43	5.04	39.56	24.91	40.57	26.46	27.88

[1] Simple average of daily prices. [2] For wheat and rye, crop year begins in June. For rice, crop year begins in August. NA-not available.

AMS, Livestock and Grain Market News branch, (202) 720–6231.

Table 1-35.—Corn: Area, yield, production, and value, United States, 2000–2009

| Year | Area planted, all purposes | Corn for grain | | | | | Corn for silage | | |
		Area harvested	Yield per harvested acre	Production	Marketing year average price per bushel	Value of production	Area harvested	Yield per harvested acre	Production
	1,000 acres	1,000 acres	Bushels	1,000 bushels	Dollars	1,000 dollars	1,000 acres	Tons	1,000 tons
2000 ...	79,551	72,440	136.9	9,915,051	1.85	18,499,002	6,082	16.8	102,156
2001 ...	75,702	68,768	138.2	9,502,580	1.97	18,878,819	6,142	16.6	101,992
2002 ...	78,894	69,330	129.3	8,966,787	2.32	20,882,448	7,122	14.4	102,293
2003 ...	78,603	70,944	142.2	10,087,292	2.42	24,472,254	6,583	16.3	107,378
2004 ...	80,929	73,631	160.3	11,805,581	2.06	24,377,913	6,101	17.6	107,293
2005 ...	81,779	75,117	147.9	11,112,187	2.00	22,194,287	5,930	18.0	106,486
2006 ...	78,327	70,638	149.1	10,531,123	3.04	32,083,011	6,487	16.2	105,129
2007 ...	93,527	86,520	150.7	13,073,875	4.20	54,666,959	6,060	17.5	106,229
2008 ...	85,982	78,570	153.9	12,091,648	4.06	49,312,615	5,965	18.7	111,619
2009 [1]	86,482	79,590	164.7	13,110,062	3.70	48,588,665	5,605	19.3	108,209

[1] Preliminary.

NASS, Crops Branch, (202) 720–2127.

Table 1-36.—Corn: Stocks on and off farms, United States, 2001–2010

| Year beginning previous December | On farms | | | | Off farms [1] | | | |
	Dec. 1	Mar. 1	Jun. 1	Sep. 1 [2]	Dec. 1	Mar. 1	Jun. 1	Sep. 1 [2]
	1,000 bushels	1,000 bushels	1,000 bushels	1,000 bushels	1,000 bushels	1,000 bushels	1,000 bushels	1,000 bushels
2001	5,550,000	3,600,000	2,230,800	753,150	2,979,634	2,442,999	1,693,158	1,145,958
2002	5,275,000	3,355,000	2,020,600	586,800	2,989,715	2,440,263	1,576,290	1,009,626
2003	4,800,000	2,940,000	1,620,200	484,900	2,837,971	2,191,873	1,364,718	601,773
2004	5,286,000	3,030,000	1,540,000	438,000	2,667,775	2,241,459	1,430,140	520,091
2005	6,144,000	4,137,000	2,462,300	820,500	3,308,488	2,619,334	1,858,513	1,293,472
2006	6,325,000	4,055,000	2,350,500	749,500	3,489,957	2,932,328	2,011,199	1,217,661
2007	5,627,000	3,330,000	1,826,600	460,100	3,305,707	2,738,250	1,706,843	843,547
2008	6,530,000	3,780,000	1,970,900	499,950	3,748,085	3,078,722	2,057,117	1,124,200
2009	6,482,000	4,085,000	2,205,400	607,500	3,590,106	2,869,145	2,056,027	1,065,811
2010 [3]	7,405,000	NA	NA	NA	3,497,460	NA	NA	NA

[1] Includes stocks at mills, elevators, warehouses, terminals, and processors. [2] Old crop only. [3] Preliminary. NA-not available.

NASS, Crops Branch, (202) 720–2127.

Table 1-37.—Corn: Area, yield, and production, by State and United States, 2007–2009

State	Area planted for all purposes			Corn for grain								
				Area harvested			Yield per harvested acre			Production		
	2007	2008	2009¹	2007	2008	2009¹	2007	2008	2009¹	2007	2008	2009¹
	1,000 acres	1,000 acres	1,000 acres	1,000 acres	1,000 acres	1,000 acres	Bush-els	Bush-els	Bush-els	1,000 bushels	1,000 bushels	1,000 bushels
AL ...	340	260	280	280	235	250	78.0	104.0	108.0	21,840	24,440	27,000
AZ ...	55	50	50	22	15	20	185.0	165.0	175.0	4,070	2,475	3,500
AR ...	610	440	430	590	430	410	169.0	155.0	148.0	99,710	66,650	60,680
CA ...	650	670	550	190	170	160	182.0	195.0	180.0	34,580	33,150	28,800
CO ..	1,200	1,250	1,100	1,060	1,010	990	140.0	137.0	153.0	148,400	138,370	151,470
CT ...	26	27	26	(²)	(²)	(²)	(²)	(²)	(²)	(²)	(²)	(²)
DE ...	195	160	170	185	152	163	99.0	125.0	145.0	18,315	19,000	23,635
FL ...	70	70	70	35	35	37	90.0	105.0	100.0	3,150	3,675	3,700
GA ..	510	370	420	450	310	370	127.0	140.0	140.0	57,150	43,400	51,800
ID	320	300	300	105	80	80	170.0	170.0	180.0	17,850	13,600	14,400
IL	13,200	12,100	12,000	13,050	11,900	11,800	175.0	179.0	174.0	2,283,750	2,130,100	2,053,200
IN	6,500	5,700	5,600	6,370	5,460	5,460	154.0	160.0	171.0	980,980	873,600	933,660
IA	14,200	13,300	13,700	13,900	12,800	13,400	171.0	171.0	182.0	2,376,900	2,188,800	2,438,800
KS ...	3,900	3,850	4,100	3,680	3,630	3,860	138.0	134.0	155.0	507,840	486,420	598,300
KY ...	1,440	1,210	1,220	1,340	1,120	1,150	128.0	136.0	165.0	171,520	152,320	189,750
LA ...	740	520	630	730	510	610	163.0	144.0	132.0	118,990	73,440	80,520
ME ...	28	29	28	(²)	(²)	(²)	(²)	(²)	(²)	(²)	(²)	(²)
MD ..	540	460	470	465	400	425	101.0	121.0	145.0	46,965	48,400	61,625
MA ..	18	19	17	(²)	(²)	(²)	(²)	(²)	(²)	(²)	(²)	(²)
MI ...	2,650	2,400	2,350	2,340	2,140	2,090	123.0	138.0	148.0	287,820	295,320	309,320
MN ..	8,400	7,700	7,600	7,850	7,200	7,150	146.0	164.0	174.0	1,146,100	1,180,800	1,244,100
MS ..	930	720	730	910	700	695	148.0	140.0	126.0	134,680	98,000	87,570
MO ..	3,450	2,800	3,000	3,270	2,650	2,920	140.0	144.0	153.0	457,800	381,600	446,760
MT ..	84	78	72	38	35	26	140.0	136.0	152.0	5,320	4,760	3,952
NE ...	9,400	8,800	9,150	9,200	8,550	8,850	160.0	163.0	178.0	1,472,000	1,393,650	1,575,300
NV ...	5	5	4	(²)	(²)	(²)	(²)	(²)	(²)	(²)	(²)	(²)
NH ...	14	15	15	(²)	(²)	(²)	(²)	(²)	(²)	(²)	(²)	(²)
NJ ...	95	85	80	82	74	70	124.0	116.0	143.0	10,168	8,584	10,010
NM ...	135	140	130	54	55	50	180.0	180.0	185.0	9,720	9,900	9,250
NY ...	1,060	1,090	1,070	550	640	595	128.0	144.0	134.0	70,400	92,160	79,730
NC ...	1,090	900	870	1,010	830	800	100.0	78.0	117.0	101,000	64,740	93,600
ND ...	2,560	2,550	1,950	2,350	2,300	1,740	116.0	124.0	115.0	272,600	285,200	200,100
OH ..	3,850	3,300	3,350	3,610	3,120	3,140	150.0	135.0	174.0	541,500	421,200	546,360
OK ...	320	370	390	270	320	320	145.0	115.0	105.0	39,150	36,800	33,600
OR ..	60	60	60	35	33	32	200.0	200.0	215.0	7,000	6,600	6,880
PA ...	1,430	1,350	1,350	980	880	920	124.0	133.0	143.0	121,520	117,040	131,560
RI	2	2	2	(²)	(²)	(²)	(²)	(²)	(²)	(²)	(²)	(²)
SC ...	400	355	335	370	315	320	97.0	65.0	111.0	35,890	20,475	35,520
SD ...	4,950	4,750	5,000	4,480	4,400	4,680	121.0	133.0	151.0	542,080	585,200	706,680
TN ...	860	690	670	790	630	590	106.0	118.0	148.0	83,740	74,340	87,320
TX ...	2,150	2,300	2,350	1,970	2,030	1,960	148.0	125.0	130.0	291,560	253,750	254,800
UT ...	70	70	65	22	23	17	150.0	157.0	155.0	3,300	3,611	2,635
VT ...	92	94	91	(²)	(²)	(²)	(²)	(²)	(²)	(²)	(²)	(²)
VA ...	540	470	480	405	340	330	86.0	108.0	131.0	34,830	36,720	43,230
WA ..	195	165	170	115	90	105	210.0	205.0	215.0	24,150	18,450	22,575
WV ..	48	43	47	27	26	30	111.0	130.0	126.0	2,997	3,380	3,780
WI ...	4,050	3,800	3,850	3,280	2,880	2,930	135.0	137.0	153.0	442,800	394,560	448,290
WY ..	95	95	90	60	52	45	129.0	134.0	140.0	7,740	6,968	6,300
US ...	93,527	85,982	86,482	86,520	78,570	79,590	150.7	153.9	164.7	13,037,875	12,091,648	13,110,062

¹ Preliminary.　² Not estimated.
NASS, Crops Branch, (202) 720–2127.

Table 1-38.—Corn: Supply and disappearance, United States, 2001–2010

Year beginning September 1	Supply				Disappearance					Ending stocks Aug. 31		
	Beginning stocks	Production	Imports	Total	Domestic use			Exports	Total disappearance	Privately held [1]	Government	Total
					Feed and residual	Food, seed, and industrial	Total					
	Million bushels	Million bushels	Million bushels	Million bushels	Million bushels	Million bushels	Million bushels	Million bushels	Million bushels	Million bushels	Million bushels	Million bushels
2001	1,899	9,503	10	11,412	5,864	2,046	7,911	1,905	9,815	1,590	6	1,596
2002	1,596	8,967	14	10,578	5,563	2,340	7,903	1,588	9,491	1,083	4	1,087
2003	1,087	10,089	14	11,190	5,795	2,537	8,332	1,900	10,232	958	0	958
2004	958	11,806	11	12,775	6,155	2,687	8,842	1,818	10,661	2,113	1	2,114
2005	2,114	11,112	9	13,235	6,115	3,019	9,134	2,134	11,268	1,967	0	1,967
2006	1,967	10,531	12	12,510	5,540	3,541	9,081	2,125	11,207	1,304	0	1,304
2007	1,304	13,038	20	14,362	5,858	4,442	10,300	2,437	12,737	1,624	0	1,624
2008	1,624	12,092	14	13,729	5,182	5,025	10,207	1,849	12,056	1,673	0	1,673
2009 [2]	1,673	13,110	8	14,791	5,525	5,900	11,425	1,980	13,405	1,386	0	1,386
2010 [3]	1,386	13,160	10	14,556	5,250	6,090	11,340	2,100	13,440	1,116	0	1,116

[1] Includes quantity under loan and farmer–owned reserve. [2] Preliminary. [3] Projected as of January 11, 2010, World Agricultural Supply and Demand Estimates. Totals may not add due to independent rounding.
ERS, Market and Trade Economics Division, (202) 694–5296.

GRAIN AND FEED

Table 1-39.—Corn: Utilization for silage, by State and United States, 2007–2009

State	Silage								
	Area harvested			Yield per acre			Production		
	2007	2008	2009 [1]	2007	2008	2009 [1]	2007	2008	2009 [1]
	1,000 acres	*1,000 acres*	*1,000 acres*	*Tons*	*Tons*	*Tons*	*1,000 tons*	*1,000 tons*	*1,000 tons*
AL	10	10	9	8.0	15.0	13.0	80	150	117
AZ	33	35	30	27.0	30.0	29.0	891	1,050	870
AR	4	4	3	15.0	14.0	15.0	60	56	45
CA	455	495	385	26.5	26.5	26.0	12,058	13,118	10,010
CO	110	120	85	22.5	21.5	23.5	2,475	2,580	1,998
CT	24	23	22	19.5	21.5	15.5	468	495	341
DE	7	6	5	10.0	13.0	15.0	70	78	75
FL	30	30	30	18.0	17.0	18.0	540	510	540
GA	40	45	30	18.0	18.0	17.0	720	810	510
ID	210	215	215	27.0	27.0	27.5	5,670	5,805	5,913
IL	100	100	100	18.0	17.0	19.0	1,800	1,700	1,900
IN	110	110	110	18.5	20.0	20.0	2,035	2,200	2,200
IA	250	200	220	19.5	20.5	22.0	4,875	4,100	4,840
KS	160	170	180	18.0	17.0	19.0	2,880	2,890	3,420
KY	85	85	60	13.5	16.0	19.5	1,148	1,360	1,170
LA	5	5	3	18.0	14.0	13.0	90	70	39
ME	25	25	25	18.0	18.0	12.5	450	450	313
MD	65	55	40	12.0	15.0	19.0	780	825	760
MA	15	15	14	20.0	19.5	15.0	300	293	210
MI	295	250	220	14.5	16.5	15.5	4,278	4,125	3,410
MN	450	400	380	13.5	16.0	20.0	6,075	6,400	7,600
MS	15	15	10	13.0	13.0	15.0	195	195	150
MO	70	50	50	15.0	14.0	16.0	1,050	700	800
MT	44	41	45	22.0	22.0	23.0	968	902	1,035
NE	170	160	210	17.0	17.0	18.0	2,890	2,720	3,780
NV	5	5	4	25.0	26.0	24.0	125	130	96
NH	13	14	15	20.5	21.5	18.0	267	301	270
NJ	11	10	9	15.0	17.0	17.5	165	170	158
NM	80	83	78	25.0	25.0	27.0	2,000	2,075	2,106
NY	505	445	470	17.0	20.0	18.0	8,585	8,900	8,460
NC	60	55	55	11.0	15.0	18.0	660	825	990
ND	180	220	170	11.0	10.0	12.0	1,980	2,200	2,040
OH	180	140	170	17.0	17.0	20.0	3,060	2,380	3,400
OK	30	30	25	19.5	16.5	14.0	585	495	350
OR	25	27	28	25.5	27.0	26.0	638	729	728
PA	430	450	420	16.5	18.5	19.5	7,095	8,325	8,190
RI	2	2	2	20.0	20.5	12.5	40	41	25
SC	12	28	10	14.0	9.0	16.0	168	252	160
SD	400	300	250	11.5	12.0	16.0	4,600	3,600	4,000
TN	55	55	50	11.0	15.0	21.0	605	825	1,050
TX	150	180	140	23.0	21.0	21.0	3,450	3,780	2,940
UT	47	47	47	21.0	23.0	23.0	987	1,081	1,081
VT	87	86	83	19.0	19.0	17.0	1,653	1,634	1,411
VA	130	125	135	14.0	16.0	18.5	1,820	2,000	2,498
WA	80	75	65	26.0	26.0	26.0	2,080	1,950	1,690
WV	20	16	16	14.0	17.0	17.5	280	272	280
WI	745	875	850	16.0	17.5	16.0	11,920	15,313	13,600
WY	31	33	32	20.0	23.0	20.0	620	759	640
US	6,060	5,965	5,605	17.5	18.7	19.3	106,229	111,619	108,209

[1] Preliminary.
NASS, Crops Branch, (202) 720–2127.

Table 1-40.—Corn for grain: Marketing year average price and value, by State and United States, 2007–2009

State	Marketing year average price per bushel			Value of production		
	2007	2008	2009 [1]	2007	2008	2009 [1]
	Dollars	*Dollars*	*Dollars*	*1,000 dollars*	*1,000 dollars*	*1,000 dollars*
AL	4.54	5.26	4.15	99,154	128,554	112,050
AZ	5.03	5.80	4.00	20,472	14,355	14,000
AR	3.80	4.42	3.75	378,898	294,593	227,550
CA	4.28	4.77	4.35	148,002	158,126	125,280
CO	3.96	4.14	3.85	587,664	572,852	583,160
DE	4.76	4.57	3.80	87,179	86,830	89,813
FL	4.00	4.50	4.00	12,600	16,538	14,800
GA	4.50	4.50	3.60	257,175	195,300	186,480
ID	4.96	4.32	4.25	88,536	58,752	61,200
IL	4.09	4.01	3.65	9,340,538	8,541,701	7,537,250
IN	4.39	4.10	3.75	4,306,502	3,581,760	3,501,225
IA	4.29	4.10	3.75	10,196,901	8,974,080	9,145,500
KS	4.13	4.12	3.60	2,097,379	2,004,050	2,153,880
KY	4.14	4.36	3.75	710,093	664,115	711,563
LA	3.80	4.45	3.55	452,162	326,808	285,846
MD	4.64	4.42	4.00	217,918	213,928	246,500
MI	4.37	3.84	3.60	1,257,773	1,134,029	1,118,880
MN	4.13	3.92	3.70	4,733,393	4,628,736	4,629,625
MS	3.68	4.63	3.70	495,622	453,740	324,009
MO	4.17	4.11	3.65	1,909,026	1,568,376	1,630,674
MT	4.76	3.80	4.15	25,323	18,088	16,401
NE	4.14	4.05	3.70	6,094,080	5,644,283	5,828,610
NJ	4.65	4.15	3.40	47,281	35,624	34,034
NM	5.20	5.30	4.00	50,544	52,470	37,000
NY	5.05	4.32	3.95	355,520	398,131	314,934
NC	4.00	4.91	3.85	404,000	317,873	360,360
ND	4.06	3.74	3.40	1,106,756	1,066,648	708,050
OH	4.29	4.21	3.70	2,323,035	1,773,252	2,021,532
OK	4.07	4.46	3.80	159,341	164,128	127,680
OR	4.36	4.15	4.10	30,520	27,390	28,208
PA	4.56	4.16	3.85	554,131	486,886	506,506
SC	3.88	4.59	3.85	139,253	93,980	136,752
SD	4.17	3.78	3.40	2,260,474	2,212,056	2,444,940
TN	3.80	4.53	3.65	318,212	336,760	318,718
TX	4.35	4.82	4.05	1,268,286	1,223,075	1,031,940
UT	4.18	4.40	4.35	13,794	15,888	11,462
VA	4.39	4.51	3.75	152,904	165,607	162,113
WA	4.50	4.56	4.50	108,675	84,132	101,588
WV	4.60	4.34	3.55	13,786	14,669	13,419
WI	4.11	3.89	3.70	1,819,908	1,534,838	1,658,673
WY	3.12	4.25	4.20	24,149	29,614	26,460
US	4.20	4.06	3.70	54,666,959	49,312,615	48,588,665

[1] Preliminary.
NASS, Crops Branch, (202) 720–2127.

GRAIN AND FEED

Table 1-41.—Corn: Area, yield, and production in specified countries, 2007/2008–2009/2010

Country	Area			Yield per hectare			Production		
	2007/ 2008	2008/ 2009	2009/ 2010	2007/ 2008	2008/ 2009	2009/ 2010	2007/ 2008	2008/ 2009	2009/ 2010
	1,000 hec- tares	1,000 hec- tares	1,000 hec- tares	Metric tons	Metric tons	Metric tons	1,000 metric tons	1,000 metric tons	1,000 metric tons
Argentina	3,412	2,500	2,630	6.45	6.00	8.56	22,017	15,000	22,500
Brazil	14,700	14,100	13,000	3.99	3.62	4.08	58,600	51,000	53,000
Canada	1,370	1,169	1,150	8.50	9.06	8.31	11,649	10,592	9,560
China, Peoples	29,478	29,864	30,400	5.17	5.56	5.10	152,300	165,900	155,000
EU-27	8,444	8,785	8,291	5.63	7.09	6.73	47,555	62,321	55,773
India	8,260	8,200	8,000	2.30	2.41	2.16	18,960	19,730	17,300
Mexico	7,330	7,318	6,230	3.22	3.31	3.42	23,600	24,226	21,300
Nigeria	4,000	4,700	4,900	1.63	1.70	1.79	6,500	7,970	8,759
South Africa, Republic	3,300	2,896	3,250	3.99	4.34	4.31	13,164	12,567	14,000
Ukraine	1,900	2,400	2,100	3.89	4.75	5.00	7,400	11,400	10,500
Others	43,326	44,458	43,767	2.32	2.47	2.47	100,693	109,983	108,315
Total foreign	125,520	126,390	123,718	3.68	3.88	3.85	462,438	490,689	476,007
United States	35,014	31,796	32,209	9.46	9.66	10.34	331,177	307,142	333,011
Total	160,534	158,186	155,927	4.94	5.04	5.19	793,615	797,831	809,018

FAS, Office of Global Analysis, (202) 720-6301.

Table 1-42.—Corn: International trade, 2007/2008–2009/2010 [1]

Country	2007/2008	2008/2009	2009/2010
	1,000 metric tons	1,000 Metric tons	1,000 metric tons
Principle exporting countries:			
Argentina	14,798	10,318	14,000
Brazil	7,791	7,136	8,500
EU-27	591	1,743	1,250
India	4,473	2,608	1,000
Paraguay	1,072	1,909	1,000
Russia	49	1,331	400
Serbia	128	1,467	1,500
South Africa, Republic of	2,162	1,671	2,500
Thailand	488	647	1,200
Ukraine	2,074	5,497	5,200
Others	3,075	3,126	3,100
Total Foreign	36,701	37,453	39,650
United States	61,913	46,965	50,167
Total	98,614	84,418	89,817
Principle importing countries:			
Algeria	1,963	2,273	2,300
Colombia	3,267	3,068	3,300
Egypt	4,151	5,031	5,000
EU-27	14,016	2,743	2,500
Iran	2,900	3,600	3,500
Japan	16,614	16,533	16,300
Korea, Republic of	9,311	7,188	8,200
Malaysia	3,181	2,447	2,500
Mexico	9,556	7,764	8,000
Taiwan	4,527	4,532	4,600
Others	28,494	26,924	28,796
Total Foreign	97,980	82,103	84,996
United States	509	82,447	85,199
Total	98,489	158,186	155,927

[1] Year beginning Oct 1.
FAS, Office of Global Analysis, (202) 720-6301. Prepared or estimated on the basis of official USDA production, supply, and distribution statistics from foreign governments.

Table 1-43.—Corn: Support operations, United States, 2000–2009

Marketing year beginning September 1	Income support pay- ment rates per bushel [1]	Program price levels per bushel		Put under loan		Acquired by CCC under loan program [5]	Owned by CCC at end of marketing year
		Loan [2]	Target [3]	Quantity	Percentage of production [4]		
	Dollars	*Dollars*	*Dollars*	*Million bushels*	*Percent*	*Million bushels*	*Million bushels*
2000/2001	0.70	1.89	NA	1,394	14.1	27	8
2001/2002	0.58	1.89	NA	1,395	14.7	0	6
2002/2003	0.28/0.00	1.98	2.60	1,367	15.2	0	4
2003/2004	0.28/0.00	1.98	2.60	1,327	13.2	1	0
2004/2005	0.28/0.29	1.95	2.63	1,366	11.6	25	0.2
2005/2006	0.28/0.35	1.95	2.63	1,064	9.6	2	1.5
2006/2007	0.28/0.00	1.95	2.63	1,108	10.5	0	0
2007/2008	0.28/0.00	1.95	2.63	1,218	9.3	0	0
2008/2009	0.28/0.00	1.95	2.63	1,074	8.9	0	0
2009/2010	0.28/0.00	1.95	2.63	....	....	....	....

[1] Payment rates for the 1998/1999 through 2001/2002 crops were calculated according to the Production Flexibility Contract (PFC) program provisions of the Federal Agriculture Improvement and Reform Act of 1996 (1996 Act) and include supplemental PFC payment rates for 1998 through 2001. Payment rates for the 2002/2003 and subsequent crops are calculated according to the Direct and Counter-cyclical program provisions, following enactment of the Farm Security and Rural Investment Act of 2002 (2002 Act). Beginning with 2002/2003, the first entry is the direct payment rate and the second entry is the counter-cyclical payment rate. [2] Starting in 2009, producers who participate in the Average Crop Revenue Election (ACRE) program get a 30 percent reduction in their loan rate, not calculated in this table. [3] Target prices were reestablished under the 2002 Act. [4] Percentage of production is on a grain basis. [5] Acquisitions for 2008/2009 as of September 1, 2009.
FSA, Feed Grains, (202) 720–7787.

Table 1-44.—Corn: United States exports, specified by country of destination, 2007/2008–2008/2009 [1]

Country of destination	2007	2008	2009
	Metric tons	*Metric tons*	*Metric tons*
Japan	14,951,458	15,121,468	15,257,547
Mexico	8,203,692	9,152,530	7,152,209
Korea, South	4,608,473	7,909,635	6,051,285
Taiwan	4,170,394	3,245,526	3,756,189
Egypt	3,743,932	2,438,333	2,272,827
Canada	2,507,834	2,627,733	1,898,720
Venezuela	527,179	1,142,314	1,294,919
Colombia	3,105,336	2,567,149	1,231,183
Dominican Republic	1,144,579	1,041,543	964,204
Rest of World	13,875,018	8,491,176	7,755,398
World Total	56,837,895	53,737,407	47,634,481

[1] Compiled from U.S. Census data. Excludes seed, popcorn.
FAS, Office of Global Analysis, (202) 720-6301.

Table 1-45.—Oats: Area, yield, production, and value, United States, 2000–2009

Year	Area		Yield per harvested acre	Production	Marketing year average price per bushel received by farmers	Value of production
	Planted [1]	Harvested				
	1,000 acres	*1,000 acres*	*Bushels*	*1,000 bushels*	*Dollars*	*1,000 dollars*
2000	4,473	2,325	64.2	149,165	1.10	175,432
2001	4,401	1,911	61.5	117,602	1.59	197,181
2002	4,995	2,058	56.4	116,002	1.81	212,078
2003	4,597	2,220	65.0	144,383	1.48	224,910
2004	4,085	1,787	64.7	115,695	1.48	178,327
2005	4,246	1,823	63.0	114,859	1.63	195,166
2006	4,166	1,564	59.8	93,522	1.87	180,899
2007	3,763	1,504	60.1	90,430	2.63	247,644
2008	3,247	1,400	63.7	89,135	3.15	269,763
2009 [2]	3,404	1,379	67.5	93,081	2.10	216,566

[1] Relates to the total area of oats sown for all purposes, including oats sown in the preceding fall. [2] Preliminary.
NASS, Crops Branch, (202) 720–2127.

Table 1-46.—Oats: Stocks on and off farms, United States, 2000–2009

Year beginning September	On farms				Off farms [1]			
	Sep. 1	Dec. 1	Mar. 1	Jun. 1	Sep. 1	Dec. 1	Mar. 1	Jun. 1
2000	101,200	86,900	55,800	32,050	49,177	57,237	54,128	40,677
2001	74,800	58,100	40,200	28,650	41,592	56,117	53,158	34,552
2002	70,500	52,500	35,000	20,600	41,212	51,284	47,879	29,233
2003	82,100	64,400	45,600	27,500	49,637	54,900	49,414	37,348
2004	74,300	60,400	43,500	25,350	41,458	44,513	38,946	32,592
2005	71,700	60,100	42,200	25,190	41,803	35,617	32,673	27,376
2006	60,800	53,000	33,900	18,400	39,284	45,889	37,158	32,198
2007	53,650	43,100	31,000	16,100	34,710	51,331	47,988	50,674
2008	52,800	42,600	30,200	17,480	66,296	72,322	65,250	66,619
2009 [2]	54,500	43,000	30,900	NA	73,875	67,629	67,126	NA

[1] Inlcudes stocks at mills, elevators, warehouses, terminals, and processors. [2] Preliminary. NA-not available.
NASS, Crops Branch, (202) 720–2127.

Table 1-47.—Oats: Supply and disappearance, United States, 2001–2010

Year beginning June 1	Supply				Disappearance					Ending stocks May 31		
	Beginning stocks	Production	Imports	Total	Domestic use			Exports	Total disappearance	Privately held [1]	Government	Total
					Feed and residual	Food, seed and industrial	Total					
	Million bushels	*Million bushels*	*Million bushels*	*Million bushels*	*Million bushels*	*Million bushels*	*Million bushels*	*Million bushels*	*Million bushels*	*Million bushels*	*Million bushels*	*Million bushels*
2001 ..	73	118	96	286	148	72	220	3	223	63	0	63
2002 ..	63	116	95	274	150	72	222	3	224	50	0	50
2003 ..	50	144	90	284	144	73	217	2	219	65	0	65
2004 ..	65	116	90	271	136	74	210	3	213	58	0	58
2005 ..	58	115	91	264	136	74	209	2	211	53	0	53
2006 ..	53	94	106	252	125	74	199	3	202	51	0	51
2007 ..	51	90	123	264	120	74	195	3	198	67	0	67
2008 ..	67	89	115	270	108	75	183	3	186	84	0	84
2009 ..	84	93	95	272	115	75	190	2	192	80	0	80
2010 [2]	80	87	80	248	115	76	191	3	194	54	0	54

[1] Includes quantity under loan and farmer-owned reserve. [2] Projected as of January 11, 2010, World Agricultural Supply and Demand Estimates. Totals may not add due to independent rounding.
ERS, Market and Trade Economics Division, (202) 694–5296.

Table 1-48.—Oats: Support operations, United States, 2000–2009

Marketing Year beginning June 1	Income support payment rates per bushel [1]	Program price levels per bushel		Put under loan		Acquired by CCC under loan program [5]	Owned by CCC at end of marketing year
		Loan [2]	Target [3]	Quantity	Percentage of production [4]		
	Dollars	Dollars	Dollars	Million bushels	Percent	Million bushels	Million bushels
2000/2001	0.06	1.16	NA	1.7	1.1	0.1	0.0
2001/2002	0.05	1.21	NA	1.7	1.5	0.0	0.0
2002/2003	0.02/0.00	1.35	1.40	2.0	1.7	0.0	0.0
2003/2004	0.02/0.00	1.35	1.40	5.2	3.6	0.4	0.0
2004/2005	0.02/0.00	1.33	1.44	3.3	2.9	0.1	0.0
2005/2006	0.02/0.00	1.33	1.44	3.0	2.6	0.0	0.0
2006/2007	0.02/0.00	1.33	1.44	1.7	1.8	0.0	0.0
2007/2008	0.02/0.00	1.33	1.44	1.2	1.3	0.0	0.0
2008/2009	0.02/0.00	1.33	1.44	1.1	1.3	....	....
2009/2010	0.02/0.00	1.33	1.44	....	....	....	....

[1] Payment rates for the 1998/1999 through 2001/2002 crops were calculated according to the Production Flexibility Contract (PFC) program provisions of the Federal Agriculture Improvement and Reform Act of 1996 (1996 Act) and include supplemental PFC payment rates for 1998 through 2001. Payment rates for the 2002/2003 and subsequent crops are calculated according to the Direct and Counter-cyclical program provisions, following enactment of the Farm Security and Rural Investment Act of 2002 (2002 Act). Beginning with 2002/2003, the first entry is the direct payment rate and the second entry is the counter-cyclical payment rate.　[2] Starting in 2009, producers who participate in the optimal Average Crop Revenue Election (ACRE) program get a 30 percent reduction in their loan rate, not calculated in this table.　[3] Target prices were reestablished under the 2002 Act.　[4] Percentage of production is on a grain basis.　[5] Acquisitions for 2008/2009 as of June 1, 2009.　NA-not applicable.
FSA, Feed Grains, (202) 720–7787.

Table 1-49.—Oats: Area, yield, and production, by State and United States, 2007–2009

State	Area planted [1]			Area harvested			Yield per harvested acre			Production		
	2007	2008	2009	2007	2008	2009	2007	2008	2009	2007	2008	2009
	1,000 acres	1,000 acres	1,000 acres	1,000 acres	1,000 acres	1,000 acres	Bushels	Bushels	Bushels	1,000 bushels	1,000 bushels	1,000 bushels
AL	45	50	50	16	15	11	58.0	50.0	50.0	928	750	550
AR [2]	....	....	10	....	....	8	....	....	80.0	....	....	640
CA	215	260	250	25	20	30	99.0	80.0	105.0	2,475	2,000	3,150
CO	75	45	60	10	7	9	55.0	70.0	65.0	550	490	585
GA	70	65	60	30	25	20	56.0	69.0	56.0	1,680	1,725	1,120
ID	70	70	80	20	20	25	61.0	69.0	78.0	1,220	1,380	1,950
IL	35	45	40	24	30	25	62.0	70.0	65.0	1,488	2,100	1,625
IN	25	15	15	8	5	7	53.0	75.0	69.0	424	375	483
IA	145	150	200	67	75	95	71.0	65.0	65.0	4,757	4,875	6,175
KS	90	60	85	35	25	35	45.0	53.0	53.0	1,575	1,325	1,855
ME	29	32	32	28	31	31	70.0	65.0	65.0	1,960	2,015	2,015
MI	70	75	70	55	60	55	56.0	66.0	63.0	3,080	3,960	3,465
MN	270	250	250	180	175	170	60.0	68.0	71.0	10,800	11,900	12,070
MO	25	15	15	8	6	9	50.0	55.0	55.0	400	330	495
MT	75	60	70	35	30	32	50.0	51.0	56.0	1,750	1,530	1,792
NE	120	95	100	35	35	30	61.0	70.0	69.0	2,135	2,450	2,070
NY	100	80	90	60	64	60	58.0	66.0	77.0	3,480	4,224	4,620
NC	50	60	50	15	30	15	55.0	80.0	70.0	825	2,400	1,050
ND	460	320	350	260	130	165	59.0	51.0	68.0	15,340	6,630	11,220
OH	75	75	65	50	50	45	62.0	70.0	75.0	3,100	3,500	3,375
OK	80	50	50	15	10	15	31.0	40.0	34.0	465	400	510
OR	60	45	45	18	18	22	78.0	100.0	100.0	1,404	1,800	2,200
PA	115	105	110	80	80	80	56.0	58.0	61.0	4,480	4,640	4,880
SC	33	33	30	14	19	15	42.0	64.0	55.0	588	1,216	825
SD	330	220	200	130	120	90	72.0	73.0	73.0	9,360	8,760	6,570
TX	710	600	600	100	100	60	40.0	50.0	47.0	4,000	5,000	2,820
UT	35	40	45	4	4	5	80.0	75.0	81.0	320	300	405
VA	16	12	12	5	4	4	60.0	70.0	54.0	300	280	216
WA	30	20	20	9	5	6	50.0	80.0	80.0	450	400	480
WI	270	270	310	160	190	195	67.0	62.0	68.0	10,720	11,780	13,260
WY	40	30	40	8	12	10	47.0	50.0	61.0	376	600	610
US	3,763	3,247	3,404	1,504	1,400	1,379	60.1	63.7	67.5	90,430	89,135	93,081

[1] Relates to the total area of oats sown for all purposes, including oats sown in the preceding fall.　[2] Estimates began in 2009.
NASS, Crops Branch, (202) 720–2127.

Table 1-50.—Oats: Marketing year average price and value of production, by State and United States, 2007–2009

State	Marketing year average price per bushel			Value of production		
	2007	2008	2009 [1]	2007	2008	2009 [1]
	Dollars	*Dollars*	*Dollars*	*1,000 dollars*	*1,000 dollars*	*1,000 dollars*
AL	2.40	2.65	2.10	2,227	1,988	1,155
AR [2]	....	....	2.30	....	....	1,472
CA	3.05	4.00	3.25	7,549	8,000	10,238
CO	3.25	3.30	2.30	1,788	1,617	1,346
GA	2.50	2.65	2.00	4,200	4,571	2,240
ID	2.40	2.95	2.45	2,928	4,071	4,778
IL	3.69	3.04	2.80	5,491	6,384	4,550
IN	3.75	3.90	2.80	1,590	1,463	1,352
IA	2.74	3.27	2.10	13,034	15,941	12,968
KS	2.73	2.94	2.05	4,300	3,896	3,803
ME	2.25	2.30	1.70	4,410	4,635	3,426
MI	2.91	3.40	2.25	8,963	13,464	7,796
MN	2.49	2.58	1.90	26,892	30,702	22,933
MO	2.85	3.90	2.70	1,140	1,287	1,337
MT	2.76	3.07	2.85	4,830	4,697	5.107
NE	3.14	3.46	2.35	6,704	8,477	4,865
NY	2.69	3.07	2.10	9,361	12,968	9,702
NC	2.50	3.10	2.70	2,063	7,440	2,835
ND	2.53	2.70	2.05	38,810	17,901	23,001
OH	2.43	3.92	2.60	7,533	13,720	8,775
OK	3.40	3.50	2.70	1,581	1,400	1,377
OR	3.89	2.74	3.05	5,462	4,932	6,710
PA	3.20	3.23	2.80	14,336	14,987	13,664
SC	2.35	2.95	2.30	1,382	3,587	1,898
SD	2.87	2.68	2.15	26,863	23,477	14,126
TX	3.47	4.00	4.70	13,880	20,000	13,254
UT	2.65	3.20	2.50	848	960	1,013
VA	2.55	2.82	2.40	765	790	518
WA	2.85	3.08	2.80	1,283	1,232	1,344
WI	2.46	2.82	2.05	26,371	33,220	27,183
WY	2.82	3.26	2.95	1,060	1,956	1,800
US	2.63	3.15	2.10	247,644	269,763	216,566

[1] Preliminary.   [2] Estimates began in 2009.
NASS, Crops Branch, (202) 720–2127.

Table 1-51.—Oats: Area, yield, and production in specified countries, 2007/2008–2009/2010

Country	Area			Yield per hectare			Production		
	2007/ 2008	2008/ 2009	2009/ 2010	2007/ 2008	2008/ 2009	2009/ 2010	2007/ 2008	2008/ 2009	2009/ 2010
	1,000 hec- tares	1,000 hec- tares	1,000 hec- tares	Metric tons	Metric tons	Metric tons	1,000 metric tons	1,000 metric tons	1,000 metric tons
Argentina	224	212	250	2.11	1.37	2.00	472	291	500
Australia	1,238	870	920	1.21	1.33	1.37	1,502	1,160	1,260
Belarus	250	180	250	2.40	3.33	3.20	600	600	800
Brazil	350	350	350	1.36	1.36	1.36	475	475	475
Canada	1,816	1,448	950	2.59	2.95	2.95	4,696	4,273	2,800
Chile	98	101	68	3.92	3.41	5.07	384	344	345
China, Peoples	500	500	500	1.20	1.20	1.20	600	600	600
EU-27	3,003	2,993	2,908	2.88	3.00	2.93	8,634	8,975	8,517
Russian Federation	3,700	3,700	3,350	1.46	1.57	1.61	5,400	5,800	5,400
Ukraine	350	450	420	1.57	2.11	1.74	550	950	730
Others	1,399	1,432	1,471	0.93	0.96	0.94	1,297	1,373	1,378
Total Foreign	12,928	12,236	11,437	1.90	2.03	1.99	24,610	24,841	22,805
United States	609	567	558	2.16	2.28	2.42	1,313	1,294	1,351
Total	13,537	12,803	11,995	1.92	2.04	2.01	25,923	26,135	24,156

FAS, Office of Global Analysis, (202) 720-6301. Prepared or estimated on the basis of official USDA production, supply, and distribution, supply, and and distribution statistics from foreign governments.

Table 1-52. Oats:[1] International trade, 2007/2008–2009/2010[2]

Country	2007/2008	2008/2009	2009/2010[3]
	1,000 metric tons	1,000 metric tons	1,000 metric tons
Principle exporting countries:			
Argentina	1	5	5
Australia	181	161	175
Brazil	5	9	5
Canada	2,386	1,942	1,450
Chile	36	26	32
EU-27	158	92	225
Russia	7	2	5
Ukraine	24	6	5
Others	15		
Total foreign	2,813	2,243	1,902
United States	42	49	31
Total	2,855	2,292	1,933
Principle importing countries:			
Argentina	1	3	5
Canada	17	16	15
China, Peoples Republic of	13	39	50
Colombia	14	2	10
Ecuador	15	9	10
Japan	68	46	60
Mexico	111	61	100
Norway	53	6	30
South Africa, Republic of	18	33	25
Switzerland	56	52	50
Others	44	35	28
Total foreign	410	302	383
United States	2,125	1,975	1,636
Total	2,535	2,277	2,019

[1] Flour and products reported in terms of grain equivalent.   [2] Year beginning July 1.   [3] Preliminary.
FAS, Office of Global Analysis, (202) 720-6301. Prepared or estimated on the basis of official USDA production, supply, and distribution statistics from foreign governments.

GRAIN AND FEED

Table 1-53.—Barley: Area, yield, production, and value, United States, 2000–2009

Year	Area		Yield per harvested acre	Production	Marketing year average price per bushel received by farmers	Value of production
	Planted [1]	Harvested				
	1,000 acres	*1,000 acres*	*Bushels*	*1,000 bushels*	*Dollars*	*1,000 dollars*
2000	5,801	5,200	61.1	317,804	2.11	647,966
2001	4,951	4,273	58.1	248,329	2.22	535,110
2002	5,008	4,123	55.0	226,906	2.72	605,635
2003	5,348	4,727	58.9	278,283	2.83	755,140
2004	4,527	4,021	69.6	279,743	2.48	698,184
2005	3,875	3,269	64.8	211,896	2.53	527,633
2006	3,452	2,951	61.1	180,165	2.85	498,691
2007	4,018	3,502	60.0	210,110	4.02	834,954
2008	4,246	3,779	63.6	240,193	5.37	1,259,357
2009 [2]	3,567	3,113	73.0	227,323	4.40	917,500

[1] Barley sown for all purposes, including barley sown in the preceding fall. [2] Preliminary.
NASS, Crops Branch, (202) 720–2127.

Table 1-54.—Barley: Stocks on and off farms, United States, 2000–2009

Year beginning September	On farms				Off farms [1]			
	Sep. 1	Dec. 1	Mar. 1	June 1	Sep. 1	Dec. 1	Mar. 1	June 1
	1,000 bushels	*1,000 bushels*	*1,000 bushels*	*1,000 bushels*	*1,000 bushels*	*1,000 bushels*	*1,000 bushels*	*1,000 bushels*
2000	151,700	111,500	58,600	28,850	142,341	117,369	103,544	77,409
2001	134,800	92,400	46,000	23,210	110,564	102,587	95,748	68,919
2002	131,300	83,400	36,730	14,860	92,419	86,601	86,710	54,480
2003	141,900	97,200	51,700	28,320	99,730	100,679	101,186	91,988
2004	175,300	130,700	79,680	41,100	114,777	115,276	111,001	87,317
2005	137,400	103,650	68,400	30,770	117,511	104,335	98,354	77,161
2006	112,850	83,650	38,310	14,580	99,939	89,171	78,756	54,300
2007	105,600	62,050	28,270	9,950	83,095	73,728	82,154	58,273
2008	127,750	77,050	44,310	27,010	81,669	95,766	84,791	61,723
2009 [2]	154,050	114,630	67,370	NA	85,414	91,759	90,029	NA

[1] Includes stocks at mills, elevators, warehouses, terminals, and processors. [2] Preliminary. NA-not available.
NASS, Crops Branch, (202) 720–2127.

Table 1-55.—Barley: Supply and disappearance, United States, 2001–2010

Year beginning June 1	Supply				Disappearance					Ending stocks May 31		
	Beginning stocks	Production	Imports	Total	Domestic use			Exports	Total disappearance	Privately held [1]	Government	Total
					Feed and residual	Food, seed, and industrial	Total					
	Million bushels	*Million bushels*	*Million bushels*	*Million bushels*	*Million bushels*	*Million bushels*	*Million bushels*	*Million bushels*	*Million bushels*	*Million bushels*	*Million bushels*	*Million bushels*
2001	106	248	24	379	104	156	260	26	286	92	0	92
2002	92	227	18	337	84	154	238	30	268	69	0	69
2003	69	278	21	368	74	155	229	19	248	120	0	120
2004	120	280	12	412	103	158	261	23	284	128	0	128
2005	128	212	5	346	48	162	210	28	238	108	0	108
2006	108	180	12	300	49	162	211	20	231	69	0	69
2007	69	210	29	308	30	169	199	41	240	68	0	68
2008	68	240	29	337	67	169	236	13	249	89	0	89
2009 [2] ...	89	227	17	333	48	164	212	6	218	115	0	115
2010 [3] ...	115	184	15	314	50	165	215	10	225	89	0	89

[1] Includes quantity under loan and farmer–owned reserve. [2] Preliminary. [3] Projected as of January 11, 2010, World Agricultural Supply and Demand Estimates. Totals may not add due to independent rounding.
ERS, Market and Trade Economics Division, (202) 694–5296.

Table 1-56.—Barley: Area, yield, and production, by State and United States, 2007–2009

State	Area planted [1]			Area harvested			Yield per harvested acre			Production		
	2007	2008	2009 [2]	2007	2008	2009 [2]	2007	2008	2009 [2]	2007	2008	2009 [2]
	1,000 acres	1,000 acres	1,000 acres	1,000 acres	1,000 acres	1,000 acres	Bushels	Bushels	Bushels	1,000 bushels	1,000 bushels	1,000 bushels
AZ	33	42	48	31	40	45	110.0	120.0	115.0	3,410	4,800	5,175
CA	85	95	90	40	60	55	64.0	55.0	54.0	2,560	3,300	2,970
CO	60	80	78	58	72	77	120.0	120.0	135.0	6,960	8,640	10,395
DE	21	25	28	19	22	26	78.0	80.0	70.0	1,482	1,760	1,820
ID	570	600	530	550	580	510	78.0	86.0	95.0	42,900	49,880	48,450
KS	20	17	14	13	10	9	52.0	37.0	51.0	676	370	459
KY³	10	8	....	3	7	....	37.0	88.0	....	111	616	....
ME	18	20	16	17	19	15	65.0	55.0	55.0	1,105	1,045	825
MD	45	45	55	30	35	48	82.0	90.0	70.0	2,460	3,150	3,360
MI	14	12	13	13	10	11	51.0	46.0	51.0	663	460	561
MN	130	125	95	110	110	80	54.0	65.0	61.0	5,940	7,150	4,880
MT	900	860	870	720	740	720	44.0	51.0	57.0	31,680	37,740	41,040
NV³	3	3	....	1	1	....	90.0	100.0	....	90	100	....
NJ³	3	3	....	2	2	....	68.0	71.0	....	136	142	....
NY	13	13	12	11	9	10	49.0	52.0	53.0	539	468	530
NC	22	21	23	14	14	19	49.0	71.0	60.0	686	994	1,140
ND	1,470	1,650	1,210	1,390	1,540	1,130	56.0	56.0	70.0	77,840	86,240	79,100
OH³	4	6	....	3	5	....	53.0	72.0	....	159	360	....
OR	63	57	40	53	42	32	53.0	50.0	60.0	2,809	2,100	1,920
PA	55	60	60	42	55	45	73.0	75.0	75.0	3,066	4,125	3,375
SD	56	63	48	29	43	22	40.0	41.0	54.0	1,160	1,763	1,188
UT	38	40	40	22	27	30	81.0	85.0	85.0	1,782	2,295	2,550
VA	48	63	67	30	36	43	71.0	85.0	74.0	2,130	3,060	3,182
WA	235	205	105	225	·195	97	62.0	57.0	64.0	13,950	11,115	6,208
WI	40	43	45	23	30	25	57.0	54.0	59.0	1,311	1,620	1,475
WY	62	90	80	53	75	64	85.0	92.0	105.0	4,505	6,900	6,720
US	4,018	4,246	3,567	3,502	3,779	3,113	60.0	63.6	73.0	210,110	240,193	227,323

[1] Includes area planted in the preceding fall. [2] Preliminary. [3] Estimates discontinued in 2009.
NASS, Crops Branch, (202) 720–2127.

Table 1-57.—Barley: Marketing year average price and value, by State and United States, 2007–2009

State	Marketing year average price per bushel			Value of production		
	2007	2008	2009 [1]	2007	2008	2009 [1]
	Dollars	Dollars	Dollars	1,000 dollars	1,000 dollars	1,000 dollars
AZ	4.00	4.80	3.70	13,640	23,040	19,148
CA	4.52	6.15	3.40	11,571	20,295	10,098
CO	3.51	5.18	5.15	24,430	44,755	53,534
DE	2.49	4.18	2.20	3,690	7,357	4,004
ID	4.02	5.86	4.80	172,458	292,297	232,560
KS	3.65	4.20	2.15	2,467	1,554	987
KY²	2.85	3.90	....	316	2,402	....
ME	2.94	3.55	2.75	3,249	3,710	2,269
MD	2.70	3.99	2.40	6,642	12,569	8,064
MI	2.50	3.25	2.80	1,658	1,495	1,571
MN	3.80	5.27	3.65	22,572	37,681	17,812
MT	4.14	5.78	4.70	131,155	218,137	192,888
NV²	4.00	5.78	....	360	578	....
NJ²	2.70	4.20	....	367	596	....
NY	2.76	4.75	3.50	1,488	2,223	1,855
NC	2.65	4.00	2.60	1,818	3,976	2,964
ND	3.91	5.18	3.55	304,354	446,723	280,805
OH²	3.10	3.90	....	493	1,404	....
OR	5.11	4.01	2.55	14,354	8,421	4,896
PA	2.90	4.50	2.80	8,891	18,563	9,450
SD	4.55	5.06	3.05	5,278	8,921	3,623
UT	3.99	4.41	2.25	7,110	10,121	5,738
VA	2.76	4.22	· 2.35	5,879	12,913	7,478
WA	5.08	3.49	2.60	70,866	38,791	16,141
WI	2.70	3.57	2.70	3,540	5,783	3,983
WY	3.62	5.08	5.60	16,308	35,052	37,632
US	4.02	5.37	4.40	834,954	1,259,357	917,500

[1] Preliminary. [2] Estimates discontinued in 2009.
NASS, Crops Branch, (202) 720–2127.

Table 1-58.—Barley: Area, yield, and production in specified countries, 2007/2008–2009/2010

Country	Area			Yield per hectare			Production		
	2007/ 2008	2008/ 2009	2009/ 2010	2007/ 2008	2008/ 2009	2009/ 2010	2007/ 2008	2008/ 2009	2009/ 2010
	1,000 hec- tares	*1,000 hec- tares*	*1,000 hec- tares*	*Metric tons*	*Metric tons*	*Metric tons*	*1,000 metric tons*	*1,000 metric tons*	*1,000 metric tons*
Australia	4,902	5,015	4,479	1.46	1.59	1.85	7,160	7,997	8,300
Canada	3,998	3,502	2,920	2.75	3.36	3.26	10,984	11,781	9,520
China	773	794	715	3.60	3.56	3.50	2,785	2,823	2,500
EU-27	13,797	14,505	13,853	4.17	4.51	4.42	57,545	65,452	61,255
Iran	1,700	1,300	1,400	1.76	1.54	1.86	3,000	2,000	2,600
Kazakhstan	1,800	2,100	2,000	1.39	0.86	1.30	2,500	1,800	2,600
Morocco	1,993	2,181	2,183	0.38	0.58	1.74	763	1,272	3,800
Russia	9,600	9,600	9,050	1.63	2.41	1.98	15,650	23,100	17,900
Turkey	3,600	3,400	3,400	1.67	1.68	1.76	6,000	5,700	6,000
Ukraine	4,100	4,150	5,000	1.46	3.04	2.36	6,000	12,600	11,800
Others	9,571	7,800	9,142	1.66	1.98	1.92	15,925	15,433	17,528
Total Foreign	55,834	54,347	54,142	2.30	2.76	2.66	128,312	149,958	143,803
United States	1,417	1,529	1,260	3.23	3.42	3.93	4,575	5,230	4,949
Total	57,251	55,876	55,402	2.32	2.78	2.69	132,887	155,188	148,752

FAS, Office of Global Analysis, (202) 720-6301. Prepared or estimated on the basis of official USDA production, supply, and distribution, supply, and and distribution statistics from foreign governments.

Table 1-59.—Barley:[1] International trade, 2007/2008–2009/2010 [2]

Country	2007/2008	2008/2009	2009/2010 [3]
	1,000 metric tons	*1,000 metric tons*	*1,000 metric tons*
Principle exporting countries:			
Argentina	911	1,018	600
Australia	3,386	3,234	3,700
Canada	3,046	1,483	1,300
EU-27	3,803	3,597	1,400
India	348	167	150
Kazakhstan, Republic of	792	291	300
Moldova	6	24	25
Russian Federation	1,046	3,444	2,600
Turkey		2	800
Ukraine	1,044	6,371	6,232
Others	198	108	55
Total foreign	14,580	19,739	17,162
United States	902	288	123
Total	15,482	20,027	17,285
Principle importing countries:			
Brazil	279	454	400
China, Peoples Republic of	1,091	1,551	1,900
Colombia	256	211	250
Iran	300	1,900	800
Israel	251	367	425
Japan	1,361	1,346	1,350
Jordan	592	546	500
Libya	166	296	400
Saudi Arabia	7,400	7,200	7,900
Syria	150	1,750	400
Others	3,281	3,099	2,097
Total foreign	15,127	18,720	16,422
United States	636	632	361
Total	15,763	19,352	16,783

[1] Flour and products reported in terms of grain equivalent.   [2] Year beginning July 1.   [3] Preliminary.
FAS, Office of Global Analysis, (202) 720-6301. Prepared or estimated on the basis of official USDA production, supply, and distribution statistics from foreign governments.

Table 1-60.—Grains and grain products: Total and per capita civilian consumption as food, United States, 1999–2008

Calendar year [1]	Wheat			Rye		Rice (milled)	
	Total consumed [2]	Per capita consumption of food products		Total consumed [2]	Per capita consumption of rye flour	Total consumed [4]	Per capita consumption
		Flour [3]	Non-milled product				
	Million bushels	Pounds	Pounds	Million bushels	Pounds	Million cwt.	Pounds
1999	920	144	2.6	3.3	0.5	59.3	21.1
2000	951	146	2.6	3.3	0.5	60.7	21.4
2001	934	141	2.5	3.3	0.5	62.6	21.8
2002	913	137	2.5	3.3	0.5	65.3	22.5
2003	919	137	2.5	3.3	0.5	65.9	22.6
2004	905	134	2.4	3.3	0.5	67.8	23.0
2005	917	134	2.4	3.3	0.5	66.1	22.2
2006	938	136	2.4	3.3	0.5	66.2	22.0
2007	948	138	2.4	3.3	0.5	65.9	21.7
2008 [9]	927	137	2.4	3.3	0.4	68.4	22.4

Calendar year [1]	Corn						Oats		Barley	
	Total consumed [5]	Per capita consumption of food products					Total consumed [6]	Per capita consumption of oat food products	Total consumed [7]	Per capita consumption of food products [8]
		Flour and meal	Hominy and grits	Syrup	Dextrose	Starch				
	Million bushels	Pounds	Pounds	Pounds	Pounds	Pounds	Million bushels	Pounds	Million bushels	Pounds
1999	984	17.3	5.8	80.0	3.5	4.7	56.8	4.4	6.4	0.7
2000	970	17.5	6.2	78.5	3.4	4.7	56.7	4.3	6.4	0.7
2001	981	17.8	6.6	78.1	3.3	4.6	59.2	4.5	6.3	0.7
2002	976	18.1	7.0	78.3	3.3	4.6	60.2	4.5	6.4	0.7
2003	986	18.3	7.4	76.2	3.1	4.6	62.4	4.6	6.5	0.7
2004	973	18.6	7.8	75.6	3.3	4.5	63.0	4.6	6.6	0.7
2005	989	18.8	8.1	74.5	3.2	4.5	62.9	4.6	6.7	0.7
2006	980	19.0	8.5	72.1	3.1	4.4	64.5	4.6	6.6	0.7
2007	958	19.1	8.9	70.0	3.0	4.4	66.0	4.7	6.7	0.7
2008 [9]	....	19.3	9.3	66.6	2.8	....	67.6	4.8	6.8	0.7

[1] Data are in marketing year; for corn, September 1-August 31; for oats and barley, June 1-May 31; and rice, August 1-July 31. Wheat, rye, syrup, and sugar are in calendar year. [2] Excludes quantities used in alcoholic beverages. [3] Includes white, whole wheat, and semolina flour. [4] Does not include shipments to U.S. territories. Excludes rice used in alcoholic beverages. Includes imports and rice used in processed foods and pet foods. [5] Includes an allowance for the quantity used as hominy and grits. This series is not adjusted for trade. [6] Oats used in oatmeal, prepared breakfast foods, infant foods, and food products. [7] Malt for food, breakfast food uses, pearl barley, and flour. [8] Malt equivalent of barley food products. [9] Preliminary. Estimates of corn syrup and sugar are unofficial estimates; industry data were not reported after April 1968.

ERS, Market & Trade Economics Division, (202) 694-5290. All figures are estimates based on data from private industry sources, the U.S. Department of Commerce, the Internal Revenue Service, and other Government agencies.

Table 1-61.—Barley: Support operations, United States, 2000–2009

Marketing year beginning June 1	Income support payment rates per bushel [1]	Program price levels per bushel		Put under loan		Acquired by CCC under loan program [5]	Owned by CCC at end of marketing year
		Loan [2]	Target [3]	Quantity	Percentage of production [4]		
	Dollars	Dollars	Dollars	Million bushels	Percent	Million bushels	Million bushels
2000/2001	0.52	1.62	NA	16.0	5.0	0.7	0.1
2001/2002	0.44	1.65	NA	10.6	4.2	0.1	0.0
2002/2003	0.24/0.00	1.88	2.21	10.4	4.6	0.0	0.0
2003/2004	0.24/0.00	1.88	2.21	17.9	6.4	0.3	0.0
2004/2005	0.24/0.15	1.85	2.24	8.3	3.0	0.3	0.0
2005/2006	0.24/0.13	1.85	2.24	12.0	5.7	0.1	0.0
2006/2007	0.24/0.00	1.85	2.24	9.3	5.1	0.0	0.0
2007/2008	0.24/0.00	1.85	2.24	4.4	2.1	0.0	0.0
2008/2009	0.24/0.00	1.85	2.24	6.8	2.9	....	....
2009/2010	0.24/0.00	1.85	2.24	....	....	....	....

[1] Payment rates for the 1998/1999 through 2001/2002 crops were calculated according to the Production Flexibility Contract (PFC) program provisions of the Federal Agriculture Improvement and Reform Act of 1996 (1996 Act) and include supplemental PFC payment rates for 1998 through 2001. Payment rates for the 2002/2003 and subsequent crops are calculated according to the Direct and Counter-cyclical program provisions, following enactment of the Farm Security and Rural Investment Act of 2002 (2002 Act). Beginning with 2002/2003, the first entry is the direct payment rate and the second entry is the counter-cyclical payment rate. [2] Starting in 2009, producers who participate in the Average Crop Revenue Election (ACRE) program get a 30 percent reduction in their loan rate, not calculated in this table. [3] Target prices were reestablished under the 2002 Act. [4] Percentage of production is on a grain basis. [5] Acquisitions for 2008/2009 as of June 1, 2009. NA-not applicable.

FSA, Feed Grains, (202) 720-7787.

Table 1-62.—Sorghum: Area, yield, production, and value, United States, 2000–2009

Year	Area planted for all purposes [1]	Sorghum for grain [2]					Sorghum for silage		
		Area harvested	Yield per harvested acre	Production	Marketing year average price per cwt [3]	Value of production [3]	Area harvested	Yield per harvested acre	Production
	1,000 acres	1,000 acres	Bushels	1,000 bushels	Dollars	1,000 dollars	1,000 acres	Tons	1,000 tons
2000	9,195	7,726	60.9	470,526	3.37	845,755	278	10.5	2,932
2001	10,248	8,579	59.9	514,040	3.46	978,783	352	11.0	3,860
2002	9,589	7,125	50.6	360,713	4.14	855,140	408	9.6	3,913
2003	9,420	7,798	52.7	411,219	4.26	964,978	343	10.4	3,558
2004	7,486	6,517	69.6	453,606	3.19	843,344	352	13.6	4,782
2005	6,454	5,736	68.5	392,739	3.33	736,629	311	13.6	4,224
2006	6,522	4,937	56.1	276,824	5.88	883,204	347	13.3	4,612
2007	7,712	6,792	73.2	497,445	7.28	1,925,312	392	13.4	5,246
2008	8,284	7,271	65.0	472,342	5.72	1,631,065	408	13.8	5,646
2009 [4]	6,633	5,520	69.4	382,983	5.90	1,242,196	254	14.5	3,680

[1] Grain and sweet sorghum for all uses, including sirup.　[2] Includes both grain sorghum for grain, and sweet sorghum for grain or seed.　[3] Based on the reported price of grain sorghum.　[4] Preliminary.
NASS, Crops Branch, (202) 720–2127.

Table 1-63.—Sorghum grain: Stocks on and off farms, United States, 2001–2010

Year beginning previous Dec.	On farms				Off farms [1]			
	Dec. 1	Mar. 1	Jun. 1	Sep. 1	Dec. 1	Mar. 1	Jun. 1	Sep. 1
	1,000 bushels	1,000 bushels	1,000 bushels	1,000 bushels	1,000 bushels	1,000 bushels	1,000 bushels	1,000 bushels
2001	74,300	40,100	19,000	8,900	187,681	127,027	57,411	32,851
2002	72,400	38,100	17,300	7,400	241,477	156,007	88,178	53,573
2003	53,600	27,500	11,150	4,500	178,252	135,423	70,744	38,530
2004	45,200	21,000	7,650	3,700	190,736	137,652	72,944	29,849
2005	78,700	33,400	16,000	5,900	203,505	170,122	97,170	51,041
2006	55,000	26,200	12,650	5,250	235,376	166,936	102,213	60,413
2007	38,100	17,100	5,380	2,150	174,094	125,122	69,490	29,903
2008	51,400	26,100	7,000	3,550	239,850	159,808	94,019	49,200
2009	54,400	32,200	12,000	4,400	243,290	173,650	90,215	50,312
2010 [2]	48,000	NA	NA	NA	202,733	NA	NA	NA

[1] Includes stocks at mills, elevators, warehouses, terminals, and processors.　[2] Preliminary.　NA-not available.
NASS, Crops Branch, (202) 720–2127.

Table 1-64.—Sorghum: Supply and disappearance, United States, 2001–2010

Year beginning September 1	Supply			Disappearance						Ending stocks Aug. 31		
	Beginning stocks	Production	Total	Domestic use			Exports	Total disappearance		Privately owned [1]	Government	Total
				Feed and residual	Food, seed and industrial	Total						
	Million bushels	Million bushels	Million bushels	Million bushels	Million bushels	Million bushels	Million bushels	Million bushels		Million bushels	Million bushels	Million bushels
2001	42	514	556	230	23	253	242	495		61	0	61
2002	61	361	422	170	24	194	184	379		43	0	43
2003	43	411	454	182	40	222	199	421		34	0	34
2004	34	454	487	191	55	246	184	430		57	0	57
2005	57	393	450	140	50	190	194	384		66	0	66
2006	66	277	343	113	45	158	153	311		32	0	32
2007	32	497	530	165	35	200	277	477		53	0	53
2008	53	472	525	233	95	328	143	471		55	0	55
2009 [2]	55	383	438	140	100	240	167	407		31	0	31
2010 [3]	31	376	407	110	100	210	160	370		37	0	37

[1] Includes quantity under loan and farmer–owned reserve.　[2] Preliminary.　[3] Projected as of January 11, 2010, World Agricultural and Supply Demand Estimates. Totals may not add due to independent rounding.
ERS, Market and Trade Economics Division, (202) 694–5296.

Table 1-65.—Sorghum: Area, yield, and production, by State and United States, 2007–2009

State	Area planted for all purposes			Sorghum for grain								
				Area harvested			Yield per harvested acre			Production		
	2007	2008	2009[1]	2007	2008	2009[1]	2007	2008	2009[1]	2007	2008	2009[1]
	1,000 acres	*1,000 acres*	*1,000 acres*	*1,000 acres*	*1,000 acres*	*1,000 acres*	*Bush- els*	*Bush- els*	*Bush- els*	*1,000 bushels*	*1,000 bushels*	*1,000 bushels*
AL[2]	12	12	....	6	6	8	40.0	53.0	85.0	240	318	....
AZ	42	57	35	20	27	37	90.0	90.0	79.0	1,800	2,430	680
AR	225	125	40	215	115	....	96.0	88.0	....	20,640	10,120	2,923
CA[2]	39	47	....	10	9	....	85.0	95.0	....	850	855	....
CO	220	230	180	150	150	150	37.0	30.0	45.0	5,550	4,500	6,750
GA	65	60	55	45	44	40	46.0	45.0	53.0	2,070	1,980	2,120
IL	80	80	40	77	76	36	81.0	103.0	82.0	6,237	7,828	2,952
KS	2,800	2,900	2,700	2,650	2,750	2,550	79.0	78.0	88.0	209,350	214,500	224,400
KY[2]	15	13	....	12	11	....	90.0	90.0	....	1,080	990	....
LA	250	120	70	245	110	65	95.0	87.0	82.0	23,275	9,570	5,330
MS	145	85	13	115	82	11	85.0	71.0	70.0	9,775	5,822	770
MO	110	90	50	100	80	43	96.0	97.0	86.0	9,600	7,760	3,698
NE	350	300	235	240	210	140	94.0	91.0	93.0	22,560	19,110	13,020
NM	105	130	85	75	80	50	40.0	43.0	46.0	3,000	3,440	2,300
NC[2]	12	16	....	8	13	....	55.0	56.0	....	440	728	....
OK	240	350	250	220	310	220	56.0	45.0	56.0	12,320	13,950	12,320
PA[2]	15	11	....	3	3	....	56.0	37.0	....	168	111	....
SC[2]	9	12	....	6	8	....	35.0	46.0	....	210	368	....
SD	210	170	180	130	115	120	60.0	64.0	61.0	7,800	7,360	7,320
TN[2]	18	26	....	15	22	....	82.0	91.0	....	1,230	2,002	....
TX	2,750	3,450	2,700	2,450	3,050	2,050	65.0	52.0	48.0	159,250	158,600	98,400
US	7,712	8,284	6,633	6,792	7,271	5,520	73.2	65.0	69.4	497,445	472,342	382,983

[1] Preliminary. [2] Estimates discontinued in 2009.
NASS, Crops Branch, (202) 720–2127.

Table 1-66.—Sorghum: Utilization for silage, by State and United States, 2007–2009

State	Silage								
	Area harvested			Yield per acre			Production		
	2007	2008	2009[1]	2007	2008	2009[1]	2007	2008	2009[1]
	1,000 acres	*1,000 acres*	*1,000 acres*	*Tons*	*Tons*	*Tons*	*1,000 tons*	*1,000 tons*	*1,000 tons*
AL[2]	3	3	....	9.0	8.0	....	27	24	....
AZ	21	30	27	19.0	19.0	20.0	399	570	540
AR	2	2	1	13.0	10.0	11.0	26	20	11
CA[2]	29	38	....	18.0	17.0	....	522	646	....
CO	15	12	7	13.0	13.0	14.0	195	156	98
GA	12	12	12	12.0	14.0	11.0	144	168	132
IL	2	3	1	12.0	15.0	11.0	24	45	11
KS	80	70	40	12.0	13.0	11.0	960	910	440
KY[2]	2	1	....	10.0	6.0	....	20	6	....
LA	1	1	1	10.0	10.0	11.0	10	10	11
MS	1	1	1	16.0	13.0	12.0	16	13	12
MO	5	4	4	13.0	9.0	9.0	65	36	36
NE	25	15	15	11.0	8.0	13.0	275	120	195
NM	20	25	18	15.0	16.0	16.0	300	400	288
NC[2]	3	2	....	10.0	11.0	....	30	22	....
OK	12	16	12	5.0	10.0	13.0	60	160	156
PA[2]	5	8	....	9.0	6.5	....	45	52	....
SC[2]	2	4	....	7.0	6.0	....	14	24	....
SD	30	30	15	10.0	10.0	10.0	300	300	150
TN[2]	2	1	....	7.0	14.0	....	14	14	....
TX	120	130	100	15.0	15.0	16.0	1,800	1,950	1,600
US	392	408	254	13.4	13.8	14.5	5,246	5,646	3,680

[1] Prelimary. [2] Estimates discontinued in 2009.
NASS, Crops Branch, (202) 720–2127.

Table 1-67.—Sorghum grain: Marketing year average price and value of production, by State and United States, 2007–2009

State	Marketing year average price per cwt			Value of production		
	2007	2008	2009 [1]	2007	2008	2009 [1]
	Dollars	*Dollars*	*Dollars*	*1,000 dollars*	*1,000 dollars*	*1,000 dollars*
AL [2]	7.00	6.10	....	941	1,086	....
AZ	8.60	9.40	7.75	8,669	12,792	2,951
AR	6.15	6.93	5.90	71,084	39,274	9,658
CA [2]	8.25	6.30	....	3,927	3,016	....
CO	7.01	5.90	5.35	21,787	14,868	20,223
GA	7.30	6.10	5.40	8,462	6,7647	6,411
IL	7.21	5.29	6.60	25,183	23,190	10,911
KS	7.23	5.61	5.70	847,616	673,8730	716,285
KY [2]	7.10	6.60	....	4,294	3,659	....
LA	6.50	6.90	5.55	84,721	38,978	16,566
MS	6.43	6.00	4.75	35,198	19,562	2,048
MO	6.68	6.38	6.10	35,912	27,725	12,632
NE	7.10	5.80	5.85	89,699	62,069	42,654
NM	7.25	6.25	5.70	12,180	12,040	7,342
NC [2]	6.95	6.90	....	1,712	2,813	....
OK	7.00	5.89	5.95	48,294	46,013	41,050
PA [2]	6.60	5.35	....	621	333	....
SC [2]	6.85	6.80	....	806	1,401	....
SD	7.06	5.62	4.90	30,838	23,163	20,086
TN [2]	6.94	6.00	....	4,780	6,727	....
TX	6.60	6.91	6.05	588,588	613,719	333,379
US	7.28	5.72	5.90	1,925,312	1,631,065	1,242,196

[1] Preliminary.   [2] Estimates discontinued in 2009.
NASS, Crops Branch, (202) 720–2127.

Table 1-68.—Sorghum grain: Support operations, United States, 2000–2009

Marketing year beginning September 1	Income support payment rates per cwt [1]	Program price levels per cwt		Put under support		Acquired by CCC under loan program [5]	Owned by CCC at end of marketing year
		Loan [2]	Target [3]	Quantity	Percentage of production [4]		
	Dollars	*Dollars*	*Dollars*	*Million cwt.*	*Percent*	*Millions cwt.*	*Million cwt.*
2000/2001	1.49	3.05	NA	8.6	3.3	0.2	0.0
2001/2002	1.24	3.05	NA	9.6	3.3	0.0	0.1
2002/2003	0.63/0.00	3.54	4.54	3.7	1.8	0.0	0.0
2003/2004	0.63/0.00	3.54	4.54	3.5	1.6	0.0	0.0
2004/2005	0.63/0.48	3.48	4.59	5.5	2.2	0.2	0.0
2005/2006	0.63/0.48	3.48	4.59	5.4	2.4	0.0	0.0
2006/2007	0.63/0.00	3.48	4.59	1.9	1.2	0.0	0.0
2007/2008	0.63/0.00	3.48	4.59	1.8	0.7	0.0	0.0
2008/2009	0.63/0.00	3.48	4.59	4.5	1.7	0.0	0.0
2009/2010	0.63/0.00	3.48	4.59	....	....	....	....

[1] Payment rates for the 1998/1999 through 2001/2002 crops were calculated according to the Production Flexibility Contract (PFC) program provisions of the Federal Agriculture Improvement and Reform Act of 1996 (1996 Act) and include supplemental PFC payment rates for 1998 through 2001. Payment rates for the 2002/2003 and subsequent crops are calculated according to the Direct and Counter-cyclical program provisions, following enactment of the Farm Security and Rural Investment Act of 2002 (2002 Act). Beginning with 2002/2003, the first entry is the direct payment rate and the second entry is the counter-cyclical payment rate.   [2] Starting in 2009, producers who participate in the Average Crop Revenue Election (ACRE) program get a 30 percent reduction in their loan rate, not calculated in this table.   [3] Target prices were reestablished under the 2002 Act.   [4] Percentage of production is on a grain basis.   [5] Acquisitions for 2008/2009 as of September 1, 2009. NA-not applicable.
FSA, Feed Grains, (202) 720–7787.

Table 1-69.—Sorghum: Area, yield, and production in specified countries, 2007/2008–2009/2010

Country	Area			Yield per hectare			Production		
	2007/ 2008	2008/ 2009	2009/ 2010	2007/ 2008	2008/ 2009	2009/ 2010	2007/ 2008	2008/ 2009	2009/ 2010
	1,000 hec- tares	1,000 hec- tares	1,000 hec- tares	Metric tons	Metric tons	Metric tons	1,000 metric tons	1,000 metric tons	1,000 metric tons
Argentina	619	450	700	4.74	3.69	5.50	2,937	1,660	3,850
Australia	942	767	545	4.02	3.51	2.94	3,790	2,690	1,600
Brazil	850	845	780	2.35	2.26	2.34	2,000	1,910	1,825
Burkina	1,608	1,620	1,620	0.94	1.16	1.04	1,507	1,875	1,684
China, Peoples	500	490	450	3.84	3.75	3.67	1,920	1,837	1,650
Ethiopia	1,533	1,553	1,550	1.73	1.69	1.34	2,659	2,619	2,084
India	7,930	7,700	7,000	1.00	0.95	0.97	7,930	7,310	6,770
Mexico	1,775	1,890	1,800	3.49	3.74	3.61	6,200	7,067	6,500
Nigeria	7,400	7,400	7,500	1.35	1.49	1.53	10,000	11,000	11,500
Sudan	6,500	6,400	6,000	0.69	0.66	0.44	4,500	4,192	2,630
Others	10,179	9,966	10,053	1.02	1.06	1.01	10,366	10,560	10,123
Total foreign	39,836	39,081	37,998	1.35	1.35	1.32	53,809	52,720	50,216
United States ...	2,749	2,942	2,234	4.60	4.08	4.35	12,636	11,998	9,728
Total	42,585	42,023	40,232	1.56	1.54	1.49	66,445	64,718	59,944

FAS, Office of Global Analysis, (202) 720-6301. Prepared or estimated on the basis of official USDA production, supply, and distribution, supply, and and distribution statistics from foreign governments.

Table 1-70.—Sorghum: International trade, 2007/2008–2009/2010

Country	2007/2008	2008/2009	2009/2010
	1,000 metric tons	1,000 metric tons	1,000 metric tons
Principle exporting countries:			
Argentina	1,223	1,113	1,200
Australia	810	1,000	800
Bolivia	9	12	10
Brazil	116		25
China, Peoples Republic of	223	32	50
EU-27	4	8	5
India	94	52	50
Nigeria	50	50	50
South Africa, Republic of	38	46	40
Thailand	24	9	25
Others	109	30	10
Total foreign	2,700	2,352	2,265
United States	7,030	3,632	4,318
Total	9,730	5,984	6,583
Principle importing countries:			
Chile	374	501	550
Columbia	66	202	200
Ethiopia	30	140	150
Israel	144	71	50
Japan	1,084	1,629	1,800
Kenya	6	9	50
Mexico	1,156	2,496	2,600
Morocco	9		100
Sudan	300	300	400
Taiwan	83	72	75
Others	6,426	690	290
Total foreign	9,678	6,110	6,265
United States	1	3	
Total	9,679	6,113	6,265

FAS, Office of Global Analysis, (202) 720-6301. Prepared or estimated on the basis of official USDA production, supply, and distribution, supply, and and distribution statistics from foreign governments.

Table 1-71.—Millet: Area, yield, and production in specified countries, 2007/2008–2009/2010

Country	Area			Yield per hectare			Production		
	2007/ 2008	2008/ 2009	2009/ 2010	2007/ 2008	2008/ 2009	2009/ 2010	2007/ 2008	2008/ 2009	2009/ 2010
	1,000 hec- tares	*1,000 hec- tares*	*1,000 hec- tares*	*metric tons*	*metric tons*	*metric tons*	*1,000 metric tons*	*1,000 metric tons*	*1,000 metric tons*
Burkina	1,185	1,200	1,200	0.82	1.05	0.85	966	1,255	1,020
China, Peoples	900	900	900	1.67	1.67	1.78	1,500	1,500	1,600
Ethiopia	399	401	400	1.35	1.33	1.23	538	533	493
India	10,800	10,000	8,500	1.15	1.14	1.01	12,410	11,370	8,590
Mali	1,586	1,600	1,600	0.74	0.78	0.75	1,175	1,242	1,200
Niger	5,200	5,200	5,200	0.54	0.71	0.52	2,782	3,700	2,678
Nigeria	5,850	5,850	5,850	1.28	1.32	1.32	7,500	7,700	7,700
Senegal	687	943	943	0.46	0.72	0.86	319	678	810
Sudan	2,250	2,250	2,250	0.34	0.28	0.21	760	630	471
Uganda	430	430	430	1.60	1.63	1.63	690	700	700
Others	4,921	4,960	4,647	0.69	0.75	0.66	3,391	3,701	3,054
Total foreign	34,208	33,734	31,920	0.94	0.98	0.89	32,031	33,009	28,316
Total	34,208	33,734	31,920	0.94	0.98	0.89	32,031	33,009	28,316

FAS, Office of Global Analysis, (202) 720-6301. Prepared or estimated on the basis of official USDA production, supply, and distribution, supply, and and distribution statistics from foreign governments.

Table 1-72.—Mixed grain: Area, yield, and production in specified countries, 2007/2008–2009/2010

Country	Area			Yield per hectare			Production		
	2007/ 2008	2008/ 2009	2009/ 2010	2007/ 2008	2008/ 2009	2009/ 2010	2007/ 2008	2008/ 2009	2009/ 2010
	1,000 hec- tares	*1,000 hec- tares*	*1,000 hec- tares*	*metric tons*	*metric tons*	*metric tons*	*1,000 metric tons*	*1,000 metric tons*	*1,000 metric tons*
Bangladesh	35	35	35	0.71	0.71	0.71	25	25	25
Canada	95	80	78	2.74	2.78	2.73	260	222	213
EU-27	4,120	4,205	4,323	3.44	3.55	3.77	14,185	14,925	16,291
Switzerland	10	10	10	5.50	5.50	5.50	55	55	55
Turkey	100	100	100	1.15	1.15	1.15	115	115	115
Others	....	....	....	....	....	....	....	....	....
Total foreign	4,360	4,430	4,546	3.36	3.46	3.67	14,640	15,342	16,699
Total	4,360	4,430	4,546	3.36	3.46	3.67	14,640	15,342	16,699

FAS, Office of Global Analysis, (202) 720-6301. Prepared or estimated on the basis of official USDA production, supply, and distribution, supply, and and distribution statistics from foreign governments.

Table 1-73.—Commercial feeds: Disappearance for feed, United States, 2001–2010

Year beginning October	Oilseed cake and meal						Animal protein			
	Soy-bean	Cotton-seed	Linseed	Peanut [1]	Sun-flower	Total	Tank-age and meat meal	Fish meal	Dried milk [2]	Total
	1,000 tons	*1,000 tons*	*1,000 tons*	*1,000 tons*	*1,000 tons*	*1,000 tons*	*1,000 tons*	*1,000 tons*	*1,000 tons*	*1,000 tons*
2001	32,568	3,340	124	151	402	36,585	1,938	274	250	2,462
2002	32,074	2,691	178	178	234	35,355	1,878	252	433	2,564
2003	31,449	2,786	197	122	340	34,894	2,320	233	374	2,928
2004	33,561	3,454	206	95	143	37,459	2,217	151	203	2,572
2005	33,195	3,355	269	117	298	37,234	2,254	199	269	2,722
2006	34,355	3,049	275	119	356	38,154	2,375	215	292	2,882
2007	33,232	2,589	210	116	343	36,490	2,398	213	250	2,861
2008	30,752	1,807	129	102	357	33,147	2,271	223	250	2,744
2009	30,200	1,784	210	92	388	32,674	2,343	200	250	2,793
2010 [3]	30,300	2,525	197	95	360	33,477	2,350	200	250	2,800

Year beginning October	Mill products [4]					Total commercial feeds
	Wheat millfeeds	Gluten feed and meal [5]	Rice millfeeds	Alfalfa meal	Total	
	1,000 tons	*1,000 tons*	*1,000 tons*	*1,000 tons*	*1,000 tons*	*1,000 tons*
2001	6,895	1,475	678	NA	9,049	48,096
2002	6,948	2,275	694	NA	9,917	47,835
2003	6,755	2,421	594	NA	9,771	47,592
2004	6,765	2,894	613	NA	10,272	50,303
2005	6,753	3,514	641	NA	10,908	50,865
2006	6,873	4,624	545	NA	12,042	53,078
2007	6,776	4,560	568	NA	11,904	51,256
2008	6,464	5,167	570	NA	12,201	48,092
2009	6,400	5,075	575	NA	12,050	47,516
2010 [3]	6,400	5,075	575	NA	12,050	48,327

[1] Year beginning August 1.   [2] Includes dried skim milk, and whey for feed, but does not include any milk products fed on farms.   [3] Preliminary.   [4] Other mill products that are not listed include screenings, hominy, and oats feed etc., for which no statistics are available.   [5] Adjusted for export data.   NA-not available.
ERS, Market and Trade Economics Division, (202) 694–5290.

Table 1-74.—High-protein feeds: Quantity for feeding, high-protein animal units, quantity per animal unit, and prices, United States, 2001–2010

| Year beginning October | Quantity for feeding [1] | | | | | | High-protein animal units | Quantity per animal unit | High protein feed prices |
| | Oilseed meal | | | Animal protein | Grain protein [3] | Total | | | |
	Soybean meal	Other oilseed meals [2]	Total						
	1,000 tons	*1,000 tons*	*1,000 tons*	*1,000 tons*	*1,000 tons*	*1,000 tons*	*Million units*	*1,000 Pounds*	*Index numbers 1992=100*
2001	35,825	3,722	39,546	2,754	879	43,179	72.1	1,198	89
2002	35,281	3,038	38,320	2,712	1,355	42,386	72.0	1,177	95
2003	34,594	3,178	37,772	3,202	1,442	42,416	70.3	1,206	131
2004	36,917	3,606	40,523	2,923	1,723	45,169	70.8	1,275	97
2005	36,515	3,724	40,239	3,047	2,092	45,378	71.6	1,267	88
2006	37,791	3,497	41,288	3,219	2,753	47,260	71.8	1,317	105
2007	36,555	3,002	39,558	3,232	2,715	45,505	71.5	1,273	170
2008	33,827	2,207	36,034	3,092	3,077	42,203	70.9	1,191	168
2009	33,220	2,267	35,487	3,149	3,022	41,658	70.2	1,186	151
2010	33,330	2,926	36,256	3,158	3,022	42,436	69.5	1,222	130

[1] In terms of 44 percent protein soybean meal equivalent. [2] Includes cottonseed, linseed, peanut meal, and sunflower meal. [3] Beginning 1974, adjusted for exports of corn gluten feed and meal.
ERS, Market and Trade Economics Division (202) 694–5290.

Table 1-75.—Feed concentrates: Fed to livestock and poultry, 2001–2010

| Year beginning October | Feed grains | | | | Wheat [2] | Rye [2] | By-product feeds [3] | Total concentrates | Grain consuming animal units | Concentrates fed per grain-consuming animal unit |
	Corn [1]	Sorghum [1]	Oats [2] and barley [2]	Total						
	Million tons	*Million tons*	*Million tons*	*Million tons*	*Million tons*	*Million tons*	*Million tons*	*Million tons*	*Millions*	*Tons*
2001	163.8	6.4	4.9	175.1	3.9	0.1	55.8	234.8	89.8	2.62
2002	155.4	4.8	5.2	165.3	7.4	0.1	55.4	228.1	88.2	2.59
2003	161.9	5.1	4.7	171.6	4.5	0.1	55.2	231.5	89.4	2.59
2004	171.8	5.4	4.3	181.4	5.3	0.1	58.4	245.3	90.1	2.72
2005	171.2	3.9	3.9	179.0	3.0	0.1	58.8	240.9	91.5	2.63
2006	155.1	3.2	4.0	162.3	5.1	0.1	60.6	228.1	92.7	2.46
2007	164.0	4.6	3.4	172.0	4.6	0.1	59.0	235.7	95.1	2.48
2008	145.1	6.5	3.4	155.0	3.6	0.1	55.2	213.9	92.7	2.31
2009	154.7	3.9	3.4	162.0	3.1	0.1	54.8	220.0	91.5	2.40
2010	147.0	3.1	4.0	154.1	6.0	0.1	55.4	215.7	91.9	2.35

[1] Marketing year beginning Sept. 1. [2] Marketing year beginning June 1. [3] Oilseed meals, animal protein feeds, mill by-products, and mineral supplements.
ERS, Market and Trade Economics Division (202) 694–5290.

Table 1-76.—Feed: Consumed per head and per unit of production, by class of livestock or poultry, with quantity expressed in equivalent feeding value of corn, 2000–2009

Year beginning October	Dairy cattle			Beef cattle				Sheep and lambs	
	Milk cows		Other dairy cattle per head	Cattle on feed per head Jan. 1 [1]	Other beef cattle per head	All beef cattle per head	Cattle and calves per 100 pounds produced [2]	Per head	Per 100 pounds produced [3]
	Per head	Per 100 pounds milk produced							
	Pounds	Pounds	Pounds	Pounds	Pounds	Pounds	Pounds	Pounds	Pounds
2000	13,088	72	6,559	9,924	5,319	6,101	1,254	1,278	1,583
2001	13,027	72	6,548	9,834	5,316	6,076	1,255	1,277	1,556
2002	12,965	70	6,538	9,745	5,313	6,020	1,239	1,276	1,533
2003	12,985	69	6,541	9,773	5,314	6,076	1,233	1,276	1,532
2004	13,268	70	6,589	10,187	5,329	6,155	1,278	1,282	1,561
2005	13,087	67	6,558	9,922	5,319	6,118	1,299	1,278	1,556
2006	12,758	64	6,503	9,441	5,302	6,031	1,266	1,272	1,567
2007	12,758	64	6,503	9,441	5,302	6,047	1,271	1,272	1,593
2008	12,379	61	6,439	8,887	5,282	5,901	1,214	1,265	1,616
2009 [4]	12,582	60	6,474	9,184	5,293	5,955	1,237	1,269	1,609

Year beginning October	Poultry								Hogs per 100 pounds produced	Horses and mules two years and over per head
	Hens and pullets		Chickens raised		Broilers produced		Turkeys raised			
	Per head Jan. 1	Per 100 eggs	Per head	Per 100 pounds live weight	Per head	Per 100 pounds produced	Per head	Per 100 pounds produced		
	Pounds	Pounds	Pounds	Pounds	Pounds	Pounds	Pounds	Pounds	Pounds	Pounds
2000	131	52	31	865	11.5	232	97	379	593	3,868
2001	129	51	31	888	11.4	230	96	368	592	3,856
2002	128	50	31	846	11.3	219	95	344	574	3,844
2003	129	50	31	917	11.3	224	95	342	581	3,847
2004	134	52	32	991	11.9	229	100	360	603	3,903
2005	131	50	31	927	11.5	214	97	353	581	3,867
2006	124	47	29	979	10.9	199	92	337	551	3,803
2007	124	46	29	1,026	10.9	199	92	331	562	3,803
2008	116	43	28	927	10.2	175	86	277	502	3,729
2009 [4]	120	45	29	1,006	10.6	192	89	300	513	3,769

[1] Feed consumed by all cattle divided by the number on feed Jan. 1. [2] Feed for all cattle, except milk cows, divided by the net live-weight production of cattle and calves. It includes the growth on dairy heifers and calves as well as all beef cattle. [3] Including wool produced. [4] Preliminary.
ERS, Market and Trade Economics Division, (202) 694–5290.

Table 1-77.—Feed: Consumed by livestock and poultry, by type of feed, with quantity expressed in equivalent feeding value of corn, 2000–2009

Year beginning October	Concentrates	Harvested roughage	Pasture	Total
	Million tons	Million tons	Million	Million tons
2000	258	87	159	504
2001	257	85	160	502
2002	250	83	162	494
2003	254	85	154	493
2004	268	86	154	509
2005	264	87	157	508
2006	253	82	162	497
2007	260	80	163	503
2008	236	83	158	477
2009 [1]	242	83	156	481

[1] Preliminary.
ERS, Market and Trade Economics Division, (202) 694–5290.

Table 1-78.—Animal units fed: Grain-consuming, roughage-consuming, and grain-and-roughage-consuming, United States, 2001–2010 [1]

Year beginning October	Grain-consuming [2]	Roughage-consuming [3]	Grain and roughage-consuming [4]
	1,000 units	1,000 units	1,000 units
2001	89,771	72,083	78,380
2002	88,236	72,045	77,765
2003	89,438	70,318	77,149
2004	90,144	70,829	77,714
2005	91,490	71,647	78,731
2006	92,749	71,753	79,289
2007	95,118	71,479	80,042
2008	92,749	70,887	78,782
2009	91,510	70,242	77,900
2010 [5]	91,923	69,461	77,563

[1] Index series based on average feeding rates for years 1969–71. In calculations for the feeding years 1969 to date, cattle numbers used are the new categories shown in the Livestock and Poultry Inventory, published by NASS, USDA. [2] Livestock and poultry numbers weighted by all concentrates consumed. [3] Livestock and poultry numbers weighted by all roughage (including pasture) consumed. [4] Livestock and poultry numbers weighted by all feed (including pasture) fed to livestock. [5] Preliminary.
ERS, Market and Trade Economics Division, (202) 694–5290.

Table 1-79.—Feed grains: Average price, selected markets and grades, 2000–2009 [1]

Calendar year	Kansas City			Minneapolis			
	Corn, No. 2 Yellow (truck)	Corn, No. 2 White (truck)	Sorghum, No. 2 Yellow (truck)	Corn, No. 2 Yellow	Barley, No. 3 or Better malting	Duluth Barley, No. 2 Feed	Oats, No. 2 White
	Dollars per bushel	Dollars per bushel	Dollars per cwt.	Dollars per bushel	Dollars per bushel	Dollars per bushel	Dollars per bushel
2000	1.93	1.94	3.19	1.79	....	1.22	1.28
2001	1.85	....	4.03	....	....	....	....
2002	2.13	2.51	4.27	2.11	2.85	1.70	....
2003	2.36	2.58	4.07	2.22	3.34	1.91	1.82
2004	2.40	2.52	4.23	2.38	2.55	1.79	1.71
2005	1.87	2.19	3.34	1.79	2.53	NA	1.84
2006	2.42	2.03	4.27	2.24	3.20	....	2.24
2007	4.61	4.43	6.05	3.38	2.02	3.95	2.98
2008	5.12	5.32	8.41	4.76	6.81	....	3.91
2009	3.60	3.90	5.57	3.46	4.26	....	2.21

Calendar year	Omaha: Corn, No. 2 Yellow (truck)	Chicago: Corn, No. 2 Yellow	Texas High Plains: Sorghum, No. 2 Yellow	Memphis		St. Louis: Corn, No. 2 Yellow (truck)
				Corn, No. 2 Yellow	Barley, No. 2 Western	
	Dollars per bushel	Dollars per bushel	Dollars per cwt.	Dollars per bushel	Dollars per bushel	Dollars per cwt.
2000	1.82	1.97	3.51	2.01	....	2.00
2001	....	1.98	....	2.03	....	....
2002	2.13	2.24	4.27	2.29	....	2.33
2003	2.24	2.34	3.94	2.42	....	2.38
2004	2.36	2.48	4.70	2.55	....	2.64
2005	1.77	1.97	3.98	2.11	....	2.01
2006	2.31	2.43	5.06	2.66	....	2.34
2007	3.54	3.67	7.10	3.71	....	3.74
2008	5.04	5.12	9.53	5.07	....	5.11
2009	3.56	3.76	6.52	3.69	....	3.78

[1] Simple average of daily prices. NA-not available.
AMS, Livestock and Grain Market News Branch, (202) 720–6231.

Table 1-80.—Feedstuffs: Average price per ton bulk, in wholesale lots, at leading markets, 2000–2009

Year beginning October	Soybean meal 44% protein Decatur	Soybean meal 48% protein Decatur	Cottonseed meal 41% protein Kansas City	Cottonseed meal 41% protein Memphis	Linseed meal 34% protein Min-neapolis	Meat meal 50% protein Kansas City	Fish meal 60% protein Gulf Coast	Wheat bran Kansas City	Wheat middlings Min-neapolis
	Dollars per ton	Dollars per ton	Dollars per ton	Dollars per ton	Dollars per ton	Dollars per ton	Dollars per ton	Dollars per ton	Dollars per ton
2000	160.03	168.10	146.50	130.70	103.10	166.50	326.40	53.37	45.81
2001	165.21	173.60	165.00	142.70	121.90	166.50	358.20	62.93	50.81
2002	153.82	167.72	160.10	136.20	119.20	166.00	460.00	59.74	58.42
2003	115.60	208.95	172.52	152.24	134.31	196.30	487.50	65.07	56.05
2004	(1)	237.30	193.58	167.68	148.09	190.63	524.97	67.82	64.19
2005	(1)	188.17	156.59	128.89	115.70	169.19	....	54.34	44.53
2006	(1)	175.60	171.84	141.87	116.12	151.43	707.27	72.68	61.12
2007	(1)	230.39	187.53	166.49	148.36	225.96	850.53	87.31	87.20
2008	(1)	331.09	298.72	265.82	227.05	326.48	866.06	134.33	136.20
2009	(1)	347.73	293.25	265.21	231.77	334.69	861.06	90.17	89.12

Year beginning October	Wheat shorts or mid-dlings Kansas City	Wheat millrun Portland	Gluten feed 21% protein Illinois Points	Hominy feed Midwest	Distillers' dried grains Lawrence-burg	Brewers' dried grains Colum-bus	Alfalfa meal Dehy-drated, 17% pro-tein Kansas City	Alfalfa meal Sun-cured Kansas City	Blackstrap molasses New Orleans
	Dollars per ton	Dollars per ton	Dollars per ton	Dollars per ton	Dollars per ton	Dollars per ton	Dollars per ton	Dollars per ton	Dollars per ton
2000	57.82	60.92	53.64	58.71	79.90	90.94	97.59	86.58	38.72
2001	62.88	63.25	60.55	55.02	80.62	94.00	139.06	130.38	63.16
2002	59.77	75.95	59.63	63.23	80.19	94.00	154.05	134.34	68.63
2003	65.27	85.49	70.15	72.66	93.13	94.95	138.61	122.48	58.00
2004	67.82	85.00	68.83	77.02	106.04	(1)	121.35	109.26	57.28
2005	54.23	74.72	68.17	50.50	75.47	(1)	135.83	110.57	NA
2006	72.53	84.51	69.51	59.84	89.04	(1)	174.13	161.77	NA
2007	....	129.30	81.34	108.64	113.38	(1)	206.53	179.50	NA
2008	134.31	185.85	....	153.50	....	(1)	236.28	205.77	NA
2009	90.24	120.48	....	100.53	114.23	(1)	224.93	189.19	NA

1 Discontinued.　NA-not available.
AMS, Livestock and Grain Market News Branch, (202) 720–6231.

Table 1-81.—Proso millet: Area, yield, production, and value, United States, 2001–2009

Year	Area		Yield per harvested acre	Production	Marketing year average price per bushel received by farmers	Value of production
	Planted	Harvested				
	1,000 acres	*1,000 acres*	*Bushels*	*1,000 bushels*	*Dollars*	*1,000 dollars*
2001	650	585	33.2	19,405	2.02	39,109
2002	520	275	13.3	3,668	7.22	26,462
2003	730	620	18.5	11,450	2.95	33,730
2004	710	595	25.3	15,065	2.83	42,611
2005	565	515	26.5	13,670	3.19	43,660
2006	580	475	21.5	10,195	4.09	41,748
2007	570	520	32.5	16,900	4.67	78,975
2008	520	460	32.3	14,880	3.23	48,017
2009 [1]	350	293	33.7	9,865	2.84	28,043

[1] Preliminary.
NASS, Crops Branch, (202) 720–2127.

Table 1-82.—Proso millet: Area, yield, and production, by State and United States, 2007–2009

State	Area planted			Area harvested		
	2007	2008	2009 [1]	2007	2008	2009 [1]
	1,000 acres	*1,000 acres*	*1,000 acres*	*1,000 acres*	*1,000 acres*	*1,000 acres*
CO	270	270	170	260	230	150
NE	145	140	95	130	130	78
SD	155	110	85	130	100	65
US	570	520	350	520	460	293

State	Yield per acre			Production		
	2007	2008	2009 [1]	2007	2008	2009 [1]
	Bushels	*Bushels*	*Bushels*	*1,000 bushels*	*1,000 bushels*	*1,000 bushels*
CO	33.0	33.0	35.0	8,580	7,590	5,250
NE	33.0	33.0	30.0	4,290	4,290	2,340
SD	31.0	30.0	35.0	4,030	3,000	2,275
US	32.5	32.3	33.7	16,900	14,880	9,865

[1] Preliminary.
NASS, Crops Branch, (202) 720–2127.

Table 1-83.—Proso millet: Marketing year average price and value, by State and United States, 2007–2009

State	Marketing year average price per bushel			Value of production		
	2007	2008	2009 [1]	2007	2008	2009 [1]
	Dollars	*Dollars*	*Dollars*	*1,000 dollars*	*1,000 dollars*	*1,000 dollars*
CO	4.50	3.25	2.90	38,610	24,668	15,225
NE	4.90	3.10	2.95	21,021	13,299	6,903
SD	4.80	3.35	2.60	19,344	10,050	5,915
US	4.67	3.23	2.84	78,975	48,017	28,043

[1] Preliminary.
NASS, Crops Branch, (202) 720–2127.

CHAPTER II

STATISTICS OF COTTON, TOBACCO, SUGAR CROPS, AND HONEY

In addition to tables on cotton, tobacco, sugar, and honey, this chapter includes tables on fibers other than cotton and syrups. Cottonseed data, however, are in the following chapter on oilseeds, fats, and oils.

Table 2-1.—Cotton: Area, yield, production, market year average price, and value, United States, 2000–2009

Year	Area		Yield per harvested acre	Production	Marketing year average price per pound received by farmers	Value of production
	Planted	Harvested				
	1,000 acres	*1,000 acres*	*Pounds*	*1,000 bales*[1]	*Cents*	*1,000 dollars*
2000	15,517.2	13,053.0	632	17,188.3	51.6	4,260,417
2001	15,768.5	13,827.7	705	20,302.8	32.0	3,121,848
2002	13,957.9	12,416.6	665	17,208.6	45.7	3,777,132
2003	13,479.6	12,003.4	730	18,255.2	63.0	5,516,761
2004	13,658.6	13,057.0	855	23,250.7	44.7	4,993,565
2005	14,245.4	13,802.6	831	23,890.2	49.7	5,695,217
2006	15,274.0	12,731.5	814	21,587.8	48.4	5,013,238
2007	10,827.2	10,489.1	879	19,206.9	61.3	5,652,907
2008	9,471.0	7,568.7	813	12,815.3	49.1	3,021,485
2009[2]	9,149.2	7,690.5	774	12,401.3	62.8	3,735,564

[1] 480-pound net weight bales. [2] Preliminary.
NASS, Crops Branch, (202) 720–2127.

Table 2-2.—Cotton: Area, yield, production, and type by State and United States, 2007–2009

State	Area planted			Area harvested			Yield per harvested acre			Production[1]		
	2007	2008	2009[2]	2007	2008	2009[2]	2007	2008	2009[2]	2007	2008	2009[2]
	1,000 acres	*1,000 acres*	*1,000 acres*	*1,000 acres*	*1,000 acres*	*1,000 acres*	*Pounds*	*Pounds*	*Pounds*	*1,000 bales*[3]	*1,000 bales*[3]	*1,000 bales*[3]
Upland:												
AL	400.0	290.0	255.0	385.0	286.0	250.0	519	787	691	416.0	469.0	360.0
AZ	170.0	135.0	145.0	168.0	133.0	144.0	1,469	1,462	1,467	514.0	405.0	440.0
AR	860.0	620.0	520.0	850.0	615.0	500.0	1,071	1,012	797	1,896.0	1,296.0	830.0
CA	195.0	120.0	71.0	194.0	117.0	70.0	1,608	1,506	1,714	650.0	367.0	250.0
FL	85.0	67.0	82.0	81.0	65.0	78.0	687	916	646	116.0	124.0	105.0
GA	1,030.0	940.0	1,000.0	995.0	920.0	990.0	801	835	882	1,660.0	1,600.0	1,820.0
KS	47.0	35.0	38.0	43.0	25.0	34.0	639	653	720	57.2	34.0	51.0
LA	335.0	300.0	230.0	330.0	234.0	225.0	1,017	576	725	699.0	281.0	340.0
MS	660.0	365.0	305.0	655.0	360.0	295.0	966	911	692	1,318.0	683.0	425.0
MO	380.0	306.0	272.0	379.0	303.0	260.0	968	1,106	960	764.0	698.0	520.0
NM	43.0	38.0	30.5	39.0	35.0	29.0	1,095	974	828	89.0	71.0	50.0
NC	500.0	430.0	375.0	490.0	428.0	370.0	767	847	986	783.0	755.0	760.0
OK	175.0	170.0	205.0	165.0	155.0	200.0	817	811	792	281.0	262.0	330.0
SC	180.0	135.0	115.0	158.0	134.0	114.0	486	881	842	160.0	246.0	200.0
TN	515.0	285.0	300.0	510.0	280.0	280.0	565	909	857	600.0	530.0	500.0
TX	4,900.0	5,000.0	5,000.0	4,700.0	3,250.0	3,650.0	843	657	644	8,250.0	4,450.0	4,900.0
VA	60.0	61.0	64.0	59.0	60.0	63.0	829	908	990	101.9	113.5	130.0
US	10,535.0	9,297.0	9,007.5	10,201.0	7,400.0	7,552.0	864	803	763	18,355.1	12,384.5	12,011.0
American Pima:												
AZ	2.5	0.8	1.7	2.5	0.8	1.7	883	480	1,129	4.6	0.8	4.0
CA	260.0	155.0	119.0	257.0	151.0	116.0	1,481	1,281	1,448	793.0	403.0	350.0
NM	4.7	2.6	3.0	4.6	1.9	3.0	856	758	688	8.2	3.0	4.3
TX	25.0	15.6	18.0	24.0	15.0	17.8	920	768	863	46.0	24.0	32.0
US	292.2	174.0	141.7	288.1	168.7	138.5	1,419	1,226	1,353	851.8	430.8	390.3
US, all	10,827.2	9,471.0	9,149.2	10,489.1	7,568.7	7,690.5	879	813	774	19,206.9	12,401.3	12,401.3

[1] Production ginned and to be ginned. [2] Preliminary. [3] 480-pound net weight bale.
NASS, Crops Branch, (202) 720–2127.

Table 2-3.—Cotton: Marketing year average price per pound, and value, by State and United States, 2007–2009

State	Marketing year average price per pound			Value of production		
	2007 [1]	2008	2009 [2]	2007 [1]	2008	2009 [2]
	Dollars	Dollars	Dollars	1,000 dollars	1,000 dollars	1,000 dollars
Upland:						
AL	0.597	0.449	0.644	119,209	101,079	111,283
AZ	0.596	0.585	0.657	147,045	113,724	138,758
AR	0.578	0.479	0.610	526,026	297,,976	243,024
CA	0.722	0.594	0.715	225,264	104,639	85,800
FL	0.580	0.504	0.663	32,294	29,998	33,415
GA	0.596	0.514	0.652	474,893	394,752	569,587
KS	0.610	0.417	0.647	16,748	6,805	15,839
LA	0.570	0.524	0.618	191,246	70,677	100,858
MS	0.576	0.481	0.619	364,401	157,691	126,276
MO	0.574	0.470	0.574	210,497	57,469	143,270
NM	0.599	0.490	0.630	25,589	16,699	15,120
NC	0.558	0.462	0.558	209,719	167,429	203,558
OK	0.610	0.415	0.650	82,277	52,190	102,960
SC	0.580	0.490	0.570	44,544	57,859	54,720
TN	0.556	0.498	0.634	160,128	126,691	152,160
TX	0.604	0.438	0.592	2,391,840	935,568	392,384
VA	0.557	0.470	0.540	27,244	25,606	33,696
US	0.593	0.478	0.605	5,248,964	2,816,852	3,522,708
American-Pima:						
AZ [3]	0.943	....	1.130	2,082	....	2,170
CA	0.990	0.992	1.140	376,834	191,892	191,520
NM [3]	0.962	....	1.100	3,786	....	2,270
TX	0.962	0.961	1.100	21,241	11,071	16,896
US	0.988	0.990	1.136	403,943	204,633	212,856
US, all	0.613	0.491	0.628	5,652,907	3,021,485	3,735,564

[1] Revised. [2] Preliminary. [3] Estimates not published to avoid disclosure of individual operations.
NASS, Crops Branch, (202) 720–2127.

Table 2-4.—Cotton, American Upland: Support operations, United States, 2000–2009

Marketing Year beginning August 1	Income support payment rates per pound [1]	Program price levels per pound		Put under Loan		Acquired by CCC under loan program	Owned by CCC at end of marketing year
		Loan *	Target [2]	Quantity	Percentage of production		
	Cents	Cents	Cents	1,000 bale	Percent	1,000 bale	1,000 bale
2000/2001	15.21	51.92	NA	8,837	52.6	69	5
2001/2002	12.66	51.92	NA	13,655	69.7	31	2
2002/2003	6.67/13.73	52.00	72.40	12,740	77.1	0	106
2003/2004	6.67/3.93	52.00	72.40	10,466	58.7	16	0
2004/2005	6.67/13.73	52.00	72.40	17,092	76.0	8	0
2005/2006	6.67/13.73	52.00	72.40	17,783	76.5	181	11
2006/2007	6.67/13.73	52.00	72.40	17,839	85.7	79	0
2007/2008	6.67/13.73	52.00	72.40	14,636	79.7	169	0
2008/2009	6.67/12.58	52.00	71.25	10,005	79.5	4	0
2009/2010 [3]	6.67/12.58	52.00	71.25	8,278	73.7	....	0

[1] Payment rates for the 2000/2001 through 2002/2003 crops were calculated according to the Production Flexibility Contract (PFC) program provisions of the Federal Agriculture Improvement and Reform Act of 1996 (1996 Act) and include supplemental PFC payment rates for 1999 through 2002. Payment rates for the 2003/2004 and subsequent crops are calculated according to the Direct and Counter-cyclical program provisions, following enactment of the Farm Security and Rural Investment Act of 2002 (2002 Act). Beginning with 2003/2004, the first entry is the direct payment rate and the second entry is the maximum counter-cyclical payment rate. [2] Target prices were reestablished under the 2002 Act. [3] As of August 30, 2010. NA-not applicable. * For Upland cotton, the loan rate is for base quality rather than average as is done for other commodities.
FSA, Fibers, (202) 720-3392.

Table 2-5.—Cotton: Area, yield, and production in specified countries, 2007/2008–2009/2010

Country	Area			Yield per hectare			Production		
	2007/ 2008	2008/ 2009	2009/ 2010	2007/ 2008	2008/ 2009	2009/ 2010	2007/ 2008	2008/ 2009	2009/ 2010
	1,000 hec- tares	1,000 hec- tares	1,000 hec- tares	Kilo- grams	Kilo- grams	Kilo- grams	1,000 metric tons	1,000 metric tons	1,000 metric tons
Australia	65	164	200	2,144.00	1,991.00	1,742.00	640	1,500	1,600
Brazil	1,077	843	836	1,488.00	1,415.00	1,498.00	7,360	5,480	5,750
China	6,200	6,050	5,300	1,299.00	1,321.00	1,335.00	37,000	36,700	32,500
Greece	350	250	235	964.00	1,002.00	834.00	1,550	1,150	900
India	9,439	9,406	10,260	554.00	523.00	499.00	24,000	22,600	23,500
Pakistan	3,000	2,900	3,000	646.00	676.00	718.00	8,900	9,000	9,900
Syria	193	176	165	1,252.00	1,330.00	1,320.00	1,110	1,075	1,000
Turkey	520	340	280	1,298.00	1,236.00	1,361.00	3,100	1,930	1,750
Turkmenistan	600	600	550	472.00	490.00	495.00	1,300	1,350	1,250
Uzbekistan, Republic of	1,430	1,420	1,300	815.00	705.00	670.00	5,350	4,600	4,000
Others	5,807	5,518	5,140	1.81	1.68	1.65	10,533	9,256	8,506
Total foreign	28,681	27,667	27,266	3.52	3.42	3.32	100,843	94,641	90,656
United States	4,245	3,063	3,047	985.00	911.00	871.00	19,207	12,815	12,188
Total	32,926	30,730	30,313	794.00	761.00	739.00	120,050	107,456	102,844

FAS, Office of Global Analysis, (202) 720-6301. Prepared or estimated on the basis of official USDA production, supply, and distribution statistics from foreign governments.

Table 2-6.—Cotton: Supply and distribution, United States, 2000–2009

Year beginning August 1	Supply			Distribution				
	Beginning of season total [1]	Ginnings in season [2]	Total supply [1]	Consumption [1]			Exports	Carryover, end of season [1]
				Upland	American Pima	Total		
	1,000 bales	1,000 bales	1,000 bales	1,000 bales	1,000 bales	1,000 bales	1,000 bales	1,000 bales
2000	4,056	16,596	20,657	8,410	118	8,528	6,425	5,930
2001	5,930	19,729	25,650	7,289	99	7,388	10,649	7,305
2002	7,305	16,683	23,989	7,022	100	7,122	11,571	5,293
2003	5,193	17,729	22,921	6,076	61	6,137	13,330	3,381
2004	3,381	22,576	25,957	5,968	60	6,028	13,593	5,411
2005	5,368	23,253	28,576	5,604	49	5,653	17,038	5,877
2006	5,878	20,998	26,872	*	*	4,745	12,631	9,221
2007	9,223	18,713	27,929	*	*	4,499	13,237	9,699
2008	9,699	12,462	22,154	*	*	3,439	12,875	6,135
2009 [3]	6,136	11,832	17,963	*	*	3,336	11,687	2,828

[1] May include small volume of foreign growths.   [2] Ginnings during the 12 months, Aug. 1–July 31. Includes an allowance for "city crop" which consists of rebaled samples and pickings from cotton damaged by fire and weather.   [3] Prliminary.
* Withheld to avoid disclosing data for individual companies.
AMS, Cotton and Tobacco Programs, (901) 384–3016. Compiled from reports of the Bureau of the Census.

Table 2-7.—Cotton, American Upland: Percentage distribution of fiber strength, United States, 2005–2009

Fiber strength [1]	Year				
	2005	2006	2007	2008	2009
17 and below	*	*	*	-	*
18	*	*	*	*	*
19	*	*	*	*	*
20	*	*	*	*	*
21	*	*	*	*	*
22	*	0.1	0.1	*	*
23	0.1	0.2	0.3	0.1	0.1
24	0.3	0.7	0.9	0.3	0.4
25	1.4	2.0	2.3	1.0	1.4
26	4.5	5.4	5.3	3.0	4.9
27	11.4	11.2	10.6	7.5	11.8
28	20.5	17.6	16.4	14.1	19.7
29	23.9	20.4	20.0	20.0	21.9
30	17.9	18.0	18.4	21.5	17.8
31	9.6	12.0	12.8	17.2	11.6
32	4.6	6.2	6.9	9.1	5.9
33	2.2	2.8	3.2	3.4	2.4
34	1.3	1.5	1.8	1.4	1.1
35	1.1	1.0	0.8	0.9	0.6
36 and above	1.1	0.8	0.2	0.6	0.3
Average	29.2	29.2	29.2	29.7	29.1

[1] Fiber strength expressed in terms of ⅛" gage (grams per tex). *Less than 0.05 percent.
AMS, Cotton and Tobacco Programs, (901) 384–3016.

Table 2-8.—Cotton, American Upland: Estimated percentage of the crop forward contracted by growers, by State and United States, 2003–2009

State	Crop of—						
	2003	2004	2005	2006	2007	2008	2009
	Percent	Percent	Percent	Percent	Percent	Percent	Percent
AL	17	19	21	15	2	10	3
AZ	33	1	-	-	4	6	-
AR	39	*	11	12	3	5	-
CA	*	*	1	*	*	-	-
FL	4	-	2	-	-	16	10
GA	21	19	10	2	*	12	16
LA	58	13	28	22	39	68	2
MS	7	2	8	14	14	6	13
MO	51	*	22	3	-	19	*
NM	3	-	-	-	-	-	-
NC	9	16	9	9	1	10	4
OK	-	-	-	-	-	-	-
SC	11	25	19	6	6	10	2
TN	71	*	13	3	-	1	*
TX	10	5	6	3	3	11	3
US	21	7	10	7	4	12	4

*Less than 0.5 percent.
AMS, Cotton and Tobacco Programs, (901) 384–3016.

Table 2-9.—Cotton, American Upland: Carryover and crop, running bales, by grade groupings, United States, 2000–2009

Year beginning August 1	White color grades					Light spotted color grades				Other color grades[1]	All grades[2]
	21 and higher	31	41	51	61 and 71	22 and higher	32	42	52 and lower		
	1,000 bales	1,000 bales	1,000 bales	1,000 bales	1,000 bales	1,000 bales	1,000 bales	1,000 bales	1,000 bales	1,000 bales	1,000 bales
Carryover:											
2000	1,274	1,007	981	123	8	68	85	192	42	50	3,830
2001	1,392	1,712	1,464	218	4	63	148	601	150	67	5,819
2002	1,234	2,325	1,976	107	2	99	238	769	54	76	6,700
2003	596	988	1,804	502	8	37	193	475	251	115	4,972
2004	435	1,573	1,106	54	1	22	47	51	7	19	3,314
2005	975	1,042	1,609	530	18	42	154	505	186	339	5,402
2006	1,642	2,178	1,466	90	1	68	92	146	62	66	5,810
2007	1,874	3,909	2,611	132	3	59	133	209	127	39	9,096
2008	2,373	4,149	2,466	123	4	77	137	200	126	38	9,692
2009	852	1,999	2,536	236	0	45	64	84	55	3	5,874

	White color grades					Light spotted color grades				Other color grades[1]	All grades[3]
	21 and higher	31	41	51	61 and 71	22 and higher	32	42	52 and lower		
Crop:											
2001	4,950	6,593	3,997	443	8	391	654	1,296	276	431	19,039
2002	2,248	3,389	5,610	1,086	29	122	594	1,627	859	488	16,053
2003	3,971	7,755	4,423	193	2	156	278	319	67	124	17,290
2004	4,063	5,228	7,079	1,955	45	180	605	1,328	567	782	21,832
2005	7,698	8,029	4,297	541	5	303	591	699	312	164	22,638
2006	3,785	8,145	6,842	397	4	146	296	425	198	24	20,262
2007	6,376	3,794	4,788	592	4	188	238	1,184	745	16	17,925
2008	2,160	4,557	4,285	391	2	169	171	161	166	12	12,075
2009	2,696	3,419	3,665	776	21	55	97	241	419	30	11,419

[1] Includes all color grades of Spotted, Tinged, Yellow Stained, and Below Grade. [2] Carryover as reported by the Bureau of the Census, Crop as reported by AMS, Cotton Program. [3] Bales classed as reported by AMS, Cotton Program. AMS, Cotton and Tobacco Programs, (901) 384–3016.

Table 2-10.—Cotton, American Upland: Carryover (2000-2009) and crop (2000-2009), running bales, by staple groupings, United States

Year beginning August 1	Staple										All staples[1]
	26 and shorter	28	29	30	31	32	33	34	35	36 and longer	
	1,000 bales	1,000 bales	1,000 bales	1,000 bales	1,000 bales	1,000 bales	1,000 bales	1,000 bales	1,000 bales	1,000 bales	1,000 bales
Carryover:											
2000	(2)	1	10	46	85	386	651	969	820	862	3,830
2001	1	4	22	88	241	558	1,209	1,385	1,341	970	5,819
2002	-	-	4	9	32	200	708	1,995	2,071	1,681	6,700
2003	-	(2)	15	35	69	214	708	1,495	1,357	1,079	4,972
2004	-	1	3	14	33	142	389	1,189	869	674	3,314
2005	-	1	4	17	77	213	543	1,128	1,615	1,803	5,402
2006	-	-	(2)	4	32	173	510	1,582	1,849	1,659	5,810
2007	-	(2)	(2)	5	62	382	924	1,873	2,236	3,613	9,096
2008	-	(2)	(2)	6	62	368	892	1,827	2,312	4,225	9,692
2009	-	-	-	3	11	61	337	816	1,423	3,224	5,874
Crop:											
2000	2	20	86	229	558	1,408	2,915	4,196	3,661	3,273	16,348
2001	(2)	1	9	53	256	974	3,084	5,592	4,947	4,123	19,039
2002	(2)	2	22	123	457	1,259	2,840	4,324	3,596	3,429	16,053
2003	(2)	1	10	57	202	624	2,205	4,873	4,805	4,512	17,290
2004	(2)	1	9	56	196	723	2,175	4,630	6,543	7,499	21,832
2005	(2)	(2)	1	16	127	650	2,460	5,892	7,261	6,232	22,638
2006	(2)	1	7	29	136	588	1,764	3,735	5,181	8,821	20,262
2007	-	(2)	2	14	113	524	1,574	4,376	8,030	8,030	17,925
2008	-	(2)	1	7	41	195	685	1,675	2,541	6,930	12,075
2009	(2)	(2)	2	11	39	120	488	1,828	3,461	5,468	11,419

[1] Carryover as reported by the Bureau of the Census, Crop as reported by AMS, Cotton Program. [2] Less than 500 bales.
AMS, Cotton and Tobacco Programs, (901) 384–3016.

Table 2-11.—Cotton, American Pima: Carryover (2005-2009) and crop, running bales (2005-2009), running bales, by grade and staple, United States,

Year beginning August 1	Grade					Staple				All grades and staples [1]
	01 and 02	03	04	05	06 and 07	42 and shorter	44	46	48 and longer	
	1,000 bales	*1,000 bales*	*1,000 bales*	*1,000 bales*	*1,000 bales*	*1,000 bales*	*1,000 bales*	*1,000 bales*	*1,000 bales*	*1,000 bales*
Carryover:										
2005	8.8	1.9	0.7	0.4	0.4	0.0	0.3	9.2	1.2	10.7
2006	54.3	11.3	1.2	0.4	0.1	0.0	2.7	24.6	40.1	67.4
2007	76.0	45.1	1.7	1.9	0.1	0.0	51.4	56.1	18.1	125.6
2008	65.5	75.0	0.8	2.9	1.6	0.0	31.7	91.7	22.7	146.2
2009	243.5	15.3	2.1	0.3	0.5	0.1	25.4	128.2	107.8	261.5
Crop:										
2005	534.7	62.2	8.0	1.4	0.3	0.2	17.4	140.5	448.5	606.6
2006	621.9	97.5	8.0	1.9	0.4	0.2	19.1	163.3	547.2	729.8
2007	784.1	29.0	7.3	1.9	0.3	0.4	51.8	400.7	369.9	822.7
2008	391.1	18.5	1.8	0.3	1.0	0.0	10.5	126.7	275.5	412.7
2009	324.1	51.9	7.7	0.9	0.0	0.0	10.8	118.2	255.5	384.6

[1] Carryover as reported by the Bureau of the Census; crop as reported by AMS, Cotton Program.
AMS, Cotton and Tobacco Programs, (901) 384–3016.

Table 2-12.—Cotton, Upland: Average staple length of Upland cotton classed, by State and United States, 2003–2009

State	Average staple length (32ds of an inch) [1]						
	2003	2004	2005	2006	2007	2008	2009
AL	34.3	34.4	34.5	33.8	33.8	34.3	34.8
AZ	35.4	35.5	35.6	36.2	35.7	36.3	36.2
AR	34.8	35.3	34.9	35.4	35.0	36.1	35.6
CA	37.2	36.3	37.1	37.4	37.2	38.1	38.0
FL	34.3	34.8	34.6	34.7	34.4	35.0	34.8
GA	34.2	34.7	34.7	34.4	34.4	34.5	34.9
KS	32.2	31.7	33.5	34.1	35.1	35.7	35.6
LA	34.6	35.4	34.4	34.2	34.8	34.5	35.1
MS	34.5	35.3	34.6	34.1	34.6	35.9	35.5
MO	35.2	35.0	35.0	36.2	34.8	36.0	35.7
NM	36.1	36.1	36.1	37.0	37.0	37.2	36.5
NC	34.5	35.1	34.9	35.2	33.9	34.8	35.0
OK	34.6	34.4	34.8	35.6	35.4	36.0	35.5
SC	34.5	35.0	34.9	35.1	33.6	35.2	35.1
TN	34.2	34.0	34.2	35.2	33.3	35.1	35.0
TX	34.3	34.5	34.7	35.8	36.0	36.3	35.6
VA	35.3	35.0	35.6	35.5	34.0	34.5	35.4
Oth Sts	(2)	(2)	(2)	(2)	(2)	(2)	(2)
US	34.7	34.9	34.8	35.2	35.3	35.7	35.5

[1] Average calculated on numerical equivalents of the staple-length designations. For example, 7/8-inch = 28, 29/32-inch = 29, etc.   [2] Not available.
AMS, Cotton and Tobacco Programs, (901) 384–3016.

Table 2-13.—Cotton: United States exports by country of destination, 2007–2009

Country of destination	Year		
	2007	2008	2009
	Metric tons	*Metric tons*	*Metric tons*
Cotton linters:			
China	15,843	8,099	61,008
Germany(*)	4,464	5,215	2,613
New Zealand(*)	0	365	1,145
Korea, South	0	5	521
Spain	0	44	397
Japan	1,694	977	384
Netherlands	17	67	312
Bangladesh	0	0	298
Rest of World	1,329	647	1,514
World Total	23,346	15,516	72,918
Cotton < 1:			
China	117,699	109,572	76,497
Turkey	9,491	3,652	24,449
Pakistan	10,428	8,730	15,287
Taiwan	10,969	5,035	13,287
Thailand	14,198	13,159	12,508
Korea, South	8,554	6,736	10,481
India	7,342	12,791	9,838
Indonesia	9,558	14,103	8,620
Bangladesh	563	3,841	5,459
Rest of World	44,530	39,051	17,471
World Total	233,332	216,670	193,895
Cotton <1 1/8:			
Turkey	483,374	272,570	264,327
China	713,456	565,152	262,032
Mexico	183,745	165,592	187,768
Vietnam	59,420	104,853	119,874
Indonesia	167,553	171,471	99,609
Thailand	88,939	120,658	77,609
Pakistan	106,300	56,596	66,142
Taiwan	57,420	66,045	47,903
Bangladesh	35,574	42,339	43,375
Rest of World	368,768	325,783	264,798
World Total	2,264,549	1,891,058	1,433,115

See end of table.

Table 2-13.—Cotton: United States exports by country of destination, 2007–2009—Continued

Country of destination	Year		
	2007	2008	2009
	1,000 metric tons	*1,000 metric tons*	*1,000 metric tons*
Pima:			
China	39,708	15,222	25,686
Pakistan	18,629	9,050	11,172
India	14,397	9,840	9,712
Thailand	6,320	6,208	7,308
Korea, South	4,102	4,535	5,208
Indonesia	12,732	9,967	5,004
Bangladesh	1,436	1,905	3,503
Japan	30,516	31,874	3,126
Mexico	202	475	2,307
Taiwan	7,841	6,827	2,174
Peru	2,865	119	2,069
Hong Kong	1,115	716	2,037
Germany(*)	2,812	1,349	1,776
Vietnam	117	587	1,183
Turkey	13,390	7,635	1,070
Egypt	845	1,148	838
Brazil	765	829	643
Phillippines	0	397	487
Malaysia	212	386	447
El Salvador	236	0	299
Belgium	887	732	298
Italy(*)	1,758	0	174
Rest of World	2,080	5,138	213
World Total	162,886	114,355	86,718
Cotton, other:			
China	597,345	779,372	840,241
Turkey	68,454	63,241	109,552
Mexico	116,722	117,857	97,466
Indonesia	51,721	80,426	75,730
Pakistan	33,693	17,808	60,479
Bangladesh	4,492	13,866	37,090
Thailand	23,614	35,697	31,053
Vietnam	4,786	16,864	24,445
India	9,016	10,713	23,346
Taiwan	20,986	8,981	22,716
Colombia	3,399	12,712	22,049
Hong Kong	4,284	4,559	17,463
Peru	5,700	6,886	17,419
Korea, South	25,087	24,490	14,329
Japan	7,065	11,045	6,577
Malaysia	728	2,952	5,717
Morocco	0	220	4,796
Ecuador	522	1,895	3,782
Italy(*)	7,015	7,288	3,616
Brazil	3,348	1,194	2,734
Chile	1,621	3,671	2,410
Singapore	0	0	1,661
Guatemala	3,678	322	1,645
Rest of World	28,188	16,453	10,625
World Total	597,345	779,372	840,241

Note: (*) Denotes a country that is a summarization of its component countries.
FAS, Office of Global Analysis, (202) 720-6301.

Table 2-14.—Cotton: International trade, 2007/2008–2009/2010 [1]

Country	2007/2008	2008/2009	2009/2010
	1,000 bales	*1,000 bales*	*1,000 bales* [2]
Principle exporting countries:			
Australia	1,219	1,201	2,100
Brazil	2,231	2,739	1,990
Burkina	775	800	775
Greece	1,299	800	875
India	7,500	2,360	6,550
Kazakhstan	530	348	425
Pakistan	269	375	750
Tajikistan	525	350	450
Turkmenistan	825	600	1,075
Uzbekistan	4,200	3,000	3,800
Others	5,971	4,256	4,862
Total foreign	25,344	16,829	23,652
United States	13,653	13,276	12,000
Total	38,997	30,105	35,652

Country	2007/2008	2008/2009	2009/2010
	1,000 bales	*1,000 bales*	*1,000 bales* [2]
Principle importing countries:			
Bangladesh	3,500	3,800	4,000
China	11,530	6,996	10,940
Indonesia	2,300	2,000	2,100
Korea, South	975	988	1,000
Mexico	1,530	1,315	1,400
Pakistan	3,907	1,950	1,500
Taiwan	964	787	975
Thailand	1,928	1,602	1,800
Turkey	3,267	2,919	4,300
Vietnam	1,208	1,251	1,700
Others	7,870	6,458	6,491
Total foreign	38,979	30,066	36,206
United States	12		1
Total	38,991	30,066	36,207

[1] Marketing year beginning Aug. 1.   [2] 480-pound net weight.
FAS, Office of Global Analysis, (202) 720-6301. Prepared or estimated on the basis of official USDA production, supply, and distribution statistics from foreign governments.

Table 2-15.—Cotton, American Upland: high, low, and season average spot prices for the base quality in the designated markets, cents per pound, 2001–2009

Season beginning August 1	Color 41, Leaf 4, Staple 34 [1]		
	Average	High	Low
	Cents	*Cents*	*Cents*
2001	33.10	41.39	25.94
2002	47.46	55.86	36.56
2003	60.15	77.66	42.45
2004	45.61	52.30	40.39
2005	48.96	53.25	43.46
2006	48.67	60.67	42.84
2007	61.50	79.16	50.34
2008	47.87	62.69	36.28
2009	67.76	78.90	50.98

[1] Prices are for mixed lots, net weight, compressed, FOB car/truck.
AMS, Cotton Program, (901) 384-3016.

Table 2-16.—Cotton and cotton linters: United States imports for consumption, by country of origin, 2006–2008

Country of origin	Year beginning August		
	2006	2007	2008
	Metric tons	*Metric tons*	*Metric tons*
Cotton, linters:			
Syria	3,699	14,035	5,384
Turkey	0	3,214	5,016
Colombia	0	30	1,281
Mexico	1,577	639	1,183
Pakistan	0	22	23
Indonesia	0	0	19
Honduras	0	0	19
Canada	0	3	15
India	0	0	11
El Salvador	0	14	0
Germany	163	0	0
Israel	18	0	0
Austria	0	0	0
World Total	5,457	17,957	12,950
Cotton:			
Indonesia	0	0	129
India	1	25	48
United Kingdom	0	0	30
Italy	11	0	0
Canada	0	0	0
Costa Rica	62	0	0
Austria	0	0	0
Turkey	279	1,306	0
Pakistan	614	596	0
China, Peoples Republic of	87	0	0
Korea; Republic of	0	0	0
Japan	0	0	0
World Total	1,053	1,928	207
Cotton:			
Turkey	0	0	632
Brazil	0	0	0
Germany	0	0	0
India	0	50	0
Egypt	473	202	0
Uganda	0	25	0
World Total	473	277	632
Cotton:			
Turkey	0	275	947
Brazil	0	0	1
Canada	0	0	0
China; Peoples Republic of	0	0	0
Egypt	4,537	1,181	0
World Total	4,537	1,457	947

FAS, Office of Global Analysis, (202) 720-6301.

Table 2-17.—Cotton, American Upland: Percentage distribution of mike readings, by specified groups, United States, 2000–2009

Year beginning August 1	Mike groups						
	26 and below	27 to 29	30 to 32	33 to 34	35 to 49	50 to 52	53 and above
	Percent	*Percent*	*Percent*	*Percent*	*Percent*	*Percent*	*Percent*
2000	0.1	0.7	2.0	2.8	85.8	7.1	1.3
2001	*	0.2	0.7	1.1	75.9	15.7	6.0
2002	*	0.3	0.7	1.1	74.2	17.7	5.8
2003	*	0.3	0.9	1.4	83.6	11.2	5.8
2004	0.4	1.5	3.4	3.7	83.8	6.4	0.8
2005	*	1.5	4.0	4.4	82.0	6.5	5.8
2006	1.1	1.8	2.7	2.3	79.2	10.8	1.8
2007	0.1	0.6	1.8	2.8	87.4	6.5	0.8
2008	0.5	1.1	2.8	3.9	77.2	9.7	1.4
2009	1.9	2.6	3.9	3.7	81.0	6.0	0.9

(*) Less than 0.05 percent.
AMS, Cotton and Tobacco Programs, (901) 384–3016.

Table 2-18.—Cotton, American Upland: Average spot prices for specified grades of staple 34 in the designated markets for mixed lots, net weight, compressed, FOB car/truck, cents per pound, 2000–2009

Year beginning August 1	White				Light Spotted			Spotted	
	Color 31 Leaf 3	Color 41 Leaf 4	Color 51 Leaf 5	Color 61 Leaf 6	Color 32 Leaf 3	Color 42 Leaf 4	Color 52 Leaf 5	Color 33 Leaf 3	Color 43 Leaf 4
	Cents	Cents	Cents	Cents	Cents	Cents	Cents	Cents	Cents
2000	52.98	51.56	47.18	43.50	51.36	48.78	45.15	47.81	43.88
2001	34.66	33.10	29.32	26.87	33.26	31.04	28.12	30.42	27.50
2002	49.72	47.46	43.38	41.40	47.53	44.94	42.22	44.99	42.04
2003	62.24	60.15	56.05	53.89	60.03	57.42	54.89	57.15	54.58
2004	48.40	45.61	41.59	39.11	45.70	43.30	40.38	42.51	40.75
2005	51.33	48.96	44.84	42.34	48.72	46.42	43.41	45.98	44.05
2006	50.83	48.67	44.56	42.12	48.39	46.25	43.24	45.75	43.81
2007	63.46	61.50	57.35	54.95	61.16	59.01	56.09	58.58	56.59
2008	49.83	47.87	43.73	41.33	47.55	45.37	42.48	44.95	42.87
2009	69.97	67.76	63.47	61.21	67.44	65.26	62.37	64.84	62.76

AMS, Cotton and Tobacco Programs, (901) 384–3016.

Table 2-19.—Cotton, American Upland: Average spot prices for specified staple lengths of Grade 41 Leaf 4, in the designated markets for mixed lots, net weight, compressed, FOB car/truck, cents per pound, 2000–2009

Year beginning August 1	Staple							
	28	29	30	31	32	33	34	35
	Cents	Cents	Cents	Cents	Cents	Cents	Cents	Cents
2000	43.14	43.14	44.59	45.90	46.10	48.24	51.56	52.82
2001	29.12	29.12	29.77	30.53	30.01	31.24	33.10	34.31
2002	43.07	43.07	43.57	44.60	44.40	45.64	47.46	49.13
2003	55.39	55.39	55.94	56.95	57.08	58.42	60.15	61.71
2004	41.54	41.54	42.13	43.28	43.32	44.07	45.61	47.02
2005	44.26	44.26	44.96	46.13	46.14	46.84	48.96	50.36
2006	43.92	43.92	44.67	45.79	45.89	46.53	48.67	49.97
2007	56.75	56.75	57.50	58.62	58.50	59.27	61.50	62.69
2008	43.12	43.12	43.87	44.99	44.82	45.62	47.87	49.04
2009	62.83	62.83	63.58	64.71	64.71	65.51	67.76	68.98

AMS, Cotton and Tobacco Programs, (901) 384–3016.

Table 2-20.—Cotton, American Upland: Season average spot prices for the base quality, by designated markets, cents per pound, 2004–2009 [1]

Market	Color 41, Leaf 4, Staple 34 [2]					
	2004	2005	2006	2007	2008	2009
	Cents	Cents	Cents	Cents	Cents	Cents
Southeast	45.91	49.65	49.90	63.95	48.97	70.13
North Delta	46.02	49.67	49.46	62.67	47.99	69.30
South Delta	46.02	49.63	49.46	62.67	47.99	69.30
East TX–OK	44.22	47.69	48.17	60.89	47.08	66.57
West Texas	44.08	47.78	48.06	60.64	48.93	66.38
Desert SW	45.66	48.26	47.08	59.07	47.03	65.40
SJ Valley	47.38	50.06	48.58	60.57	49.10	67.20
Average	45.61	48.96	48.67	61.50	47.87	67.76

[1] Year beginning August 1.   [2] Prices are for mixed lots, net weight, compressed, FOB car/truck.
AMS, Cotton and Tobacco Programs, (901) 384–3016.

Table 2-21.—Sugarbeets: Area, yield, production, and value, United States, 2000–2009 [1]

Year	Area		Yield per harvested acre	Production	Marketing year average price per ton received by farmers [2]	Value of production
	Planted	Harvested				
	1,000 acres	*1,000 acres*	*Tons*	*1,000 tons*	*Dollars*	*1,000 dollars*
2000	1,564.2	1,373.0	23.7	32,541	34.20	1,113,030
2001	1,365.3	1,241.1	20.7	25,708	39.80	1,023,054
2002	1,427.3	1,360.7	20.4	27,707	39.60	1,097,329
2003	1,365.4	1,347.8	22.8	30,710	41.40	1,270,026
2004	1,345.6	1,306.7	23.0	30,021	36.90	1,109,272
2005	1,299.8	1,242.9	22.1	27,433	43.50	1,193,151
2006	1,366.2	1,303.6	26.1	34,064	44.20	1,506,985
2007	1,268.8	1,246.8	25.5	31,834	42.00	1,337,173
2008	1,090.7	1,004.5	26.8	26,881	48.00	1,289,621
2009 [3]	1,183.2	1,145.3	25.8	29,519	NA	NA

[1] Relates to year of intended harvest except for overwintered spring planted beets in CA. [2] Prices do not include Government payments under the Sugar Act. [3] Preliminary. NA-not available.
NASS, Crops Branch, (202) 720–2127.

Table 2-22.—Sugarbeets: Area, yield, and production, by State and United States, 2007–2009 [1]

State	Area planted			Area harvested			Yield per harvested acre			Production		
	2007	2008	2009	2007	2008	2009	2007	2008	2009	2007	2008	2009
	1,000 acres	*1,000 acres*	*1,000 acres*	*1,000 acres*	*1,000 acres*	*1,000 acres*	*Tons*	*Tons*	*Tons*	*1,000 tons*	*1,000 tons*	*1,000 tons*
CA	40.0	26.0	25.1	39.1	25.3	24.6	35.5	41.6	40.0	1,388	1,052	984
CO	32.0	33.8	35.1	29.2	28.6	35.0	26.2	26.5	27.0	765	758	945
ID	169.0	131.0	164.0	167.0	116.0	163.0	34.4	31.2	34.3	5,745	3,619	5,591
MI	150.0	137.0	138.0	149.0	136.0	136.0	23.4	28.7	24.4	3,487	3,903	3,318
MN	486.0	440.0	463.0	481.0	399.0	448.0	23.8	24.7	23.5	11,448	9,855	10,528
MT	47.5	31.7	38.4	47.0	30.7	33.6	24.7	26.8	29.8	1,161	823	1,001
NE	47.5	45.2	53.0	44.3	37.3	52.6	23.5	22.6	24.5	1,041	843	1,289
ND	252.0	208.0	225.0	247.0	197.0	218.0	23.1	25.9	22.0	5,706	5,102	4,796
OR	12.0	6.7	10.6	11.0	5.9	10.5	31.9	33.1	37.6	351	195	395
WA [2]	2.0	1.6	....	2.0	1.6	....	42.0	41.9	....	84	67	....
WY	30.8	29.7	31.0	30.2	27.1	24.0	21.8	24.5	28.0	658	664	672
US	1,268.8	1,090.7	1,183.2	1,246.8	1,004.5	1,145.3	25.5	26.8	25.8	31,834	26,881	29,519

[1] Relates to year of intended harvest except for overwintered spring planted beets in CA. [2] Estimates discontinued in 2009.
NASS, Crops Branch, (202) 720–2127.

Table 2-23.—Sugarbeets: Production and value, by State and United States, 2007–2008 [1]

State	Production		Marketing year average price per ton received by farmers		Value of production	
	2007	2008	2007	2008	2007	2008
	1,000 tons	*1,000 tons*	*Dollars*	*Dollars*	*1,000 dollars*	*1,000 dollars*
CA	1,388	1,052	43.60	44.80	60,517	47,130
CO	765	758	36.00	47.80	27,540	36,232
ID	5,745	3,619	36.90	42.00	211,991	151,998
MI	3,487	3,903	36.00	44.00	125,532	171,732
MN	11,448	9,855	45.20	49.90	517,450	491,765
MT	1,161	823	39.10	50.80	45,395	41,808
NE	1,041	843	40.40	50.80	42,056	42,824
ND	5,706	5,102	46.30	51.00	264,188	260,202
OR	351	195	36.90	42.00	12,952	8,190
WA	84	67	36.90	42.00	3,100	2,814
WY	658	664	40.20	52.60	26,452	34,926
US	31,834	26,881	42.00	48.00	1,337,173	1,289,621

[1] Relates to year of intended harvest in all States except CA. In CA, relates to year of intended harvest for fall planted beets in central CA and to year of planting for overwintered beets in central and southern CA.
NASS, Crops Branch, (202) 720–2127.

Table 2-24.—Sugarcane for sugar and seed: Area, yield, production, and value, United States, 2000–2009

Year [1]	Area harvested			Yield of cane per acre			Production		
	For sugar	For seed	Total	For sugar	For seed	For sugar and seed	For sugar	For seed	Total
	1,000 acres	1,000 acres	1,000 acres	Tons	Tons	Tons	1,000 tons	1,000 tons	1,000 tons
2000	976.7	55.6	1,032.3	35.1	32.8	35.0	34,291	1,823	36,114
2001	970.3	57.5	1,027.8	33.8	31.5	33.7	32,775	1,812	34,587
2002	971.9	51.3	1,023.2	34.9	32.2	34.7	33,903	1,650	35,553
2003	930.6	61.7	992.3	34.3	31.1	34.1	31,942	1,916	33,858
2004	879.5	58.7	938.2	31.0	30.2	30.9	27,243	1,770	29,013
2005	858.2	63.7	921.9	28.8	29.5	28.9	24,728	1,878	26,606
2006	846.6	51.1	897.7	33.0	31.4	32.9	27,962	1,602	29,564
2007	827.9	51.7	879.6	34.2	32.8	34.1	28,273	1,696	29,969
2008	821.6	46.4	868.0	31.8	31.7	31.8	26,131	1,472	27,603
2009 [2]	820.7	57.0	877.7	34.5	32.8	34.4	28,283	1,868	30,151

Year [1]	Marketing year average price per ton received by farmers [3]	Value of production [4]	
		Of cane used for sugar	Of cane used for sugar and seed [4]
	Dollars	1,000 dollars	1,000 dollars
2000	26.10	895,917	941,791
2001	29.00	951,813	1,003,046
2002	28.40	961,896	1,007,142
2003	29.50	943,646	998,269
2004	28.30	771,734	821,118
2005	28.40	701,920	754,529
2006	30.40	849,157	897,601
2007	29.40	831,218	880,616
2008	29.50	771,134	814,479
2009	NA	NA	NA

[1] In Hawaii, harvest continues throughout the year and production statistics are on a calendar year basis. In other states, harvest is seasonal and the production statistics year relates to the year in which the season begins. [2] Preliminary. [3] Prices do not include Government payments under the Sugar Act. [4] Price per ton of cane for sugar used in evaluating value of production for seed. NA-not available.
NASS, Crops Branch, (202) 720–2127.

Table 2-25.—Sugarcane for sugar and seed: Production and value, by State and United States, 2007–2008

State	Sugarcane for sugar						Sugar and seed: Value of production	
	Production		Price per ton [1]		Value of production [1]		2007	2008
	2007	2008	2007	2008	2007	2008		
	1,000 tons	1,000 tons	Dollars	Dollars	1,000 dollars	1,000 dollars	1,000 dollars	1,000 dollars
FL	13,500	12,634	31.60	30.10	426,600	380,283	447,993	398,975
HI	1,493	1,422	31.90	31.10	47,627	44,224	49,892	46,463
LA	11,856	10,754	27.30	29.10	323,669	312,941	348,567	333,544
TX	1,424	1,321	23.40	25.50	33,322	33,686	34,164	35,497
US	28,273	26,131	29.40	29.50	831,218	771,134	880,616	814,479

[1] Price per ton of cane for sugar used in evaluating value of production for seed.
NASS, Crops Branch, (202) 720–2127.

Table 2-26.—Sugarcane for sugar and seed: Area, yield, and production, by State and United States, 2007–2009

State	Sugarcane for sugar and seed [1]								
	Area harvested			Yield of cane per acre [2]			Cane production [2]		
	2007	2008	2009	2007	2008	2009	2007	2008	2009
	1,000 acres	1,000 acres	1,000 acres	Tons	Tons	Tons	1,000 tons	1,000 tons	1,000 tons
For sugar:									
FL	375.0	384.0	372.0	36.0	32.9	36.1	13,500	12,634	13,429
HI	20.4	20.4	19.7	73.2	69.7	71.0	1,493	1,422	1,399
LA	390.0	380.0	390.0	30.4	28.3	31.0	11,856	10,754	12,090
TX	42.5	37.2	39.0	33.5	35.5	35.0	1,424	1,321	1,365
US	827.9	821.6	820.7	34.2	31.8	34.5	28,273	26,131	28,283
For seed:									
FL	18.0	17.0	18.0	37.6	36.5	36.3	677	621	653
HI	2.5	2.4	2.0	28.3	30.0	30.0	71	72	60
LA	30.0	25.0	35.0	30.4	28.3	31.0	912	708	1,085
TX	1.2	2.0	2.0	30.4	35.5	35.0	36	71	70
US	51.7	46.4	57.0	32.8	31.7	32.8	1,696	1,472	1,868
For sugar and seed:									
FL	393.0	401.0	390.0	36.1	33.1	36.1	14,177	13,255	14,082
HI	22.9	22.8	21.7	68.3	65.5	67.2	1,564	1,494	1,459
LA	420.0	405.0	425.0	30.4	28.3	31.0	12,768	11,462	13,175
TX	43.7	39.2	41.0	33.4	35.5	35.0	1,460	1,392	1,435
US	879.6	868.0	877.7	34.1	31.8	34.4	29,969	27,603	30,151

[1] In Hawaii, harvest continues throughout the year and production statistics are on a calendar year basis. In other states, harvest is seasonal and the production statistics year relates to the year in which the season begins.   [2] Net tons.
NASS, Crops Branch, (202) 720–2127.

Table 2-27.—Sugar, cane (raw value [1]): Refiners' raw stocks, receipts, meltings, continental United States, 2000–2009

Year	Jan. 1 stocks	Receipts [2]	Meltings
	1,000 tons	1,000 tons	1,000 tons
2000	356	5,543	5,575
2001	274	5,362	5,221
2002	351	5,607	5,681
2003	299	5,408	5,533
2004	286	5,181	5,171
2005	244	5,215	5,270
2006	217	5,543	5,405
2007	358	5,388	5,464
2008	299	5,634	5,329
2009	304	5,459	5,575

[1] Raw value is the equivalent in terms of 96° sugar.   [2] Receipts include refiners' total offshore raw sugar receipts in continental U.S. ports, whether entered through the customs or held pending availability of quota and raw cane sugar produced from sugarcane in the continental United States.
FSA, Dairy and Sweeteners Analysis, (202) 720–3451.

Table 2-28.—Sugar, cane and beet: Domestic marketings, by source of supply, continental United States, 2007–2009 [1]

Area of supply	2007	2008	2009
	1,000 tons	1,000 tons	1,000 tons
Domestic areas:			
Mainland (beet)	5,206	5,258	4,442
Mainland and Hawaii (cane)	5,520	5,431	5,769
Total domestic areas	10,726	10,689	10,211

[1] Source: U.S. Census.
FSA, Dairy and Sweeteners Analysis Division, (202) 720–3451.

Table 2-29.—Sugar, cane and beet (refined): Stocks, production or receipts, and deliveries, continental United States, 2000–2009

Item and year	Cane sugar refineries	Beet sugar factories	Importers of direct consumption sugar	Mainland cane sugar mills [1]	Total
JAN. 1 STOCKS [2]	*1,000 tons*	*1,000 tons*	*1,000 tons*	*1,000 tons*	*1,000 tons*
2000	208	1,554	0	22	1,784
2001	262	1,972	0	19	2,253
2002	288	1,812	0	19	2,119
2003	298	1,374	0	6	1,678
2004	326	1,853	0	5	2,184
2005	368	1,782	0	4	2,154
2006	328	1,429	0	7	1,764
2007	452	1,792	0	3	2,247
2008	406	1,806	0	4	2,216
2009	440	1,464	0	5	1,909
PRODUCTION OR RECEIPTS					
2000	5,681	6,014	37	32	11,764
2001	5,467	4,839	58	26	10,390
2002	5,896	4,258	109	8	10,271
2003	5,761	4,817	60	8	10,646
2004	5,389	5,305	64	16	10,774
2005	5,112	4,690	197	19	10,018
2006	5,741	4,758	576	16	11,091
2007	5,525	5,219	207	21	10,972
2008	5,460	4,937	916	28	11,341
2009	5,862	4,434	835	34	11,165
DELIVERIES [3]					
2000	5,738	5,573	37	15	11,363
2001	5,538	4,961	58	13	10,570
2002	5,768	4,596	109	15	10,488
2003	5,573	4,476	60	8	10,117
2004	5,362	5,153	64	16	10,595
2005	5,453	5,012	197	17	10,679
2006	5,587	4,419	576	19	10,601
2007	5,520	5,206	207	20	10,953
2008	5,404	5,258	916	27	11,605
2009	5,769	4,441	835	32	11,077

[1] Sugar for human consumption only. [2] Stocks include sugar in bond and in Customs custody and control. [3] Consists of all refined sugar.
FSA, Dairy and Sweeteners Analysis, (202) 720–3451.

Table 2-30.—Sugar (raw and refined): Average price per pound at specified markets, 2000–2009

Year	Cane sugar		Refined beet: Mid-west	Retail price, granulated: United States
	Raw, 96 centrifugal			
	Caribbean ports, f.o.b. and stowed, plus freight to Far East	New York, c.i.f. duty paid		
	Cents	*Cents*	*Cents*	*Cents*
2000	8.51	19.09	20.80	42.41
2001	9.12	21.11	23.31	43.42
2002	7.88	20.87	25.79	43.10
2003	7.51	21.42	26.21	42.68
2004	8.61	20.46	23.48	42.64
2005	11.35	21.28	29.54	43.54
2006	15.50	22.14	33.10	49.58
2007	11.60	20.99	25.06	51.48
2008	13.84	21.30	32.54	52.91
2009	18.72	24.93	38.10	57.03

ERS, Specialty Crops Branch, (202) 694–5247. Compiled from the following sources: (New York) Coffee, Sugar & Cocoa Exchange; the U.S. Department of Labor, Bureau of Labor Statistics; Milling and Baking News.

Table 2-31.—Sugar, centrifugal: International trade, 2006/2007–2008/2009

Country	2006/2007	2007/2008	2008/2009
	1,000 Metric tons, raw value		
Principle exporting countries:			
Australia	3,860	3,700	3,522
Brazil	20,850	19,500	21,550
Cuba	705	800	725
EU-27	2,439	1,656	1,331
Guatemala	1,500	1,333	1,654
Mexico	160	677	1,367
Saudi Arabia	540	700	670
South Africa	1,267	1,154	1,185
Thailand	4,705	4,914	5,295
United Arab Emirates ...	1,600	1,715	1,725
Others	13,430	15,202	9,713
Total Foreign	51,056	51,351	48,737
United States	383	184	123
Total	51,439	51,535	48,860

Country	2006/2007	2007/2008	2008/2009
	1,000 Metric tons, raw value		
Principle importing countries:			
Egypt	936	1,390	1,382
EU-27	3,530	2,948	3,173
India			2,786
Indonesia	1,800	2,420	2,197
Japan	1,405	1,440	1,452
Korea,South	1,518	1,648	1,550
Malaysia	1,670	1,390	1,430
Russia	2,950	3,100	3,100
Saudi Arabia	1,280	1,625	1,575
United Arab Emirates ...	1,605	1,890	1,930
Others	24,923	25,149	24,798
Total Foreign	41,617	43,000	45,373
United States	1,887	2,377	2,796
Total	43,504	45,377	48,169

FAS, Office of Global Analysis, (202) 720-6301. Prepared or estimated on the basis of official USDA production, supply, and distribution statistics from foreign governments.

Table 2-32.—Sugar, cane and beet (raw value): Production, stocks, trade, and supply available for consumption in continental United States includes Puerto Rico, 2000–2009

Year	Production	Visible stocks beginning of period	Imports	Exports	Total deliveries
	1,000 short tons	1,000 short tons	1,000 short tons	1,000 short tons	1,000 short tons
2000	8,955	3,855	1,639	109	10,091
2001	8,642	4,337	1,643	147	10,075
2002	7,504	4,525	1,574	136	9,994
2003	8,929	3,432	1,564	148	9,713
2004	8,366	4,088	1,652	280	9,901
2005	7,478	4,029	2,143	243	10,213
2006	7,754	3,357	3,195	299	10,162
2007	8,467	4,039	2,238	368	10,265
2008	7,947	4,012	2,844	168	10,912
2009	7,530	3,976	3,009	150	10,867

ERS, Specialty Crops Branch, (202) 694–5247.

Table 2-33.—Honey: United States exports and imports for consumption, by country of origin, 2007–2009

Country of origin	2007	2008	2009
	Metric tons	Metric tons	Metric tons
Exports:			
Canada	450	624	686
Yemen(*)	164	280	529
Israel(*)	855	1,253	433
Korea, South	402	332	291
Kuwait	102	109	244
Philippines	96	137	238
Taiwan	35	49	233
United Arab Emirates	85	160	208
Indonesia	119	184	207
Saudi Arabia	131	182	128
China	98	58	85
India	61	31	83
Panama	6	7	58
Thailand	115	83	49
Malaysia	307	127	47
Bahamas; The	14	14	35
Hong Kong	35	59	28
Guatemala	22	43	28
Australia(*)	50.3	82.5	26.7
Bahrain	4.5	22.3	26.1
Netherlands	7	9	24
Rest of World	271	483	251
World Total	3,781	4,572	4,381
Imports:			
Brazil	12,103	13,598	17,709
Vietnam	15,707	19,378	17,430
India	7,671	13,648	13,137
Argentina	20,379	10,043	10,899
Malaysia	1,891	4,150	9,068
Canada	13,961	17,305	8,301
Taiwan	753	3,983	5,576
Indonesia	447	1,814	5,124
Thailand	790	956	1,847
Mexico	3,192	1,412	1,625
New Zealand (*)	355	650	1,002
Mongolia	1,060	363	836
Ukraine	502	84	635
Germany(*)	447	358	277
Rest of World	26,421	17,243	1,987
World Total	105,676	104,984	95,453

Note: (*) Denotes a country that is a summarization of its component countries.
FAS, Office of Global Analysis, (202) 720-6301.

Table 2-34.—Honey: Number of colonies, yield, production, stocks, price and value, United States, 2000–2009 [1]

State	Honey pro-ducing colo-nies [2]	Yield per colony	Production [3]	Stocks Dec 15 [4]	Average price per pound [5]	Value of produc-tion
	1,000	*Pounds*	*1,000 pounds*	*1,000 pounds*	*Cents*	*1,000 dollars*
2000	2,622	84	220,286	85,244	60	132,865
2001	2,550	73	186,051	64,901	71	132,989
2002	2,574	67	171,718	39,393	133	228,338
2003	2,599	69.9	181,724	40,785	138.7	252,051
2004	2,554	71.8	183,494	61,203	108.8	199,641
2005	2,409	72.5	174,614	62,455	92.2	160,994
2006	2,394	64.7	154,910	60,484	100.5	155,685
2007	2,443	60.7	148,341	52,635	107.7	159,763
2008	2,342	69.9	163,789	51,159	142.1	232,744
2009	2,462	58.5	144,108	37,153	144.5	208,236

[1] For producers with 5 or more colonies. [2] Honey producing colonies are the maximum number of colonies from which honey was taken during the year. It is possible to take honey from colonies which did not survive the entire year. [3] Due to rounding, total colonies multiplied by total yield may not exactly equal production. [4] Stocks held by producers. [5] Average price per pound based on expanded sales.
NASS, Livestock Branch, (202) 720-3570.

Table 2-35.—Honey: Number of colonies, yield, production, stocks, price and value, by State and United States, 2009 [1]

State	Honey pro-ducing colo-nies [2]	Yield per colony	Production	Stocks Dec 15 [3]	Average price per pound [4]	Value of production [5]
	1,000	*Pounds*	*1,000 pounds*	*1,000 pounds*	*Cents*	*1,000 dollars*
AL	9	49	441	66	182	803
AZ	20	52	1,040	562	153	1,591
AR	24	57	1,368	301	139	1,902
CA	355	33	11,715	2,109	139	16,284
CO	28	53	1,484	326	140	2,078
FL	150	68	10,200	1,428	138	14,076
GA	65	41	2,665	346	147	3,918
HI	10	95	950	323	163	1,549
ID	103	46	4,738	1,706	145	6,870
IL	8	34	272	57	226	615
IN	9	32	288	101	198	570
IA	26	42	1,092	339	151	1,649
KS	9	63	567	164	189	1,072
KY	5	35	175	25	273	478
LA	37	103	3,811	610	132	5,031
ME	6	50	300	51	186	558
MI	66	60	3,960	1,505	151	5,980
MN	122	65	7,930	1,427	140	11,102
MS	14	104	1,456	87	132	1,922
MO	11	47	517	57	198	1,024
MT	146	70	10,220	3,577	145	14,819
NE	48	56	2,688	1,102	144	3,871
NV	10	52	520	57	129	671
NJ	9	32	288	46	193	556
NM	7	60	420	143	163	685
NY	47	65	3,055	978	183	5,591
NC	11	45	495	84	252	1,247
ND	450	77	34,650	7,623	137	47,471
OH	11	50	550	132	275	1,513
OR	55	34	1,870	767	149	2,786
PA	21	40	840	319	199	1,672
SD	270	66	17,820	6,237	139	24,770
TN	7	51	357	86	235	839
TX	74	63	4,662	886	138	6,434
UT	26	38	988	198	147	1,452
VT	5	49	245	69	236	578
VA	6	39	234	56	328	768
WA	62	44	2,728	1,064	149	4,065
WV	5	37	185	33	267	494
WI	63	60	3,780	1,588	151	5,708
WY	37	48	1,776	391	143	2,540
Oth Sts [6] [7] ...	15	51	768	127	280	2,150
US [7] [8]	2,462	58.5	144,108	37,153	144.5	208,236

[1] For producers with 5 or more colonies. Colonies which produced honey in more than one State were counted in each State. [2] Honey producing colonies are the maximum number of colonies from which honey was taken during the year. It is possible to take honey from colonies which did not survive the entire year. [3] Stocks held by producers. [4] Average price per pound based on expanded sales. [5] Value of production is equal to production multiplied by average price per pound. [6] CT, DE, MD, MA, NH, OK, RI, and SC not published separately to avoid disclosing data for individual operations. [7] Due to rounding, total colonies multiplied by total yield may not exactly equal production. [8] Summation of States will not equal U.S. level value of production.
NASS, Livestock Branch, (202) 720-3570.

Table 2-36.—U.S. per capita caloric sweeteners estimated deliveries for domestic food and beverage, use by calendar year 2000–2009

| Calendar year | U.S.population (July 1) | Refined sugar | Corn Sweetener | | | | Pure honey | Edible syrups | Total caloric sweeteners |
			HFCS	Glucose syrup	Dex-trose	Total			
									Millions
2000	282.3	65.5	62.7	15.8	3.4	81.8	1.1	0.4	148.9
2001	285.0	64.5	62.6	15.5	3.3	81.4	0.9	0.4	147.3
2002	287.7	63.3	62.9	15.5	3.3	81.6	1.1	0.4	146.5
2003	290.3	61.0	61.0	15.2	3.1	79.3	1.0	0.4	141.7
2004	293.0	61.7	59.9	15.6	3.3	78.9	0.9	0.4	141.9
2005	295.7	63.2	59.2	15.3	3.3	77.8	1.1	0.4	142.5
2006	298.4	62.6	58.3	13.8	3.1	75.2	1.2	0.4	139.4
2007	301.1	62.3	56.3	13.7	3.0	73.0	0.9	0.4	136.7
2008	303.8	65.4	53.2	13.4	2.8	69.3	1.0	0.5	136.2
2009	307.2	63.6	50.1	13.0	2.7	65.8	0.9	0.5	130.7

Note: Total may not add exactly, due to rounding.
ERS, Market and Trade Economics Division, Specialty Crops Branch, (202) 694–5247.

Table 2-37.—Tobacco: Area, yield, production, price, and value, United States, 2000–2009

Year	Area harvested	Yield per acre	Production [1]	Marketing year average price per pound received by farmers	Value of production
	Acres	*Pounds*	*1,000 pounds*	*Dollars*	*1,000 dollars*
2000	469,420	2,244	1,053,264	1.910	2,001,811
2001	432,490	2,292	991,293	1.956	1,938,892
2002	427,310	2,039	871,122	1.936	1,686,809
2003	411,150	1,952	802,560	1.964	1,576,436
2004	408,050	2,161	881,875	1.984	1,749,856
2005	297,080	2,171	645,015	1.642	1,059,324
2006	339,000	2,147	727,897	1.665	1,211,885
2007	356,000	2,213	787,653	1.693	1,329,235
2008	354,490	2,258	800,504	1.859	1,488,069
2009	354,140	2,325	823,290	1.820	1,498,629

[1] Production figures are on farm-sales-weight basis.
NASS, Crops Branch, (202) 720–2127.

Table 2-38.—Tobacco: Area, yield, and production, by State and United States, 2007–2009

| State | Area harvested | | | Yield per harvested acre | | | Production | | |
	2007	2008	2009 [1]	2007	2008	2009 [1]	2007	2008	2009 [1]
	Acres	*Acres*	*Acres*	*Pounds*	*Pounds*	*Pounds*	*1,000 pounds*	*1,000 pounds*	*1,000 pounds*
CT	2,900	2,600	1,800	1,733	1,352	1,283	5,025	3,516	2,310
GA	18,500	16,000	14,000	2,150	2,100	2,000	39,775	33,600	28,000
KY	89,200	87,800	88,700	2,209	2,345	2,333	197,040	205,850	206,900
MA	1,320	690	390	1,725	1,403	1,331	2,277	968	519
MO [2]	1,600	1,500		2,330	2,240		3,728	3,360	
NC	170,000	174,300	177,400	2,255	2,240	2,389	383,420	390,360	423,856
OH	3,500	3,400	3,400	2,050	2,050	2,000	7,175	6,970	6,800
PA	7,900	7,900	8,200	2,318	2,232	2,276	18,310	17,630	18,660
SC	20,500	19,000	18,500	2,250	2,100	2,100	46,125	39,900	38,850
TN	19,980	21,800	21,600	1,934	2,403	2,313	38,636	52,380	49,960
VA	20,600	19,500	20,150	2,240	2,357	2,354	46,142	45,970	47,435
US	356,000	354,490	354,140	2,213	2,258	2,325	787,653	800,504	823,290

[1] Preliminary.   [2] Estimates discontinued in 2009.
NASS, Crops Branch, (202) 720–2127.

Table 2-39.—Tobacco: Area, yield, production, stocks, supply, disappearance, and price, by types, United States including Puerto Rico, 1999–2006 (farm-sales-weight basis)[1]

Type and crop year	Area	Yield per acre	Produc- tion	Stocks[2]	Supply	Disappearance			Average price per pound to growers
						Total	Exports	Domes- tic	
	Acres	*Pounds*	*1,000 pounds*	*1,000 pounds*	*1,000 pounds*	*1,000 pounds*	*1,000 pounds*	*1,000 pounds*	*Cents*
Total flue-cured, types 11–14:[3]									
1999	303,800	2,162	656,752	1,234,280	1,888,172	698,684	261,818	436,866	173.6
2000	250,000	2,396	598,915	1,189,488	1,753,609	717,242	238,025	479,217	179.3
2001	238,100	2,432	579,091	1,036,367	1,580,790	664,912	276,007	388,905	185.8
2002	245,600	2,105	525,940	915,878	1,480,678	643,008	219,631	423,377	182.5
2003	233,400	1,957	525,941	837,670	1,345,326	522,478	215,520	306,958	184.9
2004	228,400	2,272	499,330	822,848	1,322,178	526,210	188,627	337,583	184.5
2005	175,500	2,182	382,950	795,968	1,178,918	574,900	316,537	258,363	147.4
2006	208,100	2,185	454,740	604,018	1,058,758	....	....	....	....
Total fire-cured, types 21–23:									
1999	16,420	2,319	38,075	89,390	127,465	36,246	12,979	14,312	226.4
2000	17,540	2,944	51,635	91,219	142,854	44,869	26,292	18,600	216.3
2001	14,620	3,096	45,299	97,962	143,261	38,955	16,379	22,576	214.9
2002	10,970	3,182	33,437	104,306	139,214	37,342	10,733	26,609	237.8
2003	11,250	3,067	34,508	101,872	136,380	33,788	20,259	13,259	247.5
2004	11,020	3,167	37,151	102,592	139,743	33,702	9,198	24,504	254.0
2005	11,840	3,178	37,631	104,819	142,450	40,501	30,703	9,798	236.9
2006	11,250	3,303	37,158	101,949	139,107	....	....	....	....
Virginia fire-cured, type 21:									
1999	1,600	1,670	2,672	2,669	5,341	1,897	979	918	181.9
2000	1,300	1,700	2,548	3,444	5,992	1,806	1,000	806	163.7
2001	1,200	1,835	2,202	4,168	6,388	1,567	150	1,417	175.8
2002	730	2,015	1,471	4,821	6,292	1,997	64	1,933	188.4
2003	550	1,525	839	4,295	5,134	1,359	63	1,296	160.3
2004	710	1,895	1,345	3,775	5,120	1,242	400	842	178.4
2005	550	2,300	805	3,878	4,683	....	....	....	....
2006	....	....	....	....	....	....	....	....	....
Kentucky and Ten- nessee fire-cured, types 22–23:									
1999	14,970	2,365	35,403	86,721	122,124	34,349	20,955	13,394	229.8
2000	16,240	3,023	49,087	87,775	136,862	43,086	25,292	17,794	216.3
2001	13,420	3,211	43,097	89,766	136,873	37,388	16,229	21,159	217.2
2002	10,240	3,265	33,437	90,787	132,922	35,345	10,669	24,676	214.9
2003	10,700	3,147	33,669	93,162	131,246	32,429	13,196	19,233	176.7
2004	11,020	3,249	35,806	94,315	134,623	32,460	8,798	23,662	161.6
2005	17,000	3,209	36,900	100,500	137,400	38,855	32,520	6,335	237.7
2006	14,000	3,343	36,440	98,545	134,985	....	....	....	....
Burley, type 31:[3]									
1999	300,600	1,829	555,185	901,415	1,452,573	412,531	139,262	273,269	189.9
2000	185,400	1,957	362,788	1,040,042	1,355,481	666,022	142,020	523,012	196.3
2001	164,400	2,032	334,066	689,459	1,033,119	385,238	139,802	245,436	197.3
2002	157,700	1,861	303,895	647,881	947,726	369,561	148,618	220,943	197.4
2003	152,300	1,850	303,896	578,165	849,861	309,832	173,650	136,182	197.7
2004	153,150	1,908	303,897	540,029	820,129	329,645	227,571	102,074	199.4
2005	100,150	2,031	203,383	492,623	696,006	292,640	92,264	200,376	156.4
2006	103,700	2,166	224,580	403,366	627,946	....	....	....	....
Maryland, type 32:[3]									
1998	9,800	1,568	15,370	22,543	37,913	18,855	6,228	12,627	129.1
2000	8,400	1,595	13,395	13,361	26,756	17,071	12,690	4,381	138.7
2001	3,300	1,620	5,346	9,685	15,031	6,817	4,126	2,691	155.4
2002	2,500	1,682	4,205	8,214	12,419	5,231	3,467	1,764	125.0
2003	2,400	1,748	4,195	7,188	11,383	6,413	4,621	1,792	130.0
2004	3,300	1,767	5,830	4,970	10,800	8,214	6,936	1,278	125.0
2005	1,500	2,000	3,000	2,586	5,586	4,754	3,154	1,600	135.0
2006	....	....	....	....	....	....	....	....	....

See footnotes at end of table.

Table 2-39.—Tobacco: Area, yield, production, stocks, supply, disappearance, and price, by types, United States including Puerto Rico, 1999–2006 (farm-sales-weight basis)[1]—Continued

Type and crop year	Area	Yield per acre	Produc- tion	Stocks[2]	Supply	Disappearance			Average price per pound to growers
						Total	Exports	Domes- tic	
	Acres	*Pounds*	*1,000 pounds*	*1,000 pounds*	*1,000 pounds*	*1,000 pounds*	*1,000 pounds*	*1,000 pounds*	*Cents*
Total dark air-cured, types 35–37:									
1999	5,100	3,878	11,795	24,094	35,889	9,176	1,433	7,743	203.3
2000	5,580	4,551	16,061	26,713	42,774	9,896	1,022	8,874	196.4
2001	5,070	4,347	14,103	32,878	46,981	8,614	322	8,292	182.7
2002	3,830	4,466	10,686	38,367	48,956	11,679	230	11,449	209.8
2003	4,150	2,726	11,314	37,374	48,688	11,307	120	11,187	215.4
2004	4,260	2,799	11,924	37,381	49,305	11,753	263	11,490	218.7
2005	5,100	2,778	11,530	37,575	49,105	12,624	60	12,564	....
2006	5,100	3,076	15,690	36,457	52,147	....	....	....	....
One Sucker, Green River type 35-36:									
1999	5,000	2,328	11,640	24,021	35,661	9,036	1,337	7,699	203.9
2000	5,480	2,901	15,896	26,625	42,521	9,824	1,000	8,824	197.1
2001	4,970	2,807	13,949	32,697	46,646	8,391	100	8,291	182.9
2002	3,760	2,811	10,570	38,255	48,825	11,548	100	11,448	210.1
2003	4,090	2,746	11,230	37,277	48,507	11,231	100	11,131	215.7
2004	4,190	2,816	11,798	37,276	49,074	11,548	58	11,490	218.7
2005	4,150	2,778	11,530	37,526	49,056	12,624	60	12,564	213.7
2006	5,100	3,076	15,690	36,432	52,122	....	....	....	....
Virginia sun-cured, type 37:									
1999	100	1,550	155	73	228	140	44	96	159.0
2000	100	1,650	165	88	253	72	50	22	180.0
2001	100	1,540	154	181	335	223	1	222	168.6
2002	70	1,655	116	112	131	131	1	130	177.8
2003	60	1,400	84	97	181	76	0	76	142.9
2004	70	1,770	126	105	231	205	....	205	145.8
2005	(4)	....	....	49	49	....	....	....	....
2006	....	....	....	25	25	....	....	....	....
Total continental cigar filler, types 41–44:									
1999	3,200	1,850	5,920	11,380	17,300	7,768	*	7,768	130.0
2000	2,400	2,100	5,040	9,532	14,572	2,453	*	2,453	NA
2001	2,000	2,060	4,120	12,119	16,239	3,968	*	3,968	150.0
2002	2,100	2,100	4,410	12,271	16,681	6,014	*	6,014	145.0
2003	2,400	2,200	5,280	10,667	15,947	6,218	*	6,218	140.0
2004	1,800	2,300	4,140	9,729	13,869	3,952	*	3,952	145.0
2005	1,300	2,200	2,860	9,917	12,777	1,992	*	1,992	145.0
2006	1,300	2,000	2,600	10,785	13,385	....	....	....	....
Pennsylvania seedleaf filler, type 41:									
1999	3,200	1,850	5,920	11,380	17,300	7,768	*	7,768	130
2000	2,400	2,100	5,040	9,532	14,572	2,453	*	2,453	NA
2001	2,000	2,060	4,120	12,119	16,239	3,968	*	3,968	150
2002	2,100	2,100	4,410	12,271	16,681	6,014	*	6,014	145
2003	2,400	2,200	5,280	10,667	15,947	6,218	*	6,218	140
2004	1,800	2,300	4,140	9,729	13,869	5,545	*	5,545	145
2005	1,300	2,200	2,860	9,917	12,777	1,992	*	1,992	145
2006	1,300	2,000	2,600	10,785	13,385	....	....	....	....
Puerto Rican filler, type 46:[4]									
1999	*	*	*	*	*	*	*	*	*
Total cigar binder, types 51–55:									
1999	3,680	1,899	6,987	17,781	24,768	9,321	8,057	1,264	342.7
2000	1,860	1,787	3,325	15,447	18,772	6,735	346	5,389	263.3
2001	3,650	2,039	7,441	12,037	19,478	8,954	162	8,592	367.3
2002	3,650	2,147	7,838	10,524	18,362	8,149	1,379	6,770	356.5
2003	4,190	1,824	7,641	10,213	17,854	7,641	4,141	2,500	256.1
2004	4,230	1,728	7,308	10,213	17,521	5,062	4,016	1,046	358.4
2005	2,420	1,701	4,117	12,459	16,576	4,561	3,261	1,300	....
2006	2,600	1,887	4,905	7,707	....	....	....	....	....

See footnotes at end of table.

Table 2-39.—Tobacco: Area, yield, production, stocks, supply, disappearance, and price, by types, United States including Puerto Rico, 1999–2006 (farm-sales-weight basis)[1]—Continued

Type and crop year	Area	Yield per acre	Produc-tion	Stocks[2]	Supply	Disappearance Total	Disappearance Exports	Disappearance Domes-tic	Average price per pound to growers
	Acres	*Pounds*	*1,000 pounds*	*1,000 pounds*	*1,000 pounds*	*1,000 pounds*	*1,000 pounds*	*1,000 pounds*	*Cents*
Connecticut Valley binder, types 51–52:									
1999	2,500	1,668	4,169	3,485	7,654	4,888	264	4,624	473.7
2000	900	1,189	1,070	2,766	3,836	1,522	346	1,176	491.6
2001	2,140	1,786	3,822	2,314	6,136	4,308	162	4,146	558.9
2002	2,200	1,828	4,021	1,828	5,849	3,779	1,379	2,400	536.9
2003	2,360	1,539	3,657	2,051	5,708	2,223	46	2,177	549.9
2004	2,420	1,557	3,767	1,576	5,343	4,127	3,527	600	530.9
2005	2,420	1,701	4,117	1,216	5,333	3,626	2,926	700	
2006	2,600	1,781	4,630	1,707	6,337				
Wisconsin binder, types 54–55:									
1999	1,180	2,388	2,818	14,296	17,114	4,433	1,000	3,433	149.0
2000	960	2,348	2,255	12,681	14,936	5,213	1,000	4,213	155.0
2001	1,510	2,397	3,619	9,723	13,342	4,646	200	4,646	165.0
2002	1,450	2,632	3,817	8,696	12,513	4,370	50	4,320	175.0
2003	1,820	2,388	4,255	8,143	12,398	3,761	1,261	2,500	175.0
2004	1,810	1,956	3,541	8,637	12,178	935	489	446	174.6
2005	(⁴)			11,243	11,243	935	635	300	
2006				6,000					
Southern Wisconsin, type 54:									
1999	890	2,530	2,252						149.0
2000	710	2,500	1,825						155.0
2001	1,200	2,535	3,042						165.0
2002	1,150	2,740	3,151						
2003	1,400	2,480	3,472						175.0
2004	1,400	1,960	2,744						175.0
2005	(⁴)								
2006									
Northern Wisconsin, type 55:									
1999	290	1,952	566						
2000	230	1,865	430						
2001	310	1,860	577						
2002	300	2,220	666						
2003	420	1,865	783						175.0
2004	410	1,945	797						175.0
2005									
2006									
Total cigar wrapper, types 61:									
1999	1,860	1,951	3,628	1,276	4,904	4,127	3,021	1,106	NA
2000	1,250	1,472	1,840	777	2,617	1,494	1,300	194	2,530
2001	1,270	1,605	1,757	1,123	2,880	1,093	800	293	NA
2002	960	1,201	1,153	1,787	2,940	2,232	750	1,482	2,250
2003	1,060	1,164	1,234	708	1,942	1,566	450	1,116	2,600
2004	1,180	1,540	1,817	376	2,193	1,579	779	800	2,530.0
2005	1,220	1,481	1,807	614	2,421	1,725	975	750	NA
2006	1,000	1,500	1,500	696	2,196				
Total tobacco, types 11–72:[5]									
1999	647,160	1,997	1,292,692	2,300,668	3,593,360	1,204,925	432,876	772,049	⁶182.8
2000	472,410	2,229	1,052,999	2,388,435	3,441,434	1,548,344	414,414	1,133,930	186.9
2001	432,310	2,293	991,223	1,893,090	2,884,313	1,145,915	436,142	709,773	⁶192.0
2002	427,310	2,039	871,122	1,738,398	2,609,520	1,025,502	382,976	642,526	⁶193.6
2003	411,150	1,952	802,560	1,584,018	2,386,578	857,460	407,271	450,189	196.4
2004	408,050	2,161	881,875	1,529,118	2,410,993	951,920	523,284	428,636	198.4
2005	298,080	2,171	647,278	1,459,073	2,106,351	939,012	470,128	468,884	165.9
2006	334,150	2,224	743,098	1,167,339	1,910,437				

[1] Data for this table is no longer collected. This table will be deleted next year.   [2] July 1 for flue-cured types 11-14 and cigar types 61 and 62; Oct. 1 for all other types.   [3] Flue-cured (type11-14) and Burley (type 31) supply based on actual marketing. Maryland (type 32) based on October 1 stocks.   [4] No longer produced.   [5] Includes Perique.   [6] Does not include cigar wrap-per type 61.   NA-not applicable.

ERS, Specialty Crops Branch, (202) 694–5311. Basic export data from the official reports of the Department of Commerce.

Table 2-40.—Tobacco: Stocks owned by dealers and manufacturers, by types, United States, 2005–2010 (farm-sales-weight basis)[1]

Type and year	Jan. 1	Apr. 1	July 1	Oct. 1
	1,000 pounds	*1,000 pounds*	*1,000 pounds*	*1,000 pounds*
Flue-cured, types 11–14:				
2005	892,026	876,511	795,968	639,446
2006	932,888	712,313	604,018	697,073
2007	671,018	570,171	493,248	578,776
2008	581,279	483,696	396,757	452,740
2009	546,889	436,658	360,324	448,901
2010	618,862	449,823		
Virginia fire-cured, type 21:				
2005	4,762	4,696	4,669	4,162
2006	4,340	4,277	3,795	3,404
2007	3,167	3,668	3,288	2,717
2008	3,131	3,154	2,833	2,579
2009	3,384	2,894	2,785	2,696
2010	886	1,092		
Kentucky and Tennessee fire-cured, types 22–23:				
2005	109,518	118,420	108,241	100,448
2006	105,126	116,038	105,864	98,545
2007	103,320	117,804	108,637	100,535
2008	111,458	121,405	112,796	103,306
2009	125,167	140,069	136,463	126,011
2010	96,965	102,349		
Burley, type 31:				
2005	603,237	615,334	548,345	422,965
2006	507,094	542,206	450,742	403,366
2007	422,568	426,348	361,305	296,177
2008	321,549	337,271	282,561	256,163
2009	283,223	297,075	265,545	239,152
2010	274,244	279,984		
Maryland, type 32:				
2005	4,970	5,622	4,942	3,692
2006	2,586	2,809	1,048	832
2007	375	1,190	372	1,028
2008	249	971	930	410
2009	116	970	43	30
2010	24	2,048		
One Sucker and Green River, types 35–36:[2]				
2005	40,638	43,677	40,010	37,526
2006	41,053	43,099	40,042	36,432
2007	39,818	44,456	40,765	36,775
2008	43,183	45,956	43,018	39,047
2009	49,492	53,357	53,517	51,812
2010	38,844	43,502		
Virginia sun-cured, type 37:				
2005	68	61	54	49
2006	42	37	32	25
2007	17	8	0	0
2008	0	5	5	5
2009	5	5	0	13
2010	13	12		
Pennsylvania seedleaf, type 41:				
2005	8,749	10,755	10,340	9,917
2006	9,998	11,691	12,179	10,785
2007	9,891	10,221	7,899	6,909
2008	6,375	9,953	9,210	7,932
2009	7,666	11,350	10,850	10,620
2010	7,262	13,550		
Connecticut Valley, types 51–52:				
2005	1,492	1,706	1,454	1,216
2006	1,359	1,036	1,464	1,707
2007	1,713	1,790	1,950	1,762
2008	1,730	1,398	1,837	533
2009	1,554	1,286	1,409	1,671
2010	1,062	953		
Wisconsin binder, types 54–55:				
2005	7,867	10,434	9,304	8,464
2006	7,750	9,319	8,251	7,529
2007	6,707	6,564	5,675	4,930
2008	4,826	5,378	4,497	3,777
2009	3,647	4,500	3,805	3,201
2010	2,798	4,492		
Cigar Wrapper, type 61:				
2005	1,050	916	614	982
2006	820	868	696	1,162
2007	727	966	511	1,149
2008	768	810	591	779
2009	611	278	239	742
2010	327	222		

See footnotes at end of table.

Table 2-40.—Tobacco: Stocks owned by dealers and manufacturers, by types, United States, 2005–2010 (farm-sales-weight basis) [1]—Continued

Type and year	Jan. 1	Apr. 1	July 1	Oct. 1
	1,000 pounds	*1,000 pounds*	*1,000 pounds*	*1,000 pounds*
Perique, type 72:				
2005	26	33	30	29
2006	34	36	30	29
2007	27	29	28	43
2008	43	42	22	36
2009	36	36	19	127
2010	105	93		
Other miscellaneous domestic, type 73:				
2005	1,095	804	4,294	3,878
2006	3,886	3,521	2,870	2,909
2007	3,558	1,851	2,661	2,781
2008	1,730	3,101	3,195	3,979
2009	5,351	3,998	2,871	2,024
2010	1,546	2,280		
Foreign-grown cigar-leaf, types 81–89:				
2005	99,508	93,720	85,179	88,571
2006	91,887	88,872	83,570	86,069
2007	91,323	84,390	82,627	79,698
2008	84,538	85,535	81,340	81,468
2009	99,181	103,158	93,970	91,668
2010	85,370	98,216		
Foreign-grown cigarette and smoking, types 91–99:				
2005	718,352	731,187	747,639	511,888
2006	788,543	743,270	752,381	745,812
2007	766,925	753,161	757,311	721,959
2008	711,251	719,283	711,278	670,380
2009	655,356	668,814	621,702	623,288
2010	621,793	609,022		

[1] Stocks shown have been converted to a farm-sales-weight basis—the equivalent of weight at the time of sale-thereby making these data of leaf-tobacco stocks comparable with the leaf-tobacco production.   [2] One Sucker and Green leaf combined.
AMS Cotton and Tobacco Programs, (901) 384–3016.

Table 2-41.—Tobacco products: Consumption, total and per capita (18 years of age and over) in the United States, 1997–2005 [1]

Year	Cigarettes			Large cigars [2]			Smoking, chewing, and snuff [3]		All tobacco products [3]	
	Total	Total	Per capita	Total	Total	Per capita	Total	Per capita	Total	Per capita
	Billion	*Million pounds*	*Number*	*Billion*	*Million pounds [4]*	*Number*	*Million pounds*	*Pounds*	*Million pounds*	*Pounds*
1997	480	805	2,422	3.5	58	18	88	0.65	1,004	4.66
1998	465	781	2,320	3.7	60	18	87	0.64	962	4.49
1999	435	721	2,148	3.8	63	19	87	0.64	876	4.32
2000	430	711	2,056	3.9	63	19	92	0.62	866	4.21
2001	425	696	2,026	3.9	67	20	89	0.63	863	4.25
2002	415	719	1,979	3.8	68	19	89	0.62	881	4.15
2003	400	674	1,890	4.5	74	21	131	0.62	841	3.98
2004	388	653	1,714	4.9	81	23	131	0.61	827	3.87
2005	376	634	1,716	4.9	80	22	133	0.61	809	3.69

[1] Includes consumption by overseas forces.   [2] Weighing over 3 pounds per 1,000.   [3] Unstemmed-processing weight equivalent.   [4] Includes weight of small cigars.   Note: Data for this table is no longer collected.
ERS, Market and Trade Economics Division, Specialty Crops Branch, (202) 694–5311. No adjustment made for quantities lost, destroyed, bartered, etc., under war and postwar conditions, but such adjustments probably would be small in relation to totals.

Table 2-42.—Tobacco: Price-support loan operations, United States, 2000–2005 [1]

Year	Flue-cured, types 11–14			Burley, type 31		
	Support price per pound	Placed under loan		Support price per pound	Placed under loan	
		Quantity	Percentage of production		Quantity	Percentage of production
	Cents	*Million pounds*	*Percent*	*Cents*	*Million pounds*	*Percent*
2000	164.0	27.4	4.4	180.5	19.3	4.8
2001	166.0	15.0	2.6	182.6	12.4	3.5
2002	165.6	24.8	4.8	183.5	24.3	31.0
2003	166.3	59.8	11.8	184.9	40.2	14.8
2004	169.0	94.9	18.5	187.3	48.0	16.1
2005	(2)	(2)	(2)	(2)	(2)	(2)

[1] Support operations for other kinds of tobacco not shown. Burley and flue-cured usually account for over 95 percent of tobacco loan placements.   [2] Price support and loans discontinued for 2005 and subsequent crops of tobacco by the Fair and Equitable Tobacco Return Act of 2004.
FSA, (202) 720–5291.

Table 2-43.—Tobacco products: Cigars, cigarettes, chewing and smoking tobacco, and snuff, manufactured in the United States, 2000–2009

Year	Cigars		Cigarettes		Chewing tobacco			
	Large	Small	Large [1]	Small	Firm	Moist	Twist	Looseleaf
	Millions	*Millions*	*Millions*	*Millions*	*1,000 pounds*	*1,000 pounds*	*1,000 pounds*	*1,000 pounds*
2000	2,824.5	2,468.9	0.0	593,173.0	2,048	543	829	45,978
2001	NA	NA	NA	NA	1,867	475	821	43,872
2002	3,815.8	2,478.3	0.0	484,332.1	1,782	376	787	41,515
2003	4,017.1	2,616.2	0.0	499,401.2	1,420	328	705	39,185
2004	4,341.7	3,359.8	0.0	492,749.4	1,403	271	651	37,012
2005	3,674.2	4,665.1	0.0	498,974.7	1,173	230	601	37,226
2006	4,256.2	5,291.3	0.0	483,678.0	1,098	199	551	36,406
2007	4,797.3	5,870.4	0.0	449,728.5	1,009	176	538	35,066
2008	4,984.4	6,478.0	0.0	396,115.4	909	144	500	30,935
2009	8,231.5	2,729.0	0.0	338,107.6	756	114	470	27,973
	Taxable removals and domestic invoices [2]							
2000	3,369.8	2,243.2	0.0	421,597.4	2,049	485	863	45,059
2001	NA	NA	NA	NA	1,828	429	803	43,532
2002	3,703.2	2,247.9	0.0	394,871.9	1,722	329	750	40,225
2003	4,018.5	2,298.2	0.0	376,682.4	1,417	289	714	38,020
2004	4,319.2	2,701.6	0.0	374,977.6	1,325	245	656	35,721
2005	4,441.0	3,772.1	0.0	363,260.2	1,166	201	614	35,701
2006	4,499.5	4,233.7	0.0	364,177.7	1,050	174	561	35,486
2007	4,658.7	4,791.3	0.0	347,960.2	978	150	539	32,721
2008	4,771.1	5,440.1	0.0	334,942.7	881	133	512	30,103
2009	7,944.2	2,150.6	0.0	308,117.2	736	72	457	27,002
	Tax-free removals and exports							
2000	113.7	228.6	0.0	153,633.8	31	34	0	85
2001	NA	NA	NA	NA	30	31	0	75
2002	79.6	270.5	0.0	136,582.4	28	26	0	68
2003	93.7	354.9	0.0	126,631.3	24	25	0	68
2004	114.5	658.6	0.0	111,202.4	28	19	0	55
2005	98.2	689.7	0.0	124,117.2	18	19	0	56
2006	100.0	830.7	0.0	116,649.0	21	20	0	59
2007	115.0	1,024.9	0.0	94,935.3	20	18	0	60
2008	152.6	857.0	0.0	61,698.4	19	19	0	96
2009	110.2	674.0	0.0	33,210.5	18	13	0	134

See footnotes at end of table.

Table 2-43.—Tobacco products: Cigars, cigarettes, chewing and smoking tobacco, and snuff, manufactured in the United States, 2000–2009—Continued

Year	Smoking tobacco			Snuff	Total chewing, smoking, and snuff
	Pipe	Granulated	Cigarette cut		
	1,000 pounds	*1,000 pounds*	*1,000 pounds*	*1,000 pounds*	*1,000 pounds*
2000	5,982	50	7,327	69,556	132,313
2001	5,088	0	7,674	70,893	130,690
2002	5,018	0	10,474	72,696	132,648
2003	4,744	0	12,636	74,895	133,913
2004	4,512	0	11,626	79,333	134,808
2005	4,280	0	13,109	81,951	138,570
2006	4,067	0	12,388	86,041	140,750
2007	4,117	0	12,164	90,153	143,223
2008	3,442	0	13,707	94,416	144,053
2009	5,102	0	8,394	95,528	138,337
	Taxable removals and domestic invoices[3]				
2000	4,620	50	8,398	68,605	130,129
2001	4,815	0	10,094	66,279	127,780
2002	4,643	0	11,258	71,668	130,595
2003	4,125	0	12,610	73,841	131,016
2004	3,773	0	11,675	74,718	128,113
2005	3,483	0	12,873	79,060	133,098
2006	3,149	0	12,311	83,618	136,349
2007	3,138	0	12,132	88,255	137,913
2008	2,949	0	13,735	93,112	141,425
2009	4,549	0	8,284	93,080	134,180
	Tax-free removals and exports				
2000	546	0	0	742	1,438
2001	455	0	0	765	1,356
2002	598	0	0	704	1,424
2003	624	0	0	697	1,438
2004	652	0	0	726	1,480
2005	446	0	0	785	1,324
2006	747	0	0	749	1,596
2007	942	0	0	740	1,780
2008	381	0	0	797	1,312
2009	239	0	0	745	1,149

[1] Weighing more than three pounds per thousand.   [2] Includes cigars and cigarettes imported or brought into the United States and Puerto Rico.   NA-not available.
AMS Cotton and Tobacco Programs, (901) 384–3016.

Table 2-44.—Cigarettes and cigars: Total output, domestic consumption, tax-exempt removals, and exports, United States, 1997–2005

Year	Cigarettes				Cigars[3]			
	Total output	Domestic consump-tion[1]	Tax-exempt removals[2]		Total output[4]	Domestic consump-tion[1]	Tax-exempt removals[2]	
			Total	Exports			Total	Exports
	Billion	*Billion*	*Billion*	*Billion*	*Million*	*Million*	*Million*	*Million*
1997	720	480	232	217	2,324	3,517	110	136
1998	680	465	213	201	2,751	3,655	112	158
1999	607	435	166	151	2,938	3,845	121	84
2000	595	430	154	148	2,825	3,850	114	113
2001	562	425	145	134	3,743	3,941	130	124
2002	532	415	136	127	3,819	3,833	80	123
2003	499	400	377	122	4,017	4,527	94	130
2004	494	388	376	119	4,407	4,935	114	171
2005	489	376	363	113	3,674	4,877	98	301

[1] As indicated by taxable removals and imports, and estimated inventory changes.   [2] In addition to exports, tax-exempt removals include principally shipments to forces overseas, to United States possessions, and ships' stores.   [3] Includes cigarillos but excludes small (approximately cigarette-size) cigars.   [4] Includes cigars shipped to mainland United States from Puerto Rico.   Note: Data for this table is no longer collected.
ERS, Market and Trade Economics Division, Specialty Crops Branch, (202) 694–5311. Compiled from annual and monthly reports of the Internal Revenue Service, U.S. Treasury Department, and the Commerce Department.

CHAPTER III

STATISTICS OF OILSEEDS, FATS, AND OILS

This chapter includes information on cottonseed, flaxseed, olive oil, peanuts, soybeans, margarine, and fats and oils. Most butter statistics are included in the chapter on dairy and poultry statistics. Lard data are mostly in the chapter on livestock.

Table 3-1.—Cottonseed: All cotton harvested area and cottonseed production, farm disposition, marketing year average price per ton received by farmers, and value, United States, 2000–2009

Year	Harvested area of all cotton	Cottonseed				
		Production	Farm disposition		Marketing year average price	Value of production
			Sales to oil mills	Other [1]		
	1,000 acres	*1,000 tons*	*1,000 tons*	*1,000 tons*	*Dollars/tons*	*1,000 dollars*
2000	13,053.0	6,435.6	3,452.2	2,983.4	105.00	667,800
2001	13,827.7	7,452.2	3,860.9	3,591.3	90.50	667,348
2002	12,416.6	6,183.9	3,287.9	2,896.0	101.00	616,352
2003	12,003.4	6,664.6	3,383.6	3,281.0	117.00	778,994
2004	13,057.0	8,198.1	4,501.5	3,696.6	107.00	872,796
2005	13,802.6	8,172.1	4,588.8	3,583.3	96.00	779,500
2006	12,731.5	7,347.9	3,608.3	3,739.6	111.00	814,151
2007	10,489.1	6,588.7	3,635.1	2,953.1	162.00	1,069,849
2008	7,568.7	4,300.3	2,526.5	1,773.8	223.00	962,708
2009	7,528.7	4,148.0	2,277.9	1,870.9	159.00	666,146

[1] Includes planting seed, feed, exports, inter-farm sales, shrinkage, losses, and other uses.
NASS, Crops Branch, (202) 720–2127.

Table 3-2.—Cottonseed: Production and farm disposition, by State and United States, 2007–2009

State	Production			Farm disposition				Used for planting [1]		
	2007	2008 [2]	2009 [3]	Sales to oil mills		Other [4]		2007	2008	2009 [3]
				2007	2008	2007	2008			
	1,000 tons	*1,000 tons*	*1,000 tons*	*1,000 tons*	*1,000 tons*	*1,000 tons*	*1,000 tons*	*1,000 tons*	*1,000 tons*	
AL	151.0	139.0	114.0	25.5	22.0	125.5	117.0	1.7	1.5	2.2
AZ	182.8	140.0	161.4	0.0	0.0	182.8	140.3	1.1	1.1	1.4
AR	671.0	443.0	294.0	508.0	357.0	163.0	86.0	3.4	3.6	3.5
CA	546.0	280.0	275.0	105.0	73.0	441.0	207.0	1.8	1.7	2.4
FL	32.9	32.6	34.5	28.2	28.5	4.7	4.0	0.3	0.4	0.8
GA	487.0	508.0	539.1	262.0	361.0	225.0	147.0	4.7	5.0	5.0
KS	20.0	12.7	19.0	4.0	0.0	16.0	12.7	0.2	0.2	0.2
LA	228.0	89.0	108.0	129.0	58.0	99.0	31.0	2.2	2.1	1.8
MS	467.0	230.0	134.0	408.0	204.0	59.0	26.0	2.0	2.2	2.4
MO	276.0	240.0	192.5	163.0	155.0	113.0	85.0	1.7	1.5	1.6
NM	33.5	25.0	25.4	0.0	0.0	33.5	25.0	0.3	0.2	0.3
NC	244.0	231.0	244.6	61.0	44.0	183.0	187.0	2.4	2.4	3.4
OK	106.5	90.5	108.4	92.4	87.2	14.1	3.3	1.0	1.0	1.3
SC	47.5	88.1	64.3	33.9	55.9	13.6	32.2	0.6	0.5	0.7
TN	203.0	169.0	157.9	156.0	146.0	47.0	23.0	2.0	2.0	2.5
TX	2,860.7	1,547.1	1,634.0	1,659.1	934.9	1,201.6	612.2	33.0	32.6	37.1
VA	31.8	35.0	42.7	0.0	0.0	31.8	35.0	0.6	0.6	0.7
US	6,588.7	4,300.3	4,148.0	3,635.1	2,526.5	2,953.1	1,773.8	59.0	58.6	67.0

[1] Included in 'other' farm disposition. Seed for planting is produced in crop year shown, but used in the following year. [2] Revised. [3] Preliminary. [4] Includes planting seed, feed, exports, inter-farm sales, shrinkage, losses, and other uses.
NASS, Crops Branch, (202) 720–2127.

Table 3-3.—Cottonseed: Marketing year average price per ton and value of production, by State and United States, crop of 2007–2009

State	Marketing year average price per ton			Value of production		
	2007	2008	2009 [1]	2007	2008	2009 [1]
	Dollars	Dollars	Dollars	1,000 dollars	1,000 dollars	1,000 dollars
AL	135.00	196.00	129.00	20,385	27,244	15,480
AZ	198.00	285.00	199.00	36,194	39,986	31,641
AR	150.00	237.00	174.00	100,650	104,991	49,938
CA	249.00	254.00	262.00	135,954	71,120	57,902
FL	161.00	207.00	135.00	5,297	6,748	4,185
GA	148.00	195.00	125.00	72,076	99,060	69,125
KS	125.00	194.00	119.00	2,500	2,464	2,261
LA	155.00	246.00	151.00	35,340	21,894	16,459
MS	143.00	212.00	157.00	66,781	48,760	22,765
MO	160.00	255.00	170.00	44,160	61,200	30,600
NM	183.00	289.00	180.00	6,131	7,225	3,420
NC	148.00	177.00	150.00	36,112	40,887	35,700
OK	135.00	202.00	132.00	14,378	18,281	15,576
SC	173.00	185.00	142.00	8,218	16,299	9,372
TN	180.00	228.00	174.00	36,540	38,532	28,362
TX	155.00	227.00	156.00	443,409	351,192	266,760
VA	180.00	195.00	165.00	5,724	6,825	6,600
US	162.00	223.00	159.00	1,069,849	962,708	666,146

[1] Preliminary.
NASS, Crops Branch, (202) 720–2127.

Table 3-4.—Cottonseed: Crushings, output of products and product prices, United States, 1999–2008

Year beginning August	Quantity crushed	Cottonseed products and prices			
		Oil		Cake and meal	
		Quantity	Price [1]	Quantity	Price [2]
	1,000 tons	Million pounds	Cents per pound	1,000 tibs	Dollars per ton
1999	3,064	939	21.6	1,390	127.43
2000	2,753	847	16.0	1,338	142.93
2001	2,791	876	18.0	1,294	136.16
2002	2,495	725	37.8	1,115	147.10
2003	2,643	874	32.0	1,244	183.47
2004	2,923	957	28.01	1,362	124.04
2005	3,010	951	29.47	1,372	144.27
2006	2,680	849	35.7	1,241	150.36
2007	2,706	856	73.55	1,262	253.81
2008 [3]	2,250	663	37.10	934	255.23

[1] Tanks, f.o.b. Valley Points.　[2] 41 percent protein, solvent, Memphis.　[3] Forecast.
ERS, Field Crops Branch, (202) 694–5300. Compiled from annual reports of the U.S. Department of Commerce.

Table 3-5.—Cottonseed meal: Production, 2007/2008–2009/2010

Country	2007/2008	2008/2009	2009/2010
	1,000 metric tons	1,000 metric tons	1,000 metric tons
Australia	126	142	183
Brazil	1,130	948	958
China, Peoples Republic of	4,875	4,766	4,368
India	3,480	3,380	3,429
Mexico	218	145	175
Pakistan	1,602	1,565	1,657
Syria	182	167	148
Turkey	448	310	267
Turkmenistan	205	215	200
Uzbekistan, Republic of	880	702	632
Others	1,359	1,247	1,170
Total Foreign	14,505	13,587	13,187
United States	1,145	848	798
Total	15,650	14,435	13,985

FAS, Office of Global Analysis, (202) 720-6301. Prepared or estimated on the basis of official USDA production, supply, and distribution statistics from foreign governments.

Table 3-6.—Cottonseed oil and cottonseed cake and meal: United States exports by country of destination 2007–2009

Country of destination	2007	2008	2009
	Metric tons	*Metric tons*	*Metric tons*
Cottonseed:			
Mexico	67,295	68,970	24,983
Korea, South	14,776	16,761	17,154
Japan	18,918	25,582	12,540
Spain	2,495	0	2,515
Canada	2,018	2,407	2,154
United Arab Emirates	550	2,441	651
Colombia	0	8	250
Taiwan	228	272	160
Mayotte	0	0	94
Rest of World	1,001	670	51
World Total	107,280	117,109	60,552
Cottonseed oil:			
Canada	20,682	26,509	23,698
Mexico	8,022	11,676	11,514
Australia(*)	4,124	6,274	3,119
Korea, South	3,584	4,863	2,949
Malaysia	0	92	2,371
Japan	6,635	6,788	1,847
El Salvador	0	0	102
Brazil	0	0	65
Trinidad and Tobago	21	178	57
Rest of World	295	1,537	206
World Total	301,284	350,053	212,960
Cottonseed cake & meal:			
Mexico	19,506	24,547	23,027
Germany(*)	680	807	860
Brazil	81	379	636
Korea, South	506	927	395
Belgium-Luxembourg(*)	188	94	377
Canada	259	631	296
United Kingdom	261	267	230
Austria	24	19	131
Italy(*)	54	395	84
Japan	112	325	18
Finland	44	17	9
Hungary	7	9	9
Kuwait	0	0	5
United Arab Emirates	0	0	5
Rest of World	154	97	3
World Total	21,876	28,514	26,086

Note: (*) Denotes a country that is a summarization of its component countries.
FAS, Office of Global Analysis, (202) 720-6301.

Table 3-7.—Cottonseed: Area and production in specified countries, 2007/2008–2009/2010

Continent and country	Area			Production		
	2007/2008	2008/2009	2009/2010	2007/2008	2008/2009	2009/2010
	1,000 hectares	*1,000 hectares*	*1,000 hectares*	*1,000 metric tons*	*1,000 metric tons*	*1,000 metric tons*
Australia	75	164	200	191	400	500
Brazil	1,077	843	836	2,740	2,040	2,100
China, Peoples Republic of	6,200	6,050	5,300	14,500	14,400	12,740
EU-27	415	303	291	643	475	462
India	9,439	9,370	10,260	10,400	9,600	10,100
Pakistan	3,000	2,900	3,000	3,875	4,000	4,300
Syria	193	176	165	551	480	454
Turkey	520	340	280	1,016	700	600
Turkmenistan	600	600	550	510	530	490
Uzbekistan, Republic of	1,430	1,420	1,300	2,330	1,800	1,575
Others	5,084	4,765	4,544	3,286	2,998	2,779
Total foreign	28,033	26,931	26,726	40,042	37,423	36,100
United States	4,245	3,063	3,047	5,977	3,901	3,764
Total	32,278	29,994	29,773	46,019	41,324	39,864

FAS, Office of Global Analysis, (202) 720-6301. Prepared or estimated on the basis of official USDA production, supply, and distribution statistics from foreign governments.

Table 3-8.—Cottonseed: Production, 2007/2008–2009/2010 [1]

Country	2007/2008	2008/2009	2009/2010 [2]
	1,000 metric tons	*1,000 metric tons*	*1,000 metric tons*
Australia	43	49	63
Brazil	380	318	322
China, Peoples Republic of	1,625	1,600	1,466
India	1,062	1,030	1,045
Mexico	82	70	70
Pakistan	541	528	560
Syria	62	57	51
Turkey	166	116	100
Turkmenistan	75	79	75
Uzbekistan, Republic	313	249	224
Others	498	448	420
Total Foreign	4,847	4,544	4,396
United States	389	301	277
Total	5,236	4,845	4,673

[1] Year beginning July 1.   [2] Preliminary.
FAS, Office of Global Analysis, (202) 720-6301. Prepared or estimated on the basis of official USDA production, supply, and distribution statistics from foreign governments.

Table 3-9.—Flaxseed: Area, yield, production, disposition, and value, United States, 2000–2009

Year	Area planted	Area harvested	Yield per harvested acre	Production	Marketing year average price per bushel received by farmers	Value of production
	1,000 acres	*1,000 acres*	*Bushels*	*1,000 bushels*	*Dollars*	*1,000 dollars*
2000	536	517	20.8	10,730	3.30	35,569
2001	585	578	19.8	11,455	4.29	49,004
2002	784	703	16.9	11,863	5.77	68,564
2003	595	588	17.9	10,516	5.88	61,900
2004	523	511	20.3	10,368	8.07	83,767
2005	983	955	20.6	19,695	5.94	117,070
2006	813	767	14.4	11,019	5.80	63,961
2007	354	349	16.9	5,896	13.00	76,521
2008	354	340	16.8	5,716	12.70	72,773
2009 [1]	317	314	23.6	7,423	8.750	64,817

[1] Preliminary.
NASS, Crops Branch, (202) 720–2127.

Table 3-10.—Flaxseed: Supply and disappearance, United States, 1999–2008

Year beginning June	Supply				Disappearance			
	Stocks June 1	Production	Imports	Total	Total used for seed	Exports	Crushings [1]	Total domestic disappearance [2]
	1,000 bushels	*1,000 bushels*	*1,000 bushels*	*1,000 bushels*	*1,000 bushels*	*1,000 bushels*	*1,000 bushels*	*1,000 bushels*
1999	2,158	7,864	6,629	16,651	434	201	11,500	14,884
2000	1,767	10,730	2,849	15,346	474	1,017	12,000	14,038
2001	1,308	11,455	1,904	14,667	635	2,386	10,000	13,774
2002	893	11,863	2,901	15,657	482	3,181	10,500	14,579
2003	1,078	10,516	4,580	16,174	424	2,516	11,260	14,886
2004	1,288	10,368	5,413	17,069	796	1,510	13,600	16,206
2005	863	19,695	4,256	24,814	659	3,780	16,400	21,279
2006	3,535	11,019	5,464	20,018	287	1,788	14,900	17,574
2007	2,444	5,896	8,019	16,359	287	2,221	11,700	14,847
2008 [3]	1,512	5,716	4,794	12,022	257	432	8,150	9,470

[1] From domestic and imported seed.   [2] Total supply minus exports and stocks June 1 of following year.   [3] Preliminary.
ERS, Field Crops Branch, (202) 694–5300.

Table 3-11.—Flaxseed: Area, yield, and production, by State and United States, 2007–2009

State	Area planted			Area harvested			Yield per harvested acre			Production		
	2007	2008	2009 [1]	2007	2008	2009 [1]	2007	2008	2009 [1]	2007	2008	2009 [1]
	1,000 acres	*1,000 acres*	*1,000 acres*	*1,000 acres*	*1,000 acres*	*1,000 acres*	*Bushels*	*Bushels*	*Bushels*	*1,000 bushels*	*1,000 bushels*	*1,000 bushels*
MN	4	3	3	4	3	3	18.0	23.0	21.0	72	69	63
MT	21	9	11	20	8	10	9.0	9.0	16.0	180	72	160
ND	320	335	295	317	323	293	17.5	17.0	24.0	5,548	5,491	7,032
SD	9	7	8	8	6	8	12.0	14.0	21.0	96	84	168
US	354	354	317	349	340	314	16.9	16.8	23.6	5,896	5,716	7,423

[1] Preliminary.
NASS, Crops Branch, (202) 720–2127.

Table 3-12.—Flaxseed: Marketing year average price and value of production, by State and United States, 2007–2009

State	Marketing year average price per bushel			Value of production		
	2007	2008	2009 [1]	2007	2008	2009 [1]
	Dollars	Dollars	Dollars	1,000 dollars	1,000 dollars	1,000 dollars
MN	11.50	11.80	8.90	828	814	561
MT	13.00	17.70	7.90	2,340	1,274	1,264
ND	13.00	12.70	8.75	72,124	69,736	61,530
SD	12.80	11.30	8.70	1,229	949	1,462
US	13.00	12.70	8.75	76,521	72,773	64,817

[1] Preliminary.
NASS, Crops Branch, (202) 720–2127.

Table 3-13.—Flaxseed: Support operations, United States, 2000–2010

Marketin year beginning June 1	Income support payment rates per bushels [1]	Program price levels per bushel		Put under loan		Acquired by CCC under loan program	Owned by CCC at end of marketing year
		Loan	Target [2]	Quantity	Percentage of production		
	Dollars	Dollars	Dollars	1,000 bushels	Percent	1,000 bushels	1,000 bushels
2000/2001	0.23	5.21	NA	352.6	3.3	151.8	0.0
2001/2002	NA	5.21	NA	107.6	0.9	35.7	1.3
2002/2003	0.45/0.00	5.38	5.49	157.2	1.3	1.8	0.0
2003/2004	0.45/0.00	5.38	5.49	276.8	2.6	0.0	0.0
2004/2005	0.45/0.00	5.21	5.66	157.1	1.5	0.0	0.0
2005/2006	0.45/0.00	5.21	5.66	1,455.4	7.4	0.0	0.0
2006/2007	0.45/0.00	5.21	5.66	598.2	5.4	0.0	0.0
2007/2008	0.45/0.00	5.21	5.66	131.0	2.2	0.0	0.0
2008/2009	0.45/0.00	5.21	5.66	140.5	2.5	0.0	0.0
2009/2010	0.45/0.00	5.21	5.66	....	....	....	....

[1] Oilseeds payment rates for 1999/2000 are calculated according to the provisions of the Agriculture, Rural Development, Food and Drug Administration, and Related Agencies Appropriations Act. Payment rates for 2000/2001 were calculated according to provisions of the Agricultural Risk Protection Act. Payment rates for the 2002/2003 through 2007/2008 crops are calculated according to the Direct and Counter-cyclical program provisions, of the Farm Security and Rural Investment Act of 2002 (2002 Act). Payment rates for the 2008/2009 crop are calculated according to the Direct and Counter-cyclical program provisions of the Food, Conservation, and Energy Act 2008 (2008 Act). Beginning with 2002/2003, the first entry is the direct payment rate and the second entry is the counter-cyclical payment rate.   [2] Target prices were established under the 2002 Act.
   FSA, Oilseeds, (202) 720–0967.

Table 3-14.—Flaxseed and linseed oil and meal: Average price Minneapolis, 1999–2008

Year	Average price received by farmers per bushel	Minneapolis	
		Oil, per pound [1]	Meal, per ton [2]
	Dollars	Cents	Dollars
1999	3.79	36.42	91.63
2000	3.30	35.83	93.77
2001	4.29	36.00	116.23
2002	5.77	38.10	119.62
2003	5.88	39.86	122.89
2004	8.07	42.00	158.90
2005	5.94	59.49	114.24
2006	5.80	53.99	124.69
2007	13.00	44.37	124.61
2008 [3]	12.70	70.31	191.54

[1] Raw oil in tank cars.   [2] Bulk carlots, 34 percent protein.   [3] Preliminary.
ERS, Field Crops Branch, (202) 694–5300.

Table 3-15.—Flaxseed and products: Flaxseed crushed; production, imports, and exports of linseed oil, cake, and meal; and June 1 stocks of oil, United States, 1999–2008

Year beginning June	Total flaxseed crushed	Linseed oil			Linseed cake and meal		
		Stocks June 1	Production	Exports	Production	Imports for consumption	Exports
	1,000 bushels	*Million pounds*	*Million pounds*	*Million pounds*	*1,000 tons*	*1,000 tons*	*1,000 tons*
1999	11,500	76	224	74	207	1	19
2000	12,000	49	234	73	216	5	25
2001	10,000	45	195	50	180	6	62
2002	10,500	31	205	70	189	19	31
2003	11,260	34	220	76	203	26	32
2004	13,600	20	265	107	245	23	62
2005	16,400	45	320	98	295	18	44
2006	14,900	29	291	76	268	17	10
2007	11,700	51	228	74	211	9	10
2008 [1]	8,150	26	159	66	147	10	28

[1] Preliminary.
ERS, Field Crops Branch, (202) 694–5300.

Table 3-16.—Sunflower: United States exports by country of destination 2007–2009

Country	2007	2008	2009
	Metric tons	*Metric tons*	*Metric tons*
Spain	28,071	33,346	33,684
Turkey	18,837	15,169	32,873
Canada	8,951	14,970	18,931
Mexico	8,247	8,836	14,871
Romania	8,919	13,743	14,142
Germany(*)	7,899	10,807	8,057
China	3,955	2,953	7,662
United Kingdom	6,797	8,826	6,015
Jordan	4,509	1,082	5,299
Israel(*)	2,660	2,754	2,885
United Arab Emirates	3,363	2,729	2,732
Greece	2,357	2,527	2,409
Netherlands	3,050	4,061	1,974
Egypt	0	90	1,768
Rest of World	19,634	23,957	14,471
World Total	127,248	145,848	167,773

Note: (*) Denotes a country that is a summarization of its component countries.
FAS, Office of Global Analysis, (202) 720-6301.

**Table 3-17.—Sunflower oil: United States exports by country of destination
2007–2009**

Country	2007	2008	2009
	Metric tons	*Metric tons*	*Metric tons*
Canada	50,504	72,424	77,179
Japan	4,631	4,893	7,081
Taiwan	200	120	4,966
Singapore	7,420	1,199	3,500
Mexico	2,746	3,078	2,639
Morocco	0	0	2,430
Costa Rica	0	0	399
Spain	0	0	344
France(*)	0	31	98
Italy(*)	0	0	90
Australia(*)	0	15	82
Brazil	107	204	41
Germany(*)	0	0	37
Sweden	0	0	24
Rest of World	9,211	1,695	108
World Total	74,818	83,659	99,018

Note: (*) Denotes a country that is a summarization of its component countries.
FAS, Office of Global Analysis, (202) 720-6301.

**Table 3-18.—Sunflower cake and meal: United States exports by country of destination
2007–2009**

Country	2007	2008	2009
	Metric tons	*Metric tons*	*Metric tons*
Mexico	959	1,527	912
Canada	472	383	478
Taiwan	0	0	15
Netherlands (Antilles(*))	0	0	11
United Kingdom	315	0	0
Malaysia	0	10	0
Costa Rica	3	0	0
Brazil	0	3	0
Ireland	569	0	0
Jamaica	0	7	0
Guatemala	0	7	0
Rest of World	154	97	3
World Total	2,318	1,937	1,416

Note: (*) Denotes a country that is a summarization of its component countries.
FAS, Office of Global Analysis, (202) 720-6301.

Table 3-19.—Sunflower oil: Production, 2007/2008–2009/2010

Country	2007/2008	2008/2009	2009/2010
	1,000 metric tons	*1,000 metric tons*	*1,000 metric tons*
Argentina	1,758	1,323	1,075
China, Peoples Republic of	130	315	263
EU-27	1,773	2,335	2,435
India	360	319	255
Pakistan	233	331	325
Russian Federation	2,130	2,565	2,505
Serbia	115	180	170
South Africa, Republic of	310	318	248
Turkey	544	515	571
Ukraine	1,726	2,632	2,615
Others	558	679	670
Total foreign	9,637	11,512	11,132
United States	287	295	305
Total	9,924	11,807	11,437

FAS, Office of Global Analysis, (202) 720-6301. Prepared or estimated on the basis of official USDA production, supply, and distribution statistics from foreign governments.

Table 3-20.—Peanuts: Area, yield, production, disposition, marketing year average price per pound received by farmers, and value, United States, 2000–2009

Year	Area planted	Peanuts for nuts				
		Area harvested	Yield per acre	Production [1]	Marketing year average price	Value of production
	1,000 acres	*1,000 acres*	*Pounds*	*1,000 pounds*	*Cents*	*1,000 dollars*
2000	1,536.8	1,336.0	2,444	3,265,505	27.4	896,097
2001	1,541.2	1,411.9	3,029	4,276,704	23.4	1,000,512
2002	1,353.0	1,291.7	2,571	3,321,040	18.2	599,714
2003	1,344.0	1,312.0	3,159	4,144,150	19.3	799,428
2004	1,430.0	1,394.0	3,076	4,288,200	18.9	813,551
2005	1,657.0	1,629.0	2,989	4,869,860	17.3	843,435
2006	1,243.0	1,210.0	2,863	3,464,250	17.7	612,798
2007	1,230.0	1,195.0	3,073	3,672,250	20.5	758,626
2008	1,534.0	1,507.0	3,426	5,162,400	23.0	1,193,617
2009 [2]	1,116.0	1,081.0	3,412	3,688,350	23.0	835,172

[1] Estimates comprised of quota and non-quota peanuts.   [2] Preliminary.
NASS, Crops Branch, (202) 720–2127.

Table 3-21.—Peanuts, farmer stock: Stocks, production, and quantity milled, United States, 1999–2008

Year beginning August	Stocks Aug. 1 [1]	Production harvested for nuts [1]	Imports	Total supply	Milled [1][2]
	1,000 pounds	*1,000 pounds*	*1,000 pounds*	*1,000 pounds*	*1,000 pounds*
1999	158,646	3,829,490	5,341	3,993,477	3,703,266
2000	139,210	3,265,505	7,625	3,412,340	3,254,950
2001	116,994	4,276,704	0	4,393,698	3,663,304
2002	483,702	3,321,040	251	3,804,993	3,585,900
2003	123,428	4,144,150	321	4,267,899	4,014,994
2004	234,770	4,288,200	0	4,522,970	3,675,410
2005	677,436	4,869,860	6	5,547,302	3,896,012
2006	1,402,614	3,464,250	48	4,866,912	3,914,354
2007	730,134	3,672,250	0	4,402,384	3,783,154
2008	346,948	5,162,400	194	5,509,542	3,901,712

[1] Net weight basis.   [2] Includes peanuts milled for seed.
NASS, Crops Branch, (202) 720–2127, and ERS.

Table 3-22.—Peanuts: Crushings, and oil and meal stocks, production, and foreign trade, United States, 1999–2008

Year beginning August	Peanuts crushed (shelled basis)	Peanut oil				Peanut cake and meal	
		Stocks Aug. 1 [1]	Production of crude	Imports	Exports [2]	Stocks Aug. 1 [3]	Production
	1,000 pounds	*1,000 pounds*	*1,000 pounds*	*1,000 pounds*	*1,000 pounds*	*1,000 pounds*	*1,000 pounds*
1999	536,164	6,770	228,839	12,835	17,519	2,847	291,491
2000	411,558	10,881	178,523	79,119	13,824	4,721	230,099
2001	521,173	3,812	230,791	38,665	8,386	3,800	296,874
2002	644,194	3,872	285,685	69,995	41,868	1,292	356,888
2003	402,958	27,698	172,977	126,346	27,695	7,769	226,995
2004	295,769	13,368	126,249	55,077	10,026	5,732	172,668
2005	407,817	20,225	181,085	61,926	7,466	1,965	232,868
2006	385,375	11,730	166,450	104,622	11,009	4,908	223,537
2007	372,980	19,824	158,144	75,545	12,979	5,651	211,733
2008	334,296	6,024	142,666	54,155	9,311	4,949	190,748

[1] Crude plus refined.   [2] Reported as edible peanut oil and crude peanut oil; in this tabulation added without converting.   [3] Holding at producing mills only.
NASS, Crops Branch, (202) 720–2127, ERS, and Bureau of the Census.

Table 3-23.—Cleaned peanuts (roasting stock): Supply and disposition, United States, 1999–2008

Year beginning August	Supply				Disposition		
	Stocks Aug. 1	Production	Imports	Total	Exports	Domestic disappearance	
						Total	Per capita
	1,000 pounds	*1,000 pounds*	*1,000 pounds*	*1,000 pounds*	*1,000 pounds*	*1,000 pounds*	*Pounds*
1999	73,108	235,756	5,341	314,205	53,406	200,877	0.72
2000	59,922	228,185	7,625	295,732	41,054	216,306	0.77
2001	38,372	245,783	0	284,155	39,099	179,907	0.63
2002	65,149	207,881	251	273,281	40,192	184,189	0.64
2003	48,900	254,048	321	303,269	32,202	211,104	0.73
2004	59,963	261,823	0	321,786	36,808	215,323	0.73
2005	69,655	240,023	6	309,684	36,844	215,847	0.72
2006	56,993	221,618	48	278,659	19,600	216,956	0.73
2007	42,103	257,386	0	299,489	56,401	185,293	0.61
2008	57,795	282,284	194	340,273	67,773	212,013	0.69

NASS, Crops Branch, (202) 720–2127, ERS, and Foreign trade from the Bureau of the Census.

Table 3-24.—Shelled peanuts (all grades): Supply, exports, and quantity crushed, United States, 1999–2008

Year beginning August	Supply						Exports	Crushed
	Stocks Aug. 1		Production		Imports	Total		
	Edible	Oil stock	Edible	Oil stock				
	1,000 pounds	*1,000 pounds*	*1,000 pounds*	*1,000 pounds*	*1,000 pounds*	*1,000 pounds*	*1,000 pounds*	*1,000 pounds*
1999	855,572	16,587	2,157,828	448,875	129,819	3,608,681	503,675	536,164
2000	707,554	70,103	1,939,736	337,324	147,103	3,201,820	354,419	411,558
2001	693,209	14,463	2,090,776	485,092	150,276	3,433,816	495,559	521,173
2002	680,850	16,648	1,983,016	611,627	54,117	3,346,258	337,332	644,194
2003	504,186	24,231	2,439,231	390,893	26,811	3,385,352	362,669	402,958
2004	603,504	17,686	2,357,314	246,663	25,290	3,250,457	341,015	295,769
2005	486,563	15,305	2,411,471	357,600	21,783	3,292,722	341,069	407,817
2006	510,097	21,499	2,415,495	347,243	42,888	3,337,222	437,659	385,375
2007	528,918	33,401	2,291,603	319,186	53,476	3,226,584	520,644	372,980
2008	431,593	39,508	2,442,345	253,778	61,199	3,228,423	494,445	334,296

NASS, Crops Branch, (202) 720–2127, ERS, and Foreign trade from the U.S. Bureau of the Census.

Table 3-25.—Peanuts: Shelled (raw basis) by types, used in primary products and apparent disappearance of peanuts, United States, 1999–2008

Type, and year beginning August	Shelled uses					Apparent disappearance [2]
	Peanut butter [1]	Snack	Candy	Other	Total	
	1,000 pounds	1,000 pounds	1,000 pounds	1,000 pounds	1,000 pounds	1,000 pounds
Virginia and Valencia:						
1999	73,926	100,384	23,173	3,321	200,804	
2000	102,050	100,650	19,101	3,271	225,072	
2001	106,573	97,046	26,640	3,097	233,356	
2002	77,018	75,100	26,930	4,178	183,226	
2003	88,053	68,257	23,580	1,669	181,559	
2004	112,027	70,216	25,466	1,702	209,411	
2005	123,402	81,617	25,738	1,136	231,893	
2006	113,689	75,858	29,542	1,103	220,196	
2007	125,497	71,059	27,909	979	225,445	
2008	110,737	52,925	26,342	1,766	191,770	
Runner:						
1999	690,564	278,440	315,467	15,922	1,300,393	
2000	643,229	247,739	320,304	15,884	1,227,156	
2001	702,454	250,079	303,668	13,575	1,269,776	
2002	734,844	257,258	312,192	19,552	1,323,846	
2003	805,852	333,198	328,560	13,847	1,481,457	
2004	824,876	367,671	349,437	20,708	1,562,692	
2005	849,176	361,176	335,748	10,925	1,557,025	
2006	869,014	328,167	329,806	8,263	1,535,250	
2007	878,026	344,551	279,564	9,666	1,511,807	
2008	981,546	303,730	276,212	8,043	1,569,531	
Spanish:						
1999	7,614	15,297	16,313	984	40,208	
2000	7,960	13,127	16,205	843	38,135	
2001	9,900	13,791	19,421	612	43,724	
2002	16,667	12,555	15,110	649	44,981	
2003	7,732	13,133	13,843	414	35,122	
2004	1,611	12,894	14,793	137	29,435	
2005	[3]	11,531	15,291	[3]	28,498	
2006	[3]	11,104	14,335	[3]	36,211	
2007	[3]	9,556	12,994	[3]	31,321	
2008	[3]	10,823	13,721	[3]	34,990	
All types:						
1999	772,104	394,121	354,953	20,227	1,541,405	2,701,205
2000	753,239	361,516	355,610	19,998	1,490,363	2,347,426
2001	818,927	360,916	349,729	17,284	1,546,856	2,586,177
2002	828,529	344,913	354,232	24,379	1,552,053	2,763,724
2003	901,637	414,588	365,983	15,930	1,698,138	2,737,351
2004	938,514	450,781	389,696	22,547	1,801,538	2,723,299
2005	974,223	454,324	376,777	12,092	1,817,416	2,739,343
2006	993,445	415,131	373,684	9,397	1,791,657	2,732,015
2007	1,012,263	425,166	320,467	10,676	1,768,572	2,702,007
2008	1,102,698	367,478	316,275	9,840	1,796,291	2,633,643

[1] Excludes peanut butter made by manufacturers for own use in candy. Includes peanut butter used in spreads, sandwiches, and cookies. [2] Apparent disappearance represents stocks beginning of year plus production, minus stocks at end of year. [3] Not published to avoid disclosure of individual operations.
NASS, Crops Branch, (202) 720–2127, and ERS.

Table 3-26.—Peanuts: Area, yield, and production, by State and United States, 2007–2009

State	Area planted			Peanuts for nuts								
				Area harvested			Yield per harvested acre			Production		
	2007	2008	2009 [1]	2007	2008	2009 [1]	2007	2008	2009 [1]	2007	2008	2009 [1]
	1,000 acres	1,000 acres	1,000 acres	1,000 acres	1,000 acres	1,000 acres	Pounds	Pounds	Pounds	1,000 pounds	1,000 pounds	1,000 pounds
AL	160.0	195.0	155.0	157.0	193.0	152.0	2,550	3,500	3,100	400,350	675,500	471,200
FL	130.0	150.0	115.0	119.0	140.0	105.0	2,700	3,200	3,200	321,300	448,000	336,000
GA	530.0	690.0	510.0	520.0	685.0	505.0	3,120	3,400	3,530	1,622,400	2,329,000	1,782,650
MS	19.0	22.0	21.0	18.0	21.0	18.0	3,300	3,900	3,000	59,400	81,900	54,000
NM	10.0	8.0	7.0	10.0	8.0	7.0	3,200	3,200	3,100	32,000	25,600	21,700
NC	92.0	98.0	67.0	90.0	97.0	66.0	2,900	3,700	3,700	261,000	358,900	244,200
OK	18.0	19.0	14.0	17.0	18.0	13.0	3,400	3,500	3,300	57,800	63,000	42,900
SC	59.0	71.0	50.0	56.0	68.0	48.0	3,100	3,900	3,100	173,600	265,200	148,800
TX	190.0	257.0	165.0	187.0	253.0	155.0	3,700	3,300	3,500	691,900	834,900	542,500
VA	22.0	24.0	12.0	21.0	24.0	12.0	2,500	3,350	3,700	52,500	80,400	44,400
US	1,230.0	1,534.0	1,116.0	1,195.0	1,507.0	1,081.0	3,073	3,426	3,412	3,672,250	5,162,400	3,688,350

[1] Preliminary.
NASS, Crops Branch, (202) 720–2127.

Table 3-27.—Peanuts: Marketing year average price, and value of production, by State and United States, 2007–2009

State	Marketing year average price per pound			Value of production		
	2007	2008	2009 [1]	2007	2008	2009 [1]
	Dollars	*Dollars*	*Dollars*	*1,000 dollars*	*1,000 dollars*	*1,000 dollars*
AL	0.191	0.225	0.222	76,467	151,988	104,606
FL	0.186	0.221	0.207	59,762	99,008	69,552
GA	0.200	0.225	0.219	324,480	524,025	390,400
MS	0.185	0.222	0.200	10,989	18,182	10,800
NM	0.200	0.242	0.328	6,400	6,195	7,118
NC	0.210	0.254	0.274	54,810	91,161	66,911
OK	0.240	0.255	0.205	13,872	16,065	8,795
SC	0.222	0.249	0.242	38,539	66,035	36,010
TX	0.235	0.240	0.239	162,597	200,376	129,658
VA	0.204	0.256	0.255	10,710	20,582	11,322
US	0.205	0.230	0.230	758,626	1,193,617	835,172

[1] Preliminary.
NASS, Crops Branch, (202) 720–2127.

Table 3-28.—Peanuts, farmer's stock: Support operations, United States, 2000–2010

Marketing year beginning August 1	Income support payment rates per short ton	Program price levels per short ton		Put under support		Owned by CCC at end of marketing year
		Quota [1]	Additional [2]	Quantity	Percentage of production	
	Dollars	*Dollars*	*Dollars*	*1,000 short tons*	*Percent*	*1,000 short tons*
2000/2001	NA	610.00	132.00	225	13.9	0
2001/2002	NA	610.00	132.00	468	21.9	0

Marketing year beginning August 1	Income support payment rates per short ton [3]	Program price levels per short ton		Put under support		Acquired by CCC under loan program [5]	Owned by CCC at end of marketing year
		Loan	Target [4]	Quantity	Percentage of production		
	Dollars	*Dollars*	*Dollars*	*1,000 short tons*	*Percent*	*1,000 short tons*	*1,000 short tons*
2003/2004	36.00/ 73.0	355.00	495.00	1,657	80.0	0.0	0.0
2004/2005	36.00/ 81.0	355.00	495.00	1,948	91.4	105.8	9.1
2005/2006	36.00/104.0	355.00	495.00	2,300	96.1	42.0	20.5
2006/2007	36.00/104.0	355.00	495.00	1,694	97.9	0.5	0.0
2007/2008	36.00/ 49.0	355.00	495.00	1,363	74.2	0.4	0.3
2008/2009	36.00/ 00.0	355.00	495.00	2,073	80.5	3.6	0.3
2009/2010	36.00/ 25.0	355.00	495.00	1,674	90.7	3.0	3.0

[1] Quota peanuts are peanuts grown within the farm poundage quota. [2] Additional peanuts are peanuts grown in excess of the quota. [3] Enactment of the Farm Security and Rural Investment Act of 2002 (2002 Act) repealed the peanut quota marketing program; and established payment rates for the 2002/03 and subsequent crops according to the provisions of the Direct Payment and Counter-cyclical Program. Beginning with 2002/2003, the first entry is the direct payment rate and the second entry is the counter-cyclical payment rate. [4] Target prices were established under the 2002 Act. [5] Acquisitions for 2008/2009 as of September 30, 2009. NA-not applicable.
FSA, Peanuts, (202) 720–4284.

Table 3-29.—Peanuts: Area and production in specified countries and the world, 2007/2008–2009/2010

Country	Area			Production		
	2007/ 2008	2008/ 2009	2009/ 2010	2007/ 2008	2008/ 2009	2009/ 2010
	1,000 hec- tares	*1,000 hec- tares*	*1,000 hec- tares*	*1,000 metric tons*	*1,000 metric tons*	*1,000 metric tons*
Argentina	227	325	220	800	900	750
Burma	650	650	670	1,000	1,000	1,000
Chad	500	354	354	223	468	468
China	3,945	4,246	4,260	13,027	14,286	14,700
India	6,500	6,400	5,300	6,800	6,250	4,900
Indonesia	720	750	750	1,150	1,250	1,250
Nigeria	1,245	1,245	1,245	1,550	1,550	1,550
Senegal	607	586	830	331	450	625
Sudan	1,000	1,000	1,000	850	850	850
Vietnam	254	256	260	510	534	550
Others	4,734	4,750	4,732	4,684	4,631	4,567
Total Foreign	20,382	20,562	19,621	30,925	32,169	31,210
United States	484	610	437	1,666	2,342	1,673
Total	20,866	21,172	20,058	32,591	34,511	32,883

FAS, Office of Global Analysis, (202) 720-6301. Prepared or estimated on the basis of official USDA production, supply, and distribution statistics from foreign governments.

Table 3-30.—Peanuts: Production, 2007/2008–2009/2010

Country	2007/2008	2008/2009	2009/2010
	1,000 metric tons	*1,000 metric tons*	*1,000 metric tons*
Argentina	85	105	66
Burkina	88	88	88
Burma	169	172	172
China	2,575	2,770	2,776
Congo (Kinshasa)	46	46	46
India	1,973	1,820	1,450
Nigeria	203	203	203
Senegal	57	104	99
Sudan	176	176	176
Vietnam	63	67	69
Others	390	411	407
Total Foreign	5,825	5,962	5,552
United States	108	94	87
Total	5,933	6,056	5,639

FAS, Office of Global Analysis, (202) 720-6301. Prepared or estimated on the basis of official USDA production, supply, and distribution statistics from foreign governments.

Table 3-31.—Soybeans: Area, yield, production, and value, United States, 2000–2009

Year	Area planted	Soybeans for beans				
		Area harvested	Yield per acre	Production	Marketing year average price per bushel received by farmers	Value of production
	1,000 acres	*1,000 acres*	*Bushels*	*1,000 bushels*	*Dollars*	*1,000 dollars*
2000	74,266	72,408	38.1	2,757,810	4.54	12,466,572
2001	74,075	72,975	39.6	2,890,682	4.38	12,605,717
2002	73,963	72,497	38.0	2,756,147	5.53	15,252,691
2003	73,404	72,476	33.9	2,453,845	7.34	18,015,097
2004	75,208	73,958	42.2	3,123,790	5.74	17,895,510
2005	72,032	71,251	43.1	3,068,342	5.66	17,297,137
2006	75,522	74,602	42.9	3,196,726	6.43	20,468,267
2007	64,741	64,146	41.7	2,677,117	10.10	26,974,406
2008	75,718	74,681	39.7	2,967,007	9.97	29,458,225
2009	77,451	76,372	44.0	3,359,011	9.45	31,760,452

NASS, Crops Branch, (202) 720–2127.

Table 3-32.—Soybeans: Stocks on and off farms, United States, 2001–2010

Year beginning previous December	On farms				Off farms [1]			
	Dec. 1	Mar. 1	June 1	Sep. 1 [2]	Dec. 1	Mar. 1	June 1	Sep. 1 [2]
	1,000 bushels	*1,000 bushels*	*1,000 bushels*	*1,000 bushels*	*1,000 bushels*	*1,000 bushels*	*1,000 bushels*	*1,000 bushels*
2001	1,217,000	780,000	365,000	83,500	1,022,991	623,908	343,180	164,247
2002	1,240,000	687,000	301,200	62,700	1,035,618	648,987	383,721	145,361
2003	1,172,000	636,500	272,500	58,000	943,373	565,528	329,862	120,329
2004	820,000	355,900	110,000	29,400	868,653	549,947	300,604	83,014
2005	1,300,000	795,000	356,100	99,700	1,004,640	586,364	343,174	156,038
2006	1,345,000	872,000	495,500	176,300	1,156,426	797,206	495,199	273,026
2007	1,461,000	910,000	500,000	143,000	1,240,366	876,887	592,185	430,810
2008	1,128,500	593,000	226,600	47,000	1,231,860	840,982	449,543	158,034
2009	1,189,000	656,500	226,300	35,100	1,086,432	645,289	369,859	103,098
2010 [3]	1,229,500	NA	NA	NA	1,109,050	NA	NA	NA

[1] Includes stocks at mills, elevators, warehouses, terminals, and processors.   [2] Old crop only.   [3] Preliminary.   NA-not available.
NASS, Crops Branch, (202) 720–2127.

Table 3-33.—Soybeans, soybean meal, and oil: Average price at specified markets, 1999–2008

Year [1]	Soybeans per bushel: No. 1 Yellow Chicago	Soybean oil per pound crude, tanks, f.o.b. Decatur	Soybean meal per short ton: 48 percent protein Decatur
	Dollars	*Cents*	*Dollars*
1999	4.82	15.59	167.62
2000	4.67	14.09	173.61
2001	4.74	16.46	167.72
2002	5.82	22.04	181.58
2003	8.18	29.97	256.05
2004	5.88	23.01	182.90
2005	5.64	23.41	174.17
2006	6.92	31.02	205.44
2007	12.22	52.03	335.94
2008 [2]	10.09	32.16	331.17

[1] Year beginning September for soybeans and October for oil and meal.   [2] Preliminary.
ERS, Field Crops Branch, (202) 694–5300.

Table 3-34.—Soybeans: Supply and disappearance, United States, 1999–2008

Year beginning September	Supply				
	Stocks by position			Production	Total [1]
	Farm	Terminal market, interior mill, elevator, and warehouse	Total		
	1,000 bushels	*1,000 bushels*	*1,000 bushels*	*1,000 bushels*	*1,000 bushels*
1999	145,000	203,482	348,482	2,653,758	3,006,411
2000	112,500	177,662	290,162	2,757,810	3,051,540
2001	83,500	164,247	247,747	2,890,682	3,140,749
2002	62,700	145,361	208,061	2,756,147	2,968,869
2003	58,000	120,329	178,329	2,453,665	2,637,773
2004	29,400	83,014	112,414	3,123,686	3,241,782
2005	99,700	156,038	255,738	3,068,342	3,327,452
2006	176,300	273,026	449,326	3,196,726	3,655,086
2007	143,000	430,810	573,810	2,677,117	3,260,798
2008 [2]	47,000	158,034	205,034	2,967,007	3,185,314

Year beginning September	Disappearance			
	Crushed [3]	Seed, feed and residual	Exports	Total
	1,000 bushels	*1,000 bushels*	*1,000 bushels*	*1,000 bushels*
1999	1,577,650	165,194	973,405	2,716,249
2000	1,639,670	168,252	995,871	2,803,793
2001	1,699,741	169,296	1,063,651	2,932,688
2002	1,614,787	131,380	1,044,372	2,790,540
2003	1,529,699	109,072	886,551	2,525,322
2004	1,696,081	192,806	1,097,156	2,986,044
2005	1,738,852	199,396	939,879	2,878,126
2006	1,807,706	157,074	1,116,496	3,081,276
2007	1,803,407	93,445	1,158,829	3,055,764
2008 [2]	1,661,987	101,849	1,283,269	3,047,106

[1] Includes imports.　[2] Preliminary.　[3] Reported by the U.S. Department of Commerce.
ERS, Field Crops Branch, (202) 694–5300.

Table 3-35.—Soybeans: Support operations, United States, 2000–2010

Marketin year beginning September 1	Income support payment rates per bushels [1]	Program price levels per bushel		Put under loan		Acquired by CCC under loan program [3]	Owned by CCC at end of marketing year
		Loan	Target [2]	Quantity	Percentage of production		
	Dollars	*Dollars*	*Dollars*	*Million bushels*	*Percent*	*Million bushels*	*Million bushels*
2000/2001	0.26	5.26	NA	313.0	11.3	5.7	2.0
2001/2002	NA	5.26	NA	311.8	10.8	0.0	2.7
2002/2003	0.44/0.00	5.00	5.80	384.3	13.9	0.0	0.7
2003/2004	0.44/0.00	5.00	5.80	156.6	6.4	0.1	0.0
2004/2005	0.44/0.00	5.00	5.80	426.0	13.6	0.5	0.0
2005/2006	0.44/0.00	5.00	5.80	463.7	15.1	8.7	0.1
2006/2007	0.44/0.00	5.00	5.80	397.2	12.5	0.0	0.0
2007/2008	0.44/0.00	5.00	5.80	181.5	6.8	0.0	0.0
2008/2009	0.44/0.00	5.00	5.80	189.5	6.4	0.0	0.0
2009/2010	0.44/------	5.00	5.80	....	....	....	....

[1] Oilseed payment rates for 1999/2000 are calculated according to the provisions of the Agriculture, Rural Development, Food and Drug Administration, and Related Agencies Appropriations Act. Payment rates for 2000/2001 are calculated according to the provisions of the Agricultural Risk Protection Act. Payment rates for the 2002/2003 through 2007/2008 crops are calculated according to the Direct and Counter-cyclical program provisions of the Farm Security and Rural Investment Act of 2002 (2002 Act). Payment rates for the 2008/2009 through 2009/2010 crops are calculated according to the Direct and Counter-cyclical program provisions of the Food, Conservation, and Energy Act 2008 (2008 Act). Beginning with 2002/2003, the first entry is the direct payment rate and the second entry is the counter-cyclical payment rate.　[2] Target prices were established under the 2002 Act.
FSA, Oilseeds, (202) 720–0967.

Table 3-36.—Soybeans: Area, yield, and production, by State and United States, 2007–2009

State	Area planted			Soybeans for beans								
	2007	2008	2009	Area harvested			Yield per harvested acre			Production		
				2007	2008	2009	2007	2008	2009	2007	2008	2009
	1,000 acres	*1,000 acres*	*1,000 acres*	*1,000 acres*	*1,000 acres*	*1,000 acres*	*Bush-els*	*Bush-els*	*Bush-els*	*1,000 bushels*	*1,000 bushels*	*1,000 bushels*
AL	190	360	440	185	350	430	21.0	35.0	40.0	3,885	12,250	17,200
AR	2,850	3,300	3,420	2,820	3,250	3,270	36.0	38.0	37.5	101,520	123,500	122,625
DE	160	195	185	155	193	183	26.0	27.5	42.0	4,030	5,308	7,686
FL	14	32	37	12	29	34	24.0	38.0	38.0	288	1,102	1,292
GA	295	430	470	285	415	440	30.0	31.0	36.0	8,550	12,865	15,840
IL	8,300	9,200	9,400	8,280	9,120	9,350	43.5	47.0	46.0	360,180	428,640	430,100
IN	4,800	5,450	5,450	4,790	5,430	5,440	46.0	45.0	49.0	220,340	244,350	266,560
IA	8,650	9,750	9,600	8,630	9,670	9,530	52.0	46.5	51.0	448,760	449,655	486,030
KS	2,650	3,300	3,700	2,610	3,250	3,650	33.0	37.0	44.0	86,130	120,250	160,600
KY	1,120	1,390	1,430	1,100	1,380	1,420	27.5	34.5	48.0	30,250	47,610	68,160
LA	615	1,050	1,020	600	950	940	43.0	33.0	39.0	25,800	31,350	36,660
MD	405	495	485	390	485	475	27.5	30.0	42.0	10,725	14,550	19,950
MI	1,800	1,900	2,000	1,790	1,890	1,990	40.0	37.0	40.0	71,600	69,930	79,600
MN	6,350	7,050	7,200	6,290	6,970	7,120	42.5	38.0	40.0	267,325	264,860	284,800
MS	1,460	2,000	2,160	1,440	1,960	2,030	40.5	40.0	38.0	58,320	78,400	77,140
MO	4,700	5,200	5,350	4,670	5,030	5,300	37.5	38.0	43.5	175,125	191,140	230,550
NE	3,870	4,900	4,800	3,850	4,860	4,760	51.0	46.5	54.5	196,350	225,990	259,420
NJ	82	92	89	80	90	87	31.0	30.0	42.0	2,480	2,700	3,654
NY	205	230	255	203	226	254	39.0	46.0	43.0	7,917	10,396	10,922
NC	1,440	1,690	1,800	1,380	1,670	1,750	22.0	33.0	34.0	30,360	55,110	59,500
ND	3,100	3,800	3,900	3,060	3,760	3,870	35.5	28.0	30.0	108,630	105,280	116,100
OH	4,250	4,500	4,550	4,240	4,480	4,530	47.0	36.0	49.0	199,280	161,280	221,970
OK	190	400	405	180	360	390	26.0	25.0	31.0	4,680	9,000	12,090
PA	435	435	450	430	430	445	41.0	40.0	46.0	17,630	17,200	20,470
SC	460	540	590	440	530	565	18.5	32.0	24.5	8,140	16,960	13,843
SD	3,250	4,100	4,250	3,240	4,060	4,190	42.0	34.0	42.0	136,080	138,040	175,980
TN	1,080	1,490	1,570	1,010	1,460	1,530	19.0	34.0	45.0	19,190	49,640	68,850
TX	95	230	215	92	205	190	37.5	24.5	25.0	3,450	5,023	4,750
VA	510	580	580	500	570	570	27.5	32.0	37.0	13,750	18,240	21,090
WV	15	19	20	14	18	19	33.0	41.0	41.0	462	738	779
WI	1,400	1,610	1,630	1,380	1,590	1,620	40.5	35.0	40.0	55,890	55,650	64,800
US	64,741	75,718	77,451	64,146	74,681	76,372	41.7	39.7	44.0	2,677,117	2,967,007	3,359,011

NASS, Crops Branch, (202) 720–2127.

Table 3-37.—Soybeans: Crushings, and oil and meal stocks, production, and foreign trade, United States, 1999–2008

Year beginning October	Soybeans crushed					Soybean oil			Soybean cake and meal		
	Oct.-Dec.	Jan.-Mar.	Apr.-Jun.	Jul.-Sep.	Total	Stocks Oct. 1	Produc-tion	Exports	Stocks Oct. 1	Produc-tion	Ex-ports
	1,000 bushels	*1,000 bushels*	*1,000 bushels*	*1,000 bushels*	*1,000 bushels*	*Million pounds*	*Million pounds*	*Million pounds*	*1,000 tons*	*1,000 tons*	*1,000 tons*
1999	435,943	395,117	360,423	381,273	1,572,757	1,520	17,825	1,375	330	37,591	7,619
2000	434,530	417,420	391,733	395,327	1,639,010	1,993	18,420	1,401	293	39,385	8,085
2001	452,757	443,946	414,412	382,741	1,693,856	2,767	18,898	2,519	383	40,292	8,015
2002	445,332	414,609	378,150	381,989	1,620,081	2,359	18,430	2,263	240	38,194	6,314
2003	437,589	406,889	339,334	339,214	1,523,026	1,491	17,080	936	220	36,324	5,169
2004	456,436	434,643	414,215	402,989	1,708,283	1,076	19,360	1,324	211	40,715	7,340
2005	457,566	438,307	419,161	433,020	1,748,053	1,699	20,387	1,153	172	41,244	8,048
2006	474,220	448,549	445,981	444,926	1,812,676	3,010	20,489	1,877	314	43,032	8,804
2007	484,090	460,817	441,080	393,720	1,779,707	3,085	20,580	2,911	343	42,234	9,242
2008 [1]	436,079	425,055	426,615	248,549	1,536,298	2,485	18,746	2,193	294	39,104	8,508

[1] Preliminary.
ERS, Field Crops Branch, (202) 694–5300. Data from the U.S. Department of Commerce.

Table 3-38.—Soybeans for beans: Marketing year average price and value, by State and United States, 2007–2009

State	Marketing year average price per bushel			Value of production		
	2007	2008	2009 [1]	2007	2008	2009 [1]
	Dollars	Dollars	Dollars	1,000 dollars	1,000 dollars	1,000 dollars
AL	11.40	10.30	10.40	44,289	126,175	178,880
AR	9.02	9.64	9.60	915,710	1,190,540	1,177,200
DE	11.50	9.40	9.60	46,345	49,895	73,786
FL	8.90	8.50	9.50	2,563	9,367	12,274
GA	11.90	9.50	9.50	101,745	122,218	153,900
IL	10.40	10.20	9.70	3,745,872	4,372,128	4,171,970
IN	10.20	10.20	9.55	2,247,468	2,492,370	2,545,648
IA	10.50	10.20	9.40	4,711,980	4,586,481	4,568,682
KS	10.10	9.39	9.25	869,913	1,129,148	1,485,550
KY	10.10	10.00	9.65	305,525	476,100	657,744
LA	8.43	9.52	9.60	217,494	298,452	351,936
MD	11.20	9.20	9.70	120,120	133,860	193,515
MI	9.69	9.82	9.40	693,804	686,713	748,240
MN	10.20	10.10	9.30	2,726,715	2,675,086	2,648,640
MS	8.36	9.29	9.15	487,555	728,336	705,831
MO	10.10	9.74	9.40	1,768,763	1,861,704	2,167,170
NE	9.92	9.79	9.40	1,947,792	2,212,442	2,438,548
NJ	10.10	9.75	9.45	25,048	26,325	34,530
NY	11.20	10.30	8.95	88,670	107,079	97,752
NC	10.10	9.33	9.50	306,636	514,176	571,710
ND	9.63	9.71	9.25	1,046,107	1,022,269	1,073,925
OH	9.93	10.30	9.60	1,978,850	1,661,184	2,130,912
OK	10.00	9.10	9.35	46,800	81,900	113,042
PA	10.70	10.20	9.35	188,641	175,440	191,395
SC	10.90	9.00	9.75	88,726	152,640	138,938
SD	9.60	9.65	9.05	1,306,368	1,332,086	1,592,619
TN	10.30	9.45	9.65	197,657	469,098	664,403
TX	10.40	9.25	9.25	35,880	46,463	43,938
VA	11.40	9.10	9.60	156,750	165,984	207,936
WV	11.30	9.75	9.60	5,221	7,196	7,478
WI	9.83	9.80	9.45	549,399	545,370	612,360
US	10.10	9.97	9.45	26,974,406	29,458,225	31,760,452

[1] Preliminary.
NASS, Crops Branch, (202) 720–2127.

Table 3-39.—Soybeans: Area, yield, and production in specified countries and the world, 2007/2008–2009/2010

Continent and country	Area			Production		
	2007/2008	2008/2009	2009/2010	2007/2008	2008/2009	2009/2010
	1,000 hec-tares	*1,000 hec-tares*	*1,000 hec-tares*	*1,000 metric tons*	*1,000 metric tons*	*1,000 metric tons*
Argentina	16,371	16,000	18,800	46,200	32,000	54,000
Bolivia	730	890	900	1,050	1,600	1,665
Brazil	21,300	21,700	23,500	61,000	57,800	69,000
Canada	1,172	1,195	1,380	2,696	3,336	3,500
China	8,750	9,130	8,800	14,000	15,540	14,700
India	8,800	9,600	9,600	9,470	9,100	8,750
Paraguay	2,650	2,550	2,680	6,900	4,000	7,200
Russia	709	709	792	652	744	942
Ukraine	630	550	625	650	800	1,050
Uruguay	448	650	800	806	1,170	1,600
Others	3,155	3,202	3,333	4,723	5,125	5,378
Total Foreign	64,715	66,176	71,210	148,147	131,215	167,785
United States	25,959	30,222	30,907	72,859	80,749	91,417
Total	90,674	96,398	102,117	221,006	211,964	259,202

FAS, Office of Global Analysis, (202) 720–6301. Prepared or estimated on the basis of official USDA production, supply, and distribution statistics from foreign governments.

Table 3-40.—Soybeans: United States exports by country of destination, 2007–2009

Country of destination	2007	2008	2009
	Metric tons	*Metric tons*	*Metric tons*
China	4,117,405	7,259,676	9,193,671
Mexico	1,160,791	1,783,595	1,347,726
Japan	1,096,110	1,356,365	1,083,422
Taiwan	712,546	952,219	718,946
Indonesia	401,451	589,266	617,161
Egypt	166,023	384,571	461,481
Germany(*)	333,896	753,160	437,599
Turkey	126,102	213,738	341,540
Korea, South	158,418	186,706	275,971
Thailand	122,976	34,531	211,353
Spain	83,430	269,158	174,900
Syria	71,451	147,207	167,662
Unidentified Countries	123,924	84,089	161,705
Malaysia	109,286	72,369	159,025
Rest of World	1,208,298	1,344,242	1,071,039
World Total	9,992,106	15,430,894	16,423,200

Note: (*) Denotes a country that is a summarization of its component countries.
FAS, Office of Global Analysis, (202) 720-6301.

Table 3-41.—Soybean oil: United States exports by country of destination, 2007–2009

Country of destination	2007	2008	2009
	Metric tons	Metric tons	Metric tons
Mexico	136,069	286,004	162,841
Morocco	51,049	122,036	140,281
India	11,627	22	119,698
Korea, South	46,456	66,775	55,835
Algeria	58,453	80,678	55,713
Peru	1,912	156	52,998
Canada	64,261	96,159	48,100
Vietnam	0	0	41,317
Venezuela	33,256	112,486	38,699
China	139,243	133,119	34,507
Malaysia	3	2,883	34,389
Dominican Republic	10,782	50,456	33,844
Cuba	20,074	21,916	22,396
Haiti	8,216	11,256	21,418
Rest of World	177,642	412,664	194,460
WorldTotal	759,041	1,396,607	1,056,494

FAS, Office of Global Analysis, (202) 720-6301.

Table 3-42.—Soybean cake and meal: United States exports by country of destination, 2007–2009

Country of destination	2007	2008	2009
	Metric tons	Metric tons	Metric tons
Mexico	438,872	560,847	528,795
Canada	353,925	470,503	434,641
Philippines	189,872	243,909	317,075
Venezuela	15,545	222,686	194,196
Indonesia	32,840	50,781	164,895
Japan	137,924	173,507	159,707
Dominican Republic	121,587	146,942	147,343
Korea, South	37,951	82,080	116,035
Guatemala	72,379	98,571	94,324
Vietnam	20,436	55,375	88,546
Turkey	41,553	52,202	79,171
Morocco	45,431	54,305	76,285
Peru	6,971	11,293	70,308
Thailand	7,945	46,496	65,108
Rest of World	641,376	931,424	964,165
World Total	2,164,608	3,200,921	3,500,594

FAS, Office of Global Analysis, (202) 720-6301.

OILSEEDS, FATS, AND OILS

Table 3-43.—Soybean oil: Production, 2007/2008–2009/2010

Country	2007/2008	2008/2009	2009/2010
	1,000 metric tons	*1,000 metric tons*	*1,000 metric tons*
Argentina	6,627	5,914	6,670
Brazil	6,160	6,120	6,040
China	7,045	7,314	8,427
Egypt	202	275	290
EU-27	2,667	2,314	2,250
India	1,458	1,340	1,108
Japan	528	456	462
Mexico	636	609	616
Russia	187	266	349
Taiwan	350	342	383
Others	2,335	2,295	2,566
Total Foreign	28,195	27,245	29,161
United States	9,335	8,503	8,820
Total	37,530	35,748	37,981

FAS, Office of Global Analysis, (202) 720-6301. Prepared or estimated on the basis of official USDA production, supply, and distribution statistics from foreign governments.

Table 3-44.—Soybeans, meal: International trade, 2007/2008–2009/2010

Country	2007/2008	2008/2009	2009/2010
	1,000 metric tons	*1,000 metric tons*	*1,000 metric tons*
Principle exporting countries:			
Argentina	26,816	24,025	25,380
Bolivia	851	1,000	1,050
Brazil	12,138	13,109	12,750
Canada	120	82	124
China	634	1,017	1,550
EU-27	414	467	450
India	5,285	3,808	2,600
Korea, South	9	116	50
Norway	148	153	150
Paraguay	1,117	1,167	1,064
Others	248	165	170
Total Foreign	47,780	45,109	45,338
United States	8,384	7,708	10,433
Total	56,164	52,817	55,771
Principle importing countries:			
Canada	1,511	1,251	1,200
EU-27	24,074	20,980	21,800
Indonesia	2,429	2,339	2,450
Iran	891	1,044	1,170
Japan	1,747	1,812	1,950
Korea, South	1,760	1,813	1,790
Mexico	1,401	1,518	1,450
Philippines	1,213	1,295	1,350
Thailand	1,935	2,160	2,400
Vietnam	2,296	2,521	2,800
Others	14,684	14,371	14,644
Total Foreign	53,941	51,104	53,004
United States	128	80	127
Total	54,069	51,184	53,131

FAS, Office of Global Analysis, (202) 720-6301. Prepared or estimated on the basis of official USDA production, supply, and distribution statistics from foreign governments.

Table 3-45.—Soybeans, oil: International trade, 2007/2008–2009/2010

Country	2007/2008	2008/2009	2009/2010
	1,000 metric tons	*1,000 metric tons*	*1,000 metric tons*
Principle exporting countries:			
Argentina	5,789	4,704	4,400
Bolivia	141	210	190
Brazil	2,388	1,910	1,415
China	102	83	80
EU-27	333	399	400
Iran	80	35	52
Malaysia	88	109	115
Norway	69	69	69
Paraguay	299	243	250
Russia	10	127	170
Others	270	238	304
Total Foreign	9,569	8,127	7,445
United States	1,320	995	1,474
Total	10,889	9,122	8,919
Principle importing countries:			
Algeria	383	365	375
Bangladesh	401	254	398
China	2,727	2,494	1,600
EU-27	1,040	793	450
India	733	1,060	1,500
Iran	545	260	300
Korea, South	296	266	300
Morocco	421	327	380
Peru	292	272	280
Venezuela	363	325	310
Others	3,265	2,499	2,656
Total Foreign	10,466	8,915	8,549
United States	30	41	50
Total	10,496	8,956	8,599

FAS, Office of Global Analysis, (202) 720-6301. Prepared or estimated on the basis of official USDA production, supply, and distribution statistics from foreign governments.

Table 3-46.—Soybean oil, local: International trade, 2007/2008–2009/2010

Country	2007/2008	2008/2009	2009/2010
	1,000 metric tons	*1,000 metric tons*	*1,000 metric tons*
Principle exporting countries:			
Argentina	4,987	3,707	5,100
Brazil	2,197	1,497	1,350
Total Foreign	7,184	5,204	6,450
United States			
Total	7,184	5,204	6,450
Principle importing countries:			
Brazil	9	41	50
Total Foreign	9	41	50
United States			
Total	9	41	50

FAS, Office of Global Analysis, (202) 720-6301. Prepared or estimated on the basis of official USDA production, supply, and distribution statistics from foreign governments.

Table 3-47.—Soybeans: International trade, 2007/2008–2009/2010

Country	2007/2008	2008/2009	2009/2010
	1,000 metric tons	*1,000 metric tons*	*1,000 metric tons*
Principle exporting countries:			
Argentina	13,839	5,590	11,000
Bolivia	79	132	135
Brazil	25,364	29,986	28,350
Canada	1,753	2,017	2,200
China	453	400	250
India	12	55	55
Paraguay	5,400	2,637	5,400
South Africa	71	131	130
Ukraine	190	277	175
Uruguay	781	1,115	1,750
Others	109	96	123
Total Foreign	48,051	42,436	49,568
United States	31,538	34,817	40,007
Total	79,589	77,253	89,575
Principle importing countries:			
China	37,816	41,098	49,500
Egypt	1,061	1,575	1,623
EU-27	15,123	13,213	13,000
Indonesia	1,147	1,393	1,500
Japan	4,014	3,396	3,600
Korea, South	1,232	1,167	1,200
Mexico	3,584	3,327	3,450
Taiwan	2,148	2,216	2,500
Thailand	1,753	1,510	1,590
Turkey	1,277	1,007	1,380
Others	8,694	6,905	6,910
Total Foreign	77,849	76,807	86,253
United States	269	361	408
Total	78,118	77,168	86,661

FAS, Office of Global Analysis, (202) 720-6301. Prepared or estimated on the basis of official USDA production, supply, and distribution statistics from foreign governments.

Table 3-48.—Soybeans, local: International trade, 2007/2008–2009/2010

Country	2007/2008	2008/2009	2009/2010
	1,000 metric tons	*1,000 metric tons*	*1,000 metric tons*
Principle exporting countries:			
Argentina	11,803	3,486	12,200
Brazil	24,515	28,041	29,950
Others			
Total Foreign	36,318	31,527	42,150
United States			
Total	36,318	31,527	42,150
Principle importing countries:			
Brazil	83	124	185
Others	2,947	157	
Total Foreign	3,030	281	185
United States			
Total	3,030	281	185

FAS, Office of Global Analysis, (202) 720-6301. Prepared or estimated on the basis of official USDA production, supply, and distribution statistics from foreign governments.

Table 3-49.—Sunflower, all: Area, yield, production, and value, United States, 2000–2009

Year	Area planted	Area harvested	Yield per harvested acre	Production	Price per cwt.	Value of production
	1,000 acres	*1,000 acres*	*Pounds*	*1,000 pounds*	*Dollars*	*1,000 dollars*
2000	2,840.0	2,647.0	1,339	3,544,428	6.89	246,869
2001	2,633.0	2,555.0	1,338	3,418,759	9.62	325,950
2002	2,581.0	2,167.0	1,131	2,451,247	12.10	294,595
2003	2,344.0	2,197.0	1,213	2,665,226	12.10	316,214
2004	1,873.0	1,711.0	1,198	2,049,613	13.70	272,732
2005	2,709.0	2,610.0	1,539	4,017,155	12.10	487,420
2006	1,950.0	1,770.0	1,211	2,143,613	14.50	308,832
2007	2,070.0	2,012.0	1,426	2,868,870	21.70	614,736
2008	2,516.5	2,396.0	1,429	3,422,840	21.80	704,105
2009	2,030.0	1,953.5	1,554	3,036,460	14.50	444,795

NASS, Crops Branch, (202) 720–2127.

Table 3-50.—Sunflower, oil varieties: Area, yield, production, and value, United States, 2000–2009

Year	Area planted	Area harvested	Yield per harvested acre	Production	Price per cwt.	Value of production
	1,000 acres	*1,000 acres*	*Pounds*	*1,000 pounds*	*Dollars*	*1,000 dollars*
2000	2,248.0	2,116.0	1,375	2,909,844	5.89	175,306
2001	2,117.0	2,060.0	1,361	2,803,704	9.07	254,705
2002	2,126.0	1,806.0	1,144	2,065,899	11.70	241,851
2003	1,998.0	1,874.0	1,206	2,259,666	11.30	254,076
2004	1,533.0	1,424.0	1,238	1,763,378	12.80	223,836
2005	2,104.0	2,032.0	1,564	3,177,635	10.50	340,584
2006	1,658.0	1,514.0	1,181	1,787,966	14.10	249,848
2007	1,765.5	1,719.0	1,445	2,483,585	21.40	527,925
2008	2,163.0	2,062.0	1,452	2,993,510	19.50	572,979
2009	1,698.0	1,653.0	1,563	2,584,010	13.20	345,950

NASS, Crops Branch, (202) 720–2127.

Table 3-51.—Sunflower, non-oil varieties: Area, yield, production, and value, United States, 2000–2009

Year	Area planted	Area harvested	Yield per harvested acre	Production	Price per cwt.	Value of production
	1,000 acres	*1,000 acres*	*Pounds*	*1,000 pounds*	*Dollars*	*1,000 dollars*
2000	592.0	531.0	1,195	634,584	11.20	71,563
2001	516.0	495.0	1,243	615,055	11.60	71,245
2002	455.0	361.0	1,067	385,348	13.70	52,744
2003	346.0	323.0	1,256	405,560	15.20	62,138
2004	340.0	287.0	997	286,235	17.20	48,896
2005	605.0	578.0	1,452	839,520	17.30	146,836
2006	292.0	256.0	1,389	355,647	16.80	58,984
2007	304.5	293.0	1,315	385,285	22.90	86,811
2008	353.5	334.0	1,285	429,330	31.30	131,126
2009	332.0	300.5	1,506	452,450	22.00	98,845

NASS, Crops Branch, (202) 720–2127.

Table 3-52.—Sunflower: Area, yield, production, and value by type, State and United States, 2008–2009

Type and State	Area planted 2008	2009¹	Area harvested 2008	2009¹	Yield per harvested acre 2008	2009¹
	1,000 acres	1,000 acres	1,000 acres	1,000 acres	Pounds	Pounds
Oil:						
CA²		34.0		33.5		1,200
CO	170.0	70.0	143.0	68.0	900	1,320
KS	220.0	150.0	205.0	140.0	1,240	1,580
MN	75.0	45.0	73.0	44.0	1,550	1,400
NE	45.0	27.0	43.0	26.0	1,300	1,200
ND	960.0	770.0	930.0	760.0	1,430	1,520
OK²		13.0		12.5		1,100
SD	550.0	520.0	545.0	510.0	1,780	1,800
TX	65.0	69.0	54.0	59.0	1,100	900
Other States³	78.0		69.0		1,191	
US	2,163.0	1,698.0	2,062.0	1,653.0	1,452	1,563
Non-oil:						
CA²		8.0		8.0		1,350
CO	24.0	21.0	19.0	19.0	1,300	1,700
KS	21.0	18.0	19.0	15.0	1,300	1,600
MN	40.0	26.0	39.0	20.0	1,300	1,250
NE	19.0	25.0	18.0	21.0	1,500	1,500
ND	155.0	115.0	150.0	108.0	1,210	1,500
OK²		3.0		2.5		1,500
SD	50.0	50.0	48.0	48.0	1,650	1,800
TX	36.0	66.0	33.0	59.0	1,000	1,300
Other States³	8.5		8.0		1,066	
US	353.5	332.0	334.0	300.5	1,285	1,506
Total:						
CA²		42.0		41.5		1,229
CO	194.0	91.0	162.0	87.0	947	1,403
KS	241.0	168.0	224.0	155.0	1,245	1,582
MN	115.0	71.0	112.0	64.0	1,463	1,353
NE	64.0	52.0	61.0	47.0	1,359	1,334
ND	1,115.0	885.0	1,080.0	868.0	1,399	1,518
OK²		16.0		15.0		1,167
SD	600.0	570.0	593.0	558.0	1,769	1,800
TX	101.0	135.0	87.0	118.0	1,062	1,100
Other States³	86.5		77.0		1,178	
US	2,516.5	2,030.0	2,396.0	1,953.5	1,429	1,554

Type and State	Production 2008	2009¹	Marketing year average price per cwt. 2008	2009¹	Value of production 2008	2009¹
	1,000 pounds	1,000 pounds	Dollars	Dollars	1,000 dollars	1,000 dollars
Oil:						
CA²		40,200		17.80		7,156
CO	128,700	89,760	21.90	14.30	28,185	12,836
KS	254,200	221,200	17.90	15.30	45,502	33,844
MN	113,150	61,600	24.00	18.00	27,156	11,088
NE	55,900	31,200	17.00	15.40	9,503	4,805
ND	1,329,900	1,155,200	19.90	13.10	264,650	151,331
OK²		13,750		16.70		2,296
SD	970,100	918,000	17.50	12.40	169,768	113,832
TX	59,400	53,100	18.00	16.50	10,692	8,762
Other States³	82,160		21.30		17,523	
US	2,993,510	2,584,010	19.50	13.20	572,979	345,950
Non-oil:						
CA²		10,800		21.80		2,354
CO	24,700	32,300	30.00	24.00	7,410	7,752
KS	24,700	24,000	29.70	23.90	7,336	5,736
MN	50,700	25,000	34.30	22.40	17,390	5,600
NE	27,000	31,500	31.00	24.30	8,370	7,655
ND	181,500	162,000	33.00	21.20	59,895	34,344
OK²		3,750		20.50		769
SD	79,200	86,400	26.50	21.00	20,988	18,144
TX	33,000	76,700	23.00	21.50	7,590	16,491
Other States³	8,530		25.20		2,147	
US	429,330	452,450	31.30	22.00	131,126	98,845
Total:						
CA²		51,000		18.60		9,510
CO	153,400	122,060	23.20	14.30	35,595	20,588
KS	278,900	245,200	20.90	17.90	52,838	39,580
MN	163,850	86,600	27.60	19.40	44,546	16,688
NE	82,900	62,700	21.60	19.90	17,873	12,460
ND	1,511,400	1,317,200	22.00	14.60	324,545	185,675
OK²		17,500		17.50		3,065
SD	1,049,300	1,004,400	19.70	12.80	190,756	131,976
TX	92,400	129,800	19.80	19.50	18,282	25,253
Other States³	90,690		21.70		19,670	
US	3,422,840	3,036,460	21.80	14.50	704,105	444,795

¹Preliminary.　²Beginning in 2009, CA and OK are published individually.　³For 2008, Other States include CA, IL, MI, MO, MT, OK, WI, and WY. Beginning in 2009 Other States is discontinued.
NASS, Crops Branch, (202) 720–2127.

Table 3-53.—Sunflower: Area and production in specified countries, 2007/2008–2009/2010

Country	Area			Production		
	2007/ 2008	2008/ 2009	2009/ 2010	2007/ 2008	2008/ 2009	2009/ 2010
	1,000 hectares	*1,000 hectares*	*1,000 hectares*	*1,000 metric tons*	*1,000 metric tons*	*1,000 metric tons*
Argentina	2,576	1,810	1,400	4,650	2,900	2,300
China	800	964	920	1,250	1,790	1,630
EU-27	3,339	3,749	3,936	4,799	7,080	7,027
India	1,630	1,530	1,400	1,120	1,000	820
Moldova	234	228	250	156	372	400
Pakistan	383	450	450	583	685	685
Russia	5,000	6,000	5,600	5,650	7,350	6,425
South Africa	565	636	400	872	801	500
Turkey	480	500	460	700	830	800
Ukraine	3,400	4,500	4,350	4,200	7,000	6,500
Others	1,885	2,338	2,219	1,915	2,323	2,019
Total Foreign	20,292	22,705	21,385	25,895	32,131	29,106
United States	814	970	791	1,301	1,553	1,377
Total	21,106	23,675	22,176	27,196	33,684	30,483

FAS, Office of Global Analysis, (202) 720-6301. Prepared or estimated on the basis of official USDA production, supply, and distribution statistics from foreign governments.

Table 3-54.—Peppermint oil: Area, yield, production, and value, United States, 2000–2009

Year	Area harvested	Yield per harvested acre	Production	Price per pound	Value of production
	1,000 acres	*Pounds*	*1,000 pounds*	*Dollars*	*1,000 dollars*
2000	88.5	78	6,877	10.80	74,320
2001	79.5	82	6,512	10.90	70,860
2002	78.5	89	6,958	11.90	82,560
2003	79.4	88	6,996	12.00	84,218
2004	78.7	92	7,236	11.90	86,421
2005	76.0	92	6,980	12.20	85,114
2006	77.7	91	7,105	12.70	89,911
2007	63.6	89	5,636	13.60	76,866
2008	60.0	92	5,499	15.90	87,450
2009 [1]	69.8	91	6,379	20.10	128,497

[1] Preliminary.
NASS, Crops Branch (202), 720–2127.

Table 3-55.—Spearmint oil: Area, yield, production, and value, United States, 2000–2009

Year	Area harvested	Yield per harvested acre	Production	Price per pound	Value of production
	1,000 acres	*Pounds*	*1,000 pounds*	*Dollars*	*1,000 dollars*
2000	21.7	101	2,199	9.06	19,919
2001	19.5	105	2,052	9.09	18,645
2002	18.4	109	2,010	9.11	18,308
2003	15.8	113	1,778	9.29	16,521
2004	15.8	116	1,839	9.62	17,700
2005	16.7	108	1,798	10.70	19,230
2006	18.5	110	2,038	11.30	23,044
2007	19.8	126	2,493	12.60	31,495
2008	20.4	118	2,399	14.90	35,765
2009 [1]	20.5	132	2,698	16.60	44,703

[1] Preliminary.
NASS, Crops Branch, (202) 720–2127.

Table 3-56.—Mint oil: Production and value, by State and United States, 2007–2009

State	Production			Price per pound			Value of production		
	2007	2008	2009 [1]	2007	2008	2009 [1]	2007	2008	2009 [1]
	1,000 pounds	*1,000 pounds*	*1,000 pounds*	*Dollars*	*Dollars*	*Dolllars*	*1,000 dollars*	*1,000 dollars*	*1,000 dollars*
Peppermint:									
CA [2]	....	....	360	....	....	19.50	....	....	7,020
ID	1,283	1,400	1,630	13.40	16.40	19.80	17,192	22,960	32,274
IN	374	293	432	14.80	19.60	25.80	5,535	5,743	11,146
MI	28	36	36	14.40	28.00	18.00	403	1,008	648
OR	1,640	1,672	1,806	14.00	15.80	21.10	22,960	26,418	38,107
WA	2,040	1,920	1,931	13.20	14.70	19.00	26,928	28,224	36,689
WI	271	178	184	14.20	17.40	14.20	3,848	3,097	2,613
US	5,636	5,499	6,379	13.60	15.90	20.10	76,866	87,450	128,497
Spearmint:									
ID	108	162	144	13.80	14.70	15.70	1,490	2,381	2,261
IN	78	81	86	12.40	15.20	16.80	967	1,231	1,445
MI	90	90	104	12.00	15.00	13.00	1,080	1,350	1,352
OR	268	240	266	12.90	14.90	18.40	3,457	3,576	4,894
WA Total	1,905	1,796	2,070	12.60	14.90	16.60	23,911	26,708	34,323
Native	1,106	1,158	1,318	12.30	14.80	16.00	13,604	17,138	21,088
Scotch	799	638	752	12.90	15.00	17.60	10,307	9,570	13,235
WI	44	30	28	13.40	17.30	15.30	590	519	428
US	2,493	2,399	2,698	12.60	14.90	16.60	31,495	35,765	44,703

[1] Preliminary.   [2] Estimates began in 2009.
NASS, Crops Branch, (202) 720–2127.

Table 3-57.—Olive oil: World Production in Specified Countries, 2007/2008–2009/2010

Country	2007/2008	2008/2009	2009/2010
	1,000 metric tons	*1,000 metric tons*	*1,000 metric tons*
Algeria	42	42	42
EU-27	2,170	2,250	2,250
Israel	9	9	9
Jordan	28	28	28
Lebanon	6	6	6
Libya	7	7	7
Morocco	90	85	85
Syria	115	185	165
Tunisia	200	160	140
Turkey	170	175	175
Total foreign	2,837	2,947	2,907
United States	2	2	3
Total	2,839	2,949	2,910

FAS, Office of Global Analysis, (202) 720-6301. Prepared or estimated on the basis of official USDA production, supply, and distribution statistics from foreign governments.

Table 3-58.—Margarine, actual weight: Supply and disposition, United States, 1999–2008

Year	Supply			Disposition		
	Production	Stocks, Jan. 1	Total supply	Exports	Domestic disappearance	
					Total	Per capita
	Million pounds	*Million pounds*	*Million pounds*	*Million pounds*	*Million pounds*	*Pounds*
1999	2,274	35	2,319	36	2,241	8.0
2000	2,398	42	2,453	31	2,353	8.3
2001	1,994	69	2,078	31	2,012	7.1
2002	1,900	34	1,951	28	1,894	6.6
2003	1,550	30	1,546	29	1,544	5.3
2004	1,567	24	1,603	33	1,554	5.3
2005	1,239	17	1,268	43	1,207	4.1
2006	1,454	18	1,483	42	1,389	4.6
2007	1,420	52	1,483	41	1,387	4.6
2008						

ERS, Field Crops Branch. (202) 694–5300. Totals and per capita estimates computed from unrounded numbers.

Table 3-59.—Margarine: Selected reported fats and oils used in manufacture, United States, 1999–2008

Year	Vegetable oils			Animal fats[1]	Total[2]
	Soybean oil	Cottonseed oil	Corn oil		
	Million pounds	*Million pounds*	*Million pounds*	*Million pounds*	*Million pounds*
1999	1,574	NA	NA	21	1,664
2000	1,465	NA	56	12	1,547
2001	1,298	NA	NA	6	1,394
2002	1,212	NA	NA	7	1,300
2003	1,138	NA	NA	16	1,207
2004	1,227	NA	NA	6	1,262
2005	848	NA	NA	3	896
2006	961	NA	NA	NA	1,033
2007	902	NA	NA	NA	956
2008 [3]	NA	NA	NA	NA	921

[1] Lard and edible tallow. [2] Includes small quantities of nuts, coconut, palm, and sunflower oil. NA-not available. [3] Preliminary
ERS, Field Crops Branch, (202) 694–5300. Compiled from reports of the U.S. Department of Commerce. Totals computed from unrounded numbers.

Table 3-60.—Shortening: Supply and disposition, United States, 1999–2008

Year	Supply			Disposition		
	Factory and warehouse stocks, Jan. 1	Production	Total supply	Exports and shipments	Domestic disappearance	
					Total	Per capita
	Million pounds	Million pounds	Million pounds	Million pounds	Million pounds	Pounds
1999	92	5,945	6,037	65	5,886	21.1
2000	86	6,593	6,630	69	6,482	23.0
2001	129	9,420	9,549	83	9,315	32.6
2002	151	9,685	9,836	89	9,607	33.3
2003	140	9,622	9,762	91	9,549	32.8
2004	122	9,671	9,794	90	9,576	32.6
2005	127	8,728	8,855	78	8,644	29.2
2006	133	7,544	7,677	90	7,434	24.9
2007	153	6,458	6,610	119	6,338	21.0
2008	153	5,639	5,792	143	5,504	18.1

ERS, Market and Trade Economics Division, Field Crops Branch, (202) 694–5300. Compiled from reports of the Commerce and Agriculture Departments.

Table 3-61.—Shortening: Fats and oils used in manufacture, United States, 1999–2008

Year	Vegetable oils				Animal fats		Total primary and secondary fats and oils [1]
	Cottonseed oil	Soybean oil	Coconut oil	Palm oil	Lard	Edible tallow	
	Million pounds	Million pounds	Million pounds	Million pounds	Million pounds	Million pounds	Million pounds
1999	167	5,069	(2)	(2)	241	262	5,968
2000	188	7,908	(2)	(2)	(2)	283	9,023
2001	185	8,234	(2)	(2)	(2)	(2)	9,405
2002	195	8,566	(2)	(2)	(2)	(2)	9,685
2003	167	8,304	(2)	(2)	(2)	(2)	9,237
2004	166	7,938	(2)	(2)	(2)	(2)	8,934
2005	213	7,799	(2)	(2)	(2)	(2)	8,918
2006	162	6,225	(2)	(2)	(2)	(2)	7,577
2007	166	5,271	(2)	(2)	(2)	(2)	6,521
2008 [3]	(2)	4,445	(2)	(2)	(2)	(2)	5,733

[1] Includes small quantities of corn, peanut, safflower, and sunflower oil.   [2] Not included to avoid disclosure.   [3] Preliminary.
ERS, Market and Trade Economics Division, Field Crops Branch, (202) 694–5300. Compiled from reports of the U.S. Department of Commerce. Totals computed from unrounded numbers.

Table 3-62.—Inedible tallow and grease: Supply and disposition, United States, and price per pound at Chicago, 1999–2008

Year	Supply				Disposition			Price of inedible tallow No. 1 at Chicago, per pound [1]
	Stocks Jan. 1	Production	Total	Exports	Factory consumption			
					Total	Use in soap	Use in feed	
	Million pounds	Million pounds	Million pounds	Million pounds	Million pounds	Million pounds	Million pounds	Cents
1999	437	7,076	7,582	1,943	3,728	229	2,751	13.0
2000	405	6,287	6,748	1,742	2,838	(2)	2,086	10.2
2001	331	5,931	6,326	1,346	3,030	(2)	2,187	11.5
2002	316	6,462	6,832	1,750	3,131	(2)	2,314	13.2
2003	242	6,245	6,512	1,573	3,234	(2)	2,419	18.2
2004	188	6,370	6,618	1,624	3,275	(2)	2,636	18.2
2005	271	6,558	6,887	1,430	3,918	(2)	2,825	17.4
2006	296	6,534	6,870	1,616	3,915	(2)	2,607	16.7
2007	282	6,628	6,972	1,806	3,865	(2)	2,395	25.5
2008	532	6,351	6,883	1,602	3,864	(2)	2,106	34.3

[1] Includes small quantities of corn, peanut, safflower, and sunflower.   [2] Not included to avoid disclosure.
ERS, Market and Trade Economics Division, Field Crops Branch, (202) 694–5300.

Table 3-63.—Fats and oils: Use in products for civilian consumption, total and per capita, United States, 1999–2008

Calendar year	Food products [1]							
	Butter (actual weight)		Lard (direct use) [2]		Tallow (direct food use)		Margarine (actual weight)	
	Total	Per capita	Total	Per capita	Total	Per capita	Total	Per capita
	Million lbs	*Lbs*	*Million lbs*	*Lbs*	*Million lbs*	*Lbs*	*Million lbs*	*Lbs*
1999	1,307	4.7	186	0.7	996	3.6	2,241	8.0
2000	1,266	4.5	221	0.8	1,125	4.0	2,353	8.3
2001	1,265	4.4	325	1.1	869	3.0	2,012	7.0
2002	1,272	4.4	370	1.3	974	3.4	1,889	6.6
2003	1,302	4.5	369	1.3	1,108	3.8	1,549	5.3
2004	1,324	4.5	220	0.7	1,163	4.0	1,554	5.3
2005	1,351	4.6	460	1.5	1,116	3.8	1,207	4.1
2006	1,408	4.7	499	1.6	1,160	3.9	1,389	4.6
2007	1,432	4.7	487	1.6	889	2.9	1,387	4.6
2008 [3]	1,535	5.0	317	1.0	872	2.9	1,302	4.3

Calendar year	Food products [1]							
	Baking and frying fats (shortening)		Salad and cooking oils		Other edible use		All food products (fat content)	
	Total	Per capita	Total	Per capita	Total	Per capita	Total	Per capita
	Million lbs	*Lbs*	*Million lbs*	*Lbs*	*Million lbs*	*Lbs*	*Million lbs*	*Lbs*
1999	5,886	21.1	8,030	28.8	431	1.5	18,385	65.8
2000 [4]	6,482	23.0	9,522	33.7	429	1.5	20,674	73.2
2001 [4]	9,315	32.7	10,144	35.6	408	1.4	23,682	83.0
2002 [4]	9,607	33.4	11,430	39.7	402	1.4	25,311	87.9
2003	9,549	32.8	11,683	40.2	386	1.3	25,375	87.3
2004	9,576	32.6	11,724	40.0	436	1.5	25,421	86.7
2005	8,644	29.2	12,658	42.8	480	1.6	25,404	85.8
2006	7,434	24.9	13,322	44.6	642	2.1	25,295	84.7
2007	6,338	21.0	15,159	50.2	507	1.7	25,636	85.0
2008 [3]	5,504	18.1	16,519	54.3	503	1.7	25,985	85.3

Calendar year	Industrial products							
	Soap		Fatty acids		Animal feeds		Other inedible products	
	Total	Per capita	Total	Per capita	Total	Per capita	Total	Per capita
	Million pounds	*Pounds*	*Million pounds*	*Pounds*	*Million pounds*	*Pounds*	*Million pounds*	*Pounds*
1999	565	2.0	2,028	7.3	3,200	11.5	553	2.0
2000	423	1.5	2,108	7.5	2,602	9.2	426	1.5
2001	366	1.3	2,060	7.2	2,651	9.3	476	1.7
2002	374	1.3	2,178	7.6	2,670	9.3	489	1.7
2003	304	1.0	2,235	7.7	2,751	9.4	445	1.5
2004	250	0.8	2,374	8.1	2,963	10.1	452	1.5
2005	257	0.9	2,271	7.6	3,223	10.9	498	1.7
2006	243	0.8	2,527	8.4	3,034	10.1	2,495	8.4
2007	238	0.8	2,696	8.1	3,062	10.1	4,232	14.0
2008 [3]	234	0.8	2,637	8.7	2,711	8.9	6,412	21.1

Calendar year	Industrial products							
	Paint and varnish		Resin and plastics		Lubricants and similar oil		Total Use [1]	
	Total	Per capita	Total	Per capita	Total	Per capita	Total	Per capita
	Million pounds	*Pounds*	*Million pounds*	*Pounds*	*Million pounds*	*Pounds*	*Million pounds*	*Pounds*
1999	79	0.3	180	0.6	128	0.5	6,733	24.1
2000	114	0.4	153	0.5	129	0.5	5,954	21.1
2001	99	0.3	141	0.5	119	0.4	6,344	22.2
2002	111	0.4	138	0.5	112	0.4	6,071	21.0
2003	109	0.4	141	0.5	110	0.4	6,095	20.9
2004	91	0.3	161	0.5	112	0.4	6,403	21.8
2005	104	0.4	157	0.5	364	1.2	6,873	23.1
2006	103	0.3	164	0.5	390	1.3	8,956	29.9
2007	86	0.3	179	0.6	275	0.9	10,769	35.6
2008 [3]	101	0.3	173	0.6	300	1.0	12,569	41.3

[1] Domestic disappearance data are computed by ERS. [2] Includes edible tallow direct use beginning in 1979. [3] Preliminary. [4] ERS estimates.

ERS, Market and Trade Economics Division, Field Crops Branch, (202) 694–5300.

Table 3-64.—Fats and oils: Wholesale price per pound, 2003–2008 [1]

Item and market	2003	2004	2005	2006	2007	2008
	Cents	Cents	Cents	Cents	Cents	Cents
Castor oil, No. 1, Brazilian, tanks, imported, New York	47.04	47.08	48.79	43.71	55.89	74.77
Coconut oil, crude, tanks, f.o.b. New York	25.86	38.05	32.44	29.10	41.68	58.69
Corn oil, crude, tank cars, f.o.b. Decatur	28.64	27.66	28.42	25.06	39.23	63.10
Cottonseed oil, crude, tank cars, f.o.b. Valley	36.73	28.87	30.42	28.56	43.53	68.01
Linseed oil, raw, tank cars, Minneapolis	41.80	48.50	64.55	43.11	55.27	86.76
Palm oil, U.S. ports, refined	32.02	34.09	29.43	29.73	41.94	48.09
Canola oil, Midwest	28.61	33.21	30.73	32.61	45.45	61.53
Safflower oil, tanks, New York	77.75	69.00	72.15	72.50	NA	NA
Soybean oil, crude, tank cars, f.o.b. Decatur	23.57	28.57	22.99	24.44	34.84	49.56
Sunflower oil, crude, Minneapolis	33.03	34.13	44.50	44.88	64.65	84.98
Tallow, inedible, number delivered Chicago	20.34	19.74	19.14	18.74	30.76	38.06
Tung oil, imported, drums, f.o.b. New York	75.63	85.42	98.33	92.06	89.72	143.03

[1] All prices are calendar year basis.
ERS, Market and Trade Economics Division, Field Crops Branch, (202) 694–5300. Compiled from the Chemical Marketing Reporter, the National Provisioner, the Wall Street Journal, and the U.S. Department of Labor.

CHAPTER IV
STATISTICS OF VEGETABLES AND MELONS

This chapter contains statistics on potatoes, sweet potatoes, and commercial vegetables and melons.

For potatoes and sweet potatoes, the estimates of area, production, value, and farm disposition pertain to the total crop and include quantities produced both for sale and for use on farms where grown. Potato statistics are shown on a within-year seasonal grouping of winter, spring, summer, and fall crops, by States. Some States have production in more than one seasonal group.

For processing vegetables, the estimates of area, production, and value for each of 10 crops relate to production used by commercial canners, freezers, and other processors, except dehydrators. These estimates include raw products grown by processors themselves and those grown under contract or purchased on the open market. This production and the actual area harvested are not duplicated in the fresh market estimates for the same commodities. The production of those vegetables used for processing for which regular processing estimates are not made is included in the fresh market estimates. The processed segment of production for asparagus, broccoli, and cauliflower, combined with fresh market production during the year, is published at the end of the season, separately. In 2000, estimates were added for collard greens, kale, mustard greens, turnip greens, okra, chili peppers, pumpkins, radishes, and squash. In 2002, estimates for fresh market lima beans, beets for canning, Brussels sprouts, cabbage for kraut, eggplant, escarole/endive, collard greens, kale, mustard greens, turnip greens, okra, and radishes were discontinued. Additionally, States were removed from the program for certain commodities. For details on the 2002 program changes see the following website: http:/www.usda.gov/nass/events/programchg/vegprogchgs.htm.

Seasonal Groups and Marketing Period

Prospective Area For Harvest

Winter: January, February, March

Summer: July, August, September

Spring: April, May, June

Fall: October, November, December

Annual Acreage, Yield, Production, and Value

The seasonal patterns of harvest do not correspond precisely in all States to the estimating period or periods designated. In some cases, only one seasonal group is shown for a State, but marketing may be active in earlier or later months. Because of the small volume from this earlier or later period, the crop estimate has been placed in the seasonal group where the largest portion is harvested.

In 2002, commercial vegetables for fresh market include 24 principal vegetable and melon crops in the major producing States. These estimates relate to crops which are grown primarily for sale, and they do not include vegetables and melons produced in farm and nonfarm gardens. The bulk of the production of the principal vegetable and melon crops is for consumption in the fresh state. However, quantities used by processors of artichokes, celery, garlic, onions, bell peppers, chile peppers, pumpkins, and squash are included, and separate estimates of commercial processing are not made for these crops. The commercial estimates of the principal crops include local market production from areas near consuming centers as well as production from well recognized commercial areas which specialize in producing supplies for shipment to distant markets.

For fresh market vegetables and melons, value per unit and total value are on a f.o.b. basis. For processed vegetables, value per unit and total value are at processing plant door.

Aggregate data for the years 2000, 2000 and 2001, and 2001 and 2002 lack comparability with data from other years because of program changes altering the crops included.

Table 4-1.—Vegetables, commercial: Area, production, and value of principal crops, United States, 2000-2009

Year	Area [1]	
	For fresh market [2]	For processing [3]
	Acres	*Acres*
2000	2,038,870	1,449,930
2001	2,020,220	1,333,310
2002	1,930,650	1,339,520
2003	1,902,160	1,333,770
2004	1,904,750	1,284,170
2005	1,858,420	1,266,460
2006	1,829,840	1,253,350
2007	1,784,290	1,249,230
2008	1,717,360	1,226,110
2009	1,708,720	1,263,699

Year	Production [4]	
	For fresh market [2]	For processing [3]
	1,000 Cwt	*Tons*
2000	476,223	17,031,310
2001	469,543	14,988,950
2002	462,976	17,074,350
2003	464,573	15,553,950
2004	481,674	17,655,540
2005	460,235	15,696,690
2006	460,812	15,910,370
2007	459,421	17,799,410
2008	446,664	17,479,610
2009	443,247	19,474,940

Year	Value [5]	
	For fresh market [2]	For processing [3]
	1,000 dollars	*1,000 dollars*
2000	9,089,706	1,415,628
2001	8,877,326	1,255,589
2002	9,416,299	1,334,583
2003	9,662,583	1,288,343
2004	9,153,763	1,391,853
2005	9,567,211	1,252,998
2006	10,150,783	1,343,800
2007	10,047,825	1,609,544
2008	10,368,620	1,898,377
2009	10,396,621	2,099,780

[1] Area for fresh market is area for harvest, including any partially harvested or notharvested because of low prices or other economic factors. Area for processing is area harvested. [2] Area, production, and farm value of the following crops for which regular seasonal estimates are prepared in major producing States: Artichokes, asparagus, snap beans,broccoli, cabbage, cantaloups, carrots, cauliflower, celery, sweet corn, cucumbers, garlic, honeydew melons, head lettuce, leaf lettuce, romaine lettuce, onions, green peppers, spinach, tomatoes, and watermelons. [3] Area, production, and farm value of the following 8 crops in all States: Lima beans, snap beans, carrots, sweet corn, cucumbers (pickles), green peas, spinach, and tomatoes. Production of other vegetables processed included in fresh market series of estimates. [4] Production for fresh market excludes some quantities not marketed because of low prices or other economic factors. [5] Value for all fresh market vegetables. For processing vegetables, value at processing plant door.
NASS, Crops Branch, (202) 720–2127.

Table 4-2.—Vegetables, commercial: Area of principal crops, by State and United States, 2007–2009 [1]

State	For fresh market [2]			For processing [4]		
	2007	2008	2009 [3]	2007	2008	2009 [3]
	Acres	*Acres*	*Acres*	*Acres*	*Acres*	*Acres*
AL	5,400	5,950	5,300	....	....	....
AZ	125,700	117,000	113,900	....	....	....
AR	3,000	2,300	2,400	....	....	....
CA	780,500	757,000	750,600	320,700	304,790	335,940
CO	25,700	23,200	22,500	....	....	....
CT	4,500	3,900	3,900	....	....	....
DE	5,900	6,280	6,000	30,100	30,810	30,950
FL	186,200	179,690	183,100	....	....	....
GA	112,600	107,600	110,200	6,700	5,000	4,300
ID	9,100	8,600	8,800	....	....	....
IL	20,850	19,770	19,600	38,000	35,800	41,800
IN	16,100	15,800	16,500	....	....	....
ME	1,900	1,800	1,500	....	....	....
MD	10,080	10,840	10,580	16,900	17,980	16,600
MA	5,200	5,200	4,700	....	....	....
MI [5]	56,000	53,800	54,500	59,100	51,600	....
MN [5]	....	....	....	199,830	....	204,030
MS	2,600	2,600	2,300	....	....	....
MO	3,800	2,500	2,600	13,400	10,500	11,000
NV	3,410	3,240	3,930	....	....	....
NH	1,700	1,600	1,400	....	....	....
NJ	25,700	24,900	25,000	3,700	4,150	4,100
NM	18,100	17,700	17,500	....	....	....
NY	68,720	65,500	64,100	58,300	58,300	50,000
NC	40,300	40,900	41,000	....	....	....
OH	30,580	30,990	28,610	8,500	9,000	8,800
OK	4,500	4,500	3,500	....	....	....
OR	31,400	29,800	28,300	61,100	57,300	62,250
PA	26,800	25,300	24,500	....	....	....
RI	900	800	750	....	....	....
SC	14,100	13,600	13,200	....	....	....
TN [6]	12,600	13,400	11,600	5,800	5,800	....
TX	57,450	52,300	55,900	19,100	22,100	19,600
UT	1,700	1,500	1,450	....	....	....
VT	1,100	1,000	1,000	....	....	....
VA	15,400	15,600	15,800	1,200	2,370	960
WA	41,700	39,900	40,000	130,400	115,100	129,500
WI	13,000	11,000	12,200	203,700	222,900	223,400
Oth Sts [7] ..	....	....	....	72,700	272,610	120,469
US	1,784,290	1,717,360	1,708,720	1,249,230	1,226,110	1,263,699

[1] Area for fresh market and for processing is area harvested. [2] Area of the following crops for which regular seasonal estimates are prepared in major producing States: Artichokes, asparagus, snap beans, broccoli, cabbage, cantaloups, carrots, cauliflower, celery, sweet corn, cucumbers, garlic, honeydew melons, head lettuce, leaf lettuce, romaine lettuce, onions, green peppers, Chile pepper, spinach, tomatoes, and watermelons. [3] Preliminary. [4] Includes Lima beans, snap beans, carrots, sweet corn, cucumbers (pickles), green peas, spinach, and tomatoes. Other vegetables processed (dual purpose) included in fresh market series of estimates. [5] Missing data included in Other States to avoid disclosure of individual operations. [6] Estimates for TN discontinued in 2009. [7] Processing, 2007 - AL, AR, FL, IA, ID, IN, MA, NC, PA, and SC. 2008 - AL, AR, FL, IA, ID, IN, MA, MN, NC, PA, and SC. 2009 - AL, AR, FL, IA, ID, IN, MA, MI, NC, PA, and SC.
NASS, Crops Branch, (202) 720–2127.

Table 4-3.—Vegetables, commercial: Production of principal crops, by State and United States, 2007–2009

State	For fresh market [1]			For processing [2]		
	2007	2008	2009 [3]	2007	2008	2009 [3]
	1,000 Cwt	1,000 Cwt	1,000 Cwt	Tons	Tons	Tons
AL	817	863	893	....	....	....
AZ	36,980	33,080	31,610	....	....	....
AR	567	615	344	....	....	....
CA	221,754	218,671	218,792	12,258,560	12,027,660	13,501,580
CO	7,248	6,828	7,219	....	....	....
CT	360	332	273	....	....	....
DE	1,333	1,643	1,363	83,310	88,910	107,580
FL	43,124	40,892	41,230	....	....	....
GA	22,943	22,013	22,763	29,230	22,910	18,830
ID	6,825	6,192	6,512	....	....	....
IL	6,272	5,772	4,873	180,520	159,260	195,410
IN	3,687	3,718	3,630	....	....	....
ME	152	108	90	....	....	....
MD	959	1,599	1,199	82,980	85,060	81,200
MA	416	416	306	....	....	....
MI [4]	8,347	8,396	9,100	419,100	413,350	....
MN [4]	....	....	....	959,000	....	1,180,050
MS	442	442	334	....	....	....
MO	741	675	858	35,710	30,440	28,690
NV	2,114	2,072	2,537	....	....	....
NH	102	128	77	....	....	....
NJ	3,990	4,127	3,906	13,150	20,810	19,710
NM	5,170	5,262	5,053	....	....	....
NY	14,860	15,928	12,131	221,360	241,460	208,030
NC	6,212	6,010	6,433	....	....	....
OH	5,835	4,239	5,668	193,040	153,940	220,220
OK	563	360	263	....	....	....
OR	17,368	15,783	14,677	371,400	329,970	388,250
PA	2,748	3,230	2,504	....	....	....
RI	54	68	45	....	....	....
SC	1,499	2,620	2,710	....	....	....
TN [5]	1,676	1,560	1,677	9,120	9,120	....
TX	11,438	12,700	13,082	114,780	144,720	132,340
UT	918	773	653	....	....	....
VT	72	50	45	....	....	....
VA	2,470	1,990	2,019	2,900	10,580	3,910
WA	17,143	15,922	15,836	1,114,800	902,630	1,050,910
WI	2,222	1,587	2,542	1,058,970	1,175,440	1,240,750
Oth Sts [6] ..	....	....	....	651,480	1,663,350	1,097,480
US	459,421	446,664	443,247	17,799,410	17,479,610	19,474,940

[1] Production of the following crops for which regular seasonal estimates are prepared in major producing States: Artichokes, asparagus, snap beans, broccoli, cabbage, cantaloups, carrots, cauliflower, celery, sweet corn, cucumbers, garlic, honeydew melons, head lettuce, leaf lettuce, romaine lettuce, onions, green peppers, Chile peppers, spinach, squash, tomatoes, and watermelons. [2] Includes Lima beans, snap beans, carrots, sweet corn, cucumbers (pickles), green peas, spinach, and tomatoes. Other vegetables processed (dual purpose) included in fresh market series of estimates. [3] Preliminary. [4] Missing data included in Other States to avoid disclosure of individual operations. [5] Estimates for TN discontinued in 2009. [6] 2007 - AL, AR, FL, IA, ID, IN, MA, NC, PA, and SC. 2008 - AL, AR, FL, IA, ID, IN, MA, MN, NC, PA, and SC. 2009 - AL, AR, FL, IA, ID, IN, MA, MI, NC, PA, and SC.

NASS, Crops Branch, (202) 720–2127.

Table 4-4.—Vegetables, commercial: Value of principal crops, by State and United States, 2007–2009

State	For fresh market [1]			For processing [3]		
	2007	2008	2009 [2]	2007	2008	2009 [2]
	1,000 dollars	*1,000 dollars*	*1,000 dollars*	*1,000 dollars*	*1,000 dollars*	*1,000 dollars*
AL	17,204	17,156	19,647	....	....	....
AZ	903,662	729,101	766,949	....	....	....
AR	11,094	16,850	5,712	....	....	....
CA	5,199,359	5,264,467	5,413,436	896,345	981,017	1,200,827
CO	82,269	102,186	101,552	....	....	....
CT	9,720	11,620	10,920	....	....	....
DE	17,117	26,610	21,713	16,467	19,200	19,518
FL	1,345,239	1,497,413	1,384,921	....	....	....
GA	458,558	456,451	530,804	9,740	8,607	5,920
ID	14,648	38,643	44,339	....	....	....
IL	38,872	38,813	27,991	19,729	19,277	22,621
IN	49,614	61,448	57,013	....	....	....
ME	5,092	4,644	4,230	....	....	....
MD	23,026	33,578	28,247	12,085	13,532	12,064
MA	16,224	17,888	13,158	....	....	....
MI [4]	156,949	169,990	171,540	67,728	69,240	....
MN [4]	....	....	....	113,294	....	163,797
MS	5,171	4,950	3,407	....	....	....
MO	4,594	4,995	6,178	6,524	5,530	5,214
NV	39,998	49,248	70,769	....	....	....
NH	5,304	7,808	4,543	....	....	....
NJ	106,870	123,158	132,466	2,441	4,711	4,885
NM	102,517	89,831	103,312	....	....	....
NY	321,399	384,421	277,721	39,311	61,707	47,567
NC	108,198	114,806	125,831	....	....	....
OH	147,720	152,009	183,108	24,163	27,103	29,776
OK	5,912	3,528	3,445	....	....	....
OR	81,208	123,791	124,718	51,649	60,797	62,889
PA	88,125	91,477	78,922	....	....	....
RI	1,890	2,516	1,800	....	....	....
SC	28,192	63,816	49,205	....	....	....
TN [5]	54,926	51,880	53,964	1,870	1,984	....
TX	335,671	244,894	226,296	20,868	31,736	33,460
UT	5,141	8,911	6,420	....	....	....
VT	3,672	2,800	2,430	....	....	....
VA	76,055	66,593	74,016	760	3,037	1,109
WA	140,491	261,111	225,364	100,550	123,971	132,541
WI	36,124	29,219	40,534	116,382	179,022	157,320
Oth Sts [6] ..	....	....	....	109,638	287,906	200,272
US	10,047,825	10,368,620	10,396,621	1,609,544	1,898,377	2,099,780

[1] Value of the following crops for which regular seasonal estimates are prepared in major producing States: Artichokes, asparagus, snap beans, broccoli, cabbage, cantaloups, carrots, cauliflower, celery, sweet corn, cucumbers, garlic, honeydew melons, head lettuce, leaf lettuce, romaine lettuce, onions, green peppers, Chile peppers, spinach, tomatoes, and water-melons.   [2] Preliminary.   [3] Includes Lima beans, snap beans, carrots, sweet corn, cucumbers (pickles), green peas, spinach, and tomatoes. Other vegetables processed (dual purpose) included in fresh market series of estimates.   [4] Missing data included in Other States to avoid disclosure of individual operations.   [5] Estimates for TN discontinued in 2009.   [6] 2007 - AL, AR, FL, IA, ID, IN, MA, NC, PA, and SC. 2008 - AL, AR, FL, IA, ID, IN, MA, MN, NC, PA, and SC. 2009 - AL, AR, FL, IA, ID, IN, MA, MI, NC, PA, and SC.
NASS, Crops Branch, (202) 720–2127.

Table 4-5.— Artichokes for fresh market and processing, commercial crop: Area, yield, production, value, and total value, United States, 2000-2009

Year	Area harvested	Yield per acre	Production	Value	
				Per cwt	Total
	Acres	*Cwt.*	*1,000 cwt*	*Dollars*	*1,000 dollars*
2000	8,800	115	1,012	60.30	61,021
2001	8,000	125	1,000	58.60	58,559
2002	8,200	115	943	71.50	67,425
2003	7,200	140	1,008	75.10	75,701
2004	7,500	110	825	45.10	37,208
2005	7,900	110	869	45.40	39,453
2006	8,700	135	1,175	42.00	49,350
2007	9,600	110	1,056	55.00	58,080
2008	8,800	130	1,144	47.80	54,683
2009 [1]	8,600	130	1,118	56.60	63,279

[1] Preliminary.
NASS, Crops Branch, (202) 720–2127.

Table 4-6.—Artichokes for fresh market and processing: Area, production, and value per hundredweight, California, 2007–2009

Crop	Area harvested			Production			Value per unit		
	2007	2008	2009	2007	2008	2009	2007	2008	2009
	Acres	*Acres*	*Acres*	*1,000 cwt.*	*1,000 cwt.*	*1,000 cwt.*	*Dollars per cwt.*	*Dollars per cwt.*	*Dollars per cwt.*
CA	9,600	8,800	8,600	1,056	1,144	1,118	55.00	47.80	56.60

NASS, Crops Branch, (202) 720–2127.

Table 4-7.—Asparagus for Fresh Market and Processing, commercial crop: Area, production, and value per hundredweight and per ton, by State and United States, 2007–2009

State	Area harvested [1]			Production			Value per unit		
	2007	2008	2009 [2]	2007	2008	2009 [2]	2007	2008	2009 [2]
	Acres	*Acres*	*Acres*	*1,000 cwt.*	*1,000 cwt.*	*1,000 cwt.*	*Dollars per cwt.*	*Dollars per cwt.*	*Dollars per cwt.*
CA [3]	20,000	14,500	12,500	580	421	400	121.00	125.00	130.00
MI	11,600	11,200	10,700	244	258	235	66.00	71.80	70.40
WA	7,000	6,500	6,000	301	273	264	59.10	67.10	76.90
US	38,600	32,200	29,200	1,125	952	899	92.50	94.00	98.80

State	For fresh market						For processing					
	Production			Value per unit			Production			Value per unit		
	2007	2008	2009 [2]	2007	2008	2009 [2]	2007	2008	2009 [2]	2007	2008	2009 [2]
	1,000 cwt.	*1,000 cwt.*	*1,000 cwt.*	*Dollars per cwt.*	*Dollars per cwt.*	*Dollars per cwt.*	*Tons*	*Tons*	*Tons*	*Dollars per ton*	*Dollars per ton*	*Dollars per ton*
CA [3]	580	421	400	121.00	125.00	130.00	....	....	....	....	....	....
MI	....	....	....	....	....	....	....	....	....	....	....	....
WA	....	....	....	....	....	....	....	....	....	....	....	....
Oth Sts [4]	347	297	303	61.90	70.60	78.60	9,900	11,700	9,800	1,260.00	1,360.00	1,330.00
US	927	718	703	98.90	103.00	108.00	9,900	11,700	9,800	1,260.00	1,360.00	1,330.00

[1] Asparagus for fresh market and for processing is frequently harvested from the same area; therefore it is not practical to make individual area estimates for these segments.   [2] Preliminary.   [3] Includes a small amount of processing asparagus.   [4] 2007-2009 - MI and WA.
NASS, Crops Branch, (202) 720–2127.

Table 4-8.—Asparagus, commercial crop: Area, yield, production, value per hundredweight and per ton, and total value, United States, 2000-2009

Year	Total crop					For fresh market			For processing		
	Area for harvest	Yield per acre	Production	Value[1]		Production	Value[1]		Production	Value[2]	
				Per cwt.	Total		Per cwt.	Total		Per ton	Total
	Acres	*Cwt.*	*1,000 cwt.*	*Dollars*	*1,000 dollars*	*1,000 cwt.*	*Dollars*	*1,000 dollars*	*Tons*	*Dollars*	*1,000 dollars*
2000	77,400	29	2,272	97.40	221,299	1,504	117.00	176,017	38,400	1,180.00	45,282
2001	70,150	30	2,078	110.00	228,925	1,372	140.00	192,346	35,290	1,040.00	36,579
2002	66,000	28	1,868	92.50	172,876	1,267	110.00	139,609	30,050	1,110.00	33,267
2003	58,000	32	1,843	88.40	162,901	1,194	105.00	125,086	32,450	1,170.00	37,815
2004	61,500	34	2,062	75.40	155,537	1,524	81.30	123,945	26,900	1,170.00	31,592
2005	49,300	31	1,534	78.50	120,436	1,144	87.40	99,988	19,500	1,050.00	20,448
2006	43,200	27	1,153	82.30	94,901	911	88.90	81,027	12,100	1,150.00	13,874
2007	38,600	29	1,125	92.50	104,074	927	98.90	91,642	9,900	1,260.00	12,432
2008	32,200	30	952	94.00	89,451	718	103.00	73,599	11,700	1,360.00	15,852
2009[3]	29,200	31	899	98.80	88,855	703	108.00	75,827	9,800	1,330.00	13,028

[1] Price and value at point of first sale. [2] Price and value at processing plant door. [3] Preliminary.
NASS, Crops Branch, (202) 720–2127.

Table 4-9.—Lima beans for processing, commercial crop: Area, yield, production, value per ton, and total value, United States, 2000-2009

Year	Area harvested	Yield per acre	Production	Value[1]	
				Per ton	Total
	Acres	*Tons*	*Tons*	*Dollars*	*1,000 dollars*
2000	49,750	1.18	58,890	432.00	25,442
2001	51,200	1.31	67,160	459.00	30,854
2002	51,400	1.28	65,540	430.00	28,176
2003	45,800	1.31	60,180	442.00	26,615
2004	31,600	1.43	45,180	434.00	19,612
2005	28,820	1.47	42,440	421.00	17,854
2006	43,050	1.31	56,330	398.00	22,444
2007	39,330	1.35	53,100	423.00	22,450
2008	38,270	1.28	49,150	500.00	24,584
2009[2]	34,740	1.38	48,030	519.00	24,945

[1] Price and value at processing plant door. [2] Preliminary.
NASS, Crops Branch, (202) 720–2127.

Table 4-10.—Lima beans for processing: Area, production, and value per ton, United States, 2007–2009

State	Area harvested			Production			Value per unit		
	2007	2008	2009[1]	2007	2008	2009[1]	2007	2008	2009[1]
	Acres	*Acres*	*Acres*	*Tons*	*Tons*	*Tons*	*Dollars per ton*	*Dollars per ton*	*Dollars per ton*
US[2]	39,330	38,270	34,740	53,100	49,150	48,030	423.00	500.00	519.00

[1] Preliminary. [2] 2007 - 2008 - CA, DE, IL, MD, OR, TN, WA, and WI. 2009 - CA, DE, IL, MD, OR, WA, and WI.
NASS, Crops Branch, (202) 720–2127.

Table 4-11.—Snap beans for fresh market, commercial crop: Area, yield, production, value hundredweight, and total value, United States, 2000-2009

Year	Area harvested	Yield per acre	Production	Value[1]	
				Per cwt	Total
	Acres	*Cwt.*	*1,000 cwt.*	*Dollars*	*1,000 dollars*
2000	92,600	64	5,881	42.60	250,261
2001	96,500	64	6,193	45.00	278,511
2002	98,400	61	5,965	47.60	283,813
2003	92,100	61	5,663	49.20	278,429
2004	92,200	62	5,757	45.20	260,109
2005	97,300	57	5,511	54.10	298,272
2006	93,900	66	6,213	50.00	310,420
2007	96,400	67	6,502	61.20	397,611
2008	90,400	64	5,824	52.80	307,790
2009[2]	88,400	55	4,862	53.50	259,922

[1] Price and value at point of first sale. [2] Preliminary.
NASS, Crops Branch, (202) 720–2127.

Table 4-12.—Snap beans for fresh market: Area, production, and value per hundredweight, by State and United States, 2007–2009

State	Area harvested			Production			Value per unit		
	2007	2008	2009[1]	2007	2008	2009[1]	2007	2008	2009[1]
	Acres	Acres	Acres	1,000 cwt.	1,000 cwt.	1,000 cwt.	Dollars per cwt.	Dollars per cwt.	Dollars per cwt.
CA	6,800	5,800	6,300	714	609	693	67.70	66.50	69.30
FL	38,000	33,200	32,800	3,420	2,656	2,132	70.00	58.60	52.90
GA	16,000	17,000	16,000	640	935	720	31.60	32.00	52.40
MD	1,900	2,000	1,800	67	120	99	50.00	40.00	47.00
MI	3,100	2,800	3,100	124	112	155	65.00	40.00	40.00
NJ	2,700	2,500	2,800	81	95	76	47.00	45.00	67.40
NY	7,800	6,600	6,700	437	482	268	89.80	84.10	88.00
NC	6,300	6,200	6,000	252	248	240	30.00	28.00	31.00
SC	900	800	900	45	36	50	57.00	55.00	43.00
TN	7,900	8,500	6,800	442	366	252	32.00	34.00	31.00
VA	5,000	5,000	5,200	280	165	177	39.00	38.00	25.00
US	96,400	90,400	88,400	6,502	5,824	4,862	61.20	52.80	53.50

[1] Preliminary.
NASS, Crops Branch, (202) 720–2127.

Table 4-13.—Snap beans for processing, commercial crop: Area, yield, production, value per ton, and total value, United States, 2000-2009

Year	Area harvested	Yield per acre	Production	Value[1]	
				Per ton	Total
	Acres	Tons	Tons	Dollars	1,000 dollars
2000	218,380	3.82	833,490	171.00	142,502
2001	193,980	3.55	688,140	161.00	111,114
2002	201,800	3.93	793,710	151.00	120,190
2003	189,600	3.84	727,640	157.00	114,520
2004	200,990	4.16	835,880	158.00	131,865
2005	204,620	4.00	819,250	140.00	114,648
2006	203,240	3.87	785,950	157.00	123,218
2007	198,770	3.79	753,730	168.00	126,620
2008	198,300	4.07	808,000	219.00	177,278
2009[2]	196,179	4.14	812,990	191.00	155,420

[1] Price and value at processing plant door.   [2] Preliminary.
NASS, Crops Branch, (202) 720–2127.

Table 4-14.—Snap beans for processing, commercial crop: Area, production, and value per ton, by State and United States, 2007–2009

State	Area harvested			Production			Value per unit		
	2007	2008	2009[1]	2007	2008	2009[1]	2007	2008	2009[1]
	Acres	Acres	Acres	Tons	Tons	Tons	Dollars per ton	Dollars per ton	Dollars per ton
IL	12,300	12,500	11,500	57,140	44,820	43,700	120.00	180.00	136.00
IN	5,000	4,500	4,400	15,770	13,980	15,650	200.00	212.00	209.00
MI	17,800	15,000	16,500	61,400	54,750	65,180	169.00	210.00	220.00
MN[2]	....	4,700	7,600	....	14,890	20,190	....	202.00	195.00
NY[2]	....	20,400	19,400	....	77,590	55,670	....	278.00	267.00
OR	19,100	18,600	18,950	115,010	112,140	112,600	202.00	236.00	216.00
PA	10,800	10,700	7,000	33,170	37,250	27,660	239.00	271.00	256.00
WI	69,900	80,500	81,700	264,910	326,870	353,290	118.00	189.00	149.00
Oth Sts[3]	63,870	31,400	29,129	206,330	125,710	119,050	212.00	253.00	244.00
Total	198,770	198,300	196,179	753,730	808,000	812,990	168.00	219.00	191.00

[1] Preliminary.   [2] Missing data included in Other States to avoid disclosure of individual operations.   [3] 2007 - AR, CA, DE, FL, GA, MD, MN, MO, NJ, NY, NC, TX, and VA. 2008 - AR, CA, DE, FL, GA, MD, MO, NJ, NC, TX, and VA. 2009 - AR, CA, DE, FL GA, MD, MO, NJ, NC, TX, and VA.

NASS, Crops Branch, (202) 720–2127.

Table 4-15.—Broccoli, commercial crop: Area, yield, production, value per hundredweight and per ton, and total value, United States, 2000-2009 [1]

Year	Total crop					For fresh market			For processing		
	Area for harvest	Yield per acre	Production	Value [2]		Production	Value [2]		Production	Value [3]	
				Per cwt.	Total		Per cwt.	Total		Per ton	Total
	Acres	Cwt.	1,000 cwt.	Dollars	1,000 dollars	1,000 cwt.	Dollars	1,000 dollars	Tons	Dollars	1,000 dollars
2000	144,300	141	20,315	30.50	620,606	19,502	31.20	607,958	40,670	311.00	12,648
2001	133,100	140	18,690	25.90	484,467	17,755	26.50	469,694	46,750	316.00	14,773
2002	130,400	141	18,375	30.90	567,767	17,595	31.40	552,713	39,000	386.00	15,054
2003	129,100	148	19,075	31.60	603,122	17,486	32.70	571,102	79,454	403.00	32,020
2004	128,800	148	19,085	32.10	613,179	17,331	33.20	575,389	87,680	431.00	37,790
2005	127,900	148	18,890	28.50	539,253	(4)	(4)	(4)	(4)	(4)	(4)
2006	130,900	145	19,040	33.30	634,394	18,538	33.70	624,827	25,110	381.00	9,567
2007	129,900	148	19,188	36.20	694,922	18,287	36.70	671,681	45,040	516.00	23,241
2008	126,900	158	20,086	35.90	721,307	19,412	36.20	701,884	33,720	576.00	19,423
2009 [5]	124,000	158	19,570	37.90	741,900	19,090	37.80	720,900	24,000	875.00	21,000

[1] Sprouting broccoli only. Does not include broccoli rabe nor heading (cauliflower) broccoli. [2] Price and value at point of first sale. [3] Price and value at processing plant door. [4] Not published to avoid disclosure of individual operations. [5] Preliminary.
NASS, Crops Branch, (202) 720–2127.

Table 4-16.—Broccoli, commercial crop: Area, production, and value per hundredweight, and per ton, by State and United States, 2007–2009 [1]

State	Area harvested			Production			Value per unit		
	2007	2008	2009 [2]	2007	2008	2009 [2]	2007	2008	2009 [2]
	Acres	Acres	Acres	1,000 cwt.	1,000 cwt.	1,000 cwt.	Dollars per cwt.	Dollars per cwt.	Dollars per cwt.
AZ	11,900	10,900	9,000	1,488	1,526	1,170	46.10	38.00	37.20
CA	118,000	116,000	115,000	17,700	18,560	18,400	35.40	35.70	38.00
US	129,900	126,900	124,000	19,188	20,086	19,570	36.20	35.90	37.90

State	For fresh market						For processing					
	Production			Value per unit			Production			Value per unit		
	2007	2008	2009 [2]	2007	2008	2009 [2]	2007	2008	2009 [2]	2007	2008	2009 [2]
	1,000 cwt.	1,000 cwt.	1,000 cwt.	Dollars per cwt.	Dollars per cwt.	Dollars per cwt.	Tons	Tons	Tons	Dollars per ton	Dollars per ton	Dollars per ton
AZ	1,488	1,526	1,170	46.10	38.00	37.20	....	....	....	....	....	....
CA	16,799	17,886	17,920	35.90	36.00	37.80	45,040	33,720	24,000	516.00	576.00	875.00
US	18,287	19,412	19,090	36.70	36.20	37.80	45,040	33,720	24,000	516.00	576.00	875.00

[1] Sprouting broccoli only. Does not include broccoli rabe nor heading (cauliflower) broccoli. [2] Preliminary.
NASS, Crops Branch, (202) 720–2127.

Table 4-17.—Cabbage for fresh market, commercial crop: Area, yield, production, value and total value, United States, 2000-2009

Year	Area harvested	Yield per acre	Production	Value	
				Per cwt	Total
	Acres	Cwt.	1,000 cwt.	Dollars	1,000 dollars
2000	76,890	333	25,623	12.30	309,171
2001	77,030	331	25,460	13.40	332,554
2002	75,680	320	24,227	12.90	307,856
2003	73,050	303	22,164	13.30	289,397
2004	72,850	331	24,118	13.10	311,997
2005	71,700	324	23,234	13.60	311,001
2006	69,250	338	23,411	14.10	324,365
2007	69,050	346	23,886	16.40	386,373
2008	65,760	373	24,516	14.70	355,065
2009 [1]	65,500	345	22,623	15.30	341,440

[1] Preliminary.
NASS, Crops Branch, (202) 720–2127.

Table 4-18.—Cabbage for fresh market: Area, production, and value per hundredweight, by State and United States, 2007–2009

State	Area harvested			Production			Value per unit		
	2007	2008	2009 [1]	2007	2008	2009 [1][2]	2007	2008	2009 [1]
	Acres	Acres	Acres	1,000 cwt.	1,000 cwt.	1,000 cwt.	Dollars per cwt.	Dollars per cwt.	Dollars per cwt.
AZ	2,700	2,800	2,300	864	1,134	1,000	20.10	16.10	19.00
CA	14,100	13,000	12,500	5,429	5,330	4,938	14.80	14.20	15.00
CO	2,400	2,300	2,700	912	920	1,269	10.90	10.60	11.00
FL	9,500	9,190	9,500	3,135	3,217	3,658	19.30	13.50	16.50
GA	7,200	6,500	6,300	2,304	2,080	1,890	13.00	12.40	14.80
IL [3]	450	370	....	126	118	....	12.50	9.36	....
MI	2,400	2,400	2,600	768	672	676	15.00	18.00	15.00
NJ	1,500	1,600	1,600	518	576	552	13.80	13.50	15.90
NY	11,200	9,500	9,200	5,152	5,605	3,496	17.70	15.00	17.00
NC	5,200	5,000	5,400	1,248	1,250	1,323	10.00	13.00	12.50
OH	1,200	1,200	1,000	516	266	127	16.80	15.50	17.90
PA	1,200	1,200	1,200	324	288	264	14.20	14.70	18.90
TX	6,000	7,100	7,500	1,620	2,272	2,400	25.60	20.30	13.50
VA	500	500	500	95	75	70	12.50	18.00	13.00
WI	3,500	3,100	3,200	857	713	960	16.90	14.00	16.20
US	69,050	65,760	65,500	23,886	24,516	22,623	16.40	14.70	15.30

[1] Preliminary.   [2] Includes some quantities of fall storage in NY harvested but not sold because of shrinkage and loss: 2007, 360,000 cwt; 2008, 331,000 cwt; and 2009, 290,000 cwt.   [3] Estimates discontinued in 2009.
NASS, Crops Branch, (202) 720–2127.

Table 4-19.—Cantaloups for fresh market, commercial crop: Area, yield, production, value, and total value, United States, 2000-2009

Year	Area harvested	Yield per acre	Production	Value	
				Per cwt	Total
	Acres	Cwt.	1,000 cwt.	Dollars	1,000 dollars
2000	97,800	223	21,774	17.10	371,984
2001	95,160	238	22,613	19.00	429,281
2002	89,800	250	22,443	17.70	398,302
2003	85,700	258	22,069	16.80	370,953
2004	84,750	251	21,298	14.70	313,981
2005	84,560	242	20,465	15.90	326,201
2006	79,300	246	19,498	17.20	335,526
2007	73,820	277	20,426	14.80	302,485
2008	71,730	269	19,294	18.50	356,781
2009 [1]	76,130	261	19,891	18.10	359,082

[1] Preliminary.
NASS, Crops Branch, (202) 720–2127.

Table 4-20.—Cantaloups for fresh market: Area, production, and value per hundredweight, by State and United States, 2007–2009

State	Area harvested			Production			Value per unit		
	2007	2008	2009 [1]	2007	2008	2009 [1]	2007	2008	2009 [1]
	Acres	Acres	Acres	1,000 cwt.	1,000 cwt.	1,000 cwt.	Dollars per cwt.	Dollars per cwt.	Dollars per cwt.
AZ	19,200	19,400	23,300	5,856	4,850	5,010	17.50	24.50	23.00
CA	39,000	39,000	38,400	11,700	11,700	11,712	12.70	15.30	14.20
CO	1,800	2,100	2,200	342	441	594	15.00	20.40	21.60
GA	4,800	4,500	5,000	1,392	1,238	1,375	12.70	20.60	28.50
IN	2,300	2,300	2,200	414	460	462	16.20	19.80	15.00
MD	620	630	530	50	57	45	30.00	25.00	30.00
PA	900	900	900	144	113	153	33.50	30.00	28.00
SC	1,600	1,100	1,200	96	237	300	16.00	16.00	20.00
TX	3,600	1,800	2,400	432	198	240	32.50	34.00	29.00
US	73,820	71,730	76,130	20,426	19,294	19,891	14.80	18.50	18.10

[1] Preliminary.
NASS, Crops Branch, (202) 720–2127.

STATISTICS OF VEGETABLES AND MELONS

Table 4-21.—Carrots for fresh market, commercial crop: Area, production, and value per hundredweight, by State and United States, 2007-2009

State	Area harvested			Production			Value per unit		
	2007	2008	2009[1]	2007	2008	2009[1]	2007	2008	2009[1]
	Acres	Acres	Acres	1,000 cwt.	1,000 cwt.	1,000 cwt.	Dollars per cwt.	Dollars per cwt.	Dollars per cwt.
CA	68,000	62,500	61,000	20,400	20,313	19,215	22.40	25.20	25.70
MI	2,200	2,300	2,200	660	667	594	15.80	19.20	21.30
TX	1,500	1,300	1,200	300	312	324	28.00	25.00	25.60
Oth Sts[2]	7,400	7,600	5,000	3,070	3,273	2,030	21.10	21.30	21.10
US	79,100	73,700	69,400	24,430	24,565	22,163	22.10	24.50	25.20

[1] Preliminary. [2] 2007 - AZ, CO, GA, and WA. 2008 - AZ, CO, GA, and WA. 2009 - CO, GA, and WA.
NASS, Crops Branch, (202) 720-2127.

Table 4-22.—Carrots for processing, commercial crop: Area, production, and value per ton, by State and United States, 2007-2009

State	Area harvested			Production			Value per unit		
	2007	2008	2009[1]	2007	2008	2009[1]	2007	2008	2009[1]
	Acres	Acres	Acres	Tons	Tons	Tons	Dollars per tons	Dollars per tons	Dollars per tons
CA	1,900	1,500	1,000	62,700	52,500	33,000	80.00	110.00	180.00
MI[2]	3,000	2,700	....	60,000	67,500	....	76.00	88.00	....
MN[2]	....	....	930	....	....	28,850	....	....	83.30
WA	4,000	4,700	3,200	128,000	141,000	92,800	73.00	80.00	83.00
WI	4,000	3,900	3,700	81,200	77,300	86,690	63.30	80.30	79.90
Oth Sts[3]	2,130	3,140	3,700	45,250	63,430	88,100	68.50	72.10	94.20
US	15,030	15,940	12,530	377,150	401,730	329,440	72.00	84.10	94.90

[1] Preliminary. [2] Missing data included in Other States to avoid disclosure of individual operations. [3] 2007 - 2008 MN and TX. 2009 - MI and TX.
NASS, Crops Branch, (202) 720–2127.

Table 4-23.—Cauliflower, commercial crop: Area, yield, production, value per hundredweight and per ton, and total value, United States, 2000-2009 [1]

Year	Total crop					For fresh market			For processing		
	Area for harvest	Yield per acre	Produc- tion	Value[2]		Produc- tion	Value[2]		Produc- tion	Value[3]	
				Per cwt.	Total		Per cwt.	Total		Per ton	Total
	Acres	Cwt.	1,000 cwt.	Dollars	1,000 dollars	1,000 cwt.	Dollars	1,000 dollars	Tons	Dollars	1,000 dollars
2000	43,160	165	7,120	31.00	220,817	6,350	32.10	203,770	38,480	443.00	17,047
2001	42,050	160	6,708	28.30	190,085	5,920	29.20	172,690	39,410	441.00	17,395
2002	41,000	152	6,220	31.80	197,568	5,842	32.20	188,340	18,910	488.00	9,228
2003	38,840	168	6,535	34.60	225,795	6,205	35.10	217,545	16,500	500.00	8,250
2004	37,330	172	6,416	30.50	195,558	6,088	30.80	187,709	16,420	478.00	7,849
2005	41,370	174	7,214	30.40	219,411	[4]	[4]	[4]	[4]	[4]	[4]
2006	39,350	177	6,965	31.40	219,008	6,678	32.30	215,607	14,350	237.00	3,401
2007	37,820	181	6,828	34.20	233,413	6,616	34.40	227,689	10,600	540.00	5,724
2008	36,700	181	6,648	40.40	268,531	6,485	40.70	263,912	8,160	566.00	4,619
2009[5]	35,000	186	6,501	44.10	286,612	6,334	44.40	281,351	8,350	630.00	5,261

[1] Includes heading (cauliflower) broccoli. [2] Price and value at point of first sale. [3] Price and value at processing plant door. [4] Not published to avoid disclosure of individual operations. [5] Preliminary.
NASS, Crops Branch, (202) 720–2127.

Table 4-24.—Cauliflower, commercial crop: Area, production, and value per hundredweight and per ton, by State and United States, 2007–2009[1]

State	Area harvested			Production			Value per unit		
	2007	2008	2009[2]	2007	2008	2009[2]	2007	2008	2009[2]
	Acres	Acres	Acres	1,000 cwt.	1,000 cwt.	1,000 cwt.	Dollars per cwt.	Dollars per cwt.	Dollars per cwt.
AZ	4,000	4,000	3,200	760	800	640	42.20	43.70	44.50
CA	33,500	32,300	31,400	6,030	5,814	5,809	33.20	39.90	44.00
NY	320	400	400	38	34	52	34.10	52.40	45.50
US	37,820	36,700	35,000	6,828	6,648	6,501	34.20	40.40	44.10

State	For fresh market						For processing					
	Production			Value per unit			Production			Value per unit		
	2007	2008	2009[2]	2007	2008	2009[2]	2007	2008	2009[2]	2007	2008	2009[2]
	1,000 cwt.	1,000 cwt.	1,000 cwt.	Dollars per cwt.	Dollars per cwt.	Dollars per cwt.	Tons	Tons	Tons	Dollars per ton	Dollars per ton	Dollars per ton
AZ	760	800	640	42.20	43.70	44.50						
CA	5,818	5,651	5,642	33.40	40.20	44.40	10,600	8,160	8,350	540.00	566.00	630.00
NY	38	34	52	34.10	52.40	45.50						
US	6,616	6,485	6,334	34.40	40.70	44.40	10,600	8,160	8,350	540.00	566.00	630.00

[1] Includes heading (cauliflower) broccoli.   [2] Preliminary.
NASS, Crops Branch, (202) 720–2127.

Table 4-25.—Celery, commercial crop: Area, production, and value per hundredweight, by State and United States, 2007–2009[1]

State	Area harvested			Production			Value per unit		
	2007	2008	2009[2]	2007	2008	2009[2]	2007	2008	2009[2]
	Acres	Acres	Acres	1,000 cwt.	1,000 cwt.	1,000 cwt.	Dollars per cwt.	Doll ars per cwt.	Dollars per cwt.
CA	26,500	26,500	26,600	19,080	19,080	18,630	20.70	18.60	18.80
MI	1,900	1,800	1,900	931	945	1,055	13.20	15.60	14.10
US	28,400	28,300	28,500	20,011	20,025	19,685	20.40	18.50	18.50

[1] Mostly for fresh market use, but includes some quantities used for processing.   [2] Preliminary.
NASS, Crops Branch, (202) 720–2127.

Table 4-26.—Celery, commercial crop: Area, yield, production, value per hundredweight, and total value, United States, 2000-2009[1]

Year	Area for harvest	Yield per acre	Production	Value[2]	
				Per cwt.	Total
	Acres	Cwt.	1,000 cwt.	Dollars	1,000 dollars
2000	26,200	703	18,425	18.50	341,391
2001	27,800	678	18,856	14.40	272,391
2002	27,100	691	18,737	12.80	239,846
2003	27,500	700	19,256	13.40	258,965
2004	27,900	698	19,479	14.80	288,791
2005	26,800	697	18,686	13.90	259,309
2006	27,700	694	19,230	18.20	350,454
2007	28,400	705	20,011	20.40	408,001
2008	28,300	708	20,025	18.50	369,684
2009[3]	28,500	691	19,685	18.50	364,816

[1] Mostly for fresh market use, but includes quantities used for processing.   [2] Price and value at point of first sale.   [3] Preliminary.
NASS, Crops Branch, (202) 720–2127.

Table 4-27.—Corn, sweet, commercial crop: Area, production, and value per hundredweight and per ton, by State and United States, 2007–2009

Utilization and State	Area harvested			Production			Value per unit		
	2007	2008	2009¹	2007	2008	2009¹	2007	2008	2009¹
FOR FRESH MARKET	Acres	Acres	Acres	1,000 cwt.	1,000 cwt.	1,000 cwt.	Dollars per cwt.	Dollars per cwt.	Dollars per cwt.
AL	1,100	1,200	1,400	31	38	74	24.20	29.50	30.00
CA	25,600	24,900	24,900	4,480	3,984	4,482	23.90	23.60	24.80
CO	7,500	7,300	6,800	1,050	1,241	1,088	11.70	15.60	13.70
CT	4,500	3,900	3,900	360	332	273	27.00	35.00	40.00
DE	3,200	3,480	3,500	320	383	420	25.00	30.00	27.00
FL	39,000	42,000	43,100	6,630	6,720	6,681	22.50	23.60	34.00
GA	25,000	23,000	25,000	3,125	3,910	3,250	17.50	20.90	26.20
IL	6,600	6,800	7,100	726	694	582	19.10	22.20	22.50
IN	5,400	5,400	6,100	459	378	421	22.90	31.00	40.00
ME	1,900	1,800	1,500	152	108	90	33.50	43.00	47.00
MD	3,600	4,100	3,900	216	328	261	35.00	30.00	32.00
MA	5,200	5,200	4,700	416	416	306	39.00	43.00	43.00
MI	8,700	8,500	9,100	740	723	1,001	19.80	23.50	23.60
NH	1,700	1,600	1,400	102	128	77	52.00	61.00	59.00
NJ	7,100	7,100	7,100	675	533	781	22.30	29.10	29.20
NY	22,500	22,900	21,500	2,700	2,863	2,150	22.00	25.80	27.10
NC	6,200	6,500	6,300	620	650	693	15.00	19.00	21.00
OH	16,000	15,200	11,400	1,680	1,125	1,357	23.20	31.90	30.40
OR	4,600	4,300	4,900	515	495	343	21.40	24.90	27.50
PA	14,900	14,900	14,400	924	1,237	979	32.60	35.40	36.30
RI	900	800	750	54	68	45	35.00	37.00	40.00
TX	2,400	2,400	2,400	185	161	156	20.80	22.00	22.00
VT	1,100	1,000	1,000	72	50	45	51.00	56.00	54.00
VA	3,400	3,200	2,900	299	125	102	17.00	33.20	25.00
WA	8,500	9,000	11,500	1,403	1,665	2,162	27.90	38.90	37.50
WI	7,400	6,800	7,000	570	544	602	24.50	27.40	26.60
US	234,000	233,280	233,550	28,504	28,899	28,421	22.70	25.90	29.40
FOR PROCESSING	Acres	Acres	Acres	Tons	Tons	Tons	Dollars per ton	Dollars per ton	Dollars per ton
MD	6,300	6,900		52,920	50,370		73.00	75.00	
MN	120,500	123,900	122,400	794,850	876,980	979,250	82.90	120.00	99.60
OR	21,900	18,400	24,000	212,240	179,310	240,000	88.40	126.00	117.00
WA	82,500	66,300	81,700	854,310	648,490	847,010	72.70	118.00	109.00
WI	83,000	87,600	85,700	578,720	651,570	666,630	87.50	124.00	93.50
Oth Sts²	53,400	57,500	65,700	404,390	425,770	501,190	88.30	97.00	110.00
US	367,600	360,600	379,500	2,897,430	2,832,490	3,234,080	81.80	117.00	104.00

¹ Preliminary. ² 2007 - DE, ID, IL, IA, NJ, NY, PA, and TN. 2008 - DE, ID, IL, IA, NJ, NY, PA, TN, and VA. 2009 - DE, ID, IL, IA, MD, NJ, NY, and PA.
NASS, Crops Branch, (202) 720–2127.

Table 4-28.—Corn, sweet, commercial crop: Area, yield, production, value per hundredweight and per ton, and total value, United States, 2000-2009

Year	For fresh market					For processing				
	Area for harvest	Yield per acre	Pro-duction	Value ¹		Area for harvest	Yield per acre	Produc-tion	Value ²	
				Per cwt.	Total				Per ton	Total
	Acres	Cwt.	1,000 cwt.	Dol-lars	1,000 dollars	Acres	Tons	Tons	Dol-lars	1,000 dollars
2000	239,200	109	26,027	18.50	481,016	460,400	6.86	3,160,020	73.40	232,021
2001	244,930	109	26,815	19.50	523,567	447,150	7.04	3,147,530	73.00	229,678
2002	245,730	108	26,480	19.20	509,421	417,100	7.35	3,067,690	68.00	208,703
2003	236,600	116	27,492	19.20	528,858	426,600	7.66	3,266,050	70.40	229,788
2004	234,000	116	27,126	19.20	521,358	405,800	7.31	2,968,180	72.10	213,993
2005	230,600	115	26,416	22.10	582,529	403,910	7.86	3,174,800	68.40	217,111
2006	218,300	118	25,745	23.00	590,859	384,700	8.02	3,085,550	66.80	205,965
2007	234,000	122	28,504	22.70	646,374	367,600	7.88	2,897,430	81.80	236,908
2008	233,280	124	28,899	25.90	748,632	360,600	7.85	2,832,490	117.00	340,486
2009³	233,550	122	28,421	29.40	835,833	379,500	8.52	3,234,080	104.00	335,563

¹ Price and value at point of first sale. ² Price and value at processing plant door. ³ Preliminary.
NASS, Crops Branch, (202) 720–2127.

Table 4-29.—Cucumbers for fresh market: Area, production, and value per hundredweight, by State and United States, 2007–2009

State	Area harvested			Production			Value per unit		
	2007	2008	2009 [1]	2007	2008	2009 [1]	2007	2008	2009 [1]
	Acres	Acres	Acres	1,000 cwt.	1,000 cwt.	1,000 cwt.	Dollars per cwt.	Dollars per cwt.	Dollars per cwt.
CA	3,200	3,000	3,200	592	540	464	22.20	25.90	25.20
FL	15,200	11,600	11,300	4,104	3,248	2,656	26.80	21.90	29.60
GA	9,500	10,500	10,000	1,900	2,048	2,500	26.40	32.60	23.60
MD	360	380	450	22	17	25	50.00	30.00	42.00
MI	4,900	4,100	4,300	858	759	968	17.90	18.60	19.20
NJ	3,400	3,100	3,100	646	543	403	17.80	24.10	28.00
NY	4,100	3,600	3,200	574	468	384	34.30	34.50	41.80
NC	7,000	7,000	7,200	735	735	756	16.00	18.00	17.00
SC	1,200	1,600	1,700	108	336	391	25.00	20.00	18.00
TX	1,100	1,100	1,100	121	104	117	23.00	23.00	27.00
VA	1,000	1,000	1,300	40	45	65	18.70	22.00	22.00
US	50,960	46,880	46,850	9,700	8,843	8,729	24.60	24.80	25.30

[1] Preliminary.
NASS, Crops Branch, (202) 720–2127.

Table 4-30.—Cucumbers (for pickles), commercial crop: Area, yield, production, value per ton, total value, and pickle stocks, United States, 2000-2009

Year	For processing					Pickle stocks on hand Dec. 1 [2] [3]
	Area harvested	Yield per acre	Production	Value [1]		
				Per ton	Total	
	Acres	Tons	Tons	Dollars	1,000 dollars	Tons
2000	104,710	5.86	613,160	269.00	164,956	387,544
2001	108,260	5.37	581,540	291.00	168,958	552,303
2002	117,800	5.26	619,310	273.00	169,006	300,580
2003	118,800	5.46	648,430	275.00	178,328	353,573
2004	113,500	5.23	593,880	269.00	159,643	240,644
2005	110,500	4.89	540,080	256.00	138,391	250,448
2006	103,000	4.90	505,190	305.00	153,968	444,306
2007	101,500	5.33	541,230	325.00	175,822	376,732
2008	96,600	5.87	567,100	316.00	178,998	447,969
2009 [4]	97,500	5.57	542,600	333.00	180,845	189,665

[1] Price and value at processing plant door. [2] Stocks in hands of original salters of both salt and dill pickles, sold and unsold, in tanks and barrels, on Dec. 1. [3] Includes stocks of fresh-pack pickles. [4] Preliminary.
NASS, Crops Branch, (202) 720-2127.

Table 4-31.—Cucumbers (for pickles), commercial crop: Area, production, and value per ton, by State and United States, 2007–2009

State	Area harvested			Production			Value per unit		
	2007	2008	2009 [1]	2007	2008	2009 [1]	2007	2008	2009 [1]
	Acres	Acres	Acres	Tons	Tons	Tons	Dollars per ton	Dollars per ton	Dollars per ton
FL	6,800	7,000	7,000	69,970	73,500	49,000	470.00	420.00	468.00
IN	1,700	1,700	1,500	7,310	7,310	9,620	348.00	350.00	366.00
MI	35,000	30,500	32,500	185,500	189,100	188,500	230.00	220.00	260.00
NC	10,700	9,400	9,300	42,270	42,300	42,780	305.00	285.00	285.00
OH	2,600	3,200	2,200	21,350	21,120	17,600	445.00	428.00	460.00
SC	2,400	2,100	2,000	9,600	10,500	12,000	237.00	235.00	330.00
TX	6,800	7,400	7,000	25,840	37,000	38,500	484.00	516.00	511.00
WI	6,100	7,400	6,300	49,290	40,400	36,920	187.00	207.00	223.00
Oth Sts [2]	29,400	27,900	29,700	130,100	145,870	147,680	395.00	363.00	360.00
US	101,500	96,600	97,500	541,230	567,100	542,600	325.00	316.00	333.00

[1] Preliminary. [2] 2007 - AL, CA, DE, GA, MD, MO, and WA. 2008 - AL, CA, DE, GA, MD, and MO. 2009 - AL, CA, DE, GA, MD, and MO.
NASS, Crops Branch, (202) 720-2127.

Table 4-32.—Cucumbers for fresh market, commercial crop: Area, yield, production, value, and total value, United States, 2000-2009

Year	Area harvested	Yield per acre	Production	Value	
				Per cwt	Total
	Acres	*Cwt.*	*1,000 cwt.*	*Dollars*	*1,000 dollars*
2000	52,130	209	10,873	19.90	216,704
2001	52,780	197	10,392	19.80	205,689
2002	54,900	199	10,939	19.00	207,784
2003	54,600	172	9,381	19.90	186,352
2004	56,570	177	10,005	20.20	201,654
2005	51,970	178	9,265	23.10	214,138
2006	50,740	179	9,079	25.30	229,775
2007	50,960	190	9,700	24.60	238,925
2008	46,880	189	8,843	24.80	219,073
2009 [1]	46,850	186	8,729	25.30	220,761

[1] Preliminary.
NASS, Crops Branch, (202) 720–2127.

Table 4-33.—Garlic for fresh market and processing, commercial crop: Area, yield, production, value, and total value, United States, 2000-2009

Year	Area harvested	Yield per acre	Production	Value	
				Per cwt	Total
	Acres	*Cwt.*	*1,000 cwt.*	*Dollars*	*1,000 dollars*
2000	34,800	160	5,581	27.80	154,971
2001	35,200	167	5,877	29.40	173,020
2002	32,800	172	5,650	27.60	155,673
2003	35,000	178	6,241	25.70	160,200
2004	31,600	165	5,224	26.50	138,622
2005	29,900	160	4,771	43.60	208,018
2006	26,120	165	4,312	29.50	127,067
2007	24,810	165	4,104	41.20	169,218
2008	25,440	168	4,282	43.60	186,807
2009 [1]	23,330	169	3,941	49.80	196,075

[1] Preliminary.
NASS, Crops Branch, (202) 720–2127.

Table 4-34.—Garlic for fresh market and processing: Area, production, and value per hundredweight, by State and United States, 2007–2009

State	Area harvested			Production			Value per unit		
	2007	2008	2009 [1]	2007	2008	2009 [1]	2007	2008	2009 [1]
	Acres	*Acres*	*Acres*	*1,000 cwt.*	*1,000 cwt.*	*1,000 cwt.*	*Dollars per cwt.*	*Dollars per cwt.*	*Dollars per cwt.*
CA	22,200	23,000	22,000	3,774	3,910	3,740	42.90	45.20	51.20
NV	610	640	530	42	96	89	15.00	19.00	25.00
OR	2,000	1,800	800	288	276	112	22.80	30.00	22.50
US	24,810	25,440	23,330	4,104	4,282	3,941	41.20	43.60	49.80

[1] Preliminary.
NASS, Crops Branch, (202) 720–2127.

Table 4-35.—Honeydew melons, commercial crop: Area, yield, production, value per hundredweight, and total value, United States, 2000-2009

Year	Area for harvest	Yield per acre	Production	Value [1]	
				Per cwt.	Total
	Acres	Cwt.	1,000 cwt.	Dollars	1,000 dollars
2000	26,000	197	5,116	19.20	98,244
2001	24,200	195	4,720	21.10	99,500
2002	24,400	208	5,065	18.10	91,453
2003	22,200	229	5,075	18.80	95,461
2004	19,900	240	4,781	17.60	84,345
2005	19,600	216	4,243	19.00	80,418
2006	18,300	231	4,221	18.20	76,943
2007	17,550	236	4,144	17.70	73,517
2008	17,200	215	3,690	17.80	65,636
2009 [2]	16,800	220	3,698	16.00	59,060

[1] Price and value at point of first sale.   [2] Preliminary.
NASS, Crops Branch, (202) 720–2127.

Table 4-36.—Honeydew melons, commercial crop: Area, production, and value per hundredweight, by State and United States, 2007–2009

State	Area harvested			Production			Value per unit		
	2007	2008	2009 [1]	2007	2008	2009 [1]	2007	2008	2009 [1]
	Acres	Acres	Acres	1,000 cwt.	1,000 cwt.	1,000 cwt.	Dollars per cwt.	Dollars per cwt.	Dollars per cwt.
AZ	3,300	3,400	3,400	1,106	815	800	16.60	27.60	20.60
CA	13,600	13,100	12,800	2,856	2,686	2,688	16.70	14.00	14.20
TX	650	700	600	182	189	210	41.00	29.30	21.00
US	17,550	17,200	16,800	4,144	3,690	3,698	17.70	17.80	16.00

[1] Preliminary.
NASS, Crops Branch, (202) 720–2127.

Table 4-37.—Head lettuce, commercial crop: Area, production, and value per hundredweight, by State and United States, 2007–2009

State	Area harvested			Production			Value per unit		
	2007	2008	2009 [1]	2007	2008	2009 [1]	2007	2008	2009 [1]
	Acres	Acres	Acres	1,000 cwt.	1,000 cwt.	1,000 cwt.	Dollars per cwt.	Dollars per cwt.	Dollars per cwt.
AZ	39,900	32,700	32,000	14,564	11,772	11,040	21.00	15.80	22.90
CA	119,000	116,000	114,000	42,245	41,180	42,180	22.00	21.30	21.40
Oth Sts [2][3]	2,900	....	....	665	....	....	19.10	....	....
US	161,800	148,700	146,000	57,474	52,952	53,220	21.70	20.10	21.70

[1] Preliminary.   [2] CO and NJ estimates discontinued in 2008.   [3] 2007 - CO and NJ.
NASS, Crops Branch, (202) 720–2127.

Table 4-38.—Head lettuce, commercial crop: Area, yield, production, value per hundredweight, and total value, United States, 2000-2009

Year	Area for harvest	Yield per acre	Production	Value [1]	
				Per cwt.	Total
	Acres	Cwt.	1,000 cwt.	Dollars	1,000 dollars
2000	184,900	377	69,673	17.30	1,208,140
2001	184,300	374	68,917	17.90	1,234,981
2002	184,500	369	68,140	21.10	1,435,296
2003	185,100	369	68,244	18.10	1,235,193
2004	181,000	366	66,228	16.90	1,118,970
2005	177,400	368	65,253	15.50	1,011,976
2006	178,800	350	62,494	16.90	1,054,941
2007	161,800	355	57,474	21.70	1,247,941
2008	148,700	356	52,952	20.10	1,063,132
2009 [2]	146,000	365	53,220	21.70	1,155,468

[1] Price and value at point of first sale.   [2] Preliminary.
NASS, Crops Branch, (202) 720–2127.

Table 4-39.—Lettuce, leaf for fresh market, commercial crop: Area, yield, production, value per hundredweight, and total value, United States, 2000-2009

Year	Area harvested	Yield per acre	Production	Value [1] Per cwt.	Value [1] Total
	Acres	Cwt.	1,000 cwt.	Dollars	1,000 dollars
2000	47,500	252	11,979	29.70	355,658
2001	50,500	226	11,394	27.50	313,621
2002	53,900	249	13,410	33.70	452,274
2003	57,400	233	13,370	31.50	420,546
2004	59,400	236	14,001	30.80	430,904
2005	56,900	241	13,701	33.90	463,995
2006	55,900	238	13,317	34.80	463,859
2007	54,600	224	12,240	30.50	373,692
2008	52,300	244	12,781	32.20	411,719
2009 [2]	49,100	229	11,238	36.10	405,829

[1] Price and value at point of first sale.　[2] Preliminary.
NASS, Crops Branch, (202) 720-2127.

Table 4-40.—Leaf lettuce for fresh market: Area, production, and value per hundredweight, by State and United States, 2007-2009

State	Area harvested 2007	Area harvested 2008	Area harvested 2009 [1]	Production 2007	Production 2008	Production 2009 [1]	Value per unit 2007	Value per unit 2008	Value per unit 2009 [1]
	Acres	Acres	Acres	1,000 cwt.	1,000 cwt.	1,000 cwt.	Dollars per cwt.	Dollars per cwt.	Dollars per cwt.
AZ	10,600	9,800	8,600	2,120	2,156	1,720	40.70	37.70	39.50
CA	44,000	42,500	40,500	10,120	10,625	9,518	28.40	31.10	35.50
US	54,600	52,300	49,100	12,240	12,781	11,238	30.50	32.20	36.10

[1] Preliminary.
NASS, Crops Branch, (202) 720-2127.

Table 4-41.—Lettuce, Romaine for fresh market, commercial crop: Area, yield, production, value per hundredweight, and total value, United States, 2000-2009

Year	Area harvested	Yield per acre	Production	Value [1] Per cwt.	Value [1] Total
	Acres	Cwt.	1,000 cwt.	Dollars	1,000 dollars
2000	48,850	308	15,045	19.90	299,278
2001	53,100	284	15,067	19.30	290,934
2002	58,300	318	18,564	25.20	466,896
2003	74,500	297	22,103	27.50	607,078
2004	74,200	347	25,712	19.10	492,208
2005	58,400	330	19,272	19.50	375,005
2006	86,400	307	26,500	22.40	593,866
2007	82,400	320	26,409	24.80	655,533
2008	77,400	294	22,774	21.00	479,006
2009 [2]	79,100	329	26,030	23.60	614,134

[1] Price and value at point of first sale.　[2] Preliminary.
NASS, Crops Branch, (202) 720-2127.

Table 4-42.—Romaine lettuce for fresh market: Area, production, and value per hundredweight, by State and United States, 2007–2009

State	Area harvested 2007	Area harvested 2008	Area harvested 2009 [1]	Production 2007	Production 2008	Production 2009 [1]	Value per unit 2007	Value per unit 2008	Value per unit 2009 [1]
	Acres	Acres	Acres	1,000 cwt.	1,000 cwt.	1,000 cwt.	Dollars per cwt.	Dollars per cwt.	Dollars per cwt.
AZ	17,400	16,400	16,100	4,959	5,084	5,555	35.30	20.80	23.20
CA	65,000	61,000	63,000	21,450	17,690	20,475	22.40	21.10	23.70
US	82,400	77,400	79,100	26,409	22,774	26,030	24.80	21.00	23.60

[1] Preliminary.
NASS, Crops Branch, (202) 720-2127.

Table 4-43.—Onions, commercial crop: Area, yield, production, shrinkage and loss, value per hundredweight, and total value, United States, 2000-2009 [1]

Year	Area for harvest	Yield per acre	Production [2]	Shrinkage and loss	Value [3] Per cwt.	Value [3] Total
	Acres	Cwt.	1,000 cwt.	1,000 cwt.	Dollars	1,000 dollars
2000	167,070	437	72,948	7,131	11.20	735,939
2001	164,990	424	69,961	6,564	10.70	680,350
2002	162,720	429	69,844	6,425	12.10	764,994
2003	165,990	442	73,346	5,583	13.70	928,907
2004	169,350	495	83,775	8,821	8.93	669,514
2005	163,420	446	72,875	5,008	12.40	839,773
2006	163,780	446	73,066	5,529	16.10	1,084,099
2007	160,080	497	79,638	6,295	11.10	816,061
2008	153,490	489	75,120	5,072	12.50	872,113
2009 [4]	148,560	505	74,970	5,735	12.20	843,570

[1] Mostly for fresh market use, but includes some quantities used for processing. [2] Includes storage crop onions harvested but not sold because of shrinkage and loss. [3] Price and value at point of first sale. [4] Preliminary.
NASS, Crops Branch, (202) 720–2127.

Table 4-44.—Onions, commercial crop: Area, production, shrinkage and loss, and value per hundredweight, by State and United States, 2007–2009 [1]

Season and State	Area harvested 2007	Area harvested 2008	Area harvested 2009 [2]	Production 2007	Production 2008	Production 2009 [2]	Shrinkage and loss 2007	Shrinkage and loss 2008	Shrinkage and loss 2009 [2]	Value per unit 2007	Value per unit 2008	Value per unit 2009 [2]
	Acres	Acres	Acres	1,000 cwt.	1,000 cwt.	1,000 cwt.	1,000 cwt.	1,000 cwt.	1,000 cwt.	Dollars per cwt.	Dollars per cwt.	Dollars per cwt.
Spring:												
AZ	1,100	1,500	1,600	495	555	576				12.00	10.50	11.30
CA	6,700	6,500	6,000	3,015	2,860	2,460				11.00	8.60	8.60
GA	12,000	11,500	10,500	3,240	3,680	2,520				35.90	29.50	32.90
TX	10,400	8,900	9,100	3,120	2,403	3,003				56.00	30.30	26.40
Total	30,200	28,400	27,200	9,870	9,498	8,559				33.40	22.30	22.20
Summer:												
Non-storage:												
CA	7,800	7,600	6,400	4,407	4,104	3,456				9.90	8.60	8.60
NV	2,800	2,600	3,400	2,072	1,976	2,448				19.00	24.00	28.00
NM	7,100	6,600	5,000	3,550	3,300	2,750				20.80	14.40	19.60
TX	1,000	700	600	400	280	180				29.50	33.40	38.20
WA	1,500	1,900	2,000	570	684	750				32.50	29.90	29.50
Non-storage total	20,200	19,400	17,400	10,999	10,344	9,584				17.00	15.50	18.90
Storage: [3]												
CA [4]	29,000	31,300	31,400	12,325	13,303	14,287	250	250	250	8.53	9.01	9.54
CO	7,700	7,500	6,700	3,157	2,850	3,015	470	400	300	9.70	17.60	12.60
ID	9,100	8,600	8,800	6,825	6,192	6,512	1,400	970	1,170	2.70	7.40	8.30
MI	3,800	3,600	3,800	988	1,008	1,330	200	200	270	11.10	15.20	13.50
NY	10,500	10,100	10,100	3,780	4,141	4,141	300	634	696	11.10	16.80	13.20
OR (Malheur)	12,600	12,200	11,200	9,828	8,662	8,176	1,310	870	1,060	2.51	7.38	9.08
OR (Other)	10,300	9,600	9,100	6,386	5,952	5,642	400	410	570	5.97	6.96	7.41
WA	22,500	20,000	19,000	13,725	12,000	11,970	1,800	1,200	1,200	3.15	11.10	7.60
WI	2,100	1,100	2,000	777	330	980	75	25	92	10.50	14.20	10.10
Oth Sts [5]	2,080	1,690	1,860	978	840	774	90	113	127	6.38	13.00	11.20
Storage total	109,680	105,690	103,960	58,769	55,278	56,827	6,295	5,072	5,735	5.69	9.96	9.25
Total summer	129,880	125,090	121,360	69,768	65,622	66,411				7.66	10.90	10.80
US	160,080	153,490	148,560	79,638	75,120	74,970				11.10	12.50	12.20

[1] Mostly for fresh market use, but includes some quantities used for processing. [2] Preliminary. [3] Includes some quantities of storage crop onions harvested but not sold because of shrinkage and loss. [4] Primarily for dehydrated and other processing. [5] OH and UT.
NASS, Crops Branch, (202) 720–2127.

Table 4-45.—Onions (fresh market): Foreign trade, United States, 1999–2008[1]

Year beginning July	Imports	Domestic exports
	1,000 cwt.	*1,000 cwt.*
1999	5,005	7,040
2000	5,671	7,964
2001	5,925	6,788
2002	6,322	6,838
2003	6,563	6,174
2004	6,713	6,916
2005	6,297	6,149
2006	8,656	6,236
2007	7,337	5,368
2008	6,639	6,131

[1] Includes onion sets and pearl onions.
ERS, Specialty Crops Branch, (202) 694–5253. Compiled from reports of the U.S. Department of Commerce.

Table 4-46.—Peas, green (for processing), commercial crop: Area, yield, production, value per ton, and total value, United States, 2000-2009

Year	Area harvested	Yield per acre	Production	Value[1] Per ton	Value[1] Total
	Acres	*Tons*	*Tons*	*Dollars*	*1,000 dollars*
2000	277,240	1.91	530,550	248.00	131,817
2001	211,640	1.85	390,980	264.00	103,313
2002	212,200	1.65	349,860	253.00	88,439
2003	229,000	2.02	462,240	251.00	116,077
2004	203,500	1.88	383,390	255.00	97,669
2005	210,900	1.75	370,050	270.00	99,905
2006	195,900	2.00	392,420	247.00	96,778
2007	202,000	2.07	419,080	259.00	108,702
2008	209,700	1.96	411,780	360.00	148,052
2009[2]	205,350	2.15	441,580	319.00	140,679

[1] Price and value at processing plant door. [2] Preliminary.
NASS, Crops Branch, (202) 720–2127.

Table 4-47.—Peas, green (for processing), commercial crop: Area, production, and value per ton, by State and United States, 2007–2009[1]

State	Area harvested 2007	Area harvested 2008	Area harvested 2009[2]	Production 2007	Production 2008	Production 2009[2]	Value per unit 2007	Value per unit 2008	Value per unit 2009[2]
	Acres	*Acres*	*Acres*	*Tons*	*Tons*	*Tons*	*Dollars per ton*	*Dollars per ton*	*Dollars per ton*
DE	5,400	5,600	5,450	10,260	10,080	9,930	275.00	286.00	280.00
MN	73,600	73,100	73,100	128,950	124,470	151,760	342.00	496.00	395.00
OR	17,600	18,000	17,600	38,300	33,320	31,400	199.00	263.00	255.00
WA	38,500	40,100	40,200	105,490	103,460	100,100	199.00	298.00	265.00
WI	37,300	40,100	40,800	80,950	76,060	91,760	231.00	266.00	271.00
Oth Sts[3]	29,600	32,800	28,200	55,130	64,390	56,630	261.00	367.00	329.00
US	202,000	209,700	205,350	419,080	411,780	441,580	259.00	360.00	319.00

[1] Shelled basis; 2½ pounds of peas in the shell produce approximately 1 pound of shelled peas. [2] Preliminary. [3] 2007 - IL, MD, NJ, and NY. 2008 - IL, MD, NJ, and NY. 2009 - IL, MD, NJ, and NY.
NASS, Crops Branch, (202) 720–2127.

Table 4-48.—Chile peppers for fresh market and processing: Area, production, and value per hundredweight, by State and United States, 2007–2009[1][2]

State	Area harvested			Production			Value per unit		
	2007	2008	2009[3]	2007	2008	2009[3]	2007	2008	2009[3]
	Acres	*Acres*	*Acres*	*1,000 cwt.*	*1,000 cwt.*	*1,000 cwt.*	*Dollars per cwt.*	*Dollars per cwt.*	*Dollars per cwt.*
AZ	3,600	3,000	2,900	288	224	234	45.10	48.90	44.60
CA	5,800	5,600	5,800	1,798	1,810	2,033	34.40	24.30	24.80
NM	11,000	11,100	12,500	1,620	1,962	2,303	17.70	21.60	21.50
TX	4,500	5,300	7,300	171	174	260	71.30	79.80	71.90
US	24,900	25,000	28,500	3,877	4,170	4,830	29.90	26.70	26.70

[1] Chile peppers are defined as all peppers excluding bell peppers.   [2] Estimates include both fresh and dry product combined.   [3] Preliminary.
NASS, Crops Branch, (202) 720–2127.

Table 4-49.—Chile peppers for fresh market and processing, commercial crop: Area, yield, production, value, and total value, United States, 2001-2009[1]

Year	Area harvested	Yield per acre	Production	Value	
				Per cwt	Total
	Acres	*Cwt.*	*1,000 cwt.*	*Dollars*	*1,000 dollars*
2001	32,350	100	3,244	33.30	108,090
2002	30,000	116	3,474	30.20	105,018
2003	29,000	153	4,443	23.10	102,748
2004	30,200	172	5,181	21.50	111,236
2005	32,700	156	5,108	22.90	117,048
2006	28,200	169	4,779	21.90	104,775
2007	24,900	156	3,877	29.90	115,745
2008	25,000	167	4,170	26.70	111,199
2009[2]	28,500	169	4,830	26.70	128,951

[1] Chile peppers are defined as all peppers excluding bell peppers. Estimates include both fresh and dry product combined.   [2] Preliminary.
NASS, Crops Branch, (202) 720–2127.

Table 4-50.—Bell peppers for fresh market and processing: Area, production, and value per hundredweight, by State and United States, 2007–2009

State	Area harvested			Production			Value per unit		
	2007	2008	2009[1]	2007	2008	2009[1]	2007	2008	2009[1]
	Acres	*Acres*	*Acres*	*1,000 cwt.*	*1,000 cwt.*	*1,000 cwt.*	*Dollars per cwt.*	*Dollars per cwt.*	*Dollars per cwt.*
CA	21,000	20,000	20,500	7,980	8,062	7,595	28.50	36.10	30.10
FL	19,300	17,800	18,200	4,340	4,984	4,482	42.20	50.80	44.30
GA	4,200	3,400	3,500	1,176	612	980	35.00	33.50	35.50
MI	1,500	1,600	1,600	390	400	384	33.00	30.00	30.00
NJ	3,100	3,100	3,200	930	1,116	928	31.50	29.50	33.80
NC	2,700	2,800	2,900	624	336	406	25.00	30.00	32.00
OH	2,200	2,200	2,200	660	378	825	34.80	45.60	46.00
US	54,000	50,900	52,100	16,100	15,888	15,600	33.10	40.10	35.60

[1] Preliminary.
NASS, Crops Branch, (202) 720–2127.

Table 4-51.—Bell peppers for fresh market and processing, commercial crop: Area, yield, production, value, hundredweight, and total value, United States, 2000-2009

Year	Area harvested	Yield per acre	Production	Value	
				Per cwt	Total
	Acres	Cwt.	1,000 cwt.	Dollars	1,000 dollars
2000	62,080	272	16,879	31.50	531,018
2001	57,780	285	16,494	28.70	473,557
2002	53,800	291	15,668	29.60	464,401
2003	53,300	302	16,118	30.70	494,663
2004	52,900	310	16,400	31.50	516,956
2005	56,800	282	16,036	33.30	534,703
2006	53,100	296	15,710	33.70	528,652
2007	54,000	298	16,100	33.10	532,799
2008	50,900	312	15,888	40.10	636,620
2009 [1]	52,100	299	15,600	35.60	555,643

[1] Preliminary.
NASS, Crops Branch, (202) 720–2127.

Table 4-52.—Potatoes: Area, yield, production, season average price, and value, United States, 2000–2009

Year	Area planted	Area harvested	Yield per harvested acre	Production	Season average price per cwt. received by farmers [1]	Value of production
	1,000 acres	1,000 acres	Cwt.	1,000 cwt.	Dollars	1,000 dollars
2000	1,383.1	1,347.5	381	513,544	5.08	2,590,053
2001	1,246.9	1,220.9	358	437,673	6.99	3,055,876
2002	1,299.6	1,265.9	362	458,171	6.67	3,045,310
2003	1,273.6	1,249.6	367	458,199	5.88	2,677,361
2004	1,192.4	1,166.0	391	455,806	5.65	2,565,260
2005	1,108.4	1,086.2	390	423,788	7.04	2,981,754
2006	1,139.4	1,120.2	393	440,698	7.31	3,208,632
2007	1,141.9	1,122.2	396	444,875	7.51	3,339,710
2008	1,059.6	1,046.9	396	415,055	8.42	3,770,462
2009	1,069.8	1,045.0	413	431,425	8.00	3,452,276

[1] 1999-2007 obtained by weighting State prices by quantity sold. 2008 obtained by weighting State prices by production.
NASS, Crops Branch, (202) 720–2127.

Table 4-53.—Potatoes: Production, seed used, and disposition, United States, 2000–2008

Year	Production	Total used for seed	Used on farms where produced		Sold
			For seed, feed, and household use	Shrinkage and loss	
	1,000 cwt.	1,000 cwt.	1,000 cwt.	1,000 cwt.	1,000 cwt.
2000	513,544	27,137	5,287	43,685	464,572
2001	437,673	28,625	5,386	31,227	401,060
2002	458,171	28,149	5,622	30,905	421,644
2003	458,199	26,652	5,546	35,324	417,329
2004	455,806	24,744	4,801	37,432	413,573
2005	423,788	25,616	4,797	28,572	390,419
2006	440,698	26,374	4,750	29,639	406,309
2007	444,875	24,476	4,105	29,561	411,209
2008	415,055	24,533	4,138	26,438	384,478

NASS, Crops Branch, (202) 720–2127.

Table 4-54.—Fall potatoes: Production and total stocks held by growers and local dealers, 15 Major States, 1999–2008

Crop year	Production	Total stocks						
		Dec. 1	Following year					
			Jan. 1	Feb. 1	Mar. 1	Apr. 1	May 1	June 1
	1,000 cwt.	*1,000 cwt.*	*1,000 cwt.*	*1,000 cwt.*	*1,000 cwt.*	*1,000 cwt.*	*1,000 cwt.*	*1,000 cwt.*
1999	420,567	275,100	239,910	207,150	169,620	128,410	86,915	47,220
2000	458,827	310,300	275,270	234,260	197,670	153,520	109,160	61,270
2001	387,033	258,750	224,680	192,090	158,590	119,950	81,200	42,990
2002	407,085	264,485	231,490	199,020	165,210	125,770	83,040	45,880
2003	403,566	267,900	233,590	200,230	166,280	126,110	85,000	46,020
2004	404,017	271,100	236,700	203,490	168,020	128,900	88,550	51,700
2005 [1]	375,118	253,800	220,500	189,100	155,500	115,700	75,900	41,560
2006	389,527	258,900	225,800	192,200	159,500	121,900	79,050	44,460
2007	397,753	265,500	232,300	199,300	163,400	125,500	83,960	50,420
2008	369,866	243,700	213,200	183,900	152,700	115,800	78,100	45,300

[1] Beginning in 2005 13 major States.
NASS, Crops Branch, (202) 720–2127.

Table 4-55.—Potatoes: Area, production, and marketing year price per hundredweight received by farmers, by State and United States, 2007–2009

Season and State	Area harvested			Yield			Production		
	2007	2008	2009	2007	2008	2009	2007	2008	2009
	1,000 acres	*1,000 acres*	*1,000 acres*	*Cwt.*	*Cwt.*	*Cwt.*	*1,000 cwt.*	*1,000 cwt.*	*1,000 cwt.*
Winter:									
CA	10.5	11.0	8.7	215	230	245	2,258	2,530	2,132
Total	10.5	11.0	8.7	215	230	245	2,258	2,530	2,132
Spring:									
AZ	4.0	3.5	4.0	280	300	280	1,120	1,050	1,120
CA	15.5	15.4	17.5	395	450	410	6,123	6,930	7,175
FL	27.2	27.9	28.9	287	285	266	7,807	7,952	7,700
Hastings	16.2	17.0	16.5	285	285	260	4,617	4,845	4,290
Other FL	11.0	10.9	12.4	290	285	275	3,190	3,107	3,410
NC	14.5	14.0	15.0	186	180	225	2,700	2,520	3,375
TX	9.0	8.0	8.3	230	210	235	2,070	1,680	1,951
Total	70.2	68.8	73.7	282	293	289	19,820	20,132	21,321
Summer:									
AL	1.1	1.2	....	140	170	....	154	204	....
CA	4.3	3.6	3.4	360	360	405	1,548	1,296	1,377
CO	2.7	4.4	3.9	350	370	400	945	1,628	1,560
DE	2.0	1.7	1.6	270	250	300	540	425	480
IL	6.1	5.3	5.2	400	395	385	2,440	2,094	2,002
KS	4.9	4.8	4.8	365	320	360	1,789	1,536	1,728
MD	3.0	2.5	2.3	320	300	320	960	750	736
MO	6.6	6.5	7.1	300	190	275	1,980	1,235	1,953
NJ	2.4	2.0	2.1	265	230	260	636	460	546
TX	9.8	7.4	5.4	395	395	460	3,871	2,923	2,484
VA	5.4	5.7	6.9	210	220	240	1,134	1,254	1,656
Total	48.3	45.1	42.7	331	306	340	15,997	13,805	14,452
Fall:									
CA	7.9	8.4	8.4	480	470	495	3,792	3,948	4,158
CO	59.1	56.9	55.2	355	385	400	20,981	21,907	22,080
ID	349.0	304.0	319.0	373	383	411	130,010	116,475	131,000
10 S.W. Co. ..	21.0	15.0	19.0	490	540	500	10,290	8,100	9,500
Other ID	328.0	289.0	300.0	365	375	405	119,720	108,375	121,500
ME	56.5	54.7	55.5	295	270	275	16,668	14,769	15,263
MA	2.6	2.7	3.4	320	260	260	832	702	884
MI	42.0	42.5	43.5	350	350	360	14,700	14,875	15,660
MN	49.0	48.0	45.0	440	425	460	21,560	20,400	20,700
MT	11.2	10.5	9.7	330	330	345	3,696	3,465	3,347
NE	19.8	19.4	19.9	415	425	440	8,217	8,245	8,756
NV	7.3	5.8	5.1	390	410	470	2,847	2,378	2,397
NM	5.4	5.9	6.4	370	390	400	1,998	2,301	2,560
NY	18.3	17.8	16.5	285	320	300	5,216	5,696	4,950
ND	91.0	81.0	75.0	260	280	255	23,660	22,680	19,125
OH	3.0	2.1	2.1	330	325	335	990	683	704
OR	36.5	35.3	37.0	556	529	580	20,293	18,676	21,460
Malheur Co. ..	3.0	2.8	....	455	460	....	1,365	1,288	....
Other OR	33.5	32.5	....	565	535	....	18,928	17,388	....
PA	10.0	9.5	9.5	220	265	310	2,200	2,518	2,945
RI	0.6	0.5	0.4	300	280	210	180	140	84
WA	160.0	155.0	145.0	630	600	610	100,800	93,000	88,450
WI	64.0	62.0	63.0	440	415	460	28,160	25,730	28,980
Total	993.2	922.0	919.6	410	411	428	406,800	378,588	393,503
US	1,122.2	1,046.9	1,045.0	396	396	413	444,875	415,055	431,425

NASS, Crops Branch, (202) 720–2127.

STATISTICS OF VEGETABLES AND MELONS

Table 4-56.—Fall potatoes: Total stocks held by growers and local dealers, 13 States, crop of 2007 and 2008 [1]

State	Crop of 2007						
	Dec. 1, 2007	Jan. 1, 2008	Feb. 1, 2008	Mar. 1, 2008	Apr. 1, 2008	May 1, 2008	June 1, 2008
	1,000 cwt.	*1,000 cwt.*	*1,000 cwt.*	*1,000 cwt.*	*1,000 cwt.*	*1,000 cwt.*	*1,000 cwt.*
CA	2,000	1,600	1,400	1,100	600	460	300
CO	15,200	13,200	11,200	9,500	7,500	4,500	2,500
ID	92,000	82,000	71,500	60,500	49,000	35,000	23,000
ME	12,900	11,400	9,700	8,000	6,400	4,300	2,500
MI	8,800	7,000	5,300	3,700	2,100	800	300
MN	13,400	11,600	9,900	8,200	6,200	4,300	2,400
MT	3,500	3,400	3,400	3,100	2,100	*	*
NE	5,300	4,500	3,700	3,000	2,000	1,000	*
NY	2,900	2,100	1,600	1,200	500	*	*
ND	14,500	12,500	10,000	8,000	6,000	4,200	2,800
OR	18,000	16,100	13,900	11,600	9,200	6,400	4,200
WA	57,000	50,200	43,800	35,400	27,500	19,000	11,000
WI	20,000	16,700	13,900	10,100	6,400	3,100	1,000
Other	....	....	....	....	....	900	720
13 State total	265,500	232,300	199,300	163,400	125,500	83,960	50,420
Klamath Basin [2]	4,200	3,400	2,700	2,050	1,150	570	(3)

State	Crop of 2008						
	Dec. 1, 2008	Jan. 1, 2009	Feb. 1, 2009	Mar. 1, 2009	Apr. 1, 2009	May 1, 2009	June 1, 2009
	1,000 cwt.	*1,000 cwt.*	*1,000 cwt.*	*1,000 cwt.*	*1,000 cwt.*	*1,000 cwt.*	*1,000 cwt.*
CA	1,900	1,500	1,300	1,000	800	500	400
CO	16,600	14,700	12,700	11,100	9,000	5,900	3,800
ID	85,000	75,500	66,000	56,500	45,000	32,500	21,000
ME	11,300	10,000	8,500	7,100	5,600	3,700	2,200
MI	8,300	6,600	4,800	3,300	1,800	700	*
MN	13,200	11,700	10,200	8,700	6,400	4,400	2,900
MT	3,400	3,400	3,400	3,200	2,100	*	*
NE	5,600	4,700	4,100	3,500	2,200	1,200	*
NY	2,600	2,000	1,600	1,200	700	*	*
ND	14,800	12,500	10,400	8,100	6,300	4,200	2,200
OR	16,100	14,100	12,000	9,800	7,400	5,300	3,400
WA	49,500	43,800	38,600	31,400	23,700	16,500	8,300
WI	15,400	12,700	10,300	7,800	4,800	2,200	300
Other	....	....	....	....	....	1,000	800
13 State total	243,700	213,200	183,900	152,700	115,800	78,100	45,300
Klamath Basin [2]	4,000	3,300	2,800	2,000	1,450	750	(3)

* Combined into "Other" to avoid disclosure of individual operations.
[1] Stocks are defined as the quantity (whether sold or not) remaining in storage for all purposes and uses, including seed potatoes that are not yet moved, and shrinkage, waste, and other losses that occur after the date of each estimate. [2] Estimates began in 2006. Includes potato stocks in CA and Klamath Co. OR. [3] Not published to avoid disclosure of individual operations.
NASS, Crops Branch, (202) 720–2127.

Table 4-57.—Potatoes: Utilization, United States, crop years 2001–2008

Item	2001	2002	2003	2004
SALES	1,000 cwt.	1,000 cwt.	1,000 cwt.	1,000 cwt.
Table stock	122,552	131,889	120,663	123,846
For processing:				
Chips and shoestring	54,080	51,640	55,524	51,284
Dehydration [1]	40,759	51,357	51,156	49,719
Frozen french fries	126,711	124,875	132,378	134,788
Other frozen products	23,598	28,951	24,959	23,555
Canned potatoes	2,590	2,744	3,286	2,912
Other canned products (hash, stews, soups)	1,722	2,089	1,189	1,008
Starch and flour	1,015	1,050	1,546	1,701
Total	250,475	262,706	270,038	264,967
Other sales:				
Livestock feed [2]	3,496	3,044	2,005	1,852
Seed	24,537	24,005	24,623	22,908
Total	28,033	27,049	26,628	24,760
Total sales	401,060	421,644	417,329	413,573
NON-SALES				
Seed used on farms where grown	4,088	4,144	4,002	3,604
Household use	1,298	1,478	1,544	1,197
Shrinkage and loss [3]	31,227	30,905	35,324	37,432
Total non-sales	36,613	36,527	40,870	42,233
Total production	437,673	458,171	458,199	455,806

Item	2005	2006	2007	2008
SALES	1,000 cwt.	1,000 cwt.	1,000 cwt.	1,000 cwt.
Table stock	113,626	113,335	110,860	109,351
For processing:				
Chips and shoestring	52,365	64,377	54,343	50,744
Dehydration [1]	43,437	48,809	49,021	40,714
Frozen french fries	126,545	126,083	139,624	134,394
Other frozen products	25,398	24,229	26,571	19,558
Canned potatoes	2,176	1,957	2,504	2,156
Other canned products (hash, stews, soups)	959	930	800	791
Starch and flour	1,732	1,369	4,029	5,067
Total	252,612	267,754	276,892	253,424
Other sales:				
Livestock feed	1,909	1,610	1,160	803
Seed	22,272	23,610	22,297	20,900
Total	24,181	25,220	23,457	21,703
Total sales	390,419	406,309	411,209	384,478
NON-SALES				
Seed used on farms where grown	3,600	3,520	2,986	3,315
Household use	1,197	1,230	1,119	823
Shrinkage and loss	28,572	29,639	29,561	26,438
Total non-sales	33,369	34,389	33,666	30,576
Total production	423,788	440,698	444,875	415,055

[1] Dehydrated products except starch and flour. [2] Includes 6,872 thousand cwt sold for livestock feed under Government Diversion Program for 2000. [3] Includes potatoes disposed of by the United Fresh Potato Growers of Idaho for 2004.
NASS, Crops Branch, (202) 720–2127.

Table 4-58.—Potatoes: Production, seed used, and disposition, by seasonal groups, crop of 2008

| Season and State | Production | Total used for seed | Used on farms where produced | | Sold |
			For seed, feed, and household use	Shrinkage and loss	
	1,000 cwt.	*1,000 cwt.*	*1,000 cwt.*	*1,000 cwt.*	*1,000 cwt.*
Winter:					
CA	2,530	297	....	55	2,475
Total	2,530	297	....	55	2,475
Spring:					
AZ	1,050	70	2	13	1,035
CA	6,930	450	....	735	6,195
FL [1]	7,952	704	....	187	7,765
Hastings	4,845	452	....	140	4,705
Other [1]	3,107	252	....	47	3,060
NC	2,520	261	13	....	2,507
TX	1,680	141	18	32	1,630
Total	20,132	1,626	33	967	19,132
Summer:					
AL	204	20	2	8	194
CA	1,296	108	....	64	1,232
CO	1,628	104	4	99	1,525
DE	425	20	1	8	416
IL	2,094	121	35	40	2,019
KS	1,536	120	....	61	1,475
MD	750	35	1	21	728
MO	1,235	144	....	13	1,222
NJ	460	38	3	13	444
TX	2,923	93	2	86	2,835
VA	1,254	84	2	51	1,201
Total	13,805	885	50	464	13,291
Fall:					
CA	3,948	215	9	371	3,568
CO	21,907	1,456	1,316	1,811	18,780
ID	116,475	7,520	908	7,150	108,417
ME	14,769	1,154	214	525	14,030
MA	702	74	12	25	665
MI	14,875	1,089	210	1,265	13,400
MN	20,400	931	139	1,600	18,661
MT	3,465	286	218	183	3,064
NE	8,245	500	187	733	7,325
NV	2,378	104	....	300	2,078
NM	2,301	143	....	245	2,056
NY	5,696	393	75	365	5,256
ND	22,680	1,600	190	2,280	20,210
OH	683	54	2	6	675
OR	18,676	875	100	1,150	17,426
PA	2,518	220	16	65	2,437
RI	140	12	....	3	137
WA	93,000	3,625	210	5,500	87,290
WI	25,730	1,461	250	1,375	24,105
Total	378,588	21,711	4,055	24,952	349,580
US	415,055	24,533	4,138	26,438	384,478

[1] Winter potatoes combined with spring potatoes in 2007. Spring potato "Total used for Seed" includes winter and spring potatoes.

NASS, Crops Branch, (202) 720–2127.

Table 4-59.—Potatoes, fresh & seed: United States exports by country of destination and imports by country of origin, 2007–2009

Country	2007	2008	2009
	Metric tons	*Metric tons*	*Metric tons*
Exports			
Canada	83,605	100,991	94,164
Mexico	24,052	32,145	30,989
Taiwan	6,776	4,100	3,369
Korea, South	3,742	4,126	3,185
Singapore	1,175	2,205	2,903
Hong Kong	1,083	2,103	1,993
Malaysia	2,158	3,136	1,938
Thailand	8	2,758	1,788
Philippines	1,210	1,660	1,145
Bahamas, The	534	906	761
Guatemala	197	423	657
Dominican Republic	824	1,019	633
Leeward-Windward Islands(*)	341	673	520
Uruguay	444	568	497
Panama	98	262	470
Russia	556	1,009	468
Costa Rica	4	490	397
Cayman Islands	0	144	393
Venezuela	0	0	388
El Salvador	76	208	358
Nicaragua	150	628	255
Japan	624	269	234
Jamaica	745	913	177
Rest of World	1,792	1,135	1,087
World Total	130,193	161,873	148,769
Imports			
Canada	126,698	168,807	111,954
Dominican Republic	26	20	26
France(*)	17	9	0
Jamaica	0	2	0
Cameroon	0	0	8
Colombia	23	0	0
Costa Rica	14	0	0
Egypt	0	0	2
Ghana	50	0	0
India	35	0	0
Japan	4	0	0
World Total	126,868	168,839	111,990

Note: (*) Denotes a country that is a summarization of its component countries.
FAS, Office of Global Analysis, (202) 720-6301.

Table 4-60.—Potatoes (fresh): Foreign trade, United States, 1999–2008 [1]

Year beginning July	Imports for consumption	Domestic exports
	1,000 cwt.	*1,000 cwt.*
1999	9,094	6,541
2000	5,986	6,695
2001	8,644	6,571
2002	9,265	6,287
2003	7,611	5,091
2004	7,604	5,430
2005	8,152	6,730
2006	10,681	5,798
2007	10,852	6,731
2008	10,550	6,588

[1] Includes seed.
ERS, Specialty Crops Branch, (202) 694–5253. Compiled from reports of the U.S. Department of Commerce.

Table 4-61.—Pumpkins for fresh market and processing: Area, production, and value per hundredweight, by State and United States, 2007–2009

State	Area harvested			Production			Value per unit		
	2007	2008	2009 [1]	2007	2008	2009 [1]	2007	2008	2009 [1]
	Acres	*Acres*	*Acres*	*1,000 cwt.*	*1,000 cwt.*	*1,000 cwt.*	*Dollars per cwt.*	*Dollars per cwt.*	*Dollars per cwt.*
CA	5,100	5,300	5,100	1,224	1,484	1,479	10.50	12.50	13.90
IL	13,800	12,600	12,500	5,420	4,960	4,291	4.32	4.50	3.47
MI	6,200	6,800	6,700	713	986	737	12.00	15.50	14.00
NY	6,400	5,900	6,000	1,152	1,062	750	19.70	36.20	29.00
OH	6,600	6,100	7,500	2,013	931	1,237	16.70	24.20	18.20
PA	7,800	6,700	6,300	936	1,240	819	23.90	16.10	15.50
US	45,900	43,400	44,100	11,458	10,663	9,313	10.80	12.90	11.00

[1] Preliminary.
NASS, Crops Branch, (202) 720–2127.

Table 4-62.—Pumpkins for fresh market and processing, commercial crop: Area, yield, production, value, hundredweight, and total value, United States, 2001-2009

Year	Area harvested	Yield per acre	Production	Value	
				Per cwt	Total
	Acres	*Cwt.*	*1,000 cwt.*	*Dollars*	*1,000 dollars*
2001	37,800	224	8,460	8.83	74,679
2002	41,000	208	8,509	10.80	91,712
2003	39,300	206	8,085	9.92	80,203
2004	45,000	225	10,135	9.04	91,609
2005	43,800	246	10,756	9.64	103,651
2006	43,700	240	10,484	9.98	104,623
2007	45,900	250	11,458	10.80	123,519
2008	43,400	246	10,663	12.90	137,072
2009 [1]	44,100	211	9,313	11.00	102,730

[1] Preliminary.
NASS, Crops Branch, (202) 720–2127.

Table 4-63.—Spinach for fresh market: Area, production, and value per hundredweight, by State and United States, 2007–2009

State	Area harvested			Production			Value per unit		
	2007	2008	2009 [1]	2007	2008	2009 [1]	2007	2008	2009 [1]
	Acres	Acres	Acres	1,000 cwt.	1,000 cwt.	1,000 cwt.	Dollars per cwt.	Dollars per cwt.	Dollars per cwt.
AZ	3,700	5,000	6,000	592	825	1,170	31.70	34.20	39.20
CA	23,000	25,000	28,500	4,025	4,125	4,418	32.00	34.00	38.00
NJ	1,600	1,600	1,500	160	280	203	42.60	37.20	43.20
TX	800	1,100	600	100	132	88	24.00	26.00	22.00
Oth Sts [2]	2,800	2,980	3,500	202	359	360	35.50	29.90	34.30
US	31,900	35,680	40,100	5,079	5,721	6,239	32.30	33.70	38.00

[1] Preliminary. [2] 2007 - 2009 - CO and MD.
NASS, Crops Branch, (202) 720–2127.

Table 4-64.—Spinach for fresh market, commercial crop: Area, yield, production, value per hundredweight, and total value, United States, 2000-2009

Year	Area harvested	Yield per acre	Production	Value [1]	
				Per cwt	Total
	Acres	Cwt.	1,000 cwt.	Dollars	1,000 dollars
2000	29,020	146	4,239	31.80	134,733
2001	26,450	131	3,458	32.40	112,068
2002	31,700	146	4,625	34.20	158,385
2003	33,880	150	5,089	36.90	187,711
2004	36,600	157	5,756	22.20	127,722
2005	42,500	167	7,096	22.80	161,732
2006	36,500	166	6,045	29.90	180,774
2007	31,900	159	5,079	32.30	163,952
2008	35,680	160	5,721	33.70	193,052
2009 [2]	40,100	156	6,239	38.00	236,808

[1] Price and value at point of first sale. [2] Preliminary.
NASS, Crops Branch, (202) 720–2127.

Table 4-65.—Spinach for processing, commercial crop: Area, yield, production, value per ton, and total value, United States, 2000-2009

Year	Area harvested	Yield per acre	Production	Value [1]	
				Per ton	Total
	Acres	Tons	Tons	Dollars	1,000 dollars
2000	14,720	9.28	136,650	121.00	16,579
2001	13,940	9.12	127,100	116.00	14,698
2002	11,420	9.30	106,170	114.00	12,150
2003	14,100	8.52	120,130	107.00	12,824
2004	12,400	10.50	130,220	116.00	15,088
2005	9,600	10.23	98,240	109.00	10,667
2006	9,400	7.40	69,560	127.00	8,809
2007	11,400	8.58	97,800	104.00	10,123
2008	10,200	10.15	103,540	124.00	12,831
2009 [2]	10,100	9.47	95,660	127.00	12,144

[1] Price and value at processing plant door. [2] Preliminary.
NASS, Crops Branch, (202) 720–2127.

Table 4-66.—Spinach for processing: Area, production, and value per ton, by State and United States, 2007-2009

State	Area harvested			Production			Value per unit		
	2007	2008	2009 [1]	2007	2008	2009 [1]	2007	2008	2009 [1]
	Acres	Acres	Acres	Tons	Tons	Tons	Dollars per ton	Dollars per ton	Dollars per ton
CA	8,500	7,100	7,500	59,500	67,450	62,550	118.00	126.00	118.00
Oth Sts [2]	2,900	3,100	2,600	38,300	36,090	33,110	81.00	120.00	144.00
US	11,400	10,200	10,100	97,800	103,540	95,660	104.00	124.00	127.00

[1] Preliminary.　[2] 2007-2009 - NJ and TX.
NASS, Crops Branch, (202) 720–2127.

Table 4-67.—Sweet Potatoes: Area, yield, production, season average price per hundredweight received by farmers, and value, United States, 2000–2009

Year	Area harvested	Yield per acre	Production	Market year average price [1]	Value of production
	1,000 acres	Cwt.	1,000 cwt.	Dollars	1,000 dollars
2000	94.8	145	13,780	15.30	210,351
2001	93.6	155	14,515	15.30	222,658
2002	82.3	156	12,799	16.80	214,650
2003	92.6	172	15,891	19.20	305,448
2004	92.8	174	16,112	17.50	281,559
2005	88.4	178	15,730	18.10	284,103
2006	87.3	188	16,401	18.20	298,388
2007	97.4	186	18,070	18.30	330,060
2008	97.3	190	18,443	21.20	390,572
2009	96.9	201	19,469	20.90	410,361

[1] Obtained by weighting State prices by production.
NASS, Crops Branch, (202) 720–2127.

Table 4-68.—Sweet Potatoes: Area, production, and season average price per hundredweight received by farmers, by State and United States, 2007–2009

State	Area harvested			Production			Market year average price per cwt.		
	2007	2008	2009	2007	2008	2009	2007	2008	2009
	1,000 acres	1,000 acres	1,000 acres	1,000 cwt.	1,000 cwt.	1,000 cwt.	Dollars	Dollars	Dollars
AL	2.3	2.5	2.3	276	438	391	31.90	12.10	20.50
AR [1]			2.5			463			13.20
CA	13.3	14.8	17.4	4,256	4,366	5,916	19.00	30.40	26.80
FL [1]			3.2			352			30.00
LA	15.0	11.0	12.0	3,000	1,100	1,620	19.00	18.30	17.80
MS	20.0	19.5	11.0	3,500	3,354	1,265	18.90	18.60	18.00
NJ	1.2	1.2	1.2	120	150	132	27.40	26.90	29.00
NC	43.0	46.0	46.0	6,665	8,740	9,200	16.40	18.50	18.10
SC [2]	0.5	0.5		55	55		18.00	18.00	
TX	1.8	1.5	1.3	162	210	130	19.80	14.40	17.00
VA [2]	0.3	0.3		36	30		12.50	9.75	
US	97.4	97.3	96.9	18,070	18,443	19,469	18.30	21.20	20.90

[1] Estimates began in 2009.　[2] Estimates discontinued in 2009.
NASS, Crops Branch, (202) 720–2127.

Table 4-69.—Squash for fresh market and processing: Area, production, and value per hundredweight, by State and United States, 2007–2009

State	Area harvested			Production			Value per unit		
	2007	2008	2009[1]	2007	2008	2009[1]	2007	2008	2009[1]
	Acres	*Acres*	*Acres*	*1,000 cwt.*	*1,000 cwt.*	*1,000 cwt.*	*Dollars per cwt.*	*Dollars per cwt.*	*Dollars per cwt.*
CA	5,600	5,500	5,800	1,176	1,128	1,218	24.70	29.70	26.40
FL	7,800	8,300	8,800	718	996	1,144	52.80	53.00	45.00
GA	4,800	5,000	5,300	552	500	1,060	35.00	28.50	28.20
MI	7,500	6,600	6,500	1,425	1,320	1,365	9.50	9.20	8.60
NJ	3,000	3,000	2,800	345	360	325	25.50	34.80	32.00
NY	3,500	3,800	4,500	595	760	540	38.90	42.80	42.60
NC	3,000	3,300	3,200	390	363	352	30.00	28.00	32.00
OH	1,200	1,500	1,500	276	363	465	20.10	29.50	23.00
OR	1,900	1,900	2,300	351	398	404	18.50	17.90	26.20
SC	1,000	1,100	1,000	69	125	141	33.40	36.00	32.50
TN	900	900	800	75	74	65	18.30	18.30	19.60
TX	1,400	1,500	1,400	294	300	140	50.00	42.30	45.80
US	41,600	42,400	43,900	6,266	6,687	7,219	27.80	30.50	28.20

[1] Preliminary.
NASS, Crops Branch, (202) 720-2127.

Table 4-70.—Squash for fresh market and processing, commercial crop: Area, yield, production, value and total value, United States, 2001-2009

Year	Area harvested	Yield per acre	Production	Value	
				Per cwt	Total
	Acres	*Cwt.*	*1,000 cwt.*	*Dollars*	*1,000 dollars*
2001	51,200	152	7,791	22.20	173,265
2002	52,300	168	8,792	23.10	202,975
2003	48,600	151	7,318	25.60	187,613
2004	49,300	146	7,202	28.50	205,229
2005	48,400	154	7,439	25.40	188,846
2006	48,200	165	7,946	24.20	192,459
2007	41,600	151	6,266	27.80	173,917
2008	42,400	158	6,687	30.50	204,283
2009[1]	43,900	164	7,219	28.20	203,464

[1] Preliminary.
NASS, Crops Branch, (202) 720-2127.

Table 4-71.—Taro: Area, yield, total production, price, and value, Hawaii, 2000–2009

Year	Total area	Yield per acre[1]	Production	Price per pound	Value of production
	Acres	*1,000 pounds*	*1,000 pounds*	*Dollars*	*1,000 dollars*
2000	470		7,000	0.530	3,710
2001	440		6,400	0.530	3,392
2002	430		6,100	0.540	3,294
2003	420		5,000	0.540	2,700
2004	370		5,200	0.540	2,808
2005	360		4,300	0.540	2,322
2006	380		4,500	0.570	2,565
2007	380		4,000	0.590	2,360
2008	390		4,300	0.620	2,666
2009[2]	445		4,000	0.610	2,440

[1] Yield not estimated.　[2] Preliminary.
NASS, Crops Branch, (202) 720-2127.

Table 4-72.—Tomatoes: Foreign trade, United States,1999–2008

Year beginning July	Imports			Domestic exports [2]				
	Fresh	Canned [1]	Paste	Fresh	Canned whole	Catsup and sauces	Paste	Juice [3]
	1,000 pounds	1,000 pounds	1,000 pounds	1,000 pounds	1,000 pounds	1,000 pounds	1,000 pounds	1,000 pounds
1999	1,596,470	203,391	46,790	356,676	85,641	334,125	198,029	29,417
2000	1,885,424	253,212	32,717	398,458	77,988	355,414	215,569	39,324
2001	1,708,004	473,841	40,729	375,744	78,828	367,393	206,113	14,480
2002	2,114,478	409,602	24,482	324,097	78,082	389,279	250,924	2,956
2003	1,984,044	387,998	15,681	333,895	82,375	417,271	280,244	3,588
2004	1,985,968	438,007	9,536	364,601	96,464	394,417	303,300	3,739
2005	2,290,631	441,506	22,923	319,206	105,562	459,632	277,484	4,643
2006	2,305,552	492,483	102,184	331,704	106,941	441,365	202,599	7,712
2007	2,390,516	435,720	28,806	364,052	125,605	449,250	472,149	3,265
2008	2,501,398	470,919	17,906	372,037	142,113	497,277	633,778	2,247

[1] Includes all canned tomato and tomato product imports except paste, and is on a product-weight-basis. [2] Includes exports for military-civilian feeding abroad. [3] Converted to pounds from liters.
ERS, Specialty Crops Branch, (202) 694–5253. Compiled from reports of the U.S. Department of Commerce.

Table 4-73.—Tomatoes, commercial crop: Area, yield, production, value per hundredweight and per ton, and total value, United States, 2000–2009

Year	For fresh market					For processing				
	Area harvested	Yield per acre	Production	Value [1]		Area harvested	Yield per acre	Production	Value [2]	
				Per cwt.	Total				Per ton	Total
	Acres	Cwt.	1,000 cwt.	Dollars	1,000 dollars	Acres	Tons	Tons	Dollars	1,000 dollars
2000 ...	126,790	307	38,890	30.70	1,194,710	289,600	37.49	10,858,240	59.80	649,066
2001 ...	130,840	288	37,701	30.00	1,131,421	274,860	33.65	9,248,720	59.20	547,473
2002 ...	129,020	307	39,588	31.60	1,252,801	312,200	37.38	11,670,820	58.20	679,823
2003 ...	119,700	295	35,364	37.50	1,324,757	293,920	33.41	9,819,710	58.70	576,441
2004 ...	128,400	296	37,948	37.40	1,420,160	300,620	40.80	12,266,410	58.60	719,285
2005 ...	124,000	307	38,033	41.60	1,583,897	281,940	36.15	10,193,120	60.90	620,987
2006 ...	120,200	302	36,274	43.70	1,584,708	299,400	35.44	10,611,820	66.40	704,669
2007 ...	108,100	311	33,627	34.80	1,168,693	313,600	40.37	12,659,890	71.20	901,761
2008 ...	105,250	296	31,137	45.50	1,415,297	296,500	41.50	12,305,820	79.80	982,373
2009 [3]	105,700	306	32,365	40.60	1,313,941	327,800	42.62	13,970,560	87.20	1,218,912

[1] Price and value at point of first sale. [2] Price and value at processing plant door. [3] Preliminary.
NASS, Crops Branch, (202) 720–2127.

Table 4-74.—Tomatoes, commercial crop: Area, production, and value per hundredweight and per ton, by State and United States, 2007–2009 [1]

Utilization and State	Area harvested			Production			Value per unit		
	2007	2008	2009 [2]	2007	2008	2009 [2]	2007	2008	2009 [2]
FOR FRESH MARKET	Acres	Acres	Acres	1,000 cwt.	1,000 cwt.	1,000 cwt.	Dollars per cwt.	Dollars per cwt.	Dollars per cwt.
AL	1,300	1,250	1,300	501	475	403	26.30	25.80	32.30
AR	1,100	900	1,000	149	279	64	45.00	51.00	56.00
CA	37,000	37,000	36,000	11,100	11,655	10,440	33.70	33.30	34.80
FL	34,600	31,500	33,600	13,321	10,458	12,298	31.90	59.50	42.30
GA	5,000	4,200	4,500	1,800	840	1,260	29.00	32.20	49.40
IN	1,000	900	800	150	144	120	81.00	94.00	80.00
MI	2,200	2,100	2,000	506	546	600	49.00	45.00	35.00
NJ	2,900	2,900	2,900	595	624	638	39.70	42.70	53.20
NY	2,400	2,700	2,500	432	513	350	75.20	84.00	93.50
NC	3,300	3,400	3,300	957	1,088	1,122	30.00	31.00	31.00
OH	3,000	4,600	4,600	630	1,109	1,536	59.40	55.00	44.00
PA	2,000	1,600	1,700	420	352	289	62.40	57.10	74.10
SC	2,900	2,500	1,900	238	488	333	48.50	65.00	48.00
TN	3,800	4,000	4,000	1,159	1,120	1,360	34.00	34.00	33.00
TX	1,100	1,000	800	143	130	112	42.10	61.60	57.60
VA	4,500	4,700	4,800	1,526	1,316	1,440	36.20	38.70	43.90
US	108,100	105,250	105,700	33,627	31,137	32,365	34.80	45.50	40.60
FOR PROCESSING	Acres	Acres	Acres	Tons	Tons	Tons	Dollars per ton	Dollars per ton	Dollars per ton
CA	296,000	279,000	308,000	12,082,000	11,822,000	13,314,000	70.30	78.60	86.10
IN	8,400	8,300	9,800	294,000	249,000	321,340	94.00	100.00	113.00
MI	3,300	3,400	3,400	112,200	102,000	132,600	90.00	100.00	110.00
OH	5,900	5,800	6,600	171,690	132,820	202,620	85.40	136.00	107.00
US	313,600	296,500	327,800	12,659,890	12,305,820	13,970,560	71.20	79.80	87.20

[1] Cherry, grape, tomatillo, and greenhouse tomatoes are exclued. [2] Preliminary.
NASS, Crops Branch, (202) 720–2127.

Table 4-75.—Vegetables and melons, fresh: Total reported domestic rail, truck, and air shipments, 2008

Commodity	Jan.	Feb.	Mar.	Apr.	May	June	July	Aug.	Sep.	Oct.	Nov.	Dec.	Total
	1,000 cwt.	*1,000 cwt.*	*1,000 cwt.*	*1,000 cwt.*	*1,000 cwt.*	*1,000 cwt.*	*1,000 cwt.*	*1,000 cwt.*	*1,000 cwt.*	*1,000 cwt.*	*1,000 cwt.*	*1,000 cwt.*	*1,000 cwt.*
Vegetables:													
Artichokes	47	87	75	48	149	80	89	81	73	99	72	45	945
Asparagus			124	286	269	113							792
Beans	128	160	308	316	330	203	74	118	66	199	273	205	2,380
Broccoli	651	796	665	568	788	669	686	620	601	624	605	674	7,947
Cabbage	1,237	1,111	1,527	945	908	436	490	793	801	789	884	1,111	11,032
Carrots	786	688	786	712	640	639	695	515	678	578	563	595	7,875
Cauliflower	325	387	337	263	380	290	250	259	293	278	260	343	3,665
Celery				1,232	1,068	1,218	1,083	1,098	1,145	1,269	1,732	1,495	15,585
Chinese cabbage	93	88	77	81	76	81	84	87	81	80	76	93	997
Corn, sweet	376	448	538	1,567	2,892	2,500	750	487	319	345	439	432	11,093
Cucumbers	29	41	366	600	595	641	551	867	578	441	271	173	5,153
Eggplant	47	51	77	64	121	165	65	28	46	109	63	60	896
Endive	29	22	21	17	5	5	6	5	5	6	8	28	157
Escarole	34	25	24	21	5	5	6	5	5	7	12	32	181
Greens	191	200	252	208	214	87	70	69	67	104	267	259	1,988
Lettuce, iceberg	2,312	2,341	2,500	2,640	2,790	2,610	2,793	2,754	2,526	2,775	2,287	2,500	30,828
Lettuce, other	370	345	366	319	317	254	257	255	252	304	275	393	3,707
Lettuce, romaine	1,290	1,157	1,193	1,389	1,375	1,281	1,258	1,135	1,156	1,353	1,113	1,444	15,144
Okra				3,920	3,924	3,110	3,437	3,656	4,050	4,177	3,398	3,537	43,512
Onions, dry	13	10	15	10	15	16	16	19	20	17	11	9	171
Onions, green				12	23	14	13	11	9	7	4	1	94
Parsley	43	41	39	33	24	22	22	20	22	24	24	40	354
Peppers, bell	664	542	718	764	944	1,097	1,013	927	909	857	821	569	9,825
Peppers, other	42	36	29	20	22		18	2		1	15	22	207
Potatoes, table	8,675	7,899	8,789	7,890	9,173	7,945	7,121	7,343	7,869	7,882	8,561	8,242	97,389
Potatoes, chipper	3,423	3,148	3,975	3,633	4,638	3,666	3,327	4,511	3,508	3,660	4,362	3,643	45,494
Potatoes, seed	423	572	3,493	5,987	3,625	73			15	130	223	195	14,736
Radishes	56	51	43	47	41	12	12	12	14	13	19	41	361
Spinach	134	141	161	58	45	39	41	41	41	37	56	129	923
Squash	95	161	118	154	293	261	152	175	148	149	95	132	1,933
Sweet potatoes	405	428	608	422	438	382	446	468	469	644	1,319	646	6,675
Tomatoes	1,715	1,310	1,507	2,058	1,994	1,884	2,068	2,260	1,952	1,695	1,742	2,184	22,369
Tomatoes, green-house	184	238	321	302	406	417	370	348	254	245	333	209	3,627
Tomatoes,grapes Type	191	197	203	234	237	189	100	175	127	102	172	187	2,114
Tomatoes, cherry	87	59	68	83	80	43	33	44	22	24	79	81	703
Tomatoes, plum Types	314	206	239	315	366	223	279	379	370	268	241	383	3,583
Total	24,409	22,986	29,562	37,218	39,210	30,670	27,675	29,567	28,491	29,292	30,675	30,132	374,435
Melons:													
Cantaloups				55	2,403	4,198	3,712	3,653	2,275	1,447	631	35	18,409
Honeydews					290	620	957	946	811	393	137	1	4,155
Mixed and misc. melons					34	168	209	122	11		3	2	549
Watermelons, seeded					146	1,388	1,712	1,347	918	226	30	1	5,768
Watermelons, seedless				8	573	4,598	7,731	5,880	4,824	1,208	240	37	25,099
Total				63	3,446	10,972	14,321	11,948	8,839	3,274	1,041	76	53,980
Grand total	24,409	22,986	29,562	37,281	42,656	41,642	41,996	41,515	37,330	32,566	31,716	30,208	428,415

AMS, Fruit and Vegetable Division, Market News Branch, (202) 720–3343.

Table 4-76.—Vegetables (fresh), melons, potatoes, sweet potatoes: Per capita civilian utilization (farm-weight basis), United States, 2000–2009 [1]

Year	Cabbage	Cucumbers	Tomatoes [2]	Asparagus	Broccoli	Carrots	Head Lettuce	Leaf/romaine
	Pounds	Pounds	Pounds	Pounds	Pounds	Pounds	Pounds	Pounds
2000	8.9	6.4	19.0	1.0	5.9	9.2	23.5	8.4
2001	8.8	6.3	19.2	0.9	5.4	9.4	23.0	8.0
2002	8.3	6.6	20.3	1.0	5.4	8.4	22.5	9.6
2003	7.4	6.2	19.4	1.0	5.4	8.8	22.2	10.8
2004	8.0	6.4	20.0	1.1	5.3	8.7	21.3	12.0
2005	7.8	6.2	20.2	1.1	5.3	8.7	20.9	9.7
2006	7.8	6.1	19.8	1.1	5.8	8.1	20.1	12.0
2007	8.0	6.4	19.2	1.2	5.6	8.1	18.4	11.5
2008	8.1	6.4	18.5	1.2	6.0	8.1	16.9	10.4
2009 [4]	7.3	6.6	19.3	1.3	6.1	7.4	17.1	11.0

Year	Snap beans	Garlic	Cauliflower	Celery	Sweet Corn	Onions	Spinach	Bell peppers
	Pounds	Pounds	Pounds	Pounds	Pounds	Pounds	Pounds	Pounds
2000	2.0	2.2	1.7	6.3	9.0	18.9	1.4	8.2
2001	2.2	2.4	1.5	6.4	9.2	18.5	1.1	8.1
2002	2.1	2.5	1.4	6.3	9.0	19.3	1.4	8.3
2003	2.0	2.8	1.6	6.3	9.2	19.5	1.6	8.4
2004	1.9	2.6	1.6	6.2	9.0	21.9	1.9	8.6
2005	1.8	2.4	1.8	5.9	8.7	20.9	2.3	9.2
2006	2.1	2.7	1.7	6.1	8.3	19.9	2.0	9.5
2007	2.2	2.7	1.7	6.3	9.2	21.6	1.6	9.4
2008	2.0	2.8	1.6	6.2	9.1	20.9	1.8	9.6
2009 [4]	1.6	2.5	1.5	6.1	9.0	19.3	1.9	9.4

Year	Watermelon	Cantaloupe	Honeydew melons	Others [3]	Total vegetables and melons	Potatoes	Sweet potatoes
	Pounds	Pounds	Pounds	Pounds	Pounds	Pounds	Pounds
2000	13.8	11.1	2.3	15.6	174.7	47.2	4.2
2001	15.0	11.2	2.0	14.8	173.3	46.6	4.4
2002	14.0	11.1	2.2	14.9	174.6	44.3	3.8
2003	13.5	10.8	2.2	14.8	173.8	46.8	4.7
2004	13.0	9.8	2.1	15.5	176.8	45.8	4.6
2005	13.6	9.6	1.9	15.9	173.7	41.3	4.5
2006	15.1	9.3	1.9	15.9	175.1	38.6	4.6
2007	14.4	9.6	1.8	15.3	174.2	38.7	5.1
2008	15.6	8.9	1.7	14.7	170.3	37.8	5.0
2009 [4]	15.3	9.3	1.6	14.0	167.6	37.4	5.3

[1] Fresh vegetable consumption computed for total commercial production for fresh market. Does not include production for home use. Consumption obtained by dividing the total apparent consumption by total July 1 population as reported by the Bureau of the Census. All data for calendar year.　[2] After 1996, includes an ERS estimate of domestically produced hothouse tomatoes. Hothouse imports included in all years.　[3] Includes artichokes, eggplant, radishes, brussels sprouts, squash, green limas, and escarole/endive. Beginning in 2000, also includes collards, mustard greens, turnip greens, kale, okra, and pumpkins.　[4] Preliminary.

ERS, Market and Trade Economics Division, Specialty Crops Branch, (202)694–5253.

Table 4-77.—Vegetables, canning: Per capita utilization (farm weight), United States, 2000–2009

Year	Cabbage for kraut	Asparagus	Snap beans	Carrots	Green peas
	Pounds	Pounds	Pounds	Pounds	Pounds
2000	1.4	0.2	4.0	1.1	1.5
2001	1.3	0.2	3.8	1.1	1.4
2002	1.2	0.2	3.4	1.0	1.1
2003	1.1	0.2	3.7	1.1	1.3
2004	1.1	0.2	3.7	1.1	1.2
2005	1.2	0.2	4.0	1.1	1.0
2006	1.2	0.2	3.9	1.0	1.1
2007	1.0	0.1	3.5	0.9	1.1
2008	0.9	0.2	3.3	1.0	1.1
2009 [1]	0.9	0.2	3.6	0.8	1.2

Year	Tomatoes	Corn	Pickles	Other [2]	Total [3]
	Pounds	Pounds	Pounds	Pounds	Pounds
2000	70.1	9.0	4.9	7.8	100.0
2001	65.5	8.7	3.7	7.9	93.6
2002	69.4	7.8	5.4	8.1	97.5
2003	69.9	8.3	4.5	8.0	98.0
2004	70.5	8.2	4.9	8.8	99.7
2005	73.7	8.6	3.8	8.9	102.6
2006	64.5	8.4	3.0	9.0	92.1
2007	68.7	8.9	3.7	8.5	96.4
2008	67.1	6.7	3.5	8.6	92.5
2009 [1]	70.2	7.6	5.0	8.9	98.4

[1] Preliminary. [2] Includes beets, chile peppers (all uses), green lima beans and spinach. [3] Totals may not add due to rounding.
ERS, Specialty Crops Branch, (202) 694–5253.

Table 4-78.—Watermelon for fresh market: Area, production, and value per hundredweight, by State and United States, 2007–2009

State	Area harvested			Production			Value per unit		
	2007	2008	2009 [1]	2007	2008	2009 [1]	2007	2008	2009 [1]
	Acres	Acres	Acres	1,000 cwt.	1,000 cwt.	1,000 cwt.	Dollars per cwt.	Dollars per cwt.	Dollars per cwt.
AL	3,000	3,500	2,600	285	350	416	11.50	10.80	10.60
AZ	6,500	6,400	5,500	3,348	2,880	2,695	12.20	16.20	11.80
AR	1,900	1,400	1,400	418	336	280	10.50	7.80	7.60
CA	11,400	11,300	12,400	6,498	6,554	6,944	11.30	15.90	12.40
DE	2,700	2,800	2,500	1,013	1,260	943	9.00	12.00	11.00
FL	22,800	26,100	25,800	7,456	8,613	8,179	18.80	16.30	16.60
GA	23,000	21,000	23,000	6,440	5,880	6,900	7.90	8.90	9.80
IN	7,400	7,200	7,400	2,664	2,736	2,627	7.60	9.90	9.00
MD	2,300	2,350	2,100	552	870	630	13.00	13.00	12.00
MS	2,600	2,600	2,300	442	442	334	11.70	11.20	10.20
MO	3,800	2,500	2,600	741	675	858	6.20	7.40	7.20
NC	6,600	6,700	6,700	1,386	1,340	1,541	8.00	9.00	10.00
OK	4,500	4,500	3,500	563	360	263	10.50	9.80	13.10
SC	6,500	6,500	6,500	943	1,398	1,495	8.00	10.80	9.00
TX	23,000	19,500	20,900	4,370	6,045	5,852	8.20	8.70	8.20
VA	1,000	1,200	1,100	230	264	165	12.50	11.00	9.00
US	129,000	125,550	126,300	37,349	40,003	40,122	11.30	12.50	11.50

[1] Preliminary.
NASS, Crops Branch, (202) 720–2127.

Table 4-79.—Watermelon for fresh market, commercial crop: Area, yield, production, value per hundredweight, and total value, United States, 2000-2009

Year	Area harvested	Yield per acre	Production	Value [1]	
				Per cwt	Total
	Acres	Cwt.	1,000 cwt.	Dollars	1,000 dollars
2000	164,360	228	37,494	6.40	239,893
2001	160,720	252	40,478	6.75	273,418
2002	152,500	260	39,585	8.30	328,497
2003	150,300	255	38,327	8.97	343,795
2004	141,900	260	36,880	8.49	313,129
2005	133,800	277	37,023	11.60	429,445
2006	131,000	304	39,865	10.40	414,111
2007	129,000	290	37,349	11.30	422,546
2008	125,550	319	40,003	12.50	499,633
2009 [2]	126,300	318	40,122	11.50	460,778

[1] Price and value at point of first sale. [2] Preliminary.
NASS, Crops Branch, (202) 720–2127.

Table 4-80.—Vegetables, freezing: Per capita utilization (farm weight basis), United States, 2000–2009

Year	Leafy, green, and yellow vegetables				
	Asparagus	Snap beans	Carrots	Peas	Broccoli
	Pounds	*Pounds*	*Pounds*	*Pounds*	*Pounds*
2000	0.1	1.8	2.7	2.1	2.3
2001	0.1	1.9	2.2	2.0	2.0
2002	0.1	1.8	2.1	1.7	2.1
2003	0.1	1.9	2.0	1.8	2.6
2004	0.1	1.9	2.0	1.6	2.7
2005	0.1	1.8	2.0	1.6	2.7
2006	0.1	1.9	2.1	1.6	2.3
2007	0.1	2.1	1.5	1.8	2.7
2008	0.1	2.1	1.5	1.8	2.7
2009 [1]	0.1	1.9	1.4	1.7	2.5

Year	Cauliflower	Sweet Corn	Other [2]	Total vegetables excluding potatoes	Potato products	Grand total [3]
	Pounds	*Pounds*	*Pounds*	*Pounds*	*Pounds*	*Pounds*
2000	0.6	9.1	3.1	21.8	57.5	79.3
2001	0.5	9.3	3.2	21.2	58.2	79.4
2002	0.3	9.3	4.3	21.7	55.2	76.9
2003	0.4	9.0	3.9	21.6	57.1	78.7
2004	0.4	9.1	3.8	21.5	57.4	78.9
2005	0.4	9.5	4.1	22.1	54.4	76.5
2006	0.4	9.4	4.1	21.8	53.3	75.1
2007	0.4	10.0	4.1	22.6	53.2	75.8
2008	0.4	8.4	4.0	21.1	51.5	72.6
2009 [1]	0.4	8.8	4.3	21.0	51.2	72.2

[1] Preliminary.   [2] Includes green lima beans, spinach, and miscellaneous freezing vegetables.   [3] Totals may not add due to rounding.
ERS, Specialty Crops Branch, (202) 694–5253.

Table 4-81.—Commercially produced vegetables: Per capita utilization, United States, 2000–2009 [1]

Year	Farm weight equivalent					Percentage of annual total			
	Total fresh and processed	Fresh [2]	Processed [3]			Fresh	Processed		
			Total	Canning	Freezing		Total	Canning	Freezing
	Pounds	*Pounds*	*Pounds*	*Pounds*	*Pounds*	*Percent*	*Percent*	*Percent*	*Percent*
2000	296.4	174.6	121.8	100.0	21.8	58.9	41.1	33.7	7.4
2001	288.1	173.3	114.8	93.6	21.2	60.2	39.8	32.5	7.4
2002	293.8	174.6	119.2	97.5	21.7	59.4	40.6	33.2	7.4
2003	293.4	173.8	119.6	98.0	21.6	59.2	40.8	33.4	7.4
2004	298.0	176.8	121.2	99.7	21.5	59.3	40.7	33.5	7.2
2005	298.4	173.7	124.7	102.6	22.1	58.2	41.8	34.4	7.4
2006	289.0	175.1	113.9	92.1	21.8	60.6	39.4	31.9	7.5
2007	293.2	174.2	119.0	96.4	22.6	59.4	40.6	32.9	7.7
2008	283.9	170.3	113.6	92.5	21.1	60.0	40.0	32.6	7.4
2009 [4]	287.0	167.6	119.4	98.4	21.0	58.4	41.6	34.3	7.3

[1] Excludes potatoes, sweet potatoes, pulses, dehydrating onions, and mushrooms.   [2] See table 4-76 for items included. Includes melons.   [3] See table 4-78 and 4-79 for items included.   [4] Preliminary
ERS, Market and Trade Economics Division, Specialty Crops Branch, (202) 694–5253.

Table 4-82.—Frozen Vegetables and potato products: Cold storage holdings, end of month, United States, 2008 and 2009

Month	Asparagus		Green beans, regular cut		Green beans, French cut		Green beans, total	
	2008	2009	2008	2009	2008	2009	2008	2009
	1,000 pounds	1,000 pounds	1,000 pounds	1,000 pounds	1,000 pounds	1,000 pounds	1,000 pounds	1,000 pounds
January	4,864	7,517	163,431	188,915	29,038	25,216	192,469	214,131
February	3,499	7,363	138,508	169,346	25,690	22,285	164,198	191,631
March	3,608	7,116	121,767	147,596	22,616	20,134	144,383	167,730
April	2,604	7,151	100,988	126,359	19,791	16,429	120,779	142,788
May	5,896	8,716	86,190	116,493	17,217	15,259	103,407	131,752
June	9,964	12,279	70,854	95,706	12,675	13,140	83,529	108,846
July	9,618	11,353	122,470	139,967	23,244	26,894	145,714	166,861
August	9,201	11,049	203,585	197,238	35,043	31,927	238,628	229,165
September	8,402	10,087	246,495	240,107	37,652	34,444	284,147	274,551
October	8,360	9,176	238,098	211,003	35,675	31,033	273,773	242,036
November	7,860	8,978	231,706	194,184	32,086	27,120	263,792	221,304
December	7,368	9,057	211,997	176,371	29,298	25,325	241,295	201,696

Month	Broccoli, spears		Broccoli, chopped & cut		Broccoli, total		Brussels sprouts	
	2008	2009	2008	2009	2008	2009	2008	2009
	1,000 pounds	1,000 pounds	1,000 pounds	1,000 pounds	1,000 pounds	1,000 pounds	1,000 pounds	1,000 pounds
January	27,906	34,963	37,792	47,286	65,698	82,249	19,586	19,918
February	29,637	33,786	44,108	47,702	73,745	81,488	13,922	18,183
March	33,594	40,639	45,708	47,196	79,302	87,835	12,384	17,596
April	35,877	43,762	43,565	48,901	79,442	92,663	11,383	16,087
May	36,186	46,206	44,882	49,289	81,068	95,495	10,444	14,964
June	33,587	37,560	49,873	54,864	83,460	92,424	10,392	12,898
July	36,858	38,593	51,504	55,598	88,362	94,191	9,786	12,792
August	36,413	31,603	61,181	60,086	97,594	91,689	10,421	12,890
September	37,850	32,241	53,262	48,863	91,112	81,104	10,761	11,084
October	36,578	29,266	54,280	42,010	90,858	71,276	13,742	14,261
November	33,700	28,360	47,476	38,746	81,176	67,106	17,191	16,807
December	36,649	29,456	45,513	41,547	82,162	71,003	21,049	20,789

Month	Limas, fordhook		Limas, baby		Mixed vegetables		Okra	
	2008	2009	2008	2009	2008	2009	2008	2009
	1,000 pounds	1,000 pounds	1,000 pounds	1,000 pounds	1,000 pounds	1,000 pounds	1,000 pounds	1,000 pounds
January	6,820	2,811	33,699	40,378	46,035	50,384	20,263	20,272
February	5,687	2,516	28,610	32,758	42,788	47,748	12,819	16,861
March	4,807	2,198	27,484	30,813	45,999	49,685	8,635	13,780
April	3,967	1,539	24,652	29,773	45,650	52,366	7,316	9,938
May	3,158	1,141	20,840	23,693	46,360	49,514	8,952	13,919
June	2,209	960	16,213	20,264	46,341	54,394	21,505	21,470
July	2,124	2,104	13,035	16,177	47,253	50,654	26,468	29,224
August	2,772	6,819	16,110	25,542	48,149	45,661	29,456	37,993
September	5,928	2,988	43,834	55,917	50,540	43,582	31,487	41,014
October	3,990	3,123	51,658	66,944	51,737	42,526	31,585	38,262
November	4,079	3,220	46,992	63,132	51,073	44,629	30,176	32,678
December	3,676	2,973	41,020	58,390	47,729	50,129	24,115	29,030

Month	Carrots, diced		Carrots, other		Carrots, total		Cauliflower	
	2008	2009	2008	2009	2008	2009	2008	2009
	1,000 pounds	1,000 pounds	1,000 pounds	1,000 pounds	1,000 pounds	1,000 pounds	1,000 pounds	1,000 pounds
January	117,837	144,332	126,037	142,190	243,874	286,522	27,151	23,026
February	105,301	138,387	112,980	133,716	218,281	272,103	24,919	19,979
March	96,253	126,357	97,645	128,601	193,898	254,958	23,143	23,677
April	84,776	114,136	90,073	120,692	174,849	234,828	22,606	23,675
May	82,619	103,930	88,882	113,031	171,501	216,961	19,402	21,804
June	68,556	95,309	85,955	99,242	154,511	194,551	20,334	16,266
July	53,747	80,418	77,702	93,868	131,449	174,286	20,679	17,707
August	50,522	64,462	78,155	95,427	128,677	159,889	20,443	18,241
September	57,579	77,495	92,989	124,221	150,568	201,716	21,764	18,630
October	107,118	133,105	136,851	153,144	243,969	286,249	25,805	26,836
November	160,330	158,827	158,254	165,734	318,584	324,561	28,535	26,044
December	156,398	165,191	150,459	155,247	306,857	320,438	27,564	27,066

See end of table.

STATISTICS OF VEGETABLES AND MELONS

Table 4-82.—Frozen Vegetables and potato products: Cold storage holdings, end of month, United States, 2008 and 2009—Continued

Month	Corn, cut		Corn, cob		Corn, total		Onion rings	
	2008	2009	2008	2009	2008	2009	2008	2009
	1,000 pounds	1,000 pounds	1,000 pounds	1,000 pounds	1,000 pounds	1,000 pounds	1,000 pounds	1,000 pounds
January	421,154	406,206	225,602	247,402	646,756	653,608	7,186	6,382
February	352,171	368,520	185,562	218,649	537,733	587,169	6,408	5,689
March	318,052	326,868	163,020	192,534	481,072	519,402	6,451	4,433
April	246,808	290,734	123,090	169,269	369,898	460,003	5,918	3,459
May	208,127	238,292	105,895	131,665	314,022	369,957	5,847	4,652
June	172,494	202,653	78,251	98,389	250,745	301,042	5,925	4,306
July	157,387	193,401	80,325	107,043	237,712	300,444	6,005	3,238
August	342,614	396,474	182,022	181,263	524,636	577,737	6,315	3,411
September	518,630	601,205	295,216	292,652	813,846	893,857	6,290	3,423
October	559,266	658,972	321,617	293,967	880,883	952,939	5,944	4,861
November	511,993	630,159	297,614	265,665	809,607	895,824	6,620	5,376
December	463,724	584,048	264,562	252,108	728,286	836,156	6,628	3,456

Month	Onions, other		Blackeye peas		Green peas		Peas & carrots mixed	
	2008	2009	2008	2009	2008	2009	2008	2009
	1,000 pounds	1,000 pounds	1,000 pounds	1,000 pounds	1,000 pounds	1,000 pounds	1,000 pounds	1,000 pounds
January	32,934	28,865	4,136	2,707	200,278	232,757	4,960	5,083
February	37,079	29,025	4,176	3,144	173,323	199,313	4,531	4,911
March	37,730	28,263	4,084	2,344	132,601	167,606	3,964	4,984
April	33,585	29,401	4,106	2,487	105,832	143,908	4,113	4,549
May	36,879	28,606	3,483	1,912	78,501	121,027	4,835	4,737
June	39,190	26,699	2,815	1,519	198,133	252,937	4,756	5,258
July	35,237	28,014	2,637	1,773	387,597	437,572	4,734	5,253
August	36,009	31,236	2,627	2,546	372,967	403,023	4,684	5,147
September	36,496	33,512	3,596	2,330	342,420	376,205	4,930	5,449
October	35,062	36,001	3,295	2,975	312,439	335,817	4,730	6,369
November	33,482	37,112	3,044	2,415	288,190	303,632	4,969	5,322
December	29,564	35,511	2,737	3,653	255,471	275,625	5,109	6,075

Month	Spinach		Squash		Southern greens		Other vegetables	
	2008	2009	2008	2009	2008	2009	2008	2009
	1,000 pounds	1,000 pounds	1,000 pounds	1,000 pounds	1,000 pounds	1,000 pounds	1,000 pounds	1,000 pounds
January	42,667	36,973	57,033	54,517	14,539	15,514	340,522	388,463
February	38,676	37,458	53,337	52,866	14,648	16,188	311,246	357,668
March	51,301	49,249	49,596	49,751	13,697	16,290	302,281	332,749
April	67,750	68,310	46,114	46,080	12,444	16,143	295,076	331,613
May	79,143	81,204	45,364	45,801	11,573	15,874	293,527	320,377
June	80,320	87,057	46,531	40,733	14,231	16,551	275,882	293,501
July	75,028	77,516	48,978	43,618	14,974	16,144	324,856	343,267
August	62,840	68,941	54,905	52,855	14,206	14,235	378,439	388,316
September	55,211	61,944	58,479	57,317	13,492	12,855	427,982	437,144
October	53,901	59,183	60,762	59,180	12,734	11,710	456,239	454,844
November	48,775	56,133	58,664	54,490	12,272	9,959	443,586	424,864
December	42,505	47,993	56,537	52,275	12,632	11,621	423,684	408,829

Month	Total vegetables		French fries		Other frozen potatoes		Total frozen potatoes	
	2008	2009	2008	2009	2008	2009	2008	2009
	1,000 pounds	1,000 pounds	1,000 pounds	1,000 pounds	1,000 pounds	1,000 pounds	1,000 pounds	1,000 pounds
January	2,011,470	2,172,077	878,496	936,401	210,592	234,619	1,089,088	1,171,020
February	1,769,625	1,984,061	899,273	945,098	218,218	246,997	1,117,491	1,192,095
March	1,626,420	1,830,459	880,608	989,311	207,053	237,532	1,087,661	1,226,843
April	1,438,084	1,716,761	901,982	983,296	233,053	238,055	1,135,035	1,221,351
May	1,344,202	1,572,106	856,672	967,888	218,274	235,295	1,074,946	1,203,183
June	1,366,986	1,563,955	947,666	1,011,186	242,365	233,941	1,190,031	1,245,127
July	1,632,246	1,832,188	857,916	973,797	249,587	213,733	1,107,503	1,187,530
August	2,059,079	2,186,385	881,540	900,912	245,170	193,926	1,126,710	1,094,838
September	2,461,285	2,624,709	925,985	932,227	254,818	198,022	1,180,803	1,130,249
October	2,621,466	2,724,568	957,150	948,178	243,336	214,746	1,200,486	1,162,924
November	2,558,667	2,603,565	963,069	894,220	249,024	213,850	1,212,093	1,108,070
December	2,365,988	2,471,765	874,095	847,238	224,485	196,607	1,098,580	1,043,845

NASS, Livestock Branch, (202) 720–3570.

CHAPTER V

STATISTICS OF FRUITS, TREE NUTS, AND HORTICULTURAL SPECIALTIES

For most fruits, production is estimated at two levels—total and utilized. Total production is the quantity of fruit harvested plus quantities which would have been acceptable for fresh market or processing but were not harvested or utilized because of economic and other reasons. Utilized production is the amount sold plus the quantities used on farms where grown and quantities held in storage. The difference between total and utilized production is the quantity of marketable fruit not harvested and fruit harvested but not sold or utilized because of economic and other reasons. Production relates to the crop produced on all farms, except for apples and strawberries. In accordance with Congressional enactment, the Department's estimates of apple production since 1938 have related only to commercial production. The estimates for strawberries cover production on area grown primarily for sale. Statistics on utilization of fruit by commercial processors refer to first utilization, not necessarily final utilization. For example, frozen fruit includes fruit which may later be used for preserves.

The price shown for each crop is a marketing year average price for all methods of sales. Prices for most fresh fruit are the average prices producers received at the point of first sale, commonly referred to as the "average price as sold." Since the point of first sale is not the same for all producers, prices for the various methods of sale are weighted by the proportionate quantity sold. For example, if in a given State part of the fruit crop is sold f.o.b. packed by growers, part sold as bulk fruit at the packinghouse door, and some sold retail at roadside stands, the fresh fruit average price as sold is a weighted average of the average price for each method of sale.

The annual estimates are checked and adjusted at the end of each marketing season on the basis of shipment and processing records from transportation agencies, processors, cooperative marketing associations, and other industry organizations. The estimates are reviewed (and revised if necessary) at 5-year intervals, when the Census of Agriculture data become available. The Department's available statistics are limited to the major tree fruits and nuts and to grapes, cranberries, and strawberries, and exclude some States where census data indicate production is of only minor importance.

Table 5-1.—Fruits and planted nuts: Bearing area, United States, 2000–2009

Year	Citrus fruits [1]	Major deciduous fruits [2]	Miscellaneous fruits [3]	Planted nuts [4]	Fruits and planted nuts
	1,000 acres	*1,000 acres*	*1,000 acres*	*1,000 acres*	*1,000 acres*
2000	1,089.9	1,889.0	300.2	831.0	4,110.0
2001	1,082.0	1,843.9	293.8	858.8	4,078.4
2002	1,053.9	1,833.5	299.0	885.0	4,071.4
2003	1,044.4	1,825.2	297.6	896.8	4,064.0
2004	995.9	1,795.8	297.9	923.2	4,012.8
2005	954.2	1,770.6	303.2	956.3	3,984.3
2006	886.8	1,752.5	309.1	981.2	3,929.6
2007	866.2	1,729.7	292.1	1,016.6	3,904.5
2008	851.2	1,728.0	303.1	1,064.3	3,946.6
2009	845.1	1,723.8	305.2	1,112.7	3,986.8

[1] Oranges, tangerines and mandarins, Temples, grapefruit, lemons, limes, and tangelos. Area is for the year of harvest. [2] Commercial apples, apricots, cherries, grapes, nectarines, peaches, pears, plums, and prunes. [3] Avocados, bananas, berries, cranberries, dates, figs, guavas, kiwifruit, olives, papayas, pineapples, and strawberries. [4] Almonds, hazelnuts, macadamia nuts, pistachios, and walnuts.
NASS, Crops Branch, (202) 720–2127.

Table 5-2.—Fruits: Total production in tons, United States, 2000–2009 [1]

Year	Apples, commercial crop [2]	Peaches	Pears	Grapes (fresh basis)	Sweet cherries	Tart cherries	Apricots	Figs (fresh basis)
	1,000 tons	*1,000 tons*	*1,000 tons*	*1,000 tons*	*1,000 tons*	*1,000 tons*	*1,000 tons*	*1,000 tons*
2000	5,291	1,276	993	7,688	208	144	97	56
2001	4,712	1,204	1,027	6,569	230	185	83	41
2002	4,262	1,268	890	7,339	181	31	90	53
2003	4,390	1,260	934	6,644	246	113	98	49
2004	5,206	1,307	878	6,240	283	107	101	51
2005	4,834	1,185	823	7,814	251	135	82	49
2006	4,912	1,010	842	6,377	294	131	45	43
2007	4,545	1,127	873	7,057	311	127	88	48
2008	4,817	1,135	870	7,319	248	107	82	43
2009 [3]	4,957	1,104	957	7,295	430	179	69	40

Year	Plums (CA)	Prunes (fresh basis) (CA)	Prunes & Plums (ID,MI,OR,WA)	Olives	Strawberries [4]	Pineapples [4]	Avocados [5]	Nectarines [8]
	1,000 tons	*1,000 tons*	*1,000 tons*	*1,000 tons*	*1,000 tons*	*1,000 tons*	*1,000 tons*	*1,000 tons*
2000	197	681	24	53	950	354	239	267
2001	210	420	21	134	826	323	223	275
2002	201	519	16	103	942	320	199	300
2003	209	578	16	118	1,078	300	233	273
2004	156	144	25	108	1,107	220	179	269
2005	171	296	9	142	1,161	212	312	251
2006	158	634	22	24	1,202	185	147	232
2007	152	241	12	133	1,223	NA	193	283
2008	160	368	16	67	1,266	NA	116	303
2009 [3]	112	496	19	46	1,401	NA	269	220

Year	Oranges [6]	Tangerines and Mandarins [6]	Grapefruit [6]	Lemons [6]	K-Early Citrus [6]	Limes [6]	Tangelos [6]	Temples [6]
	1,000 tons	*1,000 tons*	*1,000 tons*	*1,000 tons*	*1,000 tons*	*1,000 tons*	*1,000 tons*	*1,000 tons*
2000	12,997	458	2,763	840	5	26	99	88
2001	12,221	373	2,462	996	2	11	95	56
2002	12,374	420	2,424	801	1	7	97	70
2003	11,545	382	2,063	1,026	NA	NA	105	59
2004	12,872	417	2,165	798	NA	NA	45	63
2005	9,251	335	1,018	870	NA	NA	70	29
2006	9,020	417	1,232	980	NA	NA	63	32
2007	7,625	361	1,627	798	NA	NA	56	(7)
2008	10,076	527	1,548	619	NA	NA	68	(7)
2009 [3]	9,198	443	1,331	950	NA	NA	52	(7)

Year	Cranberries	Bananas [4]	Kiwifruit	Dates	Papayas [4]	Berries [9]	Guavas	Total
	1,000 tons	*1,000 tons*	*1,000 tons*	*1,000 tons*	*1,000 tons*	*1,000 tons*	*1,000 tons*	*1,000 tons*
2000	286	15	34	17	27	229	8	36,410
2001	266	14	26	20	28	216	8	33,277
2002	284	10	26	24	23	210	5	33,490
2003	310	11	25	18	21	227	3	32,334
2004	309	8	27	17	18	244	4	33,368
2005	312	11	37	17	16	267	4	29,964
2006	345	11	26	18	14	286	4	28,706
2007	328	13	25	16	17	286	2	27,567
2008	393	9	23	21	17	279	2	30,499
2009 [3]	346	9	26	23	16	346	1	30,335

[1] For some crops in certain years, production includes some quantities unharvested for economic reasons or excess cullage fruit. [2] Estimates of the commercial crop refer to production in orchards of 100 or more bearing-age trees. [3] Preliminary. [4] Utilized production only. [5] Year of bloom. [6] Year harvest was complete. [7] Temples included in early, midseason, and navel varieties beginning with the 2006-07 season. [8] Washington added in 2005; prior years are California only. [9] Excludes strawberries and cranberries. NA-not available.

NASS, Crops Branch, (202) 720–2127.

Table 5-3.—Apples, commercial crop: Production and season average price per pound, by State and United States, 2007–2009 [1]

State	Total production			Utilized production			Price per pound [3] for crop of—		
	2007	2008	2009 [2]	2007	2008	2009 [2]	2007	2008	2009 [2]
	Million pounds	Million pounds	Million pounds	Million pounds	Million pounds	Million pounds	Dollars	Dollars	Dollars
AZ	23.0	18.0	5.5	23.0	18.0	5.5	0.219	0.223	0.237
CA	345.0	360.0	265.0	345.0	360.0	265.0	0.263	0.305	0.235
CO	13.0	18.0	16.0	13.0	17.0	15.0	0.215	0.234	0.258
CT	23.0	19.5	19.5	22.0	19.0	18.0	0.489	0.507	0.517
GA [4]	2.0	12.0	....	2.0	12.0	....	0.500	0.373	....
ID	35.0	85.0	45.0	35.0	85.0	45.0	0.252	0.202	0.218
IL	6.0	46.2	46.0	4.9	39.6	39.6	0.688	0.464	0.518
IN	20.0	23.0	30.0	18.0	21.5	27.0	0.280	0.378	0.300
IA	2.7	4.7	4.8	1.7	3.6	4.1	0.642	0.545	0.662
KY [4]	0.6	7.7	....	0.4	6.9	....	0.518	0.540	....
ME	40.0	38.5	34.0	36.0	35.0	32.0	0.409	0.389	0.426
MD	29.0	41.5	46.5	29.0	41.0	46.0	0.196	0.189	0.154
MA	38.5	41.0	43.5	36.5	38.5	41.0	0.437	0.515	0.461
MI	770.0	590.0	1,150.0	770.0	590.0	1,050.0	0.169	0.200	0.125
MN	26.0	27.1	23.2	20.1	24.3	20.3	0.637	0.734	0.591
MO	1.5	30.2	18.5	1.4	30.0	18.0	0.213	0.253	0.266
NH	34.5	36.5	30.0	33.0	35.0	28.0	0.356	0.466	0.451
NJ	42.0	43.0	43.0	42.0	39.0	42.0	0.229	0.381	0.499
NY	1,310.0	1,270.0	1,380.0	1,300.0	1,240.0	1,360.0	0.222	0.210	0.154
NC	60.0	165.0	120.0	60.0	165.0	105.0	0.098	0.152	0.169
OH	55.6	104.0	115.5	55.6	88.9	110.0	0.435	0.423	0.352
OR	135.0	119.0	130.0	135.0	119.0	130.0	0.281	0.234	0.197
PA	470.0	440.0	510.0	467.0	430.0	483.0	0.142	0.180	0.139
RI	2.6	2.4	2.4	2.4	2.3	2.3	0.561	0.673	0.610
SC [4]	0.3	7.0	....	0.3	6.3	....	0.473	0.178	....
TN	0.1	10.0	8.0	0.1	8.7	7.8	0.400	0.344	0.327
UT	19.0	12.0	18.0	18.0	11.6	16.0	0.329	0.286	0.296
VT	38.0	44.0	40.0	33.0	41.0	37.0	0.332	0.356	0.237
VA	215.0	226.0	245.0	215.0	226.0	245.0	0.121	0.166	0.135
WA	5,200.0	5,650.0	5,400.0	5,200.0	5,650.0	5,400.0	0.342	0.228	0.273
WV	80.0	85.0	82.0	77.0	81.0	79.0	0.098	0.144	0.138
WI	52.0	57.0	43.5	49.0	54.5	36.5	0.468	0.515	0.411
US	9,089.4	9,633.3	9,914.9	9,045.4	9,539.7	9,708.1	0.288	0.232	0.231

[1] In orchards of 100 or more bearing-age trees. [2] Preliminary. [3] Fresh fruit prices are equivalent packinghouse-door returns for CA, MI, NY, and WA; prices at point of first sale for other States. Processing prices are equivalent at processing plant door. [4] Estimates discontinued in 2009.
NASS, Crops Branch, (202) 720–2127.

Table 5-4.—Apples: Production and value, United States, 2000–2009

Year	Apples, commercial crop [1]			
	Total production	Utilized production	Marketing year average price [2]	Value
	Million pounds	Million pounds	Cents per pound	1,000 dollars
2000	10,580.9	10,319.8	12.8	1,320,618
2001	9,423.0	9,209.2	15.8	1,452,344
2002	8,523.9	8,374.1	18.9	1,581,260
2003	8,780.1	8,692.0	18.8	1,634,141
2004	10,412.1	10,332.8	13.6	1,405,946
2005	9,666.9	9,567.2	17.3	1,657,947
2006	9,823.4	9,730.2	22.7	2,213,155
2007	9,089.4	9,045.4	28.8	2,608,220
2008	9,633.3	9,539.7	23.2	2,214,717
2009 [3]	9,914.9	9,708.1	22.8	2,246,584

[1] In orchards of 100 or more bearing-age trees. [2] Fresh fruit prices are equivalent packinghouse-door returns for CA, NY, MI, and WA; prices at point of first sale for other States. Processing prices are equivalent at processing plant door. [3] Preliminary.
NASS, Crops Branch, (202) 720–2127.

Table 5-5.—Apples, fresh: Area and production in specified countries, 2007/2008–2009/2010

Country	Area			Production		
	2007/2008	2008/2009	2009/2010	2007/2008	2008/2009	2009/2010
	1,000 hec- tares	1,000 hec- tares	1,000 hec- tares	1,000 metric tons	1,000 metric tons	1,000 metric tons
Argentina	26,000	26,000	26,000	980,000	933,000	800,000
Brazil	38,804	38,804	38,832	1,121,290	1,220,499	1,257,114
Chile	2,049	32,780	32,900	1,350,000	1,280,000	1,090,000
China	....	....	....	24,800,000	29,800,000	32,000,000
EU-27	480,485	502,478	496,900	10,294,980	12,575,063	12,061,600
India	....	....	....	2,001,000	1,930,000	1,935,000
Japan	39,900	39,500	38,800	840,100	910,700	845,600
Russia	320,000	190,000	188,000	1,300,000	1,115,000	1,150,000
Turkey	164,000	160,000	160,000	2,458,000	2,600,000	2,800,000
Ukraine	....	....	....	707,000	719,000	760,000
Others	95,642	103,429	103,139	2,723,021	2,844,650	2,782,445
Total Foreign	1,166,880	1,092,991	1,084,571	48,575,391	55,927,912	57,481,759
United States	....	....	....	4,122,925	4,358,750	4,417,800
Total	1,166,880	1,092,991	1,084,571	52,698,316	60,286,662	61,899,559

FAS, Office of Global Analysis, (202) 720-6301. Prepared or estimated on the basis of official USDA production, supply, and distribution statistics from foreign governments.

Table 5-6.—Apples, commercial crop: Production and utilization, United States, 2000–2009

Crop of—	Total production	Utilized production	Utilization of quantities sold				
			Fresh [1]	Processed (fresh basis)			
				Canned	Dried	Frozen	Juice, cider & other [2]
	Million pounds	Million pounds	Million pounds	Million pounds	Million pounds	Million pounds	Million pounds
2000	10,580.9	10,319.8	6,265.5	1,183.6	248.2	195.9	2,426.6
2001	9,423.0	9,209.2	5,467.5	1,257.2	221.0	248.5	2,015.0
2002	8,523.9	8,374.1	5,366.0	1,078.7	207.9	191.7	1,529.8
2003	8,780.1	8,692.0	5,453.3	1,235.1	182.2	282.8	1,538.6
2004	10,412.1	10,332.8	6,619.0	1,255.2	200.8	255.8	1,948.0
2005	9,666.9	9,567.2	6,096.9	1,163.8	191.1	259.4	1,758.3
2006	9,823.4	9,730.2	6,308.5	1,167.3	252.8	271.8	1,611.6
2007	9,089.4	9,045.4	6,077.3	1,091.2	203.7	257.7	1,257.3
2008	9,633.3	9,539.7	6,273.9	1,253.4	212.7	211.2	1,460.7
2009 [3]	9,914.9	9,708.1	6,403.9	1,218.2	206.2	241.2	1,494.7

[1] Includes "Home use." [2] Mostly crushed for vinegar, cider, and juice. For some States, small quantities canned, dried, and frozen are included. Beginning in 2004, "fresh slices" included. [3] Preliminary.
NASS, Crops Branch, (202) 720-2127.

Table 5-7.—Apples, commercial crop: Production and utilization, by State and United States, crop of 2008

State	Total production	Utilized pro- duction	Utilization				
			Fresh [1]	Processed (fresh basis)			
				Canned	Dried	Frozen	Juice, cider & other [2]
	Million pounds	Million pounds	Million pounds	Million pounds	Million pounds	Million pounds	Million pounds
CA	360.0	360.0	160.0	....	....	....	....
MI	590.0	590.0	165.0	180.0	....	....	105.0
NY	1,270.0	1,240.0	550.0	380.0	....	....	210.0
NC	165.0	165.0	49.5	69.3	....	....	37.0
OR	119.0	119.0	79.0	....	....	....	20.0
PA	440.0	430.0	160.0	189.0	....	....	46.0
VA	226.0	226.0	51.0	140.0	....	....	30.0
WA	5,650.0	5,650.0	4,550.0	220.0	....	....	560.0
WV	85.0	81.0	16.0	30.0	....	....	30.0
Oth Sts	728.3	678.7	493.4	45.1	....	211.2	310.6
US	9,609.3	9,515.7	6,273.9	1,253.4	212.7	211.2	1,348.6

[1] Includes "Home use." [2] Mostly vinegar, wine, and fresh slices for pie making.
NASS, Crops Branch, (202) 720-2127.

Table 5-8.—Fruits, fresh: United States exports by country of destination and imports by country of origin, 2007–2009

Country	2007	2008	2009
	Metric tons	*Metric tons*	*Metric tons*
Fresh fruits, deciduous:			
Canada	585,862	603,220	579,995
Mexico	297,777	377,342	289,575
Taiwan	126,147	137,792	132,723
Hong Kong	93,510	112,056	130,025
Japan	65,,308	69,424	71,290
Indonesia	42,388	57,660	67,434
Australia(*)	41,247	59,070	63,983
United Kingdom	69,500	76,202	60,129
India	36,228	23,482	47,009
United Arab Emirates	24,723	32,511	46,135
Russia	20,190	31,439	30,799
Philippines	17,420	20,569	29,530
Thailand	25,363	30,722	28,435
Korea, South	36,980	35,336	27,734
Rest of World	307,398	362,887	381,754
World Total	1,790,039	2,029,712	1,986,548
Fresh fruit, other:			
Canada	391,309	485,200	486,007
Japan	36,796	37,479	43,685
Mexico	23,774	37,145	20,492
United Kingdom	27,882	21,733	15,936
Korea, South	4,009	10,202	9,253
Taiwan	3,839	6,528	8,135
Germany(*)	2,674	4,120	7,624
Hong Kong	645	1,787	6,884
United Arab Emirates	941	1,990	5,482
Netherlands Antilles (exc. Aruba)	4,308	5,870	4,212
Bermuda	3,236	4,007	4,107
China	856	1,692	2,912
Australia(*)	3,769	3,419	2,911
Cayman Islands	7,373	6,541	2,797
Rest of World	11,524	16,014	18,672
World Total	552,934	643,729	639,110
Fresh melons:			
Canada	114,676	115,386	115,355
Mexico	4,758	8,197	7,616
Japan	5,366	6,720	7,616
China	0	48	1,032
Hong Kong	885	1,150	793
Kora,South	163	171	600
Taiwan	251	556	374
Kuwait	274	283	319
Bahamas,The	689	473	225
Bermuda	160	201	209
Jamaica	60	39	163
United Arab Emirates	9	31	77
Panama	50	48	50
Netherlands Antilles(*)	0	4	32
Rest of World	127	874	106
World Total	127,467	134,182	134,605

Note: (*) Denotes a country that is a summarization of its component countries.
FAS, Office of Global Analysis, (202) 720-6301.

Table 5-9.—Apples: Foreign trade, United States, 2000–2008 [1]

Year beginning October	Imports, fresh and dried, in terms of fresh	Domestic exports	
		Fresh	Dried, in terms of fresh [1]
	Metric tons	*Metric tons*	*Metric tons*
2000	180,616	743,644	33,308
2001	193,893	592,955	21,232
2002	231,504	522,525	26,250
2003	243,293	438,300	32,960
2004	155,095	638,905	32,512
2005	198,472	654,137	27,383
2006	255,320	652,827	32,925
2007	238,333	680,618	31,680
2008	211,423	810,123	26,836

[1] Dried converted to terms of fresh apples on following basis; 1 pound dried is equivalent to 8 pounds fresh. No re-exports reported.
ERS, Food and Specialty Crops Branch, (202) 694–5260.

Table 5-10.—Apricots: Production and value, United States, 2000–2009 [1]

Year	Total production	Utilized production	Market year average price per ton [2]	Value
	Tons	*Tons*	*Dollars*	*1,000 dollars*
2000	96,900	87,760	369.00	32,346
2001	82,460	75,430	353.00	26,598
2002	90,040	80,030	357.00	28,565
2003	97,580	97,560	356.00	34,702
2004	101,130	92,590	378.00	35,012
2005	81,650	76,645	520.00	39,880
2006	44,480	44,455	665.00	29,563
2007	88,460	88,460	477.00	42,227
2008	81,610	77,480	532.00	41,196
2009 [3]	68,720	68,690	654.00	44,912

[1] Production, price, and value for CA, UT, and WA.   [2] Fresh fruit prices are equivalent packinghouse-door returns for CA and WA. Quantities processed are priced at the equivalent processing plant door level.   [3] Preliminary.
NASS, Crops Branch, (202) 720–2127.

Table 5-11.—Apricots: Production and marketing year average price per ton, by State and United States, 2007–2009

State	Total production			Utilized production			Price [2] for crop of—		
	2007	2008	2009 [1]	2007	2008	2009 [1]	2007	2008	2009 [1]
	Tons	*Tons*	*Tons*	*Tons*	*Tons*	*Tons*	*Dollars*	*Dollars*	*Dollars*
CA	81,000	77,000	59,500	81,000	72,900	59,500	431.00	472.00	621.00
UT	260	410	320	260	380	290	815.00	468.00	862.00
WA	7,200	4,200	8,900	7,200	4,200	8,900	991.00	1,570.00	868.00
Total	88,460	81,610	68,720	88,460	77,480	68,690	477.00	532.00	654.00

[1] Preliminary.   [2] Fresh fruit prices are equivalent packinghouse-door returns for CA and WA. Quantities processed are priced at the equivalent processing plant door level.
NASS, Crops Branch, (202) 720–2127.

Table 5-12.—Apricots: Production and utilization, United States, 2000–2009 [1]

Crop of—	Total production	Utilized production	Utilization of quantities sold			
			Fresh [2]	Processed [3]		
				Canned [4]	Dried (fresh basis)	Frozen
	Tons	*Tons*	*Tons*	*Tons*	*Tons*	*Tons*
2000	96,900	87,760	26,580	32,000	8,000	10,000
2001	82,460	75,430	18,230	31,000	6,000	9,000
2002	90,040	80,030	18,290	30,500	8,000	10,500
2003	97,580	97,560	26,250	30,000	6,800	11,000
2004	101,130	92,590	23,650	(5)	11,800	9,700
2005	81,650	76,645	23,645	23,500	11,500	(5)
2006	44,480	44,455	13,755	14,900	5,500	(5)
2007	88,460	88,460	29,270	24,000	13,000	(5)
2008	81,610	77,480	25,760	22,000	14,000	(5)
2009 [6]	68,720	68,690	25,170	23,100	9,000	(5)

[1] CA, UT, and WA. [2] Includes "Home use." [3] CA only. [4] Includes some quantities frozen or otherwise processed. [5] Missing data not published to avoid disclosure of individual operations. [6] Preliminary.
NASS, Crops Branch, (202) 720–2127.

Table 5-13.—Apricots: Production and utilization, by State and United States, crop of 2009 [1]

State	Total production	Utilized production	Utilization			
			Fresh	Processed [2]		
				Canned [3]	Dried (fresh basis)	Frozen
	Tons	*Tons*	*Tons*	*Tons*	*Tons*	*Tons*
CA	59,500	59,500	17,500	23,100	9,000	(5)
UT [4]	320	290	(4)			(5)
WA [4]	8,900	8,900	(4)			(5)
US	68,720	68,690	25,170	23,100	9,000	(5)

[1] Preliminary. [2] CA only. [3] Some quantities used for juice are included in "Canned" to avoid disclosure of individual operations. [4] Missing data not published to avoid disclosure of individual operations, but included in U.S. total. [5] Missing data not published to avoid disclosure of individual operations.
NASS, Crops Branch, (202) 720–2127.

Table 5-14.—Apricots: Foreign trade, United States, 2000–2008

Year beginning October	Domestic exports				
	Fresh	Canned [1]	Dried [1]	Dried, in fruit salad [2]	Total, in terms of fresh [3]
	Metric tons	*Metric tons*	*Metric tons*	*Metric tons*	*Metric tons*
2000	7,663	769	2,251	312	21,393
2001	7,732	1,600	2,004	202	20,218
2002	7,914	1,554	3,310	402	28,113
2003	7,534	1,588	1,073	486	16,683
2004	5,514	1,581	706	715	13,952
2005	3,935	1,508	1,142	633	14,140
2006	7,061	1,201	615	364	12,951
2007	7,357	1,217	679	405	13,805
2008	5,044	1,291	599	358	10,887

[1] Net processed weight. [2] Dried apricots are 12⅓ percent of total dried fruit for salad. [3] Dried fruit converted to unprocessed dry weight by dividing by 1.07. Unprocessed dry weight converted to terms of fresh fruit on the basis that 1 pound dried equals 5.5 pounds fresh. Canned apricots converted to terms of fresh on the basis that 1 pound canned equals 0.717 pounds fresh.
ERS, Food and Specialty Crops Branch, (202) 694–5260.

Table 5-15.—Avocados: Foreign trade, United States, 1999–2008

Year beginning October	Imports
	Metric tons
1999	63,944
2000	73,070
2001	103,339
2002	136,708
2003	132,644
2004	248,313
2005	197,354
2006	338,559
2007	307,167
2008	413,931

ERS, Food and Specialty Crops Branch, (202) 694–5260.

Table 5-16.—Avocados: Production, marketing year average price per ton, and value, United States, 1999–2000 to 2008–09

Season	California [1]			Florida [1]		
	Production [2]	Price [3]	Value	Production [2]	Price [3]	Value
	Tons	*Dollars*	*1,000 dollars*	*Tons*	*Dollars*	*1,000 dollars*
1999–00	161,000	2,110	339,594	22,000	748	16,456
2000–01	213,000	1,480	315,842	26,000	584	15,184
2001–02	200,000	1,790	358,000	23,000	676	15,548
2002–03	168,000	2,170	364,560	31,000	556	17,236
2003–04	216,000	1,760	380,160	17,000	808	13,736
2004–05	151,000	1,830	276,330	28,000	516	14,448
2005–06	300,000	1,140	342,000	12,000	940	11,280
2006–07	132,000	1,890	249,480	14,000	912	12,768
2007–08	165,000	1,990	328,350	27,500	440	12,100
2008–09	88,000	2,280	200,640	27,450	480	13,176

Season	Hawaii			United States		
	Production [2]	Price [3]	Value	Production [2]	Price [3]	Value
	Tons	*Dollars*	*1,000 dollars*	*Tons*	*Dollars*	*1,000 dollars*
1999–00	300	1,200.00	360	183,300	1,950.00	356,410
2000–01	320	1,160.00	371	239,320	1,400.00	331,397
2001–02	300	1,140.00	342	223,300	1,670.00	373,890
2002–03	350	1,120.00	392	199,350	1,920.00	382,188
2003–04	380	1,240.00	471	233,380	1,690.00	394,367
2004–05	370	1,260.00	466	179,370	1,620.00	291,244
2005–06	400	1,320.00	528	312,400	1,130.00	353,808
2006–07	510	1,360.00	694	146,510	1,800.00	262,942
2007–08	580	1,360.00	789	193,080	1,770.00	341,239
2008–09	500	1,460.00	730	115,950	1,850.00	214,546

[1] Season from Nov 1 to Oct 31 (following year) for California and June 20 to Mar 1 for Florida.   [2] Production is the quantity sold or utilized.   [3] Quantities processed are priced at the equivalent processing plant door level.
NASS, Crops Branch, (202) 720–2127.

Table 5-17.—Bananas: Area, yield, utilized production, marketing year average price, and value, Hawaii, 2000–2009

Year	Area harvested	Yield per acre	Production	Price per pound	Value
	Acres	*1,000 pounds*	*1,000 pounds*	*Cents*	*1,000 dollars*
2000	1,460	19.9	29,000	36.0	10,440
2001	1,490	18.8	28,000	38.0	10,640
2002	1,330	15.0	20,000	43.0	8,600
2003	1,350	16.7	22,500	41.0	9,225
2004	1,000	16.5	16,500	49.0	8,085
2005	980	21.3	20,900	43.9	9,175
2006	1,100	20.0	22,000	49.0	10,780
2007	1,300	19.7	25,600	41.0	10,496
2008	1,100	15.8	17,400	46.0	8,004
2009 [1]	1,100	16.8	18,500	55.0	10,175

[1] Preliminary.
NASS, Crops Branch, (202) 720–2127.

Table 5-18.—Kiwifruit: Area, yield, utilized production, marketing year average price, and value, California, 2000–2009

Year	Bearing acreage	Yield [1]	Production	Price per ton	Value
	Acres	*Tons*	*Tons*	*Dollars*	*1,000 dollars*
2000	5,300	6.42	30,500	455	13,888
2001	4,900	5.27	23,000	667	15,340
2002	4,500	5.80	23,100	783	18,097
2003	4,500	5.64	24,000	853	20,472
2004	4,500	5.93	24,700	809	19,977
2005	4,500	8.27	36,200	620	22,461
2006	4,200	6.21	25,400	911	23,148
2007	4,200	5.83	23,700	950	22,517
2008	4,200	5.48	22,000	888	19,545
2009	4,200	6.19	23,900	1,470	35,048

[1] Yield based on total production.
NASS, Crops Branch, (202) 720-2127.

Table 5-19.—Cherries: Foreign trade, United States, 1999–2008

Year beginning October	Imports		Domestic exports	
	Fresh	Dried and preserved	Fresh	Canned
	Metric tons	*Metric tons*	*Metric tons*	*Metric tons*
1999	2,815	2,184	43,289	14,970
2000	3,858	2,561	42,880	20,515
2001	6,680	3,023	36,232	19,355
2002	8,548	3,062	47,829	12,519
2003	5,170	3,652	43,079	13,144
2004	7,214	3,738	43,043	14,263
2005	12,062	4,998	47,618	15,831
2006	13,940	5,303	51,182	14,637
2007	22,125	5,403	45,462	14,427
2008	19,411	5,390	64,595	12,684

ERS, Food and Specialty Crops Branch, (202) 694–5260.

Table 5-20.—Sweet cherries: Production and value, United States, 2000–2009

Year	Total production	Utilized production	Marketing year average price per ton [1]	Value
	Tons	Tons	Dollars	1,000 dollars
2000	207,900	205,420	1,340.00	274,995
2001	230,380	219,620	1,230.00	270,914
2002	181,355	177,305	1,550.00	274,471
2003	245,700	243,580	1,410.00	342,113
2004	283,060	279,160	1,570.00	437,133
2005	250,830	243,570	1,990.00	484,348
2006	294,160	287,520	1,620.00	465,225
2007	310,680	306,210	1,820.00	557,056
2008	248,060	240,720	2,390.00	574,043
2009 [2]	429,870	375,625	1,350.00	505,881

[1] Fresh fruit prices are equivalent packinghouse-door returns for Western States, and the average price as sold for other States. Quantities processed are priced at the equivalent processing plant door level.　[2] Preliminary.
NASS, Crops Branch, (202) 720–2127.

Table 5-21.—Tart cherries: Production and value, United States, 2000–2009

Year	Total production	Utilized production	Marketing year average price per pound [1]	Value
	Million pounds	Million pounds	Dollars	1,000 dollars
2000	288.5	281.4	0.187	52,488
2001	370.1	307.9	0.186	57,150
2002	62.5	62.2	0.448	27,879
2003	226.3	226.3	0.354	80,210
2004	213.0	213.0	0.328	69,941
2005	269.9	267.9	0.238	63,741
2006	262.0	248.6	0.215	53,454
2007	253.2	248.7	0.273	67,923
2008	214.4	213.2	0.377	80,344
2009 [2]	358.9	320.5	0.197	63,231

[1] Fresh fruit prices are equivalent packinghouse-door returns for Western States, and the average price as sold for other States. Quantities processed are priced at the equivalent processing plant door level.　[2] Preliminary.
NASS, Crops Branch, (202) 720–2127.

Table 5-22.—Sweet cherries: Production and season average price, by State and United States, 2007–2009

State	Total production			Utilized production			Price [2]		
	2007	2008	2009 [1]	2007	2008	2009 [1]	2007	2008	2009 [1]
	Tons	Tons	Tons	Tons	Tons	Tons	Dollars per ton	Dollars per ton	Dollars per ton
CA	85,000	86,000	78,000	81,800	82,800	75,000	1,900.00	2,350.00	2,470.00
ID	1,500	1,900	6,000	1,500	1,800	2,700	2,100.00	3,120.00	1,100.00
MI	27,300	26,500	28,700	27,300	26,300	28,600	649.00	614.00	478.00
MT	2,440	1,560	2,390	2,180	1,450	1,055	1,630.00	2,730.00	1,490.00
NY	1,190	1,050	1,240	1,180	920	940	2,980.00	3,520.00	2,440.00
OR	35,000	31,000	67,000	34,000	27,400	56,000	1,450.00	2,100.00	1,450.00
UT	1,250	50	1,540	1,250	50	1,330	1,380.00	2,440.00	2,280.00
WA	157,000	100,000	245,000	157,000	100,000	210,000	2,060.00	2,930.00	1,030.00
US	310,680	248,060	429,870	306,210	240,720	375,625	1,820.00	2,390.00	1,350.00

[1] Preliminary.　[2] Fresh fruit prices are equivalent packinghouse-door returns for CA, OR, and WA, and the average price as sold for other States. Quantities processed are priced at the equivalent processing plant door level.
NASS, Crops Branch, (202) 720–2127.

Table 5-23.—Tart cherries: Production and season average price, by State and United States, 2007–2009

State	Total production			Utilized production			Price [2]		
	2007	2008	2009 [1]	2007	2008	2009 [1]	2007	2008	2009 [1]
	Million pounds	Million pounds	Million pounds	Million pounds	Million pounds	Million pounds	Dollars per Lb.	Dollars per Lb.	Dollars per Lb.
MI	196.0	165.0	266.0	193.0	165.0	242.0	0.264	0.382	0.157
NY	11.3	9.6	11.2	11.3	9.4	10.1	0.343	0.413	0.243
OR	0.5	2.8	3.2	0.5	2.8	3.2	0.346	0.419	0.845
PA	3.5	3.9	3.9	3.5	3.9	3.9	0.398	0.425	0.250
UT	20.0	20.0	47.0	19.0	19.0	34.0	0.250	0.330	0.270
WA	11.5	12.5	16.7	11.4	12.5	16.4	0.350	0.330	0.468
WI	10.4	0.6	10.9	10.0	0.6	10.9	0.284	0.350	0.208
US	253.2	214.4	358.9	248.7	213.2	320.5	0.273	0.377	0.197

[1] Preliminary.　[2] Fresh fruit prices are equivalent packinghouse-door returns for OR and WA, and the average price as sold for other States. Quantities processed are priced at the equivalent processing plant door level.
NASS, Crops Branch, (202) 720–2127.

Table 5-24.—Sweet cherries: Production and utilization, by State and United States, crop of 2009[1]

State	Total production	Utilized production	Utilization			
			Fresh[2]	Processed		
				Canned and otherwise processed	Brined	Other[3]
	Tons	*Tons*	*Tons*	*Tons*	*Tons*	*Tons*
CA	78,000	75,000	63,000			
MI	28,700	28,600	800	1,250	17,750	8,800
OR	67,000	56,000	36,400	2,000	17,000	600
WA	245,000	210,000	178,000	5,000	11,000	16,000
Oth Sts	11,170	6,025	5,400		9,371	3,254
US	429,870	375,625	283,600	8,250	55,121	28,654

[1] Preliminary. [2] Includes "Home use." [3] Includes California canned utilization and other processed utilizations from all States.
NASS, Crops Branch, (202) 720–2127.

Table 5-25.—Tart cherries: Production and utilization, by State and United States, crop of 2009[1]

State	Total production	Utilized production	Utilization			
			Fresh[2]	Processed		
				Canned and otherwise processed[3]	Frozen	Other[3]
	Million pounds	*Million pounds*	*Million pounds*	*Million pounds*	*Million pounds*	*Million pounds*
MI	266.0	242.0	0.5	43.0	175.0	23.5
Oth Sts	92.9	78.5	0.8	9.1	57.0	11.6
US	358.9	320.5	1.3	52.1	232.0	35.1

[1] Preliminary. [2] Includes "Home use." [3] Some quantities used for juice, wine, brined, and dried.
NASS, Crops Branch, (202) 720–2127.

Table 5-26.—Sweet cherries: Production and utilization, United States, 2000–2009

Crop of—	Total production	Utilized production	Utilization of quantities sold		
			Fresh[1]	Processed	
				Other[2]	Brined
	Tons	*Tons*	*Tons*	*Tons*	*Tons*
2000	207,900	205,420	120,760	27,710	56,950
2001	230,380	219,620	145,710	25,730	48,180
2002	181,355	177,305	126,595	18,570	32,140
2003	245,700	243,580	175,570	25,960	42,050
2004	283,060	279,160	185,050	33,380	60,730
2005	250,830	243,570	167,190	30,050	46,330
2006	294,160	287,520	190,770	40,520	56,230
2007	310,680	306,210	222,560	35,490	48,160
2008	248,060	240,720	175,320	25,650	39,750
2009[3]	429,870	375,625	283,600	36,904	55,121

[1] Includes "Home use." [2] Includes canned utilization and other processed utilizations from all States. [3] Preliminary.
NASS, Crops Branch, (202) 720–2127.

Table 5-27.—Tart cherries: Production and utilization, United States, 2000–2009

Crop of—	Total production	Utilized production	Utilization of quantities sold		
			Fresh[1]	Processed	
				Other[2]	Frozen
	Million pounds	*Million pounds*	*Million pounds*	*Million pounds*	*Million pounds*
2000	288.5	281.4	1.8	135.3	144.3
2001	370.1	307.9	1.9	129.2	176.8
2002	62.5	62.2	0.8	32.5	28.9
2003	226.3	226.3	1.0	76.6	148.7
2004	213.0	213.0	1.3	61.6	150.1
2005	269.9	267.9	1.2	78.7	188.0
2006	262.0	248.6	1.4	90.4	156.8
2007	253.2	248.7	1.6	67.8	179.3
2008	214.4	213.2	1.0	58.7	145.0
2009[3]	358.9	320.5	1.3	87.2	232.0

[1] Includes "Home use." [2] Includes canned utilization and other processed utilizations from all states. [3] Preliminary.
NASS, Crops Branch, (202) 720–2127.

Table 5-28.—Citrus fruit: Utilized production and value, United States, for season of 1999–2000 to 2008–09

Season [1]	Production	Marketing year average returns per box [2]	Value	Quantities processed [3]	Production	Marketing year average returns per box [2]	Value	Quantities processed [3]
		Oranges [4]				Grapefruit		
	1,000 boxes	*Dollars*	*1,000 dollars*	*1,000 boxes*	*1,000 boxes*	*Dollars*	*1,000 dollars*	*1,000 boxes*
1999–2000	299,760	5.56	1,666,100	244,582	66,980	6.07	409,716	38,509
2000–01	280,935	5.88	1,682,790	223,232	59,750	4.69	285,065	32,600
2001–02	283,760	6.37	1,846,199	228,276	58,660	4.92	292,156	32,113
2002–03	267,040	5.79	1,564,658	206,000	50,080	5.24	269,381	26,150
2003–04	294,620	5.88	1,774,453	238,690	52,540	5.77	307,811	27,225
2004–05	216,500	6.68	1,475,381	158,338	25,640	14.93	383,041	9,556
2005–06	210,750	8.60	1,829,860	157,930	30,600	11.18	345,032	15,086
2006–07	177,280	12.56	2,216,471	142,030	39,900	7.69	311,914	20,579
2007–08	234,376	9.36	2,198,836	179,687	37,900	7.15	273,076	18,677
2008–09 [5]	212,609	9.03	1,950,452	163,651	32,825	7.01	233,539	14,659
		Lemons				Temples (FL)		
1999–2000	22,100	13.51	298,677	8,476	1,950	4.70	9,173	1,510
2000–01	26,200	9.06	237,362	12,793	1,250	4.23	5,282	907
2001–02	21,100	15.54	327,964	6,678	1,550	4.46	6,919	1,132
2002–03	27,000	10.79	291,425	12,354	1,300	4.30	5,591	995
2003–04	21,000	13.12	275,620	6,792	1,400	3.51	4,915	1,058
2004–05	22,900	13.38	306,434	9,772	650	5.10	3,314	437
2005–06	25,800	15.90	410,338	7,746	700	5.76	4,034	491
2006–07	21,000	21.40	449,417	8,105	(6)	(6)	(6)	(6)
2007–08	16,300	32.12	523,528	5,918	(6)	(6)	(6)	(6)
2008–09 [5]	25,000	13.71	342,667	10,786	(6)	(6)	(6)	(6)
		Tangerines and Mandarins [7]				Tangelos (FL)		
1999–2000	10,350	10.43	108,192	3,640	2,200	5.11	11,232	1,464
2000–01	8,450	11.26	96,789	2,517	2,100	3.90	8,193	1,358
2001–02	9,420	12.97	124,718	2,665	2,150	5.00	10,758	1,454
2002–03	8,730	13.23	117,432	1,989	2,350	4.89	11,489	1,742
2003–04	9,390	11.81	112,232	2,545	1,000	10.02	10,021	455
2004–05	7,750	16.28	127,251	1,633	1,550	5.16	8,004	1,055
2005–06	9,650	14.11	137,666	2,989	1,400	8.17	11,431	853
2006–07	8,400	18.30	156,198	2,530	1,250	11.00	13,755	822
2007–08	12,600	17.91	236,193	3,411	1,500	5.76	8,638	1,068
2008–09 [5]	10,800	18.69	209,515	2,221	1,150	4.78	5,496	646
		Limes (FL)				K-Early Citrus (FL)		
1999–2000	600	16.21	9,728	100	110	3.24	356	95
2000–01	250	17.00	4,249	30	40	4.68	187	19
2001–02	150	11.55	1,732	25	30	3.77	113	24
2002–03	(8)	(8)	(8)	(8)	(8)	(8)	(8)	(8)
2003–04	(8)	(8)	(8)	(8)	(8)	(8)	(8)	(8)
2004–05	(8)	(8)	(8)	(8)	(8)	(8)	(8)	(8)
2005–06	(8)	(8)	(8)	(8)	(8)	(8)	(8)	(8)
2006–07	(8)	(8)	(8)	(8)	(8)	(8)	(8)	(8)
2007–08	(8)	(8)	(8)	(8)	(8)	(8)	(8)	(8)
2008–09	(8)	(8)	(8)	(8)	(8)	(8)	(8)	(8)

[1] See footnote 1, table 5-29. [2] Equivalent packing-house door returns. [3] Includes quantities used for juice, concentrates, grapefruit segments, and other citrus products. In some seasons, includes appreciable quantities of oranges and lemons in CA delivered to processing plants which were not utilized, but for which growers received payment. [4] Includes small quantities of tangerines in TX and Temples in FL. [5] Preliminary. [6] Included in early, midseason, and navel orange varieties beginning with the 2006-07 season. [7] AZ and CA tangelos and tangors included. [8] Estimates discontinued.
NASS, Crops Branch, (202) 720–2127.

Table 5-29.—Citrus fruit: Utilized production and marketing year average returns per box, by State, 2007–08 and 2008–09 [1]

Crop and State	Utilized production		Market year average price [2]	
	2007–08	2008–09	2007–08	2008–09
ORANGES Early, midseason, and Navel varieties: [3]	*1,000 boxes* [4]	*1,000 boxes* [4]	*Dollars*	*Dollars*
AZ	230	150	7.91	11.62
CA	45,000	34,500	10.46	13.91
FL	83,500	84,600	8.51	7.34
TX	1,600	1,300	5.16	6.94
Total early, midseason, and Navel varieties	130,330	120,550	9.06	8.99
Valencia:				
AZ	150	100	3.53	6.18
CA	17,000	14,000	8.12	10.45
FL	86,700	77,800	10.00	8.89
TX	196	159	6.91	5.87
Total Valencia	104,046	92,059	9.72	9.09
All oranges:				
AZ	380	250	6.18	9.45
CA	62,000	48,500	9.82	12.91
FL	170,200	162,400	9.27	8.08
TX	1,796	1,459	5.35	6.82
US, all oranges	234,376	212,609	9.72	9.09
GRAPEFRUIT				
AZ	100	25	8.68	9.49
CA	5,200	5,600	9.18	10.09
FL, all	26,600	21,700	6.90	6.46
Colored seedless	17,600	15,100	7.92	7.49
White seedless	9,000	6,600	4.90	4.11
TX	6,000	5,500	6.83	6.64
US, all grapefruit	37,900	32,825	7.15	7.01
LEMONS				
AZ	1,500	3,000	33.32	9.88
CA	14,800	22,000	32.00	14.23
US, lemons	16,300	25,000	32.12	13.71
TANGELOS				
FL	1,500	1,150	5.76	4.78
TANGERINES AND MANDARINS				
AZ	400	250	21.58	13.45
CA [5]	6,700	6,700	25.17	24.33
FL	5,500	3,850	10.72	11.20
US, tangerines and mandarins	12,600	10,800	17.91	18.69

[1] The crop year begins with the bloom of the first year shown and ends with completion of harvest the following year. [2] Equivalent packinghouse-door returns. [3] Includes small quantities of tangerines in TX and Temples in FL. [4] Net lbs. per box: oranges—AZ and CA, 75; FL, 90; and TX, 85; grapefruit—AZ and CA, 67; FL, 85; TX, 80; lemons—76; tangelos, tangerines and mandarins—AZ and CA, 75. [5] Includes tangelos and tangors.
NASS, Crops Branch, (202) 720–2127.

Table 5-30.—Citrus: Area and production in specified countries, 2007/2008–2009/2010

Country	Area			Production		
	2007/ 2008	2008/ 2009	2009/ 2010	2007/ 2008	2008/ 2009	2009/ 2010
	1,000 hec- tares	*1,000 hec- tares*	*1,000 hec- tares*	*1,000 metric tons*	*1,000 metric tons*	*1,000 metric tons*
Oranges, fresh:						
Argentina	45,500	45,500	45,500	940	900	750
Brazil	729,600	727,600	724,600	16,850	17,422	16,238
China				5,450	6,000	6,350
Egypt	105,000	140,000	140,000	2,759	3,500	3,570
EU-27	282,013	284,526	285,104	6,492	6,506	6,202
Mexico	338,336	335,000	336,000	4,297	4,140	3,900
Morocco	41,000	42,000	41,000	732	790	815
South Africa	35,932	40,384	38,900	1,410	1,526	1,600
Turkey	48,000	50,000	50,000	1,427	1,560	1,580
Vietnam	....	....	....	601	600	600
Others	18,372	22,140	21,650	1,108	1,205	1,233
Total Foreign	1,643,753	1,687,150	1,682,754	42,066	444,149	42,838
United States	....	....	....	9,141	8,344	7,444
Total	1,643,753	1,687,150	1,682,754	51,207	52,493	50,282
Tangerines/Mandarines:						
Argentina	32,000	32,000	32,000	410	400	300
China	....	....	....	11,000	12,650	13,300
EU-27	157,827	160,121	161,053	2,975	3,160	3,074
Israel	5,320	4,700	4,900	145	139	150
Japan	57,810	56,210	54,710	1,193	1,018	1,100
Korea, South	19,663	19,671	19,538	746	593	700
Morocco	35,000	36,000	36,500	471	532	585
Phillippines	....	....	....	26	26	26
South Africa	....	....	....	135	135	145
Turkey	27,400	27,450	27,450	740	756	750
Others	....	....	....	....	....	....
Total Foreign	335,020	336,152	336,151	17,841	19,409	20,130
United States	....	....	....	540	449	531
Total	335,020	336,152	336,151	18,381	19,858	20,661
Lemons:						
Argentina	41,500	42,000	42,000	1,360	1,400	1,000
EU-27	81,774	74,223	73,850	1,139	1,264	1,158
Israel	1,250	1,300	1,500	35	30	55
Japan	435	470	510	5	5	5
Mexico	148,292	145,000	146,000	2,229	2.000	2,040
Morocco	3,000	3,100	3,100	15	15	15
South Africa	4,180	4,230	4,300	200	144	160
Turkey	22,880	22,880	22,900	652	680	680
Others	....	....	....	....	....	....
Total Foreign	303,311	293,203	294,160	5,635	5,538	5,113
United States	....	....	....	562	862	776
Total	303,311	293,203	294,160	6,197	6,400	5,889

See end of table.

Table 5-30.—Citrus: Area and production in specified countries, 2007/2008–2009/2010—Continued

Country	Area			Production		
	2007/ 2008	2008/ 2009	2009/ 2010	2007/ 2008	2008/ 2009	2009/ 2010
	1,000 hec- tares	1,000 hec- tares	1,000 hec- tares	1,000 metric tons	1,000 metric tons	1,000 metric tons
Grapefruit:						
Argentina	11,000	7,000	7,000	240	237	200
China; Peoples Republic of	....	....	....	2,230	2,520	2,900
EU-27	1,839	1,812	1,804	91	85	82
Israel	4,100	4,200	4,250	242	232	245
Mexico	17,412	17,500	17,600	425	390	410
South Africa	7,900	8,436	8,200	388	341	390
Turkey	4,707	4,800	4,800	167	160	180
Others	....	....	....	....	....	....
Total Foreign	46,958	43,748	43,654	3,783	3,965	4,407
United States	....	....	....	1,404	1,207	1,080
World Total	46,958	43,748	43,654	5,187	5,172	5,487

FAS, Office of Global Analysis, (202) 720-6301. Prepared or estimated on the basis of official USDA production, supply, and distribution statistics from foreign governments.

Table 5-31.—Fresh fruits, citrus: United States exports by country of destination, 2007–2009

Country of destination	2007	2008	2009
	Metric tons	Metric tons	Metric tons
Canada	197,451	266,537	243,281
Japan	307,136	282,822	233,305
Korea, South	103,375	115,515	87,050
Hong, Kong	45,983	72,271	82,925
China	20,790	38,783	40,128
Australia(*)	24,419	29,578	24,875
France(*)	25,252	30,406	22,773
Nrtherland	23,254	34,456	22,071
Malaysia	10,525	23,578	19,803
Singapore	5,177	15,334	15,460
United Arab Emirates	150	17,269	12,142
Mexico	16,525	19,521	8,974
Taiwan	7,384	13,306	8,945
New Zealand	11,040	11,066	7,948
Phillippines	1,270	3,907	5,329
Germany(*)	6,251	4,796	4,696
Belgium(!)	8,200	7,740	4,433
United Kingdom	8,054	8,025	4,268
Russia	141	2,059	3,481
Switzerland	3,191	3,249	3,166
Rest of World	15,531.40	27,028	26,452
World Total	835,814	1,021,445	876,357

Note: (*) Denotes a country that is a summarization of its component countries. (!) Denotes a country which is summarized into its obsolete country.
FAS, Office of Global Analysis, (202) 720-6301.

Table 5-32.—Fresh citrus fruits: Foreign trade, United States, 1999–2008

Year [1]	Oranges		Grapefruit		Lemons		Limes		Tangerines	
	Imports	Domes-tic ex-ports	Imports	Domes-tic ex-ports	Imports	Domes-tic ex-ports	Imports	Domes-tic ex-ports	Imports	Domes-tic ex-ports
	Metric tons	Metric tons	Metric tons	Metric tons	Metric tons	Metric tons	Metric tons	Metric tons	Metric tons	Metric tons
1999	48,885	511,852	5,769	390,958	25,160	106,249	179,394	3,752	5,674	10,983
2000	52,785	570,162	19,409	389,629	34,127	110,373	182,412	3,846	4,117	11,786
2001	56,789	499,988	27,327	396,400	36,351	99,906	179,101	3,659	4,324	12,678
2002	55,590	638,079	17,781	350,953	27,901	99,566	251,973	2,236	4,545	14,406
2003	58,041	626,060	18,983	396,229	34,461	101,603	267,027	2,364	3,593	17,030
2004	69,986	572,601	13,983	226,397	35,400	97,982	306,122	3,358	4,780	13,133
2005	73,842	550,277	17,580	253,408	35,245	97,445	306,181	2,260	3,434	14,991
2006	112,108	340,660	21,531	331,538	65,656	115,862	337,356	3,413	6,148	12,700
2007	81,033	613,155	15,216	270,363	64,214	155,668	359,020	2,772	7,681	19,321
2008	90,523	496,834	11,697	246,882	37,564	91,435	358,304	2,361	8,049	17,456

[1] Year beginning October for all commodities.
ERS, Food and Specialty Crops Branch, (202) 694–5260.

Table 5-33.—Concentrated citrus juices: Annual packs, Florida, 1999–2008

Season beginning December	Frozen concentrated juice [1]		
	Orange [2]	Grapefruit [2]	Tangerine
	1,000 gallons	1,000 gallons	1,000 gallons
1999	207,708	28,642	1,646
2000	196,055	27,207	852
2001	215,057	27,294	1,758
2002	156,845	19,375	872
2003	218,296	20,897	1,263
2004	85,998	3,058	699
2005	84,600	9,717	525
2006	79,054	15,782	446
2007	135,196	13,687	650
2008	120,800	10,740	466

[1] Net pack.   [2] Frozen orange juice reported in 42.0° Brix; Grapefruit 40.0° Brix. Includes concentrated juice for manufacture.
ERS, Food and Specialty Crops Branch, (202) 694–5260.

Table 5-34.—Dates: Area, yield, total production, marketing year average price per ton, and value, California, 2000–2009

Year	Bearing acreage	Yield per acre	Production	Price per ton	Value
	Acres	Tons	Tons	Dollars	1,000 dollars
2000	5,000	3.48	17,400	1,230	21,402
2001	4,900	4.02	19,700	1,360	26,792
2002	4,800	5.04	24,200	1,550	37,510
2003	5,200	3.46	18,000	1,380	24,840
2004	5,500	3.13	17,200	1,310	22,532
2005	5,700	3.02	17,200	1,430	24,596
2006	5,500	3.22	17,700	2,140	37,878
2007	5,300	3.08	16,300	2,290	37,327
2008	5,700	3.67	20,900	1,260	26,334
2009 [1]	6,300	3.71	23,400	1,230	28,782

[1] Preliminary.
NASS, Crops Branch, (202) 720–2127.

Table 5-35.—Dates: Foreign trade, United States, 1999–2008

Year beginning October	Imports
	Metric tons
1999	5,006
2000	2,996
2001	4,347
2002	5,253
2003	5,536
2004	5,178
2005	6,409
2006	9,446
2007	5,370
2008	

ERS, Specialty Crops and Fiber Branch, (202) 694–5260.

Table 5-36.—Cranberries: Area, yield, production, season average price per barrel, value and quantities processed, United States, 2000–2009 [1]

Year	Area harvested	Yield per acre [2]	Total production [3]	Utilized production	Price [4]	Value	Quantities processed [5]
	Acres	Barrels [6]	Barrels [6]	Barrels [6]	Dollars	1,000 dollars	Barrels [6]
2000	37,200	153.5	5,712,000	5,579,000	18.10	100,851	5,137,000
2001	35,600	149.7	5,329,000	4,783,000	23.80	113,646	4,357,000
2002	39,400	144.4	5,689,000	5,682,000	32.20	182,783	5,312,000
2003	39,300	157.6	6,193,000	6,193,000	33.90	209,834	5,842,000
2004	39,000	158.3	6,175,000	6,167,000	32.90	202,670	5,770,000
2005	38,600	161.7	6,243,000	6,243,000	35.20	219,985	5,896,000
2006	38,500	179.0	6,890,000	6,785,000	41.10	278,888	6,429,500
2007	38,100	172.0	6,554,000	6,554,000	50.70	332,092	6,194,000
2008	38,200	205.9	7,865,000	7,865,000	58.00	455,927	7,494,000
2009	38,500	179.6	6,913,000	6,913,000	49.30	340,706	6,580,000

[1] Estimates relate to MA, NJ, OR, WA, and WI.   [2] Derived from total production.   [3] Differences between utilized and total production are quantities unharvested for economic reasons or excess cullage and/or set-aside production under provisions of the Cranberry Marketing Order.   [4] Average price of utilized production. Equivalent returns at first delivery point, screened basis of utilized production.   [5] Mainly for canning.   [6] Barrels of 100 pounds.
NASS, Crops Branch, (202) 720–2127.

Table 5-37.—Cranberries: Area, yield, production, and season average price per barrel, by State and United States, 2007–2009

State	Area harvested			Yield per acre			Total production			Price per barrel [1]		
	2007	2008	2009	2007	2008	2009	2007	2008	2009	2007	2008	2009
	Acres	Acres	Acres	Bbl.[2]	Bbl.[2]	Bbl.[2]	Bbl.[2]	Bbl.[2]	Bbl.[2]	Dollars	Dollars	Dollars
MA	13,000	13,000	13,000	117.1	182.6	139.8	1,522,000	2,374,000	1,817,000	49.80	58.30	46.80
NJ	3,100	3,100	3,100	171.3	165.2	179.0	531,000	512,000	555,000	46.10	53.00	54.00
OR	2,700	2,700	2,700	183.3	148.1	159.3	495,000	400,000	430,000	59.60	91.50	36.30
WA	1,700	1,700	1,700	103.5	64.1	94.7	176,000	109,000	161,000	49.60	56.90	58.50
WI	17,600	17,000	18,000	217.6	252.5	219.4	3,830,000	4,470,000	3,950,000	50.50	55.40	50.80
US	38,100	38,200	38,500	172.0	205.9	179.6	6,554,000	7,865,000	6,913,000	50.70	58.00	49.30

[1] Average price of utilized production. Equivalent returns at first delivery point, screened basis of utilized production. [2] Barrels of 100 pounds.
NASS, Crops Branch, (202) 720–2127.

Table 5-38.—Figs: Total production, marketing year average price per ton, and value, California, 2000–2009

Year	Dried (dry basis)				Total		
	Production			Price per ton	Production (fresh basis) [1]	Price per ton	Value
	Total	Standard	Substandard				
	Tons	1,000 tons	Tons	Dollars	Tons	Dollars	1,000 dollars
2000	17,300	15,400	1,900	672	55,900	272	15,226
2001	13,000	11,700	1,300	923	41,000	366	15,012
2002	16,900	15,000	1,900	902	53,200	340	18,087
2003	15,200	13,300	1,900	828	48,500	317	15,373
2004	15,600	13,700	1,900	897	51,100	396	20,214
2005	15,100	12,500	2,600	847	49,000	404	19,807
2006	13,000	11,100	1,900	829	42,800	426	18,253
2007	14,500	11,100	3,400	873	47,800	401	19,145
2008	13,100	11,800	1,300	1,200	43,300	599	25,954
2009	13,240	12,030	1,210	1,550	39,950	763	30,465

[1] Dried figs converted to fresh basis at ratio of 3 pounds fresh to 1 pound dried.
NASS, Crops Branch, (202) 720–2127.

Table 5-39.—Figs, dried: Foreign trade, United States, 1999–2008

Year beginning October	Imports for consumption	Domestic exports
	Metric tons	Metric tons
1999	3,900	2,763
2000	3,070	2,506
2001	6,788	2,399
2002	7,627	2,962
2003	4,477	3,607
2004	6,221	4,152
2005	5,273	5,355
2006	5,487	3,019
2007	2,978	3,775
2008	6,134	5,019

ERS, Food and Specialty Crops Branch, (202) 694–5260.

Table 5-40.—Ginger Root: Area, yield, production, marketing year average price, and value, Hawaii, 1999–2000 to 2007–08

Year	Area harvested	Yield per acre	Total production	Price per pound	Value
	Acres	1,000 pounds	1,000 pounds	Cents	1,000 dollars
1999–00	270	50.0	13,500	66.0	8,910
2000–01	360	50.0	18,000	45.0	8,100
2001–02	320	45.0	14,400	30.0	4,320
2002–03	160	37.5	6,000	60.0	3,600
2003–04	150	40.0	6,000	90.0	5,400
2004–05	120	42.5	5,100	80.0	4,080
2005–06	100	43.0	4,300	70.0	3,010
2006–07	80	35.0	2,800	85.0	2,380
2007–08	60	30.0	1,800	160.0	2,880

NASS, Crops Branch, (202) 720–2127.

Table 5-41.—Grapes: Production, price, and value, United States, 2000–2009

Year	Grapes			
	Production (fresh basis)		Market year average price per ton [1]	Value
	Total	Utilized		
	Tons	Tons	Dollars	1,000 dollars
2000	7,687,970	7,687,330	403.00	3,098,427
2001	6,569,250	6,568,100	449.00	2,947,867
2002	7,338,900	7,336,810	387.00	2,841,569
2003	6,643,530	6,489,630	402.00	2,609,289
2004	6,240,030	6,229,930	483.00	3,009,945
2005	7,813,700	7,810,500	447.00	3,494,095
2006	6,377,470	6,366,170	519.00	3,304,631
2007	7,057,250	7,056,250	489.00	3,453,124
2008	7,319,260	7,305,550	458.00	3,342,966
2009 [2]	7,294,840	7,267,300	507.00	3,688,065

[1] Fresh fruit prices are equivalent packinghouse-door returns for California and Washington, and the average price as sold for other States. Quantities processed are priced at the equivalent processing plant door level.   [2] Preliminary.
NASS, Crops Branch, (202) 720–2127.

Table 5-42.—Grapes: Production and marketing year average price per ton, by State and United States, 2007–2009

State	Total production			Utilized production			Price per ton [1]		
	2007	2008	2009 [2]	2007	2008	2009 [2]	2007	2008	2009 [2]
	Tons	Tons	Tons	Tons	Tons	Tons	Dollars	Dollars	Dollars
AZ [3]	900	800		900	800			775.00	
AR [4]	500	1,700	1,900	500	1,700	1,790		1,080.00	753.00
CA:									
All types	6,230,000	6,548,000	6,544,000	6,230,000	6,548,000	6,544,000	494.00	446.00	499.00
Wine	3,288,000	3,055,000	3,743,000	3,288,000	3,055,000	3,743,000	564.00	609.00	613.00
Table [5]	791,000	973,000	874,000	791,000	973,000	874,000	787.00	405.00	465.00
Raisin [5]	2,151,000	2,520,000	1,927,000	2,151,000	2,520,000	1,927,000	278.00	265.00	295.00
GA	2,900	3,500	4,500	2,900	3,500	4,500	1,200.00	1,130.00	1,480.00
MI	100,100	73,700	96,500	100,100	73,700	78,400	280.00	369.00	352.00
MO	2,500	5,200	4,400	2,500	5,200	4,400	932.00	962.00	902.00
NY	180,000	172,000	133,000	180,000	172,000	130,000	296.00	342.00	372.00
NC	3,650	5,600	4,800	3,650	5,600	4,800	1,270.00	1,170.00	1,250.00
OH	7,600	5,660	5,740	7,600	5,350	5,180	390.00	356.00	973.00
OR	38,600	34,700	40,200	38,600	34,700	40,200	1,880.00	2,050.00	1,910.00
PA	84,000	107,200	64,000	84,000	94,200	62,000	266.00	296.00	299.00
TX	4,900	4,200	6,200	4,100	4,000	3,050	1,160.00	1,200.00	1,170.00
VA	5,600	7,000	8,600	5,400	6,800	7,980	1,400.00	1,530.00	1,600.00
WA:									
All types	396,000	350,000	381,000	396,000	350,000	381,000	440.00	573.00	551.00
Wine	127,000	145,000	156,000	127,000	145,000	156,000	954.00	1,030.00	989.00
Juice	269,000	205,000	225,000	269,000	205,000	225,000	198.00	250.00	248.00
Other States							787.00		
US	7,057,250	7,319,260	7,294,840	7,056,250	7,305,550	7,267,300	489.00	458.00	507.00

[1] Fresh fruit prices are equivalent packinghouse-door returns for CA and WA, and the average price as sold for other States. Quantities processed are priced at the equivalent processing plant door level.   [2] Preliminary.   [3] Estimates discontinued in 2009.   [4] Missing estimates included in Other States to avoid disclosure of individual operations.   [5] Fresh equivalent of dried and not dried.
NASS, Crops Branch, (202) 720–2127.

Table 5-43.—Grapes: Production and utilization, United States, 2000–2009

Crop of—	Total production [1]	Utilized production	Utilization of quantities sold				
			Fresh	Processed			
				Canned	Dried (fresh basis)	Crushed for wine	Crushed for juice, etc.[2]
	Tons	Tons	Tons	Tons	Tons	Tons	Tons
2000	7,687,970	7,687,330	906,825	32,000	2,194,600	4,129,655	424,250
2001	6,569,250	6,568,100	864,330	29,000	1,736,800	3,568,190	369,780
2002	7,338,900	7,336,810	982,340	31,000	1,907,000	3,998,970	417,500
2003	6,643,530	6,489,630	805,460	27,000	1,597,000	3,581,420	478,750
2004	6,240,030	6,229,930	882,580	25,000	1,107,000	3,818,130	397,220
2005	7,813,700	7,810,500	995,610	(4)	1,645,000	4,550,780	619,110
2006	6,377,470	6,366,170	797,590	21,000	1,424,000	3,725,380	398,200
2007	7,057,250	7,056,250	920,330	21,000	1,621,000	3,920,520	573,400
2008	7,319,260	7,305,550	985,200	25,000	1,873,000	3,943,220	479,130
2009[3]	7,294,840	7,267,300	939,010	20,000	1,497,000	4,372,990	438,300

[1] Total production includes utilized production plus production not harvested and harvested not sold: 1999—1,530 tons fresh equivalent; 2000—640 tons fresh equivalent; 2001—1,150 tons fresh equivalent; 2002—2,090 tons fresh equivalent; 2003—153,900 tons fresh equivalent; 2004—10,100 tons fresh equivalent; 2005—3,200 tons; 2006—11,300 fresh equivalent; 2007—1,000 fresh equivalent; and 2008—13,710 tons fresh equivalent.   [2] Mostly juice, but includes some quantities used for jam, jelly, etc.   [3] Preliminary.   [4] Included with fresh in 2005 to avoid disclosure of individual operations.
NASS, Crops Branch, (202) 720–2127.

Table 5-44.—Grapes: Production and utilization, by State and United States, crop of 2009 [1]

State	Total production	Utilized production	Utilization				
			Fresh	Processed			
				Canned	Dried (fresh basis)[2]	Crushed for—	
						Wine	Juice, etc.[3]
	Tons	Tons	Tons	Tons	Tons	Tons	Tons
AZ[4]	....	....	....	....	....	....	....
AR	1,900	1,790	....	....	....	....	....
CA:							
All types	6,544,000	6,544,000	932,000	20,000	1,497,000	4,095,000	....
Wine	3,743,000	3,743,000	40,000	....	....	3,703,000	....
Table	874,000	874,000	755,000	....	34,000	85,000	....
Raisin	1,927,000	1,927,000	137,000	20,000	1,463,000	307,000	....
GA	4,500	4,500	1,300	....	....	4,300	....
MI	96,500	78,400	60	....	....	4,340	72,800
MO	4,400	4,400	2,000	....	....	44,000	....
NY	133,000	130,000	430	....	....	4,370	84,000
NC	4,800	4,800	20	....	....	2,060	....
OH	5,740	5,180	200	....	....	40,200	3,100
OR	40,200	40,200	50	....	....	8,400	....
PA	64,000	62,000	....	....	....	3,000	53,400
TX	6,200	3,050	....	....	....	7,980	....
VA	8,600	7,980	....	....	....	156,000	....
WA:.							
All types	381,000	381,000	....	....	....	....	225,000
Wine	156,000	156,000	....	....	....	....	....
Juice	225,000	225,000	....	....	....	....	....
Oth Sts[5]	....	....	2,950	....	....	3,340	....
US	7,294,840	7,267,300	939,010	20,000	1,497,000	4,372,990	438,300

[1] Preliminary.   [2] Equivalent raisins produced (dried basis): 347,700 tons.   [3] Mostly juice, but includes some quantities used for jam, jelly, etc.   [4] Estimates discontinued in 2009.   [5] Grapes processed for juice are included in other states wine to avoid disclosure of individual operations.
NASS, Crops Branch, (202) 720–2127.

Table 5-45.—Grapes and raisins: Foreign trade, United States, 1999–2008

Year beginning October	Grapes		Raisins [1]	
	Imports, fresh	Domestic exports, fresh	Imports for consumption	Domestic exports
	Metric tons	*Metric tons*	*Metric tons*	*Metric tons*
1999	452,182	272,901	18,283	90,539
2000	418,012	303,396	12,571	118,838
2001	501,055	293,754	16,421	125,319
2002	564,512	307,602	15,416	121,438
2003	532,746	321,079	11,955	134,329
2004	614,599	301,552	23,301	119,892
2005	580,870	354,731	24,106	118,134
2006	626,189	294,670	29,806	122,349
2007	567,079	299,891	22,648	175,112
2008	600,908	340,616	21,467	160,481

[1] Raisins converted to sweatbox or production basis by multiplying by 1.08.
ERS, Food and Specialty Crops Branch, (202) 694–5260.

Table 5-46.—Guavas: Area, yield, utilized production, marketing year average price, and value, Hawaii, 2000–2009

Year	Area harvested	Yield per acre	Production	Price per pound	Value
	Acres	*1,000 pounds*	*1,000 pounds*	*Cents*	*1,000 dollars*
2000	680	23.4	15,900	12.9	2,051
2001	610	25.1	15,300	14.1	2,157
2002	550	17.6	9,700	15.0	1,455
2003	530	12.6	6,700	13.8	925
2004	525	15.4	8,100	14.4	1,166
2005	620	13.1	8,100	13.9	1,126
2006	365	20.3	7,400	14.2	1,051
2007	170	25.3	4,300	15.7	675
2008	160	21.9	3,500	15.8	553
2009	135	15.6	2,100	14.0	294

NASS, Crops Branch, (202) 720–2127.

Table 5-47.—Nectarines: Production, utilization, and value, United States, 2000–2009 [1]

Crop of—	Production	Utilization		Marketing year average price per ton [3]	Value
		Fresh [2]	Processed (fresh basis)		
	Tons	*Tons*	*Tons*	*Dollars*	*1,000 dollars*
2000	267,000	260,700	6,300	398.00	106,256
2001	275,000	265,400	9,600	464.00	127,642
2002	300,000	300,000	([5])	382.00	114,600
2003	273,000	273,000	([5])	436.00	119,028
2004	269,000	252,000	([5])	342.00	86,184
2005	250,500	250,500	([5])	507.00	126,942
2006	231,900	231,900	([5])	522.00	121,004
2007	283,000	283,000	([5])	340.00	96,305
2008	302,500	302,500	([5])	367.00	110,915
2009 [4]	219,800	219,800	([5])	631.00	138,611

[1] Washington added in 2005, prior years are California only. [2] Includes "Home use." [3] Processing fruit prices are equivalent returns at processing plant door. [4] Preliminary. [5] Small quantities of processed nectarines are included in fresh to avoid disclosure of individual operations.
NASS, Crops Branch, (202) 720–2127.

Table 5-48.—Olives: Total production, marketing year average price, value, and processed utilization, California, 2000–2009

Year	Production	Marketing year average price per ton	Value	Processed utilization			
				Crushed for oil	Canned	Limited	Undersized
	Tons	*Dollars*	*1,000 dollars*	*Tons*	*Tons*	*Tons*	*Tons*
2000	53,000	656	34,743	3,000	41,400	5,100	3,000
2001	134,000	672	90,096	3,000	109,700	15,300	5,500
2002	103,000	573	58,983	6,000	82,800	9,900	3,800
2003	118,000	409	48,289	7,500	96,000	10,500	3,500
2004	107,500	564	60,643	11,500	74,400	16,100	5,000
2005	142,000	564	80,097	14,000	100,000	21,200	6,300
2006	23,500	771	18,119	4,000	17,000	1,500	500
2007	132,500	654	86,694	12,000	96,000	20,000	4,000
2008	66,800	697	46,587	14,000	45,500	6,000	1,300
2009[1]	46,300	696	32,209	20,000	24,500	1,500	300

[1] Preliminary.
NASS, Crops Branch, (202) 720–2127.

Table 5-49.—Olives and olive oil: Foreign trade, United States, 1999–2008

Year beginning October	Imports			
	Olives		Olive oil	
	In brine	Dried	Edible	Inedible
	Metric tons	*Metric tons*	*Metric tons*	*Metric tons*
1999	86,724	314	189,302	0
2000	98,384	415	212,341	0
2001	100,343	367	217,649	276
2002	106,852	464	219,883	97
2003	108,375	504	244,976	26
2004	112,054	723	248,176	3
2005	105,736	423	242,186	295
2006	118,375	1,043	260,398	1,607
2007	118,085	133	262,716	1,575
2008	109,237	289	275,678	594

ERS, Food and Specialty Crops Branch, (202) 694–5260.

Table 5-50.—Peaches: Production and value, United States, 2000–2009

Year	Total production	Utilized production	Marketing year average price [1]	Value
	1,000 tons	*1,000 tons*	*Dollars per ton*	*1,000 dollars*
2000	1,275.7	1,230.5	382	470,399
2001	1,203.9	1,155.0	418	483,043
2002	1,267.5	1,217.7	400	488,011
2003	1,259.6	1,205.3	377	454,406
2004	1,307.3	1,229.9	375	461,804
2005	1,184.5	1,145.0	447	511,464
2006	1,010.3	987.2	520	513,093
2007	1,127.2	1,115.9	450	502,087
2008	1,135.3	1,113.5	490	545,854
2009 [2]	1,103.8	1,082.6	548	593,653

[1] Fresh fruit prices are equivalent packinghouse-door returns for CA and WA except equivalent returns for bulk fruit at the first delivery point for CA Clingstone, and the average price as sold for other States. Quantities processed are priced at the equivalent processing plant door level.　[2] Preliminary.
NASS, Crops Branch, (202) 720–2127.

Table 5-51.—Peaches: Production and utilization, United States, 2000–2009

Crop of—	Total production [1]	Utilized production	Fresh [2]	Processed (fresh basis)			
				Canned	Dried	Frozen	Other [3]
	Tons	*Tons*	*Tons*	*Tons*	*Tons*	*Tons*	*Tons*
2000	1,275.7	1,230.5	566.7	513.3	12.6	109.8	28.1
2001	1,203.9	1,155.0	564.7	453.2	14.6	100.4	22.2
2002	1,267.5	1,217.7	537.3	530.5	14.2	102.2	33.6
2003	1,259.6	1,205.3	542.9	498.5	10.2	111.5	42.3
2004	1,307.3	1,229.9	535.7	523.9	10.4	105.8	54.1
2005	1,184.5	1,145.0	502.0	479.7	12.7	101.3	49.4
2006	1,010.3	987.2	481.6	374.1	13.1	96.2	22.2
2007	1,127.2	1,115.9	441.2	484.8	12.7	135.4	41.8
2008	1,135.3	1,113.5	529.8	426.3	9.5	111.4	36.5
2009 [4]	1,103.8	1,082.6	502.9	463.7	7.1	92.0	16.9

[1] Includes harvested not sold and unharvested production for California Clingstone peaches.　[2] Includes "Home use."　[3] Used for jams, preserves, pickles, wine, brandy, baby food, etc. Includes small quantities frozen for some years.　[4] Preliminary.
NASS, Crops Branch, (202) 720–2127.

Table 5-52.—Peaches: Foreign trade, United States, 2000–2008

Year beginning October	Domestic exports				
	Fresh	Canned	Canned, in fruit salad [1]	Dried, in fruit salad [2][3]	Total, in terms of fresh [4]
	Metric tons	*Metric tons*	*Metric tons*	*Metric tons*	*Metric tons*
2000	129,292	13,008	4,677	532	149,931
2001	127,434	10,922	3,885	344	144,152
2002	120,802	29,850	3,478	685	157,937
2003	112,506	42,418	4,438	827	163,955
2004	103,904	32,915	7,274	1,217	150,856
2005	86,230	27,038	6,162	1,078	125,420
2006	105,559	18,359	5,100	619	132,458
2007	112,352	37,266	10,019	690	163,446
2008	98,896	20,250	9,640	610	132,172

[1] Canned peaches are 40 percent of total canned fruit for salad.　[2] Net processed weight.　[3] Dried peaches are 21 percent of total dried fruit for salad.　[4] Dried fruit converted to unprocessed dry weight by dividing by 1.08. Unprocessed dry weight converted to terms of fresh fruit on the basis that 1 pound dried equals 6.0 pounds fresh. Canned peaches converted to terms of fresh on basis that 1 pound canned equals about 1 pound fresh.
ERS, Food and Specialty Crops Branch, (202) 694–5260.

Table 5-53.—Peaches: Production and season average price per pound, by State and United States, 2007–2009

State	Total production			Utilized production			Price per ton [2]		
	2007	2008	2009 [1]	2007	2008	2009 [1]	2007	2008	2009 [1]
	Tons	Tons	Tons	Tons	Tons	Tons	Dollars	Dollars	Dollars
AL	3,000	7,000	4,500	2,500	6,000	3,500	1,050.00	1,030.00	1,250.00
AR	15	4,400	1,500	10	4,200	1,120	1,300.00	1,110.00	1,400.00
CA:									
Freestone	446,000	433,000	350,000	446,000	433,000	350,000	402.00	339.00	480.00
CO	13,000	14,000	13,000	11,000	13,000	11,000	1,550.00	1,430.00	1,660.00
CT	1,100	1,200	1,300	1,100	1,200	1,200	1,800.00	2,000.00	1,800.00
GA	13,000	28,000	32,000	12,000	25,000	30,000	819.00	773.00	930.00
ID	7,000	8,000	9,200	6,860	7,420	8,300	1,150.00	681.00	877.00
IL	100	8,730	8,210	90	7,960	7,580	1,200.00	1,160.00	1,200.00
KY [3]	20	1,700	....	20	1,600	....	2,050.00	1,630.00	....
LA [3]	600	450	....	600	450	....	1,900.00	2,310.00	....
MD	3,300	3,480	3,800	3,200	3,470	3,800	1,170.00	1,150.00	1,120.00
MA	1,650	1,650	1,800	1,600	1,650	1,750	1,800.00	2,500.00	2,400.00
MI	20,500	14,000	17,200	19,100	13,700	16,700	853.00	661.00	723.00
MO	15	6,100	4,800	15	6,100	4,800	1,730.00	1,850.00	1,500.00
NJ	32,000	34,000	35,000	28,800	26,000	33,000	1,140.00	920.00	1,020.00
NY	6,300	5,500	6,500	6,300	5,200	6,400	634.00	922.00	845.00
NC	650	5,600	4,200	650	5,250	4,150	1,130.00	1,010.00	990.00
OH	4,100	6,600	2,560	4,000	6,500	2,410	1,510.00	1,370.00	1,640.00
OK [3]	900	1,000	....	800	800	....	1,590.00	1,600.00	....
OR [3]	3,000	1,600	....	2,700	1,500	....	970.00	1,000.00	....
PA	19,400	21,200	27,900	19,400	21,200	27,800	903.00	1,020.00	1,040.00
SC	12,500	60,000	75,000	12,000	57,000	66,000	1,130.00	874.00	977.00
TN [3][4]	....	1,600	....	....	1,600	....	....	1,580.00	....
TX	7,200	7,900	4,900	5,700	6,300	3,800	1,950.00	2,100.00	1,900.00
UT	4,500	5,000	5,800	4,400	4,500	5,500	667.00	868.00	1,040.00
VA	1,600	5,200	5,800	1,550	5,100	5,010	1,040.00	1,070.00	1,250.00
WA	18,500	16,800	14,500	18,500	16,800	14,500	480.00	498.00	360.00
WV	4,200	5,600	5,300	4,000	5,000	5,290	858.00	650.00	741.00
Total above	624,150	709,310	634,770	612,895	687,500	613,610	570.00	579.00	709.00
CA:									
Clingstone	503,000	426,000	469,000	503,000	426,000	469,000	304.00	347.00	338.00
US	1,127,150	1,135,310	1,103,770	1,115,895	1,113,500	1,082,610	450.00	490.00	548.00

[1] Preliminary.   [2] Fresh fruit prices are equivalent packinghouse-door returns for CA and WA except equivalent returns for bulk fruit at the first delivery point for CA Clingstone, and the average price as sold for other States. Quantities processed are priced at the equivalent processing plant door level.   [3] Estimates discontinued in 2009.   [4] No significant commercial production in 2007 due to freeze damage.
NASS, Crops Branch, (202) 720–2127.

Table 5-54.—Peaches: Production and utilization, by State and United States, crop of 2009 [1]

State	Total production	Utilized production [2]	Utilization				
			Fresh [3]	Processed (fresh basis)			
				Canned	Dried	Frozen	Other [4]
	Tons	Tons	Tons	Tons	Tons	Tons	Tons
CA, all	819,000	819,000	....	....	....	....	....
Clingstone	469,000	469,000	....	449,000	....	....	....
Freestone	350,000	350,000	261,000	....	7,100	70,300	....
GA	32,000	30,000	28,000	....	....	....	....
NJ	35,000	33,000	....	....	....	....	....
PA	27,900	27,800	....	....	....	....	....
SC	75,000	66,000	63,000	....	....	....	....
WA	14,500	14,500	....	....	....	....	....
Oth Sts	100,370	92,310	150,920	14,740	....	21,700	....
US	1,103,770	1,082,610	502,920	463,740	7,100	92,000	16,850

[1] Preliminary.   [2] Difference between total and utilized production is harvested not sold and unharvested production.   [3] Includes "Home use."   [4] Used for jams, preserves, brandy, etc.
NASS, Crops Branch, (202) 720–2127.

Table 5-55.—Fruit: Exports, 2007–2009

Country	2007	2008	2009
	Metric tons	Metric tons	Metric tons
Fruit, processed:			
Canada	69,014	96,573	109,193
Netherlands Antilles (exc. Aruba)	24,775	44,273	42,848
China	22,280	27,717	39,444
Mexico	18,686	23,868	22,563
Japan	15,333	18,254	15,080
Australia(*)	5,491	7,180	8,902
Taiwan	4,984	6,960	7,570
Korea, South	6,448	5,716	3,655
Singapore	3,386	4,680	3,324
Phillippines	4,233	3,737	2,938
Malaysia	1,421	1,580	2,597
Thailand	12,087	24,700	2,435
New Zealand(*)	998	1,410	2,395
United Kingdom	2,695	1,756	2,262
Indonesia	941	1,589	2,199
Hong Kong	2,636	2,167	1,961
Guatemala	1,391	3,538	1,950
Germany(*)	1,265	935	1,679
Israel(*)	3,254	2,359	1,581
Saudi Arabia	795	749	1,241
Turkey	0	187	1085
Panama	875	1,722	1,082
France(*)	269	224	900
South Africa	294	669	890
Chile	48	118	868
Poland	269	388	827
Rest of World	13,834	12,907	12,080
World total	217,704	295,955	293,549
Fruit, prepared, misc:			
Canada	58,045	65,422	62,584
Korea, South	11,559	19,037	15,335
China	1,704	6,797	14,470
Mexico	12,566	12,803	11,894
Taiwan	9,120	9,333	10,326
Japan	12,926	10,710	8,822
Saudi Arabia	4,472	5,762	6,272
Hong Kong	3,221	4,074	5,377
Germany(*)	3,632	3,627	4,215
United Kingdom	16,673	3,188	3,879
Thailand	3,365	2,204	2,233
United Arab Emirates	1,397	1,482	2,134
Panama	765	785	2,129
Malaysia	2,458	1,738	2,004
Philippines	2,090	1,404	1,752
Israel(*)	1,981	2,049	1,529
Singapore	1,818	2,377	1,473
Australia(*)	2,454	1,293	1,253
Indonesia	1,399	1,183	1,130
Netherlands	5,088	4,989	1,046
Kuwait	555	633	887
Bahamas, The	1,003	1,065	825
Belgium-Luxembourg(*)	2,224	2,897	821
New Zealand(*)	322	319	701
Costa Rica	1,156	1,310	665
Jamaica	1,051	885	561
Rest of World	15,004	* 16,422	9,351
World total	178,047	183,786	173,673

FAS, Office of Global Analysis, (202) 720-6301.

Table 5-56.—Pineapples: Total area, utilized production, utilization, marketing year average price, and value, Hawaii, 2000–2009

Year	Total area	Utilized production	Utilization		Price per ton	Value
			Fresh	Processed		
	Acres	Tons	Tons	Tons	Dollars	1,000 dollars
2000	20,700	354,000	122,000	232,000	287	101,530
2001	20,100	323,000	110,000	213,000	298	96,337
2002	19,100	320,000	117,000	203,000	314	100,616
2003	16,000	300,000	130,000	170,000	338	101,470
2004	13,000	220,000	104,000	116,000	378	83,104
2005	14,000	212,000	106,000	106,000	374	79,288
2006	12,600	185,000	96,000	89,000	398	73,652
2007	(1)	(1)	(1)	(1)	(1)	(1)
2008	(1)	(1)	(1)	(1)	(1)	(1)
2009	(2)	(2)	(2)	(2)	(2)	(2)

[1] Missing data not published to avoid disclosure of individual operations.   [2] Estimates discontinued in 2009.
NASS, Crops Branch, (202) 720–2127.

Table 5-57.—Pears: Production and value, United States 2000–2009

Year	Total production	Utilized production	Marketing year average price [1]	Value
	Tons	Tons	Dollars per ton	1,000 dollars
2000	993,250	975,270	267.00	260,626
2001	1,026,930	989,430	266.00	263,431
2002	890,020	888,570	297.00	264,334
2003	934,050	928,450	294.00	273,142
2004	878,260	873,400	335.00	292,969
2005	823,320	821,670	358.00	293,863
2006	842,035	831,120	397.00	329,928
2007	872,950	871,850	416.00	363,092
2008	869,850	868,880	456.00	396,081
2009 [2]	957,220	955,820	372.00	355,192

[1] Fresh fruit prices are equivalent packinghouse-door returns for CA, OR, and WA, and the average price as sold for other States. Quantities processed are priced at the equivalent processing plant door level. [2] Preliminary.
NASS, Crops Branch, (202) 720–2127.

Table 5-58.—Pears: Production and season average price per ton, by State and United States, 2007–2009

Variety and State	Total production			Utilized production			Price per ton [2]		
	2007	2008	2009 [1]	2007	2008	2009 [1]	2007	2008	2009 [1]
	Tons	Tons	Tons	Tons	Tons	Tons	Dollars	Dollars	Dollars
CA, all	243,000	243,000	255,000	243,000	243,000	255,000	340.00	436.00	366.00
Bartlett	201,000	195,000	200,000	201,000	195,000	200,000	289.00	373.00	351.00
Other	42,000	48,000	55,000	42,000	48,000	55,000	586.00	689.00	422.00
CO [3]	1,700	1,900	....	1,700	1,900	....	976.00	602.00	....
CT [3]	1,000	800	....	1,000	800	....	1,300.00	1,340.00	....
MI	4,000	2,850	4,200	3,600	2,800	4,200	450.00	414.00	343.00
NY	11,000	10,300	11,200	10,300	9,400	9,900	497.00	504.00	490.00
OR, all	206,000	231,300	229,000	206,000	231,000	229,000	434.00	473.00	409.00
Bartlett	59,000	56,300	66,000	59,000	56,300	66,000	375.00	395.00	439.00
Other	147,000	175,000	163,000	147,000	175,000	163,000	458.00	498.00	397.00
PA	4,000	2,400	5,820	4,000	2,400	5,720	717.00	744.00	711.00
UT [3]	250	300	....	250	280	....	760.00	729.00	....
WA, all	402,000	377,000	452,000	402,000	377,000	452,000	443.00	453.00	349.00
Bartlett	163,000	166,000	186,000	163,000	166,000	186,000	374.00	368.00	314.00
Other	239,000	211,000	266,000	239,000	211,000	266,000	491.00	520.00	374.00
US	872,950	869,850	957,220	871,850	868,880	955,820	416.00	456.00	372.00

[1] Preliminary. [2] Fresh fruit prices are equivalent packinghouse-door returns for CA, OR, and WA, and the average price as sold for other States. Quantities processed are priced at the equivalent processing plant door level. [3] Estimates discontinued in 2009.
NASS, Crops Branch, (202) 720–2127.

Table 5-59.—Pears: Foreign trade, United States, 1999–2008

Year beginning October	Imports for consumption, fresh	Domestic exports				
		Fresh [1]	Canned	Dried, in fruit salad [1][2]	Canned, in fruit salad [3]	Total, in terms of fresh fruit [4]
	Metric tons	Metric tons	Metric tons	Metric tons	Metric tons	Metric tons
1999	89,827	162,629	4,655	238	6,885	175,669
2000	85,219	158,333	5,887	422	4,092	170,976
2001	79,967	175,346	6,181	273	3,400	186,649
2002	86,328	160,240	4,944	544	3,043	171,659
2003	66,923	167,084	4,952	656	3,883	180,061
2004	76,834	142,157	10,174	966	6,365	164,793
2005	85,498	143,024	12,954	856	5,392	166,769
2006	108,587	132,730	6,669	492	4,462	146,964
2007	85,855	165,436	8,028	548	8,767	185,688
2008	82,136	146,246	7,378	484	8,435	165,113

[1] Net processed weight. [2] Dried pears are 16⅔ percent of total dried fruit for salad. [3] Canned pears are 35 percent of total canned fruit for salad. [4] Dried converted to unprocessed dry weight by dividing by 1.03. Unprocessed dry weight converted to terms of fresh on the basis that 1 pound dried equals about 6.5 pounds fresh. Canned converted to terms of fresh on basis that 1 pound of canned equals about 1 pound fresh.
ERS, Food and Specialty Crops Branch, (202) 694–5260.

Table 5-60.—Fruits: Area and production in specified countries, 2007/2008–2009/2010

Country	Area			Production		
	2007/2008	2008/2009	2009/2010	2007/2008	2008/2009	2009/2010
	Hec-tares	Hec-tares	Hec-tares	Metric tons	Metric tons	Metric tons
Pears, fresh:						
Argentina	23,000	24,000	24,000	720,000	780,000	700,000
Australia	1,000	1,000	1,000	127,557	125,000	105,000
Chile	5,911	6,000	6,000	232,000	227,000	222,000
China	....	....	....	12,895,000	13,538,142	13,800,000
EU-27	130,631	130,763	130,800	2,784,137	2,435,057	2,724,500
Japan	16,340	16,000	15,880	326,400	361,700	351,500
Mexico	4,480	3,749	3,750	24,739	21,104	21,110
Russian	30,000	30,000	30,000	190,000	180,000	180,000
South Africa	10,230	10,265	10,280	342,143	347,636	348,000
Turkey	205,000	210,000	210,000	355,281	360,000	376,000
Others	2,982	2,895	2,591	52,756	47,834	43,500
Total Foreign	429,573	434,672	435,301	18,049,732	18,423,473	18,871,610
United States	....	....	....	791,000	789,2000	848,100
Total	429,574	434,672	435,301	18,840,732	19,212,673	19,719,710
Apricots, fresh:						
Austria	....	....	....	23,500	24,000	0
Chile	1,845	1,855	....	26,500	26,600	0
China	....	....	....	1,515,000	1,725,000	0
France	....	....	....	170,000	170,500	0
Germany	....	....	....	550	500	0
Greece	....	....	....	74,400	74,400	0
Italy	....	....	....	220,000	218,000	0
Poland	....	....	....	1,100	2,400	0
Russia	....	....	....	37,000	37,000	0
South Africa	....	....	....	100,000	100,000	0
Spain	19,810	19,810	....	86,700	90,000	0
Turkey	....	....	....	538,000	540,000	0
Others	....	2,620	....	2,900	....	0
Total Foreign	21,955	21,665	....	2,792,750	3,008,400	0
United States	....	....	....	73,500	78,930	0
Total	21,955	21,665	....	2,866,250	3,087,330	0
Fresh Cherries, sweet and sour:						
Australia	999	1,050	1,050	12,500	11,250	13,750
Canada	....	....	....	9,000	8,000	8,500
Chile	9,156	9,520	9,798	61,000	59,000	61,500
China	....	....	....	150,000	174,000	195,000
EU-27	142,058	136,464	136,272	486,051	614,344	634,200
Iran	....	....	....	225,000	225,000	225,000
Japan	4,490	4,490	4,493	17,000	16,700	18,000
Russia	....	....	....	100,000	100,000	100,000
Serbia	....	....	....	28,546	29,000	29,000
Turkey	....	....	....	432,000	440,000	500,000
Others	....	....	....	18,306	18,400	18,400
Total Foreign	156,703	151,524	151,610	1,539,403	1,695,694	1,803,350
United States	....	....	....	396,644	321,315	468,299
Total	156,703	151,524	151,610	1,936,047	2,017,009	2,271,649

See end of table.

Table 5-60.—Fruits: Area and production in specified countries, 2007/2008–2009/2010—Continued

Country	Area			Production		
	2007/2008	2008/2009	2009/2010	2007/2008	2008/2009	2009/2010
	Hec-tares	Hec-tares	Hec-tares	Metric tons	Metric tons	Metric tons
Fresh Peaches & nectarines:						
Australia				110,000	105,000	105,000
Canada				32,050	32,000	32,000
Chile	10,724	10,820	9,606	175,000	177,000	156,000
China				9,015,000	9,615,000	9,815,000
EU-27	231,375	230,951	230,888	4,049,534	4,083,541	4,192,111
Japan	10,100	10,000	10,000	157,300	165,000	160,000
Russia				42,800	43,000	43,000
Taiwan	2,630	2,686	2,533	28,435	29,329	30,373
Turkey				553,000	540,000	552,000
Others		77,380				
Total Foreign	254,829	331,837	253,027	14,163,119	14,790,370	15,085,484
United States				1,269,061	1,302,563	1,194,354
Total	254,829	331,837	253,027	15,432,180	16,092,933	16,279,838
Fresh Plums & prunes:						
Argentina				119,000	102,340	
Chile	6,083	6,100		135,500	135,500	
China				2,006,000	2,207,500	
France				240,000	247,200	
Italy				190,0000	184,300	
Russia				98,000	97,500	
South Africa				62,300	66,115	
Spain	21,000	21,000		191,100	200,000	
Taiwan	9,822	9,645		63,664	59,281	
Turkey				235,400	220,000	
Others	10,000	1,200		214,540	216,640	
Total Foreign	46,905	37,945		3,555,504	3,736,376	
United States				352,895	381,120	
Total	46,905	37,945		3,908,399	4,117,496	

FAS, Office of Global Analysis, (202) 720-6301. Prepared or estimated on the basis of official USDA production, supply, and distribution statistics from foreign governments. *Note: There are no data available for fresh plums & prunes 2009/2010 at this time.

Table 5-61.—Pears: Production and utilization, by State and United States, crop of 2009 [1]

State and variety	Total production	Utilized production	Utilization	
			Fresh [2]	Processed [3]
	Tons	Tons	Tons	Tons
CA, all	255,000	255,000	(5)	(5)
Bartlett	200,000	200,000	70,000	130,000
Other	55,000	55,000	(5)	(5)
CO [4]	....	....	(5)	(5)
CT [4]	....	....	(5)	(5)
MI	....	....	(5)	(5)
NY	....	....	(5)	(5)
OR, all	229,000	229,000	(5)	(5)
Bartlett	66,000	66,000	40,000	26,000
Other	163,000	163,000	(5)	(5)
PA	....	....	(5)	(5)
UT [4]	....	....	(5)	(5)
WA, all	452,000	452,000	(5)	(5)
Bartlett	186,000	186,000	67,000	119,000
Other	266,000	266,000	(5)	(5)
US	957,220	955,820	603,800	352,000

[1] Preliminary.　[2] Includes "Home use."　[3] Mostly canned, but includes small quantities dried, juiced, and other uses.　[4] Estimates discontinued in 2009.　[5] Data not published to avoid disclosure of individual operations, but included in U.S. totals.
NASS, Crops Branch, (202) 720–2127.

Table 5-62.—Pears: Production and utilization, United States, 2000–2009

Crop of—	Total production	Utilized production	Utilization of quantities sold—Fresh [1]
	Tons	Tons	Tons
2000	993,250	975,270	573,230
2001	1,026,930	989,430	568,320
2002	890,020	888,570	524,440
2003	934,050	928,450	559,950
2004	878,260	873,400	514,270
2005	823,320	821,670	504,400
2006	842,035	831,120	500,720
2007	872,950	871,850	551,960
2008	869,850	868,880	548,930
2009 [2]	957,220	955,820	603,800

[1] Includes "Home use."　[2] Preliminary.
NASS, Crops Branch, (202) 720–2127.

Table 5-63.—Papayas: Area, utilized production, utilization, marketing year average price, and value, Hawaii, 2000–2009

Year	Area harvested	Utilized production	Utilization		Price per pound	Value
			Fresh	Processed		
	Acres	1,000 pounds	1,000 pounds	1,000 pounds	Cents	dollars
2000	1,650	54,500	50,250	4,250	29.4	16,007
2001	1,950	55,000	52,000	3,000	26.5	14,598
2002	1,720	45,900	42,700	3,200	26.0	11,924
2003	1,565	42,600	40,800	1,800	30.7	13,069
2004	1,265	35,800	34,100	1,700	34.5	12,361
2005	1,480	32,900	30,700	2,200	34.2	11,241
2006	1,530	28,700	26,600	2,100	38.5	11,049
2007	1,310	33,400	31,200	2,200	39.2	13,094
2008	1,380	33,500	31,500	2,000	43.0	14,393
2009 [1]	1,325	31,500	30,300	1,200	45.0	14,186

[1] Preliminary.
NASS, Crops Branch, (202) 720–2127.

Table 5-64.—Plums, California: Production, value, and utilization, 2000–2009

Season	Total production	Utilized production	Marketing year average price per ton [1]	Value
	Tons	Tons	Dollars	1,000 dollars
2000	197,000	197,000	442.00	87,115
2001	210,000	210,000	306.00	64,362
2002	201,000	201,000	386.00	77,586
2003	209,000	209,000	418.00	87,362
2004	156,000	144,000	516.00	74,347
2005	171,000	171,000	541.00	92,463
2006	158,000	158,000	688.00	108,648
2007	152,000	152,000	665.00	101,077
2008	160,000	160,000	356.00	56,960
2009 [2]	112,000	112,000	514.00	57,568

[1] Fresh fruit prices are equivalent returns at point of first sale. Processing fruit prices are equivalent returns at processing plant door.   [2] Preliminary.
NASS, Crops Branch, (202) 720–2127.

Table 5-65.—Prunes (dried basis): Production, price and value, California, 2000–2009 [1]

Season	Total production	Utilized production	Marketing year average price per ton [2]	Value
	Tons	Tons	Dollars	1,000 dollars
2000	219,000	201,000	770.00	154,770
2001	150,000	135,000	726.00	98,010
2002	172,000	163,000	810.00	132,030
2003	181,000	168,000	772.00	129,696
2004	49,000	48,000	1,500.00	72,000
2005	97,000	94,000	1,470.00	138,180
2006	198,000	189,000	1,390.00	262,710
2007	83,000	81,000	1,450.00	117,450
2008	129,000	129,000	1,500.00	193,500
2009 [3]	166,000	166,000	1,200.00	199,200

[1] The drying ratio is approximately 3 pounds of fresh fruit to 1 pound of dried fruit.   [2] Equivalent returns at the processing plant door.   [3] Preliminary.
NASS, Crops Branch, (202) 720–2127.

Table 5-66.—Prunes and plums: Production, value, and utilization, 4-States, 2000–2009 [1]

Year	Total production	Utilized production	Marketing year average price per ton	Value	Utilization of quantities sold			
					Fresh [2]	Processed (fresh basis)		
						Dried and other	Canned	Frozen
	Tons	Tons	Dollars	1,000 dollars	Tons	Tons	Tons	Tons
2000	23,900	21,950	239.00	5,247	9,400	5,650	5,400	1,500
2001	21,200	20,000	273.00	5,459	11,000	3,250	4,470	1,280
2002	15,650	14,790	286.00	4,237	6,360	3,930	3,340	1,160
2003	16,300	14,880	353.00	5,260	7,700	2,780	3,100	1,300
2004	25,000	18,920	360.00	6,802	10,350	4,390	3,140	1,040
2005	9,100	9,050	562.00	5,085	5,500	680	2,450	420
2006	21,500	19,200	452.00	8,678	9,550	4,300	3,250	2,100
2007	12,100	10,920	454.00	4,956	6,420	1,300	2,550	650
2008	15,500	15,480	382.00	5,918	8,700	3,540	2,130	1,110
2009 [3]	18,600	17,700	345.00	6,105	10,150	5,040	1,730	780

[1] ID, MI, OR, and WA. Mostly prunes; however, estimates include small quantities of plums in all States.   [2] Includes "Home use."   [3] Preliminary.
NASS, Crops Branch, (202) 720–2127.

Table 5-67.—Prunes and plums (fresh basis): Production and season average price per ton, by State, 2007–2009

State	Total production			Utilized production			Price per ton[2]		
	2007	2008	2009[1]	2007	2008	2009[1]	2007	2008	2009[1]
	Tons	Tons	Tons	Tons	Tons	Tons	Dollars	Dollars	Dollars
ID	1,800	2,200	2,000	1,720	2,180	2,000	709.00	585.00	496.00
MI	3,100	2,300	2,900	2,000	2,300	2,000	440.00	357.00	689.00
OR	3,000	7,500	9,400	3,000	7,500	9,400	421.00	278.00	218.00
WA	4,200	3,500	4,300	4,200	3,500	4,300	380.00	497.00	393.00
Total, 4 States	12,100	15,500	18,600	10,920	15,480	17,700	454.00	382.00	345.00

[1] Preliminary. [2] Fresh fruit prices are equivalent packinghouse-door returns for OR and WA, and the average price as sold for other States. Quantities processed are priced at the equivalent processing plant door level.
NASS, Crops Branch, (202) 720–2127.

Table 5-68.—Prunes and plums: Utilization and marketing year average price per ton, by State, 2003–2009

State and season	Quantity				Price [3]			
	Fresh[1]	Dried and other[2]	Canned	Frozen	Fresh	Dried and other	Canned	Frozen
	Tons	Tons	Tons	Tons	Dollars	Dollars	Dollars	Dollars
MI:								
2003	1,100	(4)	(4)	(4)	480.00	(4)	(4)	(4)
2004	350	(4)	(4)	(4)	769.00	(4)	(4)	(4)
2005	450	(4)	(4)	(4)	760.00	(4)	(4)	(4)
2006	1,800	(4)	(4)	(4)	730.00	(4)	(4)	(4)
2007	900	(4)	(4)	(4)	765.00	(4)	(4)	(4)
2008	700	(4)	(4)	(4)	775.00	(4)	(4)	(4)
2009	1,400	(4)	(4)	(4)	890.00	(4)	(4)	(4)
OR:								
2003	1,500	(4)	(4)	(4)	385.00	(4)	(4)	(4)
2004	3,000	(4)	(4)	(4)	537.00	(4)	(4)	(4)
2005	1,200	(4)	(4)	(4)	445.00	(4)	(4)	(4)
2006	2,500	(4)	(4)	(4)	496.00	(4)	(4)	(4)
2007	1,800	(4)	(4)	(4)	501.00	(4)	(4)	(4)
2008	3,500	(4)	(4)	(4)	407.00	(4)	(4)	(4)
2009	4,200	(4)	(4)	(4)	322.00	(4)	(4)	(4)
Total 4 States:[5]								
2003	7,700	2,780	3,100	1,300	446.00	214.00	255.00	339.00
2004	10,350	4,390	3,140	1,040	468.00	237.00	196.00	289.00
2005	5,500	680	2,450	420	759.00	244.00	262.00	250.00
2006	9,550	4,300	3,250	2,100	639.00	285.00	277.00	213.00
2007	6,420	1,300	2,550	650	622.00	213.00	217.00	200.00
2008	8,700	3,540	2,130	1,110	547.00	153.00	182.00	207.00
2009	10,150	5,040	1,730	780	489.00	129.00	191.00	204.00

[1] Includes "Home use." [2] Some quantities otherwise processed are included to avoid disclosure of individual operations. [3] Prices for fresh sales are average prices as sold for ID and MI; equivalent packinghouse door returns for OR and WA. Quantities processed are priced at the equivalent processing plant door level. [4] Not published to avoid disclosure of individual operations, but is included in total. [5] Includes ID, MI, OR, and WA.
NASS, Crops Branch, (202) 720–2127.

Table 5-69.—Prunes: Foreign trade, United States, 1999–2008

Year beginning October	Imports				Domestic exports			
	Fresh prunes and plums	Other-wise pre-pared or pre-served	Dried prunes [1]	Total, in terms of fresh [2]	Fresh prunes and plums	Dried prunes [1]	Dried, in fruit salad [1][3]	Total, in terms of fresh [2]
	Metric tons	Metric tons	Metric tons	Metric tons	Metric tons	Metric tons	Metric tons	Metric tons
1999	22,893	778	510	24,915	61,354	66,304	613	235,080
2000	33,400	792	431	35,231	62,926	83,746	1,089	283,170
2001	32,459	811	969	35,704	62,802	69,660	704	245,478
2002	32,336	921	570	34,643	60,028	66,624	1,403	236,637
2003	35,959	1,039	677	38,650	45,105	73,976	1,693	241,554
2004	40,061	1,089	9,871	66,668	48,539	45,835	2,493	174,005
2005	30,606	1,197	6,875	49,531	48,004	48,903	2,208	180,695
2006	36,434	865	839	39,392	46,633	66,253	1,268	221,930
2007	29,104	678	828	31,863	50,551	61,345	1,414	213,483
2008	29,409	829	3,906	40,293	46,094	54,729	1,248	191,420

[1] Net processed weight.   [2] Exports and imports of dried prunes converted to unprocessed dry weight by dividing by 1.04. Unprocessed dry weight converted to terms of fresh fruit on the basis that 1 pound dried equals 2.7 pounds fresh. "Other-wise prepared or preserved" converted to terms of fresh fruit on the basis that 1 pound equals 0.899 pound fresh.   [3] Dried prunes in salad estimated at 43 percent of total dried fruit for salad.
ERS, Food and Specialty Crops Branch, (202) 694–5260.

Table 5-70.—Strawberries, commercial crop: Production and value per hundredweight, by State and United States, 2007–2009

Utilization, season, and State	Production			Value per unit		
	2007	2008	2009 [1]	2007	2008	2009 [1]
	1,000 cwt.	1,000 cwt.	1,000 cwt.	Dollars per cwt.	Dollars per cwt.	Dollars per cwt.
FOR FRESH MARKET:						
CA	17,159	18,605	20,040	75.70	77.30	79.00
FL	2,112	1,794	2,376	124.00	139.00	132.00
MI	41	47	43	120.00	122.00	150.00
NY	46	45	44	165.00	165.00	205.00
NC	189	208	195	95.00	100.00	105.00
OH	38	42	30	154.00	174.00	191.00
OR	27	26	29	134.00	172.00	140.00
PA	56	73	65	187.00	211.00	208.00
WA	17	22	11	146.00	172.00	158.00
WI	48	49	47	136.00	145.00	155.00
US	19,733	20,911	22,880	82.10	84.10	85.90
PROCESSING:						
CA	4,381	4,070	4,816	25.50	34.40	29.50
MI	2	2	3	54.00	56.00	55.00
OR	220	212	182	58.00	58.00	54.00
WA	117	122	132	53.70	51.70	50.00
US	4,720	4,406	5,133	27.70	36.00	30.90

[1] Preliminary.
NASS, Crops Branch, (202) 720–2127.

Table 5-71.—Strawberries, commercial crop: Area, yield, production, value per hundred weight, and total value, United States, 2000–2009

Year	Area for harvest	Yield per acre	Production	Value[2] Per cwt	Value[2] Total	Produc-tion	Value[2] Per cwt	Value[2] Total	Produc-tion	Value[2] Per cwt	Value[2] Total
	Acres	*Cwt.*	*1,000 cwt.*	*Dollars per cwt.*	*1,000 dollars*	*1,000 cwt.*	*Dollars per cwt.*	*1,000 dollars*	*1,000 cwt.*	*Dollars per cwt.*	*1,000 dollars*
2000	47,350	401	19,008	55.00	1,044,594	14,333	64.90	930,125	4,675	24.50	114,469
2001	45,700	361	16,509	64.70	1,068,582	12,597	75.80	954,413	3,912	29.20	114,169
2002	47,600	396	18,845	61.60	1,161,630	14,063	71.30	1,003,145	4,782	33.10	158,485
2003	48,400	445	21,560	63.80	1,375,142	16,424	74.90	1,230,583	5,136	28.10	144,559
2004	51,500	430	22,138	58.50	1,295,464	16,944	68.40	1,159,082	5,194	26.30	136,382
2005	52,460	443	23,227	60.10	1,396,385	18,110	68.90	1,248,407	5,117	28.90	147,978
2006	53,460	450	24,038	63.20	1,519,494	19,109	72.20	1,379,658	4,929	28.40	139,836
2007	52,180	469	24,453	71.60	1,751,108	19,733	82.10	1,620,241	4,720	27.70	130,867
2008	54,470	465	25,317	75.80	1,918,288	20,911	84.10	1,759,564	4,406	36.00	158,724
2009[3]	58,080	482	28,013	75.80	2,123,735	22,880	85.90	1,965,070	5,133	30.90	158,665

[1] Fresh market price and value at point of first sale. Processing price and value at processing plant door. [2] Mostly for fresh market, but includes some quantities used for processing in States for which processing estimates are not prepared. [3] Preliminary.
NASS, Crops Branch, (202) 720–2127.

Table 5-72.—Strawberries, commercial crop: Area harvested, production, value per hundred weight, by State and United States, 2007–2009 [1]

Season and State	Area harvested 2007	2008	2009[2]	Production 2007	2008	2009[2]	Value per unit 2007	2008	2009[2]
	Acres	*Acres*	*Acres*	*1,000 cwt.*	*1,000 cwt.*	*1,000 cwt.*	*Dollars per cwt.*	*Dollars per cwt.*	*Dollars per cwt.*
CA	35,500	37,600	39,800	21,540	22,675	24,856	65.50	69.60	69.40
FL	6,600	6,900	8,800	2,112	1,794	2,376	124.00	139.00	132.00
MI	850	800	800	43	49	46	117.00	119.00	144.00
NY	1,400	1,400	1,400	46	45	44	165.00	165.00	205.00
NC	1,500	1,600	1,500	189	208	195	95.00	100.00	105.00
OH	780	770	710	38	42	30	154.00	174.00	191.00
OR	1,900	1,800	1,700	247	238	211	66.30	70.50	65.80
PA	1,200	1,200	1,100	56	73	65	187.00	211.00	208.00
WA	1,600	1,600	1,500	134	144	143	65.40	70.10	58.30
WI	850	800	770	48	49	47	136.00	145.00	155.00
US	52,180	54,470	58,080	24,453	25,317	28,013	71.60	75.80	75.80

[1] Includes quantities used for fresh market and processing. [2] Preliminary.
NASS, Crops Branch, (202) 720–2127.

Table 5-73.—Fruits, noncitrus: Production, utilization, and value, United States, 2000–2009 [1]

Year	Utilized produc- tion	Fresh [2]	Processed						Value of uti- lized produc- tion
			Canned	Dried	Juice	Frozen	Wine	Other	
	1,000 tons	*1,000 tons*	*1,000 tons*	*1,000 tons*	*1,000 tons*	*1,000 tons*	*1,000 tons*	*1,000 tons*	*1,000 dollars*
2000	18,854	7,015	1,812	3,023	1,712	691	4,130	191	7,883,036
2001	16,740	6,488	1,859	2,290	1,462	665	3,568	169	7,918,636
2002	17,122	6,549	1,727	2,582	1,251	591	3,999	138	8,137,640
2003	16,848	6,672	1,762	2,293	1,295	716	3,582	219	8,434,610
2004	16,823	7,168	1,710	1,425	1,418	685	3,819	290	8,553,060
2005	18,272	7,188	1,575	2,101	1,555	712	4,551	277	9,805,757
2006	16,816	6,930	1,400	2,219	1,256	710	3,726	235	10,510,417
2007	17,047	7,013	1,453	2,030	1,277	748	3,921	278	11,436,449
2008	17,602	7,247	1,406	2,413	1,228	682	3,944	289	11,279,829
2009 [3]	18,129	7,555	1,426	2,154	1,254	754	4,373	267	11,795,005

[1] Includes the following crops: Apples, apricots, avocados, bananas, berries, cherries, cranberries, dates, figs, grapes, guavas, kiwifruit, nectarines, olives, papayas, peaches, pears, pineapples, plums, prunes, and strawberries. [2] Includes "Home Use," local and roadside sales. [3] Preliminary.
NASS, Crops Branch, (202) 720–2127.

Table 5-74.—Fruits, fresh: Total reported domestic rail, truck, and air shipments, 2008

Commodity	Jan.	Feb.	Mar.	Apr.	May	Jun.	Jul.	Aug.	Sep.	Oct.	Nov.	Dec.	Total
	1,000 cwt.	*1,000 cwt.*	*1,000 cwt.*	*1,000 cwt.*	*1,000 cwt.*	*1,000 cwt.*	*1,000 cwt.*	*1,000 cwt.*	*1,000 cwt.*	*1,000 cwt.*	*1,000 cwt.*	*1,000 cwt.*	*1,000 cwt.*
Citrus:													
Grapefruit	1,847	1,915	1,710	994	250	54	5	9	37	904	1,303	1,459	10,487
Lemons	32	29	24	32	36	46	48	30	24	31	43	52	427
Oranges	860	757	756	733	526	329	137	126	158	526	914	1,190	7,012
Tangelos	98	28	5	....	....	....	....	....	....	....	41	131	303
Temples	38	58	14	....	....	....	....	....	....	....			110
Total	2,875	2,787	2,509	1,759	812	429	190	165	219	1,461	2,301	2,832	18,339
Noncitrus:													
Apples	4,560	4,733	5,999	4,124	4,950	2,248	1,760	2,426	3,318	4,716	6,023	4,392	49,249
Apricots	....	....	....	....	114	91	94	21	2	....	....	....	322
Avocados	72	99	272	374	540	514	632	684	385	251	68	52	3,943
Blueberries	....	....	4	73	151	340	503	195	56	12	....	....	1,334
Cherries	....	....	....	....	816	1,112	1,187	227	....	....	....	....	3,342
Grapes	13	....	....	....	427	702	1,986	2,915	2,2854	2,497	2,059	1,459	14,912
Nectarines	....	....	....	....	309	922	1,127	1,062	660	114	....	....	4,194
Papaya	13	12	13	10	13	10	8	13	11	14	16	10	143
Peaches	....	....	....	2	559	1,117	1,545	1,550	939	261	7	....	5,980
Pears	1,035	790	858	566	615	136	345	605	916	1,310	1,474	1,052	9,702
Persimmons	....	....	....	....	....	....	....	....	....	6	15	1	22
Plums	....	....	....	....	104	584	912	840	578	210	55	13	3,296
Pomegranates	8	....	....	....	....	....	....	10	38	63	37	6	162
Prunes	....	....	....	....	....	....	....	22	23	5	....	....	50
Strawberries	318	760	1,430	2,167	2,053	1,760	1,690	1,244	1,071	941	321	253	13,978
Total	6,019	6,394	8,576	7,316	10,651	9,536	11,789	11,814	10,851	10,400	10,075	7,238	110,629
Grand total	8,894	9,181	11,085	9,075	11,463	9,965	11,979	11,979	11,070	11,861	12,376	10,070	128,968

AMS, Fruit and Vegetable Division, Market News Branch, (202) 720–3343.

Table 5-75.—Fruits, dried: Production (dry basis), California, 2000–2009

Year	Apricots	Figs [1]	Peaches [2]	Pears [3]	Prunes	Grapes [4]	Total
	Tons	Tons	Tons	Tons	Tons	Tons	Tons
2000	1,120	17,300	1,350	600	201,000	493,700	715,070
2001	820	13,000	1,450	500	135,000	417,100	567,870
2002	1,120	16,900	1,525	460	163,000	443,400	626,405
2003	900	15,200	1,070	610	168,000	351,900	537,680
2004	1,630	15,600	870	620	48,000	277,300	344,020
2005	1,360	15,100	1,160	400	94,000	357,500	469,520
2006	640	13,000	1,290	....	189,000	309,500	513,430
2007	1,970	14,500	1,365	....	81,000	360,000	458,835
2008	1,830	13,100	1,050	....	129,000	390,300	535,280
2009 [5]	1,090	13,240	850	....	166,000	332,600	513,780

[1] Standard and substandard.　[2] Freestone only.　[3] Bartlett only.　[4] Raisin and table type.　[5] Preliminary.　NA-not available.
NASS, Crops Branch, (202) 720–2127.

Table 5-76.—Raisins: Production in specified countries, 2007/2008–2009/2010 [1]

Country	2007/2008	2008/2009	2009/2010
	1,000 metric tons	1,000 metric tons	1,000 metric tons
Afghanistan	25,000	27,000	28,000
Argentina	33,000	37,000	36,000
Australia	12,000	14,000	15,000
Chile	67,350	70,000	70,000
China	150,000	135,000	150,000
EU-27	10,000	10,000	10,000
Iran	150,000	70,000	100,000
South Africa	40,200	40,300	39,000
Turkey	250,000	310,000	280,000
Uzbekistan	37,000	25,700	26,000
Others	8,500	8,500	8,500
Total Foreign	783,050	747,500	762,500
United States	321,143	337,472	300,000
Total	1,104,193	1,084,972	1,062,500

[1] Preliminary
FAS, Office of Global Analysis, (202) 720–6301. Prepared or estimated on the basis of official USDA production, supply, and distribution statistics from foreign governments.

Table 5-77.—Fruits, frozen: Commercial pack, United States, 1997–2004

Commodity	1997	1998	1999	2000	2001	2002	2003	2004
	1,000 pounds	1,000 pounds	1,000 pounds	1,000 pounds	1,000 pounds	1,000 pounds	1,000 pounds	1,000 pounds
Apples	119,180	124,866	111,944	141,820	146,145	123,232	113,836	80,506
Apricots	24,267	20,929	18,492	22,786	30,638	20,591	14,767	4,804
Cherries, RSP	([1])	([1])	([1])	([1])	([1])	6,912	40,709	40,332
Cherries, sweet	24,515	21,628	13,640	15,901	13,101	9,062	8,175	11,010
Peaches	124,220	110,491	123,942	148,083	131,694	135,884	136,204	123,378
Plums and prunes	789	1,518	986	1,331	1,380	680	1,732	1,359
Purees, noncitrus	85,333	100,239	85,535	74,663	58,924	36,052	31,359	31,253
Berries:								
Blackberries	26,272	24,734	23,895	26,857	22,884	25,074	23,938	12,962
Blueberries	122,767	90,850	96,567	102,185	98,369	39,887	52,750	38,122
Boysenberries	4,983	3,338	4,703	3,597	3,537	3,174	1,808	1,407
Loganberries	([2])	([2])	([2])	([2])	([2])	([2])	([2])	([2])
Raspberries	27,504	23,851	23,324	23,902	21,736	12,220	30,554	4,888
Strawberries	328,150	373,824	419,768	439,749	422,371	415,865	246,202	215,481
Miscellaneous fruits and berries	110,644	107,716	101,907	135,066	54,799	5,197	4,682	5,421
Total	998,624	1,003,984	1,024,703	1,135,940	1,005,578	826,918	706,716	570,918

[1] Data not available.　[2] Included in miscellaneous.
ERS, Specialty Crops and Fiber Branch, (202) 694-5260. Data from American Frozen Food Institute.

Table 5-78.—Fruits: Per capita consumption, United States, 1999–2008 [1]

Year	Fruits used fresh		
	Citrus fruit [2]	Noncitrus fruits [3]	Canned fruits [4]
	Per capita	Per capita	Per capita
	Pounds	Pounds	Pounds
1999	20.4	80.7	16.7
2000	23.5	77.7	15.3
2001	23.9	73.5	15.5
2002	23.3	75.9	14.8
2003	23.8	77.5	15.0
2004	22.6	79.8	14.8
2005	21.6	78.3	14.4
2006	21.6	79.6	13.2
2007	17.9	79.4	14.2
2008 [8]	20.6	79.8	13.3

Year	Juice [5]	Frozen fruit [6]	Dried fruits [7]
	Per capita	Per capita	Per capita
	Gallons	Pounds	Pounds
1999	9.1	4.5	2.5
2000	8.6	3.4	2.5
2001	8.3	6.5	2.4
2002	8.2	4.1	2.6
2003	8.4	5.1	2.3
2004	8.2	4.6	2.3
2005	7.7	5.7	2.3
2006	7.8	5.2	2.3
2007	7.1	5.5	2.3
2008 [8]	8.0	5.1	2.2

[1] Fresh citrus fruits, canned fruit, and fruit juices are on a crop-year basis. Dried fruits are on a pack-year basis. The per capita consumption was obtained by dividing the total consumption by total population.　[2] Oranges and temples, tangerines and tangelos, lemons, limes, and grapefruit.　[3] Apples, apricots, avocados, bananas, cherries, cranberries, grapes, kiwifruit, mangoes, peaches and nectarines, pears, pineapples, papayas, plums and prunes, and strawberries.　[4] Apples, apricots, cherries, olives, peaches, pears, pineapples, and plums and prunes.　[5] Orange, grapefruit, lemon, lime, apple, grape, pineapple, prune, and cranberry.　[6] Blackberries, blueberries, raspberries, strawberries, other berries, apples, apricots, cherries, and peaches.　[7] Apples, apricots, dates, figs, peaches, pears, prunes, and raisins. Dried data in terms of processed weight.　[8] Preliminary.
ERS, Food and Specialty Crops Branch, (202) 694–5260.

Table 5-79.—All tree nuts: Supply and utilization, United States, 1999/2000–2008/09

Market year [1]	Beginning stocks	Marketable production [2]	Imports	Total supply	Exports	Ending stocks	Domestic consumption	
							Total	Per capita Pounds
	—Million pounds (shelled)—							
1999/00	193.2	1,310.3	285.9	1,789.4	669.1	331.5	788.8	2.81
2000/01	331.5	1,127.9	293.1	1,752.5	781.0	237.7	733.8	2.58
2001/02	237.7	1,347.3	338.8	1,923.7	848.7	256.3	818.8	2.86
2002/03	256.3	1,571.3	362.4	2,190.1	927.8	310.3	952.0	3.29
2003/04	310.3	1,519.3	430.2	2,259.8	964.8	279.9	1,015.1	3.48
2004/05	279.9	1,552.4	502.7	2,335.1	1,041.6	263.0	1,030.5	3.50
2005/06	263.0	1,472.2	431.9	2,167.1	1,120.8	267.2	779.1	2.62
2006/07	267.2	1,651.0	438.3	2,356.5	1,131.0	243.1	982.4	3.27
2007/08	243.1	2,070.9	488.2	2,802.3	1,345.2	389.1	1,168.0	3.52
2008/09 [3]	389.1	2,233.1	429.1	3,051.3	1,437.2	537.3	1,076.7	3.52

[1] Marketing season begins July 1 for almonds, hazelnuts, macadamias, pecans, and other nuts; August 1 for walnuts; and September 1 for pistachios.　[2] Utilized production (NASS data) minus inedibles and noncommercial useage.　[3] Preliminary.
ERS, Food and Specialty Crops Branch, (202) 694–5260.

Table 5-80.—Nuts: Area and production in specified countries, 2007/2008–2009/2010

Country	Area			Production		
	2007/2008	2008/2009	2009/2010	2007/ 2008	2008/ 2009	2009/ 2010
	Hec- tares	Hec- tares	Hec- tares	Metric tons	Metric tons	Metric tons
Almonds:						
Australia	....	....	....	26,500	26,000	30,000
Chile	6,400	6,500	....	8,800	9,500	10,000
China				1,300	400	1,500
EU-27	711,051	710,616	710,022	88,500	79,800	88,950
India	17,000	17,000	17,000	1,000	1,200	1,200
Turkey	....	....	....	15,500	16,000	16,000
Others	....	....	....	....	....	....
Total Foreign	734,451	734,116	727,022	141,600	132,900	147,650
United States	248,882	267,093	275,186	630,490	739,350	612,350
Total	983,333	1,001,209	1,002,208	772,090	872,250	760,000
Filberts:						
Azerbaijan	....	....	....	26,000	20,800	25,000
EU-27	88,300	88,500	....	128,000	149,000	134,000
Turkey	....	....	....	550,000	780,000	470,000
Others	....	....	....			
Total Foreign	88,300	88,500	....	704,000	949,000	629,000
United States	11,370	11,412	....	33,570	29,030	34,500
Total	99,670	99,912	....	737,570	978,030	663,500
Walnuts:						
Brazil	....	....	....	2,250	2,300	2,300
Chile	7,840	9,336	....	26,000	30,000	32,000
China	945,000	995,000	1,050,000	460,000	490,000	560,000
EU-27	21,970	21,970		60,600	70,900	66,000
India	30,800	30,800	30,800	33,000	34,000	36,000
Turkey	....	....	....	90,000	85,000	88,000
Ukraine	....	....	....	70,000	80,000	80,000
Others	....	....	....			
Total Foreign	1,005,610	1,057,136	1,080,800	741,850	792,200	864,300
United States	87,400	88,220	....	297,560	395,530	376,480
Total	1,093,010	1,145,356	1,080,800	1,039,410	1,187,730	1,240,780
Pecans:						
Mexico	57,013	70,000	....	77,888	90,337	0
Others	....	....	....			....
Total Foreign	57,013	70,000	....	77,888	90,337	0
United States	....	....	....	174,770	92,560	0
Total	57,013	70,000	....	252,658	182,897	0
Macadamia:						
Australia	14,300	....	....	41,800	35,000	0
Kenya	9,500	9,500	....	12,500	13,500	0
South Africa	8,000	9,360	....	20,025	24,600	0
Others	....	....	....	19,900	....	....
Total Foreign	31,800	18,860	....	74,325	73,100	0
United States	6,070	6,070	....	16,330	16,400	0
Total	37,870	24,930	....	90,655	89,500	0

See end of table.

Table 5-80.—Nuts: Area and production in specified countries, 2007/2008–2009/2010—Continued

Country	Area			Production		
	2007/ 2008	2008/ 2009	2009/ 2010	2007/ 2008	2008/ 2009	2009/ 2010
	Hec- tares	Hec- tares	Hec- tares	Metric tons	Metric tons	Metric tons
Pistachios:						
EU-27	8,372	8,352	8,352	13,000	9,200	10,000
Iran				200,000	70,000	100,000
Syria	53,000	54,000	54,000	60,000	70,000	70,000
Turkey				40,000	85,000	40,000
Others						
Total Foreign	61,372	62,352	62,352	313,000	234,200	220,000
United States	46,130			188,694	126,100	175,000
Total	107,502	62,352	62,352	501,694	360,300	395,000

FAS, Office of Global Analysis, (202) 720–6301. Prepared or estimated on the basis of official USDA production, supply, and distribution statistics from foreign governments.

Table 5-81.—Almonds (shelled basis): Bearing acreage, yield, production, price, and value, California, 2000–2009 [1]

Year	Bearing Acreage	Yield per acre	Utilized produc- tion	Price per pound	Value
	Acres	Pounds	1,000 pounds	Dollars	1,000 dollars
2000	510,000	1,380	703,000	0.97	666,487
2001	530,000	1,570	830,000	0.91	740,012
2002	545,000	2,000	1,090,000	1.11	1,200,687
2003	550,000	1,890	1,040,000	1.57	1,600,144
2004	570,000	1,760	1,005,000	2.21	2,189,005
2005	590,000	1,550	915,000	2.81	2,525,909
2006	610,000	1,840	1,120,000	2.06	2,258,790
2007	640,000	2,170	1,390,000	1.75	2,401,875
2008	680,000	2,400	1,630,000	1.45	2,343,200
2009 [2]	720,000	1,960	1,410,000	1.65	2,293,500

[1] Price and value are based on edible portion of the crop only. Included in production are inedible quantities of no value as follows (million pounds): 1998-21.0; 1999-33.2; 2000-15.9; 2001-16.8; 2002-8.3; 2003-20.8; 2004-14.5; 2005-16.1; 2006-23.5; 2007-17.5; 2008-14.0; 2009-20.0.   [2] Preliminary.
NASS, Crops Branch, (202) 720–2127.

Table 5-82.—Almonds (shelled basis): Foreign trade, United States, 1999–2008 [1]

Year beginning October	Imports	Domestic exports
	Metric tons	Metric tons
1999	39	197,271
2000	173	225,550
2001	319	261,563
2002	750	289,589
2003	830	308,041
2004	1,233	304,711
2005	2,076	324,798
2006	1,515	342,046
2007	1,722	397,105
2008	796	440,029

[1] Imports of unshelled nuts converted to shelled basis at ratio of 1.67 to 1. Exports of unshelled nuts converted to shelled basis at ratio of 1.67 to 1.0.
ERS, Food and Specialty Crops Branch, (202) 694–5260.

Table 5-83.—Hazelnuts (in-shell basis): Bearing acreage, yield, production, price, and value, Oregon, Washington, and United States, 2000–2009

Year	Bearing Acreage	Yield per acre	Utilized production	Price per ton	Value
	Acres	Tons	Tons	Dollars	1,000 dollars
			Oregon		
2000	28,300	0.79	22,300	80	19,847
2001	29,000	1.71	49,500	701	34,700
2002	29,200	0.67	19,500	1,000	19,500
2003	28,000	1.35	37,900	1,030	39,037
2004	28,400	1.32	37,500	1,440	54,000
2005	28,300	0.98	27,600	2,240	61,824
2006	28,200	1.52	43,000	1,080	46,440
2007	28,600	1.29	37,000	2,040	75,480
2008	28,300	1.13	32,000	1,620	51,840
2009 [1]	28,700	1.64	47,000	1,690	79,430
			Washington		
2000	350	0.57	200	960	192
2001 [2]					
2002					
2003					
2004					
2005					
2006					
2007					
2008					
2009					
			United States		
2000	28,650	0.79	22,500	891	20,039
2001	29,000	1.71	49,500	701	34,700
2002	29,200	0.67	19,500	1,000	19,500
2003	28,000	1.35	37,900	1,030	39,037
2004	28,400	1.32	37,500	1,440	54,000
2005	28,300	0.98	27,600	2,240	61,824
2006	28,200	1.52	43,000	1,080	46,440
2007	28,600	1.29	37,000	2,040	75,480
2008	28,300	1.13	32,000	1,620	51,840
2009 [1]	28,700	1.64	47,000	1,690	79,430

[1] Preliminary. [2] WA discontinued.
NASS, Crops Branch, (202) 720–2127.

Table 5-84.—Hazelnuts (shelled basis): Foreign trade, United States, 1999–2008 [1]

Year beginning October	Imports	Domestic exports
	Metric tons	Metric tons
1999	5,425	6,563
2000	5,129	5,706
2001	6,736	11,110
2002	6,441	4,524
2003	4,916	11,142
2004	4,108	10,459
2005	5,383	11,183
2006	4,344	11,193
2007	4,979	12,259
2008	2,969	9,715

[1] Imports of unshelled nuts converted to shelled basis at ratio of 2.22 to 1. Exports of unshelled nuts converted to shelled basis at ratio of 2.50 to 1.
ERS, Food and Specialty Crops Branch, (202) 694–5260.

Table 5-85.—Macadamia nuts (in-shell basis): Bearing acreage, yield, production, price, and value, Hawaii, 2000–2009

Year	Bearing Acreage	Yield per acre	Utilized production	Price per pound	Value
	Acres	Pounds	1,000 pounds	Cents	1,000 dollars
2000	17,700	2,820	50,000	59.0	29,500
2001	17,800	3,150	56,000	59.0	33,040
2002	17,800	2,980	53,000	57.0	30,210
2003	17,800	2,980	53,000	61.0	32,330
2004	17,800	3,170	56,500	73.0	41,245
2005	18,000	3,000	54,000	81.0	43,740
2006	15,000	3,870	58,000	67.0	38,860
2007	15,000	2,730	41,000	60.0	24,600
2008	15,000	3,330	50,000	67.0	33,500
2009 [1]	15,000	2,800	42,000	70.0	29,400

[1] Preliminary.
NASS, Crops Branch, (202) 720–2127.

Table 5-86.—Pecans (in-shell basis): Production, price per pound, and value, United States, 2000–2009

Year	Improved varieties [1]			Native and seedling			All pecans		
	Utilized production	Price per pound	Value	Utilized production	Price per pound	Value	Utilized production	Price per pound	Value
	1,000 pounds	*Cents*	*1,000 dollars*	*1,000 pounds*	*Cents*	*1,000 dollars*	*1,000 pounds*	*Cents*	*1,000 dollars*
2000	160,550	126.0	201,575	49,300	75.4	37,193	209,850	114.0	238,768
2001	246,550	66.2	163,204	91,950	41.2	37,897	338,500	59.4	201,101
2002	130,720	107.0	139,597	42,180	60.3	25,436	172,900	95.5	165,033
2003	202,900	110.0	223,547	79,200	68.3	54,082	282,100	98.4	277,629
2004	138,970	192.0	267,215	46,830	128.0	59,709	185,800	176.0	326,924
2005 [2]	228,700	154.0	351,353	51,550	108.0	55,567	280,250	145.0	406,920
2006	152,130	173.0	262,544	55,170	109.0	59,949	207,300	156.0	322,493
2007	303,462	123.0	373,131	83,843	72.2	60,513	387,305	112.0	433,644
2008	166,660	142.0	236,300	27,420	87.9	24,097	194,080	134.00	260,397
2009	240,720	154.0	369,580	51,110	93.4	47,720	291,830	143.00	417,300

[1] Budded, grafted or topworked varieties.   [2] MO added.
NASS, Crops Branch, (202) 720–2127.

Table 5-87.—Pecans (in-shell basis): Production and marketing year average price per pound, by State and United States, 2007–2009

Item and State	Utilized production			Price per pound		
	2007	2008	2009 [1]	2007	2008	2009 [1]
IMPROVED VARIETIES [2]	*1,000 pounds*	*1,000 pounds*	*1,000 pounds*	*Dollars*	*Dollars*	*Dollars*
AL	10,000	7,400	12,800	0.890	1.270	1.200
AZ	23,000	17,500	20,000	1.600	1.450	1.720
AR	1,500	1,000	1,300	1.090	1.360	1.020
CA	4,400	3,750	3,920	1.780	1.310	1.510
FL	1,700	1,400	1,500	1.000	2.000	1.200
GA	135,000	66,000	79,000	1.100	1.470	1.330
LA	3,000	1,000	2,500	1.200	1.350	1.300
MS	2,200	900	2,300	1.150	1.300	1.150
MO	2	110	200	1.200	1.400	1.350
NM	74,000	43,000	68,000	1.300	1.450	1.760
NC [3]	160	600	....	1.510	1.300	....
OK	3,000	1,000	3,000	1.350	1.600	1.500
SC	1,500	3,000	1,200	1.160	1.210	1.300
TX	44,000	20,000	45,000	1.350	1.220	1.640
US	303,462	166,660	240,720	1.230	1.420	1.540
NATIVE AND SEEDLING						
AL	2,000	600	1,200	0.660	0.860	0.740
AR	800	500	1,200	0.760	1.020	0.930
FL	200	300	1,600	0.700	1.100	1.100
GA	15,000	4,000	11,000	0.720	1.000	0.890
KS	500	1,900	1,000	0.900	1.000	1.300
LA	11,000	4,000	6,500	0.700	0.850	0.750
MS	800	600	700	0.700	1.000	0.650
MO	3	1,020	1,610	1.000	1.050	1.200
NC [3]	40	100	....	0.810	0.600	....
OK	27,000	4,000	10,500	0.750	0.800	0.900
SC	500	400	800	0.900	0.650	0.880
TX	26,000	10,000	15,000	0.700	0.825	1.030
US	83,843	27,420	51,110	0.722	0.879	0.934
ALL PECANS						
AL	12,000	8,000	14,000	0.852	1.240	1.160
AZ	23,000	17,500	20,000	1.600	1.450	1.720
AR	2,300	1,500	2,500	0.975	1.250	0.977
CA	4,400	3,750	3,920	1.780	1.310	1.510
FL	1,900	1,700	3,100	0.968	1.840	1.150
GA	150,000	70,000	90,000	1.060	1.440	1.280
KS	500	1,900	1,000	0.900	1.000	1.300
LA	14,000	5,000	9,000	0.807	0.950	0.903
MS	3,000	1,500	3,000	1.030	1.180	1.030
MO	5	1,130	1,810	1.000	1.080	1.220
NM	74,000	43,000	68,000	1.300	1.450	1.760
NC [3]	200	700	....	1.370	1.200	....
OK	30,000	5,000	13,500	0.810	0.960	1.030
SC	2,000	3,400	2,000	1.100	1.140	1.130
TX	70,000	30,000	60,000	1.110	1.090	1.490
US	387,305	194,080	291,830	1.120	1.340	1.430

[1] Preliminary.   [2] Budded, grafted or topworked varieties.   [3] Estimates discontinued in 2009.
NASS, Crops Branch, (202) 720–2127.

Table 5-88.—Pecans (shelled basis): Foreign trade, United States, 1999–2008 [1]

Year beginning October	Imports	Domestic exports
	Metric tons	Metric tons
1999	12,152	9,238
2000	12,902	8,963
2001	14,323	11,115
2002	14,555	13,243
2003	20,953	15,275
2004	28,672	13,528
2005	30,983	16,419
2006	23,923	19,145
2007	33,689	29,409
2008	26,866	22,043

[1] Imports of unshelled nuts converted to shelled basis at ratio of 2.50 to 1. Exports of unshelled nuts converted to shelled basis at ratio of 2.50 to 1.
ERS, Food and Specialty Crops Branch, (202) 694–5260.

Table 5-89.—Pistachios (in-shell basis): Bearing acreage, yield, production, price, and value, California, 2000–2009

Year	Bearing Acreage	Yield per acre	Utilized production	Price per pound	Value
	Acres	Pounds	1,000 pounds	Dollars	1,000 dollars
2000	74,600	3,260	243,000	1.01	245,430
2001	78,000	2,060	161,000	1.01	162,610
2002	83,000	3,650	303,000	1.10	333,300
2003	88,000	1,350	119,000	1.22	145,180
2004	93,000	3,730	347,000	1.34	464,980
2005	105,000	2,700	283,000	2.05	580,150
2006	112,000	2,130	238,000	1.89	449,820
2007	115,000	3,620	416,000	1.41	586,560
2008	118,000	2,360	278,000	2.05	569,900
2009 [1]	126,000	2,820	355,000	1.67	592,850

[1] Preliminary.
NASS, Crops Branch, (202) 720–2127.

Table 5-90.—Walnuts, English (in-shell basis): Bearing acreage, yield, production, price, and value, California, 2000–2009

Year	Bearing Acreage	Yield per acre	Utilized production	Price per ton	Value
	Acres	Tons	Tons	Dollars	1,000 dollars
2000	200,000	1.20	239,000	1,240	296,360
2001	204,000	1.50	305,000	1,120	341,600
2002	210,000	1.34	282,000	1,170	329,940
2003	213,000	1.53	326,000	1,160	378,160
2004	214,000	1.52	325,000	1,390	451,750
2005	215,000	1.65	355,000	1,570	557,350
2006	216,000	1.60	346,000	1,630	563,980
2007	218,000	1.50	328,000	2,290	751,120
2008	223,000	1.96	436,000	1,280	558,080
2009 [1]	223,000	1.96	437,000	1,690	738,530

[1] Preliminary.
NASS, Crops Branch, (202) 720–2127.

Table 5-91.—Walnuts (shelled basis): Foreign trade, United States, 1999–2008 [1]

Year beginning October	Imports	Domestic exports
	Metric tons	Metric tons
1999	76	41,428
2000	523	41,918
2001	49	46,937
2002	99	49,925
2003	170	56,608
2004	331	60,541
2005	561	89,668
2006	974	69,581
2007	4,059	77,704
2008	837	96,206

[1] Imports of unshelled nuts converted to shelled basis at ratio of 2.50 to 1. Exports of unshelled nuts converted to shelled basis at ratio of 2.50 to 1.
ERS, Food and Specialty Crops Branch, (202) 694–5260.

Table 5-92.—Coffee: International trade, exports from principal producing countries, 2007/2008–2009/2010

Country of origin	2007/2008	2008/2009	2009/2010
	1,000 bags	*1,000 bags*	*1,000 bags*
Principle exporting countries:			
Brazil	29,260	27,290	31,475
Colombia	11,155	11,525	8,716
Guatemala	3,980	3,890	3,783
Honduras	3,370	3,440	3,250
India	3,609	3,820	3,120
Indonesia	6,400	5,900	7,675
Mexico	2,770	2,515	2,825
Peru	4,200	2,660	3,830
Uganda	1,840	2,160	2,740
Vietnam	18,840	16,418	16,463
Others	16,605	16,984	16,192
Total Foreign	102,029	96,602	100,069
United States			
Total	102,029	96,602	100,069
Principle importing countries:			
Algeria	1,825	2,025	2,045
Australia	900	920	950
Canada	1,950	1,940	1,940
EU-27	46,485	45,760	45,355
Japan	7,025	6,860	7,115
Korea,South	1,400	1,550	1,680
Malaysia	1,156	1,200	1080
Philippines	585	1,025	1,265
Russia	4,410	4,520	3,455
Switzerland	1,490	1,710	1,800
Others	5,703	5,341	6,052
Total Foreign	72,929	72,851	72,737
United States	23,670	23,475	23,400
Total	96,599	96,326	96,137

FAS, Office of Global Analysis, (202) 720-6301. Prepared or estimated on the basis of official USDA production, supply, and distribution statistics from foreign governments.

Table 5-93.—Coffee: Area, yield, production, marketing year average price, and value, Hawaii and Puerto Rico, 2000–2010

Year	Area	Yield per acre	Production [1]	Price per pound	Value
	Acres	*Pounds*	*1,000 pounds*	*Dollars*	*1,000 dollars*
			Hawaii		
2000–2001	6,800	1,280	8,700	2.65	23,055
2001–2002	6,300	1,270	8,000	2.45	19,600
2002–2003	5,900	1,270	7,500	3.10	23,250
2003–2004	5,900	1,410	8,300	2.90	24,070
2004–2005	5,800	965	5,600	3.55	19,880
2005–2006	6,100	1,340	8,200	4.55	37,310
2006–2007	6,300	1,170	7,400	4.30	31,820
2007–2008	6,400	1,170	7,500	4.25	31,875
2008–2009	6,300	1,380	8,700	3.40	29,580
2009–2010 [2]	6,300	1,270	8,000	3.20	25,600
			Puerto Rico		
2000–2001					
2001–2002					
2002–2003					
2003–2004	47,000	480	22,500	1.99	44,775
2004–2005	44,000	420	18,500	1.94	35,890
2005–2006	42,000	465	19,500	2.66	51,870
2006–2007	40,000	450	18,000	2.57	46,260
2007–2008	39,000	450	17,500	2.67	46,725
2008–2009	33,000	405	13,300	2.19	29,127
2009–2010 [2]	27,000	350	9,500	2.17	20,615

[1] Parchment basis.   [2] Preliminary.
NASS, Crops Branch, (202) 720–2127.

Table 5-94.—Coffee and tea: U.S. imports, 2007–2009

Country	2007	2008	2009
	1,000 metric tons	*1,000 metric tons*	*1,000 metric tons*
Coffee and coffee products:			
Colombia	701,040	847,275	593,588
Brazil	696,778	828,552	589,398
Guatemala	309,414	369,796	311,788
Mexico	282,117	318,491	271,890
Vietnam	309,365	296,002	205,449
Indonesia	209,039	240,125	153,409
Costa Rica	156,820	190,919	145,160
Peru	137,189	168,890	96,108
Germany(*)	138,939	153,269	102,282
Canada	126,184	140,935	126,892
Nicaragua	82,605	139,833	102,429
Honduras	100,983	132,232	71,197
El Salvador	92,124	121,640	75,376
Ethiopia(*)	45,040	76,227	28,289
Italy(*)	42,888	44,950	27,266
Switzerland(*)	34,680	36,885	40,765
Sweden	38,112	36,475	18,277
Kenya	26,301	36,332	24,670
Papua New Guinea	33,985	33,970	26,512
Malaysia	32,718	29,371	24,569
Tanzania	12,761	15,716	21,648
Venezuela	3,388	15,269	613
Panama	14,008	15,260	7,488
Belgium-Luxembourf(*)	11,355	10,671	3,617
Uganda	8,734	10,518	11,012
Rwanda	6,464	9,403	6,684
Rest of World	115,256	93,174	71,946
World Total	3,768,282	4,412,186	3,158,316
Tea, except herbal tea:			
China	79,546	88,802	57,233
Canada	73,068	78,999	78,593
Germany(*)	44,821	51,514	37,039
India	40,892	51,045	35,492
Argentina	39,542	45,427	44,906
United Kingdom	23,610	25,044	14,073
Japan	21,282	20,121	16,862
Sri Lanka	20,948	19,699	12,637
Mexico	23,316	19,026	8,179
Kenya	10,815	13,240	10,060
Indonesia	10,134	9,922	8,678
Chile	7,663	9,062	6,811
Taiwan	8,071	9,054	5,492
Morocco	5,365	7,580	3,908
Brazil	4,751	5,375	3,910
Korea, South	4,009	3,713	1,823
Vietnam	2,754	3,260	3,559
Malawi	4,236	3,245	3,713
Thailand	2,023	3,132	2,084
Spain	3,215	2,870	2,010
France(*)	2,089	2,556	1,290
Hong Kong	1,174	2,285	1,482
Egypt	1,746	1,695	573
Israel(*)	1,734	1,695	573
Netherlands	1,807	1,677	1,167
Singapore	1,180	1,645	1,238
Rest of World	15,632	15,707	13,573
World Total	455,416	497,708	377,185

Note: (*) Denotes a country that is a summarization of its component countries.
FAS, Office of Global Analysis, (202) 720-6301.

Table 5-95.—Specialty mushrooms: Number of growers, total production, volume of sales, price per pound, and value of sales, July 1–June 30, 2006–2007/2008–2009 [1]

Year and variety	Growers [2]	All sales	Total production [3]		
			Volume of sales [4]	Price per pound [5]	Value of sales
	Number	*1,000 pounds*	*1,000 pounds*	*Dollars*	*1,000 dollars*
2006–2007					
Shiitake	160	7,155	6,985	3.36	23,469
Oyster	62	5,265	5,055	2.41	12,185
Other	30	2,180	2,113	4.84	10,231
US [6]	191	14,600	14,153	3.24	45,885
2007–2008					
Shiitake	155	9,848	9,673	2.69	26,049
Oyster	61	4,371	4,253	2.88	12,232
Other	24	1,371	1,330	5.16	6,868
US [6]	181	15,590	15,256	2.96	45,149
2008–2009					
Shiitake	153	9,682	9,416	3.19	30,038
Oyster	67	5,383	5,056	2.46	12,442
Other	26	1,070	979	5.59	5,470
US [6]	184	16,135	15,451	3.10	47,950

[1] Specialty mushroom estimates represent growers who have at least 200 natural wood logs in production or some commercial indoor growing area, and 200 dollars in sales.   [2] Growers counted only once for US total if growing more than one specialty type mushroom. Growers growing Agaricus and Specialty are included.   [3] Total production includes all fresh market and processing sales plus amount harvested but not sold (shrinkage, cullage, dumped, etc.).   [4] Virtually all specialty mushroom sales are for fresh market.   [5] Prices for mushrooms are the average prices producers receive at the point of first sale, commonly referred to as the average price as sold. For example, if in a given State, part of the fresh mushrooms are sold F.O.B. packed by growers, part are sold bulk to brokers or repackers, and some are sold retail at roadside stands, the mushroom average price as sold is a weighted average of the average price for each method of sale.   [6] 2005-06: AR, CA, CT, DE, FL, HI, IL, IN, KS, KY, ME, MD, MA, MI, MN, MO, MT, NH, NY, NC, OH, OR, PA, SC, TN, TX, VA, WA, WV, and WI. 2006-07: AR, CA, CT, FL, HI, IL, IN, KS, KY, ME, MA, MI, MN, MO, MT, NY, NC, OH, OR, PA, SC, TN, TX, VA, VT, WA, WV, and WI. 2007-08: AR, CA, CO, CT, DE, FL, GA, HI, IL, IN, KS, KY, ME, MA, MI, MN, MO, MT, NY, NC, OH, OR, PA, SC, TN, TX, VA, VT, WA, WV, and WI.
NASS, Crops Branch, (202) 720–2127.

Table 5-96.—Agaricus mushrooms: Area, volume of sales, marketing year average price, and value of sales, United States, 1999/2000–2008/2009 [1]

Year	Area in production	Volume of sales	Price per pound	Value of sales		
				Total	Fresh market	Processing
	1,000 sq. ft.	*1,000 pounds*	*Cents*	*1,000 dollars*	*1,000 dollars*	*1,000 dollars*
1999–2000	151,487	854,394	97.0	828,551	715,943	112,608
2000–2001	143,873	846,209	97.6	825,500	736,543	88,957
2001–2002	140,822	831,107	105.0	870,573	796,522	74,051
2002–2003	141,844	836,398	102.0	855,983	778,307	77,676
2003–2004	146,510	841,162	104.0	878,405	805,200	73,205
2004–2005	143,093	838,083	103.0	862,192	796,493	65,699
2005–2006	142,550	833,677	102.0	848,836	793,538	55,298
2006–2007	145,743	813,849	112.0	915,561	840,560	75,001
2007–2008	136,011	797,348	115.0	917,607	841,753	75,854
2008–2009	134,089	801,523	113.0	909,078	840,095	68,983

[1] Marketing year begins July 1 and ends June 30 the following year.
NASS, Crops Branch, (202) 720–2127.

Table 5-97.—Cut flowers: Sales and wholesale value for operations with $100,000+ sales, Surveyed States, 2000–2009

Year	Quantity sold	Wholesale price	Value of sales at wholesale [1]	Quantity sold	Wholesale price	Value of sales at wholesale [1]
		Alstromeria			Standard carnations	
	1,000 stems	*Cents* per stem	*1,000* dollars	*1,000* stems	*Cent* per stem	*1,000* dollars
2000	25,056	23.7	5,939	40,206	16.0	6,430
2001	21,253	24.2	5,137	24,760	15.6	3,870
2002	17,153	27.2	4,674	21,643	15.8	3,416
2003	13,402	29.7	3,978	13,491	17.6	2,374
2004	12,023	31.1	3,735	9,251	17.6	1,624
2005	7,313	35.4	2,588	8,955	20.3	1,816
2006 [2]	8,595	21.9	1,885	5,428	17.6	955
2007	9,879	20.8	2,057	3,328	18.8	626
2008	10,774	17.9	1,927	3,343	17.0	567
2009	8,800	18.5	1,629	2,587	17.0	440
		Pompon chrysanthemums			Delphinium & Larkspur	
	1,000 spikes	*Cents* per spike	*1,000* dollars	*1,000* stems	*Cents* per stem	*1,000* dollars
2000	13,159	1.31	17,214	43,840	25.0	10,955
2001	12,933	1.30	16,831	45,515	24.2	11,008
2002	14,766	1.31	19,351	47,023	23.3	10,971
2003	14,002	1.30	18,196	40,945	23.9	9,797
2004	15,035	1.33	19,980	36,349	25.0	9,082
2005	12,320	1.40	17,246	34,150	23.7	8,087
2006 [2]	10,338	1.26	12,985	26,142	23.5	6,133
2007	18,059	0.76	13,810	32,158	24.4	7,842
2008	10,058	1.34	13,428	31,221	24.0	7,505
2009	7,920	1.43	11,298	21,775	27.3	5,950
		Gerbera Daisy			Gladioli	
	1,000 stems	*Cents* per stem	*1,000* dollars	*1,000* stems	*Cents* per stem	*1,000* dollars
2000	67,520	30.9	20,886	127,109	22.3	28,339
2001	72,916	30.6	22,317	112,948	21.5	24,284
2002	84,917	29.8	25,343	126,001	21.3	26,853
2003	94,046	29.9	28,164	121,465	23.3	28,325
2004	97,656	30.8	30,059	113,906	23.0	26,159
2005	104,682	30.9	32,314	105,432	22.8	24,074
2006 [2]	112,587	30.2	33,997	95,350	23.8	22,694
2007	117,403	30.6	35,939	85,471	27.0	23,081
2008	120,836	29.8	35,976	76,850	25.9	19,935
2009	106,805	30.9	33,027	95,249	24.2	23,086
		Iris			All Lilies	
	1,000 spikes	*Cents* per spike	*1,000* dollars			*1,000* dollars
2000	79,012	25.8	20,395	81,301	68.8	55,975
2001	83,594	23.4	19,549	91,267	66.7	60,876
2002	81,837	22.4	18,344	101,748	61.3	62,347
2003	89,976	22.6	20,367	112,946	65.0	73,400
2004	88,973	23.0	20,473	117,456	63.2	74,282
2005	88,803	22.5	20,021	114,188	67.4	77,009
2006 [2]	81,194	22.6	18,315	107,044	70.5	75,459
2007	90,890	22.4	20,349	111,185	67.4	74,954
2008	92,404	22.1	20,462	116,797	67.3	78,609
2009	64,114	24.3	15,550	101,151	63.0	63,765

See footnotes at end of table.

Table 5-97.—Cut flowers: Sales and wholesale value for operations with $100,000+ sales, Surveyed States, 2000–2009—Continued

Year	Quantity sold	Wholesale price	Value of sales at wholesale [1]	Quantity sold	Wholesale price	Value of sales at wholesale [1]
	Lisianthus			All orchids		
	1,000 stems	*Cents* per stem	*1,000* dollars	*1,000* stems	*Cents* per stem	*1,000* dollars
2000	15,150	45.5	6,891	11,719	68.9	8,071
2001	19,040	44.7	8,505	11,571	74.0	8,563
2002	14,530	45.1	6,551	11,113	70.2	7,796
2003	14,410	45.0	6,491	12,237	69.8	8,536
2004	12,667	46.4	5,875	11,398	68.7	7,834
2005	12,333	39.8	4,906	10,228	66.9	6,847
2006 [2]	8,518	43.1	3,670	10,332	120.3	12,428
2007	13,956	38.2	5,338	11,209	99.5	11,150
2008	15,180	35.6	5,406	7,882	98.2	7,737
2009	8,693	38.3	3,329	7,633	189.1	14,433
	All roses			Snapdragons		
	1,000 blooms	*Cents* per bloom	*1,000* dollars	*1,000* stems	*Cents* per stem	*1,000* dollars
2000	185,975	37.3	69,294	63,711	30.1	19,166
2001	160,301	37.4	59,976	60,939	27.9	16,980
2002	157,253	37.4	58,878	60,860	28.0	17,041
2003	123,483	38.1	46,997	55,392	28.2	15,639
2004	103,860	40.3	41,894	50,549	29.7	15,002
2005	99,771	39.1	38,969	47,016	27.9	13,132
2006 [2]	82,138	37.7	30,974	36,559	28.0	10,244
2007	67,701	41.5	28,110	41,887	29.1	12,202
2008	57,999	38.8	22,481	42,696	27.6	11,790
2009	42,031	42.0	17,662	37,113	26.8	9,962
	Tulips			Other cut flowers		
	1,000 spikes	*Cents* per spike	*1,000* dollars	*1,000* stems	*Cents* per stem	*1,000* dollars
2000	77,039	34.7	26,760	....	....	133,648
2001	75,769	35.5	26,864	....	....	133,343
2002	90,625	32.0	29,001	....	....	136,515
2003	92,551	33.6	31,055	....	....	129,663
2004	105,138	35.3	37,096	....	....	119,336
2005	128,978	32.7	42,121	....	....	124,832
2006 [2]	141,893	34.1	48,391	....	....	134,198
2007	157,992	35.9	56,719	....	....	133,017
2008	170,854	37.6	64,285	....	....	126,990
2009	150,850	37.8	57,033	....	....	102,063

[1] Equivalent wholesale value of all sales. [2] Beginning in 2006, program was reduced to 15 States from 36.
NASS, Crops Branch, (202) 720–2127.

Table 5-98.—Cut Cultivated Greens: Sales and wholesale value for operations with $100,000+ sales, Surveyed States, 2000–2009

Year	Quantity sold	Wholesale price	Value of sales at wholesale [1]	Quantity sold	Wholesale price	Value of sales at wholesale [1]
	Leatherleaf Ferns			Other cut cultivated greens		
	1,000 bunches	*Dollars* per bunch	*1,000* dollars	*1,000* bunches	*Dollars*	*1,000* dollars
2000	75,611	0.88	66,245	....	....	59,923
2001	63,002	0.88	55,310	....	....	57,048
2002	61,907	0.87	53,634	....	....	60,139
2003	58,305	0.84	48,868	....	....	53,197
2004	54,115	0.88	47,541	....	....	54,435
2005	49,213	1.03	50,668	....	....	56,776
2006 [2]	44,183	1.04	45,902	....	....	51,706
2007	39,437	1.00	39,543	....	....	58,527
2008	34,761	.98	33,924	....	....	57,824
2009	31,643	.94	29,796	....	....	44,160

[1] Equivalent wholesale value of all sales. [2] Beginning in 2006, program was reduced to 15 States from 36.
NASS, Crops Branch, (202) 720–2127.

Table 5-99.— Potted Flowering Plants for Indoor on Patio Use: Sales and wholesale value for operations with $100,000+ sales, Surveyed States, 2000–2009

Year	Quantity sold		Wholesale price		Value of sales at wholesale [1]
	Less than 5 inches	5 inches or more	Less than 5 inches	5 inches or more	
	1,000 pots	*1,000 pots*	*Dollars per pot*	*Dollars*	*1,000 dollars*

African violets

Year	*1,000 pots*	*1,000 pots*	*Dollars per pot*	*Dollars*	*1,000 dollars*
2000	16,043	257	1.12	3.57	18,909
2001	15,834	260	1.21	3.52	20,034
2002	15,513	621	1.24	2.52	20,816
2003	14,365	663	1.18	2.33	18,540
2004	12,089	548	1.17	2.41	15,419
2005	11,931	522	1.16	2.33	15,010
2006 [2]	5,997	434	1.20	1.92	8,046
2007	4,357	430	1.37	1.95	6,809
2008	2,946	9	1.35	3.34	3,993
2009	2,295	38	1.49	3.48	3,552

Florist azaleas

Year	*1,000 pots*	*1,000 pots*	*Dollars per pot*	*Dollars per pot*	*1,000 dollars*
2000	4,880	10,032	2.08	5.14	61,719
2001	3,987	9,974	2.58	5.32	63,333
2002	3,035	7,679	2.64	5.29	48,603
2003	2,330	6,281	3.04	5.09	39,048
2004	2,047	6,596	2.39	5.13	38,742
2005	2,410	6,443	2.27	4.86	36,750
2006 [2]	2,237	6,844	2.10	4.41	34,909
2007	1,514	5,081	2.34	4.90	28,435
2008	1,095	7,188	2.24	4.65	35,897
2009	2,361	5,453	2.31	4.91	32,208

Florist chrysanthemums

Year	*1,000 pots*	*1,000 pots*	*Dollars per pot*	*Dollars*	*1,000 dollars*
2000	8,439	19,936	1.50	3.47	81,869
2001	6,585	18,592	1.56	3.49	75,225
2002	7,096	23,948	1.58	3.09	85,128
2003	8,721	17,982	1.78	3.01	69,641
2004	11,251	15,685	1.80	3.05	68,123
2005	4,421	18,891	1.75	3.23	68,797
2006 [2]	1,299	12,693	1.82	3.03	40,815
2007	1,810	11,363	1.65	3.15	38,777
2008	1,927	9,902	1.57	3.21	34,762
2009	1,418	6,123	1.92	3.64	25,001

Easter lilies

Year	*1,000 pots*	*1,000 pots*	*Dollars per pot*	*Dollarsper pot*	*1,000 dollars*
2000	141	9,002	3.22	4.09	37,246
2001	214	9,236	3.07	4.01	37,735
2002	241	8,853	2.86	4.10	37,014
2003	244	8,580	2.03	4.19	36,434
2004	133	8,420	2.12	4.25	36,109
2005	34	8,251	3.21	4.25	35,204
2006 [2]	2	6,334	6.53	4.12	26,106
2007	....	6,546	....	4.05	26,512
2008	31	5,824	4.05	4.33	25,335
2009	112	6,138	3.31	4.39	27,316

Potted Orchids

Year	*1,000 pots*	*1,000 pots*	*Dollars per pot*	*Dollars per pot*	*1,000 dollars*
2000	4,782	4,912	6.58	11.72	89,018
2001	6,992	5,208	6.17	11.31	102,049
2002	7,835	5,430	6.62	11.02	111,735
2003	8,871	6,209	6.15	10.85	121,908
2004	11,277	6,016	6.01	10.21	129,141
2005	11,535	5,975	6.46	10.87	139,482
2006 [2]	10,140	4,615	6.90	10.96	120,521
2007	10,661	6,655	6.95	9.89	139,960
2008	10,689	5,415	7.16	9.23	126,509
2009	12,625	6,848	7.39	9.68	159,614

Poinsettias

Year	*1,000 pots*	*1,000 pots*	*Dollars per pot*	*Dollars per pot*	*1,000 dollars*
2000	15,457	50,931	1.81	4.28	246,263
2001	14,682	52,284	1.84	4.37	255,323
2002	14,837	51,707	1.86	4.36	252,983
2003	13,092	48,432	1.91	4.54	244,973
2004	11,301	48,287	2.06	4.62	246,598
2005	11,251	47,494	2.04	4.60	241,705
2006 [2]	7,762	33,743	1.98	4.61	171,012
2007	7,130	31,901	2.07	4.60	161,409
2008	6,373	28,922	2.07	4.82	152,611
2009	6,785	27,691	2.08	4.73	145,088

See footnotes at end of table.

Table 5-99.—Potted flowering for indoor or patio use: Sales and wholesale value for operations with $100,000+ sales, Surveyed States, 2000–2009—Continued

Year	Quantity sold		Wholesale price		Value of sales at wholesale [1]
	Less than 5 inches	5 inches or more	Less than 5 inches	5 inches or more	
	Potted florist roses [3]				
	1,000 pots	*1,000 pots*	*Dollars per pot*	*Dollars*	*1,000 dollars*
2000	8,784	2,844	1.77	4.19	27,499
2001	7,257	3,072	1.91	3.83	25,645
2002	6,662	2,483	2.19	5.18	27,492
2003	6,863	945	2.44	3.87	20,394
2004	6,149	646	2.27	4.68	17,004
2005	8,396	1,897	2.01	4.64	25,706
2006 [2]	6,389	1,901	1.90	4.90	21,446
2007	6,834	2,364	1.94	5.16	25,425
2008	7,252	1,064	2.74	5.34	25,569
2009	6,640	1,607	2.60	4.93	25,153
	Potted spring flowering bulbs [3]				
	1,000 pots	*1,000 pots*	*Dollars per pot*	*Dollars*	*1,000 dollars*
2000	6,408	8,775	1.59	3.33	39,392
2001	7,517	10,360	1.50	3.36	46,075
2002	7,590	12,347	1.65	3.44	55,012
2003	7,206	12,181	1.66	3.52	54,927
2004	5,563	11,928	1.61	3.61	51,992
2005	14,051	9,581	1.50	3.55	55,132
2006 [2]	13,061	8,469	1.41	3.43	47,447
2007	10,073	8,624	1.91	3.66	50,861
2008	13,317	10,958	1.69	3.56	61,532
2009	5,075	11,131	1.77	3.39	46,701
	Other flowering [3]				
	1,000 pots	*1,000 pots*	*Dollars per pot*	*Dollars per pot*	*1,000 dollars*
2000	33,585	35,766	1.73	3.90	197,684
2001	34,770	32,178	1.89	4.16	199,331
2002	37,033	31,103	1.89	4.35	205,157
2003	37,521	30,550	1.89	4.14	197,597
2004	32,315	32,315	1.87	4.24	197,407
2005	33,669	31,032	1.91	4.20	194,572
2006 [2]	23,220	27,703	2.01	3.77	151,158
2007	27,176	36,862	2.09	4.51	223,111
2008	22,529	31,482	2.10	4.91	201,962
2009	19,139	26,251	2.14	4.83	167,723

[1] Equivalent wholesale value of all sales except for potted foliage which is value of sales less cost of plant material purchased from other growers for growing on.   [2] Beginning in 2006, program was reduced to 15 States from 36.   [3] Cyclamen and kalanchoes included 2000–2003.
NASS, Crops Branch, (202) 720–2127.

Table 5-100.—Foliage Plants for Indoor or Patio Use: Sales and Wholesale value for operations with $100,000+ sales, Surveyed States, 2000–2009

Year	Foliage, Hanging Baskets		
	Quantity Sold	Wholesale price	Value of sales at wholesale
	1,000 baskets	*Dollars*	*1,000 dollars*
2000	20,983	4.20	88,113
2001	21,292	3.85	81,922
2002	19,984	4.19	83,723
2003	19,452	4.25	82,697
2004	17,160	4.47	76,627
2005	19,713	4.55	89,719
2006	13,341	4.60	61,303
2007	14,118	4.66	65,857
2008	11,003	4.90	53,949
2009	9,739	5.43	52,897
	Foliage, Pots		
	1,000 pots	*Dollars*	*1,000 dollars*
2000	....	....	472,079
2001	....	....	568,668
2002	....	....	538,837
2003	....	....	566,984
2004	....	....	608,637
2005	....	....	619,793
2006 [1]	....	....	466,609
2007	....	....	589,545
2008	....	....	456,362
2009	....	....	401,443

[1] Beginning in 2006, program was reduced to 15 States from 36.
NASS, Crops Branch, (202) 720–2127.

Table 5-101.—Annual Bedding and Garden Hanging Baskets: Sales and wholesale value for operations with $100,000+ sales, Surveyed States, 2000–2009

Year	Quantity sold	Wholesale price	Value of sales at wholesale [1]	Quantity sold	Wholesale price	Value of sales at wholesale [1]
	Begonias			Geraniums from vegetative cuttings		
	1,000 baskets	*Dollars*	*1,000 dollars*	*1,000 baskets*	*Dollars*	*1,000 dollars*
2000	2,855	5.51	15,733	4,146	7.00	29,024
2001	3,335	5.72	19,062	4,121	6.84	28,200
2002	2,536	5.88	14,919	4,431	7.00	30,997
2003	3,352	5.14	17,229	4,900	6.91	33,848
2004	2,777	5.81	16,122	5,424	7.00	37,943
2005	2,834	5.82	16,505	4,802	7.02	33,732
2006 [2]	3,199	5.97	19,091	3,285	6.75	22,186
2007	1,701	5.76	9,796	3,296	6.87	22,640
2008	3,056	7.32	22,366	3,056	7.32	22,366
2009	3,472	7.51	26,073	3,472	7.51	26,073
	Geraniums from seeds			Impatiens		
	1,000 baskets	*Dollars*	*1,000 dollars*	*1,000 baskets*	*Dollars*	*1,000 dollars*
2000	684	5.83	3,991	4,054	5.15	20,859
2001	647	5.98	3,869	3,414	5.42	18,492
2002	567	6.47	3,666	4,096	5.12	20,972
2003	688	6.11	4,201	3,638	5.30	19,267
2004	692	6.20	4,290	3,732	5.43	20,257
2005	724	5.88	4,260	3,824	5.51	21,087
2006 [2]	246	6.03	1,483	2,846	5.19	14,761
2007	316	5.63	1,778	2,597	5.22	13,548
2008	267	6.05	1,616	2,375	5.29	12,561
2009	461	5.94	2,739	2,102	5.77	12,135
	New Guinea Impatiens			Marigolds		
	1,000 baskets	*Dollars*	*1,000 dollars*	*1,000 baskets*	*Dollars*	*1,000 dollars*
2000	4,635	6.39	29,615	59	7.71	455
2001	4,663	6.34	29,572	50	6.90	345
2002	5,140	6.34	32,584	41	7.07	290
2003	4,540	6.44	29,247	23	5.96	137
2004	5,084	6.44	32,725	50	6.34	317
2005	4,558	6.62	30,169	21	5.14	108
2006 [2]	3,174	6.81	21,624	150	6.66	999
2007	2,987	6.96	20,797	184	6.08	1,118
2008	2,668	7.23	19,280	202	6.30	1,272
2009	2,569	7.08	18,196	62	6.58	408
	Pansies/Violas			Petunias		
	1,000 baskets	*Dollars*	*1,000 dollars*	*1,000 baskets*	*Dollars*	*1,000 dollars*
2000	303	6.38	1,932	2,941	5.30	15,595
2001	466	5.97	2,784	3,102	5.89	18,269
2002	600	6.09	3,651	3,558	5.89	20,950
2003	747	5.87	4,383	3,933	5.93	23,325
2004	931	6.50	6,049	4,771	5.98	28,547
2005	1,131	6.51	7,368	4,891	6.18	30,218
2006 [2]	510	5.15	2,625	3,673	5.74	21,081
2007	694	5.02	3,485	4,011	5.99	24,017
2008	695	5.98	23,752	3,969	5.98	23,752
2009	1,085	6.40	26,786	4,187	6.40	26,786
	Other flowering hanging baskets and foliar					
	1,000 baskets	*Dollars*	*1,000 dollars*			
2000	14,760	6.01	88,656			
2001	15,979	6.24	99,761			
2002	17,679	6.25	110,492			
2003	17,836	6.62	118,125			
2004	21,089	6.51	137,301			
2005	21,284	6.55	139,480			
2006 [2]	14,910	6.22	92,736			
2007	15,153	6.95	105,330			
2008	14,718	6.92	101,856			
2009	14,732	7.34	108,104			

[1] Equivalent wholesale value of all sales.   [2] Beginning in 2006, program was reduced to 15 States from 36.
NASS, Crops Branch, (202) 720–2127.

Table 5-102.—Annual bedding garden plants flats: Sales and wholesale value for operations with $100,000+ sales, Surveyed States, 2000–2009

Year	Quantity sold	Wholesale price	Value of sales at wholesale [1]	Quantity sold	Wholesale price	Value of sales at wholesale [1]
	Begonias [2]			Geraniums from vegetative cuttings [2]		
	1,000 flats	*Dollars*	*1,000 dollars*	*1,000 flats*	*Dollars*	*1,000 dollars*
2000	6,645	7.83	52,004	1,574	9.47	14,906
2001	8,272	7.76	64,193	1,003	10.82	10,849
2002	7,906	7.69	60,817	1,126	10.87	12,242
2003	7,424	7.63	56,633	914	10.11	9,239
2004	6,302	8.06	50,818	933	11.26	10,503
2005	7,043	8.06	56,757	654	11.87	7,763
2006 [3]	4,947	8.17	40,429	520	11.54	5,999
2007	4,094	8.17	33,444	417	12.32	5,138
2008	4,360	8.49	36,999	396	12.57	4,979
2009	4,154	8.72	36,234	360	15.06	5,423
	Geraniums from seeds [2]			Impatients		
	1,000 flats	*Dollars*	*1,000 dollars*	*1,000 flats*	*Dollars*	*1,000 dollars*
2000	861	9.33	8,035	15,380	7.70	118,381
2001	766	10.27	7,868	14,904	7.81	116,331
2002	837	10.30	8,623	14,650	8.20	120,133
2003	749	10.31	7,725	13,418	7.80	104,689
2004	774	10.31	7,983	12,596	7.92	99,802
2005	606	10.86	6,584	12,409	8.09	100,334
2006 [3]	398	10.99	4,373	9,884	7.77	76,771
2007	380	11.54	4,387	8,915	8.11	72,320
2008	384	11.00	4,223	8,547	8.52	72,815
2009	428	9.26	3,962	7,942	8.64	68,650
	New Guinea Impatiens			Marigolds [2]		
	1,000 flats	*Dollars*	*1,000 dollars*	*1,000 flats*	*Dollars*	*1,000 dollars*
2000	657	9.71	6,381	5,443	7.75	42,169
2001	589	11.12	6,547	6,623	8.09	53,600
2002	793	9.93	7,872	6,311	8.50	53,616
2003	628	9.81	6,160	6,386	8.19	52,298
2004	474	10.46	4,960	6,121	8.31	50,843
2005	468	10.97	5,134	6,158	8.54	52,569
2006 [3]	305	10.14	3,093	4,032	8.13	32,788
2007	243	11.09	2,696	3,694	8.39	31,001
2008	218	11.55	2,518	3,933	8.69	34,190
2009	295	11.01	3,249	3,966	8.96	35,550
	Pansies/Violas [2]			Petunias		
	1,000 flats	*Dollars*	*1,000 dollars*	*1,000 flats*	*Dollars*	*1,000 dollars*
2000	10,153	8.23	83,521	12,093	7.90	95,488
2001	13,109	7.87	103,151	11,542	8.03	92,669
2002	14,201	8.55	121,452	11,635	8.47	98,595
2003	14,179	8.35	118,358	11,583	8.22	95,161
2004	13,264	8.37	111,032	11,092	8.43	93,551
2005	13,340	8.41	112,165	10,821	8.72	94,351
2006 [3]	8,238	8.03	66,168	7,349	8.12	59,682
2007	8,047	8.33	67,050	7,023	8.52	59,808
2008	8,169	8.82	72,036	7,402	8.80	65,129
2009	7,422	9.14	67,819	7,148	9.15	65,400
	Other flowering and foliar plants, flats [2]			Vegetable type plants, flats [4]		
	1,000 flats	*Dollars*	*1,000 dollars*	*1,000 flats*	*Dollars*	*1,000 dollars*
2000	47,709	8.04	383,686	8,604	7.97	68,604
2001	43,226	7.84	339,064	8,480	8.37	70,946
2002	40,978	8.35	342,326	8,121	8.74	70,991
2003	39,880	8.49	338,557	7,594	8.64	65,629
2004	38,157	8.59	327,616	7,557	9.39	70,963
2005	34,837	8.83	307,711	7,845	9.66	75,747
2006 [3]	25,652	8.02	205,649	4,776	9.55	45,604
2007	21,350	8.70	185,788	4,135	9.39	38,822
2008	19,441	9.00	175,027	4,545	10.25	46,573
2009	17,182	9.60	164,936	4,872	10.31	50,231

[1] Equivalent wholesale value of all sales.   [2] Begonias, Marigolds, Geraniums from seed/cuttings, and Pansy/Violas included in other flowering and foliar type flats prior to 2000.   [3] Beginning in 2006, program was reduced to 15 States from 36.   [4] Does not include vegetable transplants grown for use in commercial vegetable production.
NASS, Crops Branch, (202) 720–2127.

Table 5-103.—Potted annual bedding and garden plants: Sales and wholesale value for operations with $100,000+ sales, Surveyed States, 2000–2009

Year	Quantity sold		Wholesale price		Value of sales at wholesale [1]
	Less than 5 inches	5 inches or more	Less than 5 inches	5 inches or more	
	Begonias				
	1,000 pots	*1,000 pots*	*Dollars*	*Dollars*	*1,000 dollars*
2000	12,559	2,321	0.89	1.81	15,427
2001	13,890	4,275	0.92	2.23	22,260
2002	15,969	4,493	0.89	1.98	23,142
2003	14,489	4,371	1.02	1.88	22,946
2004	19,261	4,760	0.88	1.94	26,166
2005	23,381	5,185	0.88	2.07	31,275
2006 [2]	19,939	4,387	0.85	2.28	27,004
2007	16,748	3,683	0.84	2.06	21,645
2008	16,481	4,296	0.91	2.17	24,293
2009	17,846	3,272	0.86	2.34	22,903
	Geraniums from cuttings				
	1,000 pots	*1,000 pots*	*Dollars*	*Dollars*	*1,000 dollars*
2000	44,004	18,423	1.41	2.62	110,223
2001	42,033	18,126	1.47	2.79	112,417
2002	41,293	18,116	1.50	2.76	111,819
2003	41,245	19,287	1.56	2.88	119,921
2004	40,509	21,028	1.61	2.97	127,671
2005	40,105	21,679	1.59	2.97	127,998
2006 [2]	23,991	12,246	1.58	3.29	78,244
2007	22,785	13,253	1.66	3.35	82,364
2008	23,094	13,050	1.76	3.35	84,300
2009	22,398	12,950	1.80	3.43	84,571
	Geraniums from seed				
	1,000 pots	*1,000 pots*	*Dollars*	*Dollars*	*1,000 dollars*
2000	46,834	1,295	0.83	2.08	41,756
2001	43,675	379	0.84	3.58	37,879
2002	40,451	1,365	0.84	1.65	36,273
2003	34,196	1,072	0.86	2.05	31,697
2004	36,671	887	0.90	2.01	34,785
2005	34,039	844	0.88	2.62	32,251
2006 [2]	31,377	456	0.81	2.20	26,545
2007	28,897	343	0.83	2.86	25,071
2008	25,114	367	0.87	3.31	23,186
2009	24,906	628	0.91	2.44	24,267
	Impatiens				
	1,000 pots	*1,000 pots*	*Dollars*	*Dollars*	*1,000 dollars*
2000	23,903	5,791	0.67	1.46	24,473
2001	26,839	4,340	0.71	1.76	26,736
2002	24,002	4,237	0.72	1.66	24,382
2003	26,557	4,788	0.71	1.71	26,989
2004	29,126	5,211	0.73	1.83	30,922
2005	30,819	4,457	0.73	2.08	31,867
2006 [2]	23,555	5,804	0.73	1.87	28,130
2007	19,575	5,575	0.81	1.69	25,267
2008	20,606	4,973	0.78	1.91	25,663
2009	22,145	4,677	0.75	1.94	25,731
	New guinea impatiens				
	1,000 pots	*1,000 pots*	*Dollars*	*Dollars*	*1,000 dollars*
2000	18,148	5,653	1.35	2.60	39,223
2001	16,382	6,678	1.40	2.35	38,601
2002	18,829	6,952	1.38	2.31	42,073
2003	18,135	7,051	1.43	2.54	43,790
2004	18,869	6,606	1.45	2.52	43,940
2005	19,105	7,074	1.44	2.65	46,320
2006 [2]	12,411	5,404	1.49	2.97	34,498
2007	11,885	4,735	1.58	2.83	32,180
2008	12,630	4,283	1.57	3.18	33,459
2009	11,984	4,220	1.55	3.22	32,182

See footnotes at end of table.

Table 5-103.—Potted annual bedding and garden plants: Sales and wholesale value for operations with $100,000+ sales, Surveyed States, 2000–2009—Continued

Year	Quantity sold		Wholesale price		Value of sales at wholesale [1]
	Less than 5 inches	5 inches or more	Less than 5 inches	5 inches or more	
			Marigolds		
	1,000 pots	*1,000 pots*	*Dollars*	*Dollars*	*1,000 dollars*
2000	4,994	1,111	0.78	1.33	5,368
2001	5,472	1,685	0.72	1.43	6,351
2002	7,760	1,497	0.76	1.35	7,895
2003	7,118	1,708	0.66	1.47	7,189
2004	8,583	2,732	0.72	1.54	10,417
2005	9,954	2,672	0.69	1.58	11,045
2006 [2]	7,928	2,207	0.69	1.79	9,392
2007	8,372	2,150	0.68	1.59	9,125
2008	7,600	1,704	0.72	1.61	8,185
2009	8,882	1,809	0.78	1.75	10,080
			Pansies/violas		
	1,000 pots	*1,000 pots*	*Dollars*	*Dollars*	*1,000 dollars*
2000	19,985	3,932	0.74	1.55	20,882
2001	18,756	5,175	0.78	1.55	22,648
2002	25,244	7,906	0.70	1.67	31,053
2003	27,291	7,885	0.70	1.76	33,026
2004	32,101	9,118	0.72	1.86	40,164
2005	38,362	10,729	0.73	1.88	48,188
2006 [2]	27,824	7,144	0.80	1.75	34,615
2007	28,021	6,145	0.74	2.13	33,788
2008	25,980	5,989	0.75	2.09	31,966
2009	26,980	5,531	0.74	2.06	31,290
			Petunias		
	1,000 pots	*1,000 pots*	*Dollars*	*Dollars*	*1,000 dollars*
2000	13,340	3,784	0.83	1.73	17,580
2001	14,724	5,280	0.90	2.11	24,389
2002	17,373	6,198	0.90	2.00	28,035
2003	17,268	7,237	0.92	2.12	31,190
2004	21,037	8,765	0.95	2.15	38,871
2005	22,714	9,664	0.92	2.28	43,000
2006 [2]	13,630	8,106	0.91	2.04	28,894
2007	18,551	6,935	0.98	2.37	34,558
2008	16,310	6,796	1.01	2.27	31,852
2009	17,465	7,424	1.01	2.19	33,982
			Other flowering and foliar plants		
	1,000 pots	*1,000 pots*	*Dollars*	*Dollars*	*1,000 dollars*
2000	159,357	53,236	0.96	2.35	277,692
2001	157,399	53,412	0.97	2.25	272,608
2002	169,081	56,477	1.02	2.29	301,859
2003	163,041	56,181	1.10	2.45	316,867
2004	175,070	64,210	1.05	2.47	343,192
2005	169,646	69,500	1.12	2.46	360,109
2006 [2]	110,614	57,880	1.09	2.73	278,296
2007	117,723	50,605	1.13	3.08	288,890
2008	108,251	55,789	1.20	2.65	278,250
2009	96,758	51,549	1.19	3.01	270,338
			Vegetable type plants [3]		
	1,000 pots	*1,000 pots*	*Dollars*	*Dollars*	*1,000 dollars*
2000	25,430	4,452	0.86	1.80	29,768
2001	24,930	4,625	0.91	1.85	31,309
2002	33,774	6,050	1.09	1.70	47,142
2003	42,492	6,305	0.93	1.82	51,028
2004	47,598	11,061	0.85	1.78	60,334
2005	48,033	10,491	0.84	2.17	63,270
2006 [2]	18,507	4,410	0.96	2.19	27,374
2007	27,676	3,874	1.01	2.22	36,668
2008	35,998	7,497	1.06	2.49	56,697
2009	37,888	12,111	1.05	2.53	70,596

[1] Equivalent wholesale value of all sales.   [2] Beginning in 2006, program was reduced to 15 States from 36.   [3] Does not include vegetable transplants grown for use in commercial vegetable production.
NASS, Crops Branch, (202) 720–2127.

Table 5-104.—Potted herbaceous perennial plants: Sales and wholesale value for operations with $100,000+ sales, Surveyed States, 2000–2009

Year	Quantity sold		Wholesale price		Value of sales at wholesale [1]
	Less than 5 inches	5 inches or more	Less than 5 inches	5 inches or more	
	Hardy/Garden Chrysanthemums				
	1,000 pots	*1,000 pots*	*Dollars*	*Dollars*	*1,000 dollars*
2000	17,813	48,534	1.11	1.78	106,385
2001	15,109	45,442	1.15	1.88	102,907
2002	12,705	50,295	1.00	2.03	114,524
2003	9,651	55,798	1.07	1.98	120,927
2004	14,421	55,457	1.06	2.18	136,149
2005	12,661	59,137	1.03	2.20	143,318
2006 [2]	10,545	40,303	.99	2.31	103,656
2007	9,763	35,148	1.06	2.66	103,831
2008	9,217	34,689	1.10	2.61	100,868
2009	8,949	34,876	1.09	2.84	108,901

Year	Quantity sold			Wholesale price			Value of sales at wholesale [1]
	Less than 1 gallon	1-2 gallons	2 gallons or more	Less than 1 gallon	1-2 gallons	2 gallons or more	
	Hosta						
	1,000 pots	*1,000 pots*	*1,000 pots*	*Dollars*	*Dollars*	*Dollars*	*1,000 dollars*
2000	3,358	7,269	296	2.02	3.67	8.28	35,874
2001	2,889	7,341	1,161	2.71	3.48	5.49	39,755
2002	3,827	7,834	439	2.33	3.76	7.68	41,771
2003	4,148	8,533	433	2.35	3.69	7.51	44,498
2004	3,166	10,200	408	2.20	3.52	7.41	45,876
2005	4,098	7,943	483	2.36	3.42	7.52	40,481
2006 [2]	1,724	6,661	399	2.10	3.25	6.66	27,924
2007	4,374	6,589	271	1.74	3.50	7.69	32,723
2008	4,451	6,793	173	1.91	3.43	9.10	33,391
2009	4,443	5,256	464	2.17	3.54	7.43	31,685
	Other Herbaceous Perennials						
	1,000 pots	*1,000 pots*	*1,000 pots*	*Dollars*	*Dollars*	*Dollars*	*1,000 dollars*
2000	66,995	56,181	3,904	1.47	3.04	5.84	291,734
2001	74,100	80,278	4,868	1.48	2.65	6.28	353,070
2002	98,314	90,170	8,527	1.44	2.92	5.96	455,793
2003	98,844	84,819	7,597	1.63	3.09	6.01	469,447
2004	93,765	97,291	9,011	1.67	3.00	6.33	505,553
2005	97,471	95,692	7,123	1.70	3.15	6.51	513,099
2006 [2]	49,602	85,660	5,075	1.55	3.14	6.97	381,511
2007	68,640	69,843	6,442	1.71	3.33	6.84	394,407
2008	65,928	74,574	7,533	1.59	3.43	6.65	410,867
2009	51,625	63,892	5,355	1.68	3.57	6.83	351,153

[1] Equivalent wholesale value of all sales.   [2] Beginning in 2006, program was reduced to 15 States from 36.
NASS, Crops Branch, (202) 720–2127.

Table 5-105.—Floriculture: Growing area by type of cover, all operations with $10,000+ sales, Surveyed States, 2008–2009

State	Glass greenhouses		Fiberglass and other rigid greenhouses		Film plastic (single/multi) greenhouses	
	2008	2009 [1]	2008	2009 [1]	2008	2009 [1]
	1,000 square feet	*1,000 square feet*	*1,000 square feet*	*1,000 square feet*	*1,000 square feet*	*1,000 square feet*
CA	11,870	14,676	39,235	37,598	53,249	59,485
FL	3,531	4,668	9,009	10,882	52,000	46,495
HI	*	*	*	*	2,761	2,675
IL	3,075	3,322	2,008	2,748	10,564	14,055
MD	1,301	1,590	685	1,158	4,449	5,609
MI	3,922	3,740	4,953	4,809	38,064	40,456
NJ	4,263	4,581	806	825	16,341	27,880
NY	3,844	3,783	1,071	1,070	18,558	18,189
NC	*	*	*	*	13,539	17,246
OH	8,117	8,656	2,344	2,444	18,281	19,439
OR	2,206	2,563	2,616	2,916	18,286	21,338
PA	2,044	2,101	2,356	2,023	15,331	16,116
SC	*	*	*	*	3,528	3,443
TX	1,121	2,415	3,117	5,764	37,689	30,388
WA	1,673	1,877	948	1,127	7,103	8,138
Oth Sts	6,246	6,157	2,429	2,865		
Total	53,213	60,129	71,575	76,229	309,743	330,952

State	Shade and temporary cover		Total covered area		Open ground [2]	
	2008	2009 [1]	2008	2009 [1]	2008	2009 [1]
	1,000 square feet	*1,000 square feet*	*1,000 square feet*	*1,000 square feet*	*Acres*	*Acres*
CA	33,321	30,281	137,673	142,040	12,653	14,086
FL	252,434	264,442	316,974	326,487	6,587	8,849
HI	16,894	17,404	21,455	21,486	1,219	1,275
IL	656	871	16,303	20,996	834	8,378
MD	313	1,067	6,748	9,424	491	4,146
MI	1,054	1,141	47,993	50,146	4,004	5,296
NJ	248	303	21,658	33,589	2,639	7,451
NY	531	405	24,004	23,447	1,382	2,589
NC	986	2,799	20,433	26,406	819	2,107
OH	416	487	29,158	31,026	639	3,932
OR	4,472	4,329	27,580	31,146	3,662	4,281
PA	398	459	20,129	20,699	622	1,618
SC	2,790	679	7,285	5,376	836	1,012
TX	6,813	13,487	48,740	52,054	1,328	3,323
WA	192	477	9,916	11,619	1,345	2,604
Total	321,518	338,631	756,049	805,941	39,060	70,948

[1] Totals are not comparable between years.   The area for 2009 includes nursery production area as well as Floriculture production area.   [2] Totals may not add due to rounding.   * Included in "Oth Sts" to avoid disclosure of individual operations.

NASS, Crops Branch, (202) 720-2127.

Table 5-106.—Floriculture Crops: Wholesale value of sales by category for operations with $100,000+ sales, Surveyed States, 2008–2009

State	Cut flowers		Potted flowering plants		Foliage for indoor or patio use	
	2008	2009	2008	2009	2008	2009
	1,000 dollars	*1,000 dollars*	*1,000 dollars*	*1,000 dollars*	*1,000 dollars*	*1,000 dollars*
CA	323,897	269,126	243,225	205,847	98,679	86,384
FL	9,046	8,136	120,685	130,928	351,993	311,119
HI	11,985	8,164	17,724	15,207	13,438	*
IL	1,946	1,167	15,042	26,655	3,644	1,562
MD	*	*	8,220	5,634	*	*
MI	*	*	32,872	31,042	3,085	9,807
NJ	13,300	11,058	23,324	18,449	*	*
NY	*	2,286	41,956	32,489	4,181	2,943
NC	*	*	39,635	32,218	10,375	11,745
OH	*	*	21,127	29,909	4,507	3,996
OR	10,509	10,598	21,730	18,949	440	*
PA	*	*	29,369	29,428	3,825	2,344
SC	*	*	13,468	15,169	*	*
TX	*	*	33,194	32,263	11,766	10,087
WA	23,086	26,073	6,599	8,169	349	*
15-State Program Oth Sts	23,329	22,619	....	....	4,029	14,353
Total	417,098	359,227	668,170	632,356	510,311	454,340

State	Annual bedding/garden plants[1]		Cut cultivated greens		Propagative materials		Total wholesale value of floriculture crops[2]	
	2008	2009	2008	2009	2008	2009	2008	2009
	1,000 dollars	*1,000 dollars*	*1,000 dollars*	*1,000 dollars*	*1,000 dollars*	*1,000 dollars*	*1,000 dollars*	*1,000 dollars*
CA	216,944	209,622	13,930	10,081	81,094	73,264	1,043,254	921,710
FL	97,456	88,447	69,114	55,730	80,203	71,084	738,905	674,190
HI	2,618	2,641	263	262	2,509	2,095	48,537	28,369
IL	46,644	43,622	45	*	5,706	5,983	127,427	124,531
MD	58,831	59,232	....	....	4,744	3,547	90,368	87,843
MI	186,938	180,157	*	*	76,651	84,311	375,744	384,427
NJ	64,950	63,854	*	*	22,665	16,581	170,470	156,986
NY	82,637	75,989	*	*	19,807	16,846	177,915	153,166
NC	114,416	162,274	*	....	*	*	199,751	238,346
OH	93,938	94,273	*	....	23,021	22,603	186,463	192,626
OR	51,044	58,375	7,061	*	7,064	5,713	121,525	114,432
PA	60,615	58,184	109	*	25,287	21,334	137,334	136,178
SC	12,250	10,235	*	*	*	*	98,838	65,739
TX	170,674	159,405	*	*	11,021	7,007	260,355	238,785
WA	48,599	50,366	....	*	20,087	23,527	123,731	131,609
15-State Program Oth Sts	....	....	1,226	7,883	3,977	3,765	24,226	37,017
Total	1,308,554	1,316,676	91,748	73,956	383,836	357,660	3,924,843	3,685,954

[1] Includes Annual Bedding Plants and Herbaceous Perennials. [2] State totals exclude plant category values denoted by asterisks (*). * Included in "Oth Sts" to avoid disclosure of individual operations.
NASS, Crops Branch, (202) 720-2127.

Table 5-107.—Fruit and orange juice: Cold storage holdings, end of month, United States, 2008 and 2009

| | Fresh | | | | | |
| Month | Apples, regular storage | | Apples, CA storage | | Apples, total | |
	2008	2009	2008	2009	2008	2009
	1,000 pounds	1,000 pounds	1,000 pounds	1,000 pounds	1,000 pounds	1,000 pounds
January	498,324	601,140	3,375,172	3,804,128	3,873,496	4,405,268
February	364,058	380,633	2,865,032	3,275,350	3,229,090	3,655,983
March	236,922	275,396	2,171,169	2,584,857	2,408,091	2,860,253
April	140,863	161,150	1,557,730	1,929,101	1,698,593	2,090,251
May	43,681	93,498	950,445	1,369,018	994,126	1,462,516
June	7,352	44,393	569,815	816,480	577,167	860,873
July	2,790	24,234	291,568	479,004	294,358	503,238
August	39,689	34,030	76,267	106,973	115,956	141,003
September ...	1,263,957	1,176,281	1,930,533	2,049,597	3,194,490	3,225,878
October	1,667,238	1,735,644	4,761,437	4,324,582	6,428,675	6,060,226
November	1,421,161	1,433,951	4,717,678	4,177,567	6,138,839	5,611,518
December	789,434	934,400	4,4458,611	3,959,086	5,248,045	4,893,486

| | Fresh | | | | | |
| Month | Pears, Bartlett | | Pears, other | | Pears, total | |
	2008	2009	2008	2009	2008	2009
	1,000 pounds	1,000 pounds	1,000 pounds	1,000 pounds	1,000 pounds	1,000 pounds
January	5,973	7,441	239,533	228,986	245,506	236,427
February	4,611	727	178,050	167,626	182,661	168,353
March	399	2,034	129,615	108,772	130,014	110,806
April	606	2,541	75,486	70,892	76,092	73,433
May	583	4,012	39,888	38,141	40,471	42,153
June	745	5,739	11,575	16,577	12,320	22,316
July	44,772	17,780	2,877	6,353	47,649	24,133
August	99,822	75,564	9,803	6,956	109,625	82,520
September ...	104,112	82,744	301,733	294,004	405,845	376,748
October	54,677	63,933	391,297	446,067	445,974	510,000
November	43,070	58,528	337,168	413,865	380,238	472,393
December	16,661	23,967	286,446	351,769	303,107	375,736

| | Frozen | | | | | | | |
| Month | Apples | | Apricots | | Blackberries, IQF | | Blackberries, pails & tubs | |
	2008	2009	2008	2009	2008	2009	2008	2009
	1,000 pounds	1,000 pounds	1,000 pounds	1,000 pounds	1,000 pounds	1,000 pounds	1,000 pounds	1,000 pounds
January	78,211	76,213	8,503	4,850	19,400	18,442	4,120	998
February	75,670	82,134	8,042	4,768	19,039	17,232	3,246	1,012
March	81,651	87,781	7,777	3,829	17,740	14,312	2,782	683
April	87,685	89,443	7,400	3,563	15,036	12,478	2,515	612
May	84,559	80,700	6,930	3,103	12,797	11,399	2,066	569
June	80,313	75,916	10,736	12,685	12,792	9,306	2,004	444
July	76,137	65,915	9,473	8,468	13,193	24,121	4,423	2,336
August	64,685	53,452	8,396	7,582	14,852	24,602	3,641	2,560
September ...	59,204	46,569	6,828	6,930	10,710	22,045	2,804	2,526
October	58,235	52,235	6,265	5,857	9,019	20,076	2,000	2,434
November	66,475	62,256	5,926	5,662	7,861	18,137	1,523	2,362
December	69,090	71,643	5,471	4,845	6,404	16,790	1,558	2,059

| | Frozen | | | | | | | |
| Month | Blackberries, barrels | | Blackberries, concentrate | | Blackberries, total | | Blueberries | |
	2008	2009	2008	2009	2008	2009	2008	2009
	1,000 pounds	1,000 pounds	1,000 pounds	1,000 pounds	1,000 pounds	1,000 pounds	1,000 pounds	1,000 pounds
January	3,809	2,873	26	57	27,355	22,370	100,696	141,089
February	3,247	2,508	46	45	25,578	20,797	94,652	130,964
March	2,084	2,410	67	100	22,673	17,505	78,106	115,615
April	1,574	2,116	81	50	19,206	15,256	66,604	100,820
May	1,331	1,676	63	98	16,257	13,742	58,282	87,790
June	1,145	1,215	48	116	15,989	11,081	51,960	81,404
July	3,597	13,010	110	241	36,058	39,708	69,783	106,327
August	4,933	12,601	209	336	40,136	40,099	172,145	189,893
September ...	5,148	12,437	149	321	39,871	37,329	182,478	189,628
October	4,565	12,092	240	253	35,328	34,855	174,334	173,849
November	4,351	11,711	196	183	31,335	32,393	164,703	154,647
December	4,104	8,576	67	205	26,148	27,630	153,445	141,883

See end of table.

Table 5-107.—Fruit and orange juice: Cold storage holdings, end of month, United States, 2008 and 2009—Continued

Month	Boysenberries		Cherries, Tart (RTP)		Cherries, Sweet		Grapes	
	2008	2009	2008	2009	2008	2009	2008	2009
	1,000 pounds	1,000 pounds	1,000 pounds	1,000 pounds	1,000 pounds	1,000 pounds	1,000 pounds	1,000 pounds
January	3,038	3,961	117,609	96,533	13,682	15,727	6,545	4,751
February	2,968	3,157	109,423	90,052	12,861	15,384	6,190	4,662
March	2,883	2,954	100,479	79,608	12,259	14,336	5,025	3,734
April	2,764	2,444	87,495	69,139	11,609	13,341	3,300	2,370
May	2,560	2,171	75,690	59,714	11,552	11,672	3,750	2,758
June	2,339	1,811	63,055	53,206	11,206	10,086	3,379	2,187
July	3,502	3,300	118,790	128,571	20,700	17,486	3,245	1,909
August	3,976	3,142	137,994	193,312	22,483	19,071	2,952	1,941
September	3,410	2,689	120,386	185,263	20,005	18,026	6,502	2,224
October	3,527	2,713	113,867	179,608	18,919	17,774	10,570	2,394
November	4,650	2,460	108,046	167,716	16,944	17,016	9,434	2,022
December	4,225	1,912	101,892	156,136	16,898	14,768	7,897	1,879

Month	Peaches		Raspberries, Black		Red Raspberries, IQF		Red Raspberries, pails & tubs	
	2008	2009	2008	2009	2008	2009	2008	2009
	1,000 pounds	1,000 pounds	1,000 pounds	1,000 pounds	1,000 pounds	1,000 pounds	1,000 pounds	1,000 pounds
January	68,716	69,001	1,066	1,403	13,193	13,078	4,423	6,528
February	62,681	59,525	1,064	1,161	14,852	12,548	3,641	5,631
March	52,331	45,757	910	824	10,710	12,039	2,804	5,497
April	41,348	40,755	742	752	9,019	11,326	2,000	4,652
May	34,291	34,667	723	692	7,861	10,285	1,523	4,099
June	27,422	29,049	661	602	6,404	8,645	1,558	3,465
July	37,217	32,155	1,562	2,226	23,081	26,723	8,666	12,837
August	54,995	48,217	2,458	2,109	22,479	24,521	10,373	11,818
September	87,006	59,280	2,366	2,060	21,000	23,481	10,365	10,322
October	87,140	56,959	1,616	2,006	20,534	21,601	9,541	9,332
November	82,656	57,070	1,523	1,693	18,894	19,582	8,785	8,528
December	76,356	52,278	1,544	1,612	13,693	17,216	7,600	7,432

Month	Red Raspberries, barrels		Red Raspberries, concentrate		Red Raspberries, total		Strawberries, IQF & Poly	
	2008	2009	2008	2009	2008	2009	2008	2009
	1,000 pounds	1,000 pounds	1,000 pounds	1,000 pounds	1,000 pounds	1,000 pounds	1,000 pounds	1,000 pounds
January	7,917	8,078	1,793	1,471	27,326	29,155	126,525	93,360
February	5,458	6,235	1,398	1,403	25,349	25,817	125,684	90,500
March	4,428	5,002	1,624	1,617	19,566	24,155	92,110	77,758
April	2,986	3,142	1,038	1,677	15,043	20,797	107,184	104,227
May	2,170	2,602	994	1,756	12,548	18,742	144,920	120,386
June	1,590	2,150	1,192	1,703	10,744	15,963	161,974	150,114
July	16,303	34,500	1,042	1,203	49,092	75,263	181,450	154,025
August	21,335	27,436	1,195	1,647	55,382	65,422	170,211	157,649
September	16,092	23,721	1,637	1,609	49,094	59,133	155,690	146,227
October	13,747	20,049	2,051	1,477	45,873	52,459	139,981	131,471
November	12,344	18,944	1,838	1,285	41,861	48,339	129,320	125,403
December	11,139	15,784	1,651	1,140	34,083	41,572	112,229	114,783

Month	Strawberries, pails & tubs		Strawberries, barrels & drums		Strawberries, juice stock		Strawberries, total	
	2008	2009	2008	2009	2008	2009	2008	2009
	1,000 pounds	1,000 pounds	1,000 pounds	1,000 pounds	1,000 pounds	1,000 pounds		
January	83,574	58,547	35,293	33,391	11,460	13,415	256,852	198,713
February	69,551	50,318	33,186	29,960	7,932	11,334	236,353	182,112
March	58,001	42,025	26,686	21,210	5,365	10,911	182,162	151,904
April	52,930	57,879	30,946	28,257	14,402	22,278	205,462	212,641
May	58,902	73,163	40,154	55,880	14,769	31,704	258,745	281,133
June	86,321	117,062	67,645	91,458	15,069	31,946	331,009	390,580
July	108,287	125,249	108,287	104,033	17,601	40,571	379,197	423,878
August	97,503	116,560	97,503	98,728	19,607	39,829	347,910	412,766
September	98,871	108,399	98,871	101,672	18,118	33,079	329,502	389,377
October	87,999	98,739	87,999	98,603	15,084	35,798	296,064	364,611
November	77,326	90,806	77,326	95,592	14,065	30,262	271,843	342,063
December	66,976	81,688	66,976	92,961	11,022	33,020	235,241	322,452

See end of table.

Table 5-107.—Fruit and orange juice: Cold storage holdings, end of month, United States, 2008 and 2009—Continued

Month	Other fruit		Total frozen fruit		Orange juice	
	2008	2009	2008	2009	2008	2009
	1,000 pounds	1,000 pounds	1,000 pounds	1,000 pounds	1,000 pounds	1,000 pounds
January	425,313	467,363	1,140,318	1,134,205	837,485	1,193,217
February	402,895	414,907	1,068,744	1,038,062	942,397	1,261,198
March	350,167	360,587	920,547	910,992	1,031,926	1,291,436
April	312,823	343,101	866,161	916,769	1,210,553	1,415,409
May	269,330	311,835	839,793	910,555	1,442,657	1,497,199
June	251,309	286,675	864,126	973,150	1,514,929	1,519,689
July	234,242	246,058	1,043,092	1,154,271	1,424,921	1,404,336
August	206,201	215,985	1,125,683	1,257,806	1,319,388	1,316,688
September ...	219,247	198,200	1,130,651	1,200,655	1,199,739	1,252,154
October	537,319	551,910	1,393,584	1,501,050	1,085,951	1,150,290
November	527,876	530,591	1,337,494	1,428,198	1,034,557	1,127,234
December	509,807	486,760	1,245,637	1,328,637	1,088,047	1,185,215

NASS, Livestock Branch, (202) 720–3570.

Table 5-108.—Nuts: Cold storage holdings, end of month, United States, 2007 and 2008

Month	Peanuts					
	Shelled		In-shell		Total	
	2007	2008	2007	2008	2007	2008
	1,000 pounds	1,000 pounds	1,000 pounds	1,000 pounds	1,000 pounds	1,000 pounds
January	346,696	279,796	10,647	15,063	357,343	294,859
February	345,665	297,411	16,315	14,914	361,980	312,325
March	391,038	347,955	20,310	16,681	411,348	364,636
April	390,132	347,491	15,619	20,207	405,751	367,698
May	370,406	324,145	18,771	23,096	389,177	347,241
June	337,598	318,268	19,802	19,731	357,400	337,999
July	327,967	291,747	17,651	18,228	345,618	309,975
August	320,788	247,958	10,033	16,893	330,821	264,851
September	305,441	180,968	8,618	10,980	314,059	191,948
October	252,134	208,643	9,358	16,570	261,492	225,213
November	250,598	242,894	8,975	17,305	259,573	260,199
December	251,651	266,624	12,006	19,762	263,657	286,386

Month	Pecans					
	Shelled		In-shell		Total	
	2007	2008	2007	2008	2007	2008
	1,000 pounds	1,000 pounds	1,000 pounds	1,000 pounds	1,000 pounds	1,000 pounds
January	34,717	35,279	114,912	202,709	149,629	237,988
February	36,309	40,683	110,515	228,997	146,824	269,680
March	35,362	43,942	99,022	226,176	134,384	270,118
April	40,724	47,175	75,069	204,857	115,793	252,032
May	40,909	51,691	56,666	176,390	97,575	228,081
June	41,154	51,250	45,214	154,915	86,368	206,165
July	40,515	51,845	30,534	126,774	71,049	178,619
August	37,143	52,146	17,902	91,942	55,045	144,088
September	25,683	50,499	11,091	75,071	36,774	125,570
October	19,238	47,638	18,505	43,500	37,743	91,138
November	19,761	44,352	57,307	55,487	77,068	99,839
December	23,952	36,992	113,933	85,806	137,885	122,798

NASS, Livestock Branch, (202) 720-3570.

CHAPTER VI

STATISTICS OF HAY, SEEDS, AND MINOR FIELD CROPS

Chapter VI deals with hay, pasture, seeds, and various minor field crops.

Table 6-1.—Hay, all: Area, yield, and production, by State and United States, 2007–2009

State	Area harvested			Yield per harvested acre			Production		
	2007	2008	2009 [1]	2007	2008	2009 [1]	2007	2008	2009 [1]
	1,000 acres	*1,000 acres*	*1,000 acres*	*Tons*	*Tons*	*Tons*	*1,000 tons*	*1,000 tons*	*1,000 tons*
AL	840	900	800	1.80	2.20	2.40	1,512	1,980	1,920
AZ	295	295	310	7.43	8.08	8.16	2,192	2,383	2,530
AR	1,465	1,405	1,415	2.11	2.21	2.21	3,084	3,111	3,131
CA	1,570	1,610	1,520	5.76	5.85	5.68	9,042	9,414	8,632
CO	1,570	1,570	1,600	2.84	2.54	2.99	4,459	3,981	4,778
CT	61	55	62	1.95	2.18	2.10	119	120	130
DE	15	18	17	2.07	2.56	3.00	31	46	51
FL	320	300	300	3.00	3.00	2.70	960	900	810
GA	670	720	700	1.90	2.20	2.30	1,273	1,584	1,610
ID	1,450	1,410	1,510	3.69	3.96	3.66	5,345	5,588	5,528
IL	680	620	610	2.82	3.03	3.28	1,916	1,878	2,001
IN	610	590	620	2.32	3.16	2.77	1,416	1,867	1,720
IA	1,380	1,550	1,220	3.58	3.44	3.28	4,944	5,330	4,002
KS	2,900	2,750	2,550	2.25	2.46	2.83	6,530	6,765	7,225
KY	2,680	2,640	2,520	1.53	1.95	2.50	4,104	5,160	6,290
LA	420	430	380	2.70	2.50	2.80	1,134	1,075	1,064
ME	144	138	149	1.85	1.57	1.70	266	217	253
MD	215	205	210	2.19	3.05	2.72	470	626	571
MA	79	73	81	1.87	2.11	1.81	148	154	147
MI	1,050	1,020	990	2.31	2.58	2.51	2,429	2,633	2,482
MN	1,800	1,950	2,050	2.36	2.70	2.56	4,240	5,265	5,250
MS	800	720	700	2.30	2.70	2.80	1,840	1,944	1,960
MO	4,050	4,200	3,880	1.86	2.10	2.07	7,528	8,820	8,040
MT	2,600	2,400	2,500	1.96	1.70	1.91	5,090	4,080	4,770
NE	2,650	2,570	2,700	2.33	2.42	2.31	6,185	6,232	6,235
NV	460	455	490	3.36	3.58	3.54	1,544	1,629	1,736
NH	55	53	57	1.95	1.98	1.56	107	105	89
NJ	115	115	110	1.79	2.08	2.11	206	239	232
NM	350	340	320	4.32	4.46	4.33	1,512	1,516	1,384
NY	1,360	1,320	1,360	1.99	2.04	1.82	2,700	2,691	2,472
NC	699	808	847	1.50	2.01	2.31	1,050	1,622	1,957
ND	2,680	3,220	2,960	1.89	1.28	1.77	5,063	4,118	5,240
OH	1,160	1,140	1,040	2.42	2.46	2.77	2,804	2,802	2,876
OK	3,140	2,910	3,220	2.18	1.90	1.64	6,858	5,536	5,278
OR	1,010	1,025	1,030	2.91	2.88	3.15	2,941	2,951	3,249
PA	1,800	1,750	1,550	2.33	2.18	2.36	4,200	3,810	3,655
RI	8	7	7	1.88	2.00	2.00	15	14	14
SC	330	330	350	1.70	1.90	2.40	561	627	840
SD	3,750	3,850	3,800	1.94	2.04	2.06	7,275	7,840	7,830
TN	1,775	1,870	1,915	1.51	2.11	2.21	2,685	3,945	4,236
TX	5,340	4,430	4,620	2.76	2.08	1.79	14,740	9,211	8,250
UT	700	695	690	3.69	3.78	3.71	2,585	2,629	2,562
VT	190	180	190	2.12	1.70	1.69	402	306	322
VA	1,290	1,270	1,180	1.86	2.16	2.26	2,394	2,748	2,668
WA	790	710	810	4.23	3.68	4.07	3,338	2,614	3,297
WV	600	605	625	1.54	1.85	1.85	924	1,117	1,158
WI	1,970	1,900	1,920	2.23	2.53	2.31	4,392	4,810	4,430
WY	1,120	1,030	1,270	2.10	2.17	2.00	2,348	2,237	2,537
US	61,006	60,152	59,755	2.41	2.43	2.47	146,901	146,270	147,442

[1] Preliminary.
NASS, Crops Branch, (202) 720–2127.

HAY, SEEDS, AND MINOR FIELD CROPS

Table 6-2.—Hay, all: Area, yield, production, and value, United States, 2000–2009

Year	Area harvested	Yield per acre	Production	Marketing year average price per ton received by farmers	Value of production
	1,000 acres	*Tons*	*1,000 tons*	*Dollars*	*1,000 dollars*
2000	60,355	2.54	153,603	84.60	11,556,882
2001	63,516	2.46	156,416	96.50	12,589,493
2002	63,942	2.34	149,467	92.40	12,338,010
2003	63,371	2.48	157,390	85.50	11,987,318
2004	61,944	2.55	158,122	92.00	12,198,171
2005	61,637	2.44	150,461	98.20	12,533,762
2006	60,632	2.32	140,783	110.00	13,633,837
2007	61,006	2.41	146,901	128.00	16,842,233
2008	60,152	2.43	146,270	152.00	18,638,748
2009 [1]	59,755	2.47	147,442	111.00	14,990,083

[1] Preliminary.
NASS, Crops Branch, (202) 720–2127.

Table 6-3.—Hay, alfalfa and alfalfa mixtures: Area, yield, and production, by State and United States, 2007–2009

State	Area harvested			Yield per harvested acre			Production		
	2007	2008	2009 [1]	2007	2008	2009 [1]	2007	2008	2009 [1]
	1,000 acres	*1,000 acres*	*1,000 acres*	*Tons*	*Tons*	*Tons*	*1,000 tons*	*1,000 tons*	*1,000 tons*
AZ	255	260	280	8.00	8.60	8.50	2,040	2,236	2,380
AR	15	15	15	2.60	3.50	3.40	39	53	51
CA	990	1,030	980	7.20	7.00	7.10	7,128	7,210	6,958
CO	820	820	850	3.70	3.30	3.90	3,034	2,706	3,315
CT	8	9	7	2.30	2.50	2.00	18	23	14
DE	5	6	5	2.60	3.30	3.90	13	20	20
ID	1,150	1,130	1,140	4.10	4.40	4.20	4,715	4,972	4,788
IL	380	350	340	3.70	3.90	3.90	1,406	1,365	1,326
IN	280	300	300	2.70	4.00	3.60	756	1,200	1,080
IA	1,060	1,150	920	4.00	3.80	3.60	4,240	4,370	3,312
KS	800	700	850	3.70	4.10	4.30	2,960	2,870	3,655
KY	280	240	220	1.80	2.50	3.50	504	600	770
ME	9	8	9	2.50	2.70	1.70	23	22	15
MD	40	45	40	3.00	4.30	4.50	120	194	180
MA	9	8	6	2.40	2.10	2.00	22	17	12
MI	770	770	700	2.50	2.90	2.80	1,925	2,233	1,960
MN	1,100	1,350	1,300	2.90	3.10	3.00	3,190	4,185	3,900
MO	400	350	280	2.85	3.20	3.00	1,140	1,120	840
MT	1,700	1,600	1,700	2.20	1.90	2.10	3,740	3,040	3,570
NE	1,100	970	950	3.65	3.95	3.80	4,015	3,832	3,610
NV	265	270	280	4.50	4.80	4.70	1,193	1,296	1,316
NH	5	5	7	2.40	2.80	2.00	12	14	14
NJ	20	20	25	2.70	2.90	2.80	54	58	70
NM	240	250	240	5.20	5.20	5.10	1,248	1,300	1,224
NY	420	350	350	2.40	2.70	2.30	1,008	945	805
NC	9	8	7	1.70	2.70	3.60	15	22	25
ND	1,550	1,660	1,780	2.10	1.40	1.85	3,255	2,324	3,293
OH	440	420	380	3.10	2.90	3.40	1,364	1,218	1,292
OK	340	310	320	3.70	3.60	2.90	1,258	1,116	928
OR	410	420	400	4.10	4.00	4.50	1,681	1,680	1,800
PA	600	550	500	3.00	3.00	2.90	1,800	1,650	1,450
RI	1	1	1	1.80	2.70	1.70	2	3	2
SD	2,200	2,400	2,500	2.25	2.30	2.30	4,950	5,520	5,750
TN	25	20	15	2.40	3.00	3.70	60	60	56
TX	140	130	120	5.00	4.70	5.00	700	611	600
UT	550	550	530	4.10	4.20	4.20	2,255	2,310	2,226
VT	30	30	35	2.20	1.70	2.10	66	51	74
VA	90	90	90	2.60	3.00	3.00	234	270	270
WA	440	410	490	5.20	4.40	4.90	2,288	1,804	2,401
WV	30	25	25	2.30	2.90	3.10	69	73	78
WI	1,550	1,500	1,550	2.40	2.70	2.50	3,720	4,050	3,875
WY	600	530	690	2.70	2.90	2.50	1,620	1,537	1,725
US	21,126	21,060	21,227	3.31	3.33	3.35	69,880	70,180	71,030

[1] Preliminary.
NASS, Crops Branch, (202) 720–2127.

Table 6-4.—Hay, all other: Area, yield, and production, by State and United States, 2007–2009

State	Area harvested			Yield per harvested acre			Production		
	2007	2008	2009 [1]	2007	2008	2009 [1]	2007	2008	2009 [1]
	1,000 acres	*1,000 acres*	*1,000 acres*	*Tons*	*Tons*	*Tons*	*1,000 tons*	*1,000 tons*	*1,000 tons*
AL	840	900	800	1.80	2.20	2.40	1,512	1,980	1,920
AZ	40	35	30	3.80	4.20	5.00	152	147	150
AR	1,450	1,390	1,400	2.10	2.20	2.20	3,045	3,058	3,080
CA	580	580	540	3.30	3.80	3.10	1,914	2,204	1,674
CO	750	750	750	1.90	1.70	1.95	1,425	1,275	1,463
CT	53	46	55	1.90	2.10	2.10	101	97	116
DE	10	12	12	1.80	2.20	2.60	18	26	31
FL	320	300	300	3.00	3.00	2.70	960	900	810
GA	670	720	700	1.90	2.20	2.30	1,273	1,584	1,610
ID	300	280	370	2.10	2.20	2.00	630	616	740
IL	300	270	270	1.70	1.90	2.50	510	513	675
IN	330	290	320	2.00	2.30	2.00	660	667	640
IA	320	400	300	2.20	2.40	2.30	704	960	690
KS	2,100	2,050	1,700	1.70	1.90	2.10	3,570	3,895	3,570
KY	2,400	2,400	2,300	1.50	1.90	2.40	3,600	4,560	5,520
LA	420	430	380	2.70	2.50	2.80	1,134	1,075	1,064
ME	135	130	140	1.80	1.50	1.70	243	195	238
MD	175	160	170	2.00	2.70	2.30	350	432	391
MA	70	65	75	1.80	2.10	1.80	126	137	135
MI	280	250	290	1.80	1.60	1.80	504	400	522
MN	700	600	750	1.50	1.80	1.80	1,050	1,080	1,350
MS	800	720	700	2.30	2.70	2.80	1,840	1,944	1,960
MO	3,650	3,850	3,600	1.75	2.00	2.00	6,388	7,700	7,200
MT	900	800	800	1.50	1.30	1.50	1,350	1,040	1,200
NE	1,550	1,600	1,750	1.40	1.50	1.50	2,170	2,400	2,625
NV	195	185	210	1.80	1.80	2.00	351	333	420
NH	50	48	50	1.90	1.90	1.50	95	91	75
NJ	95	95	85	1.60	1.90	1.90	152	181	162
NM	110	90	80	2.40	2.40	2.00	264	216	160
NY	940	970	1,010	1.80	1.80	1.65	1,692	1,746	1,667
NC	690	800	840	1.50	2.00	2.30	1,035	1,600	1,932
ND	1,130	1,560	1,180	1.60	1.15	1.65	1,808	1,794	1,947
OH	720	720	660	2.00	2.20	2.40	1,440	1,584	1,584
OK	2,800	2,600	2,900	2.00	1.70	1.50	5,600	4,420	4,350
OR	600	605	630	2.10	2.10	2.30	1,260	1,271	1,449
PA	1,200	1,200	1,050	2.00	1.80	2.10	2,400	2,160	2,205
RI	7	6	6	1.90	1.90	2.00	13	11	12
SC	330	330	350	1.70	1.90	2.40	561	627	840
SD	1,550	1,450	1,300	1.50	1.60	1.60	2,325	2,320	2,080
TN	1,750	1,850	1,900	1.50	2.10	2.20	2,625	3,885	4,180
TX	5,200	4,300	4,500	2.70	2.00	1.70	14,040	8,600	7,650
UT	150	145	160	2.20	2.20	2.10	330	319	336
VT	160	150	155	2.10	1.70	1.60	336	255	248
VA	1,200	1,180	1,090	1.80	2.10	2.20	2,160	2,478	2,398
WA	350	300	320	3.00	2.70	2.80	1,050	810	896
WV	570	580	600	1.50	1.80	1.80	855	1,044	1,080
WI	420	400	370	1.60	1.90	1.50	672	760	555
WY	520	500	580	1.40	1.40	1.40	728	700	812
US	39,880	39,092	38,528	1.93	1.95	1.98	77,021	76,090	76,412

[1] Preliminary.
NASS, Crops Branch, (202) 720–2127.

Table 6-5.—Hay, all: Stocks on farms, United States, 2000–2009

Crop year	Dec. 1	May 1 [1]
	1,000 tons	*1,000 tons*
2000	106,412	21,248
2001	110,384	22,458
2002	102,978	22,013
2003	111,011	25,947
2004	114,489	27,758
2005	105,181	21,345
2006	96,400	14,990
2007	104,089	21,585
2008	103,658	22,065
2009 [2]	107,222	20,913

[1] Following year.   [2] Preliminary.
NASS, Crops Branch, (202) 720–2127.

Table 6-6.—Hay, all: Marketing year average price and value of production, by State and United States, 2007–2009

State	Marketing year average price per ton, baled			Value of production		
	2007	2008	2009[1]	2007	2008	2009[1]
	Dollars	*Dollars*	*Dollars*	*1,000 dollars*	*1,000 dollars*	*1,000 dollars*
AL	97.00	98.00	77.00	146,664	194,040	147,840
AZ	150.00	184.00	122.00	328,864	438,356	309,860
AR	91.50	88.00	73.00	281,616	273,996	229,012
CA	154.00	191.00	107.00	1,405,800	1,797,032	927,496
CO	138.00	161.00	136.00	606,976	612,084	633,030
CT	189.00	226.00	209.00	22,501	27,103	27,220
DE	214.00	165.00	143.00	6,642	7,600	7,311
FL	116.00	136.00	137.00	111,360	122,400	110,970
GA	80.00	92.00	85.00	101,840	145,728	136,850
ID	142.00	198.00	114.00	752,995	1,091,772	615,370
IL	131.00	140.00	116.00	245,152	256,059	223,848
IN	151.00	144.00	127.00	213,192	268,169	218,720
IA	113.00	129.00	106.00	547,440	678,270	420,072
KS	99.50	116.00	103.00	614,770	699,460	664,275
KY	105.00	114.00	102.00	385,200	543,360	569,680
LA	79.00	87.00	102.00	89,586	93,525	108,528
ME	151.00	178.00	178.00	40,196	38,685	45,145
MD	214.00	164.00	143.00	100,470	102,488	81,591
MA	186.00	220.00	209.00	27,470	33,909	30,774
MI	124.00	153.00	142.00	299,411	401,948	352,454
MN	111.00	125.00	102.00	468,780	649,890	537,375
MS	65.00	66.00	64.00	119,600	128,304	125,440
MO	110.00	103.00	78.50	746,372	834,820	581,400
MT	78.50	116.00	95.50	400,760	471,120	455,520
NE	88.50	88.50	75.00	532,455	537,344	445,150
NV	148.00	187.00	108.00	227,108	301,590	191,632
NH	180.00	214.00	209.00	19,312	22,435	18,636
NJ	151.00	145.00	123.00	31,034	34,643	28,640
NM	164.00	186.00	152.00	244,584	280,480	208,656
NY	122.00	135.00	125.00	330,552	367,398	310,154
NC	91.50	100.00	92.50	95,880	162,910	180,578
ND	57.00	79.50	65.00	276,299	306,974	326,587
OH	136.00	143.00	124.00	382,020	406,260	351,240
OK	74.00	93.50	111.00	554,850	503,320	509,646
OR	157.00	198.00	132.00	466,353	581,756	464,832
PA	175.00	173.00	133.00	756,600	676,620	501,885
RI	189.00	225.00	211.00	2,835	3,145	2,960
SC	125.00	115.00	105.00	70,125	72,105	88,200
SD	97.00	93.00	78.00	685,650	716,400	606,395
TN	93.00	101.00	76.50	249,375	397,635	323,020
TX	135.00	119.00	124.00	1,940,700	1,006,824	929,550
UT	129.00	167.00	113.00	332,695	436,403	288,918
VT	150.00	170.00	166.00	60,390	52,122	53,318
VA	132.00	146.00	129.00	316,494	400,476	347,778
WA	149.00	222.00	139.00	498,224	581,302	456,603
WV	85.50	98.00	95.50	79,053	109,743	110,526
WI	85.00	107.00	92.50	370,056	517,690	414,448
WY	109.00	114.00	108.00	255,932	253,055	270,950
US	128.00	152.00	111.00	16,842,233	18,638,748	14,990,083

[1] Preliminary.
NASS, Crops Branch, (202) 720–2127.

Table 6-7.—Hay: Area and production, by type, United States, 2000–2009

Year	Area harvested			Production		
	Alfalfa	All other hay	All hay	Alfalfa	All other hay	All hay
	1,000 acres	*1,000 acres*	*1,000 acres*	*1,000 acres*	*1,000 acres*	*1,000 acres*
2000	23,463	36,892	60,355	81,520	72,083	153,603
2001	23,952	39,564	63,516	80,354	76,062	156,416
2002	22,923	41,019	63,942	73,014	76,453	149,467
2003	23,527	39,844	63,371	76,098	81,292	157,390
2004	21,697	40,247	61,944	75,375	82,747	158,122
2005	22,359	39,278	61,637	75,610	74,851	150,461
2006	21,138	39,494	60,632	70,548	70,235	140,783
2007	21,126	39,880	61,006	69,880	77,021	146,901
2008	21,060	39,092	60,152	70,180	76,090	146,270
2009[1]	21,227	38,528	59,755	71,030	76,412	147,442

[1] Preliminary.
NASS, Crops Branch, (202) 720–2127.

Forage production is the sum of all dry hay production and haylage/greenchop production after converting the haylage/greenchop production to a dry equivalent basis (13 percent moisture) by multiplying the green weight (weight at harvest) by 0.4943. The conversion factor (0.4943) is based on the assumption that one ton of dry hay is 0.87 ton of dry matter, one ton of haylage is 0.45 ton dry matter and one ton of greenchop is 0.25 ton dry matter. The total haylage/greenchop production is assumed to be comprised of 90 percent haylage and 10 percent greenchop. Therefore, the conversion factor used to adjust haylage/greenchop production to a dry equivalent basis = ((0.45*0.9)+(0.25*0.1))/0.87 = 0.4943. The factors assumed here may vary by State and can be adjusted. Adjustments would result in a slightly different conversion factor.

Table 6-8.—All forage: Area harvested, yield, and production, by State and 18 State total, 2007–2009 [1]

State	Area harvested			Yield		
	2007	2008	2009	2007	2008	2009
	1,000 acres	*1,000 acres*	*1,000 acres*	*Tons*	*Tons*	*Tons*
CA	1,815	1,930	1,820	5.98	6.12	6.05
ID	1,528	1,475	1,560	3.80	4.18	3.80
IL	715	650	650	2.89	3.06	3.33
IA	1,460	1,615	1,265	3.64	3.53	3.34
KS	3,030	2,810	2,605	2.29	2.47	2.86
MI	1,270	1,250	1,200	2.62	2.81	2.73
MN	2,055	2,150	2,290	2.49	2.77	2.69
MO	4,105	4,260	3,905	1.87	2.13	2.08
NE	2,665	2,585	2,715	2.38	2.47	2.35
NM	378	376	365	4.30	4.45	4.26
NY	1,850	1,830	1,830	2.64	2.73	2.60
OH	1,245	1,210	1,150	2.52	2.58	2.95
PA	2,045	1,915	1,800	2.67	2.62	2.89
SD	3,830	3,895	3,870	1.95	2.04	2.07
TX	5,495	4,550	4,740	2.78	2.13	1.81
VT	315	310	315	3.07	2.95	2.75
WA	835	770	878	4.50	3.81	4.19
WI	2,850	2,900	2,800	3.13	3.34	3.12
18 State Total	37,486	36,481	35,758	2.80	2.84	2.78

State	Production		
	2007	2008	2009
	1,000 tons	*1,000 tons*	*1,000 tons*
CA	10,854	11,808	11,020
ID	5,813	6,166	5,925
IL	2,067	1,992	2,163
IA	5,319	5,705	4,226
KS	6,928	6,945	7,440
MI	3,324	3,512	3,273
MN	5,119	5,957	6,151
MO	7,687	9,067	8,107
NE	6,342	6,381	6,370
NM	1,627	1,672	1,556
NY	4,890	4,990	4,756
OH	3,143	3,123	3,394
PA	5,456	5,015	5,207
SD	7,470	7,953	8,016
TX	15,284	9,677	8,602
VT	968	913	866
WA	3,756	2,937	3,682
WI	8,912	9,674	8,730
18 State Total	104,959	103,487	99,484

[1] All forage production is the sum of the following dry equivalents: alfalfa hay harvested as dry hay, all other hay harvested as dry hay, alfalfa haylage and greenchop, all other hay haylage and greenchop; after converting alfalfa and all other haylage and greenchop to a dry equivalent basis.
NASS, Crops Branch, (202) 720–2127.

Table 6-9.—All alfalfa forage: Area harvested, yield, and production, by State and 18 State total, 2007–2009 [1]

State	Area harvested			Yield		
	2007	2008	2009	2007	2008	2009
	1,000 acres	1,000 acres	1,000 acres	Tons	Tons	Tons
CA	1,015	1,050	1,020	7.30	7.07	7.08
ID	1,215	1,190	1,175	4.22	4.65	4.36
IL	400	370	360	3.81	3.94	3.96
IA	1,130	1,200	950	4.04	3.91	3.67
KS	830	740	890	3.73	4.05	4.26
MI	980	990	900	2.85	3.12	3.01
MN	1,300	1,515	1,500	3.03	3.17	3.14
MO	415	360	290	2.89	3.32	3.00
NE	1,110	980	955	3.73	4.03	3.86
NM	250	259	252	5.12	5.16	4.99
NY	700	690	680	3.63	3.86	3.55
OH	500	470	460	3.33	3.17	3.82
PA	745	665	685	3.71	3.97	3.92
SD	2,245	2,430	2,550	2.26	2.31	2.30
TX	160	140	132	4.63	4.61	4.79
VT	75	75	70	3.92	4.00	3.86
WA	450	425	508	5.28	4.40	4.83
WI	2,350	2,450	2,350	3.43	3.55	3.39
18 State Total	15,870	15,999	15,727	3.69	3.77	3.71

State	Production		
	2007	2008	2009
	1,000 tons	1,000 tons	1,000 tons
CA	7,405	7,424	7,225
ID	5,130	5,536	5,126
IL	1,524	1,457	1,424
IA	4,569	4,686	3,491
KS	3,098	2,994	3,791
MI	2,790	3,087	2,705
MN	3,944	4,801	4,716
MO	1,200	1,194	870
NE	4,135	3,953	3,688
NM	1,279	1,336	1,257
NY	2,543	2,664	2,412
OH	1,663	1,490	1,756
PA	2,765	2,638	2,687
SD	5,076	5,603	5,871
TX	740	645	632
VT	294	300	270
WA	2,377	1,868	2,455
WI	8,057	8,687	7,958
18 State Total	58,589	60,363	58,334

[1] All alfalfa forage production is the sum of alfalfa harvested as dry hay; and alfalfa haylage and greenchop production after converting it to a dry equivalent basis.
NASS, Crops Branch, (202) 720–2127.

Table 6-10.—All haylage and greenchop: Area harvested, yield, and production, by State and 18 State total, 2007–2009 [1]

State	Area harvested			Yield		
	2007	2008	2009	2007	2008	2009
	1,000 acres	1,000 acres	1,000 acres	Tons	Tons	Tons
CA	310	390	320	11.83	12.42	15.09
ID	88	82	80	10.77	14.25	10.04
IL	53	45	48	5.74	5.13	6.85
IA	105	120	75	7.23	6.33	6.07
KS	155	75	70	5.19	4.84	6.21
MI	270	285	315	6.70	6.24	5.08
MN	305	250	290	5.83	5.60	6.28
MO	100	100	25	3.23	5.00	5.40
NE	50	45	45	6.34	6.68	6.09
NM	28	36	45	8.32	8.75	7.71
NY	700	700	630	6.33	6.64	7.34
OH	147	124	144	4.67	5.24	7.28
PA	450	370	450	5.65	6.58	6.98
SD	93	55	70	4.25	4.15	5.39
TX	173	130	120	6.36	7.24	5.93
VT	170	170	165	6.74	7.22	6.67
WA	90	75	100	9.39	8.70	7.80
WI	1,450	1,500	1,500	6.31	6.56	5.80
18 State Total	4,737	4,552	4,492	6.59	7.09	7.02

State	Production		
	2007	2008	2009
	1,000 tons	1,000 tons	1,000 tons
CA	3,666	4,842	4,830
ID	948	1,169	803
IL	304	231	329
IA	759	760	455
KS	805	363	435
MI	1,810	1,778	1,601
MN	1,778	1,401	1,822
MO	323	500	135
NE	317	301	274
NM	233	315	347
NY	4,430	4,651	4,622
OH	686	650	1,049
PA	2,541	2,438	3,141
SD	395	228	377
TX	1,101	941	712
VT	1,145	1,229	1,100
WA	845	653	780
WI	9,145	9,840	8,700
18 State Total	31,231	32,290	31,512

[1] Includes all types of forage harvested as haylage or greenchop (green weight). Forage harvested as dry hay and corn and sorghum silage/greenchop are not included.
NASS, Crops Branch, (202) 720–2127.

Table 6-11.—Alfalfa haylage and greenchop: Area harvested, yield, and production, by State and 18 State total, 2007–2009 [1]

State	Area harvested			Yield		
	2007	2008	2009	2007	2008	2009
	1,000 acres	*1,000 acres*	*1,000 acres*	*Tons*	*Tons*	*Tons*
CA	85	90	60	6.60	4.80	9.00
ID	73	77	65	11.50	14.80	10.50
IL	36	35	24	6.60	5.30	8.30
IA	90	100	55	7.40	6.40	6.60
KS	50	50	50	5.60	5.00	5.50
MI	250	270	290	7.00	6.40	5.20
MN	250	215	250	6.10	5.80	6.60
MO	33	30	10	3.70	5.00	6.00
NE	35	35	25	6.90	7.00	6.30
NM	10	9	12	6.30	8.00	5.50
NY	450	470	440	6.90	7.40	7.39
OH	112	95	124	5.40	5.80	7.57
PA	310	270	325	6.30	7.40	7.70
SD	58	40	50	4.40	4.20	4.90
TX	23	12	12	3.50	5.66	5.35
VT	65	65	55	7.10	7.75	7.20
WA	20	20	23	9.00	6.50	4.80
WI	1,350	1,400	1,400	6.50	6.70	5.90
18 State Total	3,300	3,283	3,270	6.58	6.81	6.50

State	Production		
	2007	2008	2009
	1,000 tons	*1,000 tons*	*1,000 tons*
CA	561	432	540
ID	840	1,140	683
IL	238	186	199
IA	666	640	363
KS	280	250	275
MI	1,750	1,728	1,508
MN	1,525	1,247	1,650
MO	122	150	60
NE	242	245	158
NM	63	72	66
NY	3,105	3,478	3,252
OH	605	551	939
PA	1,953	1,998	2,503
SD	255	168	245
TX	81	68	64
VT	462	504	396
WA	180	130	110
WI	8,775	9,380	8,260
18 State Total	21,703	22,367	21,271

[1] Includes only alfalfa and alfalfa mixtures that were harvested as haylage or greenchop (green weight). Alfalfa harvested as dry hay is not included.
NASS, Crops Branch, (202) 720–2127.

Table 6-12.—Hay: Supply and disappearance, prices, and number of animal units fed annually, United States, 2000–2009 [1]

Year beginning May	Farm carryover May 1	Production	Total supply	Disappearance	Roughage-consuming animal units	Supply per animal unit	Disappearance per animal unit	Price received per ton
	Million tons	Million tons	Million tons	Million tons	Million units	Tons	Tons	Dollars
2000	28.8	153.6	182.5	161.2	72.1	2.52	2.23	84.6
2001	21.2	156.4	177.7	155.2	72.1	2.46	2.15	96.5
2002	22.5	149.5	171.9	149.9	72.0	2.39	2.08	92.4
2003	22.0	157.4	179.4	153.5	70.3	2.55	2.18	85.5
2004	25.9	158.1	184.1	156.3	70.8	2.60	2.21	92.0
2005	27.8	150.5	178.3	157.9	71.6	2.49	2.19	98.2
2006	21.3	140.8	162.1	147.1	71.8	2.26	2.05	110.0
2007	15.0	146.9	161.9	140.3	71.5	2.26	1.96	128.0
2008	21.6	146.3	167.9	145.8	70.9	2.37	2.06	152.0
2009 [2]	22.1	147.2	169.3	148.6	69.7	2.41	NA	111.0

[1] Excludes trade. [2] Preliminary. NA - Not available.
ERS, Market and Trade Economics Division, (202) 694-5296.

Table 6-13.—Field seeds: Average retail price paid by farmers for seed, Mar. 15, United States, 2000–2009 [1]

Kind of seed	2000	2001	2002	2003	2004	2005	2006	2007	2008	2009
	Price per 100 pounds									
	Dollars	Dollars	Dollars	Dollars	Dollars	Dollars	Dollars	Dollars	Dollars	Dollars
Alfalfa, uncertified varieties	165.00	158.00	157.00	178.00	163.00	177.00	181.00	201.00	246.00	262.00
Alfalfa, certified varieties	277.00	278.00	280.00	286.00	291.00	281.00	286.00	292.00	342.00	379.00
Clover, ladino	285.00	285.00	280.00	305.00	291.00	280.00	306.00	316.00	344.00	394.00
Clover, red	143.00	132.00	130.00	144.00	145.00	174.00	177.00	202.00	241.00	289.00
Lespedeza, Korean	77.50	160.00	98.00	102.00	81.50	79.30	87.00	126.00	184.00	127.00
Lespedeza, Striate, Kobe	90.00	180.00	104.00	108.00	93.60	83.10	89.50	113.00	263.00	198.00
Lespedeza, Sericea	310.00	330.00	300.00	281.00	230.00	220.00	181.00	[4]	[4]	[4]
Timothy	115.00	105.00	90.00	107.00	110.00	105.00	106.00	112.00	133.00	149.00
Orchardgrass	108.00	135.00	143.00	147.00	140.00	137.00	158.00	189.00	321.00	329.00
Blue Grass, Kentucky: Public and common	158.00	140.00	155.00	159.00	180.00	180.00	161.00	175.00	227.00	269.00
Proprietary, including Merion	214.00	220.00	225.00	228.00	217.00	235.00	224.00	232.00	251.00	326.00
Ryegrass, annual	60.50	55.50	58.00	51.30	52.60	59.30	69.60	71.80	78.80	78.90
Tall fescue	91.00	114.00	106.00	92.60	93.70	100.00	124.00	146.00	158.00	143.00
Sudangrass	53.00	53.00	56.00	55.30	55.60	57.40	50.20	56.70	62.10	72.50
Potatoes	10.45	8.50	10.90	10.80	9.69	9.30	11.80	12.00	13.10	15.60
Peanuts	81.70	82.60	82.10	55.90	56.90	56.40	57.70	60.70	83.10	76.50
Sunflower	395.00	407.00	407.00	417.00	425.00	476.00	520.00	616.00	718.00	749.00
Cottonseed, all	128.00	154.00	213.00	218.00	270.00	309.00	356.00	408.00	455.00	521.00
Biotech [2]	....	217.00	271.00	293.00	340.00	390.00	443.00	500.00	525.00	609.00
Non-biotech	....	87.00	94.00	107.00	108.00	110.00	118.00	97.10	88.50	113.00
Grain sorghum, hybrid	93.00	93.00	96.00	100.00	105.00	114.00	116.00	120.00	142.00	161.00
Rice	17.25	15.70	14.90	14.90	19.60	20.80	27.30	30.20	33.50	48.60
	Price per bushel									
	Dollars	Dollars	Dollars	Dollars	Dollars	Dollars	Dollars	Dollars	Dollars	Dollars
Corn, hybrid, all [3]	87.50	92.20	92.00	102.00	105.00	111.00	118.00	133.00	165.00	217.00
Biotech [2][3]	....	110.00	113.00	115.00	122.00	131.00	137.00	154.00	184.00	235.00
Non-biotech [3]	....	85.30	85.80	90.90	91.10	93.40	95.10	100.00	115.00	139.00
Wheat (spring)	6.10	6.20	6.50	8.77	7.00	7.30	7.60	8.40	20.50	11.80
Wheat (winter)	7.05	7.20	7.70	8.01	8.26	9.06	9.32	10.60	14.80	16.00
Oats (spring)	4.50	4.70	5.35	7.05	5.88	5.54	5.83	6.81	8.19	8.19
Barley (spring)	5.80	5.80	5.80	6.90	6.39	6.72	6.58	7.18	10.10	9.78
Soybeans for seed, all	17.10	20.70	22.50	24.20	24.10	27.60	28.90	34.80	38.80	48.30
Biotech [2]	....	23.90	27.00	28.80	30.50	34.60	34.10	36.70	40.00	49.60
Non-biotech	....	17.90	15.00	19.60	17.40	19.10	21.10	20.50	26.30	33.70
Flaxseed	7.90	7.60	7.60	9.96	9.60	14.40	8.80	9.73	19.80	13.80

[1] Beginning in 2009 program changed from April 15 to March 15. [2] Biotech varities are made to be resistant to herbicides, insects, or both. A technology fee is included within the price. [3] Price per 80,000 kernels. [4] Estimate discontinued in 2007.
NASS, Environmental, Economics, and Demographics Branch, (202) 720–6146.

Table 6-14.—Beans, dry edible (cleaned basis): Production, by classes, United States, 2007–2009 [1]

Class	2007	2008	2009
	1,000 cwt.	*1,000 cwt.*	*1,000 cwt.*
Navy (pea beans)	3,832	4,542	3,332
Great northern	1,186	1,598	999
Small white	10		71
Pinto	11,778	10,257	10,914
Red kidney, light	813	1,023	967
Red kidney, dark	663	992	850
Pink	578	557	497
Small red	537	816	703
Cranberry	124	141	84
Black	2,803	2,923	3,010
Large lima (CA)	302	317	333
Baby lima (CA)	377	239	352
Blackeye	497	394	771
Small chickpeas (Garbanzo)	129	129	299
Large chickpeas (Garbanzo)	1,386	989	1,145
Chickpeas, all (Garbanzo)	1,515	1,118	1,444
Other	571	641	1,033
Total	25,586	25,558	25,360

[1] Excludes beans grown for garden seed.
NASS, Crops Branch, (202) 720–2127.

Table 6-15.—Beans, dry edible: Area, yield, and production, by State and United States, 2007–2009 [1]

State	Area planted			Area harvested			Yield per acre (clean basis)			Production (clean basis)		
	2007	2008	2009	2007	2008	2009	2007	2008	2009	2007	2008	2009
	1,000 acres	*1,000 acres*	*1,000 acres*	*1,000 acres*	*1,000 acres*	*1,000 acres*	*Pounds*	*Pounds*	*Pounds*	*1,000 cwt.*	*1,000 cwt.*	*1,000 cwt.*
AZ[2]			15.5			15.2			2,120			322
CA	59.0	52.0	68.5	58.0	51.9	68.0	2,090	1,850	2,220	1,212	960	1,508
CO	48.0	48.0	57.0	46.0	44.0	53.0	1,600	1,500	1,600	736	660	848
ID	90.0	80.0	100.0	89.0	79.0	99.0	1,800	1,850	2,000	1,602	1,462	1,980
KS	6.5	6.0	8.5	6.0	5.5	8.0	2,300	2,100	2,800	138	116	224
MI	200.0	200.0	200.0	195.0	195.0	195.0	1,600	1,850	1,800	3,120	3,607	3,510
MN	150.0	150.0	150.0	145.0	145.0	140.0	1,800	1,950	1,800	2,610	2,828	2,520
MT	18.3	11.2	11.9	16.6	9.8	11.5	1,670	1,950	2,100	278	191	242
NE	110.0	135.0	130.0	107.0	126.0	115.0	2,260	2,290	2,140	2,418	2,885	2,461
NM	8.3	9.3	12.5	8.3	9.3	12.4	2,180	2,300	2,220	181	214	275
NY	17.0	17.0	16.0	16.5	16.8	15.6	1,500	1,930	1,240	248	324	193
ND	690.0	660.0	610.0	665.0	640.0	580.0	1,620	1,570	1,470	10,773	10,048	8,526
OR	7.7	4.8	6.4	7.6	4.7	6.3	1,970	2,000	2,330	149	94	147
SD	13.0	8.5	10.3	11.7	8.3	9.9	1,760	1,840	2,340	206	153	232
TX	17.0	24.0	37.0	16.2	21.8	33.7	1,500	1,300	1,260	243	283	425
UT[3]	1.5	1.2		1.3	1.2		400	580		5	7	
WA	60.0	50.0	60.0	60.0	50.0	60.0	1,700	1,770	1,900	1,020	885	1,140
WI	6.1	6.5	6.4	6.0	6.4	6.4	1,530	2,130	1,980	92	136	127
WY	25.0	31.5	37.5	24.0	30.5	34.0	2,310	2,310	2,000	555	705	680
US	1,527.4	1,495.0	1,537.5	1,479.2	1,445.2	1,463.0	1,730	1,768	1,733	25,586	25,558	25,360

[1] Excludes beans grown for garden seed. [2] Estimates began in 2009. [3] Estimates discontinued in 2009.
NASS, Crops Branch, (202) 720–2127.

Table 6-16.—Beans, dry edible: Area, yield, production, price, and value, United States, 2000–2009 [1]

Year	Area planted	Area harvested	Yield per acre [2]	Production [2]	Marketing year average price per 100 pounds received by farmers	Value of production
	1,000 acres	*1,000 acres*	*Pounds*	*1,000 cwt.*	*Dollars*	*1,000 dollars*
2000	1,767.7	1,616.5	1,642	26,543	15.50	416,462
2001	1,437.4	1,250.0	1,569	19,610	22.10	427,055
2002	1,929.7	1,738.9	1,743	30,312	17.10	519,341
2003	1,406.1	1,346.9	1,670	22,492	18.40	422,793
2004	1,346.3	1,212.3	1,464	17,743	25.70	451,605
2005	1,623.0	1,526.6	1,741	26,576	18.50	512,833
2006	1,622.8	1,531.6	1,577	24,155	22.10	554,154
2007	1,527.4	1,479.2	1,730	25,586	28.80	748,680
2008	1,495.0	1,445.2	1,768	25,558	34.60	910,200
2009	1,537.5	1,463.0	1,733	25,360	30.90	793,722

[1] Excludes beans grown for garden seed.   [2] Clean basis.
NASS, Crops Branch, (202) 720–2127.

Table 6-17.—Beans, dry edible (cleaned basis): Marketing year average price and value of production, by State and United States, 2007–2009 [1]

State	Marketing year average price per cwt.			Value of production		
	2007	2008	2009	2007	2008	2009
	Dollars	*Dollars*	*Dollars*	*1,000 dollars*	*1,000 dollars*	*1,000 dollars*
AZ [2]			42.00			13,524
CA	48.90	61.40	52.20	59,267	58,944	78,718
CO	31.20	35.80	31.00	22,963	23,628	26,288
ID	29.00	37.00	27.10	46,458	54,094	53,658
KS	27.60	36.20	30.00	3,809	4,199	6,720
MI	31.90	36.30	32.90	99,528	130,934	115,479
MN	29.10	41.80	29.70	75,951	118,210	74,844
MT	25.00	33.50	29.30	6,950	6,399	7,091
NE	29.90	35.70	30.70	72,298	102,995	75,553
NM	39.00	50.00	45.00	7,059	10,700	12,375
NY	41.20	56.00	46.00	10,218	18,144	8,878
ND	25.70	29.70	27.50	276,866	298,426	234,465
OR	32.50	34.90	33.50	4,843	3,281	4,925
SD	24.80	29.60	26.50	5,109	4,529	6,148
TX	29.20	35.00	34.00	7,096	9,905	14,450
UT [3]	29.10	31.00		146	217	
WA	30.80	36.90	30.70	31,416	32,657	34,998
WI	38.00	53.50	39.40	3,496	7,276	5,004
WY	27.40	36.40	30.30	15,207	25,662	20,604
US	28.80	34.60	30.90	748,680	910,200	793,722

[1] Excludes beans grown for garden seed.   [2] Estimates began in 2009.   [3] Estimates discontinued in 2009.
NASS, Crops Branch, (202) 720–2127.

Table 6-18.—Beans, dry edible: Season average wholesale price per 100 pounds, selected markets, 1999–2008

Year beginning September	F.o.b. California points			F.o.b. Northern Colorado points: Pinto	F.o.b. Western Nebraska points: Great northern	F.o.b. Southern Idaho points: Small red	F.o.b. Michigan points:		
	Baby lima	Large lima	Blackeye				Pea bean (Navy)	Black	Light red kidney
	Dollars	*Dollars*	*Dollars*	*Dollars*	*Dollars*	*Dollars*	*Dollars*	*Dollars*	*Dollars*
1999	28.57	35.90	23.40	19.75	24.25	21.92	19.16	18.37	26.79
2000	26.26	34.56	25.95	21.02	23.20	24.33	16.43	18.33	25.32
2001	33.73	41.65	29.93	31.39	23.52	33.36	25.65	37.44	34.04
2002	32.28	42.33	34.48	22.87	26.47	28.81	18.00	19.24	29.68
2003	32.34	42.40	30.12	22.19	22.22	28.53	23.53	24.64	30.29
2004	41.66	43.51	31.28	35.23	24.78	32.02	29.64	26.51	36.24
2005	38.28	47.62	44.26	23.47	24.32	27.33	24.44	29.54	27.74
2006	47.32	66.26	47.38	29.52	31.61	30.95	29.07	30.94	35.33
2007	42.39	65.64	41.51	38.66	47.65	44.52	44.48	43.47	53.71
2008	58.87	73.39	51.97	40.85	54.00	52.71	36.63	46.70	57.67

ERS, Specialty Crops Branch, (202) 694–5253. Compiled from the Bean Market Summary, Agricultural Marketing Service, U.S. Department of Agriculture, Greeley, Colorado.

Table 6-19.—Beans, dry edible: United States exports to specified countries, 2007–2009 [1]

Country	2007	2008	2009
	Metric tons	*Metric tons*	*Metric tons*
Mexico	81,379	100,074	187,133
Taiwan	1,186	4,143	45,652
United Kingdom	27,715	47,624	42,134
Canada	24,409	50,293	38,033
Dominican Republic	17,178	18,438	15,347
Tanzania	2,887	12,569	13,194
Japan	13,869	13,031	11,811
Haiti	2,553	9,771	9,318
France(*)	4,773	6,966	8,951
Venezuela	1,225	5,375	6,663
Guatemala	6,132	6,864	5,282
Cuba	87	65	5,161
Italy(*)	3,657	4,624	4,201
South Africa	736	16,597	4,014
Mozambique	1,128	2,230	3,670
Australia(*)	3,385	3,294	3,002
Colombia	34	2,704	2,946
Djibouti	0	0	2,490
Angola	12,163	15,366	2,416
Spain	2,911	5,181	2,383
Jamaica	3,086	3,149	2,319
French West Indies(*)	1,758	3,052	2,287
India	7,350	4,382	1,833
Malaysia	2,412	2,350	1,833
Rest of World	61,309	57,642	22,963
World Total	283,321	395,786	445,148

[1] Excluding seed bean exports. Compiled from U.S. Census data.   Note: (*)denotes a country that is a summarization of its component countries.
FAS, Office of Global Analysis, (202) 720-6301.

Table 6-20.—Chickpeas & lentils, dried: United States exports by class and quantity, 2007–2009 [1]

Country	2007	2008	2009
	1,000 metric tons	*1,000 metric tons*	*1,000 metric tons*
Dried chickpeas:			
India	0	373	8,248
Spain	7,259	5,775	6,848
Canada	2,540	4,512	2,743
Italy	1,114	478	1,238
Colombia	1,395	849	1,201
Pakistan	182	0	1,115
Lebanon	734	597	562
Algeria	681	1,026	458
Japan	197	2,478	324
Philippines	131	44	229
Norway(*)	735	504	196
New Zealand	481	972	187
Taiwan	151	148	181
Sri Lanka	47	0	147
Sweden	222	303	118
Malaysia	242	107	111
Hong Kong	21	8	93
Guatemala	70	0	90
Israel(!)	52	82	84
Germany(*)	87	90	66
Rest of World	3,483	1,456	141
World Total	19,823	19,700	24,372
Dried lentils:			
India	4,254	1,271	57,642
Spain	20,365	39,571	28,616
Sri Lanka	8,720	2,977	15,410
Sudan	19,269	25,241	12,316
Saudi Arabia	22	13,051	10,160
Peru	8,568	6,138	8,897
Ethiopia(*)	639	7,161	4,649
Canada	1,900	2,426	3,612
Taiwan	11	21	2,601
Mexico	3,115	1,180	2,090
Somalia	2,057	7,874	2,040
Pakistan	1,488	2,327	2,027
Turkey	878	1,968	1,739
Portugal	68	655	1,735
Italy(*)	1,838	2,628	1,668
United Arab Emirates	0	1,821	1,603
Germany(*)	932	1,260	1,577
Colombia	2,314	1,424	1,403
French Pacific Islands(*)	538	948	1,346
French Polynesia(!)	423	785	1,213
French West Indies(*)	137	4	1,197
Guadeloupe	137	4	1,027
Rest of World	27,043	28,550	5,804
World Total	103,877	149,284	170,372

[1] Excluding seed pea exports.   Note (*)denotes a country that is a summarization of its component countries.
FAS, Office of Global Analysis, (202) 720-6301.

Table 6-21.—Peas, dry: United States exports to specified countries, 2007–2009 [1]

Country	2007	2008	2009
	Metric tons	*Metric tons*	*Metric tons*
India	190,706	139,229	156,408
Kenya	33,603	10,285	43,9994
Pakistan	18,646	21,871	42,831
Canada	28,256	13,625	32,162
Ethiopia(*)	19,394	30,512	19,577
China	9,495	5,216	14,884
Philippines	15,120	15,914	14,308
United Arab Emirates	2,332	15,594	11,333
Afghanistan	2,979	5,780	11,161
Zimbabwe	322	10,668	9,840
Indonesia	3,579	2,441	9,666
Peru	7,356	6,610	9,171
Djibouti	1,664	200	8,175
Taiwan	4,506	4,475	8,007
South Africa	139	15,226	6,322
Cameroon	0	382	5,929
Bangladesh	6,089	5,370	5,436
Tanzania	5,926	0	5,110
Oman	140	16,854	5,028
Mexico	3,456	17,379	4,907
Korea, South	4,612	4,731	4,577
Nepal	2,430	255	4,342
Turkey	381	3,385	4,123
Rest of Total	92,809	133,524	35,908
World Total	453,936	480,066	473,197

[1] Excluding seed pea exports. Note (*)denotes a country that is a summarization of its component countries.
FAS, Office of Global Analysis, (202) 720-6301.

Table 6-22.—Hops: Area, yield, production, price, value, and Sept. 1 stocks, United States, 2000–2009

Year	Area harvested	Yield per acre	Production	Marketing year average price per pound received by farmers	Value of production	Stocks Sept. 1
	1,000 acres	*Pounds*	*1,000 pounds*	*Dollars per pound*	*1,000 dollars*	*1,000 pounds*
2000	36.1	1,871	67,577	1.87	126,217	48,000
2001	35.9	1,861	66,832	1.85	123,843	54,000
2002	29.3	1,990	58,337	1.91	111,546	65,000
2003	28.7	1,903	54,565	1.86	101,637	69,000
2004	27.7	1,990	55,204	1.88	103,969	65,000
2005	29.5	1,796	52,915	1.94	102,818	60,000
2006	29.4	1,964	57,672	2.05	118,008	49,000
2007	30.9	1,949	60,253	2.99	179,978	47,000
2008	40.9	1,971	80,630	4.03	325,092	47,000
2009 [1]	39.7	2,383	94,678	3.55	336,375	65,000

[1] Preliminary.
NASS, Crops Branch, (202) 720–2127.

Table 6-23.—Hops: Area, yield, and production, by State and United States, 2007–2009

State	Area harvested			Yield per acre			Production		
	2007	2008	2009 [1]	2007	2008	2009 [1]	2007	2008	2009 [1]
	1,000 acres	*1,000 acres*	*1,000 acres*	*Pounds*	*Pounds*	*Pounds*	*1,000 pounds*	*1,000 pounds*	*1,000 pounds*
ID	2,896	3,933	4,030	1,417	1,841	1,943	4,104.9	7,239.8	7,829.1
OR	5,270	6,370	6,108	1,811	1,569	1,948	9,542.8	9,997.6	11,896.7
WA	22,745	30,595	29,588	2,049	2,072	2,533	46,605.4	63,392.7	74,952.1
US	30,911	40,898	39,726	1,949	1,971	2,383	60,253.1	80,630.1	94,677.9

[1] Preliminary.
NASS, Crops Branch, (202) 720–2127.

Table 6-24.—Hops: Marketing year average price and value of production, by State and United States, 2007–2009

State	Marketing year average price per pound			Value of production		
	2007	2008	2009 [1]	2007	2008	2009 [1]
	Dollars	*Dollars*	*Dollars*	*1,000 dollars*	*1,000 dollars*	*1,000 dollars*
ID	2.77	4.00	3.75	11,371	28,959	29,359
OR	3.31	3.75	3.63	31,587	37,491	43,185
WA	2.94	4.08	3.52	137,020	258,642	263,831
Total	2.99	4.03	3.55	179,978	325,092	336,375

[1] Preliminary.
NASS, Crops Branch, (202) 720–2127.

Table 6-25.—Hops: United States exports by country of destination and imports by country of origin, 2007–2009

Country	Year beginning September		
	2007	2008	2009
	Metric tons	*Metric tons*	*Metric tons*
Belgium-Luxembourg	1502	661	1,713
Germany(*)	1,447	1,372	1,639
Mexico	413	552	1,581
Canada	1,069	2,515	1,543
United Kingdom	1,146	1,380	1,435
Brazil	724	1,065	1,301
Japan	587	760	1,148
Colombia	632	845	947
China	833	706	432
Peru	116	424	388
Australia(*)	134	162	325
Korea,South	340	173	284
Mexico	413	552	1,581
Hong Kong	215	159	270
Argentina	217	416	260
Thailand	77	162	251
Philippines	115	210	246
India	146	71	203
Ecuador	120	192	196
Dominican Republic	84	94	119
Nigeria	99	80	99
Bolivia	38	14	83
Chile	112	128	82
Rest of World	1,980	2,263	1,027
World Total	12,084	14,330	15,503

Note (*)denotes a country that is a summarization of its component countries.
FAS, Office of Global Analysis, (202) 720-6301.

CHAPTER VII

STATISTICS OF CATTLE, HOGS, AND SHEEP

This chapter contains information about most kinds of farm livestock and livestock products, with the exception of dairy and poultry. The information relates to inventories, production, disposition, prices, and income for farm animals, and to livestock slaughter, meat production, and market statistics for meat animals.

Table 7-1.—All cattle and calves: Operations, inventory, and value, United States, Jan. 1, 2001–2010

Year	Operations	Inventory	Value	
			Per head	Total
	Number	Thousands	Dollars	1,000 dollars
2001	1,049,170	97,298	725	70,510,630
2002	1,036,430	96,723	747	72,300,065
2003	1,013,570	96,100	728	69,952,520
2004	989,460	94,403	818	77,201,950
2005	982,510	95,018	916	87,023,945
2006	971,400	96,342	1,009	97,230,415
2007	965,510	96,573	922	89,063,310
2008	955,500	96,035	990	95,112,820
2009 [1]	950,500	94,521	872	82,435,620
2010 [1]		93,701	829	77,677,310

[1] Preliminary inventory estimates.   Operation estimates for 2010 not yet available.
NASS, Livestock Branch, (202) 720–3570.

Table 7-2.—All cattle and calves: Number by class, United States, Jan. 1, 2001–2010

Year	All cattle and calves [1]	Cows and heifers that have calved		500 pounds and over					Calves under 500 pounds
		Beef cows	Milk cows	Heifers			Steers	Bulls	
				Beef cow replacements	Milk cow replacements	Other			
	Thousands	Thousands	Thousands	Thousands	Thousands	Thousands	Thousands	Thousands	Thousands
2001	97,298	33,398	9,172	5,588	4,057	10,131	16,461	2,274	16,216
2002	96,723	33,134	9,106	5,571	4,055	10,057	16,804	2,244	15,753
2003	96,100	32,983	9,142	5,624	4,114	9,891	16,554	2,248	15,545
2004	94,403	32,531	8,988	5,508	4,018	9,756	16,201	2,201	15,200
2005	95,018	32,674	9,004	5,638	4,117	9,690	16,466	2,214	15,215
2006	96,342	32,703	9,104	5,864	4,298	9,788	16,988	2,258	15,339
2007	96,573	32,644	9,145	5,835	4,325	9,914	17,185	2,214	15,311
2008	96,035	32,435	9,257	5,647	4,415	9,793	17,163	2,207	15,118
2009	94,521	31,712	9,333	5,531	4,410	9,635	16,769	2,184	14,948
2010 [2]	93,701	31,376	9,081	5,436	4,516	9,714	16,440	2,190	14,949

[1] Totals may not add due to rounding.   [2] Preliminary.
NASS, Livestock Branch, (202) 720–3570.

STATISTICS OF CATTLE, HOGS, AND SHEEP

Table 7-3.—All cattle and calves: Inventory and value, by State and United States, Jan. 1, 2009–2010

State	Inventory		Value			
	2009	2010[1]	Value per head		Total value	
			2009	2010[1]	2009	2010[1]
	Thousands	Thousands	Dollars	Dollars	1,000 dollars	1,000 dollars
AL	1,260	1,280	680	670	856,800	857,600
AK	14.0	14.5	980	950	13,720	13,775
AZ	1,020	930	970	810	989,400	753,300
AR	1,810	1,890	670	700	1,212,700	1,323,000
CA	5,250	5,150	930	960	4,882,500	4,944,000
CO	2,600	2,600	880	850	2,288,000	2,210,000
CT	52	48	1,250	1,030	65,000	49,440
DE	21	20	1,110	920	23,310	18,400
FL	1,700	1,720	810	710	1,377,000	1,221,200
GA	1,110	1,060	760	720	843,600	763,200
HI	150	151	640	640	96,000	96,640
ID	2,110	2,140	1,070	990	2,257,700	2,118,600
IL	1,200	1,170	880	840	1,056,000	982,800
IN	860	870	970	870	834,200	756,900
IA	3,950	3,850	860	830	3,397,000	3,195,500
KS	6,300	6,000	800	780	5,040,000	4,680,000
KY	2,300	2,300	700	710	1,610,000	1,633,000
LA	890	840	790	730	703,100	613,200
ME	89	87	1,240	1,010	110,360	87,870
MD	185	195	1,180	940	218,300	183,300
MA	43	43	1,240	970	53,320	41,710
MI	1,070	1,100	1,150	930	1,230,500	1,023,000
MN	2,400	2,420	1,020	870	2,448,000	2,105,400
MS	960	970	710	700	681,600	679,000
MO	4,250	4,150	780	790	3,315,000	3,278,500
MT	2,600	2,550	930	960	2,418,000	2,448,000
NE	6,350	6,250	870	850	5,524,500	5,312,500
NV	450	450	960	920	432,000	414,000
NH	39	37	1,290	1,090	50,310	40,330
NJ	38	36	1,200	1,030	45,600	37,080
NM	1,540	1,550	970	870	1,493,800	1,348,500
NY	1,380	1,410	1,290	940	1,780,200	1,325,400
NC	850	820	720	680	612,000	557,600
ND	1,760	1,720	960	1,010	1,689,600	1,737,200
OH	1,280	1,280	970	850	1,241,600	1,088,000
OK	5,400	5,450	750	730	4,050,000	3,978,500
OR	1,240	1,260	850	850	1,054,000	1,071,000
PA	1,590	1,620	1,160	970	1,844,400	1,571,400
RI	5.0	4.7	1,050	950	5,250	4,465
SC	380	380	750	730	285,000	277,400
SD	3,700	3,800	940	930	3,478,000	3,534,000
TN	1,980	2,040	660	690	1,306,800	1,407,600
TX	13,600	13,300	770	740	10,472,000	9,842,000
UT	810	800	930	830	753,300	664,000
VT	270	265	1,450	1,060	391,500	280,900
VA	1,470	1,550	700	690	1,029,000	1,069,500
WA	1,080	1,040	1,040	950	1,123,200	988,000
WV	415	370	730	740	302,950	273,800
WI	3,350	3,400	1,260	1,040	4,221,000	3,536,000
WY	1,350	1,320	910	940	1,228,500	1,240,800
US	94,521.0	93,701.2	872	829	82,435,620	77,677,310

[1] Preliminary.
NASS, Livestock Branch, (202) 720–3570.

**Table 7-4.—Cattle and calves, Jan. 1: Number by class,
State and United States, 2009–2010**

State	Cows and heifers that have calved				Heifers, 500 pounds and over					
	Beef cows		Milk cows		Beef cow replacements		Milk cow replacements		Other	
	2009	2010[1]	2009	2010[1]	2009	2010[1]	2009	2010[1]	2009	2010[1]
	Thou-sands	Thou-sands	Thou-sands	Thou-sands	Thou-sands	Thou-sands	Thou-sands	Thou-sands	Thou-sands	Thou-sands
AL	668	669	12.0	11.0	95	95	5.0	5.0	40	40
AK	5.5	5.4	0.6	0.6	1.6	1.7	0.2	0.2	0.1	0.1
AZ	200	208	190	167	35	30	50	55	20.0	15.0
AR	906	937	14.0	13.0	165	179	5.0	6.0	95	85
CA	620	610	1,840	1,760	115	120	780	750	170	210
CO	720	714	130	116	120	120	60	70	530	500
CT	6.0	5.5	19.0	18.5	2.0	1.5	11.0	9.0	0.5	0.5
DE	4.0	4.0	6.5	6.0	0.6	0.7	2.3	2.7	0.6	0.4
FL	942	958	118	112	140	135	35	30	25	25
GA	536	524	74	76	73	68	23	25	31	39
HI	84.4	81.2	1.6	1.8	13.0	12.0	1.0	1.0	5.0	6.0
ID	451	440	554	550	90	90	275	295	195	195
IL	408	389	102	101	49	53	59	52	117	125
IN	213	221	167	169	36	42	66	70	55	55
IA	925	885	215	215	140	130	130	130	600	640
KS	1,505	1,434	125	116	255	240	65	70	1,520	1,480
KY	1,114	1,070	86	80	160	150	50	50	110	130
LA	515	499	25	21	87	82	7.0	7.0	17.0	16.0
ME	11.0	11.0	33	33	4.0	3.0	18.0	16.0	2.0	2.0
MD	39	41	56	54	8.0	9.0	26	28	7.0	6.0
MA	7.5	8.0	14.5	13.0	2.0	2.5	7.0	6.0	1.0	1.5
MI	92	96	353	354	27	27	148	158	50	50
MN	392	380	468	470	95	95	285	295	165	180
MS	491	503	19.0	17.0	83	92	7.0	7.0	36	37
MO	1,992	1,968	108	102	320	280	40	45	270	275
MT	1,494	1,465	16.0	15.0	385	340	10.0	7.0	225	273
NE	1,851	1,781	59	59	320	310	20	20	1,450	1,440
NV	237	237	28	28	35	35	10.0	10.0	29	28
NH	6.0	4.0	15.0	15.0	1.5	1.0	7.5	8.0	1.0	1.0
NJ	10.0	9.5	9.5	8.5	2.5	2.2	5.0	4.8	2.0	2.0
NM	484	502	336	318	85	95	140	145	95	90
NY	85	90	625	610	28	33	320	325	42	52
NC	384	367	46	43	69	69	20.0	19.0	30	24
ND	895	869	25	21	180	165	20.0	10.0	215	220
OH	304	288	276	272	70	65	115	120	75	70
OK	2,038	2,073	62	57	390	405	30	25	580	530
OR	535	546	115	114	90	95	55	55	110	115
PA	150	160	550	540	40	40	270	300	60	50
RI	1.4	1.3	1.1	1.1	0.5	0.4	0.5	0.5	0.1	0.1
SC	188	183	17.0	17.0	32	36	7.0	8.0	11.0	13.0
SD	1,626	1,618	94	92	280	285	40	35	580	640
TN	951	997	59	53	150	170	45	35	75	70
TX	5,170	5,140	430	410	790	760	220	250	1,550	1,540
UT	350	338	85	82	55	66	45	48	50	51
VT	9.0	10.0	139	134	3.5	4.0	61	56	4.5	5.0
VA	643	665	97	95	100	105	45	50	55	65
WA	271	227	244	243	55	54	110	123	95	93
WV	204	190	11.0	10.0	38	37	3.0	4.0	28	23
WI	265	260	1,255	1,260	65	60	650	670	55	60
WY	713	694	7.0	6.0	150	145	5.0	5.0	155	145
US	31,711.8	31,375.9	9,332.8	9,080.5	5,531.2	5,436.0	4,409.5	4,516.2	9,634.8	9,713.6

See footnote at end of table.

STATISTICS OF CATTLE, HOGS, AND SHEEP

Table 7-4.—Cattle and calves, Jan. 1: Number by class, State and United States, 2009–2010—Continued

State	Steers, 500 pounds and over		Bulls, 500 pounds and over		Calves under 500 pounds	
	2009	2010 [1]	2009	2010 [1]	2009	2010 [1]
	Thousands	*Thousands*	*Thousands*	*Thousands*	*Thousands*	*Thousands*
AL	55	70	45	50	340	340
AK	0.8	0.6	1.9	2.0	3.3	3.9
AZ	370	300	20	20	135	135
AR	145	155	60	60	420	455
CA	600	580	65	70	1,060	1,050
CO	770	860	50	45	220	175
CT	2.4	2.3	0.6	0.7	10.5	10.0
DE	3.0	2.6	0.4	0.3	3.6	3.3
FL	20	20	60	60	360	380
GA	48	39	35	29	290	260
HI	8.0	8.0	5.0	5.0	32	36
ID	260	255	35	35	250	280
IL	230	210	25	25	210	215
IN	104	122	16.0	19.0	203	172
IA	1,390	1,330	70	60	480	460
KS	2,020	1,890	95	90	715	680
KY	215	235	75	75	490	510
LA	24	21	34	32	181	162
ME	3.5	4.0	1.5	1.5	16.0	16.5
MD	15.0	18.0	4.0	4.0	30	35
MA	2.0	2.0	1.0	1.0	8.0	9.0
MI	185	200	15.0	15.0	200	200
MN	440	445	40	35	515	520
MS	55	50	39	39	230	225
MO	510	450	110	110	900	920
MT	265	265	90	90	115	95
NE	2,270	2,240	95	95	285	305
NV	36	37	13.0	15.0	62	60
NH	1.0	1.5	0.5	0.5	6.5	6.0
NJ	2.0	2.0	1.0	1.0	6.0	6.0
NM	170	130	30	40	200	230
NY	35	39	15.0	16.0	230	245
NC	30	36	31	32	240	230
ND	260	275	60	60	105	100
OH	190	205	30	30	220	230
OK	1,160	1,210	140	150	1,000	1,000
OR	145	150	35	35	155	150
PA	160	150	25	25	335	355
RI	0.4	0.4	0.2	0.1	0.8	0.8
SC	15.0	17.0	17.0	16.0	93	90
SD	680	760	85	85	315	285
TN	137	137	73	78	490	500
TX	2,780	2,530	370	370	2,290	2,300
UT	105	90	20	22	100	103
VT	3.0	4.0	3.0	3.0	47	49
VA	140	180	40	40	350	350
WA	143	144	22.0	18.0	140	138
WV	41	33	15.0	15.0	75	58
WI	360	360	30	30	670	700
WY	165	175	40	40	115	110
US	16,769.1	16,440.4	2,184.1	2,190.1	14,947.7	14,948.5

[1] Preliminary.
NASS, Livestock Branch, (202) 720–3570.

Table 7-5.—Cows and calf crop: Cows and heifers that have calved, Jan. 1, 2009–2010, and calves born, by State and United States, 2008–2009

State	Cows and heifers that have calved		Calves born	
	Jan. 1		2008	2009 [1]
	2009	2010 [1]		
	Thousands	Thousands	Thousands	Thousands
AL	680	680	580	590
AK	6.1	6.0	3.6	4.2
AZ	390	375	300	295
AR	920	950	780	810
CA	2,460	2,370	2,010	1,990
CO	850	830	780	770
CT	25	24	20	20
DE	10.5	10.0	8.5	7.5
FL	1,060	1,070	880	900
GA	610	600	510	500
HI	86	83	65	65
ID	1,005	990	960	930
IL	510	490	470	440
IN	380	390	315	335
IA	1,140	1,100	1,070	1,030
KS	1,630	1,550	1,430	1,390
KY	1,200	1,150	1,070	1,030
LA	540	520	420	400
ME	44	44	36	32
MD	95	95	72	74
MA	22	21	19.0	20.0
MI	445	450	375	380
MN	860	850	840	820
MS	510	520	390	420
MO	2,100	2,070	1,930	1,900
MT	1,510	1,480	1,490	1,460
NE	1,910	1,840	1,730	1,680
NV	265	265	210	210
NH	21.0	19.0	14.0	13.5
NJ	19.5	18.0	14.0	13.5
NM	820	820	590	610
NY	710	700	500	490
NC	430	410	390	370
ND	920	890	920	890
OH	580	560	460	470
OK	2,100	2,130	1,880	1,900
OR	650	660	610	610
PA	700	700	610	610
RI	2.5	2.4	2.4	2.3
SC	205	200	165	165
SD	1,720	1,710	1,660	1,660
TN	1,010	1,050	950	950
TX	5,600	5,550	4,800	4,750
UT	435	420	360	365
VT	148	144	123	122
VA	740	760	680	690
WA	515	470	430	410
WV	215	200	200	195
WI	1,520	1,520	1,350	1,370
WY	720	700	680	660
US	41,044.6	40,456.4	36,152.5	35,819.0

[1] Preliminary.
NASS, Livestock Branch, (202) 720–3570.

Table 7-6.—Cattle and calves: All cattle on feed, United States, Jan. 1, 2001–2010 [1]

Year	Inventory
	Thousands
2001	14,276
2002	14,050
2003	13,220
2004	13,913
2005	13,925
2006	14,392
2007	14,647
2008	14,827
2009	13,856
2010 [2]	13,642

[1] Cattle and calves on feed are animals for slaughter market being fed a ration of grain or other concentrates and are expected to produce a carcass that will grade select or better.　[2] Preliminary.
NASS, Livestock Branch, (202) 720–3570.

Table 7-7.—Cattle and calves: Total number on feed by State and United States, Jan. 1, 2009–2010

State	2009	2010 [1]
	1,000 Head	*1,000 Head*
AZ	358	288
AR	2.0	5.0
CA	490	465
CO	1,020	1,020
ID	230	225
IL	180	170
IN	120	122
IA	1,300	1,360
KS	2,370	2,370
KY	10.0	10.0
MD	12.0	12.0
MI	165	170
MN	280	290
MO	60	60
MT	45	26
NE	2,500	2,500
NV	6.0	9.0
NY	29	30
ND	70	90
OH	195	195
OK	350	370
OR	80	75
PA	75	75
SD	390	400
TN	4.0	5.0
TX	2,800	2,700
UT	25	25
VA	29	20
WA	160	167
WV	9.0	7.0
WI	240	240
WY	70	65
Oth Sts [2]	181.7	76.2
US	13,855.7	13,642.2

[1] Preliminary. [2] AL, AK, CT, DE, FL, GA, HI, LA, ME, MA, MS, NC, NH, NM, NJ, RI, SC, and VT.
NASS, Livestock Branch, (202) 720–3570.

Table 7-8.—Cattle: Average price per 100 pounds, by grades, at Nebraska Direct, Sioux Falls, SD and Louisville, KY, 2000–2009

Year	Nebraska Steers [1]		Nebraska Heifers [2]		Sioux Falls, SD		Louisville, KY	
					Cows [3]		Cows	
	Choice	Select	Choice	Select	Commer-cial	Breaking utility	Breaking utility	85-95% Lean
	Dollars	*Dollars*	*Dollars*	*Dollars*	*Dollars*	*Dollars*	*Dollars*	*Dollars*
2000 65-80%	69.52	....	69.55	....	49.25	44.51	....	....
2001 65-80%	67.68	....	67.81	....	52.35	46.67	....	....
2002 65-80%	66.39	....	67.39	....	44.99	40.97	....	....
2003 65-80%	82.37	....	82.06	....	53.49	49.50	....	....
2004 65-80%	84.78	....	84.40	....	60.64	57.22	....	....
2005 65-80%	86.54	....	87.35	....	61.89	57.82	....	....
2006 65-80%	85.55	....	86.58	....	58.72	54.19	....	....
2007 65-80%	91.87	....	91.86	....	60.91	56.85	47.02	45.22
2008	93.24	....	93.06	....	63.15	61.32	52.80	42.99
2009	82.70	....	82.71	....	53.94	54.32	46.61	37.95

[1] 1,100 to 1,500 pound weight range; weighted average of price range. [2] 1,000 to 1,300 pound weight range; simple average of price range. [3] All weights; simple average of price range.
AMS, Livestock and Grain Market News, (202) 720–7316.

Table 7-9.—Cattle and calves: Production, disposition, cash receipts, and gross income, United States, 2000–2009

| Year | Calf crop[1] | Death loss | | Marketings[2] | | Inshipments[3] | Farm slaughter |
		Cattle	Calves	Cattle	Calves		Cattle and calves
	1,000 head	*1,000 head*	*1,000 head*	*1,000 head*	*1,000 head*	*1,000 head*	*1,000 head*
2000	38,631	1,711	2,387	48,986	9,693	23,448	203
2001	38,300	1,722	2,487	47,102	9,183	21,813	194
2002	38,224	1,710	2,366	46,804	9,296	21,522	193
2003	37,593	1,710	2,320	47,773	9,649	22,353	191
2004	37,260	1,711	2,292	44,774	9,100	21,418	185
2005	37,106	1,718	2,335	43,665	8,888	21,010	188
2006	37,016	1,818	2,348	44,789	8,856	21,213	187
2007	36,759	1,856	2,394	45,008	8,956	21,104	188
2008	36,153	1,760	2,314	44,365	8,803	19,761	186
2009[4]	35,819	1,741	2,323	43,701	8,479	19,790	185

| Year | Production (live weight)[5] | Value of production | Cash receipts[6] | Value of home consumption | Gross income[7] | Average price per 100 pounds received by farmers | |
						Cattle	Calves
	1,000 pounds	*1,000 dollars*	*1,000 dollars*	*1,000 dollars*	*1,000 dollars*	*Dollars*	*Dollars*
2000	43,040,893	28,498,670	40,783,472	366,744	41,150,216	68.60	104.00
2001	42,581,294	29,403,098	40,540,645	362,317	40,902,962	71.30	106.00
2002	42,409,258	27,097,532	38,095,116	333,768	38,428,884	66.50	96.40
2003	42,236,472	32,111,711	45,341,098	384,713	45,725,811	79.70	102.00
2004	41,552,792	34,890,118	47,429,892	427,777	47,857,669	85.80	119.00
2005	41,246,788	36,348,156	49,283,098	459,007	49,742,175	89.70	135.00
2006	41,824,568	35,490,732	49,110,330	447,857	49,558,187	87.20	133.00
2007	41,437,021	35,973,068	49,843,322	441,051	50,284,373	89.90	119.00
2008	41,594,392	35,608,404	48,517,768	415,924	48,933,692	89.10	110.00
2009[4]	40,919,268	31,769,067	43,776,568	386,304	44,162,872	80.30	103.00

[1] Calves born during the year.　[2] Includes custom slaughter for use on farms where produced and State outshipments, but excludes interfarm sales within the State.　[3] Includes cattle shipped in from other States, but excludes cattle for immediate slaughter.　[4] Preliminary.　[5] Adjustments made for changes in inventory and for inshipments.　[6] Receipts from marketings and sale of farm slaughter.　[7] Cash receipts from sales of cattle, calves, beef, and veal plus value of cattle and calves slaughtered for home consumption.
NASS, Livestock Branch, (202) 720–3570.

Table 7-10.—Cattle: Weighted average weight and price per 100 pounds, Texas-Oklahoma, Kansas, Colorado, Nebraska, Iowa-So. Minnesota Feedlots, 2002–2009[1]

| Year | Steers SE/CH 65-80% | | | Steers SE/CH 35-65% | | |
	Price	Average Weight	Number of Head	Price	Average Weight	Number of Head
	Dollars	*Pounds*		*Dollars*	*Pounds*	
2002	66.74	1,327	270,924	67.40	1,263	1,965,036
2003	82.81	1,294	372,429	82.79	1,245	2,391,746
2004	84.65	1,319	389,144	85.03	1,242	2,336,418
2005	86.28	1,336	326,751	87.66	1,265	2,492,108
2006	85.51	1,358	317,732	85.75	1,282	2,319,896
2007	92.21	1,366	359,511	92.10	1,289	2,081,144
2008	93.37	1,360	344,633	93.16	1,294	1,757,049
2009	82.81	1,376	476,823	83.59	1,314	1,474,925

| Year | Heifers SE/CH 65-80% | | | Heifers SE/CH 35-65% | | |
	Price	Average Weight	Number of Head	Price	Average Weight	Number of Head
	Dollars	*Pounds*		*Dollars*	*Pounds*	
2002	67.53	1,229	324,078	67.42	1,142	1,692,785
2003	82.70	1,192	358,900	83.59	1,126	2,077,258
2004	84.38	1,210	403,193	84.92	1,128	2,193,273
2005	87.23	1,219	313,240	87.90	1,145	1,901,730
2006	86.57	1,245	247,902	86.06	1,165	1,781,077
2007	92.01	1,241	357,880	93.09	1,165	1,582,183
2008	92.95	1,229	342,547	93.22	1,170	1,317,211
2009	82.88	1,252	498,481	83.63	1,180	1,136,234

[1] Sales FOB feedlots and delivered. Estimated net weights after 3-4 % shrink.
AMS, Livestock and Grain Market News, (202) 720–7316.

STATISTICS OF CATTLE, HOGS, AND SHEEP

Table 7-11.—Cattle and calves: Receipts at selected public stockyards, 2000–2009 [1]

Year	Oklahoma City	Greeley	Amarillo	South St. Joseph	Sioux Falls	All others reporting	Total markets reporting [2][3]
				Cattle			
	Thousands	Thousands	Thousands	Thousands	Thousands	Thousands	Thousands
2000	497	53	95	101	272	4,454	4,847
2001	512	49	89	113	228	2,698	3,593
2002	516	43	144	30	239	2,293	3,169
2003	574	37	130	110	222	2,530	3,528
2004	456	24	99	101	211	2,121	2,942
2005	491	51	87	97	198	2,124	2,974
2006	471	58	69	114	179	1,999	2,798
2007	421	45	65	51	115	1,654	2,236
2008	464	47	43	83	90	1,985	2,622
2009	499	209	56	96	43	2,466	3,369
				Calves			
	Thousands	Thousands	Thousands	Thousands	Thousands	Thousands	Thousands
2000	....	....	....	....	....	89	90
2001	....	....	....	....	....	86	87
2002	....	3	....	....	....	106	113
2003	....	7	....	....	....	486	494
2004	....	6	....	1	....	574	581
2005	0	13	0	0	....	216	229
2006	0	11	0	0	....	223	235
2007	1	37	0	1	....	264	303
2008	1	33	0	1	....	290	325
2009	2	135	0	0	0	319	456

[1] Total rail and truck receipts unloaded at public stockyards. [2] Rounded totals of the complete figures. [3] The number of stockyards varies from 23 to 46.
AMS, Livestock & Grain Market News, (202) 720–7316. Compiled from reports received from stockyard companies.

Table 7-12.—Cattle and calves: Number slaughtered, United States, 2000–2009

Year	Cattle slaughter					Calf slaughter				
	Commercial			Farm	Total	Commercial			Farm	Total
	Federally in-spected	Other	Total [1]			Federally in-spected	Other	Total [1]		
	Thou-sands	Thou-sands	Thou-sands	Thou-sands	Thou-sands	Thou-sands	Thou-sands	Thou-sands	Thou-sands	Thou-sands
2000	35,631	615	36,246	170	36,416	1,089	43	1,132	40	1,172
2001	34,771	599	35,370	160	35,530	981	26	1,007	40	1,047
2002	35,120	614	35,735	153	35,888	1,019	26	1,045	37	1,082
2003	34,907	587	35,493	154	35,647	976	25	1,001	38	1,039
2004	32,156	573	32,728	152	32,880	823	20	842	37	879
2005	31,832	556	32,388	152	32,539	718	17	734	38	772
2006	33,145	553	33,698	150	33,849	699	13	711	37	748
2007	33,721	543	34,264	150	34,414	745	13	758	37	795
2008	33,805	560	34,365	150	34,515	942	15	957	36	993
2009	32,765	573	33,338	150	33,488	930	14	944	36	980

[1] Totals are based on unrounded numbers.
NASS, Iowa Field Office, (515) 284–4340.

Table 7-13.—Cattle and calves: Number slaughtered commercially, total and average live weight, by State and United States, 2009[1]

State	Cattle			Calves		
	Number slaughtered	Total live weight[2]	Average live weight[2]	Number slaughtered	Total live weight[2]	Average live weight[2]
	Thousands	1,000 pounds	Pounds	Thousands	1,000 pounds	Pounds
AL	7.0	6,268	900			
AK	0.7	745	1,136			
AZ	515.9	659,433	1,284	0.1	23	428
AR	9.7	9,431	978	0.2	104	461
CA	1,649.5	2,109,390	1,297	244.2	25,971	108
CO	2,356.9	3,119,244	1,324	0.1	36	485
DE-MD	43.1	56,082	1,301	2.6	735	279
FL				1.3	247	191
GA	233.0	231,537	998	0.9	414	455
HI	11.1	12,692	1,145			
ID	313.3	390,476	1,267			
IL				34.5	13,049	387
IN	42.9	45,387	1,057			
IA						
KS	6,336.9	8,174,971	1,290	0.2	70	458
KY	20.6	19,724	955	0.4	218	528
LA	6.9	5,091	736	4.4	2,259	514
MI	606.8	830,355	1,376	18.2	7,927	436
MN	837.9	1,167,105	1,400	0.2	63	417
MS	1.4	946	682	0.5	170	370
MO	90.1	108,055	1,205	0.6	240	376
MT	22.8	27,226	1,195			
NE	6,765.9	9,104,449	1,347			
NV	1.6	1,612	989			
N ENG[3]	16.2	16,765	1,036	30.6	3,018	99
NJ	34.7	39,244	1,146	67.8	28,896	427
NM	6.7	7,134	1,088	1.0	121	121
NY	31.9	37,610	1,195	141.1	14,119	103
NC	222.4	257,968	1,166	1.5	755	496
ND	15.7	19,304	1,230			
OH	113.4	135,018	1,199	104.4	27,819	267
OK	28.0	29,807	1,066	0.7	308	441
OR	40.7	47,337	1,210	0.1	26	406
PA	928.5	1,152,461	1,255	129.5	55,108	426
SC	149.5	165,068	1,115	0.3	99	369
SD						
TN	36.2	30,672	865	0.5	219	449
TX	6,619.4	8,208,471	1,243	43.7	7,560	173
UT				0.1	22	308
VA	11.1	11,485	1,031			
WA	1,043.5	1,344,331	1,294			
WV	10.6	11,514	1,089			
WI	1,681.3	2,237,558	1,344			
WY	8.0	9,306	1,162			
US[4]	33,338.2	42,966,102	1,293	944.2	233,403	250
PR	39.5	43,922	1,118	5.2	780	149

[1] Includes slaughter in federally inspected and other slaughter plants; excludes animals slaughtered on farms. Average live weight is based on unrounded numbers. Totals may not add due to rounding. [2] Excludes postmortem condemnations. [3] CT, ME, MA, NH, RI, and VT. [4] States with no data printed are still included in the U.S. total. Data are not printed to avoid disclosing individual operations.
NASS, Iowa Field Office, (515) 284-4340.

Table 7-14.—Cattle and calves: Number slaughtered under Federal inspection, and average live weight, 2000–2009

Year	Cattle		Calves	
	Number slaughtered	Average live weight[1]	Number slaughtered	Average live weight[1]
	Thousands	Pounds	Thousands	Pounds
2000	35,631	1,222	1,089	311
2001	34,771	1,224	981	318
2002	35,120	1,253	1,019	310
2003	34,907	1,234	976	316
2004	32,156	1,242	823	329
2005	31,832	1,259	718	352
2006	33,145	1,277	699	344
2007	33,721	1,275	745	304
2008	33,805	1,284	942	255
2009	32,765	1,296	930	248

[1] Excludes postmortem condemnations.
NASS, Iowa Field Office, (515) 284-4340.

Table 7-15.—Cattle and calves: Production, disposition, cash receipts, and gross income, by State and United States, 2009 [1]

| State | Marketings [2] | | Inshipments [3] | Farm slaugh-ter of cattle and calves [4] | Production (live weight) [5] | Value of production | Cash receipts [6] | Value of home con-sumption | Gross income [7] |
	Cattle	Calves							
	1,000 head	1,000 head	1,000 head	1,000 head	1,000 pounds	1,000 dollars	1,000 dollars	1,000 dollars	1,000 dollars
Al	433.0	101.0	20.0	2.0	451,778	309,937	308,286	2,006	310,292
AK	1.4	0.2	0.2	0.5	2,011	2,162	1,772	454	2,226
AZ	666.0	167.0	492.0	1.0	609,364	320,523	600,422	2,169	602,591
AR	610.0	219.0	180.0	2.0	536,083	412,649	436,771	6,485	443,256
CA	2,076.0	535.0	790.0	14.0	1,919,081	1,111,796	1,676,375	9,959	1,686,334
CO	2,205.0	100.0	1,647.0	2.0	1,803,820	1,586,292	2,605,779	7,161	2,612,940
CT	13.9	8.9	2.0	1.0	14,319	8,569	9,851	1,028	10,879
DE	5.5	2.7	1.0	0.3	6,230	5,127	5,918	551	6,469
FL	231.0	710.0	120.0	2.0	384,677	307,161	375,149	358	375,507
GA	319.0	238.0	52.0	3.0	364,983	250,160	296,777	3,438	300,215
HI	19.0	38.0		1.0	41,540	29,837	28,945	1,173	30,118
ID	1,034.0	150.0	375.0	2.0	1,045,960	797,856	961,618	7,050	968,668
IL	444.0	80.0	115.0	7.0	501,470	421,593	486,900	16,988	503,888
IN	271.0	117.0	100.0	4.0	273,484	200,311	224,421	10,965	235,386
IA	2,254.0	102.0	1,400.0	4.0	1,786,596	1,436,961	2,470,351	8,488	2,478,839
KS	5,109.0	1.5	3,600.0	4.5	3,915,772	2,964,814	5,546,577	12,301	5,558,878
KY	512.0	423.0	29.0	7.0	583,099	477,771	484,572	13,657	498,229
LA	153.0	260.0	2.0	2.0	222,360	170,208	189,092	888	189,980
ME	15.7	14.2	2.0	1.5	17,645	11,114	10,845	1,346	12,191
MD	45.0	12.0	3.0	2.0	68,770	53,364	48,137	2,497	50,634
MA	8.0	10.0	2.0	1.0	8,544	5,118	4,931	1,047	5,978
MI	296.0	37.2	61.2	4.0	417,194	284,037	288,660	8,749	297,409
MN	912.0	107.0	360.0	6.0	1,108,160	800,217	983,682	18,193	1,001,875
MS	236.0	130.0	12.0	2.0	210,273	144,991	140,141	4,194	144,335
MO	925.0	936.0	40.0	4.0	1,347,628	1,169,816	1,242,256	30,649	1,272,905
MT	1,290.0	230.0	105.0	5.0	952,347	752,678	896,144	9,115	905,259
NE	5,478.0	85.0	3,980.0	2.0	4,597,667	3,733,330	6,239,571	11,189	6,250,760
NV	178.0	88.0	75.0	2.0	182,550	148,730	192,441	2,626	195,067
NH	7.8	6.6	1.0	0.5	10,007	6,918	7,232	883	8,115
NJ	6.2	7.9	0.5	0.5	8,542	5,010	5,549	676	6,225
NM	1,145.0	390.0	1,000.0	2.0	774,218	594,541	1,007,546	3,398	1,010,944
NY	134.4	264.6	14.0	2.0	211,440	128,085	121,116	2,967	124,083
NC	260.0	116.0	10.0	2.0	312,124	205,617	213,812	10,987	224,799
ND	817.0	104.5	88.0	1.5	588,075	477,740	596,093	5,211	601,304
OH	355.0	108.0	43.0	4.0	392,698	304,910	348,159	16,877	365,036
OK	2,310.0	335.0	1,045.0	10.0	1,983,903	1,746,563	2,226,324	19,858	2,246,182
OR	367.0	171.0	20.0	10.0	529,175	420,628	405,691	8,867	414,558
PA	414.0	216.0	145.0	10.0	487,535	349,765	413,672	18,880	432,552
RI	1.1	1.2	0.1	0.1	1,269	738	780	105	885
SC	139.0	17.0	8.0	2.0	158,813	114,914	115,106	3,254	118,360
SD	1,561.0	390.0	588.0	2.0	1,471,124	1,325,948	1,550,379	14,267	1,564,646
TN	536.0	284.0	31.0	3.0	606,394	438,142	423,767	6,495	430,262
TX	6,840.0	155.0	2,500.0	15.0	6,923,911	5,481,429	6,938,721	18,622	6,957,343
UT	360.0	38.0	66.0	4.0	225,883	184,624	243,648	6,656	250,304
VT	47.0	71.0	7.0	2.0	61,014	39,438	41,265	1,283	42,548
VA	332.0	214.0	13.0	5.0	461,443	344,366	287,517	13,018	300,535
WA	564.0	8.0	170.0	9.0	582,189	472,958	600,834	6,639	607,473
WV	170.0	89.0	37.0	4.0	164,491	102,024	132,411	5,796	138,207
WI	796.0	412.0	100.0	7.0	1,124,481	708,065	726,337	18,206	744,543
WY	798.0	177.0	338.0	1.0	467,134	399,522	614,225	8,635	622,860
US	43,701.0	8,478.5	19,790.0	185.4	40,919,268	31,769,067	43,776,568	386,304	44,162,872

[1] Preliminary.　[2] Includes custom slaughter for use on farms where produced and State outshipments, but excludes interfarm sales within the State.　[3] Includes cattle shipped in from other states, but excludes cattle for immediate slaughter.　[4] Excludes custom slaughter for farmers at commercial establishments.　[5] Adjustments made for changes in inventory and for inshipments.　[6] Includes receipts from marketings and sales of farm-slaughter.　[7] Includes cash receipts from sales of cattle, calves, beef, and veal plus value of cattle and calves slaughtered for home consumption.
NASS, Livestock Branch, (202) 720–3570.

Table 7-16.—Cattle: Number slaughtered under Federal inspection and percentage distribution, by classes, 2000–2009 [1]

Year	Number						Percentage of total					
	Steers	Heif-ers	Cows			Bulls	Steers	Heif-ers	Cows			Bulls
			Dairy cows	Other cows	Total cows				Dairy cows	Other cows	Total cows	
	Thou-sands	Thou-sands	Thou-sands	Thou-sands	Thou-sands	Thou-sands	Per-cent	Per-cent	Per-cent	Per-cent	Per-cent	Per-cent
2000 ...	17,758	11,835	2,632	2,796	5,427	612	49.8	33.2	7.4	7.8	15.2	1.7
2001 ...	17,097	11,379	2,582	3,092	5,674	621	49.2	32.7	7.4	8.9	16.3	1.8
2002 ...	17,523	11,342	2,607	3,051	5,658	598	49.9	32.3	7.4	8.7	16.1	1.7
2003 ...	17,177	11,078	2,860	3,163	6,023	629	49.2	31.7	8.2	9.1	17.3	1.8
2004 ...	16,192	10,345	2,363	2,706	5,069	550	50.4	32.2	7.3	8.4	15.8	1.7
2005 ...	16,797	9,761	2,252	2,523	4,775	498	52.8	30.7	7.1	7.9	15.0	1.6
2006 ...	17,478	9,820	2,354	2,983	5,336	511	52.7	29.6	7.1	9.0	16.1	1.5
2007 ...	17,285	10,207	2,497	3,178	5,675	554	51.3	30.3	7.4	9.4	16.8	1.6
2008 ...	16,949	10,091	2,591	3,569	6,161	605	50.1	29.9	7.7	10.6	18.2	1.8
2009 ...	16,313	9,742	2,815	3,325	6,140	570	49.8	29.7	8.6	10.1	18.7	1.7

[1] Totals and percentages based on unrounded data and may not equal sum of classes due to rounding.
NASS, Iowa Field Office, (515) 284–4340.

Table 7-17.—Cattle and calves: Inventory Jan 1, 2009–2010, and number of operations, 2007, by State and United States [1]

State	January 1 Cattle inventory		Operations with cattle [3]
	2009	2010 [2]	2007
	1,000 head	1,000 head	Number
AL	1,260	1,280	24,000
AK	14.0	14.5	130
AZ	1,020	930	7,700
AR	1,810	1,890	28,000
CA	5,250	5,150	16,600
CO	2,600	2,600	14,700
CT	52	48	1,200
DE	21	20	400
FL	1,700	1,720	21,000
GA	1,110	1,060	21,000
HI	150	151	1,100
ID	2,110	2,140	10,600
IL	1,200	1,170	18,500
IN	860	870	18,500
IA	3,950	3,850	30,000
KS	6,300	6,000	30,000
KY	2,300	2,300	44,000
LA	890	840	14,100
ME	89	87	2,100
MD	185	195	3,700
MA	43	43	1,800
MI	1,070	1,100	14,500
MN	2,400	2,420	25,000
MS	960	970	17,800
MO	4,250	4,150	59,000
MT	2,600	2,550	12,300
NE	6,350	6,250	22,000
NV	450	450	1,500
NH	39	37	1,000
NJ	38	36	1,400
NM	1,540	1,550	9,500
NY	1,380	1,410	13,600
NC	850	820	19,200
ND	1,760	1,720	10,500
OH	1,280	1,280	26,000
OK	5,400	5,450	55,000
OR	1,240	1,260	16,100
PA	1,590	1,620	26,000
RI	5.0	4.7	280
SC	380	380	8,800
SD	3,700	3,800	15,700
TN	1,980	2,040	47,000
TX	13,600	13,300	152,000
UT	810	800	7,600
VT	270	265	2,500
VA	1,470	1,550	26,000
WA	1,080	1,040	12,700
WV	415	370	12,800
WI	3,350	3,400	35,000
WY	1,350	1,320	5,600
US	94,521.0	93,701.2	965,510
PR			4,500

[1] An operation is any place having one or more head of cattle on hand on December 31. [2] Preliminary. [3] State level estimates only available in conjunction with the Census of Agriculture every 5 years.
NASS, Livestock Branch, (202) 720–3570.

Table 7-18.—Cattle and calves: Average dressed weight under Federal inspection, 2000–2009 [1]

Year	Cattle					Calves
	All cattle	Steers	Heifers	Cows	Bulls	
	Pounds	Pounds	Pounds	Pounds	Pounds	Pounds
2000	745	798	733	579	892	192
2001	744	798	734	584	893	196
2002	765	823	753	590	912	190
2003	746	803	732	590	904	194
2004	756	806	740	614	893	201
2005	769	817	750	621	905	216
2006	781	833	767	622	914	207
2007	776	830	764	617	893	182
2008	778	838	772	609	888	150
2009	784	847	782	610	878	147

[1] Excludes postmortem condemnations.
NASS, Iowa Field Office, (515) 284–4340.

Table 7-19.—Cattle and calves: Number of operations and percent of inventory by size group, United States, 2008-2009 [1]

Head	Operations		Percent of inventory	
	2008	2009	2008	2009
	Number	Number	Percent	Percent
Cattle and Calves				
1-49	645,000	642,000	11.4	11.4
50-99	133,000	132,000	9.7	9.7
100-499	147,600	146,300	31.2	31.2
500-999	19,100	19,000	13.4	13.5
1,000-1,999	6,600	6,500	9.1	9.0
2,000-4,999	3,030	3,010	8.8	8.8
5,000-9,999	700	720	4.7	4.8
10,000-19,999	260	270	3.5	3.7
20,000+	210	200	8.2	7.9
Total	955,500	950,000	100.0	100.0
Beef Cows [2]				
1-49	601,000	598,000	28.5	28.3
50-99	83,000	82,000	17.1	17.1
100-499	67,200	67,200	38.0	38.0
500-999	4,340	4,350	8.7	8.8
1,000-1,999	1,125	1,110	4.5	4.5
2,000-4,999	280	280	2.1	2.1
5,000+	55	60	1.1	1.2
Total	757,000	753,000	100.0	100.0

[1] An operation is any place having one or more head of cattle on hand on December 31.   [2] Included in operations with cattle.
NASS, Livestock Branch, (202) 720–3570.

Table 7-20.—Beef cows: Inventory Jan 1, 2009–2010, and number of operations, 2007, by State and United States [1]

State	January 1 beef cow inventory		Operations with beef cows [3]
	2009	2010 [2]	2007
	1,000 head	*1,000 head*	*Number*
AL	668	669	22,000
AK	5.5	5.4	100
AZ	200	208	5,300
AR	906	937	25,000
CA	620	610	11,800
CO	720	714	11,600
CT	6.0	5.5	750
DE	4.0	4.0	250
FL	942	958	16,700
GA	536	524	17,700
HI	84.4	81.2	850
ID	451	440	7,400
IL	408	389	14,800
IN	213	221	12,700
IA	925	885	21,000
KS	1,505	1,434	26,000
KY	1,114	1,070	38,000
LA	515	499	12,400
ME	11.0	11.0	1,300
MD	39	41	2,500
MA	7.5	8.0	1,200
MI	92	96	7,800
MN	392	380	14,400
MS	491	503	16,000
MO	1,992	1,968	52,000
MT	1,494	1,465	11,100
NE	1,851	1,781	18,300
NV	237	237	1,300
NH	6.0	4.0	640
NJ	10	9.5	930
NM	484	502	8,200
NY	85	90	6,800
NC	384	367	15,000
ND	895	869	9,700
OH	304	288	17,400
OK	2,038	2,073	47,000
OR	535	546	12,900
PA	150	160	12,300
RI	1.4	1.3	230
SC	188	183	8,200
SD	1,626	1,618	13,800
TN	951	997	42,000
TX	5,170	5,140	132,000
UT	350	338	5,600
VT	9.0	10.0	1,000
VA	643	665	22,000
WA	271	227	10,100
WV	204	190	10,700
WI	265	260	14,800
WY	713	694	4,800
US	31,711.8	31,375.9	766,350
PR	....	....	2,400

[1] An operation is any place having one or more beef cows on hand on December 31. [2] Preliminary. [3] State level estimates only available in conjunction with the Census of Agriculture every 5 years.
NASS, Livestock Branch, (202) 720–3570.

Table 7-21.—Hogs and pigs: Operations, inventory and value, United States, Dec. 1, 2000–2009

Year	Operations [1]	Inventory	Value Per head	Value Total
	Number	Thousands	Dollars	1,000 dollars
2000	87,470	59,110	77.00	4,540,410
2001	81,220	59,722	77.00	4,584,078
2002	76,250	59,554	71.00	4,230,728
2003	73,720	60,453	67.00	4,024,626
2004	69,500	60,982	103.00	6,306,282
2005	67,280	61,463	95.00	5,833,763
2006	65,940	62,516	90.00	5,598,613
2007	75,450	68,177	73.00	4,986,206
2008	73,150	67,148	89.00	5,957,633
2009 [2]	71,450	65,807	83.00	5,464,612

[1] An operation is any place having one or more hogs and pigs on hand December 31. [2] Preliminary.
NASS, Livestock Branch, (202) 720–3570.

Table 7-22.—Hogs and pigs: Inventory and value, Dec. 1, 2008–2009, and number of operations, 2007, by State and United States

State	Inventory 2008	Inventory 2009 [2]	Value per head 2008	Value per head 2009 [2]	Total value 2008	Total value 2009 [2]	Operations [1] 2007
	Thousands	Thousands	Dollars	Dollars	1,000 dollars	1,000 dollars	Number
AL	175.0	140.0	88	82	15,400	11,480	750
AK	0.9	1.4	190	190	171	266	40
AZ	165.0	167.0	93	87	15,345	14,529	380
AR	280.0	200.0	95	91	26,600	18,200	1,100
CA	80.0	100.0	120	110	9,600	11,000	1,400
CO	730.0	720.0	90	88	65,700	63,360	1,200
CT	3.1	2.9	120	110	372	319	240
DE	9.0	7.5	88	82	792	615	80
FL	20.0	20.0	93	87	1,860	1,740	1,900
GA	235.0	210.0	80	76	18,800	15,960	1,100
HI	13.0	13.0	150	140	1,950	1,820	230
ID	33.0	36.0	93	87	3,069	3,132	660
IL	4,350.0	4,350.0	91	83	395,850	361,050	2,900
IN	3,550.0	3,650.0	91	84	323,050	306,600	3,400
IA	19,900.0	19,300.0	94	86	1,870,600	1,659,800	8,300
KS	1,740.0	1,810.0	77	75	133,980	135,750	1,500
KY	355.0	350.0	67	62	23,785	21,700	1,500
LA	11.0	10.0	93	87	1,023	870	720
ME	4.4	4.9	93	87	409	426	440
MD	31.0	30.0	88	82	2,728	2,460	410
MA	10.0	11.0	93	87	930	957	450
MI	1,030.0	1,080.0	99	91	101,970	98,280	2,700
MN	7,500.0	7,400.0	100	97	750,000	717,800	4,400
MS	375.0	365.0	93	87	34,875	31,755	680
MO	3,150.0	3,100.0	74	72	233,100	223,200	3,000
MT	175.0	175.0	93	87	16,275	15,225	490
NE	3,350.0	3,150.0	95	89	318,250	280,350	2,200
NV	3.5	2.9	120	110	420	319	90
NH	2.8	2.4	100	93	280	223	270
NJ	8.0	8.0	100	93	800	744	270
NM	2.0	1.5	93	87	186	131	400
NY	95.0	77.0	88	82	8,360	6,314	1,900
NC	9,700.0	9,700.0	75	70	727,500	679,000	2,800
ND	151.0	160.0	93	87	14,043	13,920	350
OH	1,940.0	2,020.0	92	85	178,480	171,700	3,700
OK	2,400.0	2,310.0	80	79	192,000	182,490	2,700
OR	20.0	17.0	93	87	1,860	1,479	1,300
PA	1,120.0	1,180.0	83	77	92,960	90,860	3,600
RI	1.8	1.7	93	87	167	148	100
SC	245.0	225.0	77	72	18,865	16,200	810
SD	1,280.0	1,170.0	97	92	124,160	107,640	960
TN	205.0	185.0	78	73	15,990	13,505	1,500
TX	1,120.0	770.0	71	69	79,520	53,130	4,500
UT	740.0	730.0	93	87	68,820	63,510	610
VT	2.8	3.0	120	110	336	330	250
VA	355.0	365.0	74	68	26,270	24,820	1,200
WA	25.0	23.0	100	93	2,500	2,139	1,500
WV	7.0	5.0	93	87	651	435	1,000
WI	360.0	360.0	78	79	28,080	28,440	3,200
WY	89.0	87.0	100	93	8,900	8,091	270
US	67,148.3	65,807.2	89	83	5,957,633	5,464,212	75,450
PR							1,500

[1] State level estimates only available in conjunction with the Census of Agriculture every 5 years. An operation is any place having one or more hogs and pigs on hand December 31. [2] Preliminary. Totals may not add due to rounding.
NASS, Livestock Branch, (202) 720–3570.

Table 7-23.—Sows farrowing and pig crop: Number, United States, 2000–2009

Year	Sows farrowing		Pig crop		
	Dec.-May	June-Nov.	Dec.-May	June-Nov.	Total
	Thousands	*Thousands*	*Thousands*	*Thousands*	*Thousands*
2000	5,683	5,726	50,086	50,656	100,742
2001	5,618	5,767	49,477	51,140	100,617
2002	5,776	5,716	50,858	50,820	101,678
2003	5,654	5,771	50,024	51,458	101,481
2004	5,706	5,791	50,747	52,039	102,787
2005	5,715	5,817	51,340	52,636	103,975
2006	5,769	5,861	52,259	53,374	105,633
2007	5,935	6,312	54,266	58,608	112,874
2008	6,123	6,103	57,019	58,011	115,030
2009 [1]	6,029	5,939	57,564	57,610	115,174

[1] Preliminary.
NASS, Livestock Branch, (202) 720–3570.

Table 7-24.—Hogs and pigs: Number for breeding and market, United States, 2000–2009

Year	All hogs and pigs	Kept for breeding	Market hogs by weight groups				
			Under 50 pounds	50 to 119 pounds	120 to 179 pounds	180 pounds and over	Total
			June 1				
	Thousands	*Thousands*	*Thousands*	*Thousands*	*Thousands*	*Thousands*	*Thousands*
2000	59,110	6,233	....	....	10,708	9,016	52,878
2001	58,525	6,178	....	....	10,531	8,971	52,347
2002	60,391	6,208	....	....	10,906	9,512	54,183
2003	59,609	6,036	....	....	10,827	9,362	53,573
2004	60,707	5,947	....	....	11,255	9,714	54,760
2005	60,744	5,988	....	....	11,143	9,813	54,756
2006	61,701	6,080	....	....	11,483	9,642	55,621
2007	63,947	6,169	....	....	11,789	9,920	57,777
2008	67,400	6,131	19,807	17,711	12,892	10,860	61,269
2009 [1]	66,809	5,968	19,554	17,838	12,604	10,847	60,842
			Dec. 1				
	Thousands	*Thousands*	*Thousands*	*Thousands*	*Thousands*	*Thousands*	*Thousands*
2000	59,110	6,267	....	....	10,841	9,663	52,843
2001	59,722	6,201	....	....	10,755	9,986	53,521
2002	59,554	6,058	....	....	10,875	10,103	53,496
2003	60,453	6,019	....	....	11,108	10,311	54,434
2004	60,982	5,980	....	....	11,185	10,401	55,002
2005	61,463	6,031	....	....	11,291	10,566	55,432
2006	62,516	6,116	....	....	11,274	10,738	56,399
2007	68,177	6,233	....	....	12,658	11,569	61,944
2008	67,148	6,062	19,428	17,396	12,731	11,533	61,087
2009 [1]	65,807	5,850	19,085	17,062	12,529	11,282	59,957

[1] Preliminary.
NASS, Livestock Branch, (202) 720–3570.

Table 7-25.—Cattle and swine: Production, 2007–2009

Country	2007	2008	2009
	1,000 head	*1,000 head*	*1,000 head*
Cattle:			
Argentina	15,900	14,900	12,300
Australia	9,369	9,079	9,213
Brazil	48,845	49,050	49,150
Canada	5,540	5,288	5,110
China	45,353	45,360	42,572
Colombia	5,750	5,670	5,675
EU-27	31,500	30,850	30,400
India	57,000	57,450	57,960
Mexico	6,732	6,754	6,775
Russia	7,310	7,100	7,010
Others	24,660	23,526	19,684
Total Foreign	257,959	255,027	245,849
United States	36,759	36,153	35,819
Total	294,718	291,180	281,668
Swine:			
Australia	5,480	4,477	4,609
Brazil	34,530	34,845	35,890
Canada	31,835	31,085	29,151
China	592,080	636,817	651,682
EU-27	265,100	258,400	254,500
Japan	17,050	16,960	17,700
Korea, South	14,422	13,792	14,916
Mexico	15,767	15,924	15,966
Russia	39,150	41,760	43,300
Ukraine	6,986	6,619	6,600
Others	84,356	5,030	....
Total Foreign	1,106,756	1,065,709	1,074,314
United States	112,873	115,030	115,115
Total	1,219,629	1,180,739	1,189,429

FAS, Office of Global Analysis, (202) 720-6301. Prepared or estimated on the basis of official USDA production, supply, and distribution statistics from foreign governments.

Table 7-26.—Hogs: Number slaughtered, United States, 2000–2009

Year	Commercial			Farm	Total
	Federally inspected	Other	Total [1]		
	Thousands	*Thousands*	*Thousands*	*Thousands*	*Thousands*
2000	96,436	1,540	97,976	130	98,106
2001	96,528	1,434	97,962	120	98,082
2002	98,915	1,348	100,263	115	100,378
2003	99,698	1,233	100,931	112	101,043
2004	102,361	1,103	103,463	110	103,573
2005	102,519	1,063	103,582	109	103,690
2006	103,689	1,048	104,737	106	104,842
2007	108,138	1,033	109,172	106	109,278
2008	115,421	1,031	116,452	106	116,558
2009	112,612	1,006	113,618	116	113,734

[1] Totals are based on unrounded numbers.
NASS, Iowa Field Office, (515) 284–4340.

Table 7-27.—Sows farrowing and pig crop: Number by State and United States, 2008–2009

State	Sows farrowing							
	Dec.–Feb.		Mar.–May		June–Aug.		Sept.–Nov.	
	2008	2009 [1]	2008	2009 [1]	2008	2009 [1]	2008	2009 [1]
	Thousands	Thousands	Thousands	Thousands	Thousands	Thousands	Thousands	Thousands
CO	84	75	81	72	76	72	71	70
IL	240	250	230	250	240	245	255	260
IN	150	140	145	135	145	135	140	135
IA	510	500	520	495	510	475	510	500
KS	86	81	80	81	81	80	76	85
MI	52	53	53	54	53	57	53	56
MN	285	285	285	285	285	275	285	285
MO	190	190	190	185	195	185	200	190
NE	190	185	185	190	185	185	185	180
NC	550	530	550	540	560	540	520	520
OH	89	88	89	88	89	86	89	87
OK	180	190	185	185	195	190	185	190
PA	43	40	41	42	38	42	40	46
SD	88	83	85	86	82	83	84	80
TX	54	46	51	48	52	46	50	31
UT	39	42	41	43	43	42	40	40
Oth Sts [2]	241	233	241	239	246	227	245	219
US	3,071	3,011	3,052	3,018	3,075	2,965	3,028	2,974

State	Pig crop							
	Dec.–Feb.		Mar.–May		June–Aug.		Sept.–Nov.	
	2008	2009 [1]	2008	2009 [1]	2008	2009 [1]	2008	2009 [1]
	Thousands	Thousands	Thousands	Thousands	Thousands	Thousands	Thousands	Thousands
CO	760	645	717	648	676	684	625	665
IL	2,220	2,375	2,151	2,375	2,268	2,328	2,397	2,457
IN	1,350	1,302	1,320	1,269	1,341	1,283	1,288	1,289
IA	4,692	4,800	4,862	4,777	4,794	4,631	4,845	4,900
KS	783	737	728	761	741	744	714	795
MI	491	514	514	521	490	547	512	549
MN	2,636	2,779	2,708	2,836	2,793	2,750	2,807	2,864
MO	1,748	1,777	1,786	1,813	1,892	1,850	1,930	1,881
NE	1,805	1,804	1,776	1,881	1,795	1,859	1,813	1,800
NC	5,060	5,062	5,143	5,238	5,404	5,238	4,966	4,992
OH	814	827	837	836	846	826	801	818
OK	1,683	1,691	1,748	1,665	1,823	1,805	1,702	1,805
PA	409	390	388	410	365	420	374	455
SD	845	818	825	869	804	830	823	808
TX	486	444	439	466	445	414	478	290
UT	378	407	410	426	426	420	400	392
Oth Sts [2]	2,228	2,180	2,279	2,221	2,337	2,148	2,296	2,073
US	28,388	28,552	28,631	29,012	29,240	28,777	28,771	28,833

[1] Preliminary. Totals may not add due to rounding.　[2] Individual State estimates not available for the 34 other States.
NASS, Livestock Branch, (202) 720–3570.

Table 7-28.—Hogs: Production, disposition, cash receipts, and gross income, United States, 1999–2008

Year	Mar-ketings[1]	Inship-ments[2]	Farm slaugh-ter[3]	Production (live weight)[4]	Value of production[5]	Cash receipts[6]	Value of home consump-tion	Gross income[7]	Average price per 100 pounds received by farmers
	1,000 head	*1,000 head*	*1,000 head*	*1,000 pounds*	*1,000 dollars*	*1,000 dollars*	*1,000 dollars*	*1,000 dollars*	*Dollars*
1999	121,138	22,634	141	25,856,590	7,770,907	8,624,295	28,381	8,652,676	30.30
2000	118,546	24,514	125	25,696,997	10,783,825	11,757,943	34,720	11,792,663	42.30
2001	119,272	26,745	119	25,866,250	11,416,397	12,394,560	35,462	12,430,022	44.40
2002	124,013	29,434	114	26,274,153	8,690,923	9,602,109	25,525	9,627,634	33.40
2003	124,363	31,542	116	26,266,840	9,668,978	10,616,050	27,738	10,643,788	37.20
2004	127,592	32,909	113	26,695,487	13,075,294	14,336,274	36,455	14,372,729	49.30
2005	129,027	33,396	107	27,368,993	13,591,029	14,970,027	34,713	15,004,740	50.20
2006	132,384	36,323	105	28,182,382	12,714,218	14,105,864	31,344	14,137,208	46.00
2007	137,519	39,433	106	29,606,420	13,468,332	14,750,490	32,148	14,782,638	46.60
2008	148,986	42,121	106	31,410,795	14,457,000	16,050,481	33,526	16,084,007	47.00

[1] Includes custom slaughter for use on farms where produced and State outshipments, but excludes interfarm sales within the State. [2] Includes hogs and pigs shipped in from other states but excludes animals for immediate slaughter. [3] Excludes custom slaughtered for farmers at commercial establishments. [4] Adjustments made for changes in inventory and for inshipments. [5] Includes allowance for higher average price of State inshipments and outshipments of feeder pigs. [6] Receipts from marketings and sale of farm slaughter includes allowance for higher average price of State outshipments of feeder pigs. [7] Cash receipts from sale of hogs, pork, and lard plus value of hogs slaughtered for home consumption.
NASS Livestock Branch, (202) 720–3570.

Table 7-29.—Hogs: Receipts at selected public stockyards and direct receipts at interior markets, 2000–2009[1]

Year	Receipts at selected public stockyards				Direct receipts in interior Iowa and Southern Minnesota[4]
	South St. Joseph	Sioux Falls	All others reporting	Total markets reporting[2][3]	
	Thousands	*Thousands*	*Thousands*	*Thousands*	*Thousands*
2000	59	298	998	1,260	36,504
2001	85	272	383	674	2,998
2002	16	243	376	546	4,486
2003	54	200	313	491	5,128
2004	44	176	197	354	19,760
2005	38	152	227	375	14,531
2006	49	158	186	305	41,723
2007	27	141	268	295	41,446
2008	21	110	295	316	17,267
2009	13	91	234	338	43,287

[1] Total rail and truck receipts. [2] Rounded total of complete figures. [3] The number of stockyards reporting varies from 25 to 55. [4] Covers receipts at 14 packing plants and 30 concentration yards.
AMS, Livestock & Grain Market News, (202) 720–7316. Compiled from reports received from stockyard companies.

Table 7-30.—Hogs and corn: Hog-corn price ratio and average price received by farmers for corn, United States, 1999-2008

Year	Hog-corn price ratio[1]	Price of corn per bushel[2]
		Dollars
1999	17.3	1.89
2000	23.3	1.86
2001	23.4	1.89
2002	15.9	2.13
2003	16.6	2.27
2004	21.1	2.47
2005	25.3	1.96
2006	20.6	2.28
2007	13.8	3.39
2008	10.0	4.78

[1] Number of bushels of corn equal in value to buy 100 pounds of live hogs at local markets, based on average prices received by farmers for hogs and corn. Annual average is a simple average of monthly ratios for the calendar year. [2] Annual average is a simple average of entire month prices for the calendar year.
NASS, Environmental, Economics, and Demographics Branch, (202) 720–6146.

Table 7-31.—Hogs: Production, disposition, cash receipts, and gross income, by State and United States, 2008

State	Mar-ketings[1]	Inship ments[2]	Farm slaughter[3]	Production (live weight)[4]	Value of produc-tion[5]	Cash receipts[6]	Value of home con-sumption	Gross income[7]
	1,000 head	1,000 head	1,000 head	1,000 pounds	1,000 dollars	1,000 dollars	1,000 dollars	1,000 dollars
AL	409.0	192.0	1.0	64,169	28,414	34,079	237	34,316
AK	1.6	0.7	0.6	438	376	306	92	398
AZ	308.0	9.0	1.0	75,380	41,713	43,057	231	43,288
AR	1,830.0	50.0	1.0	141,380	89,283	95,075	466	95,541
CA	300.0	59.0	7.0	55,839	26,177	33,217	1,141	34,358
CO	2,883.0	111.0	1.0	281,303	160,522	174,640	249	174,889
CT	4.4	0.3	0.1	709	277	297	45	342
DE	57.0	2.3	0.1	6,352	2,879	3,115	39	3,154
FL	62.0	11.0	1.0	7,403	2,813	3,226	53	3,279
GA	750.3	113.3	2.0	101,010	52,730	62,117	473	62,590
HI	18.0	....	1.0	3,395	3,299	3,359	187	3,546
ID	76.2	10.0	1.0	20,451	9,513	9,586	259	9,845
IL	9,825.0	1,132.0	3.0	1,711,029	916,599	977,431	1,050	978,481
IN	7,916.0	2,718.0	1.0	1,726,027	818,183	917,338	756	918,094
IA	38,802.0	22,800.0	11.0	9,427,603	4,039,926	4,757,993	4,028	4,762,021
KS	3,703.0	790.0	1.0	868,883	351,801	400,064	457	400,521
KY	762.0	108.0	4.0	172,731	76,419	82,713	1,577	84,290
LA	22.7	2.0	1.0	4,713	1,817	1,709	75	1,784
ME	10.2	1.2	0.3	2,458	1,009	792	234	1,026
MD	88.6	20.0	0.2	15,368	6,503	7,400	152	7,552
MA	16.1	1.2	0.6	3,405	1,308	1,259	159	1,418
MI	2,097.0	172.0	4.0	575,459	243,828	249,776	455	250,231
MN	17,430.0	7,260.0	4.0	3,776,536	1,762,668	2,053,302	2,039	2,055,341
MS	663.0	5.0	2.0	168,878	75,408	73,904	408	74,312
MO	9,072.0	2,157.0	1.0	1,746,537	764,508	870,817	759	871,576
MT	387.8	29.0	2.0	80,119	35,785	42,834	490	43,324
NE	7,498.0	600.0	1.0	1,384,543	714,631	728,702	1,314	730,016
NV	7.0	5.6	0.3	1,474	681	737	55	792
NH	4.9	2.2	0.3	784	242	332	44	376
NJ	26.4	19.0	0.1	1,499	376	940	101	1,041
NM	3.1	0.8	0.6	762	300	235	113	348
NY	155.2	11.0	2.0	27,066	9,508	9,462	439	9,901
NC	19,491.0	478.0	10.0	4,209,853	2,119,707	2,170,176	1,407	2,171,583
ND	737.9	55.0	2.0	52,741	35,474	39,709	512	40,221
OH	3,695.0	657.0	10.0	965,679	421,083	434,543	3,751	438,294
OK	7,527.0	1,062.0	1.0	1,284,545	513,574	562,400	231	562,631
OR	43.5	....	0.8	10,960	5,809	5,565	293	5,858
PA	1,730.8	296.8	5.0	399,000	161,145	173,726	1,020	174,746
RI	5.5	0.1	0.2	733	289	293	36	329
SC	494.5	47.0	1.5	91,757	44,490	48,210	991	49,201
SD	4,295.0	951.0	2.0	739,054	349,924	399,208	1,684	400,892
TN	365.0	55.0	2.0	95,671	42,010	39,848	509	40,357
TX	1,701.0	38.0	6.0	317,446	128,679	131,744	1,744	133,488
UT	1,527.0	12.0	1.0	312,262	163,240	167,601	251	167,852
VT	4.9	0.7	0.2	1,036	390	365	67	432
VA	522.0	50.0	2.0	126,031	57,344	57,796	1,021	58,817
WA	51.8	1.4	1.5	10,145	4,526	4,603	382	4,985
WV	11.9	1.0	1.0	2,626	1,193	970	357	1,327
WI	958.0	18.0	2.0	201,129	107,923	112,800	586	113,386
WY	634.8	5.0	2.0	136,424	60,704	61,110	507	61,617
US	148,986.1	42,120.6	106.4	31,410,795	14,457,000	16,050,481	33,526	16,084,007

[1] Includes custom slaughter for use on farms where produced and State outshipments, but excludes interfarm sales within the State. [2] Includes hogs and pigs shipped in from other states but excludes animals for immediate slaughter. [3] Excludes custom slaughter for farmers at commercial establishments. [4] Adjustments made for changes in inventory and for inshipments. [5] Includes allowance for higher average price of State inshipments and outshipments of feeder pigs. [6] Receipts from marketings and sale of farm-slaughter. [7] Cash receipts from sales of hogs, pork, and lard plus value of hogs slaughtered for home consumption.
NASS, Livestock Branch, (202) 720–3570.

Table 7-32.—Hogs: Number slaughtered commercially, total and average live weight, by State and United States, 2009 [1]

State	Number slaughtered	Total live weight [2]	Average live weight [2]
	Thousands	*1,000 pounds*	*Pounds*
AL	80.6	34,974	434
AK	1.0	240	249
AZ	1.5	377	256
AR	179.6	82,177	458
CA	2,648.7	643,725	243
CO	10.1	2,599	257
DE and MD	18.1	4,454	247
FL	82.5	11,809	143
GA	72.5	18,642	257
HI	16.5	3,757	227
ID	120.5	30,723	255
IL	9,614.3	2,674,387	278
IN	8,529.6	2,256,328	265
IA	32,150.2	8,682,322	270
KS			
KY			
LA	13.2	2,515	191
MI	122.6	45,705	374
MN	9,818.7	2,592,322	264
MS	121.1	36,058	298
MO	8,028.2	2,228,686	278
MT	13.3	3,227	243
NE	7,691.1	2,067,922	269
NV			
N ENG [3]	21.1	4,546	216
NJ	100.9	9,676	96
NM	1.7	419	253
NY	27.1	5,002	185
NC	12,381.5	3,218,589	260
ND	105.1	27,041	258
OH	1,073.4	293,736	274
OK	5,325.2	1,472,745	277
OR	159.8	40,490	254
PA	3,008.3	774,417	258
SC			
SD	4,356.1	1,135,646	261
TN	695.1	320,370	462
TX	423.2	105,276	249
UT	55.5	10,815	195
VA	2,256.3	600,419	266
WA	24.1	5,242	217
WV	7.2	1,784	248
WI	525.4	228,925	442
WY	4.1	1,063	261
US [4]	113,618.1	30,723,171	271
PR	44.6	8,918	200

[1] Includes slaughter in federally inspected and other slaughter plants; excludes animals slaughtered on farms. Average live weight is based on unrounded numbers. Totals may not add due to rounding. [2] Excludes postmortem condemnations. [3] CT, ME, MA, NH, RI, and VT. [4] States with no data printed are still included in US total. Data are not printed to avoid disclosing individual operations.
NASS, Iowa Field Office, (515) 284–4340.

Table 7-33.—Hogs: Number slaughtered, average dressed and live weights, Federally inspected, 2000–2009 [1]

Year	Federally inspected											
	Barrows and gilts			Sows			Boars			Total		
	Head	Percent of total	Avg. dressed weight[2]	Head	Percent of total	Avg. dressed weight[2]	Head	Percent of total	Avg. dressed weight[2]	Head	Avg. dressed weight[2]	Avg. live weight[2]
	1,000		Pounds	1,000		Pounds	1,000		Pounds	1,000	Pounds	Pounds
2000 ..	93,115	96.6	191	3,005	3.1	309	316	0.3	226	96,436	194	262
2001 ..	93,201	96.6	193	3,009	3.1	316	318	0.3	226	96,528	197	265
2002 ..	95,459	96.5	193	3,185	3.2	317	271	0.3	235	98,915	197	265
2003 ..	96,242	96.5	195	3,215	3.2	315	241	0.2	241	99,698	199	267
2004 ..	98,831	96.6	196	3,271	3.2	313	259	0.3	220	102,361	199	267
2005 ..	99,123	96.7	197	3,116	3.0	310	280	0.3	213	102,519	201	269
2006 ..	100,113	96.6	198	3,227	3.1	309	348	0.3	227	103,689	202	269
2007 ..	104,352	96.5	198	3,309	3.1	308	477	0.4	213	108,138	202	269
2008 ..	111,461	96.6	198	3,502	3.0	308	458	0.4	208	115,421	201	268
2009 ..	108,950	96.7	200	3,243	2.9	306	419	0.4	199	112,612	203	271

[1] All weights calculated using unrounded totals. Totals and percentages based on unrounded data and may not equal sum of classes due to rounding.   [2] Excludes postmortem condemnations.
NASS, Iowa Field Office, (515) 284–4340.

Table 7-34.—Hogs and pigs: Number of operations and percent of inventory by size group, United States, 2008–2009 [1]

Head	Operations		Percent of inventory	
	2008	2009	2008	2009
	Number	Number	Percent	Percent
1-99	50,680	50,400	0.9	0.9
100-499	6,740	6,100	2.5	2.3
500-999	3,490	3,200	3.5	3.3
1,000-1,999	3,950	3,550	8.0	7.5
2,000-4,999	5,370	5,250	24.0	24.0
5,000+	2,920	2,950	61.1	62.0
Total	73,150	71,450	100.0	100.0

[1] An operation is any place having one or more head of hogs and pigs on hand on December 31. Percents reflect average distributions based primarily on end of year surveys.
NASS, Livestock Branch, (202) 720–3570.

Table 7-35.—Lard: Supply and disappearance, United States, 1999–2008

Calendar year	Supply				Disappearance							Per capita domestic disappearance
	Beginning stocks[1]	Production[2]	Imports[3]	Total	Domestic	Exports[3]	Total	Direct food use	Indirect food use	Nonfood use[4]	Ending stocks	
	Million lbs	Million lbs	Million lbs	Million lbs	Million lbs	Million lbs	Million lbs	Million lbs	Million lbs	Million lbs	Million lbs	Million lbs
1999 ..	28	735	2	765	591	147	739	202	278	128	27	0.7
2000 ..	27	718	2	748	558	174	731	221	242	95	16	0.8
2001 ..	16	724	3	744	627	103	730	325	232	69	14	1.1
2002 ..	14	744	8	766	671	84	755	370	236	65	11	1.3
2003 ..	11	753	7	770	640	117	757	369	207	64	13	1.3
2004 ..	13	772	5	791	488	289	777	220	201	66	14	0.8
2005 ..	14	779	5	798	695	94	789	460	175	60	9	1.6
2006 ..	9	788	7	805	719	72	790	499	176	44	14	1.7
2007 ..	14	821	9	844	757	73	830	487	177	93	14	1.6
2008[3]	14	874	7	894	801	81	882	317	182	302	12	10

[1] Domestic disappearance data are computed by ERS.   [2] Includes edible tallow direct use beginning in 1979.   [3] Preliminary.   [4] Including paint, varnish, resin, plastic, and lubricants.
ERS, Market and Trade Economics Division, Field Crops Branch, (202) 694–5300.

Table 7-36.—Lard: United States exports by country of destination, 2007–2009

Country	2007	2008	2009 [1]
	Metric tons	Metric tons	Metric tons
Mexico	22,762	31,585	36,040
Canada	5,958	2,727	715
Trinidad and Tobago	342	569	363
Netherlands Antilles (exc. Aruba)	13	132	256
Aruba	13	92	253
Leeward-Windward Islands(*)	106	158	140
Antigua and Barbuda	102	131	115
United Kingdom	0	0	77
Barbados	2	8	60
Cuba	0	20	43
Bermuda	3	51	38
Marshal Island	0	0	31
St. Christopher-Nevis-Anguilla(*)	0	14	24
Colombia	3	11	23
China	34	7	20
Venezuela	980	8	17
Micronesia	0	0	17
Honduras	0	0	17
Cayman Island	1	0	5
Suriname	0	0	4
South Africa	0	0	3
Netherland	0	40	3
Saint Lucia	0	3	1
Nicaragua	204	0	0
Rest of World	2,649	379	0
World Total	33,053	36,796	37,861

[1] Preliminary.   Note: (*)denotes a country that is a summarization of its component countries.
FAS, Office of Global Analysis, (202) 720-6301.

Table 7-37.—Sheep and lambs: Operations, inventory, and value, United States, Jan. 1, 2001–2010

Year	Operations	Inventory	Value	
			Per head	Total
	Number	Thousands	Dollars	1,000 dollars
2001	68,600	6,908	100.00	690,489
2002	68,150	6,623	92.00	614,466
2003	67,720	6,321	104.00	656,638
2004	67,630	6,065	119.00	720,443
2005	68,460	6,135	130.00	798,209
2006	69,180	6,200	141.00	872,351
2007	83,130	6,120	134.00	818,491
2008	82,500	5,950	138.00	823,424
2009	82,000	5,747	133.00	765,194
2010 [1]		5,630	135.00	762,295

[1] Preliminary.   Inventory operations estimates for 2010 not yet available.
NASS, Livestock Branch, (202) 720–3570.

Table 7-38.—Sheep and lambs: Number by class, United States, Jan. 1, 2001–2010

Year	All sheep and lambs	Breeding sheep			
		Total [1]	Replacement lambs	1 year and over	
				Ewes	Rams
	Thousands	Thousands	Thousands	Thousands	Thousands
2001	6,908	4,952	679	4,071	202
2002	6,623	4,871	732	3,939	201
2003	6,321	4,670	703	3,773	194
2004	6,065	4,464	705	3,570	190
2005	6,135	4,520	783	3,545	192
2006	6,200	4,616	786	3,630	200
2007	6,120	4,553	735	3,620	199
2008	5,950	4,432	697	3,540	195
2009	5,747	4,247	647	3,405	196
2010 [2]	5,630	4,190	655	3,340	195

[1] Categories may not add to total due to rounding.   [2] Preliminary.
NASS, Livestock Branch, (202) 720–3570.

Table 7-39.—Lamb mutton, goat, etc. meat: U.S. exports, 2007–2009

Country	2007	2008	2009
	1,000 metric tons	*1,000 metric tons*	*1,000 metric tons*
Australia	72,434	61,422	5,963
New Zealand (exc. Cook; Nieu; & Toke-lau)(*)	20,860	24,740	1,680
Canada	552	850	55
Mexico	19	88	0
Iceland	44	69	0
World Total	93,909	87,170	7,698

Note: (*)denotes a country that is a summarization of its component countries.
FAS, Office of Global Analysis, (202) 720-6301.

Table 7-40.—Breeding sheep: Number by class, State and United States, Jan. 1, 2009–2010

State	Under one year old		One year and over			
	Replacement lambs		Ewes		Rams	
	2009	2010[1]	2009	2010[1]	2009	2010[1]
	1,000 head	*1,000 head*	*1,000 head*	*1,000 head*	*1,000 head*	*1,000 head*
AZ	9.0	14.0	75.0	75.0	6.0	6.0
CA	45.0	45.0	290.0	263.0	10.0	12.0
CO	32.0	30.0	167.0	155.0	6.0	5.0
ID	25.0	25.0	148.0	150.0	5.0	5.0
IL	12.0	11.0	35.0	41.0	3.0	3.0
IN	6.5	7.0	34.0	36.0	2.5	3.0
IA	18.0	23.0	120.0	116.0	7.0	6.0
KS	7.0	7.0	47.0	41.0	3.0	2.0
KY	6.0	4.5	25.0	23.0	2.0	1.5
MD	3.0	....	14.0	....	1.0	....
MI	10.0	12.0	47.0	46.0	3.0	3.0
MN	14.0	11.0	83.0	76.0	5.0	5.0
MO	12.0	12.0	58.0	55.0	4.0	4.0
MT	43.0	39.0	180.0	188.0	7.0	8.0
NE	8.0	8.0	46.0	47.0	3.0	3.0
NV	11.0	13.0	44.0	49.0	2.0	2.0
N ENG[2]	7.0	7.5	31.0	30.0	3.0	4.0
NM	15.0	16.0	83.0	84.0	5.0	5.0
NY	7.0	9.0	40.0	42.0	3.0	3.0
NC	5.0	3.0	17.0	15.0	2.0	2.0
ND	10.0	10.0	55.0	59.0	2.0	2.0
OH	18.0	16.0	77.0	81.0	5.0	6.0
OK	12.0	11.0	46.0	43.0	5.0	5.0
OR	22.0	25.0	121.0	121.0	7.0	8.0
PA	14.0	12.0	64.0	63.0	6.0	5.0
SD	32.0	32.0	205.0	205.0	8.0	8.0
TN	5.0	4.0	20.0	19.0	2.0	2.5
TX	100.0	105.0	520.0	510.0	40.0	35.0
UT	31.0	36.0	220.0	215.0	9.0	9.0
VA	11.0	12.0	48.0	55.0	3.0	4.0
WA	6.0	7.0	35.0	38.0	2.0	3.0
WV	4.0	4.0	23.0	20.0	1.0	1.0
WI	11.0	14.0	54.0	57.0	3.0	4.0
WY	55.0	51.0	260.0	240.0	10.0	9.0
Oth Sts[3]	20.0	19.0	73.0	82.0	10.0	11.0
US	646.5	655.0	3,405.0	3,340.0	195.5	195.0

[1] Preliminary. [2] N ENG includes CT, ME, MA, NH, RI, and VT. [3] Individual state estimates not available for states not shown, but are included in Other States. MD is included in Other States for 2010.
NASS, Livestock Branch, (202) 720–3570.

Table 7-41.—Sheep and lambs: Average price per 100 pounds at San Angelo, 2000–2009[1]

Year	Sheep			Slaughter lambs			
				Shorn		Spring	
	Good	Utility	Cull	Prime	Choice	Prime	Choice
2000	45.37	42.53	29.84	80.36	80.36	80.10	80.10
2001	44.14	45.11	30.49	70.05	70.05	69.78	69.78
2002	38.04	39.26	24.51	71.69	71.69	72.09	72.09
2003	41.33	44.65	31.32	91.90	91.90	92.13	92.14
2004	46.67	47.54	34.51	96.25	96.25	96.31	96.31
2005	54.21	56.59	41.39	97.50	97.50	97.69	97.69
2006	42.16	42.33	26.41	78.17	78.17	80.41	80.41
2007	41.06	41.31	25.16	85.18	85.18	85.36	85.36
2008	36.91	37.62	18.01	85.66	85.66	85.62	85.62
2009	40.27	40.53	19.40	90.55	90.55	90.49	90.49

[1] Simple average of monthly bulk-of-sales prices from data of the livestock reporting service. 1995 to present price reflects wooled lamb as well as the weight range of 110-130.
AMS, Livestock & Grain Market News, (202) 720–7316.

Table 7-42.—Sheep and lambs: Number of breeding and market sheep, by State and United States, Jan. 1, 2009–2010

State	Breeding sheep and lambs		Market sheep and lambs	
	2009	2010[1]	2009	2010[1]
	1,000 head	*1,000 head*	*1,000 head*	*1,000 head*
AZ	90.0	95.0	60.0	65.0
CA	345.0	320.0	315.0	290.0
CO	205.0	190.0	205.0	185.0
ID	178.0	180.0	32.0	40.0
IL	50.0	55.0	8.0	9.0
IN	43.0	46.0	7.0	6.0
IA	145.0	145.0	55.0	65.0
KS	57.0	50.0	23.0	30.0
KY	33.0	29.0	7.0	8.0
MD	18.0	....	6.0	....
MI	60.0	61.0	18.0	19.0
MN	102.0	92.0	38.0	38.0
MO	74.0	71.0	9.0	8.0
MT	230.0	235.0	25.0	20.0
NE	57.0	58.0	14.0	16.0
NV	57.0	64.0	10.0	11.0
N ENG[2]	41.0	41.5	7.0	6.0
NM	103.0	105.0	17.0	15.0
NY	50.0	54.0	12.0	12.0
NC	24.0	20.0	4.0	5.0
ND	67.0	71.0	21.0	17.0
OH	100.0	103.0	30.0	25.0
OK	63.0	59.0	17.0	16.0
OR	150.0	154.0	70.0	71.0
PA	84.0	80.0	16.0	14.0
SD	245.0	245.0	60.0	75.0
TN	27.0	25.5	7.0	6.0
TX	660.0	650.0	210.0	180.0
UT	260.0	260.0	30.0	30.0
VA	62.0	71.0	13.0	18.0
WA	43.0	48.0	10.0	12.0
WV	28.0	25.0	5.0	5.0
WI	68.0	75.0	17.0	15.0
WY	325.0	300.0	95.0	75.0
Oth Sts[3]	103.0	112.0	27.0	33.0
US	4,247.0	4,190.0	1,500.0	1,440.0

[1] Preliminary.   [2] N ENG includes CT, ME, MA, NH, RI, and VT.   [3] Individual state estimates not available for states not shown, but are included in Other States.   MD is included in Other States for 2010.
NASS, Livestock Branch, (202) 720–3570.

Table 7-43.—Lamb crop: Per 100 ewes 1+, number and percent of previous year, by State and United States, 2008–2009

State	Breeding ewes 1 year & older, Jan. 1		Lambs per 100 ewes 1+, Jan. 1		Lamb crop [1]		
	2008	2009	2008	2009 [2]	2008	2009 [2]	2009 as % of 2008
	1,000 head	*1,000 head*	*Number*	*Number*	*1,000 head*	*1,000 head*	*Percent*
AZ	76.0	75.0	62	67	47.0	50.0	106
CA	285.0	290.0	81	86	230.0	250.0	109
CO	170.0	167.0	124	108	210.0	180.0	86
ID	162.0	148.0	117	132	190.0	195.0	103
IL	35.0	35.0	129	143	45.0	50.0	111
IN	35.0	34.0	117	129	41.0	44.0	107
IA	130.0	120.0	141	142	183.0	170.0	93
KS	53.0	47.0	109	128	58.0	60.0	103
KY	23.0	25.0	117	112	27.0	28.0	104
MD	14.0	14.0	107	....	15.0	....	....
MI	48.0	47.0	135	138	65.0	65.0	100
MN	86.0	83.0	169	157	145.0	130.0	90
MO	57.0	58.0	123	117	70.0	68.0	97
MT	190.0	180.0	124	131	235.0	235.0	100
NE	52.0	46.0	135	148	70.0	68.0	97
NV	46.0	44.0	91	100	42.0	44.0	105
N ENG [3]	32.0	31.0	113	106	36.0	33.0	92
NM	89.0	83.0	73	80	65.0	66.0	102
NY	43.0	40.0	102	113	44.0	45.0	102
NC	16.0	17.0	94	82	15.0	14.0	93
ND	60.0	55.0	128	147	77.0	81.0	105
OH	75.0	77.0	136	136	102.0	105.0	103
OK	46.0	46.0	107	98	49.0	45.0	92
OR	120.0	121.0	131	129	157.0	156.0	99
PA	63.0	64.0	116	102	73.0	65.0	89
SD	220.0	205.0	120	127	265.0	260.0	98
TN	18.0	20.0	128	110	23.0	22.0	96
TX	590.0	520.0	61	74	360.0	385.0	107
UT	210.0	220.0	110	105	230.0	230.0	100
VA	50.0	48.0	120	131	60.0	63.0	105
WA	36.0	35.0	147	157	53.0	55.0	104
WV	24.0	23.0	125	113	30.0	26.0	87
WI	56.0	54.0	125	139	70.0	75.0	107
WY	260.0	260.0	100	98	260.0	255.0	98
Oth Sts [4]	70.0	73.0	97	99	68.0	72.0	106
US	3,540.0	3,405.0	105	108	3,710.0	3,690.0	99

[1] Lamb crop is defined as lambs born in the Eastern States and lambs docked or branded in the Western States. [2] Preliminary. [3] N ENG includes CT, ME, MA, NH, RI, and VT. [4] Individual state estimates not available for states not shown, but are included in Other States. MD is included in Other States for 2009 for Lamb crop and Lambs per 100 ewes 1+Jan 1.

NASS, Livestock Branch, (202) 720–3570.

STATISTICS OF CATTLE, HOGS, AND SHEEP

Table 7-44—Sheep and lambs: Production, disposition, cash receipts, and gross income, United States, 2000–2009

Year	Lamb crop [1]	Marketings [2]		Inshipments	Farm slaughter	Production (live weight) [3]
		Sheep	Lambs			
	1,000 head	*1,000 head*	*1,000 head*	*1,000 head*	*1,000 head*	*1,000 pounds*
2000	4,645	811	4,875	1,754	70	512,305
2001	4,520	740	4,838	1,589	65	501,483
2002	4,355	855	4,794	1,749	66	485,149
2003	4,035	871	4,304	1,586	72	466,621
2004	4,040	690	4,091	1,484	74	466,205
2005	4,015	677	4,093	1,496	75	472,273
2006	3,950	743	4,035	1,465	80	460,580
2007	3,895	780	3,927	1,398	85	440,286
2008	3,710	737	3,652	1,232	92	417,019
2009 [4]	3,690	615	3,507	1,045	95	413,106

Year	Value of production	Cash receipts [5]	Value of home consumption	Gross income [6]	Average price per 100 pounds received by farmers	
					Sheep	Lambs
	1,000 dollars	*1,000 dollars*	*1,000 dollars*	*1,000 dollars*	*Dollars*	*Dollars*
2000	365,183	476,131	9,532	485,663	34.30	79.80
2001	303,186	403,175	8,166	411,341	34.60	66.90
2002	313,946	429,125	8,560	437,685	27.90	73.80
2003	389,201	508,376	11,091	519,467	34.90	94.40
2004	412,691	515,156	12,463	527,619	38.80	101.00
2005	451,467	567,317	13,616	580,933	45.10	110.00
2006	367,799	478,714	11,998	490,712	35.20	95.50
2007	362,941	474,749	13,705	488,454	31.00	98.50
2008	351,287	451,081	13,892	464,973	27.20	99.60
2009 [4]	356,660	436,299	15,177	451,476	32.50	99.60

[1] Lamb crop defined as lambs born in the native States and lambs docked or branded in the Western States. [2] Includes custom slaughter for use on farms where produced and State outshipments, but excludes interfarm sales within the State. [3] Adjustments made for changes in inventory and for inshipments. [4] Preliminary. [5] Receipts from marketings and sale of farm-slaughtered meat. [6] Cash receipts from sales of sheep, lambs, and mutton and lamb plus value of sheep and lambs slaughtered for home consumption.
NASS, Livestock Branch, (202) 720–3570.

Table 7-45.—Sheep and lambs: Receipts at selected public stockyards, 2000–2009 [1]

Year	Sioux Falls	South St. Joseph	San Angelo	All others reporting	Total markets reporting [2][3]
	Thousands	*Thousands*	*Thousands*	*Thousands*	*Thousands*
2000	61	3	448	935	1,054
2001	48	4	337	913	1,021
2002	48	4	397	832	947
2003	45	3	211	658	750
2004	40	3	206	553	633
2005	41	2	199	543	622
2006	48	2	217	518	593
2007	40	2	186	457	497
2008	42	4	192	503	549
2009	28	4	144	409	585

[1] Total rail and truck receipts unloaded at public stockyards. [2] Rounded totals of complete figures. [3] The number of stockyards reporting varies from 41 to 68.
AMS, Livestock & Grain Market News, (202) 720–7316. Compiled from reports received from stockyard companies.

Table 7-46.—Sheep and lambs: Production, disposition, cash receipts, and gross income, by State and United States, 2009 [1]

State	Marketings [2]		Inshipments	Farm slaughter [3]	Production (live weight) [4]	Value of production	Cash receipts [5]	Value of home consumption	Gross income [6]
	Sheep	Lambs							
	1,000 head	*1,000 head*	*1,000 head*	*1,000 head*	*1,000 pounds*	*1,000 dollars*	*1,000 dollars*	*1,000 dollars*	*1,000 dollars*
AZ	16.0	46.0	42.0	13.0	4,149	3,566	4,644	1,138	5,782
CA	58.0	281.0	70.0	5.0	36,334	29,162	36,707	901	37,608
CO	57.5	655.0	526.0	2.5	62,387	47,480	97,914	268	98,182
ID	20.5	155.5	10.0	2.0	20,035	16,800	16,518	308	16,826
IL	3.0	35.0	5.0	1.0	3,887	3,013	2,724	344	3,068
IN	6.0	28.5	4.5	3.5	3,654	2,827	2,576	664	3,240
IA	20.7	162.0	53.0	1.3	23,048	21,296	23,379	439	23,818
KS	15.5	48.0	16.0	2.5	6,258	5,363	5,329	519	5,848
KY	7.2	19.0	1.6	0.4	2,689	2,495	2,714	78	2,792
MI	8.5	45.0	3.0	2.5	4,895	4,430	4,153	323	4,476
MN	17.0	119.0	22.0	4.0	14,527	13,729	15,339	450	15,789
MO	13.4	50.6	3.0	1.0	6,342	5,785	6,041	172	6,213
MT	26.5	180.5	9.0	2.0	21,386	18,168	18,690	353	19,043
NE	11.0	77.6	34.0	0.4	7,823	7,046	8,971	142	9,113
NV	3.0	29.0	15.0	1.0	2,760	2,280	2,808	223	3,031
N ENG [7] ...	7.0	21.0	2.2	3.0	2,485	2,707	2,351	495	2,846
NM	10.0	45.0	10.0	6.0	5,051	5,099	4,741	773	5,514
NY	7.6	32.3	8.4	2.5	3,055	2,694	2,845	321	3,166
NC	6.4	7.7	1.1	0.7	1,170	936	1,052	166	1,218
ND	5.0	64.4	5.0	0.6	6,505	5,747	6,239	123	6,362
OH	8.0	96.0	20.0	3.0	11,787	9,926	11,065	532	11,597
OK	12.5	31.5	7.0	2.0	3,731	2,975	3,192	417	3,609
OR	16.0	142.0	29.0	6.0	14,800	12,792	13,141	983	14,124
PA	13.4	45.8	0.9	1.7	5,288	5,351	5,777	241	6,018
SD	26.0	200.0	37.0	1.0	22,353	20,330	21,728	400	22,128
TN	5.7	15.4	2.0	0.9	1,956	1,741	1,968	86	2,054
TX	79.0	264.0	22.0	2.0	34,673	31,992	37,964	254	38,218
UT	26.0	186.0	15.0	4.0	19,315	17,417	17,676	672	18,348
VA	1.5	35.0	3.0	3.0	5,526	5,400	4,086	495	4,581
WA	0.4	43.5	5.0	5.1	5,548	4,908	3,429	1,436	4,865
WV	6.5	18.8	2.0	0.2	1,841	1,852	2,173	36	2,209
WI	4.5	50.5	1.0	2.0	7,019	5,848	5,265	382	5,647
WY	62.0	219.0	14.0	2.0	36,681	32,119	35,989	374	36,363
Oth Sts [8] ..	34.0	57.5	46.5	7.5	4,148	3,386	7,111	669	7,780
US	615.3	3,507.1	1,045.2	95.3	413,106	356,660	436,299	15,177	451,476

[1] Preliminary.　[2] Includes custom slaughter for use on farms where produced and State outshipments, but excludes interfarm sales within the State.　[3] Excludes custom slaughter for farmers at commercial establishments.　[4] Adjustments made for changes in inventory and for inshipments.　[5] Receipts from marketings and sale of farm-slaughter.　[6] Cash receipts from sales of sheep, lambs, and mutton and lamb plus value of sheep and lambs slaughtered for home consumption.　[7] N ENG includes CT, ME, MA, NH, RI, and VT.　[8] Individual state estimates not available for states not shown, but are included in Other States.

NASS, Livestock Branch, (202) 720–3570.

Table 7-47.—Sheep and lambs: Number slaughtered commercially, total and average live weight, by State and United States, 2009 [1]

State	Number slaughtered	Total live weight [2]	Average live weight [2]
	Thousands	*1,000 pounds*	*Pounds*
AL	0.5	38	78
AK	0.1	8	140
AZ	2.4	255	106
AR	0.7	68	104
CA	......	......	......
CO	917.8	145,350	159
DE and MD	44.0	4,227	96
FL	3.6	197	54
GA	3.7	274	75
HI	0.7	90	135
ID	3.9	518	133
IL	......	......	......
IN	40.4	4,452	110
IA	387.3	57,990	150
KS	3.5	368	105
KY	11.6	1,371	119
LA	4.2	266	64
MI	187.9	24,394	130
MN	17.8	2,374	134
MS	0.9	52	60
MO	5.9	693	118
MT	4.2	450	107
NE	1.0	138	138
NV	0.8	103	131
N ENG [3]	39.0	3,473	89
NJ	136.0	10,892	80
NM	13.5	1,900	141
NY	37.7	3,415	91
NC	9.6	725	75
ND	0.7	82	121
OH	15.2	1,648	108
OK	1.6	158	100
OR	23.1	3,189	138
PA	46.3	4,537	98
SC	1.7	138	80
SD	2.9	360	126
TN	14.4	1,046	73
TX	87.5	7,421	85
UT	36.8	5,026	137
VA	16.2	1,573	97
WA	9.0	1,410	156
WV	......	......	......
WI	10.1	1,331	131
WY	1.6	226	140
US [4]	2,515.9	341,351	136
PR	0.4	23	58

[1] Includes slaughter in federally inspected and in other slaughter plants; exludes animals slaughtered on farms. Average live weight is based on unrounded numbers. Totals may not add due to rounding.   [2] Excludes postmortem condemnations.   [3] CT, ME, MA, NH, RI, and VT.   [4] States with no data printed are still included in US total. Data are not printed to avoid disclosing individual operations.
NASS, Iowa Field Office, (515) 284–4340.

Table 7-48.—Sheep and lambs: Number slaughtered, United States, 2000–2009

Year	Commercial			Farm	Total
	Federally inspected	Other	Total [1]		
	Thousands	*Thousands*	*Thousands*	*Thousands*	*Thousands*
2000	3,308	152	3,460	67	3,527
2001	3,065	157	3,222	68	3,290
2002	3,092	194	3,286	65	3,351
2003	2,805	174	2,979	64	3,042
2004	2,676	163	2,839	67	2,906
2005	2,554	143	2,698	65	2,763
2006	2,547	151	2,699	68	2,766
2007	2,529	165	2,694	85	2,779
2008	2,394	177	2,556	92	2,647
2009	2,323	174	2,516	95	2,611

[1] Totals are based on unrounded numbers.
NASS, Iowa Field Office, (515) 284–4340.

Table 7-49.—Sheep and lambs: Number slaughtered, average dressed and live weights, percentage distribution, by class, Federally inspected, 2000–2009 [1]

Year	Federally inspected								
	Lambs and yearlings			Mature sheep			Total		
	Head	Pct. of total	Avg. dressed weight [2]	Head	Pct. of total	Avg. dressed weight [2]	Head	Avg. dressed weight [2]	Avg. live weight [2]
	1,000		*Pounds*	*1,000*		*Pounds*	*1,000*	*Pounds*	
2000	3,141	95.0	68	167	5.0	63	3,308	68	137
2001	2,921	95.3	71	144	4.7	62	3,065	70	142
2002	2,944	95.2	68	148	4.8	63	3,092	68	135
2003	2,662	94.9	68	143	5.1	66	2,805	68	136
2004	2,529	94.5	69	147	5.5	66	2,676	69	138
2005	2,425	94.9	71	129	5.1	69	2,554	70	140
2006	2,429	95.4	70	118	4.6	67	2,547	70	138
2007	2,413	95.4	69	116	4.6	67	2,529	69	138
2008	2,271	94.9	69	122	5.1	67	2,394	69	138
2009	2,165	93.2	70	158	6.8	64	2,323	70	139

[1] All percents and weights calculated using unrounded totals.   [2] Excludes postmortem condemnations.
NASS, Iowa Field Office, (515) 284–4340.

Table 7-50.—Sheep and lambs: Inventory Jan 1, 2009–2010, and number of operations, 2007, by State and United States [1]

State	January 1 Sheep inventory		Operations with sheep [2]
	2009	2010 [3]	2007
	1,000 head	*1,000 head*	*Number*
AZ	150.0	160.0	5,000
CA	660.0	610.0	4,100
CO	410.0	375.0	1,600
ID	210.0	220.0	1,200
IL	58.0	64.0	1,900
IN	50.0	52.0	2,000
IA	200.0	210.0	3,500
KS	80.0	80.0	1,200
KY	40.0	37.0	1,400
MD	24.0		800
MI	78.0	80.0	2,300
MN	140.0	130.0	2,500
MO	83.0	79.0	2,200
MT	255.0	255.0	1,500
NE	71.0	74.0	1,300
NV	67.0	75.0	250
N ENG [4]	48.0	47.5	3,000
NM	120.0	120.0	2,900
NY	62.0	66.0	1,800
NC	28.0	25.0	1,300
ND	88.0	88.0	680
OH	130.0	128.0	3,400
OK	80.0	75.0	1,900
OR	220.0	225.0	3,200
PA	100.0	94.0	3,800
SD	305.0	320.0	1,700
TN	34.0	31.5	1,300
TX	870.0	830.0	8,700
UT	290.0	290.0	1,600
VA	75.0	89.0	2,100
WA	53.0	60.0	2,400
WV	33.0	30.0	1,300
WI	85.0	90.0	2,800
WY	420.0	375.0	900
Oth Sts [5]	130.0	145.0	5,600
US	5,747.0	5,630.0	83,130
PR			600

[1] An operation is any place having one or more head of sheep on hand December 31.   [2] State level estimates only available in conjunction with the Census of Agriculture every 5 years.   [3] Preliminary.   [4] N Eng includes CT, ME, MA, NH, RI, and VT.   [5] Individual state estimates not available for states not shown, but are included in Other States. MD is included in Other States for 2010.
NASS, Livestock Branch, (202) 720–3570.

Table 7-51.—Breeding Sheep: Survey percent by size groups, United States, 2008–2009 [1]

Item	1–99 head		100–499 head		500–4,999 head		5,000+ head	
	2008	2009	2008	2009	2008	2009	2008	2009
	Percent	*Percent*	*Percent*	*Percent*	*Percent*	*Percent*	*Percent*	*Percent*
Operations	92.5	93.7	6.2	5.2	1.2	1.0	0.1	0.1
Inventory	32.6	36.2	22.7	20.8	30.2	31.3	14.5	11.7

[1] Percents reflect distributions from the January survey.
NASS, Livestock Branch, (202) 720–3570.

Table 7-52.—Wool: Number of sheep shorn, weight per fleece, production, average price per pound received by farmers, value of production, exports, imports, total new supply of apparel wool, and imports of carpet wool, United States, 1999–2008

Year	Sheep and lambs shorn [1]	Weight per fleece	Shorn wool production	Price per pound [2]	Value of production [3]
	Thousands	*Pounds*	*1,000 pounds*	*Cents*	*1,000 dollars*
1999	6,141	7.58	46,428	0.38	17,704
2000	6,032	7.55	45,551	0.33	14,948
2001	5,596	7.53	42,156	0.35	14,841
2002	5,462	7.52	41,078	0.53	21,689
2003	5,077	7.52	38,197	0.74	28,129
2004	5,066	7.42	37,581	0.80	29,954
2005	5,061	7.35	37,182	0.71	26,249
2006	4,847	7.41	35,899	0.68	24,300
2007	4,657	7.46	34,723	0.87	30,242
2008 [4]	4,434	7.43	32,963	0.99	32,486

Year	Shorn wool production	Raw wool supply (clean)				
		Domestic production [5]	Exports [6]	Imports for consumption		Total new supply [9]
				48's and Finer [7]	Not Finer than 46's [8]	
	1,000 pounds	*1,000 pounds*	*1,000 pounds*	*1,000 pounds*	*1,000 pounds*	*1,000 pounds*
1999	46,428	24,514	3,694	21,264	21,810	63,894
2000	45,551	24,051	6,629	23,902	21,099	62,423
2001	42,156	22,258	6,154	15,843	19,727	51,674
2002	41,078	21,689	8,461	10,526	14,159	37,913
2003	38,197	20,168	11,067	4,986	15,749	29,836
2004	37,581	19,843	11,168	6,204	16,455	31,334
2005	37,182	19,632	12,573	6,220	12,155	25,434
2006	35,899	18,955	17,998	7,324	9,929	18,210
2007	34,723	18,334	17,077	5,245	9,025	15,527
2008 [4]	32,963	17,404	10,307	4,551	8,631	20,279

[1] Includes sheep shorn at commercial feeding yards. [2] Price computed by weighting State average prices for all wool sold during the year by sales of shorn wool. [3] Production by States multiplied by annual average price. [4] Preliminary. [5] Conversion factor from grease basis to clean basis are as follows: Shorn wool production—52.8 percent (Stat. Bull. 616) from 1987-1997. [6] Includes carpet wool exports. [7] Prior to 1989, known as dutiable imports. [8] Prior to 1989, known as duty-free imports. In 1994 includes 24,645,306 pounds of imported raw wool not finer than 46's and 2,182,576 pounds of miscellaneous imported raw wool. [9] Production minus exports plus imports; stocks not taken into consideration.
ERS, Field Crops Branch, (202) 694–5300 and NASS. Imports and exports from reports of the U.S. Department of Commerce.

Table 7-53.—Wool: Price-support operations, United States, 2000–2009 [1]

Year	Income support payment rates per pound	Program price levels per pound		Put under loan		Acquired by CCC under loan program [2]	Owned by CCC at end of marketing year
		Graded wool loan	Nongraded loan	Quantity	Percentage of production		
	Dollars	*Dollars*	*Dollars*	*1,000 pounds*	*Percent*	*1,000 pounds*	*1,000 pounds*
2000	NA	NA	NA	NA	NA	NA	NA
2001	NA	NA	NA	NA	NA	NA	NA
2002	NA	1.00	0.40	36.0	0.09	0.0	0.0
2003	NA	1.00	0.40	23.0	0.06	0.0	0.0
2004	NA	1.00	0.40	63.0	0.17	0.0	0.0
2005	NA	1.00	0.40	76.2	0.20	5.0	0.0
2006	NA	1.00	0.40	3.2	0.01	0.0	0.0
2007	NA	1.00	0.40	3.0	0.01	0.0	0.0
2008	NA	1.00	0.40	8.6	0.03	0.0	0.0
2009	NA	1.00	0.40	28.9	0.09	0.0	0.0

[1] Nonrecourse Marketing Loan Program authorized following enactment of the Farm Security and Rural Investment Act of 2002. [2] Acquistions for 2006/2007 as of September 30, 2007. NA-not available.
FSA, Fibers, (202) 720–3008.

Table 7-54.—Wool: Mill consumption, by grades, on the woolen and worsted systems, scoured basis, United States, 1999–2008 [1][2][3]

Item	1999	2000	2001	2002	2003	2004	2005	2006	2007	2008
Apparel wool:										
Woolen system:	Mil. lb.	Mil. lb.	Mil. lb.	Mil. lb.	Mil. lb.	Mil. lb.	Mil. lb.	Mil. lb.	Mil. lb.	Mil. lb.
60's and finer	18.4	18.5	16.1	9.6	6.7	6.4	NA	NA	NA	NA
Coarser than 60's	10.8	13.4	9.8	8.5	5.3	8.1	NA	NA	NA	NA
Total	29.2	31.9	25.9	18.1	12.0	14.5	NA	NA	NA	NA
Worsted system:										
60's and finer	27.4	NA	NA	NA	NA	NA	NA	NA	NA	NA
Coarser than 60's	7.0	NA	NA	NA	NA	NA	NA	NA	NA	NA
Total	34.4	30.1	27.1	17.9	NA	NA	NA	NA	NA	NA
Total apparel:										
60's and finer	45.8	NA	NA	NA	NA	NA	NA	NA	NA	NA
Coarser than 60's	17.7	NA	NA	NA	NA	NA	NA	NA	NA	NA
Total	63.5	63.0	53.0	36.0	NA	NA	NA	NA	NA	NA
Carpet wool	13.9	15.2	13.3	6.9	6.0	6.9	NA	NA	NA	NA
Grand total mill	77.5	77.2	66.3	42.9	NA	NA	NA	NA	NA	NA

[1] Scoured wool, plus greasy wool converted to a scoured basis, using assumed average yields. Includes both pulled and shorn, foreign and domestic wool. Wool was considered as consumed (1) on the woolen system when laid in mixes and (2) on the worsted system as the sum of top and noil production.   [2] Domestic, duty-paid, and duty-free foreign.   [3] Excludes wool consumed on the cotton system and in the manufacture of felt, hat bodies, and other miscellaneous products.   NA-not available.
ERS, Field Crops Branch, (202) 694–5300. Compiled from reports of the U.S. Department of Commerce.

Table 7-55.—Wool: United States imports (for consumption), clean content, by grades, 1999–2008 [1][2]

Grade	1999	2000	2001	2002	2003	2004	2005	2006	2007	2008
48's and finer:	Mil. lb.	Mil. lb.	Mil. lb.	Mil. lb.	Mil. lb.	Mil. lb.	Mil. lb.	Mil. lb.	Mil. lb.	Mil. lb.
Finer than 58's [3]	19.9	22.1	14.2	9.2	4.5	5.1	5.6	6.5	4.7	4.0
48's–58's [4]	1.3	1.7	1.6	1.3	0.5	1.1	0.6	0.8	0.5	0.5
Total	21.3	23.9	15.8	10.5	5.0	6.2	6.2	7.3	5.2	4.6
Not Finer than 46's:										
Wool for special use [5]	1.8	2.2	2.4	1.3	1.4	2.8	3.1	1.1	0.7	0.6
Not finer than 40's [6]	6.5	4.5	6.5	3.9	5.3	4.4	2.4	2.6	2.6	1.8
Finer than 40's– 44's [7]	9.4	10.0	6.7	7.1	6.3	5.8	4.5	5.4	4.4	5.0
46's [8]	4.1	4.4	4.1	1.9	2.7	3.4	2.1	0.8	1.4	1.3
Total	21.9	21.0	19.7	14.2	15.7	16.5	12.2	9.9	9.0	8.6
Miscellaneous [9]	0	0	0	0	0	0	0	0	0	0
Grand total	43.1	45.0	35.6	24.7	20.7	22.7	18.4	17.3	14.3	13.2

[1] Natural fiber grown by sheep or lambs.   [2] Beginning 1989 the following Harmonized Tariff Schedule numbers are in the above 7 wool import groups: 5101.19.606060, 5101.19.6060, 5101.21.4000, 5101.29.4060, 0.5(5101.30.4000).   [4] 5101.11.6030, 5101.19.6030, 5101.21.4030, 5101.29.4030, 0.5(5101.30.4000).   [5] 5101.11.1000, 5101.19.1000, 5101.21.1000, 5101.29.1000.   [6] 5101.11.2000, 5101.19.2000, 5101.21.1500, 5101.29.1500, 5101.30.1000.   [7] 5101.11.4000, 5101.19.4000, 5101.21.3000, 5101.29.3000, 5101.30.1500.   [8] 5101.11.5000, 5101.19.5000, 5101.21.3500, 5101.29.3500, 5101.30.3000.   [9] 5101.21.6000, 5101.29.6000, 5101.30.6000. They include wool not carded or combed but processed beyond the scoured or carbonized condition, e.g. dyed. This wool is not identified by use or grade. In 1989 this quantity was 48,074 pounds, 1990 was 32,979 pounds, 1991 was 47,245 pounds, and 1992 was 25,728 pounds.
ERS, Field Crops Branch, (202) 694–5300. Compiled from reports of the U.S. Department of Commerce.

Table 7-56.—Wool: United States imports (for consumption), clean content, by country of origin, 1999–2008[1]

Country of origin	1999	2000	2001	2002	2003	2004	2005	2006	2007	2008
48's and finer:	Mil. lb.	Mil. lb.	Mil. lb.	Mil. lb.	Mil. lb.	Mil. lb.	Mil. lb.	Mil. lb.	Mil. lb.	Mil. lb.
Argentina	0.2	0.0	0.0	0.0	0.0	0.0	0.1	0.0	0.1	0.0
Australia	17.6	20.2	12.7	8.1	3.6	4.2	4.5	5.5	3.7	2.9
Canada	0.7	0.8	0.8	0.8	0.4	0.6	0.6	0.6	0.5	0.5
Chile	0.0	0.0	0.0	0.0	0.0	0.0	0.0	0.0	0.0	0.0
New Zealand	1.3	1.3	1.0	0.5	0.4	0.4	0.4	0.4	0.2	0.2
South Africa	1.1	0.8	0.6	0.5	0.4	0.5	0.3	0.2	0.3	0.4
United Kingdom	0.1	0.1	0.0	0.0	0.0	0.1	0.1	0.0	0.0	0.1
Uruguay	0.2	0.4	0.3	0.3	0.1	0.2	0.1	0.3	0.3	0.2
Other	0.1	0.3	0.4	0.3	0.1	0.2	0.1	0.3	0.1	0.3
Total	21.3	23.9	15.8	10.5	5.0	6.2	6.2	7.3	5.2	4.6
Not finer than 46's:										
Argentina	0.4	0.5	0.3	0.4	0.6	0.5	0.7	0.3	0.2	0.1
Australia	0.1	0.6	0.4	0.4	0.2	0.7	0.4	0.2	0.0	0.2
Canada	0.2	0.2	0.1	0.1	0.1	0.0	0.1	0.0	0.0	0.0
New Zealand	18.01	16.0	15.8	10.3	11.8	12.1	9.7	7.9	7.2	6.5
Uruguay	0.0	0.0	0.0	0.0	0.0	0.0	0.0	0.0	0.0	0.0
South Africa	0.0	0.3	0.2	0.1	0.1	0.2	0.2	0.2	0.1	0.2
United Kingdom	3.0	3.2	2.8	2.7	2.5	2.7	1.0	1.2	1.4	1.4
Other	0.2	0.2	0.1	0.2	0.4	0.3	0.1	0.1	0.1	0.2
Total	21.9	21.0	19.7	14.2	15.7	16.5	12.2	9.9	9.0	8.6
Grand total	43.1	45.0	35.6	24.7	20.7	22.7	18.4	17.3	14.3	13.2

[1] Wool not advanced in any manner or by any process of manufacture beyond washed, scoured, or carbonized condition.
ERS, Field Crops Branch, (202) 694–5300. Compiled from reports of the U.S. Department of Commerce.

Table 7-57.—Wool: Average price per pound, clean basis, delivered to United States mills, 1999–2008[1]

Year	Territory[2]		Australian 64's good topmaking (in bond, American yield)
	64's (20.60–22.04 microns)	Avg. 58's–56's (24.95–27.84 microns)	
	Cents	Cents	Cents
1999	110	70	148
2000	108	61	150
2001	121	72	166
2002	190	130	268
2003	241	164	314
2004	235	162	275
2005	186	126	257
2006	179	115	265
2007	265	157	373
2008	309	204	347

[1] Beginning January 1976 the unit designation terminology for wool prices changed to microns. For example 64's (20.60–22.04 microns) formerly was fine good French combing and staple. Two designations 56's (26.40–27.84 microns) and 58's (24.95–26.39 microns) have been averaged in the price data shown here and together were formerly the category fleece 3/8 blood good French combing and staple.   [2] Wool grown in the range areas of California, Oregon, Washington, Texas, the intermountain States (including Arizona and New Mexico), and parts of the Dakotas, Kansas, Nebraska, and Oklahoma. These wools vary considerably in shrinkage and color.
ERS, Field Crops Branch, (202) 694–5300 and AMS.

Table 7-58.—Wool: Number of sheep shorn, weight per fleece, and production, by State and United States, 2008–2009

State	Sheep and lambs shorn		Weight per fleece		Shorn wool production	
	2008	2009 [1]	2008	2009 [1]	2008	2009 [1]
	1,000 head	*1,000 head*	*Pounds*	*Pounds*	*1,000 pounds*	*1,000 pounds*
AZ	120.0	125.0	6.3	6.0	750	750
CA	470.0	440.0	7.0	6.1	3,300	2,700
CO	380.0	300.0	6.8	7.3	2,600	2,200
ID	180.0	190.0	9.5	9.4	1,710	1,786
IL	48.0	53.0	7.5	7.5	360	395
IN	40.0	40.0	6.3	6.0	250	240
IA	195.0	195.0	6.1	5.9	1,180	1,150
KS	56.0	57.0	6.9	6.5	384	370
KY	13.0	12.0	6.5	6.7	85	80
MD	12.0	....	6.8	....	82	....
MI	67.0	62.0	6.0	6.1	400	380
MN	140.0	130.0	6.6	6.4	930	830
MO	69.0	55.0	6.0	6.9	415	380
MT	230.0	230.0	9.3	9.3	2,150	2,150
NE	55.0	56.0	7.2	7.4	395	415
NV	48.0	53.0	9.9	9.7	475	515
N ENG [2]	37.0	37.0	7.1	6.9	263	255
NM	105.0	100.0	7.6	7.3	800	730
NY	40.0	40.0	6.5	6.5	260	260
NC	6.0	8.0	6.3	5.6	38	45
ND	70.0	75.0	8.3	8.5	580	640
OH	103.0	92.0	6.0	6.1	620	560
OK	40.0	30.0	5.5	5.0	220	150
OR	182.0	180.0	6.5	6.3	1,190	1,130
PA	60.0	54.0	6.7	6.5	400	350
SD	280.0	300.0	7.8	7.5	2,184	2,250
TN	18.0	17.0	5.8	5.9	105	100
TX	600.0	500.0	7.0	7.0	4,200	3,500
UT	255.0	260.0	9.2	9.0	2,350	2,350
VA	36.0	30.0	5.7	5.7	205	170
WA	36.0	48.0	7.8	7.3	280	350
WV	24.0	19.0	5.1	5.3	122	101
WI	62.0	64.0	7.1	7.3	440	470
WY	320.0	300.0	9.4	9.3	3,000	2,800
Oth Sts [3]	37.0	45.0	6.5	6.9	240	310
US	4,434.0	4,197.0	7.4	7.4	32,963	30,862

[1] Preliminary.   [2] N ENG includes CT, ME, MA, NH, RI, and VT.   [3] Individual state estimates not available for states not shown, but are included in Other States.   MD is included in Other States for 2009.
NASS, Livestock Branch, (202) 720–3570.

Table 7-59.—Wool: Price and value, by State and United States, 2008–2009

State	Price per pound		Value [1]	
	2008	2009 [2]	2008	2009 [2]
	Dollars	*Dollars*	*Dollars*	*1,000 dollars*
AZ	0.30	0.25	225	188
CA	1.10	0.85	3,630	2,295
CO	1.12	0.82	2,912	1,804
ID	1.14	0.83	1,949	1,482
IL	0.36	0.31	130	122
IN	0.18	0.23	45	55
IA	0.35	0.27	413	311
KS	0.65	0.37	250	137
KY	0.45	0.30	38	24
MD	0.55	....	45	....
MI	0.34	0.43	136	163
MN	0.36	0.29	335	241
MO	0.57	0.40	237	152
MT	1.40	1.10	3,010	2,365
NE	0.41	0.40	162	166
NV	1.35	1.10	641	567
N ENG [3]	0.55	0.55	145	140
NM	1.70	1.00	1,360	730
NY	0.30	0.30	78	78
NC	0.40	0.50	15	23
ND	0.80	0.70	464	448
OH	0.37	0.28	229	157
OK	0.40	0.40	88	60
OR	0.68	0.66	809	746
PA	0.31	0.33	124	116
SD	0.91	0.82	1,987	1,845
TN	0.53	0.43	56	43
TX	1.16	1.04	4,872	3,640
UT	1.20	0.80	2,820	1,880
VA	0.50	0.60	103	102
WA	1.35	1.35	378	473
WV	0.44	0.41	54	41
WI	0.45	0.50	198	235
WY	1.46	1.16	4,380	3,248
Oth Sts [4]	0.70	1.00	168	310
US	0.99	0.79	32,486	24,387

[1] Production multiplied by marketing year average price.　[2] Preliminary.　[3] N ENG includes CT, ME, MA, NH, RI, and VT.　[4] Individual state estimates not available for states not shown, but are included in Other States. MD is included in Other States for 2009.
NASS, Livestock Branch, (202) 720–3570.

Table 7-60.—Mohair: Price-support operations, United States, 2000–2009 [1]

Year	Income support payment rates per pound	Program price levels per pound		Put under loan		Acquired by CCC under loan program [2]	Owned by CCC at end of marketing year
		Loan	Target	Quantity	Percentage of production		
	Dollars	Dollars	Dollars	1,000 pounds	Percent	1,000 pounds	1,000 pounds
2000	NA	NA	NA	NA	NA	NA	NA
2001	NA	NA	NA	NA	NA	NA	NA
2002	NA	4.20	NA	47.0	2.16	1.0	0.0
2003	NA	4.20	NA	46.0	2.42	1.0	0.0
2004	NA	4.20	NA	37.0	1.85	0.0	0.0
2005	NA	4.20	NA	36.4	2.14	0.0	0.0
2006	NA	4.20	NA	25.0	1.67	0.0	0.0
2007	NA	4.20	NA	20.4	1.46	0.0	0.0
2008	NA	4.20	NA	8.7	0.73	0.0	0.0
2009	NA	4.20	NA	14.5	1.45	0.0	0.0

[1] Nonrecourse Marketing Loan Program authorized following enactment of the Farm Security and rural Investment Act of 2002.years.　[2] Acquistions for 2006/2007 as of September 30, 2007.　NA-not applicable.
FSA, Fibers, (202) 720–3008.

Table 7-61.—Mohair: Goats clipped, production, price, and value, by State and United States, 2008–2009

State	Goats clipped		Average clip per goat		Production		Price per pound		Value [1]	
	2008	2009 [2]	2008	2009 [2]	2008	2009 [2]	2008	2009 [2]	2008	2009 [2]
	Head	Head	Pounds	Pounds	1,000 pounds	1,000 pounds	Dollars	Dollars	1,000 dollars	1,000 dollars
AZ	21,000	14,000	5.2	5.7	110	80	1.00	0.55	110	44
CA	3,000	2,500	6.0	6.8	18	17	4.00	3.50	72	60
NM	12,500	11,500	6.1	6.5	76	75	4.00	2.50	304	188
TX	130,000	105,000	6.3	6.4	820	670	3.80	3.10	3,116	2,077
Oth Sts [3] ..	27,000	27,500	5.9	6.2	160	170	1.95	1.90	312	323
US	193,500	160,500	6.1	6.3	1,184	1,012	3.31	2.66	3,914	2,692

[1] Production multiplied by marketing year average price. U.S. value is summation of State values.　[2] Preliminary.　[3] Individual State estimates not available for States not shown but are included in Other States.
NASS, Livestock Branch, (202) 720–3570.

Table 7-62.—Angora goats: Inventory Jan 1, 2009–2010, and number of operations, 2007, by State and United States [1]

State	January 1 angora goats inventory		Operations with angora goats [3]
	2009	2010 [2]	2007
	1,000 head	*1,000 head*	*Number*
AL			60
AZ	24,000	16,000	1,500
CA	4,000	3,500	260
CO	1,000		180
FL			50
GA			110
ID			60
IL			50
IN			60
IA			80
KS			50
KY			130
MD			60
MI			150
MN	1,000	1,000	100
MS			40
MO	1,500	1,400	100
MT			90
NE			50
N ENG [4]	1,250	1,150	270
NJ			80
NM	12,500	10,500	740
NY			150
NC			180
OH	1,400	1,300	160
OK	1,000		100
OR	1,800	1,900	250
PA	1,100		230
SC			80
SD			20
TN			60
TX	120,000	95,000	780
UT			130
VA		1,400	160
WA	1,000	1,000	200
WI	1,000	1,000	180
WY			40
Oth Sts [5]	12,450	14,850	200
US	185,000	150,000	7,190

[1] An operation is any place having one or more head of angora goats on hand December 31. [2] Preliminary. [3] State level estimates only available in conjunction with the Census of Agriculture every 5 years. [4] N ENG includes CT, ME, MA, NH, RI, and VT. [5] Individual state estimates not available for states not shown, but are included in Other States. CO, OK, and PA are included in Other States for 2010. VA is included in Other States for 2009.
NASS, Livestock Branch, (202) 720–3570.

Table 7-63.—Milk goats: Inventory Jan 1, 2009–2010, and number of operations, 2007, by State and United States [1]

State	January 1 milk goats inventory		Operations with milk goats [3]
	2009	2010 [2]	2007
	1,000 head	*1,000 head*	*Number*
AL	4,500	4,200	450
AZ	2,000	2,000	270
AR	4,900	4,400	480
CA	37,000	38,000	1,400
CO	8,300	8,400	780
FL	6,000	5,000	780
GA	4,500	3,000	450
ID	2,500	2,800	380
IL	5,500	4,700	620
IN	10,500	11,800	1,000
IA	24,500	29,500	650
KS	4,500	3,800	480
KY	7,000	6,500	750
LA	1,200	1,100	170
MD	2,800	2,400	280
MI	9,100	10,900	1,100
MN	11,500	13,000	620
MS	2,800	2,900	250
MO	9,000	9,000	950
MT	1,900	2,700	240
NE	1,900	3,100	260
NV	4,300	4,600	....
N ENG [4]	10,500	13,500	1,200
NJ	1,700	2,000	230
NM	2,000	3,100	250
NY	13,400	13,000	1,100
NC	9,500	8,000	790
ND	1,000	....	....
OH	9,000	8,000	1,200
OK	8,000	8,300	850
OR	9,200	9,100	900
PA	14,500	17,000	1,300
SC	3,000	2,900	320
SD	1,500	2,000	180
TN	5,800	6,400	590
TX	20,000	20,000	2,100
UT	2,600	2,700	230
VA	4,500	5,800	620
WA	7,100	7,300	1,100
WV	2,200	2,000	440
WI	40,000	46,000	1,100
WY	1,800	1,700	240
Oth Sts [5]	1,500	2,400	300
US	335,000	355,000	27,400

[1] An operation is any place having one or more head of milk goats on hand December 31. [2] Preliminary. [3] State level estimates only available in conjunction with the Census of Agriculture every 5 years. [4] N ENG includes CT, ME, MA, NH, RI, and VT. [5] Individual state estimates not available for states not shown, but are included in Other States. ND is included in Other States for 2010.

NASS, Livestock Branch, (202) 720–3570.

Table 7-64.—Meat and other goats: Inventory Jan 1, 2009–2010, and number of operations, 2007, by State and United States [1]

State	January 1 meat and other goats inventory		Operations with Meat goats [3]
	2009	2010 [2]	2007
	1,000 head	*1,000 head*	*Number*
AL	65,000	60,000	3,800
AZ	14,000	25,000	1,700
AR	47,000	53,000	2,600
CA	95,000	93,000	4,000
CO	34,000	38,000	2,200
FL	55,000	60,000	3,600
GA	80,500	79,000	4,000
HI	8,400	9,500	280
ID	12,000	13,000	1,100
IL	24,500	22,500	2,000
IN	32,000	33,500	2,700
IA	32,500	25,000	1,800
KS	50,000	42,000	1,700
KY	79,000	79,000	4,800
LA	20,000	20,500	1,500
MD	13,600	12,500	980
MI	13,500	16,000	2,400
MN	20,000	22,000	1,600
MS	27,000	28,000	1,600
MO	86,500	84,600	3,800
MT	9,000	9,000	510
NE	27,000	20,000	930
NV	6,100	6,400	270
N ENG [4]	12,100	13,000	1,900
NM	8,000	7,200	870
NJ	18,000	21,000	1,500
NY	27,000	35,000	2,000
NC	86,000	95,000	5,100
ND	2,800	2,700	210
OH	55,500	50,000	4,100
OK	105,000	90,000	5,200
OR	28,000	30,000	2,500
PA	42,000	42,000	3,800
SC	38,000	39,100	2,800
SD	7,500	6,000	450
TN	134,000	125,000	6,500
TX	980,000	990,000	24,800
UT	14,000	13,000	840
VA	62,000	52,000	3,400
WA	27,000	22,000	2,500
WV	23,500	22,500	1,900
WI	18,000	21,000	2,300
WY	6,000	7,000	490
Oth Sts [5]	3,000	3,000	170
US	2,549,000	2,538,000	123,200

[1] An operation is any place having one or more head of meat goats on hand December 31.　[2] Preliminary.　[3] State level estimates only available in conjunction with the Census of Agriculture every 5 years.　[4] N ENG includes CT, ME, MA, NH, RI, and VT.　[5] Individual state estimates not available for states not shown, but are included in Other States.
NASS, Livestock Branch, (202) 720–3570.

Table 7-65.—All goats: Number of operations, 2007, by State and United States [1]

State	2007 [2]	State	2007 [2]
	Number		Number
AL	4,100	NV	320
AZ	3,400	ENG [3]	2,900
AR	2,800	NJ	1,100
CA	5,000	NM	2,300
CO	2,700	NY	2,700
FL	4,100	NC	5,600
GA	4,300	ND	280
HI	330	OH	4,900
ID	1,300	OK	5,700
IL	2,500	OR	3,200
IN	3,400	PA	4,800
IA	2,300	SC	3,000
KS	2,000	SD	590
KY	5,300	TN	6,800
LA	1,600	TX	26,400
MD	1,200	UT	1,100
MI	3,000	VA	4,000
MN	2,000	WA	3,200
MS	1,800	WV	2,100
MO	4,500	WI	3,200
MT	730	WY	640
NE	1,100	Oth Sts [4]	220
		US	144,510

[1] An operation is any place having one or more head of goats on hand December 31. [2] State level estimates only available in conjunction with the Census of Agriculture every 5 years. [3] N ENG includes CT, ME, MA, NH, RI, and VT. [4] Individual state estimates not available for states not shown, but are included in Other States.

NASS, Livestock Branch, (202) 720–3570.

**Table 7-66.—Red meat: Production, by class of slaughter,
United States, 2000–2009**

Year	Commercial (Beef) Federally inspected	Other *	Total [1]	Farm	Total	Commercial (Pork, excluding lard) Federally inspected	Other *	Total [1]	Farm	Total
	Beef					Pork, excluding lard				
	Million pounds	Million pounds	Million pounds	Million pounds	Million pounds	Million pounds	Million pounds	Million pounds	Million pounds	Million pounds
2000	26,405	371	26,776	111	26,887	18,672	257	18,929	24	18,953
2001	25,743	365	26,108	105	26,213	18,899	240	19,139	22	19,161
2002	26,714	377	27,091	102	27,193	19,437	227	19,664	21	19,685
2003	25,880	358	26,238	101	26,340	19,739	207	19,946	21	19,967
2004	24,189	358	24,547	102	24,649	20,325	186	20,511	20	20,531
2005	24,328	355	24,683	104	24,786	20,506	179	20,685	20	20,705
2006	25,792	360	26,152	104	26,256	20,877	177	21,054	20	21,073
2007	26,070	351	26,421	103	26,524	21,768	175	21,943	20	21,962
2008	26,200	361	26,561	102	26,663	23,170	177	23,347	20	23,367
2009	25,598	....	25,966	102	26,068	22,827	....	22,999	21	23,020

Year	Veal Federally inspected	Other *	Total [1]	Farm	Total	Lamb and Mutton Federally inspected	Other *	Total [1]	Farm	Total
	Veal					Lamb and Mutton				
	Million pounds	Million pounds	Million pounds	Million pounds	Million pounds	Million pounds	Million pounds	Million pounds	Million pounds	Million pounds
2000	205	10	215	10	225	224	8	232	4	236
2001	188	6	194	10	204	216	8	224	4	228
2002	190	6	196	9	205	209	9	218	4	222
2003	185	7	192	10	201	191	9	200	4	204
2004	162	5	167	9	176	185	9	194	5	199
2005	152	4	156	9	165	180	7	187	4	192
2006	144	3	147	9	155	177	8	185	5	190
2007	134	3	137	9	146	175	8	183	6	189
2008	140	3	143	9	152	166	8	174	6	180
2009	135	....	138	9	147	162	....	171	7	177

Year	All meat, excluding lard Federally inspected	Other *	Total [1]	Farm	Total
	All meat, excluding lard				
	Million pounds	Million pounds	Million pounds	Million pounds	Million pounds
2000	45,506	645	46,151	149	46,300
2001	45,045	619	45,664	141	45,805
2002	46,549	620	47,169	137	47,305
2003	45,995	581	46,576	136	46,712
2004	44,861	557	45,418	136	45,554
2005	45,166	545	45,711	138	45,848
2006	46,990	547	47,537	137	47,675
2007	48,147	537	48,684	137	48,820
2008	49,675	550	50,225	137	50,362
2009	48,721	....	49,274	139	49,413

[1] Totals are based on unrounded data.   * Note Other class no longer reported.
NASS, Iowa Field Office, (515) 284–4340.

Table 7-67.—Meat: United States exports and imports into the United States, carcass weight equivalent, 2001–2010 [1]

Year	Exports Beef and veal	Lamb and mutton	Pork [4]	All meat	Imports Beef & veal	Lamb and mutton	Pork [4]	All meat
	Exports				Imports			
	Million pounds	Million pounds	Million pounds	Million pounds	Million pounds	Million pounds	Million pounds	Million pounds
2001	2,269	7	1,539	3,815	3,163	146	951	4,260
2002	2,448	7	1,612	4,067	3,218	160	1,071	4,448
2003	2,518	7	1,717	4,242	3,006	168	1,185	4,359
2004	460	8	2,181	2,649	3,679	181	1,099	4,960
2005	697	9	2,666	3,373	3,599	180	1,024	4,803
2006	1,145	18	2,995	4,158	3,085	190	990	4,265
2007	1,434	9	3,141	4,585	3,052	203	968	4,223
2008	1,887	12	4,667	6,566	2,538	183	832	3,553
2009 [2]	1,869	16	4,126	6,011	2,628	171	834	3,633
2010 [3]	2,505	17	4,360	6,437	2,660	183	840	3,533

[1] Carcass weight equivalent of all meat, including the meat content of minor meats and of mixed products.   Includes shipments to U.S. Territories are included in domestic consumption.   [2] Preliminary.   [3] Forecast.   [4] The pork series has been revised to a dressed weight equivalent rather than "Pork, excluding lard."
ERS, Market and Trade Economics Division, Animal Products and Cost of Production Branch, (202) 694–5265. Data on imports and commercial exports are computed from records of the U.S. Department of Commerce, those on exports by the U.S. Department of Agriculture are separately estimated from deliveries and stocks.

Table 7-68.—Meat, beef & veal: Production, 2007–2009

Country	2007	2008	2009
	1,000 metric tons[1]	*1,000 metric tons*[1]	*1,000 metric tons*[1]
Beef and veal:			
Argentina	3,300	3,150	3,400
Australia	2,172	2,159	2,100
Brazil	9,303	9,024	8,935
Canada	1,278	1,288	1,245
China	6,134	6,132	5,764
EU-27	8,188	8,090	7,970
India	2,413	2,525	2,610
Mexico	1,600	1,667	1,700
Pakistan	1,113	1,168	1,226
Russia	1,370	1,315	1,285
Others	9,347	9,424	8,893
Total Foreign	46,218	45,942	45,128
United States	12,096	12,163	11,889
Total	58,314	58,105	57,017
Swine:			
Brazil	2,990	3,015	3,130
Canada	1,746	1,786	1,790
China, Peoples	42,878	46,205	48,890
EU-27	22,858	22,596	22,060
Japan	1,250	1,249	1,310
Korea, South	1,043	1,056	1,062
Mexico	1,152	1,161	1,162
Philippines	1,250	1,225	1,225
Russia	1,910	2,060	2,200
Vietnam	1,832	1,850	1,850
Others	5,714	5,726	5,662
Total Foreign	84,623	87,929	90,341
United States	9,962	10,599	10,439
Total	94,585	98,528	100,780

[1] Carcass weight equivalent.
FAS, Office of Global Analysis, (202) 720-6301. Prepared or estimated on the basis of official USDA production, supply, and distribution statistics from foreign governments.

Table 7-69.—Meat: U.S. exports, 2007–2009

Country	2007	2008	2009
	1,000 metric tons	*1,000 metric tons*	*1,000 metric tons*
Beef & veal;fr/ch/fz:			
Mexico	188,731	209,764	201,094
Canada	100,325	115,585	107,163
Japan	44,721	69,271	81,264
Korea,South	24,340	53,738	52,511
Vietnam	11,013	39,081	49,525
Taiwan	22,563	27,258	26,816
Hong Kong	9,293	9,235	21,086
Egypt	65	1,888	14,273
Netherlands	4,369	11,174	7,756
Russia	3	15,572	4,262
United Arab Emirates	2,924	4,051	4,007
Philippines	2,682	3,785	3,943
Dominican Republic	2,923	2,996	3,362
Bahamas, The	3,239	2,693	3,130
Saudi Arabia	1,354	2,530	2,026
Belgium-Luxembourg	570	1,081	1,823
Netherlands Antilles	699	1,250	1,800
Jamaica	1,977	2,283	1,760
Kuwait	1,720	1,984	1,675
Italy(*)	1,030	1,158	1,306
Cayman Islands	1,165	1,071	1,214
Rest of Total	13,422	20,335	17,629
World Total	438,731	596,533	607,628
Beef&veal, prep/pres:			
Canada	23,445	26,127	23,462
Hong Kong	141	259	1,736
Australia(*)	967	1,372	1,099
Mexico	670	432	676
Panama	1	89	245
Vietnam	0	574	211
Guatemala	25	70	105
Bahamas, The	71	54	102
Finland	0	77	69
Saudi Arabia	2	2	55
Philippines	39	25	52
Netherlands Antilles	35	37	51
Dominican Republic	12	15	45
United Arab Emirates	13	15	45
Russia	0	73	42
Kuwait	126	9	36
New Zealand	33	67	31
Lebanon	0	0	30
Taiwan	1	0	25
Turks and Caicos Islands	8	10	24
Rest of World	846	721	252
World Total	26,543	30,085	28,392
Pork, fr/ch/fz:			
Japan	346,159	419,089	391,757
Mexico	141,954	228,351	271,553
Canada	108,525	125,058	112,812
Hong Kong	37,260	155,081	94,226
Russia	79,095	133,967	91,007
Korea, South	82,764	94,885	83,939
Australia(*)	23,626	30,685	38,473
Taiwan	10,276	18,739	27,061
Philippines	4,004	20,875	25,465
China	62,860	108,164	15,743
Honduras	9,828	11,384	15,713
Dominican Republic	3,739	8,379	14,226
Singapore	1,759	8,542	5,512
New Zealand	4,027	4,488	5,416
Guatemala	4,123	4,353	5,085
United Kingdom	3,286	8,278	4,887
Cuba	3,284	5,216	4,750
Vietnam	1,901	10,068	3,882
UKraine	0	4,819	3,759
Colombia	3,475	2,831	3,744
Bahamas, The	2,633	2,451	3,041
El Salvador	1,196	1,333	2,137
Rest of World	30,188	59,748	25,626
World Total	965,962	1,466,784	1,249,815

See end of table.

Table 7-69.—Meat: U. S. exports, 2007–2009—Continued

Country	2007	2008	2009
	1,000 metric tons	*1,000 metric tons*	*1,000 metric tons*
Pork, hams/shldrs,cured:			
Mexico	12,196	882	38,997
Canada	12,776	16,464	20,238
Guatemala	819	1,008	1,060
Trinidad and Tobago	668	856	925
China	4,947	4,661	789
Bahamas,The	841	274	656
Panama	514	633	645
Ukraine	0	0	493
Netherlands, The	385	347	415
Japan	26	90	365
Belize	279	391	355
Colombia	329	245	230
Dominican Republic	39	178	175
Paraguay	0	0	152
India	0	0	149
Korea, South	0	444	146
Ecuador	74	13	111
Russia	1,175	284	86
Barbados	61	87	72
Leeward-Winward Islands(*)	220	48	52
Aruba(!)	0	16	51
Taiwan	0	7	50
Rest of World	446	1,155	424
World Total	35,795	28,083	66,636
Pork, bacon, cured:			
Mexico	11,152	21,338	14,201
Japan	892	1,392	3,279
Canada	1,647	85	789
Korea, South	373	334	624
China	0	20	602
Hong Kong	126	200	525
Cuba	0	0	399
Dominican Republic	240	197	364
Bahamas, The	201	155	345
Netherlands Antilles (exc.Aruba)	90	219	223
Philippines	1	2	192
Honduras	84	144	169
Taiwan	24	139	164
Afghanistan	0	0	146
New Zealand(*)	453	300	142
Singapore	59	410	140
Australia(*)	109	80	138
Venezuela	0	80	114
Leeward-Windward Islands(*)	78	95	107
Guatemala	173	284	97
Trinidad and Tobago	49	189	90
Ecuador	35	53	87
Rest of World	945	967	870
World Total	16,732	26,683	23,807
Pork, prep/pres, nt/cn:			
Canada	15,330	17,010	18,854
Russia	104	7,026	9,664
Mexico	1,479	2,605	5,552
Dominican Republic	82	195	539
Cuba	0	0	452
Korea, South	1,042	77	417
Ecuador	0	34.3	389
Japan	1,220	376	259
Guatemala	18	126	229
Bahamas, The	177	189	222
Colombia	3	47	169
Hong Kong	335	167	143
Panama	32	120	142
Marshal Island	43	84	119
Australia(*)	849	1,943	116
Netherland, The	43	74	104
El Salvador	79	226	97
Singapore	71	317	90
Nicaragua	9	8	72
Bermuda	25	33	69
Taiwan	38	164	67
New Zealand	81	75	57
Rest of World	514	1,922	386
World Total	21,573	32,818	38,208

See end of table.

Table 7-69.—Meat: U.S. exports, 2007–2009—Continued

Country	2007	2008	2009
	1,000 metric tons	*1,000 metric tons*	*1,000 metric tons*
Pork, prep/pres, canned:			
Japan	3,380	4,365	4,796
Korea, South	1,061	2,285	3,419
Philippines	2,073	2,117	3,100
Australia(*)	1,424	1,714	1,840
Canada	1,151	1,979	1,454
Mexico	661	610	1,400
Hong Kong	753	932	1,312
Panama	478	554	600
Singapore	162	355	385
Bahamas, The	178	97	272
New Zealand(*)	204	184	208
Dominican Republic	43	73	169
Haiti	12	39	116
French Pacific Islands(*)	83	78	88
Taiwan	79	43	55
Thailand	22	54	40
United Arab Emirates	56	126	31
Netherlands	2	11	27
Vietnam	8	33	27
malaysia	0	13	26
Russia	0	1,248	25
Rest of World	304	486	206
World Total	12,154	17,397	19,596
Lamb & mutton; fr/ch/fz:			
Mexico	2,199	1,207	2,956
Bermuda	401	1,037	801
Bahamas, The	143	251	478
Canada	560	775	234
Leeward-Windward Islands(*)	46	93	194
Netherlands Antilles	32	58	163
United Kingdom	1	71	144
Italy(*)	34	93	134
Dominican Republic	31	104	113
Barbados	20	20	77
Guatemala	14	52	67
British Virgin Islands	6	41	60
Saint Lucia(!)	2	1	57
Costa Rica	21	109	54
Saudi Arabia	0	52	52
New Zealand(*)	4	132	42
Antigua and Barbuda	21	38	41
Vietnam	1	7	28
Hong Kong	89	186	26
United Arab Emirates	11	34	26
Rest of World	236	266	0
World Total	3,955	5,004	6,975
Sausages & bologna:			
Japan	23,708	26,370	27,847
Canada	15,909	17,324	19,192
Hong Kong	3,257	3,424	14,675
Mexico	5,775	5,898	7,080
China	10,477	3,532	5,424
Vietnam	9,777	6,832	5,368
Korea, South	1,974	1,585	1,981
Bahamas, The	719	550	1,365
Philippines	679	836	1,232
Netherlands, The	267	475	713
Taiwan, The	291	365	539
Aruba(!)	100	266	522
Guatemala	324	394	436
Belize	321	368	390
Germany(*)	0	53	270
El Salvador	1	25	232
Singapore	41	183	197
Netherlands, The	167	209	192
Panama	141	318	170
Bermuda	62	109	143
Leeward-Islands(*)	29	47	126
Dominican Republic	149	273	121
Russia	108	215	100
Cuba	0	0	87
Rest of World	1,794	659	196
World Total	76,071	70,326	88,599

See end of table.

Table 7-69.—Meat: U.S. exports, 2007–2009—Continued

Country	2007	2008	2009
	1,000 metric tons	*1,000 metric tons*	*1,000 metric tons*
Oth meat prods, f/c/f:			
Mexico	103	33	118
Bermuda	84	105	83
Cayman Islands	64	74	42
Leeward-Windward Islands(*)	8	4	22
British Virgin Islands	6	4	20
Singapore	0	99	13
Netherlands Antilles (exc. Aruba)	49	6	17
United Arab Emirates	0	0	8
Sweden	0	14	7
Aruba	0	2	5
New Zealand	0	0	3
Honduras	0	0	2
Panama	0	0	2
Turks and Caicos Islands	0	0	2
Antigua and Barbuda	2	0	1
Saint Lucias	0	0	1
Bahamas, The	0	0	1
Germany(*)	43	84	1
Guatemala	1	80	1
Australia(!)	0	62	0
Russia	1,193	0	0
Belgium	3,525	0	0
Rest of World	911	22	0
World Total	5,980	588	318
Ot meat prod, prp/prs:			
Canada	3,871	3,839	3,915
Mexico	1,249	1,352	1,242
Bahamas, The	23	36	45
Netherlands Antilles (exc.Aruba)	3	27	34
Aruba	18	19	32
Panama	2	18	28
St. Lucia	0	0	25
Dominica Republic	5	9	13
Trinidad and Tobago	2	57	9
Bermuda	0	0	6
Honduras	5	1	4
Costa Rica	27	16	3
Colombia	27	13	3
El Salvador	0	3	3
Cayman Islands	16	9	3
St.Vincent and the Grenadines	0	0	3
Guatemala	31	9	3
Haiti	0	2	2
Jamaica	3	0	2
Antigua and Barbuda	6	7	0
Nicaragua	0	0	0
Turks and Caicos Islands	3	4	0
British Virgin Islands	2	2	0
St.Christopher-Nevis-Anguilla	4	0	0
Barbados	3	25	0
Venezuela	19	0	0
Rest of World	7,998	7,819	8,036
World Total	13,317	13,267	13,411
Variety meats, beef:			
Mexico	169,969	185,844	89,217
Egypt	86,130	78,064	70,974
Russia	197	24,995	19,384
Canada	8,703	13,245	12,747
Japan	2,023	4,843	10,092
Angola	9,517	5,712	5,082
Cote d'Ivoi	2,739	3,770	4,646
Peru	2,622	4,630	4,306
Philippines	1,170	4,395	4,013
Vietnam	848	1,296	3,734
Indonesia	0	2,659	3,664
Ukraine	1,081	2,476	3,413
Korea, South	826	3,531	3,122
Jamaica	2,450	3,494	3,050
Gabon	1,958	2,432	2,642
Saudi Arabia	1,941	1,475	2,465
Poland	3,279	491	2,224
Moldova	0	1,066	1,572
Hong Kong	278	334	1,300
South Africa	0	0	1,017
Netherlands, The	410	188	987
Congo (Brazzaville)	638	791	907
Ecudor	111	409	900
Guatemala	681	800	856
Rest of World	9,463	10,837	7,878
World Total	307,036	357,777	260,192

See end of table.

Table 7-69.—Meat: U.S. exports, 2007–2009—Continued

Country	2007	2008	2009
	1,000 metric tons	*1,000 metric tons*	*1,000 metric tons*
Variety meats, pork:			
Mexico	100,993	127,129	158,657
Hong Kong	13,270	77,051	108,359
Russia	17,432	59,030	28,610
Japan	6,211	25,180	20,130
China	37,457	38,099	17,910
Taiwan	5,263	12,045	14,943
Korea, South	13,832	33,790	14,908
Canada	8,869	8,393	13,813
Philippines	3,319	5,509	12,995
Australia(*)	4,558	7,043	7,158
Vietnam	1,002	6,636	4,156
Dominican Republic	1,607	4,099	3,313
Haiti	1,434	1,822	2,874
Singapore	146	1,663	1,905
Colombia	1,170	4,267	1,777
New Zealand(*)	495	1,012	1,266
Honduras	1,383	1,100	957
Guatemala	890	961	890
Ukraine	0	6,156	807
Chile	145	713	735
Panama	890	961	
Trinidad and Tobago	522	672	650
Rest of World	2,571	11,450	6,228
World Total	222,838	434,001	423,720
Variety meats, other:			
Mexico	7,606	3,569	7,134
Canada	3,783	2,863	3,531
Hong Kong	2,585	3,043	3,173
Japan	392	887	497
Bahamas;The	260	327	178
Dominican Republic	83	161	154
Netherlands, Thec	0	37	102
Vietnam	49	0	94
China	861	314	85
Korea, South	79	88	77
Singapore	29	0	65
Brazil	118	74	57
United Arab Emirates	27	0	49
Italy(*)	40	21	46
India	6	51	45
France(*)	13	12	38
Saudi Arabia	0	0	38
Leeward-Windward Islands(*)	44	32	35
Findland	0	0	35
Taiwan	123	0	24
Kuwait	0	0	21
Rest of World	997	2,031	165
World Total	17,113	13,534	15,667

Note: (*) denotes a country that is a summarization of its component countries. (!) denotes a country which is summarized into its obsolete country.
FAS, Office of Global Analysis, (202) 720-6301.

Table 7-70.—Meat, beef, veal, and swine: International trade, imports and exports, 2007–2009

Country	2007	2008	2009
	1,000 metric tons	*1,000 metric tons*	*1,000 metric tons*
Principle exporters, beef and veal: [1]			
Argentina	534	422	653
Australia	1,400	1,407	1,364
Brazil	2,189	1,801	1,596
Canada	457	494	480
Colombia	114	206	135
EU-27	140	204	148
India	678	672	590
New Zealand	496	533	514
Paraguay	206	233	254
Uruguay	385	361	370
Others	321	300	315
Total Foreign	6,920	6,633	6,419
United States	650	856	848
Total	7,570	7,489	7,267
Principle importers, beef and veal: [1]			
Canada	242	230	247
Chile	151	129	166
Egypt	293	166	180
EU-27	642	466	495
Japan	686	659	697
Korea, South	308	295	315
Mexico	403	408	322
Russia	1,030	1,137	895
Venezuela	186	320	250
Vietnam	90	200	250
Others	1,745	1,709	1,607
Total Foreign	5,776	5,719	5,424
United States	1,384	1,151	1,192
Total	7,160	6,870	6,616

See end of table.

Table 7-70.—Meat, beef, veal, and swine: International trade, imports and exports, 2007–2009—Continued

Country	2007	2008	2009
	1,000 metric tons	*1,000 metric tons*	*1,000 metric tons*
Principle exporters, swine: [1]			
Australia	54	48	40
Brazil	730	625	707
Canada	1,033	1,129	1,123
Chile	148	142	152
China	350	223	232
Croatia	2	3	4
EU-27	1,286	1,727	1,415
Korea, South	13	11	9
Mexico	80	91	70
Vietnam	19	11	13
Others	22	22	19
Total Foreign	3,737	4,032	3,784
United States	1,425	2,117	1,872
Total	5,162	6,149	5,656
Principle importers, swine: [1]			
Australia	141	152	176
Canada	171	194	180
China	198	431	194
Hong Kong	302	346	369
Japan	1,210	1,267	1,138
Korea, South	447	430	390
Mexico	451	535	678
Russia	894	1,053	845
Singapore	97	91	97
Ukraine	82	238	186
Others	655	802	803
Total Foreign	4,648	5,539	5,056
United States	439	377	378
Total	5,087	5,916	5,434

[1] Carcass weight equivalent.
FAS, Office of Global Analysis, (202) 720-6301. Prepared or estimated on the basis of official USDA production, supply, and distribution statistics from foreign governments.

Table 7-71.—Meats and lard: Production and consumption, United States, 2001–2010 [1]

Year	Beef			Veal			Lamb and mutton		
	Produc-tion	Consumption		Produc-tion	Consumption		Produc-tion	Consumption	
		Total	Per capita		Total	Per capita		Total	Per capita
	Million pounds	*Million pounds*	*Pounds*	*Million pounds*	*Million pounds*	*Pounds*	*Million pounds*	*Million pounds*	*Pounds*
2001	26,212	27,025	94.7	205	204	0.7	227	368	1.3
2002	27,192	27,877	96.8	205	204	0.7	223	381	1.3
2003	26,339	27,000	92.9	202	204	0.7	203	367	1.3
2004	24,650	27,750	94.6	176	177	0.6	200	373	1.3
2005	24,787	27,754	93.8	165	164	0.6	191	356	1.2
2006	26,256	28,137	94.2	156	155	0.5	190	356	1.2
2007	26,523	28,141	93.3	146	145	0.5	189	385	1.3
2008	26,664	27,300	89.6	152	150	0.5	180	343	1.1
2009 [2]	26,068	26,904	87.5	147	147	0.5	177	338	1.1
2010 [3]	25,793	26,258	84.7	145	147	0.5	171	337	1.1

Year	Pork			All meats			Lard		
	Produc-tion	Consumption		Produc-tion	Consumption		Produc-tion	Consumption	
		Total	Per capita		Total	Per capita		Total	Per capita
	Million pounds	*Million pounds*	*Pounds*	*Million pounds*	*Million pounds*	*Pounds*	*Million pounds*	*Million pounds*	*Pounds*
2001	19,160	18,511	64.9	45,804	46,108	162	NA	NA	NA
2002	19,685	19,142	66.5	47,305	47,604	165	NA	NA	NA
2003	19,966	19,443	66.9	46,710	47,013	162	NA	NA	NA
2004	20,531	19,446	66.3	45,557	47,746	163	NA	NA	NA
2005	20,705	19,093	64.5	45,848	47,366	160	NA	NA	NA
2006	21,074	19,055	63.8	47,675	47,703	160	NA	NA	NA
2007	21,962	19,763	65.5	48,817	48,434	160	NA	NA	NA
2008	23,367	19,415	63.7	50,362	47,211	155	NA	NA	NA
2009 [2]	23,020	19,839	64.5	49,412	47,227	154	NA	NA	NA
2010 [3]	22,259	18,769	60.5	48,368	45,511	147	NA	NA	NA

[1] Carcass weight equivalent or dressed weight. Beginning 1977, pork production was no longer reported as "pork, excluding lard." This series has been revised to reflect pork production in prior years on a dressed weight basis that is comparable with the method used to report beef, veal, and lamb and mutton. Edible offals are excluded. Shipments to the U.S. territories are included in domestic consumption. [2] Preliminary. [3] Forecast. NA-not available.

ERS, Animal Products, Grains and Oilseeds Branch, (202) 694–5265.

Table 7-72.—Hides and skins: United States imports by country of origin, 2007–2009

Country of origin	2007	2008	2009
	1,000 pieces	*1,000 pieces*	*1,000 pieces*
Hides and skins, mixed:			
Canada	1,243,470	1,181,715	92,966
China; Peoples Republic of	963	293,614	30,618
New Zealand (*)	36,872	110,991	1,170
Mexico	149,381	96,898	7,155
Colombia	40,718	29,090	2,689
Saudi Arabia	48,000	24,050	0
Turkey	325	20,000	0
Belgium-Luxembourg(*)	13,000	19,550	2,000
Pakistan	3,313	18,681	0
Rest of World	450,366	68,890	817
World Total	1,986,408	1,863,479	137,415
Furskins:			
Canada	1,587,568	2,081,843	425,069
Sweden	274,763	316,938	49,729
Belgium -Luxembourg(*)	592,318	308,018	2,640
Germany(*)	662,608	195,007	11,216
Netherlands	220,405	178,884	52,533
Czech Republic	500	100,000	0
Denmark	65,763	39,098	3,655
Finland	71,579	36,661	1,621
Russia	32,460	24,230	0
Rest of World	195,172	63,401	6,179
World Total	3,703,136	3,344,080	552,642

Note: (*) denotes a country that is a summarization of its component countries.
FAS, Office of Global Analysis, (202) 720-6301.

Table 7-73.—Hides, packer: Average price per hundred pounds, Central U.S., 2000–2008

Year	Steers				Heifers	
	Heavy native	Heavy Texas	Butt branded	Colorado branded	Heavy native	Branded
	Dollars	*Dollars*	*Dollars*	*Dollars*	*Dollars*	*Dollars*
2000	80.17	....	73.67	71.24	83.41	77.54
2001	85.84	....	79.79	75.90	85.52	85.44
2002	82.25	....	75.97	71.07	85.73	78.75
2003	83.83	....	78.58	73.29	88.34	80.20
2004 [1]	67.09	64.91	64.39	61.48	57.07	54.02
2005	65.64	63.50	63.53	60.90	57.89	54.20
2006	68.87	67.76	67.79	65.99	60.30	57.52
2007	72.01	70.51	70.72	67.79	65.70	61.85
2008	63.94	63.22	62.62	59.35	58.35	57.18
2009	44.79	44.05	42.98	36.11	29.22	32.60

[1] Effective 2004, price is per piece not per hundred pounds.
AMS, Livestock & Grain Market News, (202) 720–7316.

Table 7-74.—Hides: U.S. trade exports, 2007–2009

Country	2007	2008	2009
Cattle hides, whole, mixed: [1]			
China	10,438,347	9,405,531	9,878,544
Korea, South	4,106,554	3,657,030	4,023,472
Mexico	1,302,291	1,253,848	1,591,216
Taiwan	1,835,239	1,442,970	1,389,296
Hong Kong	1,826,994	2,359,094	1,375,691
Italy(*)	810,647	896,778	880,929
Thailand	987,967	611,122	716,968
Japan	1,019,654	864,134	676,542
Vietnam	559,942	549,730	416,315
Canada	91,638	114,771	345,381
India	89,932	143,010	163,180
Netherlands, The	85,492	106,073	128,837
Turkey	172,004	122,717	115,825
Israel(*)	112,377	113,266	99,962
Spain	47,911	98,515	62,489
Costa Rica	57,901	79,495	48,589
Pakistan	4,100	2,690	28,109
El Salvador	0	970	22,113
Indonesia	91,282	54,248	18,863
Egypt	0	0	18,632
Belgium-Luxembourg(*)	161,731	184,673	17,654
Rest of World	604,479	161,828	124,409
World Total	24,394,258	22,213,453	22,126,208
Sheep & lambskins, mixed: [2]			
China, Peoples Republic	1,080,550	1,017,894	1,347,624
Dominican Republic	769	50	38,546
Turkey	1,063,530	210,996	30,796
Mexico	40,653	38,381	17,469
Russian Federation	472,763	170,070	9,240
Canada	24,660	22,550	6,999
United Kingdom	0	122	4,000
Uruguay	0	0	1,151
Australia(*)	362	0	396
Belgium-Luxembourg(*)	0	0	357
Cyprus	0	236	249
Germany	209	508	245
Namibia	0	0	227
Italy(*)	244	5,656	222
Hong Kong	435	2,495	180
Japan	0	0	168
Spain	0	0	89
France(*)	232	395	85
Chile	0	0	78
Malta	0	0	66
Rest of World	416,175	26,176	0
World Total	2,895,765	1,495,440	1,458,187
Pig and hog skins, pieces: [1]			
Mexico	1,960,335	1,639,975	2,169,481
Taiwan	468,132	585,477	909,339
China, Peoples Republic	1,947,707	1,966,554	655,096
Vietnam	8,780	39,553	262,125
Hong Kong	35,830	73,153	119,653
United Kingdom	69,096	62,600	69,804
Japan	10,141	3,015	11,986
Canada	0	168	5,419
Thailand	0	30	5,200
Germany(*)	2,800	5,600	0
Haiti	0	2,000	0
Italy	1,026	0	0
Indonesia	0	828	0
Korea, South	13	0	0
Slovenia	11,014	0	0
Rest of World	13	0	0
World Total	4,514,874	4,378,923	4,208,103

See end of table.

Table 7-74.—Hides: U.S. trade exports, 2007–2009—Continued

Country	2007	2008	2009
	1,000 metric tons	*1,000 metric tons*	*1,000 metric tons*
Mink furskins, undressed: [2]			
China, Peoples Republic of	1,006,596	2,878,736	2,497,242
Canada	1,876,396	1,774,807	1,773,477
Korea, Republic of	675,926	430,848	441,704
Greece	153,588	114,481	61,814
Germany(*)	147,780	291,548	54,129
Kenya	0	0	21,659
Hong Kong	1,072,956	222,336	20,661
Poland	27,397	34,605	15,507
Italy	37,732	47,037	8,316
Russian Federation	0	13,760	6,941
United Kingdom	2,791	5,969	5,677
France(*)	64,886	59,817	5,558
Japan	4,712	5,904	4,536
Spain	10,968	0	4,217
Denmark(*)	10,828	2,308	2,140
Ukraine	1,238	10,832	1,319
Estonia	481	0	1,013
Mexico	0	692	782
Finland	17,936	4,021	631
Switzerland	2,950	10,203	200
Liechtenstein	0	0	200
Rest of World	3,402	14,232	0
World Total	5,115,613	5,911,933	4,927,523
Other furskins, whole: [2]			
Canada	1,966,303	1,403,347	1,366,028
China	269,721	426,228	627,708
Hong Kong	511,451	763,687	312,400
Poland	267,110	124,095	121,206
Germany(*)	97,408	68,919	50,514
Czech Republic	64,874	87,600	49,509
Russia	64,245	80,887	35,950
Italy(*)	50,714	74,516	32,731
Turkey	128,065	67,135	21,871
Greece	77,649	50,898	13,210
United Kingdom	19,978	35,708	11,204
Estonia	0	0	8,667
Finland	13,535	3,745	5,926
Indonesia	0	0	2,383
Australia(*)	0	147	2,351
Denmark(*)	304	0	2,203
Venezuela	0	0	1,965
India	0	0	1,644
United Arab Emirates	110	1,290	1,611
Japan	389	400	1,572
Ukraine	1,952	0	1,078
Rest of World	165,585	85,848	6,571
World Total	3,643,378	3,246,717	2,677,111
Other hides & skins, mixed: [3]			
China	1,013,122	2,618,560	4,809,237
Korea, South	478,495	861,568	1,640,881
Taiwan	577,532	643,049	655,393
Mexico	379,813	237,882	426,058
Hong Kong	290,622	596,501	370,034
Thailand	108,314	379,450	295,349
Vietnam	331,729	480,282	223,267
Italy(*)	144,413	262,064	197,173
France(*)	235,119	269,829	192,718
Singapore	110,992	94,300	144,561
Germany(*)	57,673	99,989	43,355
Brazil	771	2,782	30,759
Japan	79,887	24,595	19,333
Canada	8,252	3,967	17,804
Netherlands	44,204	14,813	10,971
Uruguay	0	8,760	6,120
Haiti	0	0	4,720
Turkey	9,556	5,772	3,613
Ecuador	0	0	3,498
Pakistan	0	0	2,366
Kazakhstan	0	0	1,980
Australia(!)	0	378	1,868
Rest of World	194,184	41,986	8,920
World Total	4,060,136	6,636,536	9,107,546

[1] Pieces. [2] Number. [3] Metric tons. Note: (*) denotes a country that is a summarization of its component countries. (!) denotes a country which is summarized into its obsolete country.
FAS, Office of Global Analysis, (202) 720-6301.

Table 7-75.—Mink: Farms, pelts produced and value of mink pelts, United States, 2000–2009

Year	Mink farms	Pelts produced	Average marketing price	Value of mink pelts
	Number	*Number*	*Dollars*	*Dollars*
2000	350	2,666,100	34.00	90,647,400
2001	329	2,565,300	33.50	85,937,550
2002	324	2,607,300	30.60	79,783,380
2003	305	2,549,000	40.10	102,214,900
2004	296	2,558,100	47.10	120,486,510
2005	275	2,637,800	60.90	160,642,020
2006	279	2,858,800	48.40	138,365,920
2007	283	2,828,200	65.70	185,812,740
2008	274	2,820,700	41.60	117,341,120
2009	278	2,855,700	65.10	185,906,070

NASS, Livestock Branch, (202) 720–3570.

Table 7-76.—Mink pelts: Number produced by color class, major States, and United States, 2009

State	Black	Demi wild	Pastel	Sapphire	Blue Iris	Mahogany	Pearl
	Number	*Number*	*Number*	*Number*	*Number*	*Number*	*Number*
ID	107,000	13,500	3,400	9,500	....	108,000	....
IL	56,000	....	....	....	....	....	....
IA	79,000	....	....	....	6,500	....	....
MI	29,000	....	....	....	....	3,400	....
MN	88,000	35,000	....	....	55,000	80,000	....
MT	....	....	....	....	....	....	....
OH	36,000	....	....	....	....	21,000	....
OR	158,000	....	....	11,000	79,000	....	....
PA	28,000	....	....	5,500	11,000	....	....
SD	....	....	....	....	....	....	....
UT	225,000	37,000	....	....	7,500	285,000	....
WA	69,000	....	....	....	....	....	....
WI	585,000	....	....	55,000	97,000	97,000	....
Oth Sts [1]	34,500	38,400	52,800	23,400	26,400	69,100	57,100
US	1,494,500	123,900	56,200	104,400	282,400	663,500	57,100

State	Lavender	Violet	White	Miscellaneous and unclassified	Total pelts [2]
	Number	*Number*	*Number*	*Number*	*Number*
ID	....	....	....	....	251,500
IL	....	....	....	....	71,100
IA	....	....	....	....	114,800
MI	....	....	5,500	....	45,300
MN	....	....	6,000	....	267,200
MT	....	....	....	....	22,500
OH	....	....	....	....	67,900
OR	....	....	....	....	270,100
PA	....	....	....	....	57,300
SD	....	....	....	....	81,500
UT	....	....	....	....	613,500
WA	....	....	....	....	93,500
WI	3,300	....	24,000	4,100	886,100
Oth Sts [1]	1,100	14,500	13,300	1,900	13,400
US	4,400	14,500	48,800	6,000	2,855,700

[1] Other States also includes some pelts from the above listed States that were not published to avoid disclosing individual operations. [2] Published color classes may not add to the State total.
NASS, Livestock Branch, (202) 720–3570.

Table 7-77.—Livestock: Number of animals slaughtered under Federal inspection and number of whole carcasses condemned, 2000–2009 [1]

Year	Cattle		Calves		Sheep and lambs	
	Total head	Condemned	Total head	Condemned	Total head	Condemned
	1,000	*1,000*	*1,000*	*1,000*	*1,000*	*1,000*
2000	35,136	188.9	1,103	22.4	3,316	5.8
2001	37,641	198.2	1,333	25.2	3,463	5.6
2002	31,404	165.9	1,034	19.5	2,922	5.4
2003	NA	NA	NA	NA	NA	NA
2004	31,515	159.7	876	15.2	2,679	4.9
2005	31,847	145.8	757	12.1	2,582	5.4
2006	32,861	143.1	682	11.1	2,534	4.7
2007	33,473	141.5	769	13.6	2,497	4.1
2008	34,220	146.8	866	24.0	2,447	5.2
2009	32,714	143.1	951	23.0	2,297	3.3

Year	Goats		Hogs		Horses [2]	
	Total head	Condemned	Total head	Condemned	Total head	Condemned
	1,000	*1,000*	*1,000*	*1,000*	*1,000*	*1,000*
2000	530	1.2	93,385	410.8	50	0.3
2001	592	1.1	96,600	449.9	62	0.2
2002	553	1.0	89,855	379.0	43	0.2
2003	NA	NA	NA	NA	NA	NA
2004	582	1.2	98,416	391.2	59	0.1
2005	553	1.1	103,849	414.8	88	0.7
2006	561	0.9	103,600	417.0	102	0.9
2007	613	0.7	105,611	404.8	58	0.4
2008	654	0.9	115,600	393.0	0	0
2009	660	0.8	113,395	333.6	0	0

[1] Data are reported by the Food Safety and Inspection Service, USDA for the fiscal year ending September 30. Condemnations include ante-mortem and post-mortem inspection. [2] Equine slaughter was discontinued during the week of September 22, 2007. NA-not available.
NASS, Iowa Field Office, (515) 284–4340.

Table 7-78.—Livestock: Inventory and value, United States, Jan. 1, 2008–2010

Class of livestock and poultry	Inventory			Value					
				Per head [2]			Total		
	2008	2009	2010 [1]	2008	2009	2010 [1]	2008	2009	2010 [1]
	Thousands	*Thousands*	*Thousands*	*Dollars*	*Dollars*	*Dollars*	*1,000 dollars*	*1,000 dollars*	*1,000 dollars*
Cattle	96,035	94,521	93,701	990.00	872.00	829.00	95,112,820	82,435,620	77,677,310
Hogs [3]	68,176.8	67,148	67,807	73.00	89.00	83.00	4,986,206	5,957,633	5,464,212
Sheep and lambs	5,950.0	5,747	5,630	138.00	133.00	135.00	823,424	765,194	762,295
Angora goats [4]	176.6	160.5	12.5	74.80	83.00	73.00	13,205	13,318	9,120
Total [5]							100,935,655	89,171,765	83,912,937
Chickens [3]	458,593	446,906	449,610	2.95	3.39	3.34	1,351,549	1,517,210	1,501,605
Total [6]							102,287,204	90,688,975	85,414,542

[1] Preliminary. [2] Based on reporters' estimates of average price per head in their localities. [3] Dec. 1 of preceding year. [4] Four state total for angora goats (AZ, CA, NM, TX). [5] Cattle, hogs, sheep, and angora goats. [6] Includes all cattle, hogs, sheep, angora goats, and chickens (excluding broilers).
NASS, Livestock Branch, (202) 720–3570.

Table 7-79.—Livestock: Average price per 100 pounds received by farmers, by State and United States, 2008–2009

State	Cows [1]		Steers and heifers		Beef cattle [2]		Calves	
	2008	2009	2008	2009	2008	2009	2008	2009
	Dollars	Dollars	Dollars	Dollars	Dollars	Dollars	Dollars	Dollars
AL	49.90	42.30	85.00	79.00	73.40	67.20	98.80	88.50
AK	90.00	90.00	120.00	120.00	105.00	105.00	120.00	120.00
AZ	52.30	45.40	95.60	84.10	74.80	62.30	106.00	102.00
AR	48.70	42.60	93.40	88.90	80.50	75.00	103.00	96.70
CA	43.70	42.80	92.90	84.60	72.90	64.80	102.00	100.00
CO	51.40	47.00	99.80	91.00	98.90	90.10	110.00	107.00
CT	46.00	39.00	65.00	58.00	61.00	54.00	100.00	95.00
DE	55.00	49.80	90.00	84.60	86.50	81.10	105.00	95.00
FL	52.10	43.90	85.80	81.50	70.70	67.30	98.90	93.70
GA	54.30	47.80	85.90	81.60	65.20	61.60	96.30	94.70
HI	29.20	29.50	58.20	59.70	51.70	53.30	85.00	88.00
ID	49.00	45.00	91.60	84.20	81.30	74.60	102.00	101.00
IL	52.80	45.20	93.60	83.90	92.80	83.10	102.00	97.40
IN	50.70	46.20	89.20	83.20	78.50	73.80	93.00	96.40
IA	52.20	47.10	92.80	84.30	92.10	83.60	108.00	103.00
KS	51.80	46.40	94.50	85.10	93.40	84.30	119.00	113.00
KY	50.50	43.90	89.20	85.60	79.80	75.70	98.70	96.00
LA	50.30	42.70	88.60	82.90	62.60	55.50	99.50	93.20
ME	46.00	39.00	70.00	63.00	65.00	58.00	100.00	95.00
MD	55.70	48.50	90.10	79.50	86.70	76.30	105.00	95.00
MA	46.00	39.00	66.00	59.00	62.00	55.00	100.00	95.00
MI	52.00	45.80	87.80	78.50	77.10	68.70	99.90	88.60
MN	54.60	47.80	90.20	81.50	82.60	74.00	112.00	111.00
MS	49.50	44.80	86.00	80.30	65.50	63.90	94.30	90.70
MO	49.40	43.70	98.40	92.70	86.20	80.60	108.00	105.00
MT	47.70	43.40	100.00	94.50	87.50	76.30	109.00	108.00
NE	54.20	48.60	95.50	85.40	94.20	84.30	117.00	113.00
NV	46.00	45.00	99.00	90.00	85.00	79.00	114.00	111.00
NH	46.00	39.00	67.00	60.00	63.00	56.00	105.00	100.00
NJ	52.00	44.00	70.00	63.00	55.00	47.00	82.00	78.00
NM	54.10	47.90	95.80	88.70	79.30	72.90	113.00	105.00
NY	47.50	47.00	86.10	81.90	52.20	53.00	99.50	90.00
NC	53.50	47.20	87.50	82.00	65.50	61.40	94.80	89.30
ND	50.80	46.80	95.60	91.10	83.30	80.40	104.00	103.00
OH	48.60	42.20	87.30	81.20	83.60	77.30	95.80	93.60
OK	53.20	46.40	97.60	90.20	94.20	86.70	111.00	105.00
OR	46.80	45.90	91.00	83.20	77.60	77.10	98.00	96.70
PA	52.70	45.40	87.00	80.20	78.10	71.10	112.00	102.00
RI	48.00	41.00	65.00	58.00	62.00	55.00	100.00	95.00
SC	50.70	45.70	80.90	80.30	73.30	72.00	91.90	89.40
SD	52.90	47.40	98.90	90.40	91.30	87.40	111.00	111.00
TN	49.90	42.80	89.30	85.50	73.60	68.30	97.10	93.80
TX	47.90	41.40	94.60	85.70	90.20	81.10	110.00	104.00
UT	43.00	42.00	94.00	83.00	90.50	80.00	105.00	104.00
VT	47.00	40.00	66.00	59.00	62.00	55.00	100.00	100.00
VA	49.80	43.00	87.50	84.50	75.30	71.40	97.50	93.70
WA	46.60	43.70	91.10	88.70	85.10	82.50	92.20	93.30
WV	49.40	43.80	78.20	77.80	60.10	58.90	88.20	85.30
WI	50.70	43.70	84.60	76.90	65.10	57.70	147.00	129.00
WY	47.50	43.30	103.00	92.70	91.20	82.60	110.00	110.00
US	50.60	44.80	94.50	85.40	89.10	80.30	110.00	103.00

See footnotes at end of table.

Table 7-79.—Livestock: Average price per 100 pounds received by farmers, by State and United States, 2008–2009—Continued

State	Hogs [3]		Lambs		Sheep	
	2008	2009	2008	2009	2008	2009
	Dollars	Dollars	Dollars	Dollars	Dollars	Dollars
AL	43.80	41.70				
AK	85.00	90.00				
AZ	56.10	50.70	96.00	96.00	34.00	35.00
AR	43.60	38.30				
CA	48.00	42.30	91.30	88.70	24.30	31.10
CO	49.30	42.00	99.70	99.20	29.20	32.60
CT	39.50	39.00				
DE	42.20	40.00				
FL	39.40	37.40				
GA	46.60	45.00				
HI	97.20	96.80				
ID	47.10	41.50	101.00	93.50	22.80	25.50
IL	47.90	44.40	94.00	102.00	28.00	31.40
IN	49.10	44.80	102.00	105.00	25.00	29.00
IA	48.10	42.30	95.80	96.50	27.50	30.70
KS	41.30	36.90	96.00	95.00	19.00	24.00
KY	44.60	39.00	103.00	109.00	35.30	40.70
LA	38.60	35.70				
ME	39.50	39.00				
MD	42.20	40.00	111.00		45.00	
MA	39.50	39.00				
MI	42.50	37.00	95.00	101.00	29.00	34.00
MN	48.90	42.60	101.00	99.90	25.60	31.70
MS	44.70	38.70				
MO	41.50	36.80	96.00	103.00	29.40	37.80
MT	47.10	40.90	101.00	101.00	19.00	31.10
NE	49.40	42.70	102.00	99.60	29.70	34.70
NV	46.30	38.80	112.00	106.00	34.00	35.00
NH [4]	39.50	39.00	125.00	130.00	48.00	50.00
NJ	38.60	33.00				
NM	40.30	34.80	99.00	109.00	30.00	34.00
NY	36.00	35.20	104.00	105.00	44.90	45.90
NC	48.50	41.70	106.00	111.00	35.00	40.00
ND	47.30	41.30	103.00	101.00	22.30	27.80
OH	44.30	40.30	99.70	106.00	28.40	33.40
OK	37.90	34.80	97.00	97.00	30.00	31.00
OR	53.00	44.70	97.40	93.40	31.20	38.80
PA	41.50	40.00	117.00	118.00	44.00	45.90
RI	39.50	39.00				
SC	47.80	43.60				
SD	46.80	41.70	105.00	104.00	19.80	28.30
TN	44.40	39.60	104.00	109.00	32.00	37.00
TX	40.50	37.60	98.20	109.00	30.20	34.70
UT	52.30	47.50	102.00	99.90	25.00	30.20
VT	39.50	39.00				
VA	45.50	43.90	110.00	114.00	33.40	37.70
WA	44.60	39.70	97.00	96.00	29.00	32.00
WV	45.00	43.10	102.00	110.00	28.80	33.00
WI	43.80	39.50	95.30	93.40	29.10	30.70
WY	42.20	38.30	104.00	100.00	23.50	32.80
Oth Sts [5]			98.00	103.00	32.00	36.00
US	47.00	41.70	99.60	99.60	27.20	32.50

[1] Beef cows and cull dairy cows sold for slaughter.   [2] Cows, steer, and heifers combined.   [3] December of preceding year through November.   [4] For lambs and sheep, CT, ME, MA, NH, RI, and VT are included in NH.   [5] Individual state estimates not available for states not shown, but are included in Other States. MD is included in Other States for 2009.
NASS, Livestock Branch, (202) 720–3570.

Table 7-80.—Frozen meat: Cold storage holdings, end of month, United States, 2008–2009

Month	Boneless beef		Beef cuts		Total beef	
	2008	2009	2008	2009	2008	2009
	1,000 pounds	1,000 pounds	1,000 pounds	1,000 pounds	1,000 pounds	1,000 pounds
January	382,386	391,691	68,396	70,956	450,782	462,647
February	359,145	369,433	77,404	65,081	436,549	434,514
March	351,252	361,064	75,456	64,814	426,708	425,878
April	350,909	346,537	65,445	64,168	416,354	410,705
May	359,928	358,042	60,437	59,833	420,365	417,875
June	362,621	375,531	65,497	59,273	428,118	434,804
July	368,464	373,924	62,354	70,883	430,818	444,807
August	365,222	358,949	75,928	61,147	441,150	420,096
September ...	376,923	363,513	77,597	65,400	454,520	428,913
October	392,795	366,840	78,184	60,876	470,979	427,716
November	402,399	371,515	79,165	59,398	481,564	430,913
December	415,048	373,133	77,589	57,147	492,637	430,280

Month	Picnics		Bellies		Butts	
	2008	2009	2008	2009	2008	2009
	1,000 pounds	1,000 pounds	1,000 pounds	1,000 pounds	1,000 pounds	1,000 pounds
January	15,039	8,992	70,647	69,166	21,407	20,940
February	18,102	12,085	79,282	75,668	25,332	25,351
March	14,389	14,062	98,896	72,940	22,978	21,844
April	12,567	12,404	100,189	79,543	28,836	19,697
May	9,918	8,661	87,428	78,801	12,563	18,406
June	7,509	8,762	74,372	76,333	11,593	15,940
July	6,901	10,191	57,964	60,238	13,429	13,498
August	10,845	8,984	31,878	48,958	15,005	14,284
September ...	9,809	7,847	21,270	38,481	14,732	10,987
October	14,374	12,038	21,696	37,127	16,363	12,895
November	11,149	11,392	33,490	44,638	20,228	16,212
December	8,924	8,889	51,593	56,764	25,122	20,066

Month	Hams					
	Bone-in		Boneless		Total	
	2008	2009	2008	2009	2008	2009
	1,000 pounds	1,000 pounds	1,000 pounds	1,000 pounds	1,000 pounds	1,000 pounds
January	42,757	44,022	46,180	43,515	88,937	87,537
February	49,768	46,059	43,116	50,146	92,884	96,205
March	51,271	29,202	43,563	41,692	94,834	70,894
April	74,730	27,219	49,241	50,115	123,791	77,334
May	80,974	40,738	50,185	54,531	131,159	95,269
June	86,240	59,584	43,798	56,941	130,038	116,525
July	96,369	63,905	48,732	64,714	145,101	128,619
August	108,420	82,847	49,276	53,074	157,696	135,921
September ...	128,433	97,383	52,998	51,773	181,431	149,156
October	108,852	81,072	45,189	48,471	154,041	129,543
November	68,950	42,969	35,772	40,408	104,722	83,377
December	36,793	20,521	37,422	29,898	74,215	50,419

Month	Loins					
	Bone-in		Boneless		Total	
	2008	2009	2008	2009	2008	2009
	1,000 pounds	1,000 pounds	1,000 pounds	1,000 pounds	1,000 pounds	1,000 pounds
January	17,582	20,608	28,068	30,749	45,650	51,357
February	18,270	20,141	29,081	23,907	47,351	44,048
March	18,082	20,774	30,458	23,192	48,540	43,966
April	19,749	20,286	26,943	22,130	46,692	42,416
May	16,009	16,079	20,876	18,191	36,885	34,270
June	15,455	13,932	19,568	18,503	35,023	32,435
July	12,450	10,713	17,718	14,788	30,168	25,501
August	10,451	8,645	15,662	12,486	26,113	21,131
September ...	9,722	8,174	14,982	15,321	24,704	23,495
October	11,294	10,247	17,494	15,056	28,788	25,303
November	15,485	14,378	25,085	22,270	40,570	36,648
December	18,043	15,077	26,131	20,807	44,174	35,884

See end of table.

Table 7-80.—Frozen meat: Cold storage holdings, end of month, United States, 2008–2009—Continued

Month	Ribs		Trimmings		Other frozen pork	
	2008	2009	2008	2009	2008	2009
	1,000 pounds	1,000 pounds	1,000 pounds	1,000 pounds	1,000 pounds	1,000 pounds
January	72,254	87,854	63,268	66,610	111,706	111,024
February	83,769	92,883	70,604	62,705	105,570	110,589
March	87,121	86,375	75,864	62,481	116,166	114,141
April	74,677	87,323	60,949	69,556	115,291	115,551
May	61,818	71,558	52,034	62,220	103,160	112,431
June	46,425	58,015	50,286	59,827	96,194	107,860
July	42,878	52,450	41,164	46,723	85,818	107,590
August	39,727	47,936	36,682	39,816	94,941	105,081
September ...	44,664	54,005	36,165	39,701	90,464	101,823
October	56,439	67,700	40,682	38,576	90,421	90,302
November	71,240	84,466	51,082	39,599	97,291	81,101
December	82,013	96,000	64,997	43,815	105,812	79,611

Month	Variety meats		Unclassified pork		Total pork	
	2008	2009	2008	2009	2008	2009
	1,000 pounds	1,000 pounds	1,000 pounds	1,000 pounds	1,000 pounds	1,000 pounds
January	31,775	25,074	54,246	78,382	574,929	606,936
February	32,480	24,368	56,456	80,575	611,830	624,477
March	31,109	23,698	67,447	83,726	657,344	594,127
April	29,544	27,176	70,727	81,290	663,443	612,290
May	25,377	24,662	59,061	78,266	579,403	584,544
June	25,366	25,812	53,317	76,405	530,123	577,914
July	27,412	24,101	54,434	70,789	505,269	539,700
August	27,400	29,540	62,382	78,497	502,669	530,148
September ...	33,264	32,250	69,483	70,936	526,168	528,681
October	33,790	30,924	71,437	71,892	528,031	516,300
November	26,153	22,145	70,755	63,238	526,680	482,816
December	25,575	22,584	73,217	57,093	555,642	471,125

Month	Veal		Lamb & mutton		Total red meat	
	2008	2009	2008	2009	2008	2009
	1,000 pounds	1,000 pounds	1,000 pounds	1,000 pounds	1,000 pounds	1,000 pounds
January	6,589	7,177	15,177	19,469	1,047,477	1,096,229
February	6,107	7,590	18,157	18,279	1,072,643	1,084,860
March	5,752	5,984	17,118	19,274	1,106,922	1,045,263
April	6,331	7,807	17,783	19,801	1,103,911	1,050,603
May	6,284	7,970	18,411	19,694	1,024,463	1,030,083
June	5,939	8,793	19,598	21,568	983,778	1,043,079
July	4,721	10,126	19,723	20,062	960,531	1,014,695
August	5,981	10,910	21,147	19,045	970,947	980,199
September ...	7,900	9,028	20,796	17,426	1,009,384	984,048
October	7,420	9,041	21,331	15,301	1,027,761	968,358
November	7,923	8,054	21,659	15,052	1,037,826	936,835
December	9,194	8,960	21,001	14,519	1,078,474	924,884

NASS, Livestock Branch, (202) 720–3570.

CHAPTER VIII

DAIRY AND POULTRY STATISTICS

Dairy statistics in this chapter include series relating to many phases of production, movement, prices, stocks, and consumption of milk and its products. Two series of number of milk cows on farms are included in this publication. One series is an inventory number of a specific classification estimated as one of the major groups making up the total cattle population on January 1. The other series identified as "milk cows" is an annual average number of milk cows during the year (excluding any not yet fresh) and is used in estimating milk production.

In comparing the several series of milk prices, it is important to note that prices received by farmers for all whole milk sold are for milk or milkfat content as actually sold, while certain prices paid by dealers for milk for fluid purposes or for specified manufacturing purposes may be quoted on a 3.5 percent butterfat basis, or for some types of manufacturing milk on the test of the milk used for that particular purpose.

Poultry and poultry products statistics include inventory numbers of chickens by classes; the production, disposition, cash receipts, and gross income from chickens and eggs; poultry and egg receipts at principal markets; commercial broiler production; turkey production, disposition, and gross income; poultry and eggs under Federal inspection; and the National Poultry Improvement Plan. Estimates relating to inventories, production, and income exclude poultry and eggs produced on places not classified as farms.

Table 8-1.—Milk cows and heifers: Number that have calved and heifers 500 pounds and over kept for milk cow replacements, United States, Jan. 1, 2001–2010

Year	Milk cows and heifers that have calved	Heifers 500 pounds and over kept for milk cow replacements
	Thousands	Thousands
2001	9,172	4,057
2002	9,106	4,055
2003	9,142	4,114
2004	8,988	4,018
2005	9,004	4,117
2006	9,104	4,298
2007	9,145	4,325
2008	9,257	4,415
2009	9,333	4,410
2010 [1]	9,081	4,516

[1] Preliminary.
NASS, Livestock Branch, (202) 720–3570.

Table 8-2.—Milk cows: Number of operations, percent of inventory and percent of milk production by size group, United States, 2008–2009 [1]

Head	Operations		Percent of inventory		Percent of production [2]	
	2008	2009	2008	2009	2008	2009
	Number	Number	Percent	Percent	Percent	Percent
1-29	21,300	20,400	1.8	1.8	1.2	1.2
30-49	11,900	11,500	5.1	4.9	3.9	3.8
50-99	17,800	17,300	13.1	13.0	11.5	11.4
100-199	8,700	8,600	12.5	12.4	11.8	11.6
200-499	3,950	3,850	12.6	12.3	13.1	12.5
500-999	1,720	1,700	12.5	12.5	12.5	12.6
1,000-1,999	900	910	13.1	13.3	15.5	15.7
2,000+	730	740	29.3	29.8	30.5	31.2
Total	67,000	65,000	100.0	100.0	100.0	100.0

[1] An operation is any place having one or more head of milk cows on hand on December 31.   [2] Percents reflect average distributions of various probability surveys conducted during the year.
NASS, Livestock Branch, (202) 720–3570.

Table 8-3.—Milk cows and heifers: Number that have calved and heifers 500 pounds and over kept for milk cow replacements, by State, Jan. 1, 2009 and 2010

State	Milk cows and heifers that have calved		Heifers 500 pounds and over kept for milk cow replacements	
	2009	2010 [1]	2009	2010 [1]
	Thousands	Thousands	Thousands	Thousands
AL	12.0	11.0	5.0	5.0
AK	0.6	0.6	0.2	0.2
AZ	190	167	50	55
AR	14.0	13.0	5.0	6.0
CA	1,840	1,760	780	750
CO	130	116	60	70
CT	19.0	18.5	11.0	9.0
DE	6.5	6.0	2.3	2.7
FL	118	112	35	30
GA	74	76	23	25
HI	1.6	1.8	1.0	1.0
ID	554	550	275	295
IL	102	101	59	52
IN	167	169	66	70
IA	215	215	130	130
KS	125	116	65	70
KY	86	80	50	50
LA	25	21	7.0	7.0
ME	33	33	18.0	16.0
MD	56	54	26	28
MA	14.5	13.0	7.0	6.0
MI	353	354	148	158
MN	468	470	285	295
MS	19.0	17.0	7.0	7.0
MO	108	102	40	45
MT	16.0	15.0	10.0	7.0
NE	59	59	20	20
NV	28	28	10.0	10.0
NH	15.0	15.0	7.5	8.0
NJ	9.5	8.5	5.0	4.8
NM	336	318	140	145
NY	625	610	320	325
NC	46	43	20.0	19.0
ND	25	21	20.0	10.0
OH	276	272	115	120
OK	62	57	30	25
OR	115	114	55	55
PA	550	540	270	300
RI	1.1	1.1	0.5	0.5
SC	17.0	17.0	7.0	8.0
SD	94	92	40	35
TN	59	53	45	35
TX	430	410	220	250
UT	85	82	45	48
VT	139	134	61	56
VA	97	95	45	50
WA	244	243	110	123
WV	11.0	10.0	3.0	4.0
WI	1,255	1,260	650	670
WY	7.0	6.0	5.0	5.0
US	9,332.8	9,080.5	4,409.5	4,516.2

[1] Preliminary.
NASS, Livestock Branch, (202) 720–3570.

Table 8-4.—Milk-feed price ratios: All milk-price; dairy feed, 16%; Milk-feed price ratios and value per 100 pounds of grain and concentrate rations fed to milk cows, United States, annual 1999–2008

Year	All milk price cwt.	16% dairy feed price cwt [1]	Milk-feed price ratio [2]
	Dollars	Dollars	Pounds
1999	14.38	9.00	3.59
2000	12.40	8.75	3.05
2001	15.04	9.20	3.39
2002	12.18	9.50	2.60
2003	12.55	10.00	2.61
2004	16.13	10.90	3.10
2005	15.09	9.85	3.24
2006	12.96	10.50	2.57
2007	19.21	12.45	2.80
2008	18.45	15.65	2.01

[1] Commercially prepared 16%dairy ration: Annual average prior to 1995, April price 1995-current.   [2] Annual ratios based on average of monthly ratios. Pounds of 16 % mixed dairy feed equal in value to one pound of whole milk. Effective January 1995, prices of commercial prepared feeds are based on current U.S. prices received for corn (51 lbs), soybeans (8 lbs), and alfalfa hay (41 lbs).
NASS, Environmental, Economics, and Demographics Branch, (202) 720–6146.

Table 8-5.—Official Dairy Herd Information test plans: Numbers of herds and cows and milk, fat, and protein production, United States, 1999–2008

Year	Herds	Cows	Cows per herd	Average production			Cows with protein informa-tion	Average protein production [1]	Average protein production [1]
				Milk	Fat	Fat			
	Number	Number	Number	Pounds	Percent	Pounds	Percent	Percent	Pounds
1999 ...	24,841	3,449,854	140.9	20,743	3.68	766	93	3.24	673
2000 ...	23,225	3,521,686	151.6	21,092	3.68	781	93	3.15	664
2001 ...	22,095	3,499,214	158.4	21,118	3.66	777	94	3.08	651
2002 ...	20,955	3,537,064	168.8	21,475	3.68	792	94	3.07	661
2003 ...	19,732	3,416,386	173.1	21,471	3.68	792	94	3.07	661
2004 ...	18,897	3,468,419	183.5	21,457	3.68	791	94	3.09	664
2005 ...	18,349	3,537,857	192.8	22,027	3.67	812	95	3.08	680
2006 ...	17,606	3,602,719	204.6	22,282	3.69	825	95	3.09	688
2007 ...	17,174	3,749,257	218.3	22,371	3.68	826	95	3.09	693
2008 ...	16,602	3,804,216	229.1	22,437	3.69	830	96	3.10	696

[1] The decline in protein production in 2000 reflects a measurement change by the dairy industry from crude to true protein beginning in May 2000. The percentage of milk that is true protein is lower than the percentage that is crude protein by an approximate difference of 0.19 percent.
ARS, Animal Improvement Programs Laboratory, (301) 504–8334, http://aipl.arsusda.gov.

Table 8-6.—Milk and milkfat production: Number of producing cows, production per cow, and total quantity produced, United States, 1999–2008

Year	Number of milk cows [1]	Production of milk and milkfat [2]				
		Per milk cow		Percentage of fat in all milk produced	Total	
		Milk	Milkfat		Milk	Milkfat
	Thousands	Pounds	Pounds	Percent	Million pounds	Million pounds
1999	9,153	17,763	652	3.67	162,589	5,970
2000	9,199	18,197	670	3.68	167,393	6,164
2001	9,103	18,162	667	3.67	165,332	6,073
2002	9,139	18,608	685	3.68	170,063	6,264
2003	9,081	18,759	688	3.67	170,348	6,247
2004	9,010	18,960	696	3.67	170,832	6,266
2005	9,050	19,550	716	3.66	176,931	6,480
2006	9,137	19,895	734	3.69	181,782	6,700
2007	9,189	20,204	744	3.68	185,654	6,832
2008	9,315	20,395	751	3.68	189,982	6,998

[1] Average number during year, excluding heifers not yet fresh.　[2] Excludes milk sucked by calves.
NASS, Livestock Branch, (202) 720–3570.

Table 8-7.—Milk: Quantities used and marketed by farmers, United States, 1999–2008

Year	Milk used on farms where produced			Milk marketed by producers	
	Fed to calves [1]	Consumed as fluid milk or cream	Total	Total [2]	Fluid grade [3]
	Million pounds	Million pounds	Million pounds	Million pounds	Percent
1999	1,107	219	1,326	161,263	98
2000	1,109	198	1,307	166,086	98
2001	1,036	173	1,209	164,123	98
2002	959	160	1,119	168,944	98
2003	959	168	1,127	169,222	98
2004	958	157	1,115	169,716	98
2005	949	146	1,095	175,836	98
2006	943	138	1,081	180,700	99
2007	952	137	1,089	184,565	99
2008	942	124	1,066	188,917	99

[1] Excludes milk sucked by calves.　[2] Milk sold to plants and dealers as whole milk and equivalent amounts of milk for cream. Includes milk produced by dealers' own herds and small amounts sold directly to consumers. Also includes milk produced by institutional herds.　[3] Percentage of milk sold that is eligible for fluid use (Grade A in most States). Includes fluid-grade milk used in manufacturing dairy products.
NASS, Livestock Branch, (202) 720–3570.

Table 8-8.—Milk and milkfat production: Number of milk cows, production per cow, and total quantity produced, by State, 2007

State	Number of milk cows [1]	Production of milk and milkfat [2]						
		Per milk cow		Percent of fat			Total	
		Milk	Milkfat	Fluid grade	Manuf. grade	All milk	Milk	Milkfat
	Thousands	Pounds	Pounds	Percent	Percent	Percent	Million pounds	Million pounds
AL	13.0	15,154	564	3.72	....	3.72	197.0	7.3
AK	0.6	14,667	505	3.44	....	3.44	8.8	0.3
AZ	181.0	23,260	835	3.59	....	3.59	4,210.0	151.1
AR	17.0	12,941	474	3.66	....	3.66	220.0	8.1
CA	1,813.0	22,440	826	3.67	4.22	3.68	40,683.0	1,497.1
CO	118.0	22,932	809	3.53	....	3.53	2,706.0	95.5
CT	19.0	19,211	709	3.69	....	3.69	365.0	13.5
DE	6.8	16,618	617	3.71	....	3.71	113.0	4.2
FL	125.0	16,832	611	3.63	....	3.63	2,104.0	76.4
GA	77.0	18,169	665	3.66	....	3.66	1,399.0	51.2
HI	2.9	12,241	416	3.40	....	3.40	35.5	1.2
ID	513.0	22,513	817	3.63	....	3.63	11,549.0	419.2
IL	103.0	18,612	696	3.76	3.66	3.74	1,917.0	71.7
IN	166.0	20,307	745	3.67	3.75	3.67	3,371.0	123.7
IA	213.0	20,085	743	3.70	3.82	3.70	4,278.0	158.3
KS	110.0	19,882	730	3.67	....	3.67	2,187.0	80.3
KY	90.0	13,889	510	3.67	....	3.67	1,250.0	45.9
LA	29.0	12,034	427	3.55	....	3.55	349.0	12.4
ME	33.0	17,788	663	3.73	....	3.73	587.0	21.9
MD	58.0	18,121	670	3.70	....	3.70	1,051.0	38.9
MA	15.0	17,000	639	3.76	....	3.76	255.0	9.6
MI	335.0	22,761	822	3.61	....	3.61	7,625.0	275.3
MN	460.0	18,817	704	3.74	3.76	3.74	8,656.0	323.7
MS	21.0	15,429	552	3.58	....	3.58	324.0	11.6
MO	112.0	14,982	550	3.67	3.78	3.67	1,678.0	61.6
MT	18.0	18,500	673	3.64	....	3.64	333.0	12.1
NE	59.0	18,220	676	3.71	....	3.71	1,075.0	39.9
NV	27.0	20,481	748	3.65	....	3.65	553.0	20.2
NH	15.0	19,333	725	3.75	....	3.75	290.0	10.9
NJ	10.0	16,800	622	3.70	....	3.70	168.0	6.2
NM	332.0	21,958	793	3.61	....	3.61	7,290.0	263.2
NY	627.0	19,303	714	3.70	....	3.70	12,103.0	447.8
NC	48.0	19,188	712	3.71	....	3.71	921.0	34.2
ND	29.0	15,310	573	3.73	3.75	3.74	444.0	16.6
OH	275.0	18,109	670	3.70	3.82	3.70	4,980.0	184.3
OK	69.0	16,580	610	3.68	....	3.68	1,144.0	42.1
OR	115.0	19,417	718	3.70	....	3.70	2,233.0	82.6
PA	550.0	19,422	719	3.70	....	3.70	10,682.0	395.2
RI	1.1	16,455	637	3.87	....	3.87	18.1	0.7
SC	18.0	17,889	701	3.92	....	3.92	322.0	12.6
SD	85.0	19,306	720	3.73	3.77	3.73	1,641.0	61.2
TN	63.0	15,857	585	3.69	....	3.69	999.0	36.9
TX	389.0	18,982	708	3.73	....	3.73	7,384.0	275.4
UT	85.0	20,376	744	3.65	....	3.65	1,732.0	63.2
VT	140.0	18,079	676	3.74	....	3.74	2,531.0	94.7
VA	100.0	17,530	643	3.67	....	3.67	1,753.0	64.3
WA	238.0	23,239	860	3.70	....	3.70	5,531.0	204.6
WV	13.0	15,000	543	3.62	....	3.62	195.0	7.1
WI	1,247.0	19,310	714	3.70	3.81	3.70	24,080.0	891.0
WY	7.1	18,831	674	3.55	3.75	3.58	133.7	4.8
US [3]	9,189.0	20,204	744	3.68	3.99	3.68	185,654.0	6,831.8
PR	89.0	8,303	276	3.32	3.32	3.32	739.0	24.5

[1] Average number during year, excluding heifers not yet fresh. U.S. total may not add due to rounding.　[2] Excludes milk sucked by calves.　[3] Sum of parts may not equal due to rounding.
NASS, Livestock Branch, (202) 720–3570.

Table 8-9.—Milk and milkfat production: Number of milk cows, production per cow, and total quantity produced, by State, 2008

State	Number of milk cows[1]	Production of milk and milkfat[2]						
		Per milk cow		Percent of fat			Total	
		Milk	Milkfat	Fluid grade	Manuf. grade	All milk	Milk	Milkfat
	Thousands	Pounds	Pounds	Percent	Percent	Percent	Million pounds	Million pounds
AL	12.0	15,333	569	3.71	....	3.71	184.0	6.8
AK	0.6	12,000	426	3.55	....	3.55	7.2	0.3
AZ	186.0	23,382	832	3.56	....	3.56	4,349.0	154.8
AR	15.0	12,400	453	3.65	....	3.65	186.0	6.8
CA	1,844.0	22,344	822	3.67	4.34	3.68	41,203.0	1,516.3
CO	128.0	22,930	805	3.51	....	3.51	2,935.0	103.0
CT	19.0	19,158	726	3.79	....	3.79	364.0	13.8
DE	6.5	16,923	633	3.74	....	3.74	110.0	4.1
FL	120.0	17,167	618	3.60	....	3.60	2,060.0	74.2
GA	76.0	17,829	654	3.67	....	3.67	1,355.0	49.7
HI	1.7	10,882	369	3.39	....	3.39	18.5	0.6
ID	549.0	22,432	808	3.60	....	3.60	12,315.0	443.3
IL	102.0	18,569	706	3.80	3.79	3.80	1,894.0	72.0
IN	167.0	19,683	734	3.73	....	3.73	3,287.0	122.6
IA	216.0	19,995	738	3.69	3.84	3.69	4,319.0	159.4
KS	117.0	20,641	760	3.68	....	3.68	2,415.0	88.9
KY	90.0	13,444	495	3.68	....	3.68	1,210.0	44.5
LA	26.0	12,269	434	3.54	....	3.54	319.0	11.3
ME	33.0	18,273	671	3.67	....	3.67	603.0	22.1
MD	56.0	18,375	684	3.72	....	3.72	1,029.0	38.3
MA	15.0	16,933	649	3.83	....	3.83	254.0	9.7
MI	350.0	22,180	807	3.64	....	3.64	7,763.0	282.6
MN	464.0	18,927	708	3.74	3.79	3.74	8,782.0	328.4
MS	20.0	14,550	502	3.45	....	3.45	291.0	10.0
MO	110.0	14,682	539	3.66	3.82	3.67	1,615.0	59.3
MT	17.0	18,412	663	3.60	....	3.60	313.0	11.3
NE	58.0	18,672	691	3.70	....	3.70	1,083.0	40.1
NV	27.0	20,704	762	3.68	....	3.68	559.0	20.6
NH	15.0	19,933	753	3.78	....	3.78	299.0	11.3
NJ	10.0	16,900	630	3.73	....	3.73	169.0	6.3
NM	338.0	23,269	833	3.58	....	3.58	7,865.0	281.6
NY	626.0	19,859	739	3.72	....	3.72	12,432.0	462.5
NC	47.0	18,979	710	3.74	....	3.74	892.0	33.4
ND	26.0	16,077	606	3.77	3.78	3.77	418.0	15.8
OH	280.0	18,321	691	3.77	3.90	3.77	5,130.0	193.4
OK	64.0	16,578	597	3.60	....	3.60	1,061.0	38.2
OR	114.0	19,772	726	3.67	....	3.67	2,254.0	82.7
PA	549.0	19,262	717	3.72	....	3.72	10,575.0	393.4
RI	1.1	18,091	706	3.90	....	3.90	19.9	0.8
SC	18.0	17,889	689	3.85	....	3.85	322.0	12.4
SD	90.0	19,956	746	3.74	3.75	3.74	1,796.0	67.2
TN	59.0	16,068	595	3.70	....	3.70	948.0	35.1
TX	418.0	20,134	749	3.72	....	3.72	8,416.0	313.1
UT	85.0	20,894	761	3.64	....	3.64	1,776.0	64.6
VT	140.0	18,400	692	3.76	....	3.76	2,576.0	96.9
VA	98.0	17,612	646	3.67	....	3.67	1,726.0	63.3
WA	244.0	23,344	859	3.68	....	3.68	5,696.0	209.6
WV	12.0	15,083	548	3.63	....	3.63	181.0	6.6
WI	1,252.0	19,546	727	3.72	3.83	3.72	24,472.0	910.4
WY	7.0	19,386	677	3.44	3.86	3.49	135.7	4.7
US[3]	9,315.0	20,395	751	3.68	4.01	3.68	189,982.0	6,998.1

[1] Average number during year, excluding heifers not yet fresh. U.S. total may not add due to rounding.   [2] Excludes milk sucked by calves.   [3] Sum of parts may not equal due to rounding.
NASS, Livestock Branch, (202) 720–3570.

Table 8-10.—Milk: Quantities used and marketed by producers, by State, 2008

State	Milk used where produced			Milk marketed by producers	
	Fed to calves [1]	Used for milk, cream, and butter	Total	Total quantity [2]	Fluid grade [3]
	Million pounds	*Million pounds*	*Million pounds*	*Million pounds*	*Percent*
AL	0.5	0.5	1.0	183.0	100
AK	1.1	0.2	1.3	5.9	100
AZ	12.0	1.0	13.0	4,336.0	100
AR	1.7	0.3	2.0	184.0	100
CA	32.0	5.0	37.0	41,166.0	98
CO	30.0	3.0	33.0	2,902.0	100
CT	2.5	0.5	3.0	361.0	100
DE	0.9	0.1	1.0	109.0	100
FL	5.0	1.0	6.0	2,054.0	100
GA	11.0	1.0	12.0	1,343.0	100
HI	0.4	0.2	0.6	17.9	100
ID	30.0	1.0	31.0	12,284.0	100
IL	9.0	2.0	11.0	1,883.0	98
IN	21.0	4.0	25.0	3,262.0	99
IA	20.0	2.0	22.0	4,297.0	99
KS	10.0	1.0	11.0	2,404.0	100
KY	20.0	2.0	22.0	1,188.0	100
LA	6.0	2.0	8.0	311.0	100
ME	4.0	1.0	5.0	598.0	100
MD	6.0	1.0	7.0	1,022.0	100
MA	1.5	0.5	2.0	252.0	100
MI	23.0	2.0	25.0	7,738.0	100
MN	95.0	5.0	100.0	8,682.0	98
MS	1.0	1.0	2.0	289.0	100
MO	18.0	5.0	23.0	1,592.0	96
MT	4.0	4.0	8.0	305.0	100
NE	8.0	1.0	9.0	1,074.0	99
NV	5.0	1.0	6.0	553.0	100
NH	2.5	0.5	3.0	296.0	100
NJ	1.5	0.5	2.0	167.0	100
NM	62.0	13.0	75.0	7,790.0	100
NY	30.0	2.0	32.0	12,400.0	100
NC	7.0	3.0	10.0	882.0	100
ND	10.0	1.0	11.0	407.0	80
OH	25.0	5.0	30.0	5,100.0	96
OK	10.0	1.0	11.0	1,050.0	100
OR	19.0	1.0	20.0	2,234.0	100
PA	48.0	16.0	64.0	10,511.0	100
RI	0.1		0.1	19.8	100
SC	2.0	1.0	3.0	319.0	100
SD	7.0	1.0	8.0	1,788.0	97
TN	3.0	1.0	4.0	944.0	100
TX	25.0	2.0	27.0	8,389.0	100
UT	10.0	1.0	11.0	1,765.0	100
VT	15.5	2.5	18.0	2,558.0	100
VA	6.0	2.0	8.0	1,718.0	100
WA	13.0	1.0	14.0	5,682.0	100
WV	1.0	1.0	2.0	179.0	100
WI	264.0	20.0	284.0	24,188.0	97
WY	1.3	0.2	1.5	134.2	87
US [4]	942.0	124.0	1,066.0	188,917.0	99

[1] Excludes milk sucked by calves.   [2] Milk sold to plants and dealers as whole milk and equivalent amounts of milk for cream. Includes milk produced by dealers' own herds and small amounts sold directly to consumers. Also includes milk produced by institutional herds.   [3] Percentage of milk sold that is eligible for fluid use (grade A for fluid use in most States). Includes fluid-grade milk used in manufacturing dairy products.   [4] May not add due to rounding.

NASS, Livestock Branch, (202) 720–3570.

Table 8-11.—Milk production: Marketings, income, and value, by State, 2008

State	Milk utilized	Average returns per cwt. [1]			Re-turns per lb milkfat	Cash receipts from mar-ketings	Used for milk, cream, and butter where produced		Gross producer income [3]	Value of milk pro-duced [2][4]
		Fluid grade	Manuf. grade	All milk			Milk utilized	Value [2]		
	Million pounds	Dollars	Dollars	Dol-lars	Dol-lars	1,000 dollars	Million pounds	1,000 dollars	1,000 dollars	1,000 dollars
AL	183.0	21.70		21.70	5.85	39,711	0.5	109	39,820	39,928
AK	5.9	23.60		23.60	6.65	1,392	0.2	47	1,439	1,699
AZ	4,336.0	17.60		17.60	4.94	763,136	1.0	176	763,312	765,424
AR	184.0	19.90		19.90	5.45	36,616	0.3	60	36,676	37,014
CA	41,166.0	16.78	18.56	16.82	4.57	6,924,121	5.0	841	6,924,962	6,930,345
CO	2,902.0	18.40		18.40	5.24	533,968	3.0	552	534,520	540,040
CT	361.0	20.20		20.20	5.33	72,922	0.5	101	73,023	73,528
DE	109.0	18.70		18.70	5.00	20,383	0.1	19	20,402	20,570
FL	2,054.0	22.60		22.60	6.28	464,204	1.0	226	464,430	465,560
GA	1,343.0	20.90		20.90	5.69	280,687	1.0	209	280,896	283,195
HI	17.9	30.50		30.50	9.00	5,460	0.2	61	5,521	5,643
ID	12,284.0	17.10		17.10	4.75	2,100,564	1.0	171	2,100,735	2,105,865
IL	1,883.0	19.60	19.20	19.60	5.16	369,068	2.0	392	369,460	371,224
IN	3,262.0	19.60		19.60	5.25	639,352	4.0	784	640,136	644,252
IA	4,297.0	18.50	17.30	18.50	5.01	794,945	2.0	370	795,315	799,015
KS	2,404.0	18.90		18.90	5.14	454,356	1.0	189	454,545	456,435
KY	1,188.0	20.00		20.00	5.43	237,600	2.0	400	238,000	242,000
LA	311.0	20.70		20.70	5.85	64,377	2.0	414	64,791	66,033
ME	598.0	20.70		20.70	5.64	123,786	1.0	207	123,993	124,821
MD	1,022.0	19.00		19.00	5.11	194,180	1.0	190	194,370	195,510
MA	252.0	20.20		20.20	5.27	50,904	0.5	101	51,005	51,308
MI	7,738.0	19.20		19.20	5.27	1,485,696	2.0	384	1,486,080	1,490,496
MN	8,682.0	19.10	17.40	19.10	5.11	1,658,262	5.0	955	1,659,217	1,677,362
MS	289.0	20.80		20.80	6.03	60,112	1.0	208	60,320	60,528
MO	1,592.0	19.00	16.60	18.90	5.15	300,888	5.0	945	301,833	305,235
MT	305.0	19.00		19.00	5.28	57,950	4.0	760	58,710	59,470
NE	1,074.0	18.90		18.90	5.11	202,986	1.0	189	203,175	204,687
NV	553.0	16.90		16.90	4.59	93,457	1.0	169	93,626	94,471
NH	296.0	19.90		19.90	5.26	58,904	0.5	100	59,004	59,501
NJ	167.0	18.50		18.50	4.96	30,895	0.5	93	30,988	31,265
NM	7,790.0	17.50		17.50	4.89	1,363,250	13.0	2,275	1,365,525	1,376,375
NY	12,400.0	19.20		19.20	5.16	2,380,800	2.0	384	2,381,184	2,386,944
NC	882.0	21.50		21.50	5.75	189,630	3.0	645	190,275	191,780
ND	407.0	18.80	17.00	18.50	4.91	75,295	1.0	185	75,480	77,330
OH	5,100.0	19.80	17.10	19.70	5.23	1,004,700	5.0	985	1,005,685	1,010,610
OK	1,050.0	20.20		20.20	5.61	212,100	1.0	202	212,302	214,322
OR	2,234.0	18.30		18.30	4.99	408,822	1.0	183	409,005	412,482
PA	10,511.0	20.00		20.00	5.38	2,102,200	16.0	3,200	2,105,400	2,115,000
RI	19.8	20.10		20.10	5.15	3,980			3,980	4,000
SC	319.0	21.50		21.50	5.58	68,585	1.0	215	68,800	69,230
SD	1,788.0	19.10	17.10	19.10	5.11	341,508	1.0	191	341,699	343,036
TN	944.0	20.20		20.20	5.46	190,688	1.0	202	190,890	191,496
TX	8,389.0	18.70		18.70	5.03	1,568,743	2.0	374	1,569,117	1,573,792
UT	1,765.0	18.10		18.10	4.97	319,465	1.0	181	319,646	321,456
VT	2,558.0	19.50		19.50	5.19	498,810	2.5	488	499,298	502,320
VA	1,718.0	21.60		21.60	5.89	371,088	2.0	432	371,520	372,816
WA	5,682.0	17.60		17.60	4.78	1,000,032	1.0	176	1,000,208	1,002,496
WV	179.0	18.80		18.80	5.18	33,652	1.0	188	33,840	34,028
WI	24,188.0	18.90	17.90	18.90	5.08	4,571,532	20.0	3,780	4,575,312	4,625,208
WY	134.2	17.40	17.50	17.40	4.99	23,351	0.2	35	23,386	23,612
US [5]	188,917.0	18.45	17.91	18.45	5.01	34,849,113	124.0	27,743	34,872,856	35,050,757

[1] Cash receipts divided by milk or milkfat in combined marketings.　[2] Value at averaged returns per 100 pounds of milk in combined marketings of milk and cream.　[3] Cash receipts from marketings of milk and cream plus value of milk used for home consumption.　[4] Includes value of milk fed to calves.　[5] May not add due to rounding.
NASS, Livestock Branch, (202) 720–3570.

Table 8-12.—Federal milk order markets: Measures of growth, 1999–2008 [1]

Year	Number of markets [2]	Population of Federal milk marketing areas	Number of handlers [2]	Number of producers [3]	Receipts of producer milk	Producer milk used in Class I	Percentage of producer milk used in Class I
	Number	Thousands	Number	Number	Million pounds	Million pounds	Percent
1999	31	212,118	487	69,008	104,479	45,216	43.3
2000	11	228,899	346	69,590	116,920	45,989	39.3
2001	11	231,487	350	66,423	120,223	45,887	38.2
2002	11	234,256	338	63,856	125,546	46,043	36.7
2003	11	236,180	331	58,110	110,581	45,843	41.5
2004	10	234,825	306	52,341	103,048	44,939	43.6
2005	10	238,428	302	53,036	114,682	44,570	38.9
2006	10	239,142	314	52,725	120,618	45,304	37.6
2007	10	241,000	312	49,782	114,407	45,226	39.5
2008	10	242,988	333	47,859	115,867	44,989	38.8

Year	Prices at 3.5 percent butterfat content per hundredweight [4]		Receipts as percentage of milk sold to plants and dealers		Daily deliveries of milk per producer	Gross value of receipts of producer milk [5]	
	Class I	Blend	Fluid grade	All milk		Per producer	All producer
	Dollars	Dollars	Percent	Percent	Pounds	Dollars	1,000 dollars
1999	16.24	14.09	67	65	4,148	216,794	14,960,544
2000	14.24	12.11	72	70	4,590	207,913	14,468,892
2001	16.96	14.90	75	73	4,959	275,642	18,308,968
2002	13.69	11.91	77	76	5,387	239,520	15,294,802
2003	14.10	12.12	67	65	5,178	242,066	14,066,672
2004	17.56	15.74	62	61	5,352	324,119	16,965,368
2005	17.13	15.07	66	65	5,904	334,626	17,747,577
2006	14.59	12.86	68	67	6,264	303,429	15,998,288
2007	20.81	19.19	63	62	6,297	452,097	22,507,219
2008	20.78	18.24	62	61	6,613	455,149	21,782,959

[1] Over this period, handlers elected periodically not to pool substantial volumes of milk that normally would have been pooled under Federal orders. This decision resulted from disadvantageous blend/class price relationships and qualification circumstances. This fact should be kept in mind if year-to-year comparisons are made using the various "producer deliveries" measures of growth. [2] End of year. [3] Average for year. [4] Prices are weighted averages. [5] Based on blend (uniform) price adjusted for butterfat content, and in later years, other milk components of producer milk.
AMS, Dairy Programs, (202) 720-7461.

Table 8-13.—Milk production: Marketings, income and value, United States, 1999–2008

Year	Combined marketings of milk and cream				Used for milk, cream, and butter on farms where produced		Gross farm income from dairy products [4]	Farm value of all milk produced [3][5]
	Milk utilized	Average returns [2]		Cash receipts from marketings	Milk utilized	Value [3]		
		Per 100 pounds milk	Per pound milkfat					
	Million pounds	Dollars	Dollars	1,000 dollars	Million pounds	1,000 dollars	1,000 dollars	1,000 dollars
1999	161,263	14.38	3.92	23,189,113	219	32,021	23,221,134	23,381,760
2000	166,086	12.40	3.37	20,586,629	198	24,777	20,611,406	20,749,871
2001	164,123	15.04	4.10	24,685,667	173	26,269	24,711,936	24,869,285
2002	168,944	12.18	3.31	20,582,238	160	19,816	20,602,054	20,720,482
2003	169,222	12.55	3.42	21,231,059	168	21,676	21,252,735	21,375,314
2004	169,716	16.13	4.40	27,366,835	157	25,915	27,392,750	27,550,637
2005	175,836	15.19	4.15	26,704,863	146	22,787	26,727,650	26,874,301
2006	180,700	12.96	3.51	23,412,552	138	18,591	23,431,143	23,556,102
2007	184,565	19.21	5.22	35,453,399	137	27,073	35,480,472	35,665,894
2008 [1]	188,917	18.45	5.01	34,849,113	124	23,743	34,872,856	35,050,757

[1] Preliminary. [2] Cash receipts divided by milk or milkfat represented in combined marketings. [3] Valued at average returns per 100 pounds of milk in combined marketings of milk and cream. [4] Cash receipts from marketings of milk and cream plus value of milk used for home consumption. [5] Includes value of milk fed to calves.
NASS, Livestock Branch, (202) 720–3570.

Table 8-14.—Dairy products: Quantities manufactured, United States, 2004–2008

Product	2004	2005	2006	2007	2008 [1]
	1,000 pounds	1,000 pounds	1,000 pounds	1,000 pounds	1,000 pounds
Butter	1,246,678	1,347,361	1,448,482	1,532,717	1,644,076
All American cheese	3,738,826	3,808,102	3,912,606	3,877,214	4,108,565
Cheddar cheese	3,004,477	3,045,971	3,124,754	3,056,668	3,186,454
Swiss cheese	281,288	300,131	314,389	313,689	293,968
Muenster cheese	72,812	77,882	95,563	103,605	117,241
Brick cheese	8,109	8,905	8,623	7,434	6,887
Limburger cheese	872	784	833	744	
Cream and Neufchatel cheese	699,119	714,791	752,029	772,770	763,595
Hispanic cheese	142,432	167,344	181,514	190,580	194,268
Mozzarella	2,916,558	3,019,052	3,144,562	3,329,540	3,222,765
All Italian varieties of cheese	3,661,590	3,803,040	3,988,502	4,198,800	4,120,831
All other varieties of cheese	268,102	268,343	280,272	311,949	307,473
Total of all cheese	8,873,150	9,149,322	9,534,331	9,776,785	9,912,828
Cottage cheese:					
Curd [2]	463,960	468,642	457,686	458,459	428,092
Creamed [2]	382,386	376,716	367,478	348,583	324,980
Lowfat [2]	396,431	407,904	407,845	425,447	389,195
Sweetened condensed milk:					
Bulk goods:					
Skimmed	30,669	23,881	23,347	29,102	29,106
Unskimmed	76,132	66,597	68,038	78,928	83,100
Unsweetened condensed milk:					
Bulk goods:					
Skimmed	903,794	1,058,114	1,218,313	1,638,894	1,509,246
Unskimmed	116,856	124,263	114,737	106,034	133,149
Evaporated and condensed milk:					
Case goods:					
Skimmed	19,089	20,370	21,364	18,542	18,313
Unskimmed	529,909	527,264	495,535	497,104	534,378
Condensed or evaporated buttermilk	49,646	77,961	93,692	55,754	64,115
Dry buttermilk	54,979	66,482	66,904	81,386	72,494
Dry whole milk	41,587	32,435	30,524	31,746	50,137
Nonfat dry milk	1,412,381	1,210,313	1,224,072	1,298,480	1,519,173
Skim milk powders [3]		322,733	289,074	200,649	373,830
Dry skim milk (animal feed)	5,243	5,572	5,935	4,863	8,283
Dry whey	1,034,898	1,040,692	1,100,346	1,133,861	1,081,910
Yogurt plain & fruit flavored	2,707,313	3,058,328	3,294,587	3,476,255	3,570,355
	1,000 gallons	1,000 gallons	1,000 gallons	1,000 gallons	1,000 gallons
Ice cream, regular [4]	919,919	959,941	965,781	956,121	930,708
Ice cream, lowfat [5]	387,106	360,237	372,200	381,946	383,828
Ice cream, nonfat	22,667	20,807	14,909	14,378	15,437
Sherbet (does not include water ices)	54,913	56,390	52,077	62,674	57,718
Frozen yogurt	64,544	66,132	67,450	74,722	78,580

[1] Preliminary.　[2] Cottage cheese curd includes pot and bakers' cheese. Creamed cottage cheese contains not less than 4 percent milkfat. Lowfat cottage cheese contains less than 4 percent milkfat.　[3] Includes protein standardized and blends.　[4] Contains minimum milkfat content of 10 percent and not less than 4.5 pounds per gallon.　[5] Includes freezer-made milkshake in most States. Contains less than 10 percent milkfat required for ice cream.
NASS, Livestock Branch, (202) 720–3570.

Table 8-15.—Dairy products: Average price per pound for specified products, 2004–2008

Item and market	2004	2005	2006	2007	2008
	Dollars	Dollars	Dollars	Dollars	Dollars
Butter, Chicago Mercantile Exchange:					
Grade AA:					
High [1]	2.3650	1.7400	1.4100	1.5625	1.7650
Low [1]	1.3900	1.3300	1.1450	1.2000	1.1100
Butter, National Agricultural Statistics Service, Grade AA: [2]	1.8239	1.5405	1.2193	1.3441	1.4356
Cheese, Cheddar, Chicago Mercantile Exchange, Barrels:					
High [1]	2.1700	1.7250	1.4400	2.1600	2.2500
Low	1.2350	1.3050	1.1050	1.2725	1.1300
Cheese, Cheddar, Chicago Mercantile Exchange, 40-lb blocks:					
High [1]	2.2000	1.7575	1.4275	2.2025	2.2850
Low [1]	1.3000	1.3575	1.1225	1.2875	1.1325
Cheese, Cheddar, National Agricultural Statistics Service, Barrels: [2]	1.6216	1.4621	1.2305	1.7267	1.8836
Cheese, Cheddar, National Agricultural Statistics Service, 40-lb blocks: [2]	1.6325	1.4821	1.2318	1.7172	1.8801
Nonfat dry milk, National Agricultural Statistics Service:					
Low/medium heat [2]	0.8405	0.9409	0.8874	1.6927	1.2256
Dry whey, National Agricultural Statistics Service:					
Edible nonhygroscopic [2]	0.2319	0.2782	0.3285	0.6004	0.2504

[1] Figures are the high and low prices for any trading day during the year.　[2] Prices used in Federal milk order price formulas. Averages were computed by Agricultural Marketing Service.
AMS, Dairy Programs, (202) 720-7461.

DAIRY AND POULTRY STATISTICS

Table 8-16.—Dairy Products: Factory production of specified items, by States, 2007 and 2008

State	Butter		Total American cheese [1]		Total cheese [2]	
	2007	2008 [3]	2007	2008 [3]	2007	2008 [3]
	1,000 pounds	*1,000 pounds*	*1,000 pounds*	*1,000 pounds*	*1,000 pounds*	*1,000 pounds*
CA	498,949	555,502	798,257	667,486	2,287,248	2,113,236
ID				674,391	789,085	829,406
IL					76,764	70,606
IA				136,741	147,241	167,450
MN			561,762		610,339	641,748
NJ						30,455
NM					602,324	602,890
NY	15,404				738,046	715,923
OH			26,587	14,318	194,466	182,147
OR			109,881	160,951		
PA	70,462	76,693			393,877	402,058
SD					209,249	230,189
UT					104,114	108,485
VT					155,562	149,267
WI	373,027	361,041	771,883	822,509	2,462,043	2,524,125
Other	574,875	650,840	1,608,844	1,632,169	1,006,427	1,144,843
US	1,532,717	1,644,076	3,877,214	4,108,565	9,776,785	9,912,828

State	Ice cream, regular, hard		Nonfat dry milk for human food	
	2007	2008 [3]	2007	2008 [3]
	gallons	*gallons*	*Pounds*	*Pounds*
CA	129,746	132,786	724,831	819,182
FL	17,495	12,443		
IN	78,280	77,339		
MN				
MO	28,014	23,445		
NY	27,894	31,390		
NC	14,719	15,872		
OH	24,026	23,612		
OR	15,430	14,506		
PA	33,421	34,134		
TN	15,513	15,961		
TX	61,502	59,126		
UT	26,702	26,831		
Other	418,227	408,934	573,649	699,991
US	890,969	876,379	1,298,480	1,519,173

[1] Includes Colby, washed curd, high and low moisture Jack, and Monterey. [2] Includes full-skim American cheese; excludes cottage cheese. [3] Preliminary.
NASS, Livestock Branch, (202) 720–3570.

Table 8-17.—Fluid milk and cream: Total and per capita consumption, United States, 1999–2008 [1]

Year	Consumption	
	Total	Per capita
	Billion pounds	*Pounds*
1999	59.5	213
2000	59.1	210
2001	59.0	208
2002	59.4	207
2003	60.2	208
2004	60.5	207
2005	60.7	206
2006	61.8	208
2007	62.0	206
2008	62.0	204

[1] Sales of beverage, cream, and specialty fluid products plus farm household use.
ERS, Animal Products and Cost of Production Branch, (202) 694–5265.

Table 8-18.—Milk cows, milk, and fat in cream: Average prices received by farmers, United States, 1999–2008

Year	Milk cows, per head [1]	Milk per 100 pounds [2]					
		Eligible for fluid market [3]		Of manufacturing grade		All milk wholesale	
		Price per 100 lb.	Fat test	Price per 100 lb.	Fat test	Price per 100 lb.	Fat test
	Dollars	*Dollars*	*Percent*	*Dollars*	*Percent*	*Dollars*	*Percent*
1999	1,280.00	14.42	3.67	12.84	3.79	14.38	3.67
2000	1,340.00	12.44	3.68	10.52	3.79	12.40	3.68
2001	1,500.00	15.08	3.67	13.44	3.78	15.04	3.67
2002	1,600.00	12.20	3.68	10.89	3.80	12.18	3.68
2003	1,340.00	12.55	3.66	11.72	3.80	12.55	3.67
2004	1,580.00	16.13	3.67	15.45	3.82	16.13	3.67
2005	1,770.00	15.19	3.66	14.42	3.84	15.19	3.66
2006	1,730.00	12.96	3.68	12.19	3.93	12.96	3.69
2007	1,830.00	19.22	3.68	18.31	3.99	19.21	3.68
2008	1,950.00	18.45	3.68	17.91	4.01	18.45	3.68

[1] Simple average of quarterly prices, by States, weighted by the number of milk cows on farms Jan. 1 of the current year. [2] Average price at average fat test for all milk sold at wholesale to plants and dealers, based on reports from milk-market administrators, cooperative milk-market associations, whole-milk distributors, and milk-products manufacturing plants, f.o.b. plant or receiving station (whichever is the customary place for determining prices) before hauling costs are deducted and including all premiums. [3] Includes fluid milk surplus diverted to manufacturing.
NASS, Livestock Branch, (202) 720–3570.

Table 8-19.—Dairy products: Manufacturers' average selling price [1] of specified products, United States, 1999–2008

Year	Dry skim milk for animal feed, per pound, f.o.b. factory	Dry whole milk, per pound, f.o.b. factory
	Cents	*Cents*
1999	51.70	125.10
2000	49.40	120.00
2001	73.20	134.70
2002	62.40	116.50
2003	46.50	108.40
2004	43.30	131.30
2005	45.20	132.30
2006	48.20	123.10
2007	70.50	183.30
2008	53.70	160.60

[1] Includes milk sold in bulk and in package.
NASS, Livestock Branch, (202) 720–3570.

Table 8-20.—Dairy products: Manufacturers' stocks, end of month, United States, 2007 and 2008

Month	Evaporated and sweetened condensed whole milk (case goods)		Dry whole milk		Nonfat dry milk (human food)	
	2007	2008	2007	2008	2007	2008
	1,000 pounds	*1,000 pounds*	*1,000 pounds*	*1,000 pounds*	*1,000 pounds*	*1,000 pounds*
January	39,843	39,100	2,263	2,933	108,023	112,970
February	36,169	31,191	2,522	2,498	96,818	121,107
March	40,974	34,730	2,378	5,290	112,867	136,993
April	43,869	41,592	2,388	4,260	123,022	134,861
May	45,715	50,585	3,386	4,947	115,381	131,607
June	44,708	49,744	3,120	4,954	118,173	135,979
July	41,422	48,005	3,682	5,050	122,193	134,875
August	44,876	45,482	1,850	3,675	102,993	115,205
September	41,568	48,704	2,223	4,535	89,844	85,541
October	39,519	54,491	1,637	4,941	100,242	121,071
November	39,975	45,081	2,258	3,739	99,994	133,702
December	38,466	45,673	4,039	3,315	108,930	155,262

NASS, Livestock Branch, (202) 720–3570.

Table 8-21.—Milk markets under Federal order program: Whole milk and fat-reduced milk products sold for fluid consumption within defined marketing areas, 2007 [1]

Federal milk order marketing area	Whole milk products [2]		Fat-reduced milk products [3]		Total fluid milk products	
	Quantity	Butterfat content	Quantity	Butterfat content	Quantity	Butterfat content
	Million pounds	*Percent*	*Million pounds*	*Percent*	*Million pounds*	*Percent*
Northeast	3,398	3.27	6,086	1.12	9,484	1.89
Appalachian	1,200	3.29	2,430	1.32	3,631	1.97
Southeast	1,856	3.31	3,267	1.35	5,123	2.06
Florida	1,128	3.33	1,771	1.24	2,899	2.05
Mideast	1,412	3.30	4,727	1.35	6,140	1.80
Upper Midwest	692	3.32	3,765	1.13	4,457	1.47
Central	1,078	3.31	3,614	1.27	4,692	1.74
Southwest	1,785	3.32	2,556	1.39	4,341	2.18
Arizona	355	3.33	853	1.34	1,208	1.93
Pacific Northwest	437	3.43	1,696	1.30	2,133	1.74
Combined areas	13,343	3.31	30,766	1.26	44,109	1.88

[1] In-area sales include total sales in each of the areas by handlers regulated under the respective order, by handlers regulated under other orders, by partially regulated handlers, by exempt handlers, and by producer-handlers. Sales routes of handlers may extend outside defined marketing areas; therefore, some handlers' in-area sales are partially estimated.　[2] Plain, organic, flavored, and miscellaneous whole milk products, and eggnog.　[3] Plain, fortified, organic, and flavored reduced fat milk (2%), low fat milk (1%), and fat-free milk (skim), and miscellaneous fat-reduced milk products, and buttermilk.

AMS, Dairy Programs, (202) 720-7461.

Table 8-22.—Milk markets under Federal order program: Whole milk and fat-reduced milk products sold for fluid consumption within defined marketing areas, 2008 [1]

Federal milk order marketing area	Whole milk products [2]		Fat-reduced milk products [3]		Total fluid milk products	
	Quantity	Butterfat content	Quantity	Butterfat content	Quantity	Butterfat content
	Million pounds	*Percent*	*Million pounds*	*Percent*	*Million pounds*	*Percent*
Northeast	3,248	3.29	6,238	1.12	9,486	1.86
Appalachian	1,138	3.31	2,453	1.31	3,590	1.94
Southeast	1,788	3.32	3,357	1.36	5,145	2.04
Florida	1,072	3.30	1,766	1.24	2,839	2.02
Mideast	1,363	3.29	4,723	1.34	6,086	1.78
Upper Midwest	657	3.32	3,775	1.13	4,432	1.46
Central	1,042	3.31	3,654	1.26	4,696	1.72
Southwest	1,733	3.31	2,636	1.39	4,369	2.15
Arizona	321	3.32	870	1.35	1,191	1.88
Pacific Northwest	437	3.38	1,755	1.31	2,191	1.72
Combined areas	12,799	3.30	31,226	1.26	44,025	1.85

[1] In-area sales include total sales in each of the areas by handlers regulated under the respective order, by handlers regulated under other orders, by partially regulated handlers, by exempt handlers, and by producer-handlers. Sales routes of handlers may extend outside defined marketing areas; therefore, some handlers' in-area sales are partially estimated.　[2] Plain, organic, flavored, and miscellaneous whole milk products, and eggnog.　[3] Plain, fortified, organic, and flavored reduced fat milk (2%), low fat milk (1%), and fat-free milk (skim), and miscellaneous fat-reduced milk products, and buttermilk.

AMS, Dairy Programs, (202) 720-7461.

Table 8-23.—Supply and utilization, United States, 2006–2007

Product	Product pounds		Butterfat		Solids nonfat	
	2006	2007	2006	2007	2006	2007
	Million pounds					
Supply:						
Milk production	181,782	185,654	6,700	6,832	15,949	16,284
Net imports of ingredients	176	191	6	7	15	17
Net change in storage cream	....	....	....	....	....	....
Total supply	181,958	185,845	6,706	6,839	15,964	16,301
Utilization:						
Total butter [1]	1,448	1,533	1,175	1,243	14	15
Cheese:						
American	3,913	3,877	1,287	1,275	1,168	1,158
Other	5,612	5,900	1,404	1,473	1,475	1,546
Net cheese [2]	....	....	2,686	2,745	2,218	2,241
Total selected whey products [3]	2,565	2,608	32	33	2,311	2,342
Canned milk:						
Evaporated and condensed						
Whole and skim	507	516	42	43	110	110
Bulk milk:						
Condensed whole sweetened	68	79	6	7	14	17
Condensed whole unsweetened	117	106	9	8	21	19
Other condensed skim and condensed or evaporated buttermilk	1,336	1,724	4	4	394	511
Total evaporated and condensed	2,028	2,424	61	62	539	657
Dry whole milk	31	32	8	9	22	22
Nonfat dry milk	1,244	1,298	10	10	1,176	1,248
Skim Milk Powder	....	201	....	2	....	193
Dry buttermilk	67	81	4	5	61	74
Total selectd dry products	....	....	22	26	1,259	1,537
Total yogurt [4]	3,301	3,476	81	85	367	388
Total sour cream	1,121	1,135	235	238	43	43
Cottage cheese:						
Creamed	369	349	17	16	61	58
Low-fat	409	425	6	6	72	75
Total cottage cheese	....	....	23	22	133	133
Ice cream and other frozen dairy products.						
Ice cream:						
Regular, total	4,346	4,303	522	516	435	430
Lowfat, total	1,675	1,719	100	103	184	189
Nonfat, total	67	65	1	1	9	9
Sherbet, total	312	376	6	8	6	8
Frozen yogurt	405	448	7	8	36	40
Other frozen dairy products	52	55	3	3	4	4
Net frozen products [2]	....	....	551	559	469	459
Fluid milk [5]	55,064	55,087	1,046	1,043	4,900	4,908
Cream products [6]	2,354	2,531	474	510	164	176
Net fluid products [2]	....	....	1,520	1,553	5,061	5,079
Other unpublished dairy products [7]	1,212	1,125	44	65	837	776
Other food products [8]	1,553	1,659	57	61	134	144
Used where produced.						
Fed to calves	993	952	37	35	87	83
Consumed on farms	150	137	6	5	13	12
Total used by producers	1,143	1,089	43	40	100	95
Residual [9]	....	....	173	107	2,479	2,392
Residual as a percent of supply	....	....	2.6	1.6	15.5	14.7

[1] Including whey cream butter.　[2] Adjustment made for duplication the use of dairy products in the manufacturing process of other dairy products.　[3] Excluding whey cream butter.　[4] Excludes frozen yogurt.　[5] Total sales in U.S. (Source: USDA-AMS).　[6] Includes half and half and light/heavy cream.　[7] Includes anhydrous milkfat, butter oil, butterine, and other products.　[8] Food products other than dairy (Source: USDA-ERS).　[9] Residual, includes minor miscellaneous uses and any inaccuracies in production, utilization estimates, or milk equivalent conversions. Includes plant and shipping losses.
NASS, Livestock Branch, (202) 720–3570.

Table 8-24.—Milk markets under Federal order program: Uniform and Class I milk prices at 3.5 percent fat test, number of producers, producer milk receipts, producer milk used in Class I, Class I percentage, daily milk deliveries per producer, average fat test of producer milk receipts, by markets, 2007

Federal milk order marketing area	Class I price per cwt. [1]	Uniform price per cwt. [1][2]	Average number of producers	Receipts of producer milk	Producer milk used in Cl. I	Class I utilization	Daily milk delivery per producer	Average fat test
	Dollars	Dollars	Number	Million pounds	Million pounds	Percent	Pounds	Percent
Northeast [3][4]	21.39	19.92	13,877	23,040	10,496	45.6	4,548	3.70
Appalachian [4][5]	21.19	20.36	2,793	5,865	4,120	70.2	5,763	3.69
Southeast [4][6]	21.20	20.09	3,104	7,521	4,772	63.5	6,652	3.67
Florida [7]	22.01	21.29	385	3,206	2,604	81.2	23,888	3.63
Mideast [4][8]	20.12	18.75	7,848	16,268	6,571	40.4	5,685	3.67
Upper Midwest [4][9]	19.94	18.41	16,108	26,490	4,508	17.0	4,495	3.71
Central [4][10]	20.12	18.67	4,088	11,193	4,345	38.8	7,499	3.68
Southwest [4][11]	21.09	19.35	778	9,990	4,161	41.6	34,883	3.67
Arizona [12]	20.47	18.95	93	3,799	1,392	36.7	112,500	3.60
Pacific Northwest [4][13]	20.04	18.62	710	7,036	2,256	32.1	27,175	3.70
All markets combined	20.81	19.19	49,782	114,407	45,226	39.5	6,297	3.68

[1] Prices are for milk of 3.5 percent butterfat content and for the principal pricing point of the market. See footnotes 3-14.　[2] For those orders that use the component pricing system for paying producers (orders 1, 30, 32, 33, 124,and 126), the figures are the statistical uniform price (the sum of the producer price differential and the Class III price). For those orders that use the skim milk/butterfat pricing system for paying producers (orders 5, 6, 7, and 131), the figures are the uniform price (the sum of the uniform butterfat price times 3.5 and the uniform skim milk price times 0.965).　[3] Suffolk Co. (Boston), MA.　[4][5] Mecklenburg Co. (Charlotte), NC.　[6] Fulton Co. (Atlanta), GA.　[7] Hillsborough Co. (Tampa), FL.　[8] Cuyahoga Co. (Cleveland), OH.　[9] Cook Co. (Chicago), IL.　[10] Jackson Co. (Kansas City), MO.　[11] Dallas Co. (Dallas), TX.　[12] Maricopa Co. (Phoenix), AZ.　[13] King Co. (Seattle), WA.
AMS, Dairy Programs, (202) 720-7461.

Table 8-25.—Milk markets under Federal order program: Uniform and Class I milk prices at 3.5 percent fat test, number of producers, producer milk receipts, producer milk used in Class I, Class I percentage, daily milk deliveries per producer, average fat test of producer milk receipts, by markets, 2008

Federal milk order marketing area	Class I price per cwt. [1]	Uniform price per cwt. [1][2]	Average number of producers	Receipts of producer milk	Producer milk used in Cl. I	Class I utilization	Daily milk delivery per producer	Average fat test
	Dollars	Dollars	Number	Million pounds	Million pounds	Percent	Pounds	Percent
Northeast [3][4]	21.21	18.63	13,584	23,895	10,385	43.5	4,807	3.73
Appalachian [4][5]	21.28	19.87	2,732	5,882	4,133	70.3	5,898	3.67
Southeast [4][6]	21.54	20.15	3,003	6,923	4,701	67.9	6,298	3.67
Florida [7]	22.89	21.84	386	3,130	2,605	83.2	22,567	3.60
Mideast [4][8]	19.94	17.92	7,476	15,707	6,588	41.9	5,741	3.70
Upper Midwest [4][9]	19.76	17.59	15,449	28,041	4,445	15.9	4,951	3.72
Central [4][10]	19.97	17.39	3,930	11,564	4,249	36.7	8,033	3.65
Southwest [4][11]	20.96	18.40	589	9,687	4,207	43.4	45,057	3.63
Arizona [12]	20.31	17.43	98	4,156	1,399	33.7	116,204	3.56
Pacific Northwest [4][13]	19.86	16.99	612	6,882	2,276	33.1	30,739	3.67
All markets combined	20.78	18.24	47,859	115,867	44,989	38.8	6,613	3.69

[1] Prices are for milk of 3.5 percent butterfat content and for the principal pricing point of the market. See footnotes 3-13.　[2] For those orders that use the component pricing system for paying producers (orders 1, 30, 32, 33, 124,and 126), the figures are the statistical uniform price (the sum of the producer price differential and the Class III price). For those orders that use the skim milk/butterfat pricing system for paying producers (orders 5, 6, 7, and 131), the figures are the uniform price (the sum of the uniform butterfat price times 3.5 and the uniform skim milk price times 0.965).　[3] Suffolk Co. (Boston), MA.　[4] Due to the disadvantageous intraorder class and uniform price relationships in some months in these markets, handlers elected not to pool milk that normally would have been pooled under these orders.　[5] Mecklenburg Co. (Charlotte), NC.　[6] Fulton Co. (Atlanta), GA.　[7] Hillsborough Co. (Tampa), FL.　[8] Cuyahoga Co. (Cleveland), OH.　[9] Cook Co. (Chicago), IL.　[10] Jackson Co. (Kansas City), MO.　[11] Dallas Co. (Dallas), TX.　[12] Maricopa Co. (Phoenix), AZ.　[13] King CO. (Seattle), WA.
AMS, Dairy Programs, (202) 720-7461.

Table 8-26.—Dairy products: Total disappearance, and total and per capita consumption, United States, 1999–2008 [1]

Year	Butter			Cheese [2]			Condensed and evaporated milk [3]		
	Total disappearance	Consumption		Total disappearance	Consumption		Total disappearance	Consumption	
		Total	Per capita		Total	Per capita		Total	Per capita
	Million pounds	Million pounds	Pounds	Million pounds	Million pounds	Pounds	Million pounds	Million pounds	Pounds
1999	1,314	1,307	4.7	8,219	8,086	29.0	648	573	2.1
2000	1,289	1,269	4.5	8,581	8,415	29.8	596	560	2.0
2001	1,275	1,265	4.4	8,742	8,578	30.1	610	564	2.0
2002	1,288	1,281	4.5	8,949	8,792	30.5	706	661	2.3
2003	1,332	1,303	4.5	9,026	8,882	30.6	810	749	2.6
2004	1,352	1,332	4.5	9,366	9,181	31.3	737	641	2.2
2005	1,370	1,352	4.6	9,550	9,387	31.7	722	651	2.2
2006	1,436	1,412	4.7	9,914	9,747	32.6	703	649	2.2
2007	1,519	1,430	5.0	10,217	9,999	32.7	687	599	2.0
2008	1,710	1,513	5.0	10,217	9,952	33.2	757	683	2.2

Year	Ice cream (product weight)			Dry whole milk			Nonfat dry milk (human food)		
	Total disappearance	Consumption		Total disappearance	Consumption		Total disappearance	Consumption	
		Total	Per capita		Total	Per capita		Total	Per capita
	Million pounds	Million pounds	Pounds	Million pounds	Million pounds	Pounds	Million pounds	Million pounds	Pounds
1999	4,667	4,667	16.7	124	111	0.40	1,275	787	2.8
2000	4,702	4,702	16.7	119	80	0.28	1,073	742	2.6
2001	4,656	4,656	16.3	50	0	0.00	1,156	927	3.3
2002	4,824	4,824	16.8	54	25	0.09	1,362	886	3.1
2003	4,766	4,766	16.4	49	27	0.09	1,758	982	3.4
2004	4,415	4,415	15.1	51	2	0.01	1,833	1,258	4.3
2005	4,608	4,608	15.6	41	0	0.00	1,872	1,251	4.2
2006	4,420	4,420	14.8	40	0	0.00	1,595	952	3.2
2007	4,303	4,303	14.3	38	0	0.00	1,443	860	2.9
2008	4,188	4,188	13.8	55	23	0.08	1,810	932	3.1

[1] Total disappearance is based on production, imports, and change in stocks during the year. Production statistics for these commodities appear in other tables in this chapter. The total apparent consumption was obtained by subtracting ending stocks, shipments, and exports, from the total supply. The per capita consumption for each year was obtained by dividing the total apparent consumption by the number of persons. If the apparent total consumption is negative, value is set at zero.　[2] Includes all kinds of cheese except cottage and full-skim American.　[3] The evaporated milk is unskimmed, unsweetened, case goods. The condensed milk is unsweetened, unskimmed, bulk goods; and sweetened condensed milk, unskimmed, case and bulk goods.
ERS, Animal Products and Cost of Production Branch, (202) 694–5265.

Table 8-27.—Dairy products: Dec. 31 stocks, United States, 1999–2008

Year	Butter [1][2]	Cheese [1][3]	Canned milk [1]	Dry whole milk	Nonfat dry milk for human consumption [1]
	1,000 pounds	1,000 pounds	1,000 pounds	1,000 pounds	1,000 pounds
1999	25,082	622,197	35,690	5,749	284,542
2000	24,115	708,597	41,228	4,390	662,182
2001	55,915	663,251	40,739	2,894	900,158
2002	157,820	732,551	54,428	3,244	1,145,689
2003	99,613	742,173	38,506	1,981	981,160
2004	44,988	709,715	36,363	1,556	511,549
2005	58,649	758,161	44,418	2,270	183,311
2006	108,605	817,437	31,176	1,713	106,886
2007	155,162	798,307	37,441	3,604	165,774
2008	118,962	851,960	41,974	4,955	247,330

[1] Includes Government holdings.　[2] Includes butter equivalent of butteroil held by CCC.　[3] Excludes cottage and full-skim American cheese. Includes process American cheese held by CCC.
ERS, Animal Products and Cost of Production Branch, (202) 694–5265.

Table 8-28.—Butter: Production, 2007/2009

Country	2007	2008	2009
	1,000 metric tons	*1,000 metric tons*	*1,000 metric tons*
Australia	117	111	132
Brazil	82	84	76
Canada	79	84	82
EU-27	2,053	2,040	2,030
India	3,360	3,690	3,910
Japan	75	72	81
Mexico	214	180	163
New Zealand	442	422	428
Russia	300	305	280
Ukraine	100	85	72
Others	2,132	61	51
Total Foreign	8,954	7,135	7,305
United States	695	746	700
Total	9,649	7,881	8,005

FAS, Office of Global Analysis, (202) 720-6301.

Table 8-29.—Cheese: Production, 2007/2009

Country	2007	2008	2009
	1,000 metric tons	*1,000 metric tons*	*1,000 metric tons*
Argentina	520	525	534
Australia	360	344	325
Brazil	580	607	314
Canada	308	285	280
EU-27	6,760	6,800	6,810
Egypt	43	47	45
Mexico	184	188	240
New Zealand	350	292	300
Russian Federation	435	430	445
Ukraine	244	249	250
Others	7,221	40	41
Total Foreign	17,005	9,807	9,884
United States	4,435	4,506	4,575
Total	21,440	14,313	14,459

FAS, Office of Global Analysis, (202) 720-6301.

Table 8-30.—Dairy products: United States imports by country of origin, 2007–2009

Commodity and country of origin	2007	2008	2009
	Metric tons	*Metric tons*	*Metric tons*
Licensed cheese items 1:			
New Zealand (*)	8,518	5,619	7,622
Netherlands	5,979	5,519	4,167
Australia(*)	2,730	2,346	2,507
Denmark(*)	2,891	2,470	2,145
United Kingdom	1,484	1,901	1,488
Canada	882	1,082	1,019
Ireland	631	761	948
Germany(*)	565	580	584
Chile	40	8	422
Italy(*)	467	331	271
Jamaica	15	48	227
Finland	120	215	143
France(*)	142	136	137
Bahrain	0	0	126
Rest of World	304	370	381
World Total	24,767	21,387	22,187
Licensed cheese items 2:			
Italy(*)	17,574	17,180	13,746
New Zealand (*)	18,134	11,216	12,374
France(*)	16,094	14,133	12,242
Finland	10,225	8,601	8,344
Norway(*)	7,721	7,625	7,142
Australia(*)	7,035	2,117	6,502
Argentina	3,674	6,155	6,285
Denmark(*)	8,798	6,667	5,567
Ireland	4,774	4,068	5,209
Switzerland(*)	6,624	5,967	5,092
Netherlands	5,688	4,905	4,376
Mexico	2,950	3,345	3,063
Canada	3,212	2,984	2,722
Nicaragua	2,021	2,396	2,640
Rest of World	15,918	12,630	11,361
World Total	130,440	109,990	107,262
Licensed dairy, misc mixed:			
Mexico	22,076	23,127	23,484
Canada	6,477	17,584	13,467
Chile	3,997	8,269	6,408
New Zealand (*)	9,128	1,430	4,688
Australia(*)	5,096	1,893	4,321
Netherlands	3,446	5,008	3,800
Spain	4	109	931
Denmark(*)	446	296	454
Korea, South	336	367	412
Argentina	283	139	360
Israel(*)	28	19	325
Belgium-Luxembourg	420	423	314
Indonesia	112	208	236
Costa Rica	137	188	122
Rest of World	2,332	2,518	946
World Total	54,319	61,576	60,265

See end of table.

Table 8-30.—Dairy products: United States imports by country of origin, 2007–2009—Continued

Commodity and country of origin	2007	2008	2009
	Metric tons	*Metric tons*	*Metric tons*
Non-lcnsd dairy, misc mixed:			
Mexico	12,169	12,582	15,320
New Zealand (*)	5,795	4,916	6,287
Canada	13,236	1,936	4,805
Chile	2,283	2,760	4,442
Denmark(*)	968	1,000	1,891
Netherlands	576	650	1,305
Argentina	20	0	1,065
Peru	0	0	1,028
Greece	8,735,849	4,723	592
Australia(*)	749	1,307	568
Norway(*)	0	341	563
Israel(*)	43	102	524
Spain	642	542	478
Brazil	2,779	2335	374
Rest of World	2,395	1,741	1,203
World Total	50,388	34,936	40,444
Non-lcnsd cheese:			
Italy(*)	18,510	15,867	13,882
France(*)	5,098	3,932	4,002
Argentina	2,981	5,651	3,217
Bulgaria	2,917	3,367	2,837
Spain	2,920	3,245	2,816
Greece	2,677	2,084	2,508
Lithuania	0	427	660
United Kingdom	627	623	448
Romania	194	189	434
Turkey	466	365	381
Israel(*)	487	273	320
Cyprus	254	215	301
Uruguay	444	159	138
Macedonia	42	98	98
Rest of World	4,707	2,473	416
World Total	42,325	38,969	32,458
Casein:			
New Zealand (*)	44,188	36,530	32,548
Ireland	7,609	8,807	8,060
Netherlands	5,222	16,675	7,394
Argentina	3,626	11,548	5,773
India	12,334	11,959	5,165
Poland	3,609	12,214	4,206
Australia(*)	3,393	4,352	3,029
Germany(*)	1,751	1,661	1,875
Denmark(*)	687	3,563	1,016
France(*)	3,635	6,289	937
Belgium-Luxembourg	427	2,127	696
Ukraine	1,057	771	550
Uruguay	0	0	366
Belarus	10	637	317
Rest of World	2,581	1,646	287
World Total	90,129	118,777	72,219

See end of table.

Table 8-30.—Dairy products: United States imports by country of origin, 2007–2009—Continued

Commodity and country of origin	2007	2008	2009
	Metric tons	Metric tons	Metric tons
Lactose:			
Netherlands	993	2,440	3,267
Canada	2,969	3,446	3,074
Germany(*)	1,329	1,083	1,495
China	41	53	203
New Zealand (*)	44	67	127
France(*)	8	3	31
Belarus	0	15	21
Ireland	0	0	2
Taiwan	11	1	2
Lebanon	0	0	2
Russia	12	19	1
Italy(*)	2	4	1
United Kingdom	1	3	
Switzerland(*)	2	0	0
Rest of World	297	221	0
World Total	5,709	7,353	8,226

Note: (*) denotes a country that is a summarization of its component countries.
FAS, Office of Global Analysis, (202) 720-6301.

Table 8-31.—Dairy products: United States imports by type of product, 2007–2009

Commodity and country of origin	2007	2008	2009
	Metric tons	Metric tons	Metric tons
Butter:			
New Zealand (*)	6,945	9,106	9,618
Australia(*)	2,513	890	2,629
Ireland	1,284	1,838	1,139
France(*)	490	481	477
India	218	189	227
Denmark(*)	343	211	218
Canada	1,001	127	159
Poland	203	89	114
United Kingdom	157	114	93
Colombia	104	232	76
Costa Rica	0	0	59
Netherlands	5	79	43
Italy(*)	17	131	32
Argentina	435	87	32
Rest of World	2,278	767	173
World Total	15,993	14,340	15,090

Note: (*) denotes a country that is a summarization of its component countries.
FAS, Office of Global Analysis, (202) 720-6301.

DAIRY AND POULTRY STATISTICS

Table 8-32.—Dairy products: United States exports by country of destination, 2007–2009

Commodity and country of destination	2007	2008	2009
	Metric tons	Metric tons	Metric tons
Condensed & evap milk:			
Mexico	18,572	8,647	6,725
Taiwan	1,335	1,417	1,211
Panama	225	80	1,106
Canada	2,206	3,486	891
Dominican Republic	40	39	785
Russia	0	182	602
Bahamas, The	422	480	482
Philippines	649	245	276
Vietnam	47	591	234
Honduras	32	113	199
Saudi Arabia	0	29	168
Guyana	688	93	163
Hong Kong	68	30	156
Colombia	12	33	151
New Zealand(*)	62	11	99
Korea, South	150	49	89
Libya	0	0	85
Indonesia	0	0	84
Egypt	0	15	84
Singapore	8	1	82
Bermuda	71	56	59
Micronesia	0	0	49
Cayman Islands	407	267	48
Rest of World	2,955	6,373	370
World Total	27,770	19,731	14,132
Non-fat dry milk:			
Mexico	74,672	117,706	104,871
Philippines	33,666	49,872	27,473
Indonesia	31,794	45,176	21,357
Vietnam	12,743	14,966	16,172
Algeria	12,328	19,950	12,501
Malaysia	19,617	22,258	9,150
Egypt	8,460	15,775	8,019
Pakistan	2,000	3,862	7,785
Thailand	12,378	13,291	6,979
China	5,187	12,303	5,678
Japan	321	9,893	5,447
Dominican Republic	1,696	4,128	4,316
Canada	3,903	4,219	2,895
Singapore	2,090	3,687	1,634
Bangladesh	5,137	1,530	1,561
Morocco	290	1,927	1,408
Israel(*)	4,330	3,115	1,222
Turkey	0	849	1,041
Peru	736	3,284	829
Guatemala	567	1,049	695
Chile	402	4,099	673
Jamaica	0	1,607	663
Honduras	17	3,249	495
Rest of World	26,551	35,303	5,197
World Total	257,893	391,165	247,634
Dry whole milk & cream:			
Mexico	2,884	15,685	11,401
Japan	3,013	1,063	2,123
Turkey	0	0	1,224
Canada	398	1,313	1,151
Dominican Republic	323	3,370	653
Brazil	0	0	529
Lebanon	0	0	524
Egypt	334	648	518
Colombia	107	932	411
Taiwan	1,829	816	410
Syria	0	0	388
Guatemala	497	1,017	383
Pakistan	0	0	340
Vietnam	505	168	321
China	1,533	1,075	308
Bahamas, The	361	506	197
Cayman Islands	9	178	193
Israel(*)	10	360	192
Singapore	15	614	169
Korea, south	497	324	167
Peru	392	183	163
El Salvador	508	713	153
Saudi Arabia	1.8	19.3	142
Rest of World	7,874	11,803	1,180
World Total	20,825	40,459	23,075

See end of table.

Table 8-32.—Dairy products: United States exports by country of destination, 2007–2009—Continued

Commodity and country of destination	2007	2008	2009
	Metric tons	*Metric tons*	*Metric tons*
Fluid milk and cream:			
Mexico	29,684,519	21,587,068	22,987,133
Canada	14,043,591	21,724,721	21,403,413
Bahamas, The	1,337,891	1,431,359	1,652,173
Cayman Islands	211,520	665,245	1,070,840
Leeward-Windward Islands	124,157	263,058	387,004
Singapore	110,472	297,259	162,887
Saudia Arabia	0	299,548	182,524
Taiwan	118,065	304,060	341,715
Hong Kong	114,724	186,667	363,734
Malaysia	66,000	217,131	205,524
Pakistan	0	77,400	154,800
Netherlands Antilles	0	76,986	158,958
China, Peoples Republic of	15,534	69,959	149,421
Rest of World	449,596	1,517,770	1,730,166
World Total	46,279,285	48,718,231	51,154,515
Butter and milkfat:			
Saudi Arabia	3,178	8,515	9,641
Egypt	1,540	8,468	4,518
Mexico	4,851	6,444	3,746
Moroccoo	4,258	9,468	1,284
Canada	1,295	3,972	1,153
South Africa	0	637	952
Russia	3,137	16,258	922
Korea, South	142	1,136	742
Dominican Republic	415	564	556
Israel(*)	162	493	548
Philippines	16	368	548
Indonesia	25	634	486
Guatemala	236	705	407
Honduras	108	385	316
Georgia	40	145	293
China	189	279	267
Jordan	102	262	244
Belgium-Luxembourg	3,398	1,786	230
Haiti	86	109	178
Panama	270	589	176
Lebanon	107	250	155
Congo (Brazzaville)	0	0	144
Rest of World	18,958	32,047	1,883
World Total	40,629	90,229	29,168
Ice cream:			
Mexico	13,813	13,575	12,979
Canada	3,674	3,430	3,138
Bahamas, The	650	759	926
Trinidad and Tobago	136	370	705
Jamaica	620	605	662
Netherlands, The	397	554	532
Korea, South	271	360	477
Sweden	1	131	444
Singapore	639	658	382
Bermuda	226	394	382
Australia(*)	235	224	370
Leeward-Windward Islands(*)	229	342	369
United Arab Emirates	41	253	333
Dominican Republic	59	123	315
Russia	559	532	306
Netherlands Antilles	183	334	294
Aruba(!)	214	220	238
Germany(*)	44	5	214
Japan	79	184	207
Cayman Islands	218	274	196
Senegal	0	51	173
Israel(*)	207	228	168
Rest of World	3,938	2,481	2,620
World Total	25,777	25,330	25,516

See end of table.

Table 8-32.—Dairy products: United States exports by country of destination, 2007–2009—Continued

Commodity and country of destination	2007	2008	2009
	Metric tons	*Metric tons*	*Metric tons*
Cheese and curd:			
Mexico	33,588	36,955	40,124
Korea, South	8,326	13,024	10,857
Canada	8,852	11,215	10,249
Japan	10,062	9,375	7,023
Taiwan	1,606	2,126	2,370
Saudi Arabia	2,096	6,075	2,288
Dominican Republic	1,646	1,992	2,175
China	1,205	2,029	1,936
Guatemala	953	1,444	1,888
Philippines	1,078	2,632	1,660
Honduras	1,348	1,400	1,572
Bahamas, The	901	1,178	1,552
Panama	700	2,515	1,296
Egypt	1,463	4,026	1,268
Rest of World	25,715	35,217	22,153
World Total	99,540	131,202	108,410
Whey, mixed:			
China	60,423	72,661	93,685
Mexico	94,011	59,330	66,269
Canada	66,438	35,859	36,494
Japan	38,823	31,436	29,726
Malaysia	18,632	17,587	16,846
Korea, South	28,125	13,880	15,102
Philippines	17,130	10,669	14,285
Indonesia	11,922	14,272	13,514
Vietnam	4,973	10,423	13,121
Thailand	10,104	12,767	8,748
Taiwan	4,791	7,354	6,916
Morocco	1,751	9,604	6,662
Brazil	3,766	4,694	2,884
El Salvador	2,297	1,720	2,765
Rest of World	61,646	48,179	31,448
World Total	424,832	350,434	358,465

Note: (*) denotes a country that is a summarization of its component countries.
FAS, Office of Global Analysis, (202) 720-6301.

Table 8-33.—Dairy products: Price-support operations, United States, 2000–2010

Marketing year beginning October 1	Manufacturing milk per cwt.		Product purchase price per pound [1]		
	Support price at national average milkfat test	Average producer received price	Butter [2]	Cheddar cheese [3]	Nonfat milk, spray process [4]
	Dollars	*Dollars*	*Cents*	*Cents*	*Cents*
2000–2001	9.90	12.82	[6]65.49	[6]113.14	[6]100.32
	9.90	12.82	[7]85.48	113.14	[7]90.00
2001–2002	9.90	11.46	85.48	113.14	90.00
2002–2003	9.90	11.10	85.48	113.14	90.00
	9.90	11.10	[8]105.00	113.14	[8]80.00
2003–2004	9.90	14.95	105.00	113.14	80.00
2004–2005	9.90	14.76	105.00	113.14	80.00
2005–2006	9.90	12.57	105.00	113.14	80.00
2006–2007	9.90	16.62	105.00	113.14	80.00
2007–2008	9.90	18.83	105.00	113.14	80.00
2008–2009	9.35	12.58	105.00	113.00	80.00
2009–2010	10.80	....	[9]105.00	131.00	92.00
	9.35	[5]14.60	[10]105.00	113.00	80.00

[1] Announced purchase prices for products in bulk containers. [2] U.S. Grade A or higher, salted, 25-kg blocks. [3] U.S. Grade A or higher, standard moisture basis 40-pound blocks. [4] U.S. Extra Grade, not more than 3.5 percent moisture content. Prices quoted are for product in 25-kg bags. [5] Estimated value of milk used in manufactured products. [6] Effective January 31, 2001. [7] Effective June 13, 2001. [8] Effective December 1, 2002. [9] Effective August 1, 2009 through October 31, 2009. [10] Effective November 1, 2009.
FSA, Dairy, (202) 690–0050

Table 8-34.—Chickens: Inventory number and value, United States, Dec. 1, 2000–2009 [1]

Year	Layers 1 year old and older	Layers 20 weeks old but less than 1 year	Total layers	Pullets			Other chickens	All chickens	Value per head	Total value
				13 weeks to 20 weeks old	Under 13 weeks old	Total				
	Thousands	*Thousands*	*Thousands*	*Thousands*	*Thousands*	*Thousands*	*Thousands*	*Thousands*	*Dollars*	*1,000 dollars*
2000	153,439	180,154	333,593	38,395	56,764	....	8,088	436,840	2.44	1,064,171
2001	153,817	186,500	340,317	42,907	52,749	....	8,126	444,099	2.41	1,069,335
2002	153,884	186,325	340,209	39,865	55,424	....	8,353	443,851	2.38	1,055,316
2003	(3)	(3)	341,099	(3)	(3)	100,583	8,477	450,159	2.48	1,116,052
2004	(3)	(3)	344,371	(3)	(3)	101,794	8,287	454,452	2.48	1,125,672
2005	(3)	(3)	349,764	(3)	(3)	97,544	8,264	455,572	2.52	1,149,736
2006	(3)	(3)	352,316	(3)	(3)	97,459	8,038	457,813	2.60	1,189,978
2007	(3)	(3)	346,613	(3)	(3)	103,816	8,164	458,593	2.95	1,351,649
2008	(3)	(3)	339,859	(3)	(3)	99,458	7,589	446,906	3.39	1,517,210
2009 [2]	(3)	(3)	339,526	(3)	(3)	101,588	8,496	449,610	3.34	1,501,605

[1] Does not include commercial broilers. [2] Preliminary. [3] Not available due to program change.
NASS Livestock Branch, (202) 720-3570.

Table 8-35.—Chickens: Layers, pullets, and other chickens, by State and United States, December 1, 2008 and 2009 [1]

State	Total layers		Total pullets		Other Chickens	
	2008	2009	2008	2009	2008	2009
	Thousands	Thousands	Thousands	Thousands	Thousands	Thousands
AL	9,382	9,029	5,291	5,446	1,042	1,208
AR	12,530	12,214	6,879	6,270	1,340	1,450
CA	20,172	19,686	4,042	3,932	6	5
CO	3,872	3,820	774	784	65	65
CT	2,838	2,637	568	539	10	6
FL	10,376	10,215	2,151	1,600	61	34
GA	18,418	17,922	7,556	7,510	1,216	1,814
HI	334	319	39	60	0	0
IL	4,906	4,801	329	200	20	20
IN	24,394	23,411	6,484	8,067	72	72
IA	53,370	54,025	11,984	11,877	75	70
KY	4,516	4,661	1,982	1,777	231	245
LA	1,785	1,782	467	689	113	159
ME	3,626	3,712	1,381	1,112	4	5
MD	2,369	2,292	236	184	15	12
MA	117	119	11	8	1	1
MI	9,638	10,384	1,890	2,157	1	2
MN	10,027	10,397	3,373	3,628	47	54
MS	6,076	6,065	3,618	3,653	532	535
MO	6,714	6,923	2,720	2,515	174	142
MT	350	375	159	165	1	0
NE	9,752	9,388	1,753	2,385	0	0
NY	4,320	4,080	1,424	1,040	14	7
NC	12,380	13,266	6,229	5,880	1,000	982
OH	27,063	27,577	6,155	7,737	32	40
OK	3,356	3,261	1,128	1,039	304	242
OR	2,540	2,147	410	631	7	10
PA	21,833	23,298	3,945	4,640	105	111
SC	4,398	4,470	1,673	1,754	150	152
SD	2,410	2,440	427	370	0	0
TN	1,470	1,506	1,133	1,091	160	211
TX	19,280	17,732	6,518	5,810	490	460
UT	3,403	3,372	509	607	0	0
VT	217	213	30	23	3	2
VA	2,873	3,267	1,131	1,509	180	240
WA	5,826	6,188	963	1,222	0	0
WV	970	731	659	349	48	68
WI	4,664	4,878	1,314	1,205	30	31
WY	11	11	2	2	0	0
Oth Sts [2]	7,283	6,912	2,121	2,121	40	41
US	339,859	339,526	99,458	101,588	7,589	8,496

[1] Totals may not add due to rounding.   [2] AK, AZ, DE, ID, KS, NV, NM, ND, NH, NJ, and RI combined to avoid disclosing data for individual operations.
NASS, Livestock Branch, (202) 720–3570.

Table 8-36.—Chicken inventory: Number, value per head, and total value, by State and United States, December 1, 2008 and 2009 [1] [2]

State	Number		Value per bird		Total value	
	2008	2009	2008	2009	2008	2009
	1,000 Head	*1,000 Head*	*Dollars*	*Dollars*	*1,000 Dollars*	*1,000 Dollars*
AL	15,715	15,683	5.60	6.70	88,004	105,076
AR	20,749	19,934	6.40	5.80	132,794	115,617
CA	24,220	23,623	3.30	2.60	79,926	61,420
CO	4,711	4,669	2.10	2.30	9,893	10,739
CT	3,416	3,182	2.80	2.80	9,565	8,910
FL	12,588	11,849	2.20	2.10	27,694	24,883
GA	27,190	27,246	6.20	5.60	168,578	152,578
HI	373	379	1.60	1.90	597	720
IL	5,255	5,021	1.00	1.00	5,255	5,021
IN	30,950	31,550	1.70	1.80	52,615	56,790
IA	65,429	65,972	2.40	2.40	157,030	158,333
KY	6,729	6,683	5.30	3.80	35,664	25,395
LA	2,365	2,630	3.40	3.50	8,041	9,205
ME	5,011	4,829	3.00	2.50	15,033	12,073
MD	2,620	2,488	3.80	3.60	9,956	8,957
MA	129	128	4.80	5.50	619	704
MI	11,529	12,543	1.90	2.00	21,905	25,086
MN	13,447	14,079	2.00	2.20	26,894	30,974
MS	10,226	10,253	5.30	5.20	54,198	53,316
MO	9,608	9,580	3.20	3.70	30,746	35,446
MT	510	540	3.80	4.10	1,938	2,214
NE	11,505	11,773	2.20	2.20	25,311	25,901
NY	5,758	5,127	2.10	1.50	12,092	7,691
NC	19,609	20,128	8.30	8.50	162,755	171,088
OH	33,250	35,354	1.80	2.00	59,850	70,708
OK	4,788	4,542	5.80	5.50	27,770	24,981
OR	2,957	2,788	2.10	2.80	6,210	7,806
PA	25,883	28,049	2.60	2.60	67,296	72,927
SC	6,221	6,376	3.80	3.40	23,640	21,678
SD	2,837	2,810	3.60	3.40	10,213	9,554
TN	2,763	2,808	7.60	6.60	20,999	18,533
TX	26,288	24,002	2.50	2.90	65,720	69,606
UT	3,912	3,979	2.30	1.80	8,998	7,162
VT	250	238	3.20	3.10	800	738
VA	4,184	5,016	3.90	5.30	16,318	26,585
WA	6,789	7,410	3.60	2.20	24,440	16,302
WV	1,677	1,148	5.20	5.50	8,720	6,314
WI	6,008	6,114	2.50	2.50	15,020	15,285
WY	13	13	3.10	3.10	40	40
Oth Sts [3]	9,444	9,074	2.55	2.78	24,073	25,249
US	446,906	449,610	3.39	3.34	1,517,210	1,501,605

[1] Excludes commercial broilers. [2] Totals may not add due to rounding. [3] AK, AZ, DE, ID, KS, NV, NH, NJ, NM, ND, and RI combined to avoid disclosing data for individual operations.
NASS, Livestock Branch, (202) 720-3570.

Table 8-37.—Broiler meat: Production in specified countries, 2007–2009

Continent and country	2007	2008	2009
	1,000 metric tons	*1,000 metric tons*	*1,000 metric tons*
Argentina	1,320	1,430	1,500
Brazil	10,305	11,033	11,023
China	11,291	11,840	12,100
EU-27	8,320	8,594	8,660
India	2,240	2,490	2,550
Iran	1,423	1,450	1,525
Japan	1,250	1,255	1,255
Mexico	2,683	2,853	2,789
Russian Federation	1,350	1,550	1,772
Thailand	1,050	1,170	1,200
Others	10,839	11,255	11,451
Total foreign	58,071	54,921	55,825
United States	16,225	16,561	15,935
Grand total	68,297	71,482	71,760

FAS, Office of Global Analysis, (202) 720-6301.

Table 8-38.—Chickens: Lost, sold for slaughter, price and value of sales, by State and United States, 2008[1] [2]

State	Number lost[3]	Number sold[4]	Pounds sold[4]	Price per pound[4]	Value of sales[4]
	1,000 head	*1,000 head*	*1,000 pounds*	*Dollars*	*1,000 dollars*
AL	2,642	11,291	86,941	0.099	8,607
AR	5,873	15,078	111,577	0.096	10,711
CA	2,511	8,872	29,278	0.001	29
CO	1,151	1,646	8,230	0.059	486
CT	804	1,362	4,631	0.004	19
FL	4,189	4,448	16,902	0.023	389
GA	7,132	15,417	109,461	0.093	10,180
HI	57	70	287	0.206	59
IL	586	1,855	6,864	0.019	130
IN	3,739	9,729	33,079	0.007	232
IA	26,148	6,311	21,457	0.005	107
KY	1,349	3,027	21,794	0.094	2,049
LA	309	2,061	12,366	0.078	965
ME	842	2,337	7,712	0.001	8
MD	268	1,758	6,505	0.021	137
MA	12	115	380	0.003	1
MI	887	2,423	7,996	0.001	8
MN	2,938	3,941	14,582	0.019	277
MS	2,408	6,588	52,045	0.101	5,257
MO	923	5,099	23,965	0.051	1,222
MT	72	108	410	0.001	0
NE	1,784	6,248	20,618	0.001	21
NY	596	3,060	10,098	0.002	20
NC	2,791	14,306	94,420	0.086	8,120
OH	5,977	8,863	30,134	0.006	181
OK	911	3,191	22,975	0.093	2,137
OR	1,654	444	1,510	0.004	6
PA	1,854	12,791	46,048	0.014	645
SC	1,167	3,387	18,629	0.070	1,304
SD	750	716	2,363	0.001	2
TN	503	2,214	16,826	0.098	1,649
TX	10,456	5,434	42,385	0.100	4,239
UT	932	1,747	5,765	0.001	6
VT	25	81	389	0.053	21
VA	522	2,921	17,818	0.080	1,425
WA	3,661	384	1,267	0.001	1
WV	244	1,408	10,982	0.100	1,098
WI	1,019	1,695	7,289	0.042	306
WY	1	7	23	0.001	0
Oth Sts[5]	2,145	3,140	11,044	0.010	110
US[6]	101,832	175,573	937,045	0.066	62,164

[1] Estimates cover the 12-month period, December 1, previous year through November 30 and exclude broilers. [2] Preliminary. [3] Includes rendered, died, destroyed, composted, or disappeared for any reason (excluding sold for slaughter) during the 12-month period. [4] Sold for slaughter. [5] AK, AZ, DE, ID, KS, NV, NH, NJ, NM, ND, and RI combined to avoid disclosing data for individual operations. [6] Totals may not add due to rounding.
NASS, Livestock Branch, (202) 720–3570.

Table 8-39.—Mature chickens: Lost, sold for slaughter, price, and value, United States, 1999–2008[1]

Year	Number		Pounds (live weight) sold[3]	Price per pound live weight[3]	Value of sales[3]
	Lost[2]	Sold[3]			
	1,000 head	*1,000 head*	*1,000 pounds*	*Dollars*	*1,000 dollars*
1999	54,951	214,063	1,059,153	0.071	75,217
2000	50,907	218,411	1,112,604	0.057	63,988
2001	56,146	202,482	1,032,115	0.045	47,249
2002	55,330	199,931	1,039,118	0.048	49,931
2003	86,933	189,660	984,853	0.049	47,997
2004	100,752	191,971	999,066	0.058	57,709
2005	93,445	193,938	1,005,838	0.065	65,072
2006	101,611	173,883	924,993	0.059	54,141
2007	101,152	168,283	912,875	0.056	51,498
2008	101,832	175,573	937,045	0.066	62,164

[1] Estimates cover the 12-month period, December 1, previous year through November 30 and exclude broilers. [2] Includes rendered, died, destroyed, composted, or disappeared for any reason (excluding sold for slaughter) during the 12-month period. [3] Sold for slaughter.
NASS, Livestock Branch, (202) 720–3570.

Table 8-40.—Broilers: Production and value, United States, 1999–2008 [1] [2] [3]

Year	Number produced	Pounds produced	Price per pound [4]	Value of production
	Thousands	1,000 pounds	Cents	1,000 dollars
1999	8,146,410	40,829,000	37.1	15,128,509
2000	8,283,700	41,626,100	33.6	13,989,424
2001	8,389,770	42,452,400	39.3	16,696,089
2002	8,591,080	44,058,700	30.5	13,437,345
2003	8,492,850	43,958,200	34.6	15,214,947
2004	8,740,650	45,796,250	44.6	20,446,086
2005	8,872,000	47,855,600	43.6	20,877,916
2006	8,867,800	48,829,900	36.3	17,739,234
2007	8,906,700	49,330,700	43.6	21,513,536
2008	9,009,300	50,441,600	46.0	23,203,136

[1] December 1, previous year through November 30, current year.   [2] Broiler production including other domestic meat-type strains.   [3] Excludes States producing less than 500,000 broilers.   [4] Live weight equivalent prices, derived from ready-to-cook (RTC) prices, minus processing costs, then multiplied by a dressing percentage.
NASS, Livestock Branch, (202) 720–3570.

Table 8-41.—Chickens: Supply, distribution, and per capita consumption, ready-to-cook basis, United States, 2001–2010

Year	Production			Commercial storage at beginning of year	Exports	Commercial storage at end of year	Consumption	
	Commercial broilers	Other chickens	Total [1]				Total [1] [2]	Per capita
	Million pounds	Million pounds	Million pounds	Million pounds	Million pounds	Million pounds	Million pounds	Pounds
2001	30,938	515	31,453	807	5,737	720	25,820	91
2002	31,895	547	32,441	720	4,940	768	27,468	95
2003	32,399	502	32,901	768	5,015	600	28,069	97
2004	33,699	504	34,203	600	4,997	705	29,129	99
2005	34,986	516	35,502	705	5,332	913	29,997	101
2006	35,120	504	35,624	913	5,365	738	30,484	102
2007	35,772	498	36,270	738	6,072	721	30,280	100
2008 [3]	36,511	559	37,070	721	7,110	748	30,018	99
2009 [4]	35,131	500	35,631	748	6,935	618	28,915	94
2010	35,936	489	36,425	618	5,925	708	30,500	98

[1] Totals may not add due to rounding.   [2] Shipments to territories now included in total consumption.   [3] Preliminary.   [4] Forecast.
ERS Markets and Trade Economics Division, Animal Products and Cost of Production Branch, (202) 694-5265.

Table 8-42.—Poultry: Feed-price ratios, United States, 1999–2008

Year	Ratios [1]		
	Egg-feed	Broiler-feed	Turkey-feed
	Pounds	Pounds	Pounds
1999	9.8	7.2	8.6
2000	10.5	6.6	8.7
2001	9.9	7.7	8.2
2002	8.6	5.3	6.8
2003	10.6	5.4	5.9
2004	8.3	5.9	6.2
2005	7.0	7.0	7.8
2006	7.5	5.7	7.8
2007	10.2	5.0	6.0
2008	8.6	3.7	4.6

[1] Number of pounds of poultry feed equivalent in value at local market prices to 1 dozen market eggs, or 1 pound of broiler or 1 pound of turkey live weight. Simple average of monthly feed-price ratios. Egg feed= corn (75 lbs) and soybeans (25 lbs); broiler feed= corn (58 lbs); soybeans (42 lbs); turkey feed= corn (51 lbs), soybeans (28 lbs), and wheat (21 lbs). Monthly equivalent prices of commercial prepared feeds are based on current U.S. prices received for corn, soybeans, and wheat.
NASS, Environmental, Economics, and Demographics Branch, (202) 720–6146.

Table 8-43.—Broilers: Production, price, and value, by State, 2007 and 2008 [1]

State	2007				2008			
	Production		Price per pound [2]	Value of production	Production		Price per pound [3]	Value of production
	Number	Weight			Number	Weight		
	Thou-sands	1,000 pounds	Dollars	1,000 dollars	Thou-sands	1,000 pounds	Dollars	1,000 dollars
AL	1,022,700	5,624,900	0.430	2,418,707	1,062,900	5,846,000	....	2,689,160
AR	1,175,900	6,232,300	0.420	2,617,566	1,160,000	6,380,000	....	2,934,800
DE	245,800	1,597,700	0.460	734,942	242,900	1,578,900	....	726,294
FL	73,300	417,800	0.430	179,654	63,800	376,400	....	173,144
GA	1,398,800	7,413,600	0.430	3,187,848	1,409,200	7,468,800	....	3,435,648
KY	303,300	1,668,200	0.430	717,326	306,100	1,652,900	....	760,334
MD	294,800	1,591,900	0.460	732,274	298,600	1,612,400	....	741,704
MN	46,600	242,300	0.430	104,189	44,900	238,000	....	109,480
MS	824,000	4,614,400	0.430	1,984,192	840,700	4,876,100	....	2,243,006
NE	4,800	27,400	0.450	12,330	....	....	....	....
NC	781,200	5,390,300	0.460	2,479,538	796,100	5,493,100	....	2,526,826
OH	49,800	273,900	0.430	117,777	57,500	327,800	....	150,788
OK	243,000	1,287,900	0.420	540,918	237,800	1,260,300	....	579,738
PA	151,200	846,700	0.450	381,015	160,900	933,200	....	429,272
SC	234,900	1,479,900	0.450	665,955	236,900	1,516,200	....	697,452
TN	205,900	1,050,100	0.430	451,543	199,700	1,018,500	....	468,510
TX	616,300	3,266,400	0.430	1,404,552	641,000	3,461,400	....	1,592,244
VA	250,200	1,301,000	0.430	559,430	250,300	1,251,500	....	575,690
WV	88,900	355,600	0.450	160,020	85,700	351,400	....	161,644
WI	47,300	203,400	0.450	91,530	51,700	217,100	....	99,866
Other States [4]	848,000	4,445,000	0.444	1,972,230	862,600	4,581,600	....	2,107,536
Total [5]	8,906,700	49,330,700	0.436	21,513,536	9,009,300	50,441,600	....	23,203,136

[1] Broilers are young chickens of the meat-type strains, raised for the purpose of meat production. Estimates cover the 12-month period, December 1, previous year through November 30. [2] Live weight equivalent prices, derived from ready-to-cook (RTC) prices minus processing costs, then multiplied by a dressing percentage. [3] No longer published. [4] 2007: CA, IN, IA, LA, MI, MO, NY, OR, and WA. 2008: CA, CT, IN, IA, LA, MI, MO, NE, NY, OR, and WA. [5] Excludes States producing less than 500,000 broilers.
NASS, Livestock Branch, (202) 720–3570.

Table 8-44.—Chicks hatched by commercial hatcheries: Number, average price, and value, United States, 1999–2008

Year	Chicks hatched			Average price of baby chicks per 100		Value of chick production
	Broiler-type	Egg-type	All	Broiler-type	Egg-type	
	Thousands	Thousands	Thousands	Dollars	Dollars	1,000 dollars
1999	8,715,423	451,721	9,167,144	20.30	52.60	1,886,007
2000	8,846,185	430,412	9,276,597	20.50	48.00	1,913,453
2001	9,021,116	452,673	9,473,789	20.60	53.90	1,982,613
2002	9,079,092	421,549	9,500,641	21.10	52.00	2,025,371
2003	9,080,614	416,003	9,496,617	21.10	50.50	2,025,209
2004	9,337,577	437,391	9,774,968	20.60	53.90	2,041,418
2005	9,483,918	437,066	9,920,984	20.60	53.50	2,071,207
2006	9,414,070	427,373	9,841,443	22.90	66.50	2,297,743
2007	9,590,018	467,763	10,057,781	25.60	69.40	2,605,945
2008	9,467,507	468,169	9,935,676	26.30	75.60	2,666,769

NASS, Environmental, Economics, and Demographics Branch, (202) 720–6146 and Livestock Branch, (202) 720-3570.

Table 8-45.—Poultry: Slaughtered under Federal inspection, by class, United States, 2007–2009

Class	Number inspected			Pounds inspected (live weight)		
	2007	2008	2009	2007	2008	2009
	Thousands	*Thousands*	*Thousands*	*Thousands*	*Thousands*	*Thousands*
Young chickens	8,903,071	8,921,070	8,520,225	49,089,999	49,780,767	47,613,116
Mature chickens	132,549	154,042	138,635	795,192	891,117	795,511
Total chickens	9,035,620	9,075,112	8,658,860	49,885,191	50,671,884	48,408,627
Young turkeys	262,748	269,165	243,949	7,452,773	7,803,131	7,056,127
Old turkeys	2,178	2,100	1,819	57,092	55,886	47,920
Total turkeys	264,926	271,265	245,768	7,509,865	7,859,017	7,104,047
Ducks	27,311	24,149	22,767	183,932	161,881	154,203
Other poultry [1]	....	....	....	3,672	4,251	4,577
Total poultry	....	....	....	57,582,660	58,697,033	55,671,454

Class	Pounds certified (ready-to-cook)		
	2007	2008	2009
	Thousands	*Thousands*	*Thousands*
Young chickens	36,159,067	36,906,310	35,510,493
Mature chickens	498,160	559,249	500,137
Total chickens	36,657,227	37,465,559	36,010,630
Young turkeys	5,907,685	6,204,960	5,625,599
Old turkeys	43,107	42,234	36,800
Total turkeys	5,950,792	6,247,194	5,662,399
Ducks	131,638	116,290	110,207
Other poultry [2]	2,155	2,537	2,796
Total poultry	42,741,812	43,831,580	41,786,032

Class	Pounds condemned					
	Ante-mortem (live weight)			Post-mortem (Carcass and parts)		
	2007	2008	2009	2007	2008	2009
	Thousands	*Thousands*	*Thousands*	*Thousands*	*Thousands*	*Thousands*
Young chickens	168,065	153,868	115,940	392,960	377,709	312,457
Mature chickens	14,465	14.674	12,796	36,858	40,601	36,963
Total chickens	182,530	168,542	128,736	429,818	418,310	349,420
Young turkeys	21,582	19,829	16,858	102,066	100,190	84,942
Old turkeys	719	615	442	2,981	2,712	1,732
Total turkeys	22,301	20,444	17,300	105,047	102,902	86,674
Ducks	452	470	491	3,867	3,345	3,039
Other poultry	19	13	5	55	24	26
Total poultry	205,302	189,469	146,532	538,787	524,581	439,159

[1] Includes geese, guineas, ostriches, emus, rheas, and squab.   [2] Includes geese, guineas, and squab.
NASS, Livestock Branch, (202) 720–3570.

Table 8-46.—Turkeys: Supply, distribution, and per capita consumption, ready-to-cook basis, United States, 2001–2010

Year	Production	Commercial storage at beginning of year	Exports	Commercial storage at end of year	Consumption Total [1][2]	Consumption Per capita
	Million pounds	Million pounds	Million pounds	Million pounds	Million pounds	Pounds
2001	5,489	241	487	241	5,004	17.5
2002	5,638	241	439	333	5,108	17.7
2003	5,576	333	484	354	5,074	17.4
2004	5,383	354	442	288	5,010	17.1
2005	5,432	288	570	206	4,952	16.7
2006	5,607	206	547	218	5,060	16.9
2007	5,873	218	547	261	5,294	17.5
2008	6,165	261	676	396	5,361	17.6
2009[3]	5,588	396	535	262	5,201	16.9
2010[4]	5,403	262	520	250	4,907	15.8

[1] Totals may not add due to rounding.   [2] Shipments to territories now included in consumption.   [3] Preliminary.
[4] Forecast.
ERS Markets and Trade Economics Division, Animal Products and Cost of Production Branch, (202) 694–5265.

Table 8-47.—Turkeys: Production, 2007–2009

Country	2007	2008	2009
	1,000 metric tons	1,000 metric tons	1,000 metric tons
China	5	5	5
Others	2,453	2,017	
Total foreign	2,458	2,022	5
United States	2,664	2,796	

FAS, Office of Global Analysis, (202) 720-6301.

Table 8-48.—Turkeys: Production, and value, United States, 1999–2008

Year	Number raised	Pounds (live weight) produced	Price per pound live weight	Value of production
	Thousands	*1,000 pounds*	*Cents*	*1,000 dollars*
1999	270,192	6,877,399	40.8	2,806,630
2000	270,466	6,959,833	40.6	2,828,489
2001	272,660	7,173,111	39.0	2,796,821
2002	275,477	7,494,861	36.5	2,732,481
2003	269,556	7,230,650	34.6	2,503,540
2004	255,987	6,949,311	41.5	2,887,170
2005	249,666	6,991,599	44.5	3,107,875
2006	256,334	7,223,675	48.0	3,467,534
2007	266,828	7,566,315	52.3	3,954,472
2008	273,088	7,922,087	56.5	4,477,054

NASS, Livestock Branch, (202) 720–3570.

Table 8-49.—Turkeys: Production and value, by State, 2008 [1]

State	Number raised [2]	Pounds produced	Price per pound [3]	Value of production
	1,000 head	*1,000 pounds*	*Dollars*	*1,000 dollars*
AR	31,000	610,700	0.57	348,099
CA	16,000	435,200	0.58	252,416
IN	14,500	519,100	0.59	306,269
IA	9,000	360,000	0.56	201,600
MN	48,000	1,305,600	0.57	744,192
MO	21,000	651,000	0.57	371,070
NC	40,000	1,208,000	0.54	652,320
OH	6,000	230,400	0.58	133,632
PA	11,500	216,200	0.64	138,368
SC	12,500	477,500	0.56	267,400
SD	4,700	189,410	0.47	89,023
UT	4,100	104,960	0.58	60,877
VA	18,000	484,200	0.56	271,152
WV	3,800	102,220	0.56	57,243
Oth Sts [4]	32,988	1,027,597	0.57	583,393
US	273,088	7,922,087	0.565	4,477,054

[1] Revised.   [2] Based on turkeys placed Sep. 1, 2007, through Aug. 31, 2008. Excludes young turkeys lost.   [3] Live weight equivalent price to producers is used when actual live weight price is not available for the State.   [4] Other States include State estimates not shown and States suppressed due to disclosure.
NASS, Livestock Branch, (202) 720–3570.

Table 8-50.—Turkeys: Net poults placements, United States, Monthly, 2007 and 2008 [1]

Month	Total all breeds		Percent of Previous Year
	2007	2008	
	Thousands	*Thousands*	*Percent*
Jan	26,068	26,183	100
Feb	24,249	25,011	103
Mar	26,433	26,394	100
Apr	26,034	25,880	99
May	27,266	27,084	99
June	26,282	25,557	97
July	27,258	26,551	97
Aug	26,523	24,363	92
Sept	24,073	22,515	94
Oct	25,222	22,165	88
Nov	24,755	20,702	84
Dec	24,091	23,179	96
Total	308,254	295,584	96

[1] Includes imports and excludes exports.
NASS, Livestock Branch, (202) 720–3570.

Table 8-51.—Turkeys: Poults hatched by commercial hatcheries, U.S. and regions, Monthly, 2007 and 2008

Month	United States			2008			
	2007	2008	Percent of Previous Year	East North Central	West North Central	North and South Atlantic [1]	South Central and West [1]
	Thou-sands	Thou-sands	Per-cent	Thou-sands	Thou-sands	Thou-sands	Thou-sands
All breeds:							
Jan	26,301	27,151	103	3,939	10,272	9,591	3,349
Feb	24,749	25,784	104	4,007	10,019	8,603	3,155
Mar	26,789	26,853	100	3,982	10,165	8,793	3,913
Apr	26,362	26,522	101	3,848	9,503	9,602	3,569
May	27,405	27,281	100	4,081	10,264	9,178	3,758
June	26,657	26,196	98	3,913	9,871	8,870	3,542
July	27,302	27,635	101	4,094	10,120	9,358	4,063
Aug	26,924	25,327	94	3,777	10,272	7,936	3,342
Sept	24,259	23,719	98	3,919	9,638	8,019	2,143
Oct	25,778	23,274	90	3,619	9,438	8,211	2,006
Nov	25,761	22,005	85	3,537	8,815	7,734	1,919
Dec	24,945	24,289	97	3,602	9,506	8,601	2,580
Total	313,232	306,036	98	46,318	117,883	104,496	37,339

[1] Regions combined to avoid disclosing individual operators.
NASS, Livestock Branch, (202) 720–3570.

Table 8-52.—Eggs: Supply, distribution, and per capita consumption, United States, 2001–2010 [1]

Year	Total egg production	Storage at beginning of the year [1]	Imports [2]	Exports [2]	Eggs used for hatching	Consumption		
						Storage at end of the year [2]	Total [3]	Per capita
	Million dozen	Million dozen	Million dozen	Million dozen	Million dozen	Million dozen	Million dozen	Number
2001	7,187	11	9	190	964	10.4	6,043	254
2002	7,270	10	15	174	961	10.3	6,150	256
2003	7,299	10	13	146	959	13.7	6,204	256
2004	7,450	14	13	168	988	14.5	6,306	258
2005	7,538	15	9	203	997	16.0	6,345	257
2006	7,650	16	9	202	992	12.5	6,468	260
2007	7,587	13	14	250	1,016	11.1	6,335	252
2008	7,501	11	15	206	996	17.2	6,307	248
2009 [4]	7,534	17	11	242	955	18.0	6,347	248
2010 [5]	7,567	18	12	235	972	18.0	6,372	247

[1] Calendar years.   [2] Shell eggs and the approximate shell-egg equivalent of egg product.   [3] Shipments to territories now included in total consumption.   [4] Preliminary.   [5] Forecast.
ERS Markets and Trade Economics Division, Animal Products and Cost of Production Branch, (202) 694–5265.

Table 8-53.—Eggs, shell: Average price per dozen on consumer Grade A cartoned white eggs to volume buyers, store-door delivery, New York, 2000–2009

Year	Large
	Cents
2000	68.90
2001	67.14
2002	67.06
2003	87.91
2004	82.18
2005	65.51
2006	71.76
2007	114.36
2008	128.32
2009	102.97

AMS, Poultry Programs, Market News and Analysis Branch, (202) 720–6911.

Table 8-54.—All layers and egg production: Annual average number of layers, eggs per layer, and total production, by State and United States, 2008 and 2009 [1] [2]

State	Average number of layers during year		Eggs per layer during year [3]		Total Egg Production	
	2008	2009	2008	2009	2008	2009
	Thousands	Thousands	Number	Number	Millions	Millions
AL	9,416	9,223	228	228	2,150	2,106
AR	13,371	12,343	235	238	3,139	2,935
CA	20,262	19,652	260	270	5,271	5,304
CO	3,816	3,755	286	296	1,090	1,110
CT	2,860	2,742	273	280	780	767
FL	10,385	9,847	265	271	2,749	2,670
GA	18,539	17,685	247	253	4,576	4,467
HI	343	330	214	211	73.3	69.5
IL	4,827	4,600	271	272	1,309	1,253
IN	24,117	23,630	271	273	6,523	6,460
IA	53,523	53,801	269	269	14,407	14,475
KY	4,486	4,476	254	253	1,140	1,130
LA	1,848	1,737	259	262	479	455
ME	3,910	3,527	263	260	1,028	916
MD	2,521	2,163	264	256	666	554
MA	116	111	310	324	36	36
MI	9,221	9,839	288	283	2,653	2,784
MN	10,028	10,186	276	273	2,767	2,777
MS	6,290	5,911	240	244	1,511	1,440
MO	6,854	7,144	275	276	1,885	1,973
MT	355	351	324	305	115	107
NE	9,731	9,620	285	286	2,777	2,749
NY	3,975	4,078	294	292	1,167	1,192
NC	12,427	12,636	247	249	3,063	3,148
OH	26,233	27,054	273	273	7,168	7,392
OK	3,357	3,328	231	231	774	769
OR	2,568	2,413	300	303	769	732
PA	21,531	22,591	288	288	6,189	6,509
SC	4,509	4,473	254	252	1,143	1,128
SD	2,548	2,428	271	287	691	696
TN	1,523	1,479	230	219	351	323
TX	18,626	18,576	266	268	4,954	4,985
UT	3,389	3,350	270	274	914	918
VT	212	210	260	271	55	57
VA	3,048	3,039	238	244	726	741
WA	5,584	6,107	275	279	1,533	1,705
WV	1,086	947	228	224	247	212
WI	4,551	4,858	268	278	1,220	1,350
WY	11	11	218	218	2.4	2.4
Oth Sts [4]	7,133	7,128	274	275	1,951	1,963
US	339,131	337,376	266	268	90,040	90,359

[1] Annual estimates cover the period December 1 previous year through November 30.   [2] Totals may not add due to rounding.   [3] Total egg production divided by average number of layers on hand.   [4] AK, AZ, DE, ID, KS, NV, NH, NJ, NM, ND, and RI combined to avoid disclosing data for individual operations.
NASS, Livestock Branch, (202) 720–3570.

Table 8-55.—Eggs: Broken under Federal inspection, United States, 2008 and 2009

Item	Quantity	
	2008	2009
	1,000 dozen	1,000 dozen
Shell eggs broken	2,047,776	1,993,663
	1,000 pounds	1,000 pounds
Edible product from shell eggs broken		
Whole	1,709,414	1,654,938
White	631,157	624,557
Yolk	334,529	320,334
Total	2,675,100	2,599,829
Inedible product from shell eggs broken	231,649	219,536

NASS, Livestock Branch, (202) 720–3570.

Table 8-56.—Eggs: Number, rate of lay, production, and value, United States, 2000–2009 [1]

Year	Layers average number during year	Rate of lay per layer during year [2]	Eggs, total produced	Price per dozen [3]	Value of production
	Thousands	Number	Millions	Dollars	1,000 dollars
2000	329,067	257	84,717	0.617	4,358,648
2001	336,330	256	86,093	0.622	4,460,701
2002	339,293	257	87,252	0.589	4,284,930
2003	338,579	259	87,516	0.731	5,333,736
2004	342,395	261	89,198	0.713	5,303,038
2005	345,027	262	90,343	0.540	4,066,669
2006	349,700	263	91,788	0.583	4,460,211
2007	346,498	263	91,101	0.885	6,718,853
2008	339,131	266	90,040	1.090	8,215,999
2009 [4]	337,376	268	90,359	NA	N/A

[1] Annual estimates cover the period December 1 previous year through November 30.　[2] Total egg production divided by average number of layers on hand.　[3] Average mid-month price of all eggs sold by producers including hatching eggs.　[4] Preliminary.　NA-not available.
NASS, Livestock Branch, (202) 720–3570.

Table 8-57.—Eggs: Production, price, and value, by State and United States, 2007 and 2008 [1] [2] [3] [4]

State	Eggs produced		Price per dozen		Value of production	
	2007	2008	2007	2008	2007	2008
	Millions	Millions	Dollars	Dollars	1,000 dollars	1,000 dollars
AL	2,090	2,150	1.800	1.670	313,003	298,550
AR	3,236	3,139	1.350	1.600	364,490	418,062
CA	5,290	5,271	0.786	1.000	346,426	440,438
CO	1,059	1,090	0.839	1.070	74,074	96,842
CT	814	780	0.766	0.925	51,938	60,116
FL	2,885	2,749	0.776	1.020	186,471	234,515
GA	4,792	4,576	1.100	1.480	437,491	564,244
HI	81.8	73.3	1.030	1.420	7,428	8,678
ID	155	....	1.030	....	13,354	....
IL	1,357	1,309	0.770	1.000	87,034	109,290
IN	6,673	6,523	0.760	0.985	422,640	535,571
IA	13,868	14,407	0.714	0.931	824,806	1,117,850
KY	1,170	1,140	1.230	1.370	120,075	130,387
LA	490	479	1.040	1.190	42,333	47,694
ME	1,013	1,028	0.949	1.220	80,093	104,433
MD	702	666	0.841	1.130	49,170	62,682
MA	52	36	0.990	1.240	4,288	3,718
MI	2,563	2,653	0.727	0.957	155,371	211,524
MN	2,880	2,767	0.805	1.030	193,219	237,237
MS	1,523	1,511	1.350	1.420	171,379	179,075
MO	1,881	1,885	0.817	1.050	128,026	165,703
MT	107	115	0.904	0.998	8,059	9,567
NE	2,984	2,777	0.744	0.973	185,092	225,242
NH	48	....	1.090	....	4,373	....
NJ	435	....	0.921	....	33,373	....
NY	1,127	1,167	0.787	0.996	73,945	96,871
NC	2,960	3,063	1.330	1.470	328,664	373,944
OH	7,151	7,168	0.811	0.980	483,441	585,477
OK	759	774	1.120	1.250	71,107	80,888
OR	725	769	0.784	1.010	47,379	64,775
PA	6,392	6,189	0.731	0.946	389,119	488,056
SC	1,215	1,143	0.917	1.090	92,809	104,178
SD	843	691	0.732	0.968	51,420	55,752
TN	385	351	1.450	1.460	46,602	42,815
TX	4,995	4,954	0.897	1.120	373,500	462,283
UT	954	914	0.662	0.951	52,618	72,422
VT	59	55	0.869	1.150	4,271	5,252
VA	794	726	1.190	1.270	78,991	77,103
WA	1,520	1,533	0.832	1.070	105,372	136,448
WV	272	247	1.440	1.480	32,723	30,390
WI	1,370	1,220	0.782	1.010	89,263	102,910
WY	3.5	2.4	0.830	0.965	242	193
Oth Sts [5]	1,432	1,951	0.782	1.070	93,381	174,824
US	91,101	90,040	0.885	1.090	6,718,853	8,215,999

[1] Revised data will be published in the "Poultry Production and Value" report.　[2] Annual estimates cover the period December 1, previous year through November 30.　[3] Totals may not add due to rounding.　[4] Includes hatching and market (table) eggs.　[5] 2007: AK, AZ, DE, KS, ND, NM, NV, and RI; 2008: AK, AZ, DE, ID, KS, NH, NJ, ND, NM, NV, and RI combined to avoid disclosing individual operations.
NASS, Livestock Branch, (202) 720–3570.

Table 8-58.—Poultry and poultry products: Cold storage holdings, end of month, United States, 2008 and 2009

Month	Frozen eggs							
	Whites		Yolks		Whole & mixed		Unclassified	
	2008	2009	2008	2009	2008	2009	2008	2009
	1,000 pounds	1,000 pounds	1,000 pounds	1,000 pounds	1,000 pounds	1,000 pounds	1,000 pounds	1,000 pounds
January	2,619	2,812	1,093	1,051	4,814	9,511	3,481	9,184
February	2,952	2,796	962	862	7,097	9,259	5,721	9,148
March	2,886	2,735	953	1,041	6,397	8,152	6,066	8,415
April	2,662	2,927	955	1,010	6,895	7,410	5,522	6,894
May	2,885	2,927	920	1,090	5,396	9,385	3,242	8,312
June	3,056	2,959	901	1,122	8,215	8,578	3,967	8,996
July	3,021	3,158	1,109	1,044	8,419	8,987	8,548	9,389
August	3,146	2,910	1,269	1,309	8,082	9,570	8,303	8,769
September ...	3,083	2,634	1,129	1,483	8,892	8,856	8,786	8,617
October	3,272	3,510	962	1,273	8,898	9,120	9,226	9,009
November	2,873	1,823	1,049	730	8,762	5,871	8,577	12,734
December	3,087	2,505	1,209	1,017	9,003	6,971	9,339	13,151

Month	Frozen eggs, total		Frozen chicken					
	2008	2009	Broilers (Whole)		Hens		Breast and breast meat	
			2008	2009	2008	2009	2008 [1]	2009
	1,000 pounds	1,000 pounds	1,000 pounds	1,000 pounds	1,000 pounds	1,000 pounds	1,000 pounds	1,000 pounds
January	12,007	22,558	20,719	19,468	1,571	4,084	147,223	137,092
February	16,732	22,065	20,240	21,435	1,969	5,132	146,777	129,986
March	16,302	20,343	22,540	20,696	1,774	3,878	142,394	128,039
April	16,034	18,241	26,848	21,739	2,628	4,217	130,403	125,146
May	12,443	21,714	21,576	19,896	2,958	4,342	122,965	119,732
June	16,139	21,655	22,250	19,075	4,805	4,979	124,295	112,051
July	21,097	22,578	25,405	19,704	5,434	5,046	117,935	109,079
August	20,800	22,558	29,452	19,120	6,491	5,167	110,569	98,425
September ...	21,890	21,590	25,618	19,431	6,963	4,975	125,209	105,595
October	22,358	22,912	23,910	19,587	5,686	4,310	139,605	111,015
November	21,261	21,158	24,584	18,761	4,736	2,820	144,507	129,011
December	22,638	23,644	27,213	16,401	3,192	2,215	144,955	127,714

Month	Frozen chicken							
	Drumsticks		Leg quarters		Legs		Thigh and thigh quarters	
	2008 [1]	2009	2008 [1]	2009	2008 [1]	2009	2008 [1]	2009
	1,000 pounds	1,000 pounds	1,000 pounds	1,000 pounds	1,000 pounds	1,000 pounds	1,000 pounds	1,000 pounds
January	18,021	12,210	86,202	93,099	6,533	10,663	15,278	10,525
February	16,643	13,715	83,423	78,923	9,860	9,521	16,876	9,441
March	14,805	11,008	84,795	80,730	11,350	9,372	12,240	8,058
April	21,752	12,594	79,441	74,713	9,276	9,767	12,798	7,969
May	16,255	12,163	98,401	69,765	9,639	7,540	13,582	8,955
June	19,652	11,702	80,366	79,832	8,613	4,000	10,634	8,084
July	18,551	10,088	83,989	87,093	6,185	4,768	10,923	9,615
August	18,847	10,163	97,432	77,291	7,798	4,674	10,371	12,300
September ...	22,141	9,439	113,050	64,305	4,242	5,429	9,556	12,347
October	21,328	10,743	117,606	70,552	7,081	4,304	11,913	8,471
November	21,296	13,620	119,971	73,123	12,474	5,664	12,083	9,613
December	14,627	11,886	102,978	73,932	10,350	6,926	15,722	7,787

Month	Frozen chicken							
	Thigh meat		Wings		Paws and feet		Other chicken	
	2008 [1]	2009	2008 [1]	2009	2008 [1]	2009	2008 [1]	2009
	1,000 pounds	1,000 pounds	1,000 pounds	1,000 pounds	1,000 pounds	1,000 pounds	1,000 pounds	1,000 pounds
January	12,702	14,358	35,811	26,712	19,220	13,814	409,725	352,408
February	11,071	12,050	36,254	26,660	16,110	12,005	401,608	328,891
March	12,796	14,612	35,054	22,201	16,349	11,746	410,668	325,135
April	12,373	13,814	35,392	29,283	22,710	13,049	399,254	345,952
May	13,878	14,524	44,272	28,939	27,568	14,955	384,040	341,125
June	14,699	19,350	39,888	30,692	24,776	16,280	418,798	347,841
July	15,379	19,252	38,958	34,053	24,853	14,672	391,260	368,039
August	15,402	19,780	38,549	39,249	29,031	15,289	391,463	345,026
September ...	15,932	17,403	39,041	39,615	23,892	16,329	373,617	339,489
October	16,421	17,609	42,170	41,070	21,958	18,249	380,252	330,084
November	19,616	19,241	44,842	40,680	17,912	16,158	389,079	330,635
December	15,984	22,799	38,084	34,493	17,259	15,159	375,078	313,978

See footnotes at end of table.

Table 8-58.—Poultry and poultry products: Cold storage holdings, end of month, United States, 2008 and 2009—Continued

Month	Frozen chicken, total		Frozen turkey					
			Toms		Hens		Total whole	
	2008	2009	2008	2009	2008	2009	2008	2009
	1,000 pounds	*1,000 pounds*	*1,000 pounds*	*1,000 pounds*	*1,000 pounds*	*1,000 pounds*	*1,000 pounds*	*1,000 pounds*
January	773,005	694,433	54,376	101,102	43,537	92,540	97,913	193,642
February	760,831	647,759	79,694	110,627	75,580	105,840	155,274	216,467
March	764,765	635,475	96,027	135,651	84,151	117,368	180,178	253,019
April	752,875	658,243	110,289	151,958	102,539	131,789	212,828	283,747
May	755,134	641,936	118,515	161,112	114,220	137,237	232,705	298,349
June	768,776	653,886	138,164	171,748	119,720	141,242	257,884	312,990
July	738,872	681,409	160,615	195,844	129,538	154,583	290,153	350,427
August	755,405	646,484	179,144	214,856	139,947	162,072	319,091	376,928
September ...	759,321	634,357	193,104	210,355	141,345	150,851	334,449	361,206
October	787,930	635,994	177,613	167,473	133,136	130,899	310,749	298,372
November	811,100	659,326	67,205	37,151	64,201	39,044	131,406	76,195
December	765,442	633,290	68,890	38,218	70,462	38,161	139,352	76,379

| Month | Total | | Frozen ducks | | Total frozen poultry | | N/A | |
	2008	2009	2008	2009	2008	2009	XX	XX
	1,000 pounds	*1,000 pounds*	*1,000 pounds*	*1,000 pounds*	*1,000 pounds*	*1,000 pounds*		
January	327,590	446,197	5,088	3,416	1,105,683	1,144,046	....	....
February	416,694	462,395	5,228	3,456	1,182,753	1,113,610	....	....
March	428,133	513,384	4,227	4,299	1,197,125	1,153,158	....	....
April	491,283	571,708	3,347	5,406	1,247,505	1,235,357	....	....
May	522,421	585,745	2,432	4,962	1,279,987	1,232,643	....	....
June	562,693	594,742	1,684	5,129	1,333,153	1,253,757	....	....
July	620,692	641,060	2,485	6,282	1,362,049	1,328,751	....	....
August	629,236	653,546	2,237	5,693	1,386,878	1,305,723	....	....
September ...	621,475	613,891	1,571	5,261	1,382,367	1,253,509	....	....
October	577,984	517,463	1,774	5,240	1,367,688	1,158,697	....	....
November	360,379	244,480	1,951	4,507	1,173,430	908,313	....	....
December	396,144	261,838	1,993	3,588	1,163,579	898,716	....	....

NASS, Livestock Branch, (202) 720–3570.

Table 8-59.—Dairy products: Cold storage holdings, end of month, United States, 2008 and 2009

Month	Butter		American cheese	
	2008	2009	2008	2009
	1,000 pounds	*1,000 pounds*	*1,000 pounds*	*1,000 pounds*
January	188,072	176,526	494,376	533,402
February	210,422	204,927	513,073	541,739
March	224,804	212,477	526,001	548,568
April	251,533	240,044	543,068	577,391
May	269,474	253,310	568,439	586,053
June	258,360	262,854	581,842	602,049
July	246,132	262,782	577,636	605,022
August	213,744	259,578	567,661	598,710
September	186,878	227,924	549,635	596,191
October	149,391	190,624	540,140	579,808
November	119,946	142,661	526,941	583,056
December	118,962	133,022	538,105	584,981

Month	Swiss cheese		Other natural cheese	
	2008	2009	2008	2009
	1,000 pounds	*1,000 pounds*	*1,000 pounds*	*1,000 pounds*
January	27,939	23,148	259,048	325,826
February	25,891	23,322	262,028	327,480
March	24,533	23,235	273,775	343,386
April	24,375	22,855	288,420	338,668
May	26,917	22,102	285,933	362,165
June	25,264	23,067	295,365	362,283
July	24,712	23,270	300,449	371,879
August	25,142	23,414	287,473	375,496
September	24,129	23,110	260,413	364,640
October	21,994	23,537	266,839	365,840
November	22,723	24,254	268,969	354,435
December	22,589	24,791	291,266	356,986

Month	Total Natural cheese	
	2008	2009
	1,000 pounds	*1,000 pounds*
January	781,363	882,376
February	800,992	892,541
March	824,309	915,189
April	855,863	938,914
May	881,289	970,320
June	902,471	987,399
July	902,797	1,000,171
August	880,276	997,620
September	834,177	983,941
October	828,973	969,185
November	818,633	961,745
December	851,960	966,758

NASS, Livestock Branch, (202) 720–3570.

CHAPTER IX

FARM RESOURCES, INCOME, AND EXPENSES

The statistics in this chapter deal with farms, farm resources, farm income, and expenses. Many of the series are estimates developed in connection with economic research activities of the Department.

Table 9-1.—Economic trends: Data relating to agriculture, United States, 2000–2009

Year	Prices paid by farmers [1]			Farm income [2]		
	Total including interest, taxes, and wage rates	Production items	Prices received by farmers [1]	Gross farm income [3]	Production expenses	Net farm income
	Index numbers 1990–92=100	Index numbers 1990–92=100	Index numbers 1990–92=100	Billion dollars	Billion dollars	Billion dollars
2000	119	115	96	241.7	191.0	50.7
2001	123	120	102	249.9	195.0	54.9
2002	124	119	98	230.6	191.4	39.1
2003	128	124	106	258.7	197.7	61.0
2004	134	132	118	294.9	207.5	87.4
2005	142	140	114	298.5	219.7	78.8
2006	150	150	115	290.2	232.7	57.4
2007	161	162	136	339.5	269.2	70.3
2008	183	188	149	379.6	293.0	86.6
2009 [4]	179	183	131	343.2	281.0	62.2

Year	National income [5]	Personal income [5]	Industrial production [6]	Consumer prices all items [7]	Producer prices consumer foods [7]
	Billion dollars	Billion dollars	Index numbers 2007=100	Index numbers 1982–84=100	Index numbers 1982=100
2000	8,938.9	8,559.4	92.0	172.2	137.2
2001	9,185.2	8,883.3	88.9	177.1	141.3
2002	9,408.5	9,060.1	89.1	179.9	140.1
2003	9,840.2	9,378.1	90.2	184.0	146.0
2004	10,534.0	9,937.2	92.3	188.9	152.7
2005	11,273.8	10,485.9	95.3	195.3	155.7
2006	12,031.2	11,268.1	97.4	201.6	156.7
2007	12,396.4	11,912.3	100.0	207.3	167.0
2008	12,557.8	12,391.1	96.7	215.3	178.3
2009 [4]	12,225.0	12,174.9	87.7	214.5	175.5

[1] U.S. Department of Agriculture - NASS. [2] U.S. Department of Agriculture - ERS. [3] Includes cash receipts from farm marketings, government payments, nonmoney income (gross rental value of dwelling and value of home consumption), other income (machine hire custom work and recreational income), and value of change in farm inventories. [4] Forecast. [5] Department of Commerce, Bureau of Economic Analysis. [6] Federal Reserve Board. [7] U.S. Department of Labor, Bureau of Labor Statistics.

ERS, Farm and Rural Business Branch, (202) 694–5592. E- mail contact is Roger Strickland at rogers@ers.usda.gov. For National Income, Personal Income, Industrial Production and Consumer Price Indexes, Contact David Torgerson at (202) 694-5334. E-mail contact is dtorg@ers.usda.gov.

Table 9-2.—Farms: Number, land in farms, and average size of farm, United States, 2000–2009 [1]

Year	Farms [2] [3]	Land in farms	Average size farm
	Number	1,000 acres	Acres
2000	2,166,780	945,080	436
2001	2,148,630	942,070	438
2002	2,135,360	940,300	440
2003	2,126,860	936,750	440
2004	2,112,970	932,260	441
2005	2,098,690	927,940	442
2006	2,088,790	925,790	443
2007	2,204,950	921,460	418
2008	2,200,100	919,910	418
2009 [4]	2,200,010	919,800	418

[1] The farm definition was changed in 1993 to include maple syrup, short rotation woody crops, and places with 5 or more horses. [2] A farm is any establishment from which $1,000 or more of agricultural products were sold or would normally be sold during the year. [3] Includes some accounting for individual farms on reservation land in AZ and NM from 1998 forward. [4] Preliminary.

NASS, Environmental, Economics, and Demographics Branch, (202) 720–6146.

Table 9-3.—Farms: Percent of farms, land in farms, and average size, by economic sales class, United States, 2008–2009

Economic sales class	Percent of total				Average size farm	
	Farms		Land		2008	2009 [1]
	2008	2009 [1]	2008	2009 [1]		
	Percent	Percent	Percent	Percent	Acres	Acres
$1,000–$2,499	27.8	28.0	4.0	3.7	60	55
$2,500–$4,999	14.4	14.4	3.3	3.2	96	93
$5,000–$9,999	13.3	13.4	4.3	4.6	135	143
$10,000–$24,999	12.1	11.8	6.8	6.5	235	230
$25,000–$49,999	8.0	8.0	7.4	7.8	387	408
$50,000–$99,999	7.4	7.4	10.3	10.4	582	588
$100,000–$249,999	6.8	6.8	16.4	16.1	1,010	991
$250,000–$499,999	4.5	4.5	15.6	16.2	1,452	1,506
$500,000–$999,999	3.4	3.4	15.4	15.4	1,897	1,895
$1,000,000+	2.3	2.3	16.5	16.1	3,004	2,936
Total	100.0	100.0	100.0	100.0	418	418

[1] Preliminary.
NASS, Environmental, Economics, and Demographics Branch, (202) 720–6146.

Table 9-4.—Number of farms: Economic sales class by region and United States, 2007–2009

Region and year	Economic Sales Class					Total
	$1,000–$9,999	$10,000–$99,999	$100,000–$249,999	$250,000–$499,999	$500,000 & over	
	Number	Number	Number	Number	Number	Number
NE: [1]						
2007	83,350	37,210	11,120	5,830	5,460	142,970
2008	83,190	37,360	10,920	5,720	5,980	143,170
2009	83,400	37,350	11,070	5,720	5,630	143,170
NC: [2]						
2007	366,300	237,800	86,500	57,400	58,300	806,300
2008	361,800	235,200	84,550	58,700	62,450	802,700
2009	361,200	233,800	84,400	59,600	62,800	802,800
South: [3]						
2007	598,500	242,950	30,410	20,870	36,670	929,400
2008	598,100	241,630	30,760	20,750	37,210	928,450
2009	605,200	235,130	30,820	20,960	35,970	928,080
West: [4]						
2007	180,410	90,710	22,270	13,130	19,760	326,280
2008	179,010	90,310	22,770	13,330	20,360	325,780
2009	178,400	91,700	22,200	13,340	20,320	325,960
US:						
2007	1,228,560	608,670	150,300	97,230	120,190	2,204,950
2008	1,222,100	604,500	149,000	98,500	126,000	2,200,100
2009	1,228,200	597,980	149,490	99,620	124,720	2,200,010
PR:						
2007	7,300	3,100	500	250	250	11,400
2008	7,200	3,000	500	250	250	11,200
2009	6,800	2,800	400	200	200	10,400

[1] CT, ME, MA, NH, NJ, NY, PA, RI, and VT. [2] IL, IN, IA, KS, MI, MN, MO, NE, ND, OH, SD, WI. [3] AL, AR, DE, FL, GA, KY, LA, MD, MS, NC, OK, SC, TN, TX, VA, WV. [4] AK, AZ, CA, CO, HI, ID, MT, NV, NM, OR, UT, WA, WY.
NASS, Environmental, Economics, and Demographics Branch, (202) 720–6146.

Table 9-5.—Land in farms: Economic sales class by region and United States, 2007–2009

Region and year	Economic Sales Class					Total
	$1,000-$9,999	$10,000-$99,999	$100,000-$249,999	$250,000-$499,999	$500,000 & over	
	1,000 Acres	*1,000 Acres*	*1,000 Acres*	*1,000 Acres*	*1,000 Acres*	*1,000 Acres*
NE: [1]						
2007	5,620	5,110	2,687	2,295	4,068	19,780
2008	5,400	5,010	2,600	2,245	4,355	19,610
2009	5,500	5,100	2,610	2,215	4,185	19,610
NC: [2]						
2007	28,150	63,400	59,650	66,300	125,200	342,700
2008	27,400	62,400	57,900	64,700	130,100	342,500
2009	26,150	62,100	57,200	65,650	131,300	342,400
South: [3]						
2007	56,180	88,500	37,440	29,830	68,020	279,970
2008	56,300	87,500	37,300	29,700	68,700	279,500
2009	56,350	88,150	34,740	35,170	64,800	279,300
West: [4]						
2007	17,920	71,210	53,630	47,300	88,950	279,010
2008	17,600	70,400	53,500	46,900	89,900	278,300
2009	17,300	71,550	53,450	46,400	89,700	278,400
US:						
2007	107,870	228,220	153,407	145,725	286,238	921,460
2008	106,700	225,310	151,300	143,545	293,055	919,910
2009	105,480	226,900	148,000	149,435	289,985	919,800
PR:						
2007	130	170	70	40	80	490
2008	120	160	70	40	80	470
2009	120	155	65	40	70	450

[1] CT, ME, MA, NH, NJ, NY, PA, RI, and VT.   [2] IL, IN, IA, KS, MI, MN, MO, NE, ND, OH, SD, WI.   [3] AL, AR, DE, FL, GA, KY, LA, MD, MS, NC, OK, SC, TN, TX, VA, WV.   [4] AK, AZ, CA, CO, HI, ID, MT, NV, NM, OR, UT, WA, WY.

NASS, Environmental, Economics, and Demographics Branch, (202) 720–6146.

Table 9-6.—Land in farms: Classification by tenure of operator, United States, 1925–2008

Year	Land in farms	Tenure of operator			
		Full owners	Part owners	Managers	All tenants
	Acres	*Percent*	*Percent*	*Percent*	*Percent*
1925	924,319,352	45.4	21.3	4.7	28.7
1930 [1]	990,111,984	37.6	24.9	6.4	31.0
1935	1,054,515,111	37.1	25.2	5.8	31.9
1940 [1]	1,065,113,774	35.9	28.2	6.5	29.4
1945	1,141,615,364	36.1	32.5	9.3	22.0
1950 [1]	1,161,419,720	36.1	36.4	9.2	18.3
1954	1,158,191,511	34.2	40.7	8.6	16.5
1959 [1]	1,123,507,574	31.0	44.0	9.8	14.8
1964 [1]	1,110,187,000	28.7	48.0	10.2	13.1
1969 [1]	1,062,892,501	35.3	51.8	....	13.0
1974 [1]	1,017,030,357	35.3	52.6	....	12.0
1978 [1]	1,014,777,234	32.7	55.3	....	12.0
1982 [1]	986,796,579	34.7	53.8	....	11.5
1987 [1]	964,470,625	32.9	53.9	....	13.2
1992 [1]	945,531,506	31.3	55.7	....	13.0
1997 [2]	932,475,414	26.7	62.2	....	11.2
1998 [2]	900,415,615	28.6	60.2	....	11.2
1999 [2]	870,720,495	25.6	61.6	....	12.8
2000 [2]	994,997,682	26.4	62.3	....	11.4
2001 [2]	959,163,331	24.7	61.2	....	14.2
2002 [2]	954,302,543	29.4	56.6	....	14.0
2003 [2]	926,985,610	28.9	59.7	....	11.4
2004 [2]	990,395,334	30.3	56.0	....	13.7
2005 [2]	916,304,251	27.6	60.2	....	12.3
2006 [2]	900,882,842	28.8	59.8	....	11.5
2007 [2]	861,754,674	29.0	57.9	....	13.1
2008 [2]	894,010,108	34.0	55.3	....	10.7

[1] Includes Alaska and Hawaii.   [2] Excludes Alaska and Hawaii.
ERS, Resource and Rural Economics Division, (202) 694–5575. Data for 1910–1992 is from the Census of Agriculture, U.S. Department of Commerce. Data for 1997-2008 is from ERS Agricultural Resource Management Survey.

Table 9-7.—Farms: Classification by tenure of operator, United States, 1925–2008

Year	Farms	Tenure of operator			
		Full owners	Part owners	Managers	All tenants
	Number	Percent	Percent	Percent	Percent
1925	6,371,640	52.0	8.7	0.6	38.6
1930 [1]	6,295,103	46.3	10.4	0.9	42.4
1935	6,812,350	47.1	10.1	0.7	42.1
1940 [1]	6,102,417	50.6	10.1	0.6	38.8
1945	5,859,169	56.4	11.3	0.7	31.7
1950 [1]	5,388,437	57.4	15.3	0.4	26.9
1954	4,783,021	57.4	18.2	0.4	24.0
1959 [1]	3,710,503	57.1	21.9	0.6	20.5
1964 [1]	3,157,857	57.6	24.8	0.6	17.1
1969 [1]	2,730,250	62.5	24.6	....	12.9
1974 [1]	2,314,013	61.5	27.2	....	11.3
1978 [1]	2,257,775	57.5	30.2	....	12.3
1982 [1]	2,240,976	59.2	29.3	....	11.6
1987 [1]	2,087,759	59.3	29.2	....	11.5
1992 [1]	1,925,300	57.7	31.0	....	11.3
1997 [2]	2,049,384	55.3	35.4	....	9.3
1998 [2]	2,054,709	56.5	33.9	....	9.6
1999 [2]	2,186,950	58.3	33.9	....	7.8
2000 [2]	2,166,060	57.7	34.1	....	8.2
2001 [2]	2,149,683	57.2	34.9	....	8.0
2002 [2]	2,152,412	65.9	26.7	....	7.3
2003 [2]	2,121,107	62.1	31.7	....	6.1
2004 [2]	2,107,925	61.8	32.1	....	6.1
2005 [2]	2,094,876	62.3	31.1	....	6.6
2006 [2]	2,083,674	62.7	31.0	....	6.3
2007 [2]	2,069,371	64.9	28.6	....	6.5
2008 [2]	2,191,844	65.7	28.3	....	6.0

[1] Includes Alaska and Hawaii.　[2] Excludes Alaska and Hawaii.
ERS, Resource and Rural Economics Division, (202) 694–5575. Data for 1920-1992 is from the Census of Agriculture, U.S. Department of Commerce. Data for 1997-2008 is from ERS Agricultural Resource Management Survey.

Table 9-8.—Farmland Rented: Classification by Tenants and Part Owners, United States, 1920–2007

Year	Land in farms	Tenure of operator [1]			Percentage of land rented
		Tenants	Part-owners	Total	
	Million acres	Million acres	Million acres	Million acres	Percent
1920	958.7	[2] 265.0	[3] 54.7	319.7	33.3
1925	924.3	264.9	96.3	361.2	39.1
1930	990.1	307.3	125.2	432.5	43.7
1935	1,054.5	336.8	134.3	471.1	44.7
1940	1,065.1	313.2	155.9	469.1	44.0
1945	1,141.6	251.6	178.9	430.5	37.7
1950	1,161.4	212.2	196.2	408.4	35.2
1954	1,158.2	192.6	212.3	404.9	35.0
1959	1,123.0	166.8	234.1	400.9	35.7
1964	1,110.2	144.9	248.1	[4] 393.0	35.4
1969	1,063.3	137.6	241.8	379.4	35.7
1974	1,017.0	122.3	258.4	380.7	37.4
1978	1,029.7	124.1	282.2	406.2	39.4
1982	986.2	113.6	269.9	383.5	38.9
1987	964.5	126.9	275.4	402.3	41.7
1992	945.5	122.7	282.2	404.9	42.8
1997	931.8	108.1	270.0	378.1	40.6
2002 [5]	938.3	86.5	266.8	353.3	37.7
2007 [5]	922.1	81.8	269.0	350.8	38.0

[1] Columns 3, 4, and 5 refer only to land rented from others and operated, so subleased land is not included. Acres of land rented are comparable in the same year, but definitions change over time. Basic sources are 1969 Census of Agriculture, table 5, p.14; 1974 Census of Agriculture, table 3, pp.1-6; 1978 Census of Agriculture, vol. 1, part 51, table 5, pp. 124-127; 1982 Census of Agriculture, vol. 1, part 51, table 48, p. 49; 1987 Census of Agriculture. 1 part 51, table 48, p. 49; 1992 Census of Agriculture vol. 1, part 51, table 46, p. 53; 1997 Census of Agriculture, vol. 1, part 51, chapter 1, table 46, p. 57; 2002 Census of Agriculture, vol. 1, part 51, chapter 1, table 61, p. 214; 2007 Census of Agriculture, vol. 1, part 51, chapter 1, table 65, p. 262; and earlier census volumes as noted.　[2] 1920 Census of Agriculture, vol. VI, part 1, table 5, p. 19.　[3] Assumes same proportion of owner and part-owner as in 1910.　[4] 1964 Census of Agriculture, vol. II, chapter 8, p.757.　[5] The 2002 Census of Agriculture introduced new methodology to account for all farms in the United States. All 2002 and 2007 published census items were weighted for undercoverage.　Strictly speaking, 2002 and 2007 data are not fully comparable with data from earlier years.
ERS, Resource and Rural Economics Division, (202) 694–5572. Data from the Census of Agriculture, National Agricultural Statistics Service and Economic Research Service.

Table 9-9.—Farms: Number and land in farms, by State and United States, 2008–2009

State	Farms [1]		Land in farms		Average per acre	
	2008	2009 [2]	2008	2009 [2]	2008	2009 [2]
	Number	Number	1,000 acres	1,000 acres	Acres	Acres
AL	48,500	48,500	8,950	9,000	185	186
AK	680	680	890	880	1,309	1,294
AZ	15,600	15,500	26,100	26,100	1,673	1,684
AR	49,300	49,100	13,700	13,600	278	277
CA	81,500	81,500	25,400	25,400	312	312
CO	36,500	36,200	31,300	31,300	858	865
CT	4,900	4,900	400	400	82	82
DE	2,500	2,480	500	490	200	198
FL	47,500	47,500	9,250	9,250	195	195
GA	47,800	47,600	10,400	10,300	218	216
HI	7,500	7,500	1,110	1,120	148	149
ID	25,200	25,500	11,400	11,400	452	447
IL	75,900	75,800	26,700	26,700	352	352
IN	61,000	61,500	14,800	14,800	243	241
IA	92,600	92,600	30,800	30,800	333	333
KS	65,500	65,500	46,200	46,200	705	705
KY	85,300	85,500	14,000	14,000	164	164
LA	30,000	30,000	8,050	8,050	268	268
ME	8,100	8,100	1,350	1,350	167	167
MD	12,850	12,800	2,050	2,050	160	160
MA	7,700	7,700	520	520	68	68
MI	55,000	54,800	10,000	10,000	182	182
MN	81,000	81,000	26,900	26,900	332	332
MS	42,000	42,300	11,000	11,050	262	261
MO	108,000	108,000	29,100	29,100	269	269
MT	29,500	29,800	60,800	60,800	2,061	2,040
NE	47,400	47,200	45,600	45,600	962	966
NV	3,100	3,080	5,900	5,900	1,903	1,916
NH	4,150	4,150	470	470	113	113
NJ	10,300	10,300	730	730	71	71
NM	20,600	20,500	43,000	43,000	2,087	2,098
NY	36,600	36,600	7,100	7,100	194	194
NC	52,500	52,400	8,600	8,600	164	164
ND	32,000	32,000	39,600	39,600	1,238	1,238
OH	75,000	74,900	13,900	13,800	185	184
OK	86,600	86,500	35,100	35,100	405	406
OR	38,600	38,600	16,400	16,400	425	425
PA	63,200	63,200	7,750	7,750	123	123
RI	1,220	1,220	70	70	57	57
SC	26,900	27,000	4,900	4,900	182	181
SD	31,300	31,500	43,700	43,700	1,396	1,387
TN	79,000	78,700	10,900	10,900	138	139
TX	247,500	247,500	130,400	130,400	527	527
UT	16,500	16,600	11,100	11,100	673	669
VT	7,000	7,000	1,220	1,220	174	174
VA	47,000	47,000	8,000	8,000	170	170
WA	39,500	39,500	14,800	14,800	375	375
WV	23,200	23,200	3,700	3,700	159	159
WI	78,000	78,000	15,200	15,200	195	195
WY	11,000	11,000	30,100	30,200	2,736	2,745
US	2,200,100	2,200,010	919,910	919,800	418	418
PR	11,200	10,400	470	450	42	43

[1] A farm is any establishment from which $1,000 or more of agricultural products were sold or would normally be sold during the year. [2] Preliminary.
NASS, Environmental, Economics, and Demographics Branch, (202) 720–6146.

Table 9-10.—Land: Utilization, by State and United States, 2007

State	Cropland			Forest land[2]	Urban land
	Used for crops [1]	Idle	Used only for pasture		
	1,000 acres	*1,000 acres*	*1,000 acres*	*1,000 acres*	*1,000 acres*
AL	2,070	406	627	22,580	1,140
AK	31	48	7	90,447	167
AZ	818	173	0	16,780	1,099
AR	7,409	94	736	18,480	589
CA	8,084	762	809	26,983	5,166
CO	8,110	4,115	1,242	20,036	831
CT	111	13	13	1,472	1,153
DE	404	10	9	376	200
DC	....	0	0	0	39
FL	2,098	90	571	15,552	4,052
GA	3,665	364	590	24,247	2,465
HI	72	49	23	1,552	226
ID	4,680	1,184	511	16,203	269
IL	22,778	970	309	4,363	2,341
IN	12,201	262	284	4,533	1,450
IA	24,277	1,608	845	2,824	520
KS	24,649	5,893	1,292	2,028	562
KY	5,395	676	1,550	11,248	799
LA	3,340	460	635	14,116	1,088
ME	341	85	37	17,163	209
MD	1,208	71	0	2,372	1,189
MA	139	14	16	2,647	1,837
MI	7,145	553	316	19,023	2,189
MN	19,857	1,745	740	15,113	965
MS	4,411	376	769	19,536	607
MO	13,301	1,372	1,887	14,674	1,186
MT	12,631	7,010	1,678	19,790	171
NE	19,495	2,240	896	1,174	298
NV	516	35	185	9,436	356
NH	76	10	17	4,674	359
NJ	415	24	40	1,476	1,816
NM	1,174	633	648	14,977	493
NY	3,594	266	280	16,015	2,571
NC	4,258	247	339	17,916	2,357
ND	23,290	4,125	817	534	95
OH	10,141	431	352	7,644	2,603
OK	9,169	1,336	2,781	6,234	733
OR	3,560	1.196	677	27,813	675
PA	4,325	513	427	16,018	2,799
RI	18	2	2	326	254
SC	1,536	201	264	12,641	1,230
SD	17,026	1,928	1,311	1,552	109
TN	4,406	410	1,203	13,613	1,594
TX	21,515	5,709	7,938	11,859	4,646
UT	1,137	407	403	16,058	450
VT	413	27	47	4,482	96
VA	2,570	194	487	15,309	1,555
WA	5,733	2,789	372	18,873	1,397
WV	678	51	192	11,797	372
WI	9,069	738	395	16,042	1,063
WY	1,657	494	420	5,997	110
US	334,996	52,409	35,989	656,596	60,542

[1] Cropland harvested, crop failure, and cultivated summer fallow.   [2] Excludes reserved and other forest land duplicated in parks and other special uses of land. Includes forested grazing land.

ERS, Resource Economics Division, (202) 694–5528. Estimates based on reports and records of the U.S. Departments of Agriculture and Commerce, and public land administering and conservation agencies. Estimates developed for years coinciding with a Census of Agriculture.

Table 9-11.—Land in farms:[1] Irrigated land, by State and United States, 1969–2007

State	1969	1974	1978[2]	1982	1987	1992	1997	2002	2007
	1,000 acres	*1,000 acres*	*1,000 acres*	*1,000 acres*	*1,000 acres*	*1,000 acres*	*1,000 acres*	*1,000 acres*	*1,000 acres*
AL	11	14	59	66	84	82	80	109	113
AK	1	1	1	1	2	2	3	3	4
AZ	1,178	1,153	1,196	1,098	914	956	1,075	932	876
AR	1,010	949	1,683	2,022	2,406	2,702	3,785	4,150	4,461
CA	7,240	7,749	8,506	8,461	7,596	7,571	8,887	8,709	8,016
CO	2,895	2,874	3,431	3,201	3,014	3,170	3,374	2,591	2,868
CT	9	7	7	7	7	6	8	10	10
DE	20	20	34	44	61	62	75	97	105
FL	1,365	1,559	1,980	1,585	1,623	1,783	1,874	1,815	1,552
GA	79	112	463	575	640	725	773	871	1,018
HI	146	142	159	146	149	134	77	69	59
ID	2,761	2,859	3,475	3,450	3,219	3,260	3,544	3,289	3,300
IL	51	54	130	166	208	328	352	391	474
IN	34	33	75	132	170	241	256	313	397
IA	21	39	101	91	92	116	133	142	190
KS	1,522	2,010	2,686	2,675	2,463	2,680	2,696	2,678	2,763
KY	20	11	14	23	38	28	60	37	59
LA	702	702	681	694	647	898	961	939	954
ME	6	6	7	6	6	10	22	20	21
MD	22	23	28	39	51	57	69	81	93
MA	19	19	17	17	20	20	27	24	23
MI	77	97	226	286	315	366	407	456	500
MN	36	78	272	315	354	370	403	455	506
MS	150	162	309	431	637	883	1,110	1,176	1,369
MO	156	150	320	403	535	709	921	1,033	1,200
MT	1,841	1,759	2,070	2,023	1,997	1,978	2,102	1,976	2,013
NE	2,857	3,967	5,683	6,039	5,682	6,312	7,066	7,625	8,559
NV	753	778	881	830	779	556	764	747	691
NH	2	2	2	1	3	2	3	2	2
NJ	72	89	77	83	91	80	94	97	95
NM	823	867	891	807	718	738	852	845	830
NY	55	55	56	52	51	47	74	75	68
NC	59	51	90	81	138	113	156	264	232
ND	63	71	141	163	168	187	183	203	236
OH	22	22	25	28	32	29	35	41	38
OK	524	515	602	492	478	512	509	518	535
OR	1,519	1,561	1,881	1,808	1,648	1,622	1,963	1,908	1,845
PA	19	18	15	18	30	23	40	43	38
RI	2	2	3	2	4	3	3	4	4
SC	15	10	32	81	81	76	89	96	132
SD	148	152	335	376	362	371	367	401	374
TN	12	10	13	18	38	37	47	61	81
TX	6,888	6,594	6,947	5,576	4,271	4,912	5,764	5,075	5,010
UT	1,025	970	1,169	1,082	1,161	1,143	1,218	1,091	1,134
VT	(3)	1	1	1	2	2	3	2	2
VA	37	28	42	43	79	62	86	99	82
WA	1,224	1,309	1,639	1,638	1,519	1,641	1,787	1,823	1,736
WV	3	2	1	1	3	3	4	2	2
WI	106	128	235	259	285	331	358	386	377
WY	1,523	1,460	1,662	1,565	1,518	1,465	1,750	1,542	1,551
US	39,122	41,243	50,350	49,003	46,386	49,404	56,289	55,316	56,599
PR	91	70	54	42	36	46	35	43	40
VI	(3)	(3)	(4)	(4)	(4)	(4)	(4)	(3)	(3)
Total	39,213	41,313	50,350	49,002	46,386	49,404	55,058	55,360	56,639

[1] Data may not add because of rounding.　[2] Data for 1978 not directly comparable with earlier censuses as it includes estimates from the direct enumeration sample for farms not represented on the mail list.　[3] Less than 500 acres.　[4] Not available.　Note: Data from the Census of Agriculture, U.S. Department of Commerce. Beginning in 1997 Census of Agriculture, U.S. Department of Agriculture.

ERS, Resource Economics Division, (202) 694–5528.

Table 9-12.—Farm real estate: Value of farmland and buildings, by State and United States, 2004–2008 [1]

State	Total value of land and buildings				
	2004	2005	2006	2007	2008
	Million dollars	Million dollars	Million dollars	Million dollars	Million dollars
AL	16,465	18,040	18,690	19,800	20,585
AZ[2]	10,846	13,709	16,311	16,926	18,156
AR	23,288	26,270	28,000	31,136	33,154
CA	100,320	130,795	137,752	151,384	163,576
CO	23,636	27,990	31,926	35,482	35,995
CT	3,952	4,368	4,840	5,207	5,080
DE	3,233	4,420	5,304	5,457	5,150
FL	28,704	45,840	49,476	51,150	52,170
GA	25,164	33,598	41,800	44,805	44,720
ID	15,444	19,305	25,520	28,520	28,500
IL	69,632	86,670	96,571	107,736	121,485
IN	40,975	44,700	48,425	53,872	60,680
IA	69,080	82,368	89,919	103,796	121,660
KS	32,760	38,200	41,282	45,374	47,124
KY	27,522	34,300	37,380	38,360	39,900
LA	12,400	13,600	14,240	15,795	16,503
ME	2,562	2,706	2,849	3,011	2,970
MD	11,536	15,450	17,304	17,510	16,400
MA	5,158	5,460	6,084	6,188	6,396
MI	29,290	31,000	34,037	37,600	39,000
MN	49,046	56,032	63,180	72,630	79,893
MS	16,133	18,122	19,470	21,670	22,880
MO	46,332	51,625	55,963	62,930	66,930
MT	24,160	29,088	43,776	50,547	54,720
NE	37,098	41,587	47,071	51,984	60,648
NV[2]	2,556	3,198	4,150	4,777	4,871
NH	1,530	1,739	1,950	2,256	2,303
NJ	8,532	9,500	10,656	11,461	11,169
NM[2]	11,592	13,942	16,684	18,505	19,829
NY	13,257	13,965	14,685	15,696	16,685
NC	29,250	33,616	35,728	37,238	38,270
ND	18,124	20,094	22,852	25,740	30,492
OH	42,195	44,588	48,222	51,800	55,878
OK	25,725	31,050	33,756	37,908	40,365
OR	21,420	23,046	24,402	28,208	31,160
PA	27,258	31,044	34,164	38,766	39,680
RI	763	896	1,071	1,148	1,176
SC	10,422	11,542	12,470	13,818	14,455
SD	21,462	25,783	29,716	33,649	40,204
TN	28,405	31,248	32,967	35,750	37,605
TX	109,200	129,800	154,343	179,952	202,120
UT[2]	8,514	9,607	11,293	12,479	13,291
VT	2,688	2,900	3,100	3,370	3,538
VA	26,040	31,540	36,326	39,690	40,000
WA	22,952	24,450	25,800	27,565	29,896
WV	5,400	6,840	7,200	8,510	9,250
WI	38,285	42,966	47,430	55,328	58,520
WY	10,560	10,982	12,640	14,798	16,856
US (48 States)	1,210,866	1,429,586	1,598,775	1,777,282	1,921,888

[1] Total value of land and buildings is derived by multiplying average value per acre of farm real estate by the land in farms.   [2] Value of all land and buildings adjusted to include American Indian reservation land value.
NASS, Environmental, Economics, and Demographics Branch, (202) 720–6146.

Table 9-13.—Land utilization, United States, selected years, 1949–2007

Major land uses	1949	1959	1969	1978	1987	1992	1997	2002	2007
	Million acres	Million acres	Million acres	Million acres	Million acres	Million acres	Million acres	Million acres	Million acres
Cropland used for crops [1]	383	359	333	369	331	338	349	340	335
Idle cropland	26	34	51	26	68	56	39	40	52
Cropland used for pasture [2]	69	66	88	76	65	67	68	62	36
Grassland pasture [3]	632	633	604	587	591	591	580	587	....
Forest land [4]	760	745	723	703	648	648	641	651	657
Special uses [5]	87	115	143	158	279	281	286	297	....
Urban areas [6]	18	27	31	45	57	59	66	60	61
Other land [7]	298	293	291	301	227	224	236	228	....
Total land area [8]	2,273	2,271	2,264	2,264	2,265	2,263	2,263	2,264	....

[1] Cropland harvested, crop failure, and cultivated summer fallow. [2] The 2007 estimate declined due to a change in the methodology for determining cropland used for pasture for non-respondents. [3] Grassland and other nonforest pasture and range. [4] Excludes reserved and other forest land duplicated in parks and special uses of land. Includes forested grazing land. [5] Includes rural transportation areas, Federal and State areas used primarily for recreation and wildlife purposes, military areas, farmsteads, and farm roads and lanes. [6] The 2002 urban acreage estimate is not directly comparable to estimates in prior years due to a change in the definition of urban areas in the 2000 Census of Population and Housing. The apparent change in "urban" acreage between 1997 and 2002 reflects a definitional change, rather than a decline in acreage. [7] Miscellaneous areas such as marshes, open swamps, bare rock areas, deserts, and other uses not inventoried. [8] Remeasurement and increases in reservoirs account for changes in total land areas except for the major increase in 1949 when data for Alaska and Hawaii were added. ERS, Rural and Resource Economics Division, (202) 694–5626. Estimates based on reports and records of the U.S. Department of Agriculture and Commerce, and public land administering and conservation agencies. Estimates developed for years coinciding with a Census of Agriculture. See http://www.ers.usda.gov/data/majorlanduses for data and more information.

Table 9-14.—Farm real estate: Average value per acre of land and buildings, by State and United States, Mar. 1, 1970, and Jan. 1, 2005–2009

State	Mar. 1, 1970	Jan. 1, 2005	Jan. 1, 2006	Jan. 1, 2007	Jan. 1, 2008	Jan. 1, 2009
	Dollars	Dollars	Dollars	Dollars	Dollars	Dollars
AL	200	2,050	2,100	2,200	2,300	2,150
AZ [1]	70	2,330	3,050	3,200	3,500	3,500
AR	260	1,850	2,000	2,240	2,420	2,390
CA	479	5,050	5,360	5,960	6,440	6,600
CO	95	900	1,020	1,130	1,150	1,100
CT	921	11,200	12,100	12,700	12,700	12,000
DE	499	8,500	10,400	10,700	10,300	8,900
FL	355	4,790	5,230	5,500	5,640	5,150
GA	234	3,140	3,800	4,350	4,300	4,100
ID	177	1,650	2,200	2,480	2,500	2,200
IL	490	3,210	3,590	4,020	4,550	4,530
IN	406	3,000	3,250	3,640	4,100	4,020
IA	392	2,640	2,910	3,370	3,950	3,850
KS	159	810	870	980	1,020	1,030
KY	253	2,450	2,670	2,740	2,850	2,850
LA	321	1,700	1,780	1,950	2,050	1,970
ME	161	1,990	2,110	2,230	2,200	2,100
MD	640	7,500	8,400	8,500	8,000	7,500
MA	565	10,500	11,700	11,900	12,300	12,000
MI	326	3,070	3,370	3,760	3,900	3,750
MN	226	2,060	2,340	2,700	2,970	2,870
MS	234	1,640	1,770	1,970	2,080	2,000
MO	224	1,750	1,910	2,170	2,300	2,200
MT	60	480	720	830	900	700
NE	154	910	1,030	1,140	1,330	1,340
NV [1]	53	620	830	980	1,000	1,000
NH	239	3,780	4,240	4,800	4,900	4,800
NJ	1,092	12,500	14,400	15,700	15,300	13,800
NM [1]	42	330	410	460	500	480
NY	273	1,900	2,020	2,180	2,350	2,400
NC	333	3,820	4,060	4,330	4,450	4,250
ND	94	510	580	650	770	780
OH	399	3,140	3,420	3,700	4,020	3,880
OK	173	900	970	1,080	1,150	1,170
OR	150	1,380	1,470	1,720	1,900	1,800
PA	373	3,980	4,380	4,970	5,120	5,100
RI	734	12,800	15,300	16,400	16,800	15,300
SC	261	2,370	2,550	2,820	2,950	2,900
SD	84	590	680	770	920	890
TN	268	2,790	2,970	3,250	3,450	3,300
TX	148	1,000	1,190	1,380	1,550	1,550
UT [1]	92	1,250	1,510	1,730	1,850	1,800
VT	224	2,320	2,480	2,740	2,900	2,800
VA	286	3,800	4,430	4,900	5,000	4,800
WA	224	1,630	1,720	1,850	2,020	2,000
WV	136	1,900	2,000	2,300	2,500	2,400
WI	232	2,790	3,100	3,640	3,850	3,750
WY	41	340	400	490	560	520
US (48 States) [2]	196	1,610	1,830	2,010	2,170	2,110

[1] Excludes American Indian Reservation Land. [2] Excludes Alaska and Hawaii.
NASS, Environmental, Economics, and Demographics Branch, (202) 720–6146.

Table 9-15.—Land values, cropland and pasture: By State and Untied States, 2008–2009

State	2008				2009			
	Cropland [1]	Irrigated cropland	Non-irrigated cropland	Pasture [2]	Cropland [1]	Irrigated cropland	Non-irrigated cropland	Pasture [2]
	Dollars	Dollars	Dollars	Dollars	Dollars	Dollars	Dollars	Dollars
AL	2,650	....	....	1,800	2,500	....	....	1,700
AZ	11,500	11,500	1,650	950	10,000	10,000	1,600	900
AR	1,770	1,920	3,570	2,200	1,860	2,100	3,400	2,200
CA	9,880	12,300	890	3,020	9,480	11,600	840	2,900
CO	1,310	3,100	....	710	1,300	3,150	....	670
CT	....	....	....	....	....	....	....	....
DE	9,800	....	6,300	....	8,500	....	5,900	....
FL	6,980	7,790	4,750	5,930	6,430	7,000	4,200	5,300
GA	4,540	3,600	1,360	7,450	4,050	3,500	1,300	6,000
ID	2,800	4,500	....	1,610	2,610	4,000	....	1,280
IL	4,850	....	....	2,550	4,670	....	....	2,400
IN	4,140	....	....	2,510	3,950	....	....	2,430
IA	4,260	....	980	2,070	4,050	....	1,000	1,880
KS	1,020	1,450	....	750	1,050	1,500	....	750
KY	3,100	0	1,890	2,570	3,150	....	1,800	2,420
LA	1,830	1,560	....	2,060	1,740	1,500	....	2,100
ME	0	....	....	....	....	....	....	....
MD	7,800	....	....	....	7,300	....	....	....
MA	....	....	....	....	....	....	....	....
MI	3,480	....	....	2,630	3,370	....	....	2,550
MN	2,700	....	1,770	1,480	2,610	....	1,800	1,400
MS	1,810	1,980	2,470	2,200	1,810	1,830	2,500	2,050
MO	2,500	2,980	580	1,800	2,540	3,050	600	1,700
MT	811	3,300	1,750	760	787	2,800	1,850	530
NE	2,050	2,650	....	480	2,180	2,700	....	450
NV	2,740	2,740	....	650	....	2,700	....	....
NH	....	....	....	....	....	....	....	....
NJ	15,600	....	410	16,500	14,000	....	400	14,900
NM	1,630	5,360	....	310	1,810	5,490	....	280
NY	2,150	....	....	1,100	2,200	....	....	1,050
NC	3,850	....	....	4,870	3,770	....	....	4,600
ND	810	....	....	350	800	....	....	350
OH	4,140	....	1,100	3,200	3,900	....	1,120	3,050
OK	1,110	1,400	1,900	1,000	1,130	....	1,850	1,010
OR	2,380	3,740	....	740	2,340	3,680	....	700
PA	6,000	....	....	3,100	5,700	....	....	2,600
RI	....	....	....	....	....	....	....	....
SC	2,610	....	1,390	3,000	2,500	....	1,390	2,900
SD	1,400	1,830	....	470	1,400	1,850	....	430
TN	3,400	....	1,480	3,880	3,270	....	1,450	3,650
TX	1,500	1,680	1,150	1,400	1,480	1,700	1,080	1,360
UT	2,700	5,260	....	940	2,810	5,200	....	870
VT	....	....	....	....	....	....	....	....
VA	5,350	....	1,210	4,830	5,000	....	1,150	4,800
WA	1,830	4,200	....	770	1,790	4,200	....	820
WV	3,800	....	....	1,950	3,500	....	....	1,900
WI	3,600	....	750	2,130	3,650	....	700	2,050
WY	1,180	2,090	....	480	1,180	1,900	....	410
Oth Sts	7,930	....	....	6,370	7,570	....	....	6,060
US	2,760	....	....	1,090	2,670	....	....	1,070

[1] Other cropland States include CT, ME, MA, NH, RI, and VT.   [2] Other pasture States include CT, DE, ME, MA, NH, RI, and VT.

NASS, Environmental, Economics, and Demographics Branch, (202) 720–6146.

Table 9-16.—Cash rents, cropland and pasture: By State, 2008–2009

State	2008				2009			
	Cropland	Irrigated cropland	Non-irrigated cropland	Pasture	Cropland	Irrigated cropland	Non-irrigated cropland	Pasture
	Dollars	Dollars	Dollars	Dollars	Dollars	Dollars	Dollars	Dollars
AL	43.00			19.50	48.00			19.50
AZ		180.00				170.00		
AR	79.50	97.00	60.00		82.50	100.00	52.00	
CA		360.00		15.00		350.00		14.00
CO	41.00	110.00	24.00	5.50	62.50	110.00	24.00	5.50
CT								
DE	72.00				70.00			
FL			40.00	25.00			45.00	18.00
GA	65.00	125.00	48.00	27.00	76.50	143.00	49.00	25.00
ID	98.50	145.00	55.00		130.00	160.00	57.00	
IL	163.00			37.00	163.00			35.00
IN	135.00				139.00			
IA	170.00			42.00	175.00			43.00
KS	47.50	92.00	42.50	15.50	48.00	89.00	43.50	15.50
KY	78.50				93.50			
LA	73.00	97.00	67.00	27.00	73.50	85.00	66.00	16.00
ME								
MD	66.50				68.50			
MA								
MI	78.00				81.00			
MN	109.00			21.00	113.00			22.50
MS	80.50	103.00	73.00	18.50	82.50	100.00	67.00	16.00
MO			80.00	26.00			90.00	25.00
MT	24.50	66.00	20.50	6.50	28.50	69.00	21.50	4.70
NE	121.00	158.00	97.00	15.50	128.00	163.00	97.00	16.00
NV								
NH								
NJ	52.50				55.00			
NM				2.70				2.40
NY	41.50				44.00			
NC	57.50			29.00	61.00			24.00
ND	42.50			13.50	45.50			14.00
OH	100.00				101.00			
OK			28.00	10.50			28.00	10.50
OR	107.00	195.00	75.00		141.00	195.00	88.00	
PA	55.00			32.00	55.00			32.00
RI								
SC	32.50				34.00			
SD			64.00	15.90			71.50	17.00
TN	68.50			22.00		73.00		20.00
TX	31.50	80.00	24.00	6.50	34.50	77.00	25.00	6.20
UT		65.00		4.80		70.00		4.80
VT								
VA	40.50			21.00	43.50			19.00
WA		250.00				245.00		
WV	28.00			36.00		28.00		36.00
WI	85.00			36.00	87.00			36.00
WY				4.00				4.00

NASS, Environmental, Economics, and Demographics Branch, (202) 720–6146.

Table 9-17.—Farm assets and claims: Comparative balance sheet of the farming sector, excluding operator households, United States, Dec. 31, 2000–2009

Item	2000	2001	2002	2003	2004
ASSETS Physical assets:	Billion dollars	Billion dollars	Billion dollars	Billion dollars	Billion dollars
Real estate	946.4	996.2	1,045.7	1,111.8	1,340.6
Non-real estate:					
Livestock and poultry [1]	76.8	78.5	75.6	78.5	79.4
Machinery and motor vehicles [2]	90.1	92.8	93.6	95.9	101.9
Crops [3]	27.9	25.2	23.1	24.4	24.4
Purchased inputs [4]	4.9	4.2	5.6	5.6	5.7
Financial	57.1	58.9	60.4	62.4	65.5
Total	1,203.2	1,255.9	1,304.0	1,378.8	1,617.6
CLAIMS Liabilities:					
Real estate	84.7	88.5	95.4	105.1	96.9
Non-real estate debt to	....	....	....	....	86.1
Reporting institutions	62.9	67.3	74.1	68.4	NA
Nonreporting creditors	21.9	21.3	21.4	22.6	NA
Total liabilities [5]	163.9	170.7	177.2	196.1	183.0
Equity	1,039.3	1,085.3	1,126.8	1,182.7	1,434.6
Ratio:					
Debt/equity [6]	15.8	15.7	15.7	16.6	12.8
Debt/assets [6]	13.6	13.6	13.6	14.2	11.3

Item	2005	2006	2007	2008	2009 [7]
ASSETS Physical assets:	Billion dollars	Billion dollars	Billion dollars	Billion dollars	Billion dollars
Real estate	1,487.0	1,625.8	1,751.4	1,703.0	1,727.3
Non-real estate:					
Livestock and poultry [1]	81.1	80.7	80.6	80.6	79.8
Machinery and motor vehicles [2]	113.1	114.2	114.7	115.8	12.3
Crops	24.3	22.7	22.7	27.6	32.9
Purchased inputs [4]	6.5	6.5	7.0	7.2	7.2
Financial	67.5	73.7	78.8	81.6	84.1
Total	1,779.4	1,923.6	2,055.3	2,015.7	2,043.5
CLAIMS Liabilities:					
Real estate debt	104.8	108.0	112.7	133.6	134.5
Non-real estate debt to—	91.6	95.5	101.4	109.1	110.8
Reporting institutions	NA	NA	NA	NA	NA
Nonreporting creditors	NA	NA	NA	NA	NA
Total liabilities [5]	196.4	203.6	214.1	242.7	245.4
Equity	1,583.0	1,720.0	1,841.2	1,773.0	1,798.1
Ratio:					
Debt/equity [6]	12.4	11.8	11.6	13.7	13.6
Debt/assets [6]	11.0	10.6	10.4	12.0	12.0

[1] The U.S. total exceeds the sum of the states because NASS does not release state data for some minor producing states due to disclosure issues. Horses and mules are excluded. [2] Includes only farm share value for trucks and autos. [3] All non-CCC crops held on farms plus the value above loan rate for crops held under CCC. [4] Data for the value of purchased inputs are unavailable before 1984. [5] Excludes debt for nonfarm purposes. [6] Percents. [7] Preliminary estimate subject to revision.

ERS, Farm & Rural Business Branch. Information contacts: for assets, Ken Erickson,(202) 694-5565, e-mail: erickson@ers.usda.gov and for debt, Bob Williams, (202) 694-5053, e-mail: williams@ers.usda.gov.

Table 9-18.—Farm labor: Number of workers on farms and average wage rates, United States, 2001–2009

Year	Self-employed and unpaid workers [1]	Ag service workers [2]	Hired workers [2] [3]	Hired workers [2] [3]
	Number	Number	Number	Wage rates
2001.				
Jan	(5)	165	691	8.66
Apr	(5)	215	804	8.31
July	(5)	335	1,039	8.29
Oct	(5)	262	991	8.59
Annual average	2,049.8	(4)	873.3	8.45
2002.				
Jan	(5)	183	707	8.97
Apr	(5)	189	890	8.83
July	(5)	256	1,006	8.57
Oct	(5)	271	940	8.95
Annual average	(5)	(4)	885.7	8.81
2003.				
Jan	(5)	160	729	9.34
Apr	(5)	157	781	9.16
July	(5)	320	943	8.88
Oct	(5)	306	891	9.05
Annual average	(5)	(4)	836	9.08
2004.				
Jan	(5)	185	662	9.41
Apr	(5)	257	827	9.23
July	(5)	343	961	9.04
Oct	(5)	324	851	9.32
Annual average	(5)	(4)	825.2	9.23
2005.				
Jan	(5)	185	589	9.78
Apr	(5)	247	753	9.35
July	(5)	408	936	9.38
Oct	(5)	294	842	9.61
Annual average	(5)	(4)	780.0	9.51
2006.				
Jan	(5)	180	614	10.10
Apr	(5)	241	720	9.78
July	(5)	320	876	9.72
Oct	(5)	286	800	9.96
Annual average	(5)	(4)	751.8	9.87
2007.				
Jan	(5)	(6)	(6)	(6)
Apr	(5)	253	736	10.20
July	(5)	363	843	9.99
Oct	(5)	329	817	10.38
Annual average	(5)	(4)	746.5	10.23
2008.				
Jan	(5)	179	594	10.81
Apr	(5)	219	700	10.57
July	(5)	345	828	10.34
Oct	(5)	316	804	10.70
Annual average	(5)	(4)	731.5	10.59
2009.				
Jan	(5)	190	595	10.93
Apr	(5)	223	680	10.84
July	(5)	381	875	10.66
Oct	(5)	285	807	10.91
Annual average	(5)	(4)	739.3	10.82

[1] Includes farm operators and partners doing 1 or more hours of farm work and other unpaid workers working 15 hours or more during the survey week without cash wages. [2] Includes all persons doing farm work for pay during the survey week. [3] Excludes agricultural service workers. [4] Annual average not computed. [5] Discontinued. [6] January 2007 Farm Labor survey cancelled.

NASS, Economic, Environmental and Demographics Branch, (202) 720–6146.

Table 9-19.—Farm labor: Number of hired workers on farms and average wage rates, by regions and United States, 2009 [1] [2]

State and region [3]	Workers on farms	Farm wage rates			
	Hired	Type of worker			
		Field	Livestock	Field and livestock	All hired workers [4]
	Thousands	*Dollars per hour*	*Dollars per hour*	*Dollars per hour*	*Dollars per hour*
Jan. 11–17, 2009					
Northeast I	26	10.66	10.03	10.25	10.91
Northeast II	21	10.80	9.69	10.30	11.40
Appalachian I	22	10.08	9.91	10.00	10.80
Appalachian II	17	8.92	10.52	9.80	10.99
Southeast	30	9.65	9.65	9.65	10.05
FL	38	8.95	9.45	9.00	10.16
Lake	43	10.83	9.96	10.20	11.02
Cornbelt I	26	11.85	10.85	11.30	11.90
Cornbelt II	20	11.06	11.27	11.20	11.40
Delta	27	9.43	9.83	9.55	10.04
N. Plains	20	10.29	10.53	10.42	10.90
S. Plains	65	10.06	10.24	10.15	10.70
Mountain I	15	11.49	10.65	10.85	11.50
Mountain II	15	9.37	8.90	9.05	10.32
Mountain III	20	9.27	10.58	9.75	10.30
Pacific	52	10.35	9.48	10.25	11.40
CA	132	9.80	10.95	10.09	11.15
HI	6	10.70	13.50	10.93	12.69
US (49 States)	595	9.96	10.27	10.08	10.93
Apr. 12–18 2009					
Northeast I	32	10.65	9.92	10.35	11.17
Northeast II	22	10.60	8.86	9.95	11.10
Appalachian I	28	10.16	10.40	10.25	11.00
Appalachian II	21	9.23	9.62	9.40	10.10
Southeast	36	8.80	9.84	9.00	9.22
FL	46	9.15	9.35	9.17	10.33
Lake	56	10.92	10.83	10.88	11.55
Cornbelt I	32	10.93	10.44	10.72	11.70
Cornbelt II	22	10.35	12.45	11.15	11.80
Delta	33	9.30	8.84	9.20	9.60
N. Plains	26	10.78	10.08	10.35	10.80
S. Plains	62	9.67	9.93	9.80	10.40
Mountain I	20	10.23	10.01	10.10	10.45
Mountain II	19	9.68	10.09	9.90	10.60
Mountain III	20	9.16	9.44	9.29	10.10
Pacific	61	10.67	12.09	10.80	11.55
CA	138	9.96	10.85	10.14	11.07
HI	6	11.30	13.20	11.40	13.36
US (49 States)	680	9.99	10.25	10.07	10.84

See footnotes at end of table.

Table 9-19.—Farm labor: Number of hired workers on farms and average wage rates, by regions and United States, 2009 [1] [2]—Continued

State and region [3]	Workers on farms	Farm wage rates			
	Hired	Type of worker			
		Field	Livestock	Field and livestock	All hired workers [4]
	Thousands	*Dollars per hour*	*Dollars per hour*	*Dollars per hour*	*Dollars per hour*
July 12–18, 2009					
Northeast I	37	10.32	9.86	10.15	10.92
Northeast II	38	9.83	9.09	9.65	10.20
Appalachian I	40	9.18	10.40	9.50	9.88
Appalachian II	29	9.36	9.76	9.55	10.34
Southeast	40	8.92	8.84	8.90	9.27
FL	36	9.14	9.50	9.20	10.51
Lake	76	10.55	10.27	10.45	11.04
Cornbelt I	56	9.78	9.45	9.70	10.30
Cornbelt II	28	10.88	10.02	10.55	11.01
Delta	34	8.69	9.15	8.85	9.42
No. Plains	40	10.67	10.20	10.50	10.73
So. Plains	61	8.94	9.72	9.30	9.88
Mountain I	29	9.67	9.33	9.50	10.00
Mountain II	20	10.16	8.54	9.60	10.21
Mountain III	18	10.09	10.94	10.38	11.19
Pacific	117	10.93	11.77	11.00	11.43
CA	170	10.10	11.30	10.30	11.08
HI	6	11.60	13.60	11.81	13.97
US (49 States)	875	10.04	10.05	10.04	10.66
October 11–17, 2009					
Northeast I	41	9.96	9.93	9.95	10.41
Northeast II	37	10.28	9.50	10.05	10.88
Appalachian I	31	8.67	9.15	8.85	9.55
Appalachian II	24	9.85	10.38	10.10	11.05
Southeast	32	8.87	9.33	9.00	9.53
FL	46	9.30	10.10	9.40	10.65
Lake	67	11.03	10.10	10.65	11.24
Cornbelt I	47	11.30	10.51	11.00	11.17
Cornbelt II	30	10.46	11.23	10.75	10.85
Delta	33	8.86	9.16	8.95	9.50
No. Plains	34	11.77	10.23	11.20	11.80
So. Plains	56	9.17	10.64	9.85	10.17
Mountain I	25	10.45	8.89	9.70	10.05
Mountain II	22	11.01	11.70	11.25	11.55
Mountain III	20	9.10	10.06	9.50	10.30
Pacific	99	11.07	10.42	11.00	11.82
CA	157	10.25	11.05	10.40	11.25
HI	6	11.55	13.40	11.69	13.60
US (49 States)	807	10.25	10.23	10.24	10.91

[1] Excludes Agricultural Service Workers.   [2] Includes all persons doing work for pay during the survey week.   [3] Regions consist of the following: Northeast I: CT, ME, MA, NH, NY, RI, VT; Northeast II: DE, MD, NJ, PA; Appalachian I: NC, VA; Appalachian II: KY, TN, WV; Southeast: AL, GA, SC; Lake: MI, MN, WI; Cornbelt I: IL, IN, OH; Cornbelt II: IA, MO; Delta: AR, LA, MS; No. Plains: KS, NE, ND, SD; So. Plains: OK, TX; Mountain I: ID, MT, WY; Mountain II: CO, NV, UT; Mountain III: AZ, NM; Pacific: OR, WA.   [4] Includes field, livestock, supervisors, and other workers doing work for pay during the survey week.

NASS, Economic, Environmental and Demographics Branch, (202) 720–6146.

Table 9-20.—Farm production and output: Index numbers of total output, and production of livestock, crops, and secondary output, by groups, United States, 1999–2008

[1996=100]

Year	Total farm output	Livestock and products			
		All livestock and products [1]	Meat animals [2]	Dairy products [3]	Poultry and eggs [4]
1999	1.0730	1.0781	1.0657	1.0581	1.0882
2000	1.0748	1.0746	1.0341	1.0889	1.1087
2001	1.0756	1.0725	1.0315	1.0760	1.1267
2002	1.0551	1.0854	1.0194	1.1076	1.1718
2003	1.0793	1.1018	1.0316	1.1094	1.2372
2004	1.1252	1.0787	0.9913	1.1127	1.1870
2005	1.1094	1.1009	0.9842	1.1528	1.2260
2006	1.1216	1.1298	1.0133	1.1847	1.2629
2007	1.1408	1.1298	1.0237	1.2100	1.2726
2008	1.1309	1.1341	1.0329	1.2385	1.2930

Year	Crops				
	All crops	Food Grains	Feed crops	Oil crops [5]	Vegetables and melons
1999	1.0484	1.0438	1.0170	1.1070	1.1030
2000	1.0657	1.0031	1.0316	1.1348	1.0720
2001	1.0632	0.9243	1.0117	1.2074	1.0349
2002	1.0196	0.7929	0.9547	1.1329	1.0661
2003	1.0614	1.0562	1.0476	1.0334	1.0426
2004	1.1625	1.0195	1.1543	1.2865	1.0860
2005	1.1179	0.9907	1.0874	1.2921	1.0539
2006	1.1070	0.8545	1.0208	1.3007	1.0307
2007	1.1546	0.9521	1.1953	1.1083	1.0399
2008	1.1321	1.1300	1.1398	1.2176	1.0013

Year	Crops		Farm-related output [5]
	Fruits and nuts	Other crops	
1999	1.1028	1.0829	1.2768
2000	1.2016	1.1049	1.1762
2001	1.1722	1.1162	1.2295
2002	1.1788	1.1446	1.1651
2003	1.1892	1.1692	1.0855
2004	1.2253	1.2327	1.1776
2005	1.1118	1.2145	1.0991
2006	1.2159	1.3241	1.1825
2007	1.2690	1.3273	1.0873
2008	1.2894	1.1477	1.1009

[1] Includes wool, mohair, horses, mules, honey, beeswax, bees, goats, rabbits, aquaculture, and fur animals. These items are not included in the separate groups of livestock and products shown.　[2] Cattle and calves, sheep and lambs, and hogs.　[3] Butter, butterfat, wholesale milk, retail milk, and milk consumed on farms.　[4] Chicken eggs, commercial broilers, chickens, and turkeys.　[5] These activities are defined as activities closely linked to agriculture for which information on production and input use cannot be separately observed.

ERS, Agricultural Structure and Productivity Branch, (202) 694–5460, (202) 694–5601.

Table 9-21.—Hired farmworkers: Number of Workers and Median Weekly Earnings, 2007–2009 [1]

Characteristics	Workers			Median Weekly Earnings [2]		
	2007	2008	2009	2007	2008	2009
	Thousands	Thousands	Thousands	Dollars	Dollars	Dollars
All workers	738	795	762	340	360	384
15–19 years old	101	136	90	115	92	170
20–24 years old	102	97	99	320	350	330
25–34 years old	171	157	180	384	400	400
35–44 years old	149	149	134	360	415	420
45–54 years old	120	155	136	420	415	500
55 years old and older	94	101	121	369	403	400
Male	615	662	642	346	384	400
Female	123	133	120	320	318	300
White [3]	380	416	373	360	400	400
Black and other races [3]	39	42	33	308	296	400
Hispanic	319	337	355	330	360	364
Schooling completed						
Less than 5th grade	65	59	65	320	360	350
5th-8th grade	140	171	145	325	350	368
9th-12th grade (no diploma) ..	180	196	163	275	280	300
High school diploma	207	220	229	380	440	410
Beyond high school	146	149	159	427	480	500
Full-time (35 or more hours per week) [4]	604	627	602	380	400	413
Part-time (less than 35 hours per week) [4]	133	163	154	124	101	169

[1] Represents average number of persons 15 years old and over in the civilian noninstitutional population who were employed per week as hired farmworkers. Based on the Current Population Survey microdata earnings file. [2] "Median weekly earnings" is the value that divides the earnings into two equal parts, one part having earnings above the median and the other part having earnings below the median. "Earnings" refers to the weekly earnings the farmworker usually earns at a farmwork job, before deductions, and includes any overtime pay or commissions. [3] Excludes persons of Hispanic origin. [4] The sum of full-time and part-time workers will not equal the total because usual hours worked varies for some individuals.
ERS, Farm and Rural Household Well-Being Branch, (202) 694–5416.

Table 9-22.—Crops: Area, United States, 2000–2009

Year	Principal crops				Commercial vegetables, harvested area	Fruits and nuts, bearing area [6]
	Area harvested			Area planted total [3]		
	Feed grains [1]	Food grains [2]	Total [3]			
	1,000 acres	1,000 acres	1,000 acres	1,000 acres	1,000 acres	1,000 acres
2000 [4]	87,691	56,398	307,955	328,685	3,488.8	4,114.9
2001 [4]	83,531	52,037	303,560	324,584	3,353.5	4,083.3
2002 [5]	82,636	49,294	299,146	327,283	3,270.2	4,071.4
2003 [5]	85,689	56,379	307,400	325,693	3,235.9	4,064.0
2004 [5]	85,956	53,594	304,521	322,317	3,188.9	4,012.8
2005 [5]	85,945	53,747	303,566	317,640	3,124.9	3,984.3
2006 [5]	80,090	49,895	294,453	315,645	3,083.2	3,929.6
2007 [5]	98,318	53,999	304,376	320,369	3,033.5	3,904.5
2008 [5]	91,020	58,764	308,810	324,997	2,943.5	3,946.6
2009 [5]	89,602	53,183	301,603	319,296	2,972.4	3,986.8

[1] Corn for grain, oats, barley, and sorghum for grain. [2] Wheat, rye, and rice. [3] Crops included in area planted and area harvested are corn, sorghum, oats, barley, winter wheat, rye, durum wheat, other spring wheat, rice, soybeans, peanuts, sunflower, cotton, dry edible beans, potatoes, canola, proso millet, and sugarbeets. Harvested acreage for all hay, tobacco, and sugarcane are used in computing total area planted. [4] For the 2000 crop year many changes occurred to the National Vegetable Estimation Program. Nine new commodities were added to the program. Additionally, States were added or dropped from the seasonal program. Some States were discontinued for the seasonal forecasts but remained in the program on an annual basis. When comparing 2001 and 2000 data to 1999 data, comparable States should be used. [5] For the 2002 crop year, many changes occured to the National Vegetable Estimation Program. Ten fresh market commodities and two processing commodities were removed from the program. States were removed from the program for certain commodities. When comparing 2000 and 2001 data to 2002 data, comparable States should be used. If you need assistance with these comparisons, please contact Debbie Flippin at (202) 720-2157. For details on the 2002 program changes see the following website: http://www.usda.gov/nass/events/programchg/vegprogchngs.htm. [6] Includes the following fruits and nuts: Citrus fruits—oranges, tangerines, Temples, grapefruit, lemons, limes, tangelos, and K-Early Citrus (area is for the year of harvest); limes and K-Early citrus were discontinued as of the 2002-03 crop; deciduous fruits—commercial apples, peaches, pears, grapes, cherries, plums, prunes, apricots, bananas, nectarines, figs, kiwifruit, olives, avocados, papayas, dates, berries, guavas, cranberries, pineapples and strawberries; nuts—almonds, hazelnuts, macadamias, pistachios, and walnuts.
NASS, Crops Branch, (202) 720–2127.

Table 9-23.—Crops: Area harvested and yield, United States, 2008–2009 [1]

Crop	Area harvested		Yield per harvested acre		
	2008	2009 [2]	Unit	2008	2009 [2]
	1,000 acres	*1,000 acres*			
Grains & Hay:					
Barley [3]	3,779.0	3,113.0	Bushel	63.6	73.0
Corn for Grain	78,570.0	79,630.0	Bushel	153.9	165.2
Corn for Silage	5,965.0	5,605.0	Ton	18.7	19.3
Hay, All	60,152.0	59,755.0	Ton	2.43	2.47
Alfalfa	21,060.0	21,227.0	Ton	3.33	3.35
All Other	39,092.0	38,528.0	Ton	1.95	1.98
Oats [3]	1,400.0	1,379.0	Bushel	63.7	67.5
Proso Millet	460.0	293.0	Bushel	32.3	33.7
Rice	2,976.0	3,103.0	Pound	6846	7,085
Rye [3]	269.0	252.0	Bushel	29.7	27.8
Sorghum for Grain	7,271.0	5,520.0	Bushel	65.0	69.4
Sorghum for Silage	408.0	254.0	Ton	13.8	14.5
Wheat, All [3]	55,699.0	49,868.0	Bushel	44.9	44.4
Winter [3]	39,608.0	34,485.0	Bushel	47.1	44.2
Durum	2,574.0	2,428.0	Bushel	32.6	44.9
Other Spring	13,517.0	12,955.0	Bushel	40.5	45.1
Oilseeds:					
Canola	989.0	814.0	Pound	1,461	1,811
Cottonseed	....	....	Ton	....	....
Flaxseed	340.0	314.0	Bushel	16.8	23.6
Mustard Seed	71.5	49.8	Pound	577	991
Peanuts	1,507.0	1,081.0	Pound	3,416	3,412
Rapeseed	0.2	0.9	Pound	1,500	1,700
Safflower	195.0	165.5	Pound	1,592	1,462
Soybeans for Beans	74,681.0	76,407.0	Bushel	39.6	44.0
Sunflower	2,396.0	1,953.5	Pound	1,429	1,554
Cotton, Tobacco & Sugar Crops:					
Cotton, All	7,568.7	7,690.5	Pound	813	774
Upland	7,400.0	7,552.0	Pound	803	763
Amer-Pima	168.7	138.5	Pound	1,226	1,353
Sugarbeets	1,004.5	1,145.3	Ton	26.8	25.8
Sugarcane	868.5	877.7	Ton	31.8	34.5
Tobacco	354.5	354.1	Pound	2,258	2,325
Dry Beans, Peas & Lentils:					
Austrian Winter Peas	8.0	13.7	Pound	1,300	1,328
Dry Edible Beans	1,445.2	1,463.0	Pound	1,768	1,733
Chickpeas, all	82.1	93.9	Pound	1,362	1,538
Large	71.2	72.2	Pound	1,389	1,586
Small	10.9	21.7	Pound	1,183	1,378
Dry Edible Peas	847.3	837.9	Pound	1,448	2,045
Lentils	261.0	407.0	Pound	917	1,440
Wrinkled Seed Peas	....	....	NA	....	....
Potatoes & Misc.:					
Coffee (HI)	6.3	6.3	Pound	1,300	1,270
Coffee (PR)	33.0	27.0	Pound	405	350
Ginger Root (HI) [4]	0.1	....	Pound	30,000	....
Hops	40.9	39.7	Pound	1,971	2,383
Maple syrup	....	....	NA	....	....
Mushrooms	....	....	NA	....	....
Peppermint Oil	60.0	69.8	Pound	92	91
Potatoes, All	1,046.9	1,045.0	Cwt	396	413
Winter	11.0	8.7	Cwt	230	245
Spring	68.8	73.7	Cwt	293	289
Summer	45.1	43.0	Cwt	306	336
Fall	922.0	919.6	Cwt	411	428
Spearmint Oil	20.4	20.5	Pound	118	132
Sweet Potatoes	97.3	97.7	Cwt	190	201
Taro (HI) [5]	0.4	0.4	Pound		

[1] Missing data are not available.　[2] Preliminary.　[3] Includes area seeded in preceding fall.　[4] Estimates discountinued in 2009.　[5] Acreage is total acres in crop, not harvested acreage. Yield is not estimated.
NASS, Crops Branch, (202) 720–2127.

Table 9-24.—Crops: Production and value, United States, 2008–2009 [1]

Crop	Unit	Production		Value of production	
		2008	2009 [2]	2008	2009 [2]
		Thou-sands	Thou-sands	1,000 dollars	1,000 dollars
Grains & Hay:					
Barley [3]	Bushel	240,193	227,323	1,259,357	917,500
Corn for Grain	Bushel	12,091,648	13,151,062	49,312,615	48,588,665
Corn for Silage	Ton	111,619	108,209	....	....
Hay, All	Ton	146,270	147,442	18,638,748	14,990,083
Alfalfa	Ton	70,180	71,030	10,747,161	7,997,221
All Other	Ton	76,090	76,412	7,891,587	6,992,862
Oats [3]	Bushel	89,135	93,081	269,763	216,566
Proso Millet	Bushel	14,880	9,865	48,017	28,043
Rice	Cwt	203,733	219,850	3,603,460	3,145,521
Rye [3]	Bushel	7,979	6,993	50,452	33,427
Sorghum for Grain	Bushel	472,342	382,983	1,631,065	1,242,196
Sorghum for Silage	Ton	5,646	3,680	....	....
Wheat, All [3]	Bushel	2,499,164	2,216,171	16,625,759	10,626,176
Winter [3]	Bushel	1,867,333	1,522,718	11,936,139	7,060,386
Durum	Bushel	83,827	109,042	731,445	613,103
Other Spring	Bushel	548,004	584,411	3,958,175	2,952,687
Oilseeds:					
Canola	Pound	1,445,064	1,474,130	270,988	237,669
Cottonseed	Ton	4,300.3	4,178.0	962,708	666,146
Flaxseed	Bushel	5,716	7,423	72,773	64,817
Mustard Seed	Pound	41,255	49,364	18,089	15,000
Peanuts	Pound	5,162,400	3,688,350	1,193,617	835,172
Rapeseed	Pound	300	1,530	76	403
Safflower	Pound	310,433	241,970	76,922	43,248
Soybeans for Beans	Bushel	2,967,007	3,361,028	29,458,225	31,760,452
Sunflower	Pound	3,422,840	3,036,460	704,105	444,795
Cotton, Tobacco & Sugar Crops:					
Cotton, All	Bale	12,815.3	12,401.3	3,021,485	3,735,564
Upland	Bale	12,384.5	12,011.0	2,816,852	3,522,708
Amer-Pima	Bale	430.8	390.3	204,633	212,856
Sugarbeets	Ton	26,881	29,519	1,289,621	....
Sugarcane	Ton	27,603	30,265	814,479	....
Tobacco	Pound	800,504	823,000	1,488,069	1,498,629
Dry Beans, Peas & Lentils:					
Austrian Winter Peas	Cwt	104	182	2,286	4,088
Dry Edible Beans	Cwt	25,558	25,360	910,200	793,722
Chickpeas, All	Cwt	1,118	1,444	34,972	39,306
Large	Cwt	989	1,145	31,647	33,305
Small	Cwt	129	297	3,325	6,001
Dry Edible Peas	Cwt	12,270	17,137	166,945	154,118
Lentils	Cwt	2,393	5,859	80,943	153,359
Wrinkled Seed Peas	Cwt	580	874	17,912	23,493
Potatoes & Misc.:					
Coffee (HI)	Pound	8,700	8,000	29,580	25,600
Coffee (PR)	Pound	13,300	9,500	46,725	29,127
Ginger Root (HI) [4]	Pound	1,800	....	2,880	....
Hops	Pound	80,630.1	94,677.9	325,092	336,375
Maple syrup	Gallon	1,912	2,327	77,477	....
Mushrooms	Pound	812,604	816,974	962,756	957,028
Peppermint Oil	Pound	5,499	6,379	87,450	128,497
Potatoes, All	Cwt	415,055	431,425	3,770,462	3,452,276
Winter	Cwt	2,530	2,132	50,600	38,589
Spring	Cwt	20,132	21,321	301,161	340,399
Summer	Cwt	13,805	14,469	198,056	204,634
Fall	Cwt	378,586	393,503	3,770,462	3,452,276
Spearmint Oil	Pound	2,399	2,698	35,765	44,703
Sweet Potatoes	Cwt	18,443	19,647	390,572	410,361
Taro (HI)	Pound	4,300	4,000	2,666	2,440

[1] Missing data are not available.   [2] Preliminary.   [3] Includes area seeded in preceding fall.   [4] Estimates discontinued in 2009.

NASS, Crops Branch, (202) 720–2127.

Table 9-25.—Fruits and nuts: Bearing acreage and yield, United States, 2008–2009 [1]

Crop	Bearing acreage		Yield per bearing acre		
	2008	2009 [2]	Unit	2008	2009 [2]
	Acres	*Acres*			
Apples, commercial crop	350,590	347,800	Ton	13.80	14.30
Apricots	12,450	12,350	Ton	6.56	5.56
Avocados	72,830	66,270	Ton	1.59	4.05
Bananas [3]	1,100	1,100	Ton	7.90	8.40
Blackberries (OR) [3][4]	6,700	7,000	Ton	3.37	3.92
Blueberries			Ton		
Cultivated [3]	60,180	63,770	Ton	2.90	2.86
Wild (ME) [5]			Ton		
Boysenberries [3]	770	660	Ton	2.08	2.67
Loganberries (OR) [3]	30		Ton	1.00	
Raspberries [3]			Ton		
Black (OR)	1,500	1,100	Ton	1.04	1.43
Red	11,100	11,100	Ton	2.71	3.21
All (CA)	5,400	5,400	Ton	8.00	9.50
Cherries, sweet	82,610	83,310	Ton	3.00	5.04
Cherries, tart	34,750	35,550	Ton	3.09	5.05
Cranberries	38,200	38,500	Ton	10.30	8.98
Dates (CA)	5,700	6,300	Ton	3.67	3.71
Figs (CA)	9,400	9,300	Ton	4.61	4.30
Grapes	935,950	942,900	Ton	7.82	7.74
Guava (HI) [3]	160	135	Ton	11.00	7.80
Kiwifruit (CA)	4,200	4,200	Ton	5.48	6.19
Nectarines	32,300	30,300	Ton	9.37	7.25
Olives (CA)	30,000	31,00	Ton	2.23	1.49
Papayas (HI) [3]	1,380	1,325	Ton	12.20	11.90
Peaches	124,000	118,830	Ton	9.16	9.29
Pears	58,620	57,000	Ton	14.80	16.80
Pineapples (HI)			Ton		
Plums (CA)	29,500	26,600	Ton	5.42	4.21
Prunes, dried (CA)	64,000	64,000	Ton	5.75	7.75
Prunes and plums, fresh basis (excluding CA)	3,240	3,190	Ton	4.78	5.83
Strawberries [3]	54,470	58,080	Ton	23.30	24.10
Oranges [6]	663,100	656,300	Boxes	353.00	324.00
Grapefruit [6]	83,400	80,400	Boxes	454.00	408.00
Lemons [6]	59,000	59,000	Boxes	276.00	424.00
Tangerines and Mandarins [6]	40,500	44,200	Boxes	311.00	244.00
Tangelos (FL) [6]	5,200	5,200	Boxes	288.00	221.00
Almonds (CA) [7]	680,000	720,000	Ton	2.07	1.64
Hazelnuts (OR) [7]	28,300	28,700	Ton	1.13	1.64
Macadamia (HI) [7]	15,000	15,000	Ton	1.67	1.40
Pecans [5]			Ton		
Pistachios (CA) [7]	118,000	126,000	Ton	1.18	1.41
Walnuts (CA) [7]	223,000	223,000	Ton	1.96	1.96

[1] Missing data are not available. [2] Preliminary. [3] Harvested acreage. Yield based on utilized production. [4] Cultivated. Estimates discontinued in 2009. [5] Bearing acreage and yield not estimated. [6] Crop year begins with bloom in one year and ends with completion of harvest the following year. Citrus production is for the the year of harvest. [7] Yield based on in-shell basis. Shelling ratios are: 2008, 0.578; 2009, 0.597.

NASS, Crops Branch, (202) 720–2127.

Table 9-26.—Fruits and nuts: Production and value, United States, 2008–2009 [1]

Crop	Unit [2]	Total production		Value of production	
		2008	2009 [3]	2008	2009 [3]
		Thousand	Thousand	1,000 dollars	1,000 dollars
Apples, commercial crop	Ton	9,633.3	9,914.9	2,214,717	2,246,584
Apricots	Ton	81.6	68.7	41,196	44,912
Avocados	Ton	116.0	268.7	214,546	345,388
Bananas [4]	Ton	8.7	9.3	8,004	10,175
Blackberries (OR)	Ton	23.6	28.1	27,773	30,842
Blueberries	Ton				
Cultivated	Ton	174.5	183.9	536,992	507,520
Wild (ME)	Ton	45.0	44.1	54,850	31,945
Boysenberries	Ton	1.7	1.7	2,334	2,102
Loganberries (OR)	Ton	(5)		63	(NA)
Raspberries	Ton				
Black (OR)	Ton	1.6	1.7	5,815	1,181
Red	Ton	30.0	35.6	100,177	64,110
All (CA)	Ton	43.2	51.3	259,200	297,315
Cherries, sweet	Ton	248.1	429.9	574,043	505,881
Cherries, tart	Ton	107.2	179.5	80,344	63,231
Cranberries	Ton	393.3	345.7	455,927	340,706
Dates (CA)	Ton	20.9	23.4	26,334	28,782
Figs (CA)	Ton	43.3	40.0	25,954	30,465
Grapes	Ton	7,319.3	7,294.8	3,342,966	3,688,065
Guava (HI) [4]	Ton	1.8	1.1	553	294
Kiwifruit (CA)	Ton	23.0	26.0	19,545	35,048
Nectarines	Ton	302.5	219.8	110,915	138,611
Olives (CA)	Ton	66.8	46.3	46,587	32,209
Papayas (HI) [4]	Ton	16.8	15.8	14,393	14,186
Peaches	Ton	1,135.3	1,103.8	545,854	593,653
Pears	Ton	869.9	957.2	396,081	355,192
Plums (CA)	Ton	160.0	112.0	56,960	57,568
Prunes, dried (CA)	Ton	368.0	496.0	193,500	199,200
Prunes and plums, fresh basis (excluding CA)	Ton	15.5	18.6	5,918	6,105
Strawberries [4]	Ton	1,265.9	1,400.7	1,918,288	2,123,735
Oranges [4] [6]	Ton	10,076.0	9,128.0	2,198,836	1,950,452
Grapefruit [4] [6]	Ton	1,548.0	1,304.0	273,076	233,539
Lemons [4] [6]	Ton	619.0	912.0	523,528	342,667
Tangelos (FL) [4] [6]	Ton	68.0	52.0	8,638	5,496
Tangerines and mandarins [4] [6]	Ton	527.0	443.0	236,193	209,515
Almonds (CA) [4] [7] [8]	Ton	1,410.0	1,180.9	2,343,200	2,293,500
Hazelnuts (OR) [4]	Ton	32.0	47.0	51,840	79,430
Macadamia (HI) [4]	Ton	25.0	21.0	33,500	29,400
Pecans [4]	Ton	97.0	145.9	260,397	417,300
Pistachios (CA) [4]	Ton	139.0	177.5	569,900	592,850
Walnuts (CA) [4]	Ton	436.0	437.0	558,080	738,530

[1] Missing data are not available.　[2] Ton refers to the 2,000 lb. short ton.　[3] Preliminary.　[4] Only utilized production estimated.　[5] Production less than 0.1 tons in 2007 and 2008.　Estimates discontinued in 2009.　[6] Value of production is packinghouse-door equivalent.　[7] Production is shelled basis, shelling ratios are: 2008, 0.578; 2009, 0.597.　[8] Value based on the edible portion of the crop only.　Production includes inedible quantities of no value as follows: 2008, 14.0 million pounds; 2009, 20.0 million pounds.

NASS, Crops Branch, (202) 720–2127.

FARM RESOURCES, INCOME, AND EXPENSES

Table 9-27.—Vegetables: Area harvested and yield, United States, 2008–2009

Crop	Area harvested		Yield per harvested acre		
	2008	2009 [1]	Unit	2008	2009 [1]
	Acres	*Acres*			
Commercial Vegetables:					
Fresh Market					
Artichokes [2]	8,800	8,600	Cwt	130	130
Asparagus [2]	32,200	29,200	Cwt	30	31
Beans, snap	90,400	88,400	Cwt	64	55
Broccoli [2]	126,900	124,000	Cwt	158	158
Cabbage	65,760	65,500	Cwt	373	345
Cantaloups	71,730	76,130	Cwt	269	261
Carrots	73,700	69,400	Cwt	333	319
Cauliflower [2]	36,700	35,000	Cwt	181	186
Celery [2]	28,300	28,500	Cwt	708	691
Corn, sweet	233,280	233,550	Cwt	124	122
Cucumbers	46,880	46,850	Cwt	189	186
Garlic [2]	25,440	23,330	Cwt	168	169
Honeydew melons	17,200	16,800	Cwt	215	220
Lettuce, head	148,700	146,000	Cwt	356	365
Lettuce, leaf	52,300	49,100	Cwt	244	229
Lettuce, Romaine	77,400	79,100	Cwt	294	329
Onions [2]	153,490	148,560	Cwt	489	505
Peppers, bell [2]	50,900	52,100	Cwt	312	299
Peppers, Chile [2]	25,000	28,500	Cwt	167	169
Pumpkins [2]	43,400	44,100	Cwt	246	211
Spinach	35,680	40,100	Cwt	160	156
Squash [2]	42,400	43,900	Cwt	158	164
Tomatoes	105,250	105,700	Cwt	296	306
Watermelons	125,550	126,300	Cwt	319	318
Processing:					
Beans, lima	38,270	34,740	Ton	1.28	1.38
Beans, snap	198,300	196,179	Ton	4.07	4.14
Carrots	15,940	12,530	Ton	25.20	26.29
Corn, sweet	360,600	379,500	Ton	7.85	8.52
Cucumbers for pickles	96,600	97,500	Ton	5.87	5.57
Peas, green	209,700	205,350	Ton	1.96	2.15
Spinach	10,200	10,100	Ton	10.15	9.47
Tomatoes	296,500	327,800	Ton	41.50	42.62

[1] Preliminary.   [2] Includes processing total for dual usage crops.
NASS, Crops Branch, (202) 720–2127.

Table 9-28.—Vegetables: Production and value, United States, 2008–2009

Crop	Unit	Production 2008	Production 2009 [1]	Value of production 2008	Value of production 2009 [1]
		Thousands	Thousands	1,000 dollars	1,000 dollars
Commercial Vegetables:					
Fresh Market					
Artichokes [2]	Cwt	1,144	1,118	54,683	63,279
Asparagus [2]	Cwt	952	899	89,451	88,855
Beans, snap	Cwt	5,824	4,862	307,790	259,922
Broccoli [2]	Cwt	20,086	19,570	721,307	741,900
Cabbage	Cwt	24,516	22,623	355,065	341,440
Cantaloups	Cwt	19,294	19,891	356,781	359,082
Carrots	Cwt	24,565	22,163	602,054	557,670
Cauliflower [2]	Cwt	6,648	6,501	268,531	286,612
Celery [2]	Cwt	20,025	19,685	369,684	364,816
Corn, sweet	Cwt	28,899	28,421	748,632	835,833
Cucumbers	Cwt	8,843	8,729	219,073	220,761
Garlic [2]	Cwt	4,282	3,941	186,807	196,075
Honeydew melons	Cwt	3,690	3,698	65,636	59,060
Lettuce, head	Cwt	52,952	53,220	1,063,132	1,155,468
Lettuce, leaf	Cwt	12,781	11,238	411,719	405,829
Lettuce, Romaine	Cwt	22,774	26,030	479,006	614,134
Onions [2]	Cwt	75,120	74,970	872,113	843,570
Peppers, bell [2]	Cwt	15,888	15,600	636,620	555,643
Peppers, Chile [2]	Cwt	4,170	4,830	111,199	128,951
Pumpkins [2]	Cwt	10,663	9,313	137,072	102,730
Spinach	Cwt	5,721	6,239	193,052	236,808
Squash [2]	Cwt	6,687	7,219	204,283	203,464
Tomatoes	Cwt	31,137	32,365	1,415,297	1,313,941
Watermelons	Cwt	40,003	40,122	499,633	460,778
Processing:					
Beans, lima	Ton	49,150	48,030	24,584	24,945
Beans, snap	Ton	808,000	812,990	177,278	155,420
Carrots	Ton	401,730	329,440	33,775	31,272
Corn, sweet	Ton	2,832,490	3,234,080	340,486	335,563
Cucumbers for pickles	Ton	567,100	542,600	178,998	180,845
Peas, green	Ton	411,780	441,580	148,052	140,679
Spinach	Ton	103,540	95,660	12,831	12,144
Tomatoes	Ton	12,305,820	13,970,560	982,373	1,218,912

[1] Preliminary.   [2] Includes processing total for dual usage crops.
NASS, Crops Branch, (202) 720–2127.

Table 9-29.—Total farm input: Index numbers of farm input, by major subgroups, United States, 1999–2008

[1996=100]

Year	Total farm input	Capital					Labor		
		All	Durable equip-ment	Service buildings	Inven-tories	Land	All	Hired labor	Self-em-ployed
1999	1.0528	0.9912	0.9828	0.9393	1.0844	0.9970	0.9322	1.1150	0.8358
2000	1.0061	0.9846	0.9794	0.9239	1.0856	0.9910	0.8425	0.9410	0.7906
2001	0.9988	0.9803	0.9775	0.9128	1.1032	0.9837	0.8416	0.9526	0.7831
2002	0.9942	0.9789	0.9891	0.8952	1.1133	0.9762	0.8499	0.9625	0.7905
2003	0.9770	0.9746	1.0007	0.8807	1.0774	0.9694	0.8210	0.9444	0.7560
2004	0.9619	0.9735	1.0261	0.8655	1.0523	0.9630	0.7911	0.8685	0.7501
2005	0.9740	0.9802	1.0693	0.8525	1.1413	0.9568	0.7859	0.8653	0.7439
2006	0.9681	0.9810	1.0941	0.8377	1.1384	0.9505	0.7387	0.8277	0.6917
2007	1.0154	0.9736	1.0932	0.8214	1.1113	0.9439	0.7554	0.8996	0.6799
2008	0.9491	0.9710	1.1114	0.8068	1.1166	0.9369	0.7349	0.8571	0.6708

Year	Materials				
	All	Farm origin	Energy	Chemicals	Purchased services
1999	1.1333	1.1639	1.0487	1.0372	1.1683
2000	1.0893	1.1416	1.0288	1.0389	1.0734
2001	1.0774	1.1056	0.0033	1.0224	1.0977
2002	1.0648	1.0994	1.0918	1.9894	1.0404
2003	1.0485	1.1364	0.9123	0.9379	1.0100
2004	1.0338	1.1180	0.9831	0.9586	0.9810
2005	1.0581	1.1297	0.9080	0.9991	1.0346
2006	1.0722	1.1426	0.8654	1.0181	1.0491
2007	1.1579	1.1841	1.0041	1.1513	1.1521
2008	1.0406	1.1038	0.8841	0.9204	1.0707

ERS, Agricultural Structure and Productivity Branch, (202) 694–5460, (202) 694–5601.

Table 9-30.—Farm food products: Marketing costs, United States, 1999–2008

Year	Labor [1]	Packaging materials	Intercity transpor-tation, rail and truck	Fuels and electricity	Corporate profits before taxes	Other [2]	Total marketing bill [3]
	Billion dollars	Billion dollars	Billion dollars	Billion dollars	Billion dollars	Billion dollars	Billion dollars
1999	241.5	50.9	25.2	22.0	29.2	134.3	503.1
2000	252.9	53.5	26.4	23.1	31.1	150.8	537.8
2001	263.8	55.0	27.5	24.1	32.0	155.1	557.5
2002	273.1	56.8	28.4	24.9	33.0	160.7	576.9
2003	285.9	59.5	29.7	26.1	34.6	168.2	604.0
2004	303.7	63.1	31.6	27.6	35.5	171.9	633.4
2005	319.8	66.5	33.2	31.6	37.4	184.4	672.9
2006	341.0	70.5	35.2	33.5	39.7	197.6	717.5
2007	347.4	71.8	35.9	34.1	40.4	201.4	731.0
2008 [4]	364.3	75.3	37.6	37.4	38.9	213.1	766.6

[1] Includes employee wages or salaries, and their health and welfare benefits. Also includes imputed earnings of propri-etors, partners, and family workers not receiving stated remuneration. [2] Includes depreciation, rent, advertising and pro-motion, interest, taxes, licenses, insurance, professional services, local for-hire transportation, food service in schools, col-leges, hospitals, and other institutions, and miscellaneous items. [3] The marketing bill is the difference between the farm value or payments to farmers for foodstuffs and consumer expenditures for these foods both at foodstores and away from home eating places. Thus, it covers processing, wholesaling, transportation, and retailing costs and profits. [4] Preliminary.
ERS, Food Markets Branch, (202) 694–5375.

Table 9-31.—U.S. farm foods: Marketing bill, farm value, and consumer expenditures, 1999–2008 [1]

Year	Total marketing bill	Farm value	Expenditures for farm foods
	Billion dollars	Billion dollars	Billion dollars
1999	503.1	122.2	625.3
2000	537.8	123.3	661.1
2001	557.5	130.0	687.5
2002	576.9	132.5	709.4
2003	604.0	140.2	744.2
2004	633.4	155.5	788.9
2005	672.9	157.8	830.7
2006	717.5	163.2	880.7
2007	731.0	194.3	925.2
2008 [2]	766.6	192.3	958.9

[1] The total marketing bill is the difference between total expenditures for domestic farm-originated food products and the farm value or payment farmers received for the equivalent farm products. It relates only to food purchased by consumers that is not imported or exported.   [2] Preliminary.
ERS, Food Markets Branch, (202) 694–5375.

Table 9-32.—Livestock and livestock products: Production and value, United States, 2006–2009

Product	Production [1]			Value of production		
	2006	2007	2008 [2]	2006	2007	2008 [2]
	1,000 pounds	1,000 pounds	1,000 pounds	1,000 dollars	1,000 dollars	1,000 dollars
Cattle and calves	41,824,568	41,437,021	41,594,392	35,490,732	35,973,068	35,608,404
Sheep and lambs	460,580	440,286	417,019	367,799	362,941	351,287
Hogs	28,182,382	29,606,420	31,410,795	12,714,218	13,468,332	14,457,000
Broilers [3]	48,829,900	49,330,700	50,441,600	17,739,234	21,513,536	23,203,136
Mature chickens	924,993	912,875	937,045	54,141	51,498	62,164
Turkeys [4]	7,223,675	7,561,579	7,922,087	3,467,534	3,951,772	4,477,054
Milk	181,782,000	185,654,000	189,982,000	23,556,102	35,665,894	35,050,757
	Millions	Millions	Millions			
Eggs	91,788	91,101	90,040	4,460,211	6,718,853	8,215,999

Product	Production			Value of production		
	2007	2008	2009	2007	2008	2009
	1,000 pounds	1,000 pounds	1,000 pounds	1,000 dollars	1,000 dollars	1,000 dollars
Catfish [5][6]	583,752	530,539	491,867	454,593	409,998	372,567
Trout [5][6]	69,467	54,643	51,580	87,856	79971	76,913
Honey	148,341	163,789	144,108	159,763	232,744	208,236
Wool (shorn)	34,723	32,963	30,862	30,242	32,486	24,387
Mohair	1,357	1,184	1,012	4,944	3,914	2,692

[1] For cattle, sheep, and hogs, the quantity of net production is the live weight actually produced during the year, adjustments having been made for animals shipped in and changes in inventory. Estimates for broilers and eggs cover the 12-month period Dec. 1, previous year through Nov. 30.   [2] Preliminary, except for wool shorn and mohair.   [3] Young chickens of meat–type strains raised for meat production.   [4] 28 State total for 2006 and a 50 state total for 2007 and 2008.   [5] Value of fish sold, excludes eggs.   [6] Live weight.
NASS, Livestock Branch, (202) 720–3570.

Table 9-33.—Agricultural productivity: Index numbers (1996=100) of farm output per unit of input, United States, 1999–2008

Year	Productivity [1]
1999	1.0192
2000	1.0683
2001	1.0768
2002	1.0613
2003	1.1047
2004	1.1697
2005	1.1390
2006	1.1585
2007	1.1235
2008	1.1917

[1] Productivity is the output-input ratio.
ERS, Agricultural Structure and Productivity Branch (202) 694–5601, (202) 694–5460.

Table 9-34.—Price components: Market basket of farm-originated food products by food group, United States, 1999–2008 [1]

Year	Market basket of food products				Bakery and cereal products			
	Retail cost [2]	Farm value [3]	Farm to retail spread [4]	Farm value share of retail cost	Retail cost	Farm value	Farm to retail spread	Farm value share of retail cost
	Index 1982–84=100	Index 1982–84=100	Index 1982–84=100	Percent	Index 1982–84=100	Index 1982–84=100	Index 1982–84=100	Percent
1999	167	98	205	21	185	83	199	6
2000	171	97	210	20	188	75	204	5
2001	177	106	215	21	194	79	210	5
2002	180	104	221	20	198	86	214	5
2003	185	110	226	21	203	94	218	6
2004	194	124	232	22	206	104	220	6
2005	198	122	239	22	209	96	225	6
2006	202	120	246	21	213	111	227	6
2007	211	142	248	24	222	150	232	8
2008 [5]	225	147	267	23	245	191	252	10

Year	Meat products				Fruits and vegetables, fresh			
	Index 1982–84=100	Index 1982–84=100	Index 1982–84=100	Percent	Index 1982–84=100	Index 1982–84=100	Index 1982–84=100	Percent
1999	142	82	205	29	252	136	308	18
2000	150	88	214	30	252	131	310	17
2001	159	97	223	31	261	138	321	17
2002	160	103	220	32	272	150	331	18
2003	169	108	231	33	280	157	339	19
2004	183	117	251	32	290	174	347	19
2005	188	121	255	33	301	159	370	17
2006	189	117	263	31	317	176	385	18
2007	195	125	267	32	331	181	403	18
2008 [5]	202	124	281	31	346	181	426	17

Year	Dairy products				Fats and oils			
	Index 1982–84=100	Index 1982–84=100	Index 1982–84=100	Percent	Index 1982–84=100	Index 1982–84=100	Index 1982–84=100	Percent
1999	160	108	207	32	148	89	170	16
2000	161	99	218	30	147	81	172	15
2001	167	119	212	34	156	77	185	13
2002	168	98	233	28	155	92	179	16
2003	168	99	231	28	157	113	174	19
2004	180	126	230	34	168	128	182	21
2005	182	119	241	31	168	108	190	17
2006	181	102	254	27	168	102	192	16
2007	195	153	233	38	173	151	181	24
2008 [5]	210	145	270	33	197	207	193	28

Year	Poultry				Fruits and vegetables, processed			
	Index 1982–84=100	Index 1982–84=100	Index 1982–84=100	Percent	Index 1982–84=100	Index 1982–84=100	Index 1982–84=100	Percent
1999	158	119	203	40	155	114	168	17
2000	160	117	209	39	154	106	168	17
2001	165	126	209	41	159	108	175	16
2002	167	102	242	33	166	111	184	16
2003	169	113	234	36	172	108	192	15
2004	182	143	226	42	183	125	201	16
2005	185	139	238	40	192	138	209	17
2006	182	129	244	38	201	140	220	17
2007	191	155	233	43	209	151	227	17
2008 [5]	201	155	253	41	229	165	248	17

[1] The market basket consists of foods that mainly originate on U.S. farms bought in foodstores in a base period, currently 1982–84. [2] Indexes of retail cost are components of the Consumer Price Index published by the Bureau of Labor Statistics. [3] Gross return or payment to farmers for the farm products equivalent to foods in the market basket. [4] The spread between the retail cost and farm value is an estimate of the gross margin received by marketing firms for assembling, processing, transporting, and distributing the products. [5] Preliminary.

ERS, Food Markets Branch (202) 694–5375.

Table 9-35.—Farm product prices: Marketing year average prices received by farmers; Parity prices for January, United States, 2007–2008

Commodity and unit	Marketing year average price [1]		Parity price [3]	
	2007	2008 [2]	2007	2008
	Dollars	*Dollars*	*Dollars*	*Dollars*
Basic commodities:				
Cotton:				
American Upland .... pound	0.593	.478	1.98	2.05
Extra long staple .... pound	0.988	.99	2.87	3.22
Wheat .... bushel	6.48	6.78	10.50	14.60
Rice .... cwt	12.80	16.80	28.30	33.70
Corn .... bushel	4.20	4.06	7.76	9.08
Peanuts .... pound	0.205	0.230	0.627	0.708
Tobacco:				
Flue-cured, types 11–14 .... pound	1.527	1.757	4.85	5.27
Virginia, fire-cured, type 21 [23] .... pound	....	....	5.14	6.20
Kentucky - Tennessee, fire-cured, types 22–23 [23] .... pound	....	....	6.45	7.79
Types 21-23 .... pound	2.401	2.460	6.45	7.27
Burley, type 31 .... pound	1.601	1.669	5.24	5.54
Maryland, type 32 [4][21] .... pound	....	....	3.96	4.54
Dark air-cured, types 35–36 .... pound	2.197	2.246	5.70	6.54
Sun-cured, type 37 .... pound	....	....	4.75	5.74
Pa., seedleaf, type 41 .... pound	1.650	1.700	4.04	4.74
Cigar binder type 51-52 .... pound	6.472	....	15.10	15.90
Puerto Rican filler, type 46 .... pound	....	....	3.13	3.78
Cigar filler and types 42–44, 54–55 .... pound	....	....	4.49	5.42
Designated nonbasic commodities:				
All milk, sold to plants .... cwt	19.21	18.45	38.80	44.70
Honey, all .... pound	1.077	1.410	2.49	3.10
Wool and mohair:				
Wool [5] .... pound	0.870	....	1.66	2.07
Mohair [6] .... pound	3.78	....	7.26	8.37
Other nonbasic commodities:				
Field crops and miscellaneous:				
Barley .... bushel	4.02	5.37	7.32	10.20
Beans, dry edible .... cwt	28.80	34.60	55.00	70.30
Cottonseed .... ton	162.00	223.00	299.00	369.00
Crude pine gum .... barrel	....	....	281.00	339.00
Flaxseed .... bushel	13.00	12.70	17.00	23.70
Hay, all, baled .... ton	128.00	152.00	257.00	320.00
Hops .... pound	2.99	4.03	5.06	7.32
Oats .... bushel	2.63	3.15	4.35	5.96
Peas, dry edible .... cwt	....	....	38.22	....
Peppermint oil .... pounds	13.60	15.90	32.96	40.50
Popcorn, shelled basis .... cwt	....	....	36.60	....
Potatoes .... cwt	7.51	8.42	16.50	20.30
Rye .... bushel	5.01	6.32	8.37	11.60
Sorghum grain .... cwt	7.28	5.72	13.30	16.00
Soybeans .... bushel	10.10	9.97	17.20	22.40
Spearmint oil .... pound	12.60	14.90	28.30	34.70
Sweetpotatoes .... cwt	18.30	21.20	46.70	55.40
Tobacco:				
Cigar wrapper, type 61 .... pound	24.50	28.50	66.70	74.90
Fruits:				
Citrus (equiv. on-tree): [7]				
Grapefruit .... box	7.69	7.15	12.60	15.80
Lemons .... box	21.40	32.12	23.10	33.70
Oranges .... box	12.56	9.36	12.10	16.10
Tangerines .... box	18.30	17.81	27.30	33.90
Temples, Florida [22] .... box	....	....	7.52	9.08
Deciduous and other:				
Apples:				
For fresh consumption [8] .... pound	0.383	0.301	0.609	0.806
For processing [9] .... ton	190.00	198.00	338.00	408.00
Apricots:				
For fresh consumption [10] .... ton	742.00	775.00	2,100.00	2,640.00
Dried, California (dried basis) [9] .... ton	1,510.00	2,364.00	5,340.00	6,180.00
For processing (excl dried) [9] .... ton	325.00	350.00	770.00	913.00
Avocados [10] .... ton	1,770.00	1,850.00	4,670.00	5,150.00

See footnotes at end of table.

Table 9-35.—Farm product prices: Marketing year average prices received by farmers; Parity prices for January, United States, 2007–2008—Continued

Commodity and unit		Marketing year average price [1]		Parity price [3]	
		2007	2008 [2]	2007	2008
		Dollars	Dollars	Dollars	Dollars
Deciduous and other—Con.					
Berries for processing:					
Blackberries (Oregon)	pound	0.448	0.486	12.50	15.10
Boysenberries (California & Oregon)	pound	0.602	0.729	1.78	2.15
Gooseberries	pound	....	....	0.750	0.903
Loganberries (Oregon)	pound	0.978	1.050	1.22	1.46
Raspberries, black (Oregon)	pound	0.982	1.860	2.14	2.59
Raspberries, red (Oregon & Washington)	pound	0.538	1.670	1.71	2.05
Cherries:					
Sweet	ton	1,820.00	2,390.00	3,980.00	5,000.00
Tart	pound	0.273	0.377	0.694	0.854
Cranberries [11]	barrel	50.70	58.00	101.00	116.00
Dates, California [10]	ton	2,290.00	1,260.00	3,800.00	4,810.00
Figs, California	ton	401.00	599.00	....	....
Grapes:					
For all sales	ton	489.00	458.00		
Raisin varieties dried, California (dried basis) [9]	ton	1,040.00	1,170.00	2,450.00	2,690.00
Other dried grapes	ton	567.00	534.00	1,450.00	1,630.00
Kiwi	ton	950.00	888.00	1,830.00	2,320.00
Nectarines (California):					
For fresh consumption [18]	ton	340.00	367.00	1,230.00	1,350.00
For processing [19]	ton	....	....	80.10	103.00
Olives (California): [12]					
For all sales	ton	654.00	697.00	....	....
Crushed for oil	ton	775.00	670.00	724.00	1,410.00
For all sales (excl crushed)	ton	643.00	....	1,656.00	1,790.00
For canning	ton	700.00	767.00	1,900.00	2,310.00
Papayas	pound	0.392	0.430	1.010	1.120
Peaches:					
For all sales	ton	450.00	490.00	....	....
For fresh consumption [8]	ton	706.00	684.00	1,690.00	2,030.00
Dried, California (dried basis) [9]	ton	530.00	452.00	1,560.00	1,580.00
For processing California (excl dried):					
Clingstone [12]	ton	304.00	347.00	690.00	823.00
Freestone [9]	ton	235.00	239.00	564.00	659.00
Pears:					
For all sales	ton	416.00	456.00	....	....
For fresh consumption [8]	ton	534.00	589.00	1,080.00	1,340.00
Dried, California (dried basis) [9]	ton	1,800.00	1,400.00	3,600.00	4,220.00
For processing (excl dried) [9]	ton	489.00	530.00	544.00	857.00
Plums (California):					
For all sales [10]	ton	665.00	356.00		
For fresh consumption [18]	ton	....	....	1,280.00	1,490.00
For processing [19]	ton	....	....	167.00	203.00
Prunes, dried (California) [9]	ton	1,450.00	1,500.00	2,810.00	3,560.00
Prunes and plums (excl California):					
For fresh consumption [13]	ton	622.00	547.00	1,270.00	1,540.00
For processing (excl dried) [9]	ton	213.00	171.00	572.00	647.00
Strawberries:					
For fresh consumption [14]	pound	0.821	0.841	1.98	2.30
For processing [9]	pound	0.277	0.362	0.810	0.918
Sugar crops:					
Sugarbeets	ton	42.00	48.00	110.00	125.00
Sugarcane for sugar	ton	29.40	29.50	77.90	87.40
Tree nuts: [15]					
Almonds	pound	1.75	1.45	4.33	4.93
Hazelnuts	ton	2,040.00	1,620.00	3,110.00	4,170.00
Pecans, all	pound	1.120	1.340	3.15	7,560.00
Pistachios	pound	1.41	2.05	3.64	4.64
Walnuts	ton	2,290.00	1,280.00	3,520.00	4,200.00

See footnotes at end of table.

Table 9-35.—Farm product prices: Marketing year average prices received by farmers; Parity prices for January, United States, 2007–2008—Continued

Commodity and unit		Marketing year average price [1]		Parity price [3]	
		2007 [2]	2008 [2]	2007	2008
		Dollars	Dollars	Dollars	Dollars
Vegetables for fresh market: [14]					
Artichokes, California	cwt	55.00	47.80	92.00	111.00
Asparagus	cwt	98.90	103.00	305.00	320.00
Broccoli	cwt	36.70	36.20	84.30	101.00
Cabbage	cwt	16.40	14.70	27.90	33.70
Cantaloups	cwt	14.80	18.50	43.10	52.00
Carrots [16]	cwt	22.10	24.50	48.10	62.00
Cauliflower [16]	cwt	34.40	40.70	90.00	105.00
Celery [16]	cwt	20.40	18.50	40.40	51.00
Cucumbers	cwt	24.60	24.80	46.10	55.70
Eggplant [20]	cwt	....	....	48.70	58.80
Escarole/Endive [20]	cwt	....	....	62.70	75.70
Garlic	cwt	41.20	43.60	59.20	71.50
Green peppers [16]	cwt	33.10	40.10	72.80	87.90
Honeydew melons	cwt	17.70	17.80	54.00	56.40
Lettuce	cwt	21.70	20.10	47.50	57.40
Onions [16]	cwt	11.10	12.50	33.00	37.10
Snap beans	cwt	61.20	52.80	91.20	110.00
Spinach	cwt	32.30	33.70	79.90	96.40
Sweet corn	cwt	22.70	25.90	54.00	67.10
Tomatoes	cwt	34.80	45.50	96.50	114.00
Watermelons	cwt	11.30	12.50	17.10	20.70
Vegetables for processing: [9]					
Asparagus	ton	1,260.00	1,360.00	3,203.00	3,610.00
Beets [20]	ton	....	....	158.00	190.00
Cabbage [20]	ton	....	....	122.00	147.00
Cucumbers	ton	325.00	316.00	....	....
Green peas	ton	259.00	360.00	732.00	837.00
Lima beans	ton	423.00	500.00	1,355.00	1,640.00
Snap beans	ton	168.00	219.00	453.00	513.00
Spinach	ton	104.00	124.00	293.00	354.00
Sweet corn	ton	81.80	117.00	199.00	244.00
Tomatoes	ton	71.20	79.80	172.00	202.00
Livestock and livestock products:					
All beef cattle	cwt	89.90	89.10	205.00	249.00
Cows	cwt	47.90	50.60	....	....
Steers and heifers	cwt	95.40	94.50	....	....
Calves	cwt	119.00	110.00	295.00	352.00
Beeswax	pound	....	....	5.90	7.10
Chickens:					
Excluding broilers, live	pound	0.056	0.066	....	....
Broilers, live [19]	pound	0.436	0.458	....	....
All Eggs	dozen	0.885	1.090	1.78	2.22
Hogs	cwt	46.60	47.00	118.00	135.00
Lambs	cwt	98.50	99.60	241.00	281.00
Milk cows [17]	head	1,830	1,950.00	....	....
Sheep	cwt	31.00	27.20	99.10	106.00
Turkeys, live	pound	0.523	0.565	1.14	1.37

[1] Unless otherwise noted, these prices are for marketing year average or calendar year average computed by weighing State prices by quantities sold, or by production for those commodities for which the production is sold. [2] Preliminary. [3] Parity prices are for January of the year shown as published in the January issue of Agricultural Prices. [4] Previous year. [5] Average local market price for wool sold excluding incentive payment. [6] Average local market price for mohair sold excluding incentive payment. Texas only prior to 1988. [7] Crop year begins with bloom in one year and ends with completion of harvest the following year. Prices refer to the year harvest begins. Thus the prices shown for 1996 relate to the citrus crop designated as 1996–97 in the production reports. [8] Equivalent packinghouse-door returns for California, Oregon (pears only), Washington, and New York (apples only), and prices as sold for other States. [9] Equivalent returns at processing plant-door. [10] Equivalent returns at packinghouse-door. [11] Weighted average of co-op and independent sales. Co-op prices represent pool proceeds excluding returns from non-cranberry products and before deductions for capital stock and other retains. [12] Equivalent per unit returns for bulk fruit at first delivery point. [13] Average price as sold. [14] FOB shipping point when available. Weighted average of prices at points of first sale. [15] Prices are in-shell basis except almonds which are shelled basis. [16] Includes some processing. [17] Simple average of States weighted by estimated Jan. 1 head for U.S. average. [18] Prices for fresh and processing breakdown no longer published to avoid disclosure of individual operations. [19] Live weight equivalent price. [20] Discontinued. [21] Price not published to avoid disclosure of individual firms. [22] Included in Oranges beginning in 2007. [23] Estimates discontinued in 2006.
NASS, Environmental, Economics, and Demographics Branch (202) 720–6146.

Table 9-36.—Producer prices: Index numbers, by groups of commodities, United States, 2000–2009

[1982=100]

Year	Total finished goods	Consumer foods	Total consumer goods	Total intermediate materials	Total crude materials
2000	138.0	137.2	138.2	129.2	120.6
2001	140.7	141.3	141.5	129.7	121.0
2002	138.9	140.1	139.4	127.8	108.1
2003	143.3	145.9	145.3	133.7	135.3
2004	148.5	152.7	151.7	142.6	159.0
2005	155.7	155.7	160.4	154.0	182.2
2006	160.4	156.7	166.0	164.0	184.8
2007	166.6	167.0	173.5	170.7	207.1
2008	177.1	178.3	186.3	188.3	251.8
2009 [1]	172.5	175.5	179.1	172.5	175.2

[1] Final.
ERS, Food Marketing Branch, (202) 694–5349. Compiled from reports of the U.S. Department of Labor.

Table 9-37.—Prices received by farmers: Index numbers by groups of commodities and parity ratio, United States, 2000–2009 [1]

[1910–14=100]

Year	Food grains	Feed grains and hay	Cotton	Tobacco	Oil-bearing crops	Fruit & nuts [2]	Commercial vegetables	Other crops
2000	272	308	421	1,614	467	681	808	541
2001	290	325	328	1,614	437	761	888	554
2002	331	356	284	1,641	480	734	914	561
2003	344	370	437	1,515	585	741	980	555
2004	379	391	460	1,419	733	856	898	556
2005	351	338	361	1,417	579	894	932	558
2006	425	388	402	1,377	550	1,074	974	572
2007	590	541	423	1,392	748	1,103	1,128	582
2008	820	734	515	1,409	1,107	1,036	1,076	604
2009 [4]	589	581	414	1,568	972	918	1,153	608

Year	Potatoes, and dry edible beans	All crops	Meat animals	Dairy products	Poultry and eggs	Livestock and livestock products	All farm products	Parity ratio [3]
2000	472	473	955	757	299	744	611	39
2001	497	490	989	920	323	812	650	40
2002	652	517	884	744	265	692	620	38
2003	527	547	1,045	770	310	788	674	40
2004	514	571	1,181	988	371	932	751	42
2005	554	546	1,201	931	347	910	726	38
2006	634	593	1,180	793	312	850	730	37
2007	637	706	1,204	1,177	393	994	862	40
2008	797	836	1,195	1,128	424	1,000	946	39
2009 [4]	779	742	1,074	790	390	857	828	35

[1] These indexes are computed using the price estimates of averages for all classes and grades for individual commodities being sold in local farm markets. In computing the group indexes, prices of individual commodities have been compared with 1990–92 weighted average prices. The resulting ratios are seasonally weighted by average quantities sold for the most recent 5-year period. For example, 1994 indexes use quantities sold for the period 1988-92. Then, the 1990–92 indexes are adjusted to a 1910–14 reference.   [2] Fresh market for noncitrus, and fresh market and processing for citrus.   [3] Ratio of Index of Prices Received to the Index of Prices Paid by Farmers for Commodities and Services, Interest, Taxes, and Farm Wage Rates.   [4] Preliminary.
NASS, Environmental, Economics, and Demographics Branch, (202) 720–6146.

Table 9-38.—Prices received by farmers: Index numbers by groups of commodities and ratio, United States, 2000–2009 [1]

(1990–92=100)

Year	Food grains	Feed grains and hay	Cotton	Tobacco	Oilseeds	Fruit & Nuts [2]	Commercial vegetables	Other Crops
2000	85	86	82	107	85	98	121	110
2001	91	91	64	107	80	109	133	112
2002	104	100	56	108	88	105	137	114
2003	109	104	85	100	107	106	137	113
2004	120	110	90	94	134	123	126	113
2005	111	95	70	94	106	128	130	113
2006	134	109	78	91	100	154	136	116
2007	186	152	82	92	137	158	158	118
2008	259	206	100	93	202	148	151	123
2009 [4]	186	163	81	104	177	132	161	124

Year	Potatoes and dry edible beans	All crops	Meat animals	Dairy products	Poultry and eggs	Livestock and live-stock products	All farm products	Ratio [3]
2000	93	96	94	94	106	97	96	80
2001	98	99	97	115	115	106	102	83
2002	129	105	87	93	94	90	98	79
2003	104	110	103	96	110	103	106	84
2004	102	115	116	123	132	122	118	88
2005	109	110	118	116	123	119	114	81
2006	125	120	116	99	111	111	115	77
2007	126	142	118	146	140	130	136	85
2008	157	169	117	140	151	130	149	82
2009 [4]	154	150	105	98	139	112	131	73

[1] These indexes are computed using the price estimates of averages for all classes and grades for individual commodities being sold in local farm markets. In computing the group indexes, prices of individual commodities have been compared with 1990–92 weighted average prices. The resulting ratios are seasonally weighted by average quantities sold for the most recent previous 5–year period. For example, 1994 indexes use quantities sold for the period 1988–92. [2] Fresh market for noncitrus, and fresh market and processing for citrus. [3] Ratio of Index of Prices Received (1990–92=100) to Index of Prices Paid by Farmers for Commodities & Services, Interest, Taxes, and Wage Rates (1990–92=100). [4] Preliminary.
NASS, Environmental, Economics, and Demographics Branch, (202) 720–6146.

Table 9-39.—Prices paid by farmers: Index numbers, by groups of commodities, United States, 2000–2009

(1990–92=100)

	Production indexes								
Year	Production (all commodities)	Feed	Livestock & Poultry	Seeds	Fertilizer	Agricultural chemicals	Fuels	Supplies and Repairs	Autos and trucks
2000	115	102	110	124	110	120	129	120	119
2001	120	109	111	132	123	121	121	124	118
2002	119	112	102	142	108	119	115	127	116
2003	124	114	109	154	124	121	140	130	111
2004	132	121	128	158	140	121	165	134	114
2005	140	117	138	168	164	123	216	140	114
2006	148	124	134	182	176	128	239	145	112
2007	160	149	131	204	216	129	264	149	111
2008	190	194	124	259	392	139	344	154	108
2009 [3]	182	187	115	299	275	150	229	157	110

	Production indexes - continued							Production, interest, taxes, and wage rates	Family living	Commodities, interest, taxes, and wage rates [2]
Year	Farm machinery	Building Materials	Farm services	Rent	Interest	Taxes	Wage rates [1]			
2000	139	121	118	110	113	123	140	117	128	119
2001	144	121	120	117	104	128	146	121	131	123
2002	148	122	120	120	100	130	153	121	133	124
2003	151	124	125	123	94	129	157	125	136	128
2004	162	134	127	126	97	133	160	133	140	134
2005	173	142	133	129	111	155	165	141	145	142
2006	182	152	139	141	132	177	171	150	150	150
2007	191	155	146	147	142	200	177	162	154	161
2008	209	165	146	165	147	209	183	188	160	183
2009 [3]	222	163	159	178	139	238	187	183	159	179

[1] Simple average of seasonally adjusted quarterly indexes. [2] Family Living component included. [3] Preliminary.
NASS, Environmental, Economics, and Demographics Branch, (202) 720–6146.

Table 9-40.—Prices paid by farmers: Index numbers, by groups of commodities, United States, 2000–2009[1]

[1910–14=100]

| Year | Family living | Production indexes | | | | | | | |
		Production (all commodities)	Feed	Livestock and poultry	Seed	Fertilizer	Agricultural chemicals	Fuels	Supplies and repairs
2000	1,636	1,117	497	1,400	1,228	404	741	993	855
2001	1,682	1,158	530	1,419	1,306	451	745	933	879
2002	1,709	1,154	547	1,306	1,402	394	738	866	900
2003	1,747	1,203	554	1,394	1,521	454	747	1,083	921
2004	1,794	1,284	590	1,641	1,561	514	746	1,271	949
2005	1,855	1,361	571	1,759	1,661	601	762	1,668	995
2006	1,915	1,434	607	1,706	1,802	644	792	1,845	1,029
2007	1,969	1,552	725	1,671	2,024	790	801	2,038	1,060
2008	2,045	1,839	945	1,587	2,563	1,436	859	2,653	1,091
2009[3]	2,038	1,766	910	1,472	2,960	1,010	925	1,762	1,113

| Year | Production indexes—Continued | | | | Interest | Taxes | Wage rates | Production, interest, taxes, and wage rates | Commodities, interest, taxes, and wage rates[2] |
	Autos and trucks	Farm machinery	Building materials	Farm services and rent					
2000	3,160	3,490	1,647	1,367	2,834	3,281	5,236	1,578	1,588
2001	3,141	3,602	1,646	1,418	2,600	3,421	5,468	1,627	1,637
2002	3,082	3,704	1,654	1,429	2,495	3,494	5,705	1,628	1,642
2003	2,962	3,789	1,679	1,446	2,360	3,450	5,885	1,685	1,696
2004	3,022	4,062	1,817	1,511	2,440	3,571	5,977	1,786	1,788
2005	3,031	4,329	1,930	1,569	2,772	4,150	6,158	1,900	1,891
2006	2,991	4,556	2,059	1,664	3,328	4,729	6,390	2,019	1,999
2007	2,949	4,794	2,104	1,746	3,560	5,356	6,618	2,178	2,138
2008	2,882	5,231	2,245	1,818	3,676	5,598	6,852	2,520	2,433
2009[3]	2,917	5,576	2,218	1,971	3,478	6,370	7,001	2,452	2,378

[1] Based on Consumer Price Index-Urban of Bureau of Labor Statistics. [2] The index known as the Parity Index is the Index of Prices Paid by Farmers for Commodities and Services, Interest, Taxes, and Wage Rates expressed on the 1910–14=100 base. [3] Preliminary.
NASS, Environmental, Economics, and Demographics Branch, (202) 720–6146.

Table 9-41.—Prices paid by farmers: April prices, by commodities, United States, 2007–2009[1]

Commodity	Unit	2007	2008	2009
		Dollars	Dollars	Dollars
Fuels and energy:				
Diesel fuel[2][3]	Gal	2.430	3.619	1,688
Gasoline, service station, unleaded[4]	Gal	2.625	3.277	1,941
Gasoline, service station, bulk delivery[4]	Gal	2.638	3.331	1,972
L. P. gas, bulk delivery[2]	Gal	1.727	2.281	1,737
Feeds:				
Alfalfa Meal	Cwt	18.20	21.10	23.90
Alfalfa Pellets	Cwt	18.70	21.30	24.10
Bran	Cwt	17.40	19.10	23.20
Beef Cattle Concentrate.				
32-36% Protein	Ton	370	433	488
Corn Meal	Cwt	12.30	13.70	13.40
Cottonseed Meal, 41%	Cwt	20.00	23.00	25.60
Dairy Feed				
14% Protein	Ton	239	301	285
16% Protein	Ton	249	313	293
18% Protein	Ton	267	324	297
20% Protein	Ton	258	320	295
32% Protein Conc.	Ton	378	469	458
Hog Feed				
14-18% Protein	Ton	305	345	328
38-42% Protein Conc.	Ton	398	490	493
Molasses, Liquid	Cwt	19.20	17.50	22.20
Poultry Feed:.				
Broiler Grower	Ton	306	387	464
Chick Starter	Ton	327	416	501
Laying Feed	Ton	278	371	391
Turkey Grower	Ton	337	434	466
Soybean Meal, 44%	Cwt	18.20	23.40	23.20
Soybean Meal, >44%	Cwt	16.90	22.40	22.00
Stock Salt	50 Lb	5.02	5.45	5.93
Trace Mineral Blocks	50 Lb	6.15	6.60	7.09

See footnotes at end of table.

Table 9-41.—Prices paid by farmers: April prices, by commodities, United States, 2007–2009 [1]—Continued

Commodity	Unit	2007	2008	2009
		Dollars	Dollars	Dollars
Fertilizer: [5]				
0-15-40	Ton	326	684	763
0-18-36	Ton	314	647	640
0-20-20	Ton	331	573	683
3-10-30	Ton	278	502	654
5-10-10	Ton	246	386	424
5-10-15	Ton	299	430	511
5-10-30	Ton	306	505	613
5-20-20	Ton	328	597	570
6- 6- 6	Ton	266	349	414
6- 6-18	Ton	314	436	591
6-12-12	Ton	293	418	435
6-24-24	Ton	371	691	737
8- 8- 8	Ton	278	410	469
8-20- 5	Ton	318	560	617
8-32-16	Ton	380	746	675
9-23-30	Ton	363	720	763
10- 3- 3	Ton	(9)	617	550
10- 6- 4	Ton	267	404	380
10-10-10	Ton	298	449	483
10-20-10	Ton	329	616	536
10-20-20	Ton	365	632	632
10-34- 0	Ton	358	650	787
11-52- 0	Ton	445	902	645
13-13-13	Ton	337	541	539
15-15-15	Ton	380	588	602
16- 0-13	Ton	290	434	468
16- 4- 8	Ton	352	506	568
16- 6-12	Ton	306	483	464
16-16-16	Ton	403	675	775
16-20- 0	Ton	369	619	528
17-17-17	Ton	389	620	607
18-46- 0 (DAP)	Ton	442	850	638
19-19-19	Ton	413	693	665
24- 8- 0	Ton	302	431	388
Ammonium Nitrate	Ton	382	509	438
Anhydrous Ammonia	Ton	523	755	680
Aqua Ammonia	Ton	164	241	228
Limestone, Spread on field	Ton	23.00	25.60	27.40
Muriate of Potash, 60–62% K2O	Ton	280	561	853
Nitrate of Soda	Ton	358	503	568
Nitrogen Solutions.				
28% N	Ton	276	376	357
30% N	Ton	277	401	320
32% N	Ton	308	426	409
Sulfate of Ammonia	Ton	288	391	378
Superphosphate, 44-46% P2O5	Ton	418	800	639
Urea, 44-46% Nitrogen	Ton	453	552	486
Farm Machinery:				
Baler, Pick-Up, Automatic Tie, P.T.O.				
Square Conventional, Under 200 Lb Bales	Each	19,000	20,100	21,600
Round, 1200-1500 Lb Bale	Each	21,900	23,300	25,100
Round, 1900-2200 Lb Bale	Each	31,000	32,600	35,400
Chisel Plow, Maxiumum 1 Foot Depth				
Tillage, Chisel or Sweep Type, Drawn.				
Mounted, 16-20 Foot	Each	18,000	22,200	22,500
Mounted, 21-25 Foot	Each	25,200	28,800	31,200
Combine, Self Propelled with Grain head				
Extra-large capacity	Each	255,000	276,000	304,000
Large capacity	Each	213,000	230,000	253,000
Corn Head for combine				
6 Row	Each	31,800	35,900	38,200
8 Row	Each	41,900	46,000	49,000
Cotton Picker, Self Propelled, with sprindle,				
4-Row	Each	272,000	279,000	288,000
Cultivator, Row Crop				
6-Row	Each	7,980	8,760	8,850
8-Row	Each	11,700	12,000	12,700
12-Row, Flexible	Each	18,200	17,400	18,600
Disk Harrow, Tandem, Drawn [6]				
15-17 Foot	Each	18,300	18,900	22,400
18-20 foot	Each	23,400	24,100	27,800
21-25 foot	Each	27,800	30,100	34,100

See footnotes at end of table.

Table 9-41.—Prices paid by farmers: April prices, by commodities, United States, 2007–2009 [1]—Continued

Commodity	Unit	2007	2008	2009
		Dollars	*Dollars*	*Dollars*
Elevator, Portable, Without Power Unit,				
Auger Type, 8 Inch Diameter, 60 Foot	Each	5,430	5,500	6,780
Feed Grinder-Mixer, Trailer Mtd., P.T.O.	Each	20,700	25,100	25,100
Field Cultivator, Mounted or Drawn				
17-19 Foot	Each	14,000	18,500	20,700
20-25 Foot, Flexible	Each	22,100	23,600	27,800
Forage Harvester, P.T.O., Shear Bar,				
With Pick-Up Attachment	Each	36,700	38,100	39,900
With Row Crop Unit, 2-Row	Each	37,300	40,900	44,900
Forage Harvester, Self-propelled, Shear Bar				
With 4–6 row	Each	294,000	294,000	331,000
Front-End Loader, Hydraulic, Tractor Mounted				
1800-2500 Lb. Capacity, 60 Inch Bucket	Each	5,770	6,000	6,380
Grain Drill, Most Common Spacing				
Plain, 15-17 Openers	Each	18,500	21,600	22,200
Press, 23-25 Openers	Each	26,100	26,900	32,400
With Fertilizer Attachment, 20-24 Openers	Each	23,500	22,700	28,800
Min/No-Till W/Fert. Attach., 15 Foot	Each	34,400	35,000	40,000
Hayrake, Side-Delivery, or Wheel Rake,				
Traction Drive, 8-12 Foot Working Width	Each	6,750	6,760	7,640
Hay Tedder, 15-18 Foot	Each	5,740	6,150	7,290
Manure Spreader, Conveyor Type, P.T.O.,				
2-Wheel, with Tires.				
141-190 Bushel Capacity	Each	8,440	8,520	9,740
225-300 Bushel Capacity	Each	12,800	13,700	15,400
Mower-Conditioner, P.T.O., Pull Type, with				
8-10 Foot, Sickle (Cutter) Bar or Disc	Each	17,200	18,400	19,700
14-16 Foot, Sickle (Cutter) Bar or Disc	Each	26,200	28,300	29,500
Mower, Mounted or Drawn,				
7-8 ft Sickle (Cutter) Bar	Each	6,120	6,560	6,780
13-14 Foot, Sickle (Cutter) Bar or Disc	Each	15,400	17,200	16,600
Planter, Row Crop				
With Fertilizer Attachment, 4-Row	Each	18,400	19,500	22,100
With Fertilizer Attachment, 8-Row	Each	33,500	38,000	40,200
With Fertilizer Attachment, 24-Row	Each	118,000	132,000	144,000
12-Row Conservation (No-Till Cond), w/Fert	Each	62,700	67,900	72,900
Rotary Hoe, 20-25 Foot	Each	9,280	12,200	11,200
Rotary Cutter, 7-8 Foot	Each	3,610	4,010	4,430
Sprayer, Field Crop, Power, Boom Type				
(Excl. Self-Propelled and Orchard).				
Tractor Mounted, w/ 300 Gal. Tank	Each	7,530	8,280	8,460
Trailer Type, w/ 500-700 Gal. Spray Tank	Each	16,800	19,100	21,100
Tractor, 2-Wheel Drive				
30-39 P.T.O. horsepower	Each	18,400	18,700	18,500
50-59 P.T.O. horsepower	Each	24,300	25,000	24,500
70-89 P.T.O. horsepower	Each	38,900	39,300	39,000
110 - 129 P.T.O. horsepower	Each	74,000	76,100	77,700
140 - 159 P.T.O. horsepower	Each	100,000	104,000	111,000
190 - 220 P.T.O. horsepower	Each	138,000	144,000	157,000
Tractor, 4-Wheel Drive				
200 - 280 P.T.O. horsepower	Each	154,000	176,000	195,000
281 - 350 Engine horsepower	Each	176,000	187,000	202,000
51-500 Engine horsepower	Each	....	244,000	249,000
Wagon, Gravity Unload, W/Box and Running				
Gear, and Tires,				
200-400 Bushel Capacity				
Without Side Extensions	Each	6,040	5,900	6,760
Wagon, Running Gear, W/O Box				
8-10 Ton Capacity	Each	2,270	2,480	2,600
Windrower, Self-Propelled,				
14-16 Foot	Each	78,700	83,500	93,300
Agricultural Chemicals: [7]				
Fungicides:				
Basic Copper Sulfate, 53% WP	Lb	2.30	2.87	([9])
Calcium Polysulfide (Lime Sulfur) Liq.Conc	Gal	8.06	11.60	10.30
Captan 50% WP	Lb	4.59	5.51	6.43
Chlorothalonil (Bravo), 6#/Gal EC	Gal	47.00	48.20	59.80
Copper Hydroxide (Kocide 101), 77% WP	Lb	3.58	3.90	4.64
Dodine (Cyprex), 65% WP	Lb	12.90	12.70	12.40
Fenarimol (Rubigan), 1#/Gal EC	Gal	330	340	379
Ferbam (Carbamate), 76% WP	Lb	4.26	3.96	4.38
Fosethyl-AL (Aliette), 80% WP	Lb	13.60	13.70	16.10
Mancozeb (Dithane 80% WP,Manzate 75% DF)	Lb	3.09	3.04	4.69
Maneb, 80% WP, 75% DF	Lb	3.00	3.64	5.14
Myclobutanil (Systhane, Nova, Rally), 40% WP	Lb	69.10	68.30	73.40
Oxytetraycline (Mycoshield), 17% WP	Lb	29.20	29.80	28.80
Triadimefon (Bayleton), 50% WP	Lb	81.20	83.40	101.00
Ziram, 76% WP	Lb	3.08	3.35	3.94

See footnotes at end of table.

Table 9-41.—Prices paid by farmers: April prices, by commodities, United States, 2007–2009 [1]—Continued

Commodity	Unit	2007	2008	2009
		Dollars	*Dollars*	*Dollars*
Herbicides:				
2,4-D, 4#/Gal EC	Gal	15.90	17.20	19.30
Acetochlor (Harness, Surpass),				
6.4–7#/Gal EC	Gal	69.20	71.70	75.50
Alachlor (Lasso), 4#/Gal EC	Gal	26.70	28.30	29.70
Atrazine(AAtrex), 4#/Gal L	Gal	12.20	15.30	20.80
Bentazon (Basagran), 4#/Gal EC	Gal	90.90	90.90	102.00
Butylate (Sutan), 6.7#/Gal EC	Gal	35.70	35.80	34.50
Chlorimuron-ethyl (Classic), 25% DF	Oz	14.40	15.20	15.20
Chlorsulfuron (Glean), 75%	Oz	16.60	18.30	22.00
DCPA (Dacthal), 75% WP	Lb	16.80	16.90	19.10
Dicamba (Banvel), 4#/Gal EC	Gal	82.40	77.60	82.60
Diuron (Karmex, Diurex), 80% WP	Lb	4.98	4.99	6.36
EPTC (Eptan), 7E-(Eradicane),6.7#/Gal EC	Gal	34.40	37.10	45.10
Glyphosate (Roundup), 4#/Gal EC	Gal	28.90	40.50	42.80
Linuron (Lorox, Linex), 50% DF	Lb	16.50	18.40	20.80
MCPA, 4#/Gal, EC	Gal	18.50	19.10	21.80
Metribuzin (Lexone or Sencor), 75% DF	Lb	17.10	17.90	18.20
Napropamide (Devrinol), 50% WP	Lb	10.20	10.20	10.90
Paraquat (Gramoxone Extra), 2.5#/Gal EC	Lb	34.80	33.90	(9)
Pendimethalin (Prowl),3.3#/Gal EC	Gal	28.20	29.70	37.40
Sethoxydim (Poast), 1.5#/Gal EC	Gal	73.60	72.80	82.90
Simazine (Princep), 4#/Gal EC	Gal	17.70	20.30	27.20
Terbacil (Sinbar), 80% WP	Lb	36.90	37.60	40.50
Trifluralin (Treflan), 4#/Gal EC	Gal	20.40	20.90	24.40
Insecticides:				
Acephate (Orthene), 75% SP	Lb	12.70	12.30	13.90
Aldicarb (Temik), 15% G	Lb	3.57	3.59	3.96
Azinphos-methyl (Guthion), 50% WP	Lb	11.70	11.60	13.50
Bt (Dipel 2X), WP	Lb	12.40	12.00	13.20
Carbaryl, (Sevin), 80% S, SP or WP	Lb	6.43	7.12	7.80
Carbofuran (Furadan), 4F	Gal	79.50	81.80	88.10
Chlorpyrifos (Lorsban), 4#/Gal EC	Gal	37.80	37.40	43.50
Cyfluthrin (Baythroid) 2#/Gal EC	Gal	364	320	326
Diazinon, 4#/Gal EC	Gal	43.90	44.00	(9)
Dicrotophos (Bidrin), 8#/Gal EC	Gal	101	104	107
Dimethoate (Cygon), 2.67#/Gal EC	Gal	40.20	40.90	47.40
Disulfoton (Di-Syston), 8#/Gal EC	Gal	120	121	129
Endosulfon (Thiodan, Phaser), 3#/Gal EC	Gal	29.70	29.50	31.90
Esfenvalerate (Asana XL),0.66#/Gal EC	Gal	102	99.20	101
Malathion, 5#/Gal EC	Gal	31.80	35.10	39.20
Methidathion (Supracide), 25% WP	Lb	8.40	9.22	9.09
Methyl Parathion, 4#/Gal EC	Gal	33.00	33.70	(9)
Oil, Superior Oil, Supreme, Volck	Gal	7.99	9.68	10.50
Oxamyl (Vydate-L), 2# L	Gal	76.30	76.60	91.80
Oxydemeton-Methyl (Metasystox-R).				
2#/Gal EC	Gal	103	102	121
Phorate (Thimet), 20% G	Lb	3.02	3.02	2.95
Phosmet (Imidan, Prolate), 50% WP	Lb	9.05	8.92	10.20
Propargite (Comite, Omite), 30% WP	Lb	8.67	9.18	9.26
Synthetic Pyrethroids,.				
(Pounce 2.0, Ambush 3.2 #/Gal) EC	Gal	105	103	98.50
Terbufos (Counter), 15% G	Lb	2.32	2.53	2.46
Zeta–Cyermethrin (Fury), 1.5#/Gal EC	Gal	205	196	207
Other:				
Gibberellic Acid,(Ry3Up,Pro-Gibb)4.0% L	Gal	142	170	139
Nad Napthalene Acetamide, 8.4 WP		72.10	67.30	68.00

[1] Prices paid by famers are collected, for the most part, from retail establishments located in smaller cities and towns in rural areas. Prior to 1995, recorded prices reflected a modified annual average based on frequency item was surveyed during the year. Recorded item values, 1995-99, are the U.S. April average price. [2] Includes Federal, State, and local per gallon taxes where applicable. [3] Excludes Federal excise tax. [4] Includes Federal, State, and local per gallon taxes. [5] Excludes cost of application, except for limestone. [6] With hydraulic lift, transport wheels, and tires. [7] Active Ingredient, (Common Names),and Formulation abbreviations: EC-Emulsifiable Concentrate, DF-Dry Flowable, DG-Dry Granular, G-Granular, L-Liquid, S-Solution, P-Soluble Powder, and WP-Wettable Powder. [8] Insufficient data. [9] Discontinued in 2009.

NASS, Environmental, Economics, and Demographics Branch, (202) 720–6146.

Table 9-42.—Agricultural commodities: Support prices per unit, United States, 2000–2009 [1]

Commodity	Unit	2000	2001	2002	2003	2004
		Dollars	Dollars	Dollars	Dollars	Dollars
Basic commodities:						
Corn:						
Target price	Bushel	NA	NA	2.60	2.60	2.63
Loan rate	do	1.89	1.89	1.98	1.98	1.95
Cotton:						
American upland:						
Target price	Cwt	NA	NA	72.40	72.40	72.40
Loan rate	do	51.92	51.92	52.00	52.00	52.00
Extra-long staple:						
Target price	do	NA	NA	NA	NA	NA
Loan rate	do	79.65	79.65	79.77	79.77	79.77
Peanuts:						
Target price	Short tons	NA	NA	495.00	495.00	495.00
Loan rate	do	NA	NA	355.00	355.00	355.00
Quota rate	do	610.00	610.00	NA	NA	NA
Additional rate	do	132.00	132.00	NA	NA	NA
Rice:						
Target price	Cwt.	NA	NA	10.50	10.50	10.50
Loan rate	do	6.50	6.50	6.50	6.50	6.50
Wheat:						
Target price	Bushel	NA	NA	3.86	3.86	3.92
Loan rate	do	2.58	2.58	2.80	2.80	2.75
Tobacco:						
Flue-cured, types 11-14	Pound	1.640	1.660	1.656	1.663	1.690
Fire-cured, type 21	do	1.559	1.572	1.603	1.636	1.636
Fire-cured, types 22-23	do	1.716	1.736	1.767	1.817	1.863
Burley, type 31	do	1.805	1.826	1.835	1.849	1.873
Dark air-cured, types 35-36	do	1.481	1.499	1.526	1.571	1.612
Virginia sun-cured, type 37	do	1.380	1.392	1.429	1.458	1.458
Ohio filler and Wisconsin binder, types 42-44 and 53-55	do	1.238	1.252	1.286	1.323	1.357
Barley:						
Target price	Bushel	NA	NA	2.21	2.21	2.24
Loan rate	do	1.62	1.65	1.88	1.88	1.85
Sorghum grain:						
Target price	Cwt.	NA	NA	4.54	4.54	4.59
Loan rate	do	3.05	3.05	3.54	3.54	3.48
Oats:						
Target price	Bushel	NA	NA	1.40	1.40	1.44
Loan rate	do	1.16	1.21	1.35	1.35	1.33
Minor oilseeds: [2]						
Target price	Cwt.	NA	NA	9.80	9.80	10.10
Loan rate	do	9.30	9.30	9.60	9.60	9.30
Soybeans:						
Target price	Bushel	NA	NA	5.80	5.80	5.80
Loan rate	do	5.26	5.26	5.00	5.00	5.00
Dry Peas:						
Target price	Cwt.	NA	NA	NA	NA	NA
Loan rate	do	NA	NA	6.33	6.33	6.22
Small chick peas:						
Target price	Cwt.	NA	NA	NA	NA	NA
Loan rate	do	NA	NA	7.56	7.56	7.43
Large chick peas:						
Target price	Cwt.	NA	NA	NA	NA	NA
Loan rate	do	NA	NA	NA	NA	NA
Lentils:						
Target price	do	NA	NA	NA	NA	NA
Loan rate	do	NA	NA	11.94	11.94	11.72
Sugar, raw cane:						
Loan rate	Pound	0.180	0.180	0.180	0.180	0.180
Sugar, refined beet:						
Loan rate	do	0.229	0.229	0.229	0.229	0.229
Honey, extracted:						
Loan rate	Pound	0.65	NA	0.60	0.60	0.60
Mohair:						
Loan rate	do	[3] 2.00	NA	4.20	4.20	4.20
Wool, graded:						
Loan rate	Pound	NA	NA	1.00	1.00	1.00
Wool, nongraded:						
Loan rate	Pound	NA	NA	0.40	0.40	0.40
Milk for manufacturing:						
Support price	Cwt	9.90	9.90	9.90	9.90	9.90

See footnotes at end of table.

Table 9-42.—Agricultural commodities: Support prices per unit, United States, 2000–2009 [1]—Continued

Commodity	Unit	2005	2006	2007	2008	2009
		Dollars	Dollars	Dollars	Dollars	Dollars
Basic commodities:						
Corn:						
Target price	Bushel	2.63	2.63	2.63	2.63	2.63
Loan rate	do	1.95	1.95	1.95	1.95	1.95
Cotton:						
American upland:						
Target price	Cwt	72.40	72.40	72.40	71.25	72.25
Loan rate	do	52.00	52.00	52.00	52.00	52.00
Extra-long staple:						
Target price	do	NA	NA	NA	NA	NA
Loan rate	do	79.77	79.77	79.77	79.77	79.77
Peanuts:						
Target price	Short tons	495.00	495.00	495.00	495.00	495.00
Loan rate	do	355.00	355.00	355.00	355.00	355.00
Quota rate	do	NA	NA	NA	NA	NA
Additional rate	do	NA	NA	NA	NA	NA
Rice:						
Target price	Cwt.	10.50	10.50	10.50	10.50	10.50
Loan rate	do	6.50	6.50	6.50	6.50	6.50
Wheat:						
Target price	Bushel	3.92	3.92	3.92	3.92	3.92
Loan rate	do	2.75	2.75	2.75	2.75	2.75
Tobacco:						
Flue-cured, types 11-14	Pound	NA	NA	NA	NA	NA
Fire-cured, type 21	do	NA	NA	NA	NA	NA
Fire-cured, types 22-23	do	NA	NA	NA	NA	NA
Burley, type 31	do	NA	NA	NA	NA	NA
Dark air-cured, types 35-36	do	NA	NA	NA	NA	NA
Virginia sun-cured, type 37	do	NA	NA	NA	NA	NA
Ohio filler and Wisconsin binder, types 42-44 and 53-55	do	NA	NA	NA	NA	NA
Barley:						
Target price	Bushel	2.24	2.24	2.24	2.24	2.24
Loan rate	do	1.85	1.85	1.85	1.85	1.85
Sorghum grain: [4]						
Target price	Cwt.	4.59	4.59	4.59	4.59	4.59
Loan rate	do	3.48	3.48	3.48	3.48	3.48
Oats:						
Target price	Bushel	1.44	1.44	1.44	1.44	1.44
Loan rate	do	1.33	1.33	1.33	1.33	1.33
Minor oilseeds: [2]						
Target price	Cwt.	10.10	10.10	10.10	10.10	10.10
Loan rate	do	9.30	9.30	9.30	9.30	9.30
Soybeans:						
Target price	Bushel	5.80	5.80	5.80	5.80	5.80
Loan rate	do	5.00	5.00	5.00	5.00	5.00
Dry Peas:						
Target price	Cwt.	NA	NA	NA	NA	NA
Loan rate	do	6.22	6.22	6.22	6.22	6.22
Small chick peas:						
Target price	Cwt.	NA	NA	NA	NA	NA
Loan rate	do	7.43	7.43	7.43	7.43	7.43
Large chick peas:						
Target price	Cwt.	NA	NA	NA	NA	12.81
Loan rate	do	NA	NA	NA	NA	11.28
Lentils:						
Target price	do	NA	NA	NA	NA	12.81
Loan rate	do	11.72	11.72	11.72	11.72	11.28
Sugar, raw cane:						
Loan rate	Pound	0.180	0.180	0.180	0.180	0.1825
Sugar, refined beet:						
Loan rate	do	0.229	0.229	0.229	0.229	0.2345
Honey, extracted:						
Loan rate	Pound	0.60	0.60	0.60	0.60	0.60
Mohair:						
Loan rate	do	4.20	4.20	4.20	4.20	4.20
Wool, graded:						
Loan rate	Pound	1.00	1.00	1.00	1.00	1.00
Wool,nongraded:						
Loan rate	Pound	0.40	0.40	0.40	0.40	0.40
Milk for manufacturing:						
Support price	Cwt	9.90	9.90	[5]9.35	[6]11.00	9.35

[1] National averages during the marketing years for the individual crops.   [2] Includes flaxseed, sunflower seed (oil and other), safflower, rapeseed, canola, mustard seed, crambe, and sesame.   [3] Recourse loans.   [4] Effective January 1, 1999, the milk for manufacturing support price became $9.90 per cwt.   [5] Effective support price calculated from product prices specified in 2008 Farm Bill, effective January 1, 2008.   NA-not applicable.
FSA, Economic Policy and Analysis Staff, (202) 720-0967.

Table 9-43.—Farm income: Cash receipts by commodity groups and selected commodities, United States, 2001–2010 [1]

Commodity	2001	2002	2003	2004	2005
	1,000 dollars	*1,000 dollars*	*1,000 dollars*	*1,000 dollars*	
All commodities	200,030,343	194,924,491	215,971,148	237,853,261	240,897,821
Livestock and products	106,696,873	93,956,698	105,671,501	123,472,726	124,931,103
Cattle and calves	40,540,660	38,095,143	45,341,079	47,429,896	49,283,094
Hogs	12,394,562	9,602,110	10,616,057	14,336,266	14,970,027
Sheep and lambs	396,586	420,633	502,900	508,405	560,047
Dairy products	24,685,667	20,582,238	21,231,059	27,366,854	26,704,863
Broilers	16,694,530	13,437,700	15,214,956	20,446,109	20,877,923
Farm chickens	46,516	49,850	47,997	57,709	65,072
Chicken eggs	4,398,045	4,232,449	5,333,753	5,303,038	4,066,669
Turkeys	2,735,961	2,643,273	2,440,460	2,819,712	3,025,891
Miscellaneous livestock	4,046,390	4,114,299	4,230,610	4,357,872	4,579,063

Commodity	2006	2007	2008	2009	2010
All commodities	240,623,888	288,545,936	318,329,989	283,406,168	291,571,152
Livestock and products	118,498,682	138,478,570	141,525,698	119,751,629	127,418,380
Cattle and calves	49,110,334	49,843,326	48,517,775	43,776,568	46,610,666
Hogs	14,105,864	14,750,486	16,050,489	14,395,118	15,341,879
Sheep and lambs	471,896	466,670	443,021	426,829	452,828
Dairy products	23,412,552	35,453,399	34,849,113	24,342,440	27,783,308
Broilers	17,852,894	21,513,538	23,203,136	21,812,789	21,803,525
Farm chickens	54,141	51,498	62,199	65,089	65,032
Chicken eggs	4,460,211	6,718,853	8,215,995	6,155,825	6,426,516
Turkeys	3,451,528	3,929,008	4,477,244	3,573,285	3,668,080
Miscellaneous livestock	4,754,102	4,851,248	4,832,806	4,347,194	4,410,056

Commodity	2001	2002	2003	2004	2005
Crops	93,333,470	100,967,793	110,299,647	114,380,535	115,966,718
Food grains	6,385,012	6,787,802	7,965,136	8,937,840	8,611,410
Feed crops	21,454,849	24,040,729	24,746,752	27,405,592	24,589,872
Cotton	3,639,446	3,418,096	6,419,910	4,825,881	6,402,504
Tobacco	1,894,764	1,743,429	1,602,392	1,577,423	1,097,081
Oil crops	13,337,838	15,049,103	17,988,338	17,862,280	18,387,789
Vegetables	15,433,371	17,140,215	17,152,918	16,563,699	17,291,250
Fruits/nuts	11,904,046	12,570,807	13,480,486	15,126,283	17,137,528
All other crops	19,284,144	20,217,612	20,943,715	22,081,537	22,449,284

Commodity	2006	2007	2008	2009	2010
Crops	122,125,206	150,067,366	176,804,291	163,654,539	164,152,771
Food grains	9,089,720	13,559,548	18,708,372	14,383,800	12,923,392
Feed crops	29,386,073	42,321,639	58,925,844	50,176,020	47,782,232
Cotton	5,545,956	6,457,260	5,227,915	3,488,956	4,166,072
Tobacco	1,156,674	1,284,098	1,450,572	1,485,238	1,490,844
Oil crops	18,545,958	24,603,108	28,688,901	31,912,060	32,387,744
Vegetables	18,074,300	19,320,989	21,017,475	20,593,169	20,091,640
Fruits/nuts	17,254,266	18,651,692	19,247,377	18,965,451	19,187,698
All other crops	23,072,266	23,869,032	23,537,835	22,649,845	26,123,149

[1] USDA estimates and publishes individual cash receipt values only for major commodities and major producing States. The U.S. receipts for individual commodities, computed as the sum of the reported States, may understate the value of sales for some commodities, with the balance included in the appropriate category labeled "other" or "miscellaneous." The degree of underestimation in some of the minor commodities can be substantial.

ERS, Farm and Rural Business Branch, (202) 694–5592.

Table 9-44.—Farm income: United States, 2002–2009 [1]

Item	2002	2003	2004	2005
	Billion dollars	Billion dollars	Billion dollars	Billion dollars
Total gross farm income	232.6	258.7	294.9	298.5
Value of Production [2]	220.2	242.2	281.9	274.1
Crops	98.3	108.6	125.1	114.4
Livestock and products	93.5	105.0	124.3	126.5
Services and forestry	28.5	28.6	32.5	33.2
Direct government payments	12.4	16.5	13.0	24.4
Total production expenses	193.1	197.7	207.5	219.7
Net farm income	39.6	61.0	87.4	78.8
Gross cash income	222.2	246.8	266.5	279.7
Cash expenses	33.2	174.7	182.9	193.1
Net cash income	170.8	72.1	83.7	86.7

Item	2006	2007	2008	2009
Total gross farm income	290.2	339.5	379.6	343.2
Value of production [2]	274.4	327.6	367.3	330.9
Crops	118.7	151.1	185.1	169.0
Livestock and product	119.3	138.4	140.3	119.2
Services and forestry	36.4	38.1	42.0	42.7
Direct government payments	15.8	11.9	12.2	12.3
Total production expenses	232.7	269.2	293.0	281.0
Net farm income	57.4	70.3	86.6	62.2
Gross cash income	273.2	318.0	352.0	317.6
Cash expenses	204.8	240.3	261.6	248.5
Net cash income	68.4	77.7	90.4	69.1

[1] Component values and additional details may be found in the value-added and cash income tables on the internet at http://www.ers.usda.gov/data/farmincome/finfidmu.htm. [2] Includes cash receipts, value of change in inventories, and home consumption. In the value-added table, value of production is synonymous with final output.
ERS, Farm and Rural Business Branch, (202) 694–5592.

Table 9-45.—Expenses: Farm production expenses, United States, 2002–2009

Item	2002	2003	2004	2005
	Thousand dollars	Thousand dollars	Thousand dollars	Thousand dollars
Total production expenses	193,055,516	197,739,218	207,453,128	219,741,952
Feed purchased	24,929,575	27,526,351	29,729,126	28,026,427
Livestock and poultry purchased	14,413,409	16,705,638	18,152,347	18,657,654
Seed purchased	8,924,511	9,423,203	9,621,835	10,421,614
Fertilizer and lime	9,619,305	10,022,392	11,424,587	12,828,950
Pesticides	8,316,338	8,416,859	8,616,932	8,818,161
Fuel and oil	6,603,708	6,839,358	8,210,573	10,294,184
Electricity	3,911,407	3,479,169	3,394,084	3,458,636
Other [1]	44,311,296	45,247,708	45,702,899	49,136,891
Interest	12,789,428	11,047,958	10,741,611	12,620,077
Contract and hired labor expenses	21,846,751	21,993,738	23,312,555	23,554,256
Net rent to nonoperator landlords [2]	9,636,417	8,787,507	8,458,236	8,982,428
Capital consumption	20,945,959	21,442,495	23,080,760	24,933,474
Property taxes	6,807,412	6,806,842	7,007,583	8,009,200

Item	2006	2007	2008	2009
Total production expenses	232,734,904	269,222,343	292,968,485	281,006,382
Feed purchased	31,423,477	41,923,656	46,929,786	45,027,927
Livestock and poultry purchased	18,638,653	18,830,252	17,744,310	16,477,444
Seed purchased	11,020,213	12,620,280	15,120,073	15,520,357
Fertilizer and lime	13,331,256	17,732,093	22,533,546	20,135,755
Pesticides	9,018,293	10,517,497	11,718,091	11,520,052
Fuel and oil	11,314,911	13,792,999	16,243,399	12,715,826
Electricity	3,796,466	4,282,844	4,544,009	4,590,239
Other [1]	52,691,659	60,833,440	64,032,637	60,795,971
Interest	14,392,506	15,116,726	15,419,630	15,154,296
Contract and hired labor expenses	24,226,518	28,638,631	29,689,714	28,704,575
Net rent to nonoperator landlords [2]	7,631,762	7,592,368	9,589,811	9,834,140
Capital consumption	26,238,232	27,028,167	28,689,535	30,116,259
Property taxes	9,010,958	10,313,390	10,713,944	10,413,541

[1] Includes repair and maintenance, machine hire and custom work, marketing, storage and transportation, insurance premiums, and miscellaneous other expenses. [2] Includes landlord capital consumption.
ERS, Farm and Rural Business Branch, (202) 694–5592.

Table 9-46.—Farm Operator Households: Average Income, United States, 2006–2010 [1]

Item	2006	2007	2008	2009	2010 [1]
	Dollars per farm operator household				
Net earnings of the household from farming activities	8,541	11,364	9,764	6,866	9,043
Off-farm income of the household	72,502	77,432	70,032	70,302	72,627
Earned income	51,674	58,933	50,761	50,852	52,903
Off-farm wages and salaries	38,481	48,947	42,606	43,852	NA
Off-farm business income	13,193	9,986	8,155	7,000	NA
Unearned income	20,827	18,499	19,271	19,450	19,724
Average household income of farm operators	81,043	88,796	79,796	77,169	81,670
Median income to farm operator households	56,274	54,428	51,431	52,235	NA
	Dollars per U.S. household				
U.S. Average household of income	66,570	67,609	68,424	67,976	NA
U.S. median household income	48,201	50,233	50,303	49,777	NA
	Percent				
Average farm operator household income as percent of U.S. average household income	121.7	131.3	116.6	113.5	NA
Median farm operator household income as percent of U.S. median household income	116.7	108.4	102.2	104.9	NA
Percent of farm household income from farming	10.5	12.8	12.2	8.9	11.1

[1] Forecast.
ERS, Farm and Rural Household Well-Being Branch, (202) 694-5583.

Table 9-47.—Grazing fees: Rates for cattle by selected States and regions, 2008–2009

State	Monthly lease rates for private non-irrigated grazing land [1]					
	Animal unit [2]		Cow-calf		Per head	
	2008	2009	2008	2009	2008	2009
	Dollars per month	Dollars per month	Dollars per month	Dollars per month	Dollars per month	Dollars per month
AZ	8.50	[7]	[7]	[7]	11.00	10.00
CA	17.80	16.70	22.10	21.00	18.50	17.50
CO	14.50	14.70	16.00	16.30	14.50	15.20
ID	12.60	12.60	16.30	15.90	14.10	14.00
KS	14.00	13.50	17.00	16.50	13.50	14.00
MT	18.10	18.00	20.00	20.20	19.80	18.90
NE	25.00	24.80	29.70	29.30	28.00	26.50
NV	13.50	11.00	14.70	12.00	14.00	12.00
NM	11.00	10.00	11.50	13.00	12.00	12.00
ND	15.80	16.00	17.80	17.70	15.50	16.20
OK	9.00	9.00	13.00	11.50	9.50	10.50
OR	14.00	14.60	16.80	17.80	14.60	15.50
SD	21.70	22.90	25.20	25.60	23.00	23.00
TX	10.70	10.50	11.00	11.50	11.00	11.20
UT	13.00	13.00	15.90	16.30	15.50	15.30
WA	11.50	11.00	13.40	13.00	14.10	12.80
WY	15.70	16.00	18.40	18.70	16.40	16.70
17-State [3]	14.70	14.60	17.00	17.00	15.60	15.50
16-State [4]	16.20	16.10	19.30	19.00	17.20	17.10
11-State [5]	15.00	14.70	17.40	17.40	16.20	15.80
9-State [6]	14.50	14.50	16.80	16.70	15.20	15.30

[1] The average rates are estimates (rates over $10.00 are rounded to the nearest dime) based on survey indications of monthly lease rates for private, non-irrigated grazing land from the January Cattle Survey. [2] Includes animal unit plus cow-calf rates. Cow-calf rate converted to animal unit (AUM) using (1 aum=cow-calf *0.833). [3] Seventeen Western States: All States listed. [4] Sixteen Western States: All States, except Texas. [5] Eleven Western States: AZ, CA, CO, ID, MT, NV, NM, OR, UT, WA, and WY. [6] Nine Great Plains States: CO, KS, NE, NM, ND, OK, SD, TX, and WY. [7] Insufficient data.
NASS, Environmental, Economics, and Demographics Branch, (202) 720–6146.

CHAPTER X

INSURANCE, CREDIT, AND COOPERATIVES

The statistics in this chapter deal with taxes, insurance, agricultural credit, and farm cooperatives. Some of the series were developed in connection with research activities of the Department, while others, such as data from agricultural credit agencies, are primarily records of operations.

Table 10-1.—Crop losses: Average percentage of indemnities attributed to specific hazards, by crops, 1948–2010

Crop	Year	Drought heat (excess)	Hail	Precip. (excess poor drainage)	Frost freeze, (other cold damage)	Flood	Cyclone, tornado, wind, hot wind	Insects	Disease	All others
		Percent	Percent	Percent	Percent	Percent	Percent	Percent	Percent	Percent
Adjusted gross revenue	2001-2008	19	5	18	26	0	1	0	0	31
Adjusted gross revenue-lite	1981-2008	3	4	2	67	0	0	0	0	24
Alfalfa seed	2002-2009	20	8	6	14	0	40	1	1	10
All other citrus trees	2000-2009	0	0	0	4	0	2	0	0	93
All other grapefruit	2001-2009	0	0	0	10	0	90	0	0	0
Almonds	1981-2009	2	4	55	33	0	6	0	0	0
Apples	1963-2009	10	25	7	50	0	2	0	1	4
Avocado trees	1996-2009	0	0	6	2	0	92	0	0	0
Avocados	1998-2010	9	0	0	67	0	16	1	0	7
Barley	1956-2009	40	17	26	5	0	2	2	4	4
Blueberries	1995-2009	9	4	27	58	0	1	0	0	1
Burley tobacco	1997-2009	24	7	34	5	6	5	0	16	1
Cabbage	1999-2009	21	3	18	7	1	11	35	2	2
Canola	1995-2009	24	14	38	12	0	8	2	1	2
Carambola trees	2001-2001	0	0	100	0	0	0	0	0	0
Cherries	1963-2009	4	6	27	48	0	5	0	0	9
Chile peppers	2000-2008	1	23	12	15	6	21	8	5	9
Cigar binder tobacco	1997-2009	2	20	30	2	0	1	0	45	1
Cigar filler tobacco	1998-2006	86	0	6	0	0	0	1	7	0
Cigar wrapper tobacco	1997-2008	0	0	53	3	0	0	0	43	0
Citrus	1989-1997	18	5	1	74	0	2	0	0	0
Citrus I	1998-2009	0	0	0	1	0	99	0	0	0
Citrus II	2000-2009	0	0	0	7	0	92	0	0	0
Citrus III	2001-2009	0	0	0	5	0	95	0	0	0
Citrus IV	1998-2009	6	0	0	11	0	83	0	0	0
Citrus trees	1990-1997	0	0	0	100	0	0	0	0	0
Citrus treesl	2008-2008	0	0	100	0	0	0	0	0	0
Citrus treesll	2008-2008	0	0	100	0	0	0	0	0	0
Citrus trees IV	2004-2008	0	0	100	0	0	0	0	0	0
Citrus V	1999-2009	0	1	0	7	0	91	0	0	0
Citrus VI	2005-2006	0	0	0	0	0	100	0	0	0
Citrus VII	1998-2009	0	4	0	4	0	92	0	0	0
Citrus VIII	2009-2009	0	48	0	52	0	92	0	0	0
Clams	2001-2009	0	0	0	6	0	6	0	0	88
Corn	1948-2009	238	133	117	150	101	180	27	67	241
Cotton	1948-2009	21	13	12	11	5	20	3	1	12
Cotton ex long staple	1984-2009	11	12	20	13	0	8	10	0	26
Crambe	1999-2003	22	12	28	9	0	23	0	5	0
Cranberries	1984-2009	13	12	10	54	1	0	5	1	4
Cultivated wild rice	1999-2009	10	20	4	6	2	33	2	2	23
Dark air tobacco	1997-2009	38	5	28	2	1	10	0	16	0
Dry beans	1948-2009	18	32	21	21	1	2	0	2	1
Dry peas	1963-2009	58	23	9	7	0	0	1	1	0
Early & midseason oranges	1998-2010	0	8	3	56	0	32	0	0	0
Figs	1988-2008	12	0	47	32	0	2	0	0	7
Fire cured tobacco	1997-2009	41	8	19	11	1	3	0	8	10
Flax	1948-2009	41	7	46	3	0	1	1	0	0
Flue cured tobacco	1997-2009	25	13	12	5	1	20	0	24	1
Forage production	1979-2009	43	4	13	32	0	0	1	0	6
Forage seeding	1978-2009	48	19	0	32	0	1	0	0	0
Fresh apricots	1997-2009	3	34	16	45	0	1	0	0	0
Fresh freestone peaches	1997-2009	5	22	23	49	0	1	0	0	0
Fresh market beans	2000-2006	0	0	72	15	0	8	0	4	0
Fresh market sweet corn	1985-2009	9	1	42	18	0	29	0	0	1
Fresh market tomatoes	1984-2009	6	13	45	14	0	10	2	10	0
Fresh nectarines	1997-2009	12	43	28	17	0	1	0	0	0
Fresh plum	1990-1997	0	59	8	32	0	1	0	0	0
Grain sorghum	1959-2009	32	8	22	15	2	13	2	0	6
Grapefruit	2000-2009	21	2	1	49	0	24	0	0	1
Grapefruit trees	2000-2009	0	0	1	1	0	0	0	0	98

See end of table.

Table 10-1.—Crop losses: Average percentage of indemnities attributed to specific hazards, by crops, 1948–2009—Continued

Crop	Year	Drought heat (excess)	Hail	Precip. (excess poor drainage)	Frost freeze, (other cold damage)	Flood	Cyclone, tornado, wind, hot wind	Insects	Disease	All others
		Percent	Percent	Percent	Percent	Percent	Percent	Percent	Percent	Percent
Grapes	1967-2009	20	3	16	59	0	0	0	0	0
Green peas	1962-2009	42	4	47	4	0	0	0	1	0
Hybrid corn seed	1983-2009	46	5	34	2	0	10	0	3	0
Hybrid sorghum seed	1988-2009	17	16	3	44	0	17	0	0	1
Income protection corn	1996-1996	3	0	93	3	0	0	0	0	0
Income protection cotton	1996-1996	96	0	4	0	0	0	0	0	0
Income protection wheat	1996-1996	9	0	90	0	1	0	0	0	0
Late oranges	1998-2009	0	5	3	43	0	50	0	0	0
Lemon trees	2005-2005	0	0	0	0	0	100	0	0	0
Lemons	1997-2009	2	0	0	97	0	0	0	0	0
Lime trees	1998-2005	0	0	0	1	0	0	0	0	99
Macadamia nuts	1996-2009	26	3	0	0	0	0	18	0	53
Macadamia trees	2000-2005	0	61	61	0	0	39	0	0	0
Mandarins	1997-2009	29	0	1	70	0	1	0	0	0
Mango trees	1997-2005	0	0	0	37	0	63	0	0	0
Maryland tobacco	1997-2004	60	8	7	4	0	6	0	14	0
Millet	1996-2009	82	13	5	0	0	1	0	0	0
Minneola tangelos	1998-2009	6	1	2	90	0	1	0	0	0
Mint	2000-2009	31	1	18	45	1	3	0	0	0
Mustard	1999-2009	70	24	1	3	0	1	0	0	3
Navel oranges	1998-2010	46	2	4	44	1	1	0	0	3
Nursery	1990-1999	24	0	7	11	7	42	3	5	0
Nursery (fg&c)	2001-2010	1	11	28	16	5	36	1	1	1
Oats	1956-2009	47	14	31	5	0	1	1	0	0
Onions	1988-2009	14	16	46	3	0	3	0	14	2
Orange trees	1996-2009	0	0	1	1	0	1	0	0	97
Oranges	1997-1997	30	0	0	37	0	21	0	0	11
Orlando tangelos	1998-2008	0	0	0	100	0	0	0	0	0
Papaya	2007-2007	0	0	0	0	0	20	80	0	0
Pastures	....	0	0	0	0	0	0	0	0	100
Peaches	1957-2009	3	36	3	55	0	0	0	0	2
Peanuts	1962-2009	43	0	21	6	0	5	0	21	3
Pears	1989-2009	1	29	3	66	0	0	0	0	1
Pecans	1998-2009	41	2	12	17	0	25	0	1	2
Peppers	1984-2009	0	6	62	25	0	5	0	1	0
Plums	1998-2009	14	26	21	34	0	6	0	0	0
Popcorn	1984-2009	55	11	23	4	1	2	1	2	0
Potatoes	1962-2009	23	6	24	23	0	1	0	19	0
Prevented planting endorse	1990-1994	31	0	11	0	53	0	0	0	5
Processing apricots	1997-2009	1	5	50	25	0	18	0	0	0
Processing beans	1988-2009	47	3	43	2	0	1	0	3	0
Processing cling peaches	1997-2009	20	9	34	34	0	2	0	0	0
Processing cucumbers	2000-2005	45	1	47	2	0	0	0	4	0
Processing freestone	1998-2009	9	8	8	72	0	2	1	0	0
Prunes	1986-2009	29	1	10	49	0	10	0	0	1
Raisins	1961-2008	0	0	100	0	0	0	0	0	0
Rangeland	1999-2008	0	0	0	0	0	0	0	0	100
Raspberry and blackberry	2002-2006	40	0	22	27	0	12	0	0	0
Revenue coverage corn	1996-1996	20	44	3	5	26	0	0	1	0
Revenue coverage soybeans	1996-1996	1	24	55	5	13	0	0	3	0
Rice	1960-2009	17	0	47	8	5	9	0	4	10
Rio red & star ruby	1998-2010	0	3	5	43	0	45	1	0	3
Ruby red grapefruit	1998-2009	0	8	11	33	48	0	0	0	0
Rye	1980-2009	22	6	54	17	0	1	0	0	0
Safflower	1964-2009	43	5	17	16	0	17	1	1	0
Silage sorghum	1965-2009	94	0	0	0	0	5	0	0	0
Soybeans	1955-2009	25	14	19	13	9	6	1	3	10
Special citrus	1992-1994	6	12	0	0	82	0	0	0	0
Stonefruit	1989-1996	1	28	44	19	0	2	0	0	6
Strawberries	2000-2008	9	0	67	6	0	0	0	17	0
Sugar beets	1965-2009	13	8	27	24	2	11	1	12	2
Sugarcane	1967-2009	21	0	13	17	0	4	2	16	27
Sunflowers	1976-2009	26	16	25	10	0	5	4	7	6
Sweet corn	1978-2009	48	0	29	20	0	2	0	1	1
Sweet oranges	1998-2009	12	0	3	82	0	2	0	0	0
Sweetpotatoes	1998-2009	39	0	43	0	0	0	13	4	0
Table grapes	1984-2009	31	4	26	37	0	0	0	0	0
Tangelos	1997-1997	3	0	0	97	0	0	0	0	0
Tobacco	1989-1996	17	20	20	1	2	18	0	20	2
Tomatoes	1963-2009	34	1	56	0	4	1	1	2	1
Valencia oranges	1998-2009	38	2	1	55	0	3	0	0	1
Walnuts	1984-2009	29	4	49	17	0	2	0	0	0
Watermelons	1999-2000	8	7	38	1	0	14	0	29	2
Wheat	1948-2009	99	46	50	132	20	51	15	35	42
Winter squash	1999-2005	10	13	75	0	0	2	0	0	1

GRP crops do not have any specific cause of loss.
RMA, Program Automation Branch, (816) 926-7910.

Table 10-2.—Crop insurance programs: Coverage, amount of premiums and indemnities, by crops, United States, 2007–2009 [1]

Commodity and year	Coverage					Indemnities		
	County pro-grams	Insured units [2]	Area in-sured [3]	Maximum insured production	Amount of premium	Number	Area in-demnified [3]	Amount
	Number	Number	1,000 acres	1,000 dollars	1,000 dollars		1,000 acres	1,000 dollars
Adjusted gross revenue:								
2007	1,155	427	0	78,841	3,629	42	0	1,807
2008	1,604	386	0	81,275	3,791	48	0	3,048
2009	1,706	383		98,968	4,436			0
Adjusted gross revenue-lite:								
2007	230	519	0	246,804	10,191	44		6,260
2008	230	471	0	244,196	9,841	65		7,761
2009	230	395	0	301,448	10,800		0	
Alfalfa seed:								
2007	12	292	19	9,508	747	43	2	474
2008	12	225	16	8,973	698	44	2	498
2009	12	323	28	17,290	1,415	18	1	309
All other citrus trees:								
2007	28	927	0	51,299	1,567	4	0	6
2008	28	720	0	52,561	976	0		0
2009	28	1,007	0	59,490	1,095	8	0	207
All other grape-fruit:								
2007	3	2	0	13	2	0	0	0
2008	3	2	0	15	2	0	0	0
2009	3	2	0	13	2	1		4
Almonds:								
2007	16	4,622	450	687,509	34,054	95	8	2,787
2008	16	4,754	478	756,678	34,514	149	12	3,597
2009	16	4,936	521	935,061	40,393	457	32	21,037
Apples:								
2007	367	5,424	235	512,018	39,508	1,371	34	50,119
2008	366	5,741	237	584,066	46,538	1,502	41	68,804
2009	366	5,984	240	694,519	60,105	1,016	21	41,791
Avocado trees:								
2007	1	192	0	14,406	460	0	0	0
2008	1	187	0	14,603	514	0	0	0
2009	1	207	0	21,192	807	1	0	9
Avocados:								
2007	6	1,101	33	53,777	6,437	384	10	6,410
2008	6	1,122	35	59,711	6,389	300	6	6,109
2009	6	1,110	34	61,615	6,929	53	1	834
Banana trees:								
2008	4	2	0	429	36	0		0
2009	4	2	0	469	34	0		0
Banana:								
2007	4	6	0	795	22	0		0
2008	4	5	0	370	16	0		0
2009	4	6	0	737	20	0		0
Barley:								
2007	1,760	34,201	2,839	315,672	40,497	8,359	1,122	35,833
2008	1,703	34,694	2,986	562,750	77,561	6,877	898	45,312
2009	1,771	27,917	2,524	398,283	51,587	4,446	348	19,069
Blueberries:								
2007	61	776	41	49,319	3,572	220	6	5,595
2008	61	863	44	63,863	4,561	53	1	754
2009	68	965	50	91,368	6,441	110	2	2,542
Burley tobacco:								
2007	277	8,327	66	153,387	15,632	2,365	23	30,093
2008	273	8,398	70	180,579	19,951	2,470	25	33,853
2009	284	9,190	82	227,177	24,546	556	5	7,663
Cabbage:								
2007	27	328	14	13,863	963	46	1	851
2008	27	309	14	14,325	1,056	49	1	1,042
2009	32	291	13	15,414	1,065	28	0	422
Canola:								
2007	257	12,579	1,167	173,992	28,234	3,946	445	29,825
2008	247	10,205	964	281,872	47,652	3,630	429	39,591
2009	259	10,120	984	181,815	29,251	3,985	331	35,672
Carambola trees:								
2007	1	9	0	258	7	0		0
2008	1	11	0	280	9	0		0
2009	1	11	0	331	10	0		0
Cherries:								
2007	21	2,415	47	80,604	7,209	296	5	4,403
2008	21	2,517	49	87,194	7,739	493	8	11,051
2009	35	2,490	51	278,141	23,243	674	11	25,383
Chili peppers:								
2007	3	40	4	1,801	121	1	0	10
2008	3	36	4	1,440	79	4	0	64
2009	3	36	4	2,009	123	0		0
Cigar binder to-bacco:								
2007	16	577	4	22,958	3,066	106	1	2,470
2008	16	540	4	26,427	4,049	243	2	10,536
2009	16	506	4	25,324	4,505	216	2	10,061

See footnotes at end of table.

Table 10-2.—Crop insurance programs: Coverage, amount of premiums and indemnities, by crops, United States, 2007–2009 [1]—Continued

Commodity and year	Coverage				Amount of premium	Indemnities		
	County pro-grams	Insured units [2]	Area in-sured [3]	Maximum insured production		Number	Area in-demnified [3]	Amount
	Number	Number	1,000 acres	1,000 dollars	1,000 dollars		1,000 acres	1,000 dollars
Cigar filler to-bacco:								
2007	3	21	0	282	8	0		0
2008	3	19	0	246	6	0		0
2009	3	17	0	303	7	0		0
Cigar wrapper to-bacco:								
2007	5	33	1	19,691	1,715	1	0	2
2008	5	26	1	13,607	1,078	0		0
2009	5	34	1	12,938	993	14	0	1,596
Citrus I:								
2007	29	2,778	203	113,457	2,762	0		0
2008	29	2,724	199	122,805	3,131	0		0
2009	29	2,870	200	131,887	3,281	10	0	70
Citrus II:								
2007	29	2,368	238	160,282	4,587	1	0	95
2008	29	2,377	234	182,619	5,764	0		0
2009	29	2,518	236	187,570	5,685	79	5	1,552
Citrus III:								
2007	29	80	1	489	14	0		0
2008	29	109	3	1,674	44	0		0
2009	29	140	4	2,091	54	4	0	19
Citrus IV:								
2007	29	1,220	22	15,173	512	1	0	84
2008	29	1,166	21	16,107	610	5	0	17
2009	29	750	12	9,580	376	11		86
Citrus trees I:								
2007	3	463	5	15,301	665	0		0
2008	3	442	5	15,221	631	5	0	38
2009	3	398	5	14,620	579	0		0
Citrus trees II:								
2007	3	116	1	3,560	163	0		0
2008	3	120	1	3,754	156	1	0	12
2009	3	112	1	3,924	154	0		0
Citrus trees III:								
2007	3	4	0	131	6	0	0	0
2008	3	4	0	131	6	0	0	0
2009	3	3	0	72	4	0		0
Citrus trees IV:								
2007	3	742	13	36,937	2,036	0		0
2008	3	722	14	36,911	1,972	3	0	7
2009	3	696	13	37,357	1,923	0		0
Citrus trees V:								
2007	3	112	2	4,971	330	0		0
2008	3	105	2	4,623	271	0		0
2009	3	94	2	4,387	235	0		0
Citrus V:								
2007	29	431	10	18,322	725	1	0	33
2008	29	411	10	17,153	713	1	0	5
2009	29	409	9	14,739	634	63	1	1,093
Citrus VI:								
2007	5	1	0	7	0	0		0
2008	5	2	1	292	5	0		0
2009	5	1	0	6	0	0		0
Citrus VII:								
2007	29	1,330	88	61,598	2,236	2	0	30
2008	29	1,266	83	64,739	2,562	5	0	98
2009	29	1,131	74	62,741	2,476	55	5	1,492
Clams:								
2007	13	153	0	26,824	977	19		502
2008	13	124	0	31,132	1,078	11		407
2009	13	115	0	28,554	726	21		1,510
Coffee								
2007	4	3	0	10,801	29	0		0
2008	4	4	0	9,715	27	4		0
2009	4	12	0	13,298	40	0		0
Corn:								
2007	10,170	984,164	74,967	31,444,356	3,109,880	155,956	12,711	1,095,746
2008	10,150	932,152	69,328	37,540,440	3,804,656	316,217	31,042	3,059,372
2009	10,151	781,730	71,828	31,044,410	3,392,677	88,197	7,495	790,746
Cotton ELS:								
2007	31	878	276	136,851	6,954	95	13	5,942
2008	32	547	172	82,912	3,945	103	12	5,878
2009	32	730	207	155,574	14,505	220	69	43,971
Cotton:								
2007	1,900	140,522	9,940	2,014,535	305,844	20,969	1,690	155,395
2008	1,898	130,521	8,808	2,346,000	397,761	61,774	7,145	565,805
2009	1,868	114,082	8,628	2,059,366	332,187	23,465	3,357	310,448
Cranberries:								
2007	30	630	31	81,504	2,750	83	2	1,768
2008	30	646	32	95,623	3,134	16	0	464
2009	30	655	31	147,889	4,611	50	1	2,182

See footnotes at end of table.

Table 10-2.—Crop insurance programs: Coverage, amount of premiums and indemnities, by crops, United States, 2007–2009 [1]—Continued

Commodity and year	Coverage				Amount of premium	Indemnities		
	County pro-grams	Insured units [2]	Area in-sured [3]	Maximum insured production		Number	Area in-demnified [3]	Amount
	Number	*Number*	*1,000 acres*	*1,000 dollars*	*1,000 dollars*		*1,000 acres*	*1,000 dollars*
Cultivated wild rice:								
2007	10	71	27	12,603	735	9	2	432
2008	11	123	33	17,743	1,021	10	2	463
2009	11	95	26	18,153	878	6	1	378
Dark air tobacco:								
2007	37	492	2	8,121	311	46	0	276
2008	37	794	5	16,928	660	30	0	405
2009	37	736	4	14,113	492	18	0	216
Dry beans:								
2007	321	19,950	1,371	291,901	45,452	4,003	295	23,812
2008	314	18,201	1,276	410,222	60,698	2,736	219	24,378
2009	283	18,974	1,345	428,410	65,760	4,483	352	49,343
Dry Peas:								
2007	128	10,610	995	77,028	10,225	1,472	163	4,839
2008	138	10,227	998	164,648	21,278	3,611	478	26,067
2009	139	11,937	1,204	200,415	27,472	1,758	189	11,554
Early and Midseason or-anges:								
2007	3	272	5	2,254	132	6	0	7
2008	3	248	4	2,328	149	11	0	63
2009	3	255	4	2,350	158	45	1	98
Figs:								
2007	4	90	5	3,655	242	3	0	107
2008	4	73	6	3,509	195	0	....	0
2009	4	70	6	3,674	181	1	0	22
Fired cured to-bacco:								
2007	43	1,035	8	32,183	1,473	155	1	1,815
2008	43	1,301	13	47,740	2,201	57	0	892
2009	43	1,238	11	46,385	2,089	51	0	584
Flax:								
2007	151	4,676	336	27,653	3,725	1,454	123	3,326
2008	151	4,367	329	62,386	8,315	1,376	133	8,401
2009	153	4,776	356	53,052	8,031	1,591	103	8,900
Flue cured to-bacco:								
2007	173	12,608	213	526,488	28,661	3,285	56	68,354
2008	173	12,382	216	568,871	32,946	3,538	67	80,820
2009	173	12,625	220	678,197	38,044	2,686	48	70,022
Forage prod.:								
2007	790	43,118	3,688	335,724	33,549	6,606	578	20,572
2008	792	38,349	3,369	317,809	31,774	6,620	580	13,576
2009	798	57,651	4,502	577,584	46,888	4,072	360	13,351
Forage seeding:								
2007	583	3,612	146	19,121	2,702	1,022	46	3,470
2008	586	3,259	125	20,390	2,787	655	32	2,683
2009	592	4,414	174	28,343	3,622	91	5	487
Fresh apricots:								
2007	29	174	3	5,819	637	10	0	96
2008	29	168	3	7,242	759	39	1	1,169
2009	29	168	3	7,172	813	31	1	747
Fresh freestone peaches:								
2007	25	1,030	24	21,969	1,282	49	1	264
2008	25	1,007	25	27,960	1,607	67	0	294
2009	25	890	23	26,311	1,099	67	1	443
Fresh market sweet corn:								
2007	232	1,131	61	37,675	3,781	104	3	1,322
2008	232	1,002	56	35,583	3,567	96	5	1,976
2009	232	940	50	33,711	3,362	102	4	1,085
Fresh market to-matoes:								
2007	59	771	57	159,524	16,377	247	11	22,584
2008	59	742	56	161,700	16,360	110	3	5,378
2009	59	720	52	166,835	20,600	237	11	28,100
Fresh nectarines:								
2007	24	1,068	23	24,845	1,673	50	1	513
2008	24	1,023	23	29,513	2,034	64	1	390
2009	24	916	20	26,411	1,766	59	1	491
Grain sorghum:								
2007	3,193	102,886	5,595	762,536	151,488	13,624	1,270	57,651
2008	3,111	99,553	5,399	974,943	199,931	28,113	2,980	154,310
2009	3,150	80,627	4,493	662,822	142,533	17,539	2,044	103,183
Grapefruit trees:								
2007	28	1,078	0	128,397	3,660	34	155	2,596
2008	28	804	0	111,555	1,756	0	....	0
2009	28	832	0	131,326	2,004	12	89	468

See footnotes at end of table.

Table 10-2.—Crop insurance programs: Coverage, amount of premiums and indemnities, by crops, United States, 2007–2009 [1]—Continued

Commodity and year	Coverage					Indemnities		
	County programs	Insured units [2]	Area insured [3]	Maximum insured production	Amount of premium	Number	Area indemnified [3]	Amount
	Number	Number	1,000 acres	1,000 dollars	1,000 dollars		1,000 acres	1,000 dollars
Grapefruit:								
2007	8	86	5	6,152	344	24	1	647
2008	8	90	5	7,238	367	12	1	195
2009	11	160	6	8,695	448	9	0	110
Grapes:								
2007	97	13,514	569	680,325	40,410	591	14	6,876
2008	102	13,532	568	679,647	39,971	1,674	41	21,032
2009	102	13,833	561	783,829	47,053	1,303	26	15,653
Green peas:								
2007	166	2,862	169	49,027	5,711	639	39	3,714
2008	166	2,878	173	76,793	8,880	580	39	6,042
2009	166	2,766	171	65,448	6,911	394	27	4,149
Hybrid corn seed:								
2007	385	6,732	409	254,575	19,899	471	37	6,697
2008	367	7,427	452	379,273	30,351	540	44	9,139
2009	367	5,475	329	225,912	16,446	239	19	4,181
Hybrid sorghum seed:								
2007	23	651	39	13,698	2,030	54	6	668
2008	23	613	41	20,387	3,265	95	9	1,206
2009	23	682	50	20,000	2,858	11	1	167
Late oranges:								
2007	3	83	1	553	87	1	0	4
2008	3	84	1	666	109	5	0	4
2009	3	86	1	717	119	7	0	34
Lemon trees:								
2007	4	2	0	1,067	28	0	....	0
2008	4	2	0	1,153	14	0	....	0
2009	4	2	0	1,448	17	0	....	0
Lemons:								
2007	15	685	42	72,464	3,573	119	4	3,740
2008	15	698	41	73,680	3,592	107	8	5,794
2009	15	736	43	78,794	3,768	8	0	161
Lime trees:								
2007	3	3	0	6	0	0	....	0
2008	3	3	0	51	1	0	....	0
2009	3	3	0	62	2	0	....	0
Macadamia nuts:								
2007	3	126	13	21,991	410	1	0	5
2008	3	120	13	22,422	464	35	4	1,778
2009	3	110	12	22,686	464	15	2	675
Macadamia trees:								
2007	3	130	13	47,648	510	0	....	0
2008	3	122	13	49,919	534	0	....	0
2009	3	111	12	45,175	480	0	....	0
Mandarins:								
2007	8	153	7	10,891	827	58	2	1,440
2008	11	190	9	16,341	1,319	5	0	332
2009	11	269	12	23,251	1,915	33	2	1,732
Mango trees:								
2007	1	21	0	365	9	0	....	0
2008	1	23	0	412	14	0	....	0
2009	1	24	0	467	16	0	....	0
Maryland tobacco:								
2007	6	1	0	2	0	0	....	0
Millet:								
2007	68	5,944	449	22,807	5,684	1,004	116	1,984
2008	68	5,769	407	25,003	6,539	873	95	2,047
2009	68	4,216	302	19,415	4,770	870	100	2,837
Minneola tangelos:								
2007	8	160	6	6,910	545	76	2	2,866
2008	10	175	5	7,207	571	8	0	130
2009	10	195	6	7,475	574	20	0	218
Mint:								
2007	9	200	16	6,571	345	63	3	394
2008	25	237	19	9,594	514	43	1	215
2009	31	296	28	21,413	1,095	44	1	227
Mustard:								
2007	46	253	40	4,136	596	130	25	1,054
2008	45	372	59	15,056	2,373	251	50	5,715
2009	45	279	36	5,798	1,010	42	5	415
Naval oranges:								
2007	16	2,770	112	162,922	9,655	1,151	52	46,917
2008	16	2,783	118	178,366	10,664	227	8	4,211
2009	16	2,863	122	183,315	11,631	487	17	8,058
Nursery:								
2007	3,088	9,859	0	4,010,257	89,184	416	....	21,448
2008	3,088	9,488	0	4,036,583	89,811	151	....	4,179
2009	2,855	7,287	0	3,186,669	64,836	239	....	18,735

See footnotes at end of table.

Table 10-2.—Crop insurance programs: Coverage, amount of premiums and indemnities, by crops, United States, 2007–2009 [1]—Continued

Commodity and year	Coverage				Amount of premium	Indemnities		
	County programs	Insured units [2]	Area insured [3]	Maximum insured production		Number	Area indemnified [3]	Amount
	Number	*Number*	*1,000 acres*	*1,000 dollars*	*1,000 dollars*		*1,000 acres*	*1,000 dollars*
Oats:								
2007	1,648	15,961	736	42,021	7,895	2,887	151	3,844
2008	1,593	12,563	536	50,463	9,264	2,801	163	7,518
2009	1,593	14,537	610	50,694	9,135	2,099	113	4,925
Onions:								
2007	110	2,110	91	132,965	18,194	582	18	17,738
2008	108	1,728	81	129,738	17,331	361	11	13,518
2009	108	1,833	86	138,858	21,025	484	15	21,021
Orange trees:								
2007	28	5,264	0	987,098	29,561	29	405	816
2008	28	3,921	0	962,634	15,882	0		0
2009	28	5,875	0	1,165,912	18,885	25	145	1,385
Orlando tangelos:								
2007	5	6	0	40	3	0		0
2008	5	6	0	46	3	1	0	3
2009	5	6	0	48	3	0		0
Papaya:								
2007	4	13	0	271	15	1	0	23
2008	4	9	0	276	10	0		0
2009	4	7	0	238	5	0		0
Papaya tree:								
2007	4	9	0	69	3	0		0
2008	4	7	0	93	2	0		0
2009	4	5	0	108	2	0		0
Pasture:								
2007		332	57,026	28,461	387,449	70,522	18,528	8,870
2008		441	59,638	29,402	377,286	68,951	32,660	13,850
2009		1,007	99,209	40,847	525,130	94,106	25,600	10,047
Peaches:								
2007	276	1,473	38	58,212	10,549	808	25	33,301
2008	276	1,500	39	58,114	11,300	468	16	12,458
2009	277	1,512	37	57,775	12,087	487	13	14,849
Peanuts:								
2007	371	22,876	1,085	401,493	39,376	6,138	271	46,985
2008	373	27,492	1,357	582,472	59,247	3,343	171	30,812
2009	389	20,955	1,020	392,661	39,548	2,447	113	16,042
Pears:								
2007	27	2,047	35	50,736	1,874	65	1	709
2008	29	2,051	35	56,219	2,076	106	1	819
2009	29	2,133	33	62,965	1,416	27	0	221
Pecans:								
2007	138	1,581	147	102,893	10,076	195	26	6,264
2008	142	1,627	159	116,021	10,699	558	53	10,805
2009	144	1,695	165	126,788	11,497	86	7	1,837
Peppers:								
2007	13	180	10	33,229	6,211	30	1	2,141
2008	13	170	11	33,391	5,968	25	2	2,981
2009	13	157	11	31,912	5,417	27	1	1,850
Plums:								
2007	7	1,307	22	23,639	2,524	260	4	1,998
2008	7	1,222	22	33,206	3,628	147	2	1,599
2009	10	1,103	19	29,875	3,255	265	3	2,328
Popcorn:								
2007	326	1,598	129	46,887	4,005	170	13	1,290
2008	254	1,727	141	73,477	6,186	225	16	2,424
2009	254	1,891	161	76,962	6,333	134	9	1,821
Potatoes:								
2007	342	7,522	903	847,949	71,738	1,074	85	36,143
2008	331	7,122	836	903,495	72,987	711	45	20,679
2009	331	6,964	854	1,053,283	86,301	514	35	18,902
Proc. apricots:								
2007	13	97	4	3,911	446	5	0	188
2008	13	89	4	4,104	468	15	1	493
2009	13	85	4	4,147	519	14	1	395
Processing beans:								
2007	154	1,779	108	26,818	2,938	446	25	2,513
2008	155	1,730	103	41,334	4,682	292	16	3,160
2009	155	1,707	108	43,595	4,445	273	15	3,077
Processing cling peaches:								
2007	10	1,214	19	28,843	1,257	72	1	470
2008	10	1,168	18	28,559	1,212	143	2	1,478
2009	10	1,233	19	36,707	1,528	25	0	252
Proc. freestone:								
2007	8	92	3	3,187	183	2	0	4
2008	8	96	3	3,298	184	3	0	21
2009	8	101	3	4,118	198	4	0	56
Prunes:								
2007	14	1,089	63	85,625	11,069	752	41	33,048
2008	14	1,091	62	82,222	12,100	363	20	12,711
2009	14	1,059	60	76,790	12,625	24	1	727
Raisins:								
2007	7	2,469	241	144,489	10,665	597	36	3,823
2008	7	2,252	249	146,359	10,648	22	1	98

See footnotes at end of table.

Table 10-2.—Crop insurance programs: Coverage, amount of premiums and indemnities, by crops, United States, 2007-2009 [1]—Continued

Commodity and year	Coverage					Indemnities		
	County programs	Insured units [2]	Area insured [3]	Maximum insured production	Amount of premium	Number	Area indemnified [3]	Amount
	Number	*Number*	*1,000 acres*	*1,000 dollars*	*1,000 dollars*		*1,000 acres*	*1,000 dollars*
Rangeland:								
2007	49	4,581	30,070	129,632	14,799	0		0
2008	49	3,833	24,601	102,521	11,930	0	0	0
2009								
Rice:								
2007	352	14,372	1,912	525,719	25,025	512	71	8,917
2008	352	15,438	2,119	700,765	32,542	714	110	15,859
2009	352	15,363	2,418	1,059,887	59,199	1,903	272	36,592
Rio Red & Star Ruby:								
2007	3	417	12	7,930	1,296	3	0	14
2008	3	379	12	8,681	1,503	5	0	11
2009	3	403	12	10,588	1,896	102	2	820
Ruby red grapefruit:								
2007	3	80	2	682	91	1	0	1
2008	3	77	2	767	114	0		0
2009	3	79	1	858	133	16	0	37
Rye:								
2007	48	478	37	1,979	323	212	20	732
2008	48	314	27	1,554	271	96	8	209
2009	48	439	33	2,364	422	124	11	349
Safflower:								
2007	79	1,095	113	6,830	1,140	203	24	457
2008	79	876	113	13,398	2,011	225	29	1,642
2009	79	813	114	15,221	2,327	75	9	509
Silage sorghum:								
2007	39	181	10	2,034	200	30	3	227
2008	39	172	9	1,677	180	81	7	377
2009	39	221	12	2,611	388	12	1	56
Soybeans:								
2007	7,657	783,032	50,775	11,529,457	1,066,139	134,247	9,167	602,692
2008	7,628	899,458	61,183	22,219,411	2,609,595	420,161	38,785	2,873,835
2009	7,639	775,825	64,344	17,021,378	1,979,298	70,478	5,438	353,559
Strawberries:								
2007	21	327	18	71,719	2,784	3	0	36
2008	21	311	16	65,061	2,393	1	0	9
2009								
Sugarbeets:								
2007	179	15,501	1,120	654,029	42,153	2,449	178	20,157
2008	154	13,746	973	580,155	36,595	2,363	192	37,736
2009	154	14,533	1,068	707,305	43,150	2,007	177	28,746
Sugarcane:								
2007	31	5,047	687	184,850	6,971	152	9	1,048
2008	31	4,800	734	195,693	4,972	175	8	920
2009	52	4,840	744	200,710	5,175	26	1	186
Sunflowers:								
2007	537	20,589	1,867	250,167	44,394	5,515	526	30,719
2008	525	22,440	2,119	581,968	120,779	9,978	1,225	119,315
2009	593	18,309	1,888	325,572	67,176	5,197	451	41,584
Sweet corn:								
2007	178	3,384	260	68,713	4,212	237	14	1,263
2008	171	3,468	265	108,294	6,380	244	14	2,297
2009	171	3,419	280	117,708	5,987	318	22	6,380
Sweet oranges:								
2007	6	49	0	527	38	13	0	108
2008	6	47	0	499	34	4	0	30
2009	6	48	0	553	38	2	0	12
Sweet potatoes:								
2007	22	55	8	9,697	866	3	0	347
2008	22	55	8	8,662	754	14	2	681
2009	22	48	6	5,246	584	5	0	176

See footnotes at end of table.

Table 10-2.—Crop insurance programs: Coverage, amount of premiums and indemnities, by crops, United States, 2007–2009[1]—Continued

Commodity and year	Coverage					Indemnities		
	County pro-grams	Insured units[2]	Area in-sured[3]	Maximum insured production	Amount of premium	Number	Area in-demnified[3]	Amount
	Number	Number	1,000 acres	1,000 dollars	1,000 dollars		1,000 acres	1,000 dollars
Table grapes:								
2007	12	1,194	85	177,306	9,146	102	3	3,979
2008	12	1,209	85	186,689	9,575	113	4	5,988
2009	12	1,194	83	197,675	10,653	55	1	1,498
Tomatoes:								
2007	89	2,925	289	297,993	14,776	125	9	2,706
2008	78	2,863	272	340,369	16,162	134	8	2,737
2009	78	3,120	295	497,913	13,205	188	12	8,031
Valencia oranges:								
2007	13	1,397	38	49,259	3,373	608	19	18,071
2008	13	1,376	37	50,711	3,417	95	4	1,961
2009	13	1,405	37	47,342	3,172	113	3	1,377
Walnuts:								
2007	26	1,312	98	89,639	3,492	68	2	681
2008	26	1,376	110	111,538	4,321	67	3	1,267
2009	26	1,523	122	159,927	6,027	43	2	671
Wheat:								
2007	7,497	586,399	46,982	5,383,985	897,349	169,083	17,230	862,988
2008	7,432	602,430	48,823	8,738,669	1,592,696	136,575	17,922	1,146,305
2009	7,461	588,532	48,712	9,910,320	1,839,094	241,302	28,283	1,881,279

[1] Data for 2007 and earlier is as of 2008 publication date. [2] Number of farms on which the insured crop was planted including duplication where both the landlord and tenant are insured. Insured farms on which no insured crop was planted are not included. [3] The insured's share of the planted area on the farm.
RMA, Program Automation Branch, (816) 926–7910.

Table 10-3.—Farm real estate debt: Amount outstanding by lender, United States, Dec. 31, 2001–2010[1]

Year	Farm Credit System	Farm Service Agency[2]	Commercial banks	Life insurance companies[3]	Individuals and others[4]	Total farm mort-gage debt
	1,000 dollars	1,000 dollars	1,000 dollars	1,000 dollars	1,000 dollars	1,000 dollars
2001	32,855	3,347	31,082	11,205	10,051	88,541
2002	37,815	3,181	33,060	11,421	9,946	95,423
2003	37,662	2,485	32,937	11,371	9,684	94,138
2004	37,078	2,395	34,630	10,726	10,598	95,653
2005	41,173	2,453	37,904	11,307	11,682	104,768
2006	43,448	2,374	40,149	12,001	9,790	108,048
2007	46,793	2,281	41,884	12,750	8,657	112,682
2008	57,124	2,313	49,705	14,736	9,552	133,582
2009[5]	58,423	2,343	50,338	14,246	8,695	134,514
2010[6]	NA	NA	NA	NA	NA	130,076

[1] Includes operator households. Includes regular mortgages, purchase-money mortgages, and sales contracts. [2] Includes farm ownership loans, soil and water loans to individuals, rural and labor housing loans, association loans for grazing, Indian tribe land acquisition loans, and one-half of economic emergency loans. [3] Compiled by American Council of Life Insurance. [4] Estimated by ERS. [5] Preliminary. [6] Forecast. NA-not available.
ERS, Farm Sector Performance Branch, (202) 694–5586.

Table 10-4.—Nonreal estate farm debt: Amount outstanding, by lender, United States, Dec. 31, 2001–2010[1]

Year	Farm Credit System	Farm Service Agency	Commericial banks	Individuals and others	Total nonreal es-tates debt
	Million dollars	Million dollars	Million dollars	Million dollars	Million dollars
2001	20,000	4,151	45,025	12,947	82,123
2002	20,491	3,973	44,344	12,993	81,801
2003	20,165	3,646	43,571	13,625	81,006
2004	22,040	3,244	45,849	15,132	86,265
2005	24,279	3,008	48,405	15,917	91,609
2006	27,811	2,736	51,253	13,733	95,533
2007	31,622	2,808	54,129	12,823	101,382
2008	37,290	2,652	57,313	11,841	109,096
2009[2]	39,883	2,823	57,027	11,113	110,846
2010[3]	NA	NA	NA	NA	104,905

[1] Includes operator households. [2] Preliminary. [3] Forecast. NA-not available.
ERS, Farm Sector Performance Branch, (202) 694–5586.

Table 10-5.—Farm Service Agency: Loans made to individuals and associations for farming purposes, and amount outstanding, United States and Territories, 2000–2009 [1]

| Year | Loans to individuals | | | | | | |
| | Farm ownership | | | Soil and water | | | Recreation |
	New borrowers	Loans made	Outstanding Jan. 1	New borrowers	Loans made	Outstanding Jan. 1	Outstanding Jan. 1
	Number	*1,000 dollars*	*1,000 dollars*	*Number*	*1,000 dollars*	*1,000 dollars*	*1,000 dollars*
2000	4,552	1,106,492	6,755,110	0	0	66,602	2,221
2001	3,704	1,015,634	7,287,728	0	0	52,883	1,784
2002	4,107	1,279,027	7,495,449	0	0	46,284	1,447
2003	4,174	1,399,740	7,749,043	0	0	38,484	1,263
2004	3,625	1,241,454	7,884,284	0	0	31,820	994
2005	4,199	1,298,943	8,190,313	0	0	27,341	875
2006	3,878	1,223,725	8,343,554	0	0	21,451	714
2007	3,865	1,268,809	8,518,399	0	0	18,477	674
2008	4,335	1,552,303	8,876,232	0	0	13,954	514
2009	5,048	1,832,709	9,800,441	0	0	12,469	267

| Year | Loans to individuals | | | | | |
| | Operating | | | Emergency | | |
	New borrowers	Loans made	Outstanding Jan. 1	New borrowers	Loans made	Outstanding Jan. 1
	Number	*1,000 dollars*	*1,000 dollars*	*Number*	*1,000 dollars*	*1,000 dollars*
2000	12,979	2,464,802	6,570,523	1,557	150,852	1,915,780
2001	10,732	2,152,814	6,823,828	962	90,026	1,712,807
2002	10,476	2,217,735	6,639,837	501	57,608	1,523,438
2003	10,577	2,121,150	6,728,636	920	95,698	1,405,430
2004	9,157	1,832,093	6,405,468	430	29,789	1,437,464
2005	8,891	1,723,953	6,404,277	235	23,569	1,150,557
2006	9,623	1,849,894	6,131,132	494	51,525	975,594
2007	8,673	1,789,590	5,732,012	691	74,898	920,453
2008	8,207	1,710,441	5,731,149	385	44,994	792,120
2009	11,778	2,611,248	6,500,532	177	30,401	726,370

| Year | Loans to associations | | | | | Economic opportunity individual loans | Economic emergency loans |
| | Indian tribe land acquisition | | | Grazing association | Irrigation, drainage, and soil conservation | | |
	New borrowers	Loans made	Outstanding Jan. 1	Outstanding Jan. 1	Outstanding Jan. 1	Outstanding Jan. 1	Outstanding Jan. 1
	Number	*1,000 dollars*	*1,000 dolllars*	*1,000 dollars*	*1,000 dollars*	*1,000 dollars*	*1,000 dollars*
2000	1	673	57,117	15,660	5,449	10	545,423
2001	1	590	62,738	12,785	5,177	8	427,176
2002	1	74	60,777	10,849	3,729	8	364,377
2003	1	110	55,421	8,947	3,330	7	315,601
2004	2	1,586	53,476	6,232	1,623	8	249,603
2005	0	0	55,205	4,883	1,471	8	249,039
2006	0	360	52,134	3,613	1,263	8	198,266
2007	0	0	47,914	3,317	1,184	7	173,095
2008	0	0	43,764	2,945	1,045	6	135,303
2009	0	0	38,510	2,590	860	6	117,942

[1] Includes loans made directly by FmHA and those guaranteed by the Agency. Amounts of loans made represent obligations and include loans to new borrowers and subsequent loans to borrowers who received an initial loan in a prior year. Amounts outstanding are loan advances less principal repayments for loans made directly by the Agency.
FSA, Loan Making Division, (202) 690–4006.

Table 10-6.—Farmers' marketing, farm supply, and related service cooperatives: Number, memberships, and business, United States, 1999–2008

Year [1]	Cooperatives [2]				Estimated memberships [4]			
	Marketing	Farm supply	Related service [3]	Total	Marketing	Farm supply	Related service [3]	Total
	Number	*Number*	*Number*	*Number*	*1,000 members*	*1,000 members*	*1,000 members*	*1,000 members*
1999	1,749	1,313	404	3,466	1,283	1,731	159	3,173
2000	1,672	1,277	397	3,346	1,243	1,718	124	3,085
2001	1,606	1,234	389	3,229	1,160	1,746	128	3,034
2002	1,559	1,201	380	3,140	1,049	1,637	107	2,794
2003	1,551	1,156	379	3,086	1,054	1,590	113	2,758
2004	1,460	1,161	362	2,983	978	1,600	100	2,677
2005	1,412	1,128	356	2,896	932	1,538	101	2,572
2006	1,280	1,090	305	2,675	914	1,522	133	2,570
2007	1,233	1,061	300	2,594	810	1,526	124	2,460
2008 [5]	1,191	1,003	279	2,473	772	1,500	119	2,391

Year [1]	Marketing volume		Farm supply volume		Service [8]	Total marketing and farm supply volume and service receipts	
	Gross [6]	Net [7]	Gross [6]	Net [7]		Gross [6]	Net [7]
	Million dollars	*Million dollars*	*Million dollars*	*Million dollars*	*Million dollars*	*Million dollars*	*Million dollars*
1999	80,506	71,982	30,879	23,177	3,905	115,291	99,064
2000	80,400	72,065	36,809	24,085	3,510	120,719	99,659
2001	83,954	75,042	36,141	24,756	3,471	123,566	103,269
2002	76,618	69,656	31,519	23,679	3,416	111,553	96,750
2003	77,242	71,002	35,498	25,499	4,118	116,858	100,619
2004	82,654	77,207	36,997	26,992	3,733	123,384	107,932
2005	78,024	74,152	39,302	28,426	4,330	121,657	106,909
2006	76,480	70,441	45,872	35,922	4,118	126,470	110,482
2007	93,120	85,352	49,322	38,296	4,133	146,575	127,781
2008 [5]	116,831	109,776	70,229	50,677	4,814	191,874	165,267

[1] Reports of cooperatives are included for the calendar year.   [2] Includes independent local cooperatives, centralized cooperatives, federations of cooperatives and cooperatives with mixed organizational structures. Cooperatives are classified according to their major activity. If, for example, more than 50 percent of a cooperative's business is derived from marketing activities, it is included as a marketing cooperative.   [3] Includes cooperatives whose major activity is providing services related to marketing and farm supply activities.   [4] Includes members (those entitled to vote for directors) but does not include nonvoting patrons. (Some duplication exists because some farmers belong to more than one cooperative.)   [5] Preliminary.   [6] Estimated gross business includes all business reported between cooperatives, such as the wholesale business of farm supply cooperatives with other cooperatives or terminal market sales for local cooperatives.   [7] Estimated net business represents the value at the first level at which cooperatives transact business for farmers. Figures are adjusted for duplication resulting from intercooperative business.   [8] Receipts for services related to marketing and purchasing activities, but not included in the volumes reported for these activities.
Rural Development, Cooperative Programs, (202) 690–1415.

Table 10-7.—Farmers' cooperatives: Business volume of marketing, farm supply, and related service cooperatives, United States, 2007 and 2008 (preliminary)

Item	Gross business		Net business [1]	
	2007	2008	2007	2008
	Billion dollars	*Billion dollars*	*Billion dollars*	*Billion dollars*
Products marketed:				
Beans and peas (dry edible)	0.124	0.202	0.118	0.191
Cotton and cotton products	3.114	3.507	1.786	3.332
Dairy products	37.67	40.127	35.49	37.817
Fish	0.206	0.208	0.206	0.208
Fruits and vegetables	6.862	7.464	4.293	5.233
Grain and oilseeds [2]	28.838	47.313	28.349	46.222
Livestock and livestock products	3.383	3.353	3.215	3.24
Nuts	0.804	0.832	0.804	0.832
Poultry products	1.36	1.457	1.36	1.457
Rice	1.116	1.395	1.116	1.395
Sugar products	4.904	4.895	3.998	4.092
Tobacco	0.105	0.135	0.105	0.135
Wool and mohair	0.006	0.006	0.006	0.006
Other [3]	4.629	5.936	4.506	5.615
Total farm products	93.12	116.831	85.352	109.776
Supplies purchased:				
Crop protectants	3.717	6.702	3.14	4.497
Feed	8.645	10.849	7.575	9.277
Fertilizer	7.686	12.76	6.844	9.941
Petroleum	21.616	30.267	14.662	19.247
Seed	2.815	3.54	1.819	2.396
Other supplies [4]	4.844	6.112	4.255	5.32
Total farm supplies	49.322	70.229	38.296	50.677
Receipts for services: [5]				
Trucking, cotton ginning, storage, grinding, locker plants, miscellaneous	4.133	4.814	4.133	4.814
Total business	146.575	191.874	127.781	165.267

[1] Represents value at the first level at which cooperatives transact business for farmers. [2] Excludes oilseed meal and oil. Oilseed meal is included in feed sales while oil sales are included in other products sales. [3] Includes coffee, forest products, hay, hops, seed marketed for growers, nursery stock, other farm products not separately classified, and sales of farm products not received directly from member-patrons. Also includes manufactured food products and resale items marketed by cooperatives. [4] Includes automotive supplies, building materials, chicks, containers, farm machinery and equipment, hardware, meats and groceries, and other supplies not separately classified. [5] Services related to marketing or purchasing but not included in the volume reported for those activities, plus other operating and non-operating income and losses and extraordinary items.

Rural Development, Cooperative Programs, (202) 690–1415.

Table 10-8.—Farmers' cooperatives: Number of cooperatives, memberships, and business volume of marketing, farm supply, and related service cooperatives, by State and United States, 2007–2008 (preliminary)

State	Cooperatives headquartered in State		Memberships in State [1]		Net business [1]	
	2007	2008	2007	2008	2007	2008
	Number	Number	Thousand number	Thousand number	Million dollars	Million dollars
AL	57	55	22.30	23.80	0.55	0.68
AZ	9	9	0.40	1.90	0.68	0.96
AR	49	39	46.50	46.80	1.91	2.81
CA	143	128	39.30	39.00	7.41	10.44
CO	33	31	28.80	29.20	0.94	1.49
FL	33	33	19.10	19.30	1.73	1.88
GA	13	13	7.30	9.30	0.53	0.97
HI	16	15	0.80	0.60	0.06	0.02
ID	33	30	9.90	16.10	0.84	1.18
IL	144	131	138.40	137.50	6.53	11.69
IN	39	35	45.40	37.30	2.91	4.39
IA	127	116	118.00	127.90	11.19	17.38
KS	109	96	117.80	109.60	5.89	6.78
KY	39	36	105.90	164.00	0.54	0.99
LA	42	40	7.90	11.70	0.96	1.40
MD	11	14	21.90	69.90	0.40	0.78
MA	10	8	5.00	3.10	0.55	0.22
MI	52	47	16.40	15.90	2.08	3.34
MN	231	216	192.10	142.60	12.97	20.06
MS	69	61	112.10	96.90	0.79	1.21
MO	61	61	130.10	128.10	4.40	7.94
MT	54	52	14.70	26.30	1.24	2.11
NE	73	67	76.00	83.80	5.36	9.17
NM	9	7	1.50	0.06	0.24	0.35
NY	72	56	8.60	6.70	1.96	2.30
NC	16	14	94.00	71.60	0.52	0.79
ND	200	180	67.60	87.30	4.04	7.18
OH	57	49	53.00	41.90	3.10	5.27
OK	67	61	52.60	45.40	1.20	1.62
OR	31	33	28.00	28.10	2.71	3.74
PA	45	42	4.40	6.10	0.89	1.86
SD	100	84	65.10	70.60	3.17	5.82
TN	73	68	128.50	130.80	0.83	1.24
TX	204	206	91.90	86.90	3.72	5.14
UT	12	12	16.70	7.50	0.79	1.11
VA	52	53	405.60	144.50	0.91	1.57
WA	64	64	21.50	25.10	3.60	5.03
WV	25	24	47.10	79.90	0.10	0.18
WI	130	120	182.60	155.60	8.78	9.79
WY	10	10	3.80	4.10	.40	0.53
Oth Sts [2]	48	57	19.60	56.20	1.44	1.90
US	2,675	2,473	2,568.10	2,389.30	108.83	163.82
Foreign [3]	2,675	....	1.50	1.70	0.87	1.95
Total	2,594	2,473	2,569.60	2,390.90	109.70	165.27

[1] Represents value at the first level at which cooperatives transact business for farmers. Net business volume by State is collected in odd years, 2008 was estimated using factors from 2007. Totals may not add due to rounding.   [2] Dollar volume or membership is not shown to avoid disclosing operations of individual cooperatives.   [3] Sales outside the United States, sales to domestic military installations, and sales of certain products not received directly from member-patrons.
Rural Development, Cooperative Programs, (202) 690–1415.

Table 10-9.—Rural Utilities Service: Long-term electric financing approved by purpose, by State and United States as of December 31, 2008

State	Borrowers	Total financing approved						Loan estimates	
		RUS loans[1]	Non-RUS financing		Financing approved by purpose			Miles of line	Consumers
			With RUS guarantee[2]	Without RUS guarantee[3]	Distribution	Generation and transmission[3]	Consumer facilities		
	Number	1,000 dollars	1,000 dollars	1,000 dollars	1,000 dollars	1,000 dollars	1,000 dollars	Number	Number
AL	27	904,941	1,601,806	280,177	1,200,363	1,585,203	1,359	68,989	643,759
AK	17	864,376	444,260	173,024	619,601	860,970	1,089	10,880	192,511
AZ	15	423,646	694,549	136,416	556,104	698,016	490	22,447	228,227
AR	20	1,086,615	1,566,841	813,414	1,419,954	2,042,444	4,471	76,395	567,479
CA	10	87,554	35,585	7,216	91,002	39,296	56	6,696	78,260
CO	25	1,377,577	3,253,709	794,887	1,458,155	3,967,943	74	76,544	496,613
CT	....								
DE	1	85,430	8,000	31,266	123,832	861	3	6,417	82,813
FL	18	1,467,692	1,999,458	1,153,744	2,026,948	2,591,428	2,519	73,672	983,931
GA	51	2,575,148	8,352,367	1,852,266	5,000,308	7,776,514	2,959	181,915	2,072,619
HI	1	215,000	32,960	8,240	256,200	....	....	820	30,500
ID	10	196,308	20,134	36,612	224,792	27,066	1,197	12,859	66,958
IL	29	628,023	1,262,165	188,935	652,228	1,426,683	212	55,409	254,144
IN	46	526,088	2,278,037	707,487	675,099	2,835,796	717	58,154	485,458
IA	46	844,419	893,648	164,548	764,802	1,137,415	397	66,852	213,606
KS	31	751,420	861,167	186,035	856,655	941,552	415	76,598	266,896
KY	26	1,766,221	4,642,465	749,746	2,400,117	4,757,156	1,159	92,461	930,176
LA	20	775,273	2,817,601	445,986	1,009,833	3,028,850	177	54,279	522,260
ME	4	41,737	10,102	20,396	48,701	23,490	44	2,171	20,680
MD	2	368,228	18,355	140,356	411,001	115,937	....	14,679	179,833
MA	....								
MI	10	660,087	969,579	92,692	770,340	951,545	473	39,495	309,601
MN	46	1,687,889	2,280,786	556,717	2,033,837	2,487,197	4,357	122,440	697,881
MS	29	979,323	1,171,895	361,096	1,171,176	1,340,444	694	88,076	717,224
MO	48	1,863,811	2,035,909	600,270	1,923,617	2,575,551	821	124,887	752,268
MT	26	440,528	53,856	66,866	468,332	92,689	229	46,894	139,561
NE	35	483,130	38,616	42,711	448,825	115,070	562	76,004	177,923
NV	8	75,008	1,241	10,441	65,315	21,127	248	6,265	24,889
NH	1	81,213	143,839	8,696	87,335	146,380	32	4,616	64,601
NJ	2	18,173	....	5,377	22,250	1,295	4	1,000	13,334
NM	17	653,437	135,898	78,251	781,857	82,719	3,010	45,989	240,232
NY	6	46,782	24,035	20,114	81,561	9,286	85	5,260	29,027
NC	33	1,985,135	2,267,251	560,907	2,458,909	2,349,749	4,635	102,202	1,066,519
ND	23	1,179,286	3,291,707	908,122	875,676	4,501,454	1,986	69,306	156,223
OH	27	652,958	1,350,235	472,666	856,144	1,619,496	218	48,643	373,579
OK	29	1,187,976	1,128,841	330,438	1,361,705	1,282,874	2,676	100,456	521,214
OR	18	309,338	87,877	93,022	345,682	144,318	237	24,691	143,034
PA	13	450,772	624,333	132,794	582,679	624,981	239	27,960	228,343
RI	1	....	3,940	....	334	3,606	....	4	160
SC	28	1,884,073	1,035,710	379,965	2,461,963	835,554	2,231	76,015	787,286
SD	30	809,257	246,434	155,806	904,472	306,144	881	66,821	180,454
TN	33	782,429	290,902	190,919	1,237,038	26,989	223	92,123	1,131,947
TX	99	2,425,390	3,897,260	1,184,403	2,920,657	4,583,873	2,524	270,570	1,489,821
UT	6	72,455	1,031,811	216,171	62,920	1,257,393	124	5,860	25,033
VT	3	75,437	37,070	7,310	57,949	60,964	903	2,956	24,819
VA	19	855,656	3,477,702	251,378	4,286,268	297,979	488	49,207	455,508
WA	23	251,065	7,234	40,181	277,358	20,870	252	21,052	111,183
WV	1	26,236	....	1,059	26,443	847	5	931	8,023
WI	26	572,695	1,195,635	464,963	518,476	1,713,019	1,798	46,684	245,285
WY	13	376,419	53,969	27,216	366,585	90,917	101	31,441	95,792
AS	1	....	3,000	....	....	3,000	....	....	....
MH	1	....	11,857	....	....	11,857	....	161	3,426
PW	....								
PR	1	300,981	....	31,424	292,851	39,554	....	16,633·	624,343
VI	1	430	....	....	234	197	....	85	912
US[4]	1,056	36,173,064	57,691,630	15,182,725	47,544,485	61,455,560	47,374	2,572,964	19,156,528

[1] Includes $629,161,727 discounted principal from 229 prepaid borrowers.　[2] Includes RUS Section 313A loan guarantees.　[3] Includes loans obtained by RUS borrowers' affiliates specifically organized to facillitate non-RUS finanacing.　[4] Includes figures not shown elsewhere in this table for two borrowers whose loans hve been foreclosed. The amount of these loans was $37,237.

Rural Development, Planning and Policy Branch, (202)692-0347

Table 10-10.—Rural Utilities Service: Composite revenues and patronage capital, average number of consumers and megawatt-hour sales reported by RUS electric borrowers operating distribution systems—calendar years 2006–2008

Item	2006		2007		2008	
	Amount	Per-cent of total	Amount	Per-cent of total	Amount	Per-cent of total
Number of borrowers reporting	591	....	587	....	582	....
Average number of consumers served:						
Residential service (farm & non-farm) ...	10,575,065	88.8	10,759,914	88.6	10,895,596	88.4
Commercial & industrial, small	1,155,189	9.7	1,195,721	9.8	1,232,497	10.0
Commercial & industrial, large	8,342	0.1	8,913	0.1	9,516	0.1
Irrigation	102,291	0.9	100,837	0.8	107,446	0.9
Other electric service	71,913	0.6	75,743	0.6	77,638	0.6
To others for resale	183	*	182	*	180	*
Total	11,912,983	100.0	12,1415,310	100.0	12,322,873	100.0
Megawatt-hour sales:						
Residential service (farm & non-farm) ...	147,094,045	56.7	152,625,537	56.9	153,272,106	56.4
Commercial & industrial, small	47,629,362	18.4	50,155,786	18.7	51,550,691	19.0
Commercial & industrial, large	56,356,671	21.7	57,378,913	21.4	58,457,988	21.5
Irrigation	4,341,541	1.7	3,973,629	1.5	4,599,981	1.7
Other electric service	2,441,836	0.9	2,564,353	1.0	2,577,064	0.9
To others for resale	1,405,315	0.5	1,494,788	0.6	1,425,311	0.5
Total	259,268,767	100.0	268,193,030	100.0	271,883,142	100.0
	1,000 dollars		*1,000 dollars*		*1,000 dollars*	
Revenue and patronage capital:						
Residential service (farm & non-farm) ...	13,691,670	62.7	14,559,034	62.4	15,663,365	61.5
Commercial & industrial, small	4,114,666	18.8	4,441,489	19.0	4,923,976	19.3
Commercial & industrial, large	2,961,960	13.6	3,175,983	13.6	3,565,697	14.0
Irrigation	373,035	1.7	358,722	1.5	457,053	1.8
Other electric service	234,591	1.1	253,511	1.1	275,456	1.1
To others for resale	66,509	0.3	79,477	0.3	82,474	0.3
Total from sales of electric energy	21,442,432	98.2	22,868,217	97.9	24,968,021	98.0
Other operating revenue	396,872	1.8	480,761	2.1	508,840	2.0
Total operating revenue	21,839,304	100.0	23,348,978	100.0	25,476,661	100.0

* Less than 0.05 percent.
Rural Development, Planning and Policy Branch, (202) 692-0347

Table 10-11.—Rural Utilities Service: Annual revenues and expenses reported by electric borrowers, United States, 1999–2008

Year	Operating revenue	Operating expense	Interest expense	Depreciation and amortization expense	Net margins	Total utility plant
	1,000 dollars	1,000 dollars	1,000 dollars	1,000 dollars	1,000 dollars	1,000 dollars
1999	23,823,791	19,536,422	1,832,553	1,746,681	1,112,665	62,684,354
2000	25,628,917	21,161,991	1,905,043	1,819,616	1,164,076	66,353,227
2001	26,458,243	21,867,226	1,909,833	1,895,495	1,219,287	69,630,602
2002	27,458,144	22,568,763	1,867,431	1,992,415	1,382,964	72,481,696
2003	31,821,409	26,393,809	2,153,155	2,314,811	1,303,510	84,991,605
2004	30,649,839	25,646,721	1,919,835	2,181,541	1,340,317	79,508,979
2005	34,330,831	29,164,368	2,075,557	2,271,565	1,441,751	83,405,976
2006	36,765,064	31,213,044	2,247,071	2,375,325	1,747,997	88,112,547
2007 [1]	38,423,386	32,659,447	2,311,524	2,369,896	1,989,271	90,936,276
2008 [1]	42,076,007	36,036,802	2,372,806	2,461,400	2,064,929	97,169,511

[1] Revised.
Rural Development, Planning and Policy Branch, (202) 692–0347.

Table 10-12.—Loans to farmers' cooperative organizations: Outstanding amounts held by the agricultural credit banks classified by type of loan, United States, Jan. 1, 2001-2010

Year	Operating capital loans	Facility loans
	1,000 dollars	1,000 dollars
2001	7,293,142	11,348,179
2002	7,660,584	11,311,516
2003	8,907,313	12,317,966
2004	12,373,082	12,400,364
2005	11,549,929	12,406,423
2006	12,293,156	14,004,128
2007	10,956,633	22,119,209
2008	13,263,702	27,226,859
2009	11,769,457	32,780,415
2010	10,743,647	33,430,718

FCA, Office of Management Services, (703) 883–4073.

CHAPTER XI

STABILIZATION AND PRICE-SUPPORT PROGRAMS

The statistics in this chapter relate to activities of the Commodity Credit Corporation (CCC), cropland diversion and production adjustment programs, and marketing agreement and order programs for fruits and vegetables. Statistics for Federal Milk Marketing Order programs are contained in chapter VIII.

Table 11-1.—Commodity Credit Corporation: Price-supported commodities owned as of Dec. 31, 2000–2009[1] (Inventory quantity)

Year	Barley	Butter and butter oil	Cheese and products	Corn	Cotton extra long staple	Cotton seed cotton upland	Sorghum and products	Milk and products
	Million bushels	Million pounds	Million pounds	Million bushels	1,000 bales	1,000 bales	Million bushels	Million pounds
2000	[3]	0	0	36	[2]	[5]	[3]	602
2001	[3]	0	5	24	[2]	[6]	[2]	844
2002	[3]	0	4	18	[3]	[6]	1	1,201
2003	0	0	17	16	[2]	[6]	0	1,456
2004	0	0	7	12	[2]	[6]	0	605
2005	0	0	7	12	[2]	[6]	0	605
2006	[2]	[2]	[2]	1	[2]	937	[2]	40
2007	[2]	0	[2]	1	[2]	1,017	2	38
2008	[2]	0	0	30	[2]	2	22	11
2009	[2]	[3]	13	9	1	1	0	35

Year	Oils and oilseeds	Oats and products	Rice and products[4]	Peanut and products	Soybeans	Beans, dry edible	Wheat
	Million cwt.	Million bushels	Million cwt.	Million pounds	Million bushels	Million bushels	Million bushels
2000	[2]	[2]	[2]	[6]	10	[6]	109
2001	[2]	[2]	[3]	[6]	4	[6]	118
2002	0	0	[3]	[6]	3	[6]	93
2003	0	[2]	[3]	[6]	[3]	[6]	78
2004	[2]	[2]	[3]	[6]	[2]	[6]	81
2005	[2]	[2]	[3]	[6]	[2]	[6]	81
2006	[2]	[2]	[3]	0	1	1	43
2007	2	[2]	18	62	8	[3]	24
2008	[3]	0	9	20	[3]	2	[3]
2009	0	0	1	9	17	5	48

Year	Blended Foods	Poultry	Meat	Fish	Vegetable Oil Products	Value of all commodities owned[6]
	Million pounds	Million pounds	Million pounds	Million pounds	Million pounds	Million dollars
2000	[6]	[6]	[6]	[6]	[6]	790
2001	[6]	[6]	[6]	[6]	[6]	844
2002	[6]	[6]	[6]	[6]	[6]	656
2003	[6]	[6]	[6]	[6]	[6]	219
2004	[6]	[6]	[6]	[6]	[6]	116
2005	[6]	[6]	[6]	[6]	[6]	116
2006	[2]	0	0	0	24	109
2007	0	0	[2]	1	19	185
2008	10	18	14	8	45	1,348
2009	2	19	8	12	26	1,089

[1] Commodities which were owned by CCC in some years but not shown in this table are as follows: tobacco, honey, sugar and products, dry whole peas, potatoes, and wool and mohair. [2] Less than 50,000 units. [3] Less than 500,000 units. [4] Total value of all commodities owned by CCC, including price-supported commodities not shown and commodities acquired under programs other than price-support programs, less, reserve for losses on inventory. [5] Less than 500 units. [6] Prior years data were not available.

CCC, Financial Management Division, (703) 305–1363.

Table 11-2.—Commodity Credit Corporation: Loans pledge made, by quantity and face amount United States and Territories, by crop year 2006–2009 [1]

Commodity	Unit	2006		2007	
		Quantity pledged	Face amount	Quantity pledged	Face amount
			1,000 dollars		*1,000 dollars*
Barley	1,000 bushels	6,733	12,229	3,570	6,532
Corn	1,000 bushels	31,873	65,968	56,674	117,648
Cotton, ELS & Upland [2]	1,000 bales	1,109	291,922	469	121,078
Seed cotton, ELS & Upland	1,000 pounds	0	0	0	0
Sugar Cane and Beet	1,000 pounds	0	0	0	0
Flaxseed	1,000 cwt.	97	890	24	219
Honey	1,000 pounds	7,444	4,466	6,641	3,985
Oats	1,000 bushels	1,249	1,618	776	972
Peanuts	1,000 pounds	60,622	10,602	34,956	6,099
Rice	1,000 cwt.	28,466	186,899	21,843	143,401
Wool	1,000 cwt.	3	1	3	1
Sorghum grain	1,000 bushels	1,159	2,385	417	798
Soybeans	1,000 bushels	3,563	17,878	2,844	14,325
Tobacco	1,000 pounds	0	0	0	0
Wheat	1,000 bushels	68,021	184,691	28,734	77,244
Sunflower Seed	1,000 cwt.	7	62	0	0
Canola Seed	1,000 cwt.	337	3,134	124	1,157
Safflower Seed	1,000 cwt.	0	0	2	13
Mustard Seed	1,000 cwt.	2	22	1	11
Sunflower Seed (non-oil)	1,000 cwt.	14	129	5	44
Crambe Oilseed	1,000 cwt.	0	0	0	0
Mohair	1,000 pounds	18	76	14	59
Chickpeas	1,000 cwt.	0	0	0	0
Dry Whole Peas	1,000 cwt.	636	3,899	361	2,207
Lentil Dry	1,000 cwt.	412	4,685	75	838

Commodity	Unit	2008		2009 [3]	
		Quantity pledged	Face amount	Quantity pledged	Face amount
			1,000 dollars		*1,000 dollars*
Barley	1,000 bushels	2,205	3,943	3,376	5,405
Corn	1,000 bushels	46,573	91,966	80,876	147,579
Oats	1,000 bushels	673	868	883	1,074
Sorghum grain	1,000 bushels	1,103	2,277	1,091	2,008
Wheat	1,000 bushels	36,027	93,590	47,126	123,780
Rice	1,000 cwt.	5,609	36,629	7,676	29,918
Tobacco	1,000 pounds			0	
Seed cotton, ELS & Upland	1,000 pounds	0	0	0	0
Cotton, ELS & Upland	1,000 bales	1,348	731,772	182	56,043
Canola Seed	1,000 cwt.	31	277	87	721
Crambe Oilseed	1,000 cwt.			0	
Flaxseed	1,000 cwt.	6	56	17	156
Mustard Seed	1,000 cwt.	2	14	11	104
Safflower Seed	1,000 cwt.			3	21
Sunflower Seed	1,000 cwt.			214	1,897
Sunflower Seed (non-oil)	1,000 cwt.	5	50	8	73
Mohair	1,000 pounds	9	37	13	54
Peanuts	1,000 pounds	92,291	15,842	152,599	27,153
Soybeans	1,000 bushels	2,220	11,028	3,287	13,924
Honey	1,000 pounds	5,293	3,176	4,039	2,369
Chickpeas	1,000 cwt.	2	12	5	34
Dry Whole Peas	1,000 cwt.	91	562	234	1,206
Lentil Dry	1,000 cwt.	22	257	52	458
Sugar Cane and Beet	1,000 pounds			0	
Wool	1,000 pounds	6	2	29	67

[1] Includes loans pledge directly by Commodity Credit Corporation.　[2] Includes extra long staple cotton and upland cotton.　[3] Loan pledges are made through 2009.
CCC, Financial Management Division, (703) 305–1363.

Table 11-3.—Commodity Credit Corporation: Loan transactions for fiscal year 2009, by commodities [1]

Commodity	Unit	Loans outstanding Oct. 1, 2008 [2]	New loans made	Loans repayments	Collateral acquired in settlement	Loans written off and transferred to accounts receivable [3]	Loans outstanding Sept. 30, 2009 Value	Loans outstanding Sept. 30, 2009 Quantity collateral remaining pledged
		1,000 dollars	1,000 dollars	1,000 dollars	1,000 dollars	1,000 dollars	1,000 dollars	1,000 units
Basic commodities:								
Corn	Bushel	91,966	2,052,035	(1,994,649)	0	(1,773)	147,579	80,876
Cotton	Bale	368,962	2,715,140	(2,126,011)	(45,518)	(856,399)	56,174	182
Seed cotton	Pound	0	3,044	(2,592)	0	(452)	0	....
Peanuts	Pound	15,842	718,163	(695,534)	(1,257)	(10,261)	26,953	152,599
Rice	Cwt	36,629	444,733	(451,239)	0	(205)	29,918	7,676
Tobacco [4]	Pound	0	0	0	0	0	0	0
Wheat	Bushel	93,590	236,243	(205,514)	0	(538)	123,780	47,126
Total [5]		606,989	6,169,358	5,475,539)	(46,775)	(869,628)	384,404	XXXXX
Designated nonbasic commodities:								
Barley	Bushel	3,943	12,616	(11,103)	0	(51)	5,405	3,376
Sorghum	Bushel	2,277	16,129	(16,354)	0	(45)	2,008	1,091
Honey	Pound	3,176	6,846	(7,639)	0	(14)	2,369	4,039
Oats	Bushel	868	1,527	(1,321)	0	(0)	1,074	883
Rye	Bushel	0	0	0	0	0	0	0
Raw:. sugar, cane	Pound	0	405,714	(405,714)	0	0	0	0
Refined:. sugar, cane	Pound	0	0	0	0	0	0	0
Raw:. sugar beet	Pound	0	149,381	(149,381)	0	0	0	0
Refined:. sugar beet	Pound	0	563,273	(563,273)	0	0	0	0
FlaxSeed	Cwt.	56	733	(633)	0	0	156	17
Sunflower. seed	Cwt.	0	8,545	(6,636)	0	(11)	1,897	214
Canola. seed	Cwt.	277	8,292	(7,847)	0	(0)	721	87
Safflower. seed	Cwt.	0	28	(28)	0	0	0	0
Rapeseed	Cwt.	0	0	0	0	0	0	0
Mustard. seed	Cwt.	14	104	(14)	0	0	104	11
Crambe. oilseed	Cwt.	0	0	0	0	0	0	0
Sunflower seed,. non oil	Cwt.	50	2,620	(2,597)	0	0	73	8
Total [5]		10,661	1,175,808	(1,172,541)	0	(121)	13,807	XXXXX
Other nonbasic commodities:								
Soybeans	Bushel	11,028	941,978	(938,152)	0	(931)	13,924	3,287
Mohair	Pound	37	54	(29)	0	(7)	54	13
Chickpeas	Pound	12	59	(37)	0	0	34	5
Lentils	Pound	257	876	(675)	0	0	458	52
Dry Whole Peas.	Pound	562	2,708	(2,064)	0	0	1,206	234
Wool	Pound	2	67	(0)	0 (2)	67	29	
Total		11,898	945,743	(940,958)	0	(940)	15,743	XXXXX
Other loans: Farm Storage facility.		0	0	0	0	0	0	0
Bollweevil		0	0	0	0	0	0	0
Total [5]		0	0	0	0	0	0	XXXXX
Grand total [6].		629,549	8,290,909	(7,589,038)	(46,775)	(870,690)	413,954	XXXXX

[1] Loans made directly by Commodity Credit Corporation.　[2] Book value of outstanding loans; includes face amounts and any charges paid.　[3] Includes transfers to accounts receivable.　[4] Charge offs represents pre-No Net Cost Tobacco loans - 1981 and prior crop loans.　[5] Totals do not include allowance for losses.　[6] Table may not add due to rounding.
CCC, Financial Management Division, (703) 305–1363.

Table 11-4.—Commodity Credit Corporation: Selected inventory transactions, programs and commodity, fiscal year 2009

Program and commodity	Unit	Quantity					
		Inventory Oct. 1, 2008	Purchases	Collateral acquired from loans	Other addition deduction	Gain/loss, sales and other dispositions [1]	Inventory Sept. 30, 2009
		Thousands	Thousands	Thousands	Thousands	Thousands	Thousands
Feed grains:.							
Barley	Bushel	0	6	0	0	6	0
Corn	Bushel	0	1,829	0	0	1,829	0
Corn products	Pound	22	316	0	0	320,415	17,573
Grain sorghum	Bushel	28	29,114	0	0	29,032	110
Sorghum grits	Pound	0	0	0	0	0	0
Oats	Bushel	0	0	0	0	0	0
Total feed grains		xxx	xxx	xxx	xxx	xxx	xxx
Wheat (A)	Bushel	0	31,532	0	532	32,065	0
Wheat flour	Pound	389	240,696	0	0	223,446	17,639
Wheat products, other	Pound	0	292,741	0	0	292,741	0
Rice, milled	Cwt	2	1,652	0	30	1,680	4
Rice, rough	Cwt	0	0	0	0	0	0
Cotton, extra long staple	Bale	0	0	0	0	0	0
Upland cotton	Bale	0	10,814	166	(166)	10,814	0
Tobacco Products	Pound	0	0	0	0	0	0
Dairy products:							
Butter	Pound	0	4,639	0	0	0	4,639
Cheese	Pound	0	7,056	0	8,335	9,041	0
Milk, dried	Pound	0	0	0	0	0	0
Milk, UHT	Pound	0	2,877	0	0	2,877	0
Dry whole milk	Pound	0	0	0	0	0	0
Non fat dry milk	Pound	0	269,553	0	(20,259)	25,568	223,725
Total dairy products		xxx	xxx	xxx	xxx	xxx	xxx
Oils and oilseeds:							
Crambe oilseed	Cwt.	0	0	0	0	0	0
Canola seed	Cwt.	0	0	0	0	0	0
Sunflower seed	Cwt.	0	0	0	0	0	0
Sunflower seed, non-oil	Cwt.	0	0	0	0	0	0
Peanuts farmers'stock	Pound	0	0	0	0	0	0
Peanut	Pound	0	298,691	7,109	(7,109)	298,691	0
Peanut butter	Pound	0	0	0	4,714	4,714	0
Soybeans	Bushel	0	186	0	0	186	0
Soybean products	Pound	0	107,140	0	0	107,140	0
Dry edible beans	Cwt.	2	1,115	0	0	1,105	12
Flaxseed	Cwt.	0	0	0	0	0	0
Blended foods	Pound	7,829	321,343	0	0	315,905	13,268
Total grains and seeds		xxx	xxx	xxx	xxx	xxx	xxx
Dry whole peas	Pound	95	4,202	0	0	4,131	166
Sugar, cane and beet	Pound	0	0	0	0	0	0
Vegetable oil products	Pound	1,830	318,472	0	9,744	324,521	5,525
Potatoes	Pound	0	265	0	0	265	0
Wool	Pound	0	0	0	0	0	0
Mohair	Pound	0	5	0	0	5	0
Other (B)		0	0	0	57,207	57,207	0
Total inventory operations		xxx	xxx	xxx	xxx	xxx	xxx

See footnotes at end of table.

Table 11-4.—Commodity Credit Corporation: Selected inventory transactions, programs and commodity, fiscal year 2009—Continued

Program and commodity	Unit	Value					
		Inventory Oct. 1, 2008	Purchases	Collateral acquired from loans	Other addition deduction	Sales and other dispositions [1]	Inventory Sept. 30, 2009
		1,000 dollars	1,000 dollars	1,000 dollars	1,000 dollars	1,000 dollars	
Feed grains:.							
Barley	Bushel	0	9	0	0	9	0
Corn	Bushel	0	10,826	0	0	10,826	0
Corn products ..	Pound	4,407	57,412	0	0	58,711	3,108
Grain sorghum	Bushel	190	128,180	0	0	127,782	588
Sorghum grits ...	Pound	0	0	0	0	0	0
Oats	Bushel	0	0	0	0	0	0
Total feed grains	....	xxx	xxx	xxx	xxx	xxx	xxx
Wheat (A)	Bushel	0	217,459	4	2,840	220,303	....
Wheat flour	Pound	89	47,737	0	0	44,291	3,534
Wheat products, other	Pound	0	51,097	0	0	51,097	0
Rice, milled	Cwt	72	48,735	0	994	49,692	109
Rice, rough	Cwt	0	0	0	0	0	0
Cotton, extra long staple	Bale	0	0	0	0	0	0
Upland cotton	Bale	0	2,744,294	47,344	(47,344)	2,744,294	0
Tobacco Products	Pound	0	0	0	0	0	0
Dairy products:							
Butter	Pound	0	4,871	0	0	4,871	0
Cheese	Pound	0	992	0	11,860	12,851	0
Milk, dried	Pound	0	0	0	0	0	0
Milk, UHT	Pound	0	0	0	0	0	0
Dry whole milk	Pound	0	0	0	0	0	0
Non fat dry milk	Pound	0	221,031	0	(6,724)	29,809	184,499
Total dairy products	....	xxx	xxx	xxx	xxx	xxx	xxx
Oils and oilseeds:							
Crambe oilseed	Cwt.	0	0	0	0	0	0
Canola seed	Cwt.	0	0	0	0	0	0
Sunflower seed	Cwt.	0	0	0	0	0	0
Sunflower seed, non-oil	Cwt.	0	0	0	0	0	0
Peanuts farmers'stock	Pound	0	0	0	0	0	0
Peanut	Pound	0	52,940	1,351	(1,351)	52,940	0
Peanut butter ...	Pound	0	0	0	3,660	3,660	0
Soybeans	Bushel	0	2,341	0	0	2,341	0
Soybean products	Pound	0	19,788	0	0	19,788	0
Dry edible beans	Cwt.	68	48,258	0	0	47,785	541
Flaxseed	Cwt.	0	0	0	0	0	0
Blended foods ..	Pound	2,086	76,491	0	0	75,292	3,285
Total grains and seeds	....	xxx	xxx	xxx	xxx	xxx	xxx
Dry whole peas	Pound	2,281	116,975	0	0	116,129	3,127
Sugar, cane and beet	Pound	0	0	0	0	0	0
Vegetable oil products	Pound	15,641	202,266	0	5,170	205,636	3,365
Potatoes	Pound	0	0	0	0	0	0
Wool	Pound	0	0	0	0	0	0
Mohair	Pound	0	19	0	0	19	0
Other (B)	....	0	0	0	38,326	38,326	0
Total inventory operations	....	xxx	xxx	xxx	xxx	xxx	xxx

[1] Includes sales, commodity donations, transfers to other government agencies and inventory adjustment.   (A)Excludes wheat set aside for Food Security Wheat Reserve (FSWR).   (B) Includes beans, dry edible, and fish, canned salmon.
FSA, Financial Management Division, (703) 305–1363.

Table 11-5.—Commodity Credit Corporation: Cost value of export and domestic commodity dispositions, by type of disposition, fiscal year 2009 [1]

(In Thousands)

Commodity	Domestic				
	Dollar sales	Transfers to other Government agencies	Donations [1]	Inventory adjustments and other recoveries (domestic)	Total domestic
Feed grains:					
Barley	9	0	0	0	9
Corn	0	0	0	0	0
Corn products	0	0	0	0	0
Grain sorghum	0	0	0	0	0
Sorghum grits	0	0	0	0	0
Oats	0	0	0	0	0
Tobacco Products	0	0	0	0	0
Bulgur	0	0	0	0	0
Wheat	0	0	2,840	4	2,844
Wheat flour	0	0	0	0	0
Wheat product, Other	0	0	0	0	0
Rice, milled	0	0	994	0	994
Rice, rough	0	0	0	0	0
Rice, brown and Textured soy	0	0	0	0	0
Cotton, extra long staple & upland	2,744,294	0	0	0	2,744,294
Veg dehyd vegetable soup	0	0	0	0	0
Dairy products:					
Butter oil	0	0	0	0	0
Butter	4,870	0	0	1	4,871
Cheese Products	0	0	12,851	0	12,851
Nonfat dry milk	816	0	26,692	1,014	28,522
Milk, dried. UT high temp	0	0	1,287	0	1,287
Oils and oilseeds:.					
Peanuts	52,940	0	0	0	0
Peanut butter	0	0	0	0	52,940
Peanuts farmer's stock & products	0	0	0	0	0
Soya flour	0	0	0	0	0
Flaxseed	0	0	0	0	0
Sunflower Seed (oil & non-oil)	0	0	0	0	0
Soybeans & Soybean products	0	0	0	0	0
Fruit fresh apples	0	0	0	0	0
Blended foods	0	0	0	0	0
Potatoes	0	0	0	0	0
Grains and seeds:.					
Feed for Government facilities	0	0	0	0	0
Foundation seeds	0	0	0	0	0
Lentils dry	0	0	0	0	0
Vegetable Seeds	0	0	0	0	0
Canola seed	0	0	0	0	0
Crambe oil seed	0	0	0	0	0
Peas, dried whole	0	0	0	0	0
Dry edible beans	0	0	0	0	0
Honey	0	0	0	0	0
Sugar	0	0	0	0	0
Vegetable oil products	0	0	5,170	0	5,170
Mohair	19	0	0	0	19
Meat (and products)	0	0	0	0	0
Veg. canned tomato sauce	0	0	0	0	0
Wool	0	0	0	0	0
Other(rice products, fish, canned salmon)	0	0	38,326	0	38,326
Total [2]	2,802,948	0	88,160	1,019	2,892,126

See footnotes at end of table.

Table 11-5.—Commodity Credit Corporation: Cost value of export and domestic commodity dispositions, by type of disposition, fiscal year 2009 [1]—Continued

(In Thousands)

Commodity	Export				Total export and domestic
	Dollar sales	Public law 480 Title II/III	Donations [1]	Total export	
Feed grains:					
Barley	0	0	0	0	9
Corn	848	656	9,322	10,826	10,826
Corn products	1,464	57,250	(3)	58,711	58,711
Grain sorghum	0	127,782	0	127,782	127,782
Sorghum grits	0	0	0	0	0
Oats	0	0	0	0	0
Tobacco Products	0	0	0	0	0
Bulgur	0	52,455	(1,358)	51,097	51,097
Wheat	0	175,858	41,601	217,459	220,303
Wheat flour	2,609	36,445	5,237	44,291	44,291
Wheat product, Other	0	0	0	0	0
Rice, milled	602	48,096	0	48,698	49,692
Rice, rough	0	0	0	0	0
Rice,brown and Textured soy	0	0	0	0	0
Cotton, extra long staple & up-land	0	0	0	0	2,744,294
Veg dehyd vegetable soup	0	0	0	0	0
Dairy products:					
Butter oil	0	0	0	0	0
Butter	0	0	0	0	4,871
Cheese Products	0	0	0	0	12,851
Nonfat dry milk	0	0	0	0	28,522
Milk, dried. UT high temp	0	0	0	0	1,287
Oils and oilseeds:.					
Peanuts	0	0	0	....	52,940
Peanut butter	0	0	3,660	3,600	3,660
Peanuts farmer's stock & products	0	0	0	....	0
Soya flour	0	48	0	....	48
Flaxseed	0	0	0	....	0
Sunflower Seed (oil & non-oil) ...	0	0	0	....	0
Soybeans & Soybean products	4,368	0	17,714	22,082	22,082
Fruit fresh apples	0	....	....	....	0
Blended foods	1,896	71,338	2,057	75,292	75,292
Potatoes	0	203	0	203	203
Grains and seeds:.					
Feed for Government facilities ...	0	0	0	0	0
Foundation seeds	0	0	0	0	0
Lentils dry	0	45,817	609	46,426	46,426
Vegetable Seeds	0	0	0	0	0
Canola seed	0	0	0	0	0
Crambe oil seed	0	0	0	0	0
Peas, dried whole	1,708	69,256	(1,260)	69,703	69,703
Dry edible beans	1,176	41,240	5,369	47,785	47,785
Honey	0	0	0	0	0
Sugar	0	0	0	0	0
Vegetable oil products	3,304	169,880	18,929	192,113	197,283
Mohair	0	0	0	0	19
Meat (and products)	0	0	0	0	0
Veg. canned tomato sauce	0	0	0	0	0
Wool	0	0	0	0	0
Other(rice products, fish, canned salmon)	0	0	0	0	38,326
Total [2]	17,975	896,323	101,876	1,016,174	3,908,300

[1] Includes donations under section 202,407,416, Section 210, P.L. 85-540, miscellaneous donations under various other authorizations.　[2] Totals not accurate to rounding.　[3] Loans through Sept. 30, 2009.

FSA, Financial Management Division, (703) 305-1363.

Table 11-6.—Commodity Credit Corporation: Investment in price-support operations, March and June[1], 2000–2009

Year Month	Inventory investment	Loan investment	Total investment
	Million dollars	Million dollars	Million dollars
2000:			
March	1,635	5,628	7,263
June	2,299	3,663	5,962
2002:			
March	876	5,324	6,199
June	920	2,724	3,644
2003:			
March	541	5,429	5,970
June	587	3,281	3,868
2004:			
March	135	4,972	5,106
June	92	2,840	2,931
2005:			
March	592	4,705	5,297
June	95	2,956	3,051
2006:			
March	84	5,503	5,587
June	93	3,016	3,109
2007:			
March	72	7,031	7,103
June	48	2,902	2,950
2008:			
March	6,530	6,093	12,623
June	3,996	3,346	7,343
2009:			
March	8,682	5,083	13,765
June	5,371	2,440	7,811

[1] Reflects total CCC loans and inventories investment.
FSA, Financial Management Division, (703) 305–1363.

Table 11-7.—Farm Service Agency programs: Payments to producers, by program and commodity, United States, calendar year 2008–2009

Program and commodity	2008	2009
	1,000 dollars	1,000 dollars
Acreage grazing payments	0	0
Additional interest	2	10
Agricultural management assistance	0	0
American indian - livestock feed	0	0
Aquaculture block grant	0	0
Auto conservation reserve program (crp) - cost shares	81,640	77,138
Avg crop revenue election program	0	471,814
Bioenergy program	0	0
Biomass crop assistance	0	10,387
Cottonseed payment program	0	0
Crop disaster - north carolina	0	0
Crop disaster - virginia	0	0
Crop disaster program	0	0
Crop disaster program - 2005	648	210
Crop hurricane damage program	0	0
Crp annual rental	1,745,708	1,703,525
Crp incentives	64,411	75,233
Dairy economic loss assistance	0	267,577
Dairy indemnity	45	664
Dairy market loss assistance	0	0
Direct and counter cyclical program	5,834,357	5,461,972
Extra long staple special provision program	28,581	13,050
Emergency conservation program	25,539	89,132
Environment quality incentives	0	0
Feed indemnity program	0	0
Finalty rule	2	0
Fl hurricane citrus disaster	418	73
Forestry conservation reserve	7,558	9,583
Fl nursery disaster	0	0
Fl vegetable disaster	0	0
Florida sugarcane program	0	0
Grasslands reserve program	6,028	6,088
Hard white winter wheat	0	0
Hurricane indemnity program	0	0
Interest payments	863	567
Lamb meat adjustment assistance	0	0
Livestock assistance grant	0	0

See footnotes at end of table.

Table 11-7.—Farm Service Agency programs: Payments to producers, by program and commodity, United States, calendar year 2008–2009—Continued

Program and commodity	2008	2009
	1,000 dollars	*1,000 dollars*
Livestock assistance program	0	0
Livestock compensation program	4	0
Livestock emergency assistance	52	0
Livestock forage program	0	118,543
Livestock indemnity program	240	48,827
Loan deficiency	85,516	157,577
Louisiana sugarcane program	0	0
Market gains	25,132	252,196
Marketing loss assistance	29	0
Milk income loss contract transitional	0	0
Milk income loss contract	0	0
Milk income loss ii	327	883,906
Noninsured assistance program	69,231	69,731
Peanut quota buyout program	0	0
Soil/water conservation assistance	0	0
Speciality crop - nursery	60	53
Speciality crop - tropical fruit	0	0
Specialty crop - citrus	949	0
Specialty crop - fruit/vegetable	9	80
Storage forgiven	6,659	32,217
Sugar beet disaster program	0	0
Texas sugarcane storage & transportation	0	0
Tobacco quota holder-interest	0	4
Trade adjustment assistance	24	16
Tree assistance program	992	81
Tree indemnity program	3	0
Ttpp tobacco producer	287,188	287,028
Upland cotton assistance	0	92,134
Wetlands reserve	0	0
01-02 crop disaster assistance	69	269
05 - 07 crop disaster assistance	1,685,536	58,636
05 - 07 dairy disaster prog	12,526	13
05 - 07 livestock compensation	295,141	572
05 - 07 livestock indemnity prog	39,392	348
Grand Total	10,304,879	10,189,254

FSA Budget/Corporate Programs Branch, (202) 720–5148.

Table 11-8.—Farm Service Agency programs: Payments received, by States, calendar year 2008–2009

State	Payments	
	2008	2009
	1,000 dollars	*1,000 dollars*
AL	139,030	132,572
AK	1,087	1,261
AZ	68,312	94,956
AR	346,358	404,525
CA	351,143	446,425
CO	222,492	162,638
CT	3,864	4,923
DE	12,703	6,773
DC	0	0
FL	70,709	47,801
GA	333,762	319,531
HI	2,228	5,928
ID	119,399	118,472
IL	617,499	551,075
IN	292,113	284,104
IA	752,224	737,859
KS	612,407	449,531
KY	202,082	182,150
LA	204,039	231,359
ME	4,207	10,371
MD	42,249	36,127
MA	6,140	5,058
MI	139,518	158,913
MN	507,557	497,346
MS	270,181	318,208
MO	380,195	342,468
MT	262,142	228,751
NE	477,085	389,512
NV	5,165	4,753
NH	1,227	3,842
NJ	8,318	6,598
NM	50,764	62,066
NY	52,552	137,084
NC	301,372	253,677
ND	484,307	422,017
OH	243,432	256,206
OK	272,427	192,661
OR	75,967	70,276
PA	69,517	144,965
RI	417	319
SC	92,268	92,707
SD	384,956	241,265
TN	159,557	169,940
TX	1,024,983	1,053,055
UT	24,372	25,861
VT	7,993	32,973
VA	88,355	75,572
WA	174,631	170,211
WV	6,659	5,313
WI	198,487	385,173
WY	29,198	17,921
KCCO	105,072	189,279
PR	2,022	6,859
VI	14	26
GU	100	0
MI	21	0
AS	0	0
Total [1]	10,304,879	10,189,254

[1] Total may not add due to rounding.
FSA, Budget, Corporate Programs Branch, (202) 720–5148.

Table 11-9.—Commodity Credit Corporation: Loans made in fiscal year 2009 for crop year 2008, by States and Territories [1]

State or Territory	Barley	Corn	Cotton	Flaxseed	Honey	Oats
	1,000 dollars	*1,000 dollars*	*1,000 dollars*	*1,000 dollars*	*1,000 dollars*	*1,000 dollars*
Alabama		1,419	82,548			1
Alaska	0	0				
Arizona	50	721	988	0	10	
Arkansas		10,046	230,010		79	
California	95	36	396,413		295	
Colorado	104	14,200				
Connecticut		46				
Delaware	90	1,370				
Florida		137	3,131		149	
Georgia		8,323	67,050		110	51
Hawaii	0					
Idaho	1,980				283	
Illinois		239,242				
Indiana		181,794				
Iowa	10	448,539			676	137
Kansas	10	28,958			41	14
Kentucky	3	26,307				
Louisiana	0	11,044	17,458		16	
Maine	77					315
Maryland	98	4,644				
Massachusetts		305				
Michigan		65,747			222	12
Minnesota	976	376,825		21	310	266
Mississippi		21,323	569,430			
Missouri		45,122	28,091			
Montana	2,563	42	0	15	1,645	2
Nebraska	21	203,852			339	45
Nevada						
New Hampshire						
New Jersey		509				
New Mexico		2,022	1,800			
New York	55	26,391		0	36	123
North Carolina	89	6,020	144,438			25
North Dakota	3,660	41,342		644	525	103
Ohio		73,286				6
Oklahoma		1,590	4,301			
Oregon	59	23			176	
Pennsylvania	44	11,276				82
Rhode Island		0	0			
South Carolina		4,084	5,911			118
South Dakota	33	92,938			1,630	178
Tennessee		9,824	121,742			
Texas		16,451	1,035,486		64	
Utah	15	93			91	
Vermont		16				
Virginia	401	4,596	6,314			
Washington	1,948	1,769			86	
West Virginia		680				
Wisconsin	18	66,054			64	27
Wyoming		994				
Adjustments	217	2,035	29	53		22
Peanut Associations						
Total [2]	12,616	2,052,035	2,715,140	733	6,847	1,527

See footnotes at end of table.

Table 11-9.—Commodity Credit Corporation: Loans made in fiscal year 2009 for crop year 2008, by States and Territories [1]—Continued

State or Territory	Oilseeds	Peanuts	Rice	Seed cottton	Sorghum	Soybeans
	1,000 dollars	*1,000 dollars*	*1,000 dollars*	*1,000 dollars*	*1,000 dollars*	*1,000 dollars*
Alabama		23,271				1,444
Alaska						
Arizona					299	
Arkansas			279,602		424	27,386
California			91,904			
Colorado	545				168	52
Connecticut						
Delaware						404
Florida		22,972				
Georgia	254	470,080			6	859
Hawaii						
Idaho	121					
Illinois	91				287	86,865
Indiana					397	105,460
Iowa						212,902
Kansas	134				7,207	18,296
Kentucky		47			4	13,062
Louisiana	0		18,848		112	182
Maine						
Maryland						1,954
Massachusetts		4,886				
Michigan			21,859			27,014
Minnesota	1,652					140,288
Mississippi					205	15,540
Missouri	0		7,917		591	51,498
Montana	206					
Nebraska	204				797	52,595
Nevada						
New Hampshire						
New Jersey						156
New Mexico		3,908				
New York	13					8,904
North Carolina		38,822				9,230
North Dakota	10,998					18,584
Ohio						60,878
Oklahoma		4,618			510	385
Oregon						
Pennsylvania						4,752
Rhode Island						
South Carolina		10,005				7,981
South Dakota	4,890				460	42,184
Tennessee			467		41	7,023
Texas		104,009	23,943	3,044	4,620	110
Utah						
Vermont						
Virginia		35,443				4,144
Washington						
West Virginia						501
Wisconsin	123					20,238
Wyoming						
Adjustments	358	102	193			1,107
Peanut Associations						
Total [2]	19,589	718,163	444,733	3,044	16,128	941,978

See footnotes at end of table.

Table 11-9.—Commodity Credit Corporation: Loans made in fiscal year 2009 for crop year 2008, by States and Territories [1]—Continued

State or Territory	Sugar	Tobacco	Wheat	Mohair	Dry whole peas	Wool
	1,000 dollars	*1,000 dollars*	*1,000 dollars*	*1,000 dollars*	*1,000 dollars*	*1,000 dollars*
Alabama			365			
Alaska						
Arizona			517			
Arkansas			2,143			
California			823			
Colorado			10,389			
Connecticut	139,349					
Delaware			53			
Florida	214,712		39			
Georgia			174			
Hawaii						
Idaho	207,839		21,938		60	
Illinois			1,308			
Indiana			1,025			
Iowa			38			
Kansas			16,525			
Kentucky			2,718			
Louisiana	178,604		336			
Maine						
Maryland			213			
Massachusetts						
Michigan	186,566		2,248			
Minnesota	170,649		21,966			
Mississippi			568			
Missouri			2,665			
Montana			40,469		1,168	
Nebraska			4,778			
Nevada						
New Hampshire						
New Jersey			20			
New Mexico						
New York			1,670			
North Carolina			963			
North Dakota			35,504		1,749	
Ohio			1,656			
Oklahoma			5,325			
Oregon			4,509			
Pennsylvania			640			
Rhode Island						
South Carolina			1,609			
South Dakota			17,335		360	
Tennessee			1,123			
Texas	12,398		682	54		
Utah			1,656			6
Vermont						
Virginia			1,342			
Washington	8,251		28,668		300	
West Virginia			50			
Wisconsin			788			
Wyoming			719			61
Adjustments			686		6	
Peanut Associations			0			
Total [2]	1,118,368	0	236,243	54	3,643	67

[1] Loans made directly by Commodity Credit Corporation.   As much as possible, loans have been distributed according to the location of producers receiving the loans.   Direct loans to cooperative associations for the benefit of members have been distributed according to the location of the association.   [2] Table may not add due to rounding.
CCC, Financial Management Division, (703) 305-1363.

Table 11-10.—Fruit, vegetable, and tree nut marketing agreements and orders and peanut program, 2007–2008 [1]

Active Programs	Estimated number of producers	Farm value
	Number	*1,000 dollars*
Citrus fruits (2007-08 season):		
Florida oranges, grapefruit, tangerines, and tangelos	8,000	253,092
Texas oranges and grapefruit	190	29,876
Deciduous fruits (2007 season):		
California fresh peaches	676	147,906
California nectarines	676	89,039
California olives	1,000	86,694
California desert grapes	50	89,000
California kiwifruit	220	22,517
Florida avocados	300	12,100
Washington apricots	300	7,137
Washington sweet cherries	1,700	309,600
Washington and Oregon pears [2]	1,550	268,324
Tart cherries (7 States) [3]	900	67,923
Washington and Oregon fresh prunes [4]	215	2,496
Cranberries (10 States) [5]	1,100	307,233
Dried fruits (2007 season):		
California dates	85	37,327
California dried prunes	800	117,450
California raisins	3,000	374,760
Vegetables (2007-08 season):		
Florida tomatoes	100	619,412
Idaho and Eastern Oregon onions	290	36,028
South Texas onions	84	174,720
Georgia onions (Vidalia)	86	63,167
Walla Walla onions	28	8,354
Potatoes (2007-08 season):		
Colorado	180	166,489
Idaho and eastern Oregon	975	233,307
Southeastern States (Virginia - North Carolina)	73	8,313
Washington	267	114,588
Nuts (2007 season):		
California almonds	6,200	2,401,875
California Pistashios	740	586,560
California walnuts	4,000	751,120
Oregon and Washington Hazelnuts	715	75,480
Peanuts [6]	10,002	758,626
Spearmint oil (2007 season) [7]	123	28,858
(Total 32 programs) [8]		8,249,371

[1] Preliminary. [2] Includes fresh and processed pears. [3] The tart cherry order covers the States of Michigan, New York, Pennsylvania, Oregon, Utah, Washington, and Wisconsin. [4] Farm value is available only for fresh and processed combined. [5] Massachusetts, Rhode Island, Connecticut, New Jersey, Wisconsin, Michigan, Minnesota, Oregon, Washington, and Long Island in New York. (Only top 5 are reported). [6] The Farm Security and Rural Investment Act of 2002 terminated the Peanut Administrative committee (which locally administered marketing agreement No. 146). As a result, the agreement was terminated and new quality standards for all domestic and imported peanuts were established. [7] The marketing order regulates the handling of spearmint oil produced in the States of Washington, Idaho, Oregon, and designated parts of Nevada and Utah. The farm value is the sum of values for Idaho, Oregon, and Washington, the only significant producing States in the marketing order area. [8] Total number of producers cannot be determined from totals for individual commodities; some producers produce more than one commodity.

AMS, Fruit and Vegetable Programs, (202) 720–2615.

CHAPTER XII

AGRICULTURAL CONSERVATION AND FORESTRY STATISTICS

Statistics in this chapter concern conservation of various natural resources, particularly soil, water, timber, wetlands, wildlife, and improvement of water quality. Forestry statistics include area of private and public-owned forest land, timber production, imports and exports, pulpwood consumption and paper and board production, area burned over by forest fires, livestock grazing, and recreational use of national forest lands.

Conservation Practices on Active CRP Contracts

Practice code	Practice	Acres
CP1	Introduced grasses and legumes	2,678,975
CP2	Native grasses	6,400,185
CP3	Tree planting	1,001,175
CP4	Wildlife habitat with woody vegetation	2,114,826
CP5	Field windbreaks	94,344
CP6	Diversions	534
CP7	Erosion control structures	380
CP8	Grass waterways	135,018
CP9	Shallow water areas for wildlife	43,683
CP10	Existing grasses and legumes 1/	11,625,126
CP11	Existing trees	951,783
CP12	Wildlife food plots	77,701
CP15	Contour grass strips	69,852
CP16	Shelterbelts	37,114
CP17	Living snow fences	6,088
CP18	Salinity reducing vegetation	238,355
CP21	Filter strips (grass)	1,026,875
CP22	Riparian buffers (trees)	870,700
CP23	Wetland restoration	1,572,590
CP24	Cross wind trap strips	458
CP25	Rare and declining habitat	1,201,867
CP26	Sediment retention	36
CP27	Farmable wetland pilot (wetland)	63,422
CP28	Farmable wetland pilot (upland)	156,890
CP29	Wildlife habitat buffer (marginal pasture)	115,667
CP30	Wetland buffer (marginal pasture)	30,828
CP31	Bottomland hardwood	57,881
CP32	Hardwood trees	8,569
CP33	Upland bird habitat buffers	225,009
CP34	Flood control structure	71
CP36	Longleaf pine	84,411
CP37	Duck nesting habitat	99,188
CP38	State acres for wildlife enhancement	296,321
CP39	FWP--Constucted wetlands	17
CP40	FWP--Aquaculture wetlands	4,046
CP41	FWP--Flooded praire wetlands	3,746
	Total	31,293,735

[1] Includes both introduced grasses and legumes and native grasses.
FSA, Conservation and Environmental Programs Division, (530)792-5594.

CRP enrollment: By sign up and initial contract year [1], as of January 2010

Sign up	Before 2001	2001	2002	2003	2004	2005	2006
1-17	13,908,531						
18	3,310,652						
19	155,395						
20		2,195,642					
21	101,906	12,395					
22	32,271	166,463					
23		213,880	238,984				
24			280,905	149,474			
25				198,288	53,914		
26					1,613,294	161,412	
27				11,419	170,162		
28					151,012	99,435	
29							1,007,881
30						197,067	195,154
31							198,714
32							
33							
35							
36							
37							
38							
All	17,508,755	2,588,381	519,889	359,181	1,988,382	457,913	1,401,749

Table 12-1.—Conservation Reserve Program (CRP): Enrollment by practice, under contract, January 2010

(CP 1 and CP 2)

State	CP 1 Establishment of permanent introduced grasses and legumes			CP 2 Establishment of permanent native grasses		
	Total acres treated	Total cost share	Avg cost share per acre treated [1]	Total acres treated	Total cost share	Cost share per acre treated [1]
Alabama	2,845.2	110,170	61.34	3,372.3	237,698	84.97
Alaska	4,849.6	164,198	81.92	0.0	0	....
Arkansas	2,859.7	165,206	63.94	4,005.8	372,752	96.10
California	5,280.2	323,484	79.60	1,273.6	303,949	270.51
Colorado	30,796.5	1,167,739	48.93	585,993.6	31,700,859	63.65
Connecticut	59.9	2,880	300.00	34.3	1,230	119.42
Delaware	25.3	1,639	64.78	23.3	1,967	84.41
Florida	92.2	17,865	193.76	150.5	335	67.00
Georgia	59.3	2,956	68.91	245.9	24,377	99.18
Hawaii	*	*	*	*	*	*
Idaho	85,196.4	2,849,808	40.80	38,611.7	2,471,078	69.04
Illinois	159,102.7	6,402,367	55.14	45,537.6	3,199,537	88.87
Indiana	30,746.1	1,451,192	65.15	26,721.4	1,605,313	80.05
Iowa	175,098.9	4,326,004	54.77	137,884.2	7,900,653	77.65
Kansas	11,424.5	243,611	48.26	796,426.7	30,402,661	52.95
Kentucky	69,352.9	4,204,973	75.88	41,029.3	3,215,228	98.55
Louisiana	179.4	10,635	63.99	3,199.8	250,875	83.40
Maine	1,319.1	110,525	103.74	50.3	8,428	174.49
Maryland	10,916.7	885,874	147.81	3,059.5	422,253	207.73
Massachusetts	0.0	0	....	0.0	0	....
Michigan	27,198.0	1,936,151	83.15	25,625.1	2,724,028	120.86
Minnesota	199,815.6	8,548,627	50.90	105,957.4	5,884,533	82.28
Mississippi	2,990.4	159,130	59.71	903.1	37,938	85.25
Missouri	299,335.4	11,341,067	50.57	162,242.5	8,649,322	74.16
Montana	584,284.6	12,368,860	25.87	790,273.5	25,244,597	41.20
Nebraska	23,541.4	526,647	31.30	372,520.1	17,563,852	60.55
Nevada	*	*	*	*	*	*
New Hampshire	0.0	0	....	0.0	0	....
New Jersey	1,069.8	117,170	154.76	218.1	14,929	228.97
New Mexico	1,861.0	89,793	48.25	165,398.4	4,936,997	46.51
New York	7,451.0	792,635	114.17	832.4	95,044	130.48
North Carolina	1,354.1	87,967	77.49	1,207.3	104,113	100.78
North Dakota	296,700.0	6,014,924	24.47	65,098.5	4,020,649	69.58
Ohio	20,467.3	821,752	61.25	64,162.0	5,352,613	97.98
Oklahoma	7,330.1	227,125	38.04	371,479.0	18,004,278	55.43
Oregon	118,895.3	3,904,219	42.08	86,111.4	5,863,664	75.63
Pennsylvania	116,280.6	17,280,432	152.23	41,298.9	6,904,282	176.35
Puerto Rico	0.0	0	....	0.0	0	....
Rhode Island	*	*	*	*	*	*
South Carolina	139.4	12,476	93.81	89.2	8,033	110.50
South Dakota	51,708.8	1,763,264	36.86	151,183.2	7,218,965	55.38
Tennessee	22,064.0	1,239,738	65.29	42,886.2	3,365,475	83.78
Texas	64,771.9	1,756,826	33.81	1,492,990.1	69,946,303	56.93
Utah	45,787.3	1,768,715	42.34	14,246.9	751,479	53.25
Vermont	5.3	397	128.06	0.0	0	....
Virginia	2,605.4	97,882	65.95	1,831.5	84,861	86.13
Washington	123,463.8	4,843,078	55.80	710,538.7	58,201,341	95.95
West Virginia	98.3	6,797	69.14	22.4	1,053	47.01
Wisconsin	29,833.8	1,617,545	65.41	41,087.5	2,810,840	89.77
Wyoming	39,717.7	1,150,080	31.75	4,362.3	175,676	40.99
United States, total	2,678,975	100,914,424	48.17	6,400,185	330,084,058	64.17

[1] Not including acres which receive no cost share. * Data withheld to avoid disclosure of individual operations. Note: Total acres treated may not add due to rounding.
FSA, Conservation and Environmental Programs Division, (530) 792-5594

Table 12-2.—Conservation Reserve Program (CRP): Enrollment by practice, under contract, January 2010

(CP 3 and CP 4)

State	CP 3 Tree planting			CP 4 Permanent wildlife habitat		
	Total acres treated	Total cost share	Avg cost share per acre treated [1]	Total acres treated	Total cost share	Avg cost share per acre treated [1]
Alabama	134,179.4	11,233,790	100.09	8,239.9	102,161	56.00
Alaska	0.0	0	....	0.0	0	....
Arkansas	42,429.5	4,244,824	115.34	2,606.7	195,366	94.88
California	62.1	850	119.72	725.5	8,625	539.06
Colorado	97.3	94,011	1,181.04	369,798.5	40,303,809	122.93
Connecticut	0.0	0	....	0.0	0	....
Delaware	3,384.6	1,180,060	354.36	1,308.4	390,523	330.84
Florida	17,514.5	1,163,790	75.15	2,193.1	70,704	60.80
Georgia	148,346.2	15,135,889	107.13	5,118.2	132,249	93.53
Hawaii	*	*	*	*	*	*
Idaho	4,382.9	487,806	138.23	118,336.5	4,037,578	44.41
Illinois	51,165.4	5,084,288	124.84	123,220.3	8,879,343	114.93
Indiana	17,985.5	2,387,168	160.33	13,079.1	1,346,092	151.13
Iowa	14,944.6	2,452,172	200.47	217,200.4	6,310,121	94.34
Kansas	559.3	66,297	142.02	14,681.4	552,681	59.71
Kentucky	6,013.1	732,343	136.76	595.4	391,817	107.12
Louisiana	135,848.8	12,988,888	100.53	33,778.0	3,409,470	123.14
Maine	172.4	14,041	123.60	659.6	5,709	64.00
Maryland	1,191.1	294,358	344.96	1,771.6	231,450	208.19
Massachusetts	0.0	0	....	0.0	0	....
Michigan	6,923.7	760,751	141.78	18,490.3	806,952	100.33
Minnesota	32,302.5	3,593,823	135.22	275,810.6	7,891,857	71.98
Mississippi	207,648.6	12,201,546	72.94	6,507.7	211,837	123.30
Missouri	16,465.4	1,451,734	127.49	5,989.7	616,478	138.70
Montana	189.4	27,586	179.60	30,017.2	801,869	52.46
Nebraska	1,352.5	95,1695	140.65	46,133.0	2,602,750	69.61
Nevada	*	*	*	*	*	*
New Hampshire	0.0	0	....	0.0	0	....
New Jersey	100.1	7,642	146.97	5.8	2,975	512.93
New Mexico	0.0	0	....	0.0	0	....
New York	1,139.8	163,861	209.35	454.6	47,288	185.44
North Carolina	21,854.0	2,258,721	129.10	1,883.2	156,924	118.77
North Dakota	377.2	38,421	175.04	426,908.3	13,585,560	48.43
Ohio	8,255.8	1,005,302	153.62	41,296.9	4,924,511	153.74
Oklahoma	416.4	45,633	123.20	2,257.2	149,616	70.93
Oregon	1,862.0	53,945	132.67	12,378.8	564,753	67.92
Pennsylvania	1,315.3	1,190,658	986.38	4,669.7	1,829,605	415.10
Puerto Rico	39.0	4,531	137.30	26.0	1	0.17
Rhode Island	*	*	*	*	*	*
South Carolina	45,151.6	2,893,986	74.70	5,480.9	109,640	69.50
South Dakota	283.7	68,718	308.01	70,654.8	6,117,766	101.72
Tennessee	17,469.0	1,446,400	110.71	12,195.3	549,134	90.48
Texas	2,598.6	35,688	52.76	40,868.5	2,739,503	92.75
Utah	0.0	0	....	762.6	2,287	3.69
Vermont	0.0	0	....	0.0	0	....
Virginia	6,430.9	629,192	115.34	683.8	82,837	164.46
Washington	1,066.6	215,301	278.71	162,414.9	12,435,404	108.04
West Virginia	14.5	3,370	391.86	0.0	0	....
Wisconsin	49,632.5	8,068,888	178.87	7,744.4	616,760	136.73
Wyoming	9.3	10,200	1,569.23	27,733.0	416,759	38.35
United States, total	1,001,175	93,832,167	109.21	2,114,826	123,282,866	91.15

[1] Not including acres which receive no cost share.   * Data withheld to avoid disclosure of individual operations.   Note: Total acres treated may not add due to rounding.

FSA, Conservation and Environmental Programs Division, (530) 792-5594

Table 12-3.—Conservation Reserve Program (CRP): Enrollment by practice, under contract, January 2010
(CP 5, CP 6, and CP 7)

State	CP 5 Establishment of field windbreaks			CP 6 Diversions I			CP 7 Erosion control structures		
	Total acres reated	Total cost share	Avg cost share per acre treated [1]	Total acres treated	Total cost share	Avg cost share per acre treated [1]	Total acres treated	Total cost share	Avg cost share per acre treated [1]
AL	0.0	0	....	0.0	0	....	0.0	0	....
AK	0.0	0	....	0.0	0	....	0.0	0	....
AR	0.0	0	....	0.0	0	....	2.0	729	364.50
CA	0.0	0	....	0.0	0	....	0.0	0	....
CO	1,645.3	1,346,132	953.89	0.0	0	....	209.9	1,400	254.55
CT	0.0	0	....	0.0	0	....	0.0	0	....
DE	0.0	0	....	0.0	0	....	0.0	0	....
FL	0.0	0	....	0.0	0	....	0.0	0	....
GA	0.0	0	....	0.0	0	....	0.0	0	....
HI	*	*	*	*	*	*	*	*	*
ID	575.6	1,624,034	2,978.48	0.0	0	....	4.0	4,500	1,125.00
IL	2,755.6	630,653	241.58	15.3	0	....	8.1	12,450	3,036.59
IN	2,342.1	472,106	209.44	0.0	0	....	3.8	3,600	1,161.29
IA	6,776.6	1,864,196	291.62	5.0	750	150.00	12.6	10,509	2,060.59
KS	1,975.7	857,545	525.04	5.7	2,975	632.98	4.0	1,659	414.75
KY	.5	1,010	2,020.00	0.0	0	....	4.0	4,305	1,076.25
LA	0.0	0	....	5.0	476	95.20	2.0	200	100.00
ME	0.0	0	....	0.0	0	....	0.0	0	....
MD	0.0	0	0	0.0	0	....	0.0	0	....
MA	0.0	0	....	0.0	0	....	0.0	0	....
MI	2,752.9	869,930	334.78	2.5	2,250	1,500.00	0.0	0	....
MN	9,717.7	3,120,785	341.28	0.0	0	....	0.3	0	....
MS	0.0	0	....	0.0	0	....	1.0	0	....
MO	111.2	25,784	262.30	296.0	21,411	72.98	93.0	54,802	659.47
MT	254.7	107,312	473.16	0.0	0	....	0.0	0	....
NE	31,700.0	17,475,044	685.84	0.0	0	....	9.9	0	....
NV	*	*	*	*	*	*	*	*	*
NH	0.0	0	....	0.0	0	....	0.0	0	....
NJ	9.5	37,414	3,938.32	0.0	0	....	0.0	0	....
NM	0.0	0	....	0.0	0	....	0.0	0	....
NY	12.8	10,100	789.06	0.0	0	....	1.0	3,500	3,500.00
NC	22.8	2,297	105.37	0.0	0	....	0.0	0	....
ND	5,380.1	2,747,163	533.61	0.6	143	238.33	0.0	0	....
OH	3,466.7	1,229,288	375.49	0.0	0	....	0.0	0	....
OK	51.0	11,658	438.27	57.3	13,288	238.14	20.0	1,741	87.05
OR	3.6	525	145.83	0.0	0	....	0.0	0	....
PA	0.0	0	0	0.0	0	....	0.0	0	....
PR	0.0	0	....	0.0	0	....	0.0	0	....
RI	*	*	*	*	*	*	*	*	*
SC	42.9	3,878	114.40	0.0	0	....	0.0	0	....
SD	24,201.5	18,052,076	811.87	0.0	0	....	0.0	0	....
TN	0.0	0	....	0.0	0	....	3.0	2,558	852.67
TX	43.1	47,898	1,111.32	0.0	0	....	0.0	0	....
UT	4.4	9,311	2,116.14	0.0	0	....	0.0	0	....
VT	0.0	0	....	0.0	0	....	0.0	0	....
VA	3.4	500	1,250.00	0.0	0	....	0.0	0	....
WA	6.2	3,250	524.19	0.0	0	....	0.0	0	....
WV	0.0	0	....	0.0	0	....	0.0	0	....
WI	201.1	57,466	306.32	0.5	600	1,200.00	1.6	12,400	7,750.00
WY	287.1	310,033	1,330.04	146.2	0	....	0.0	0	....
US	94,344.1	50,935,388	608.97	534	41,893	114.31	380	114,353	802.48

[1] Not including acres which receive no cost share.　* Data withheld to avoid disclosure of individual operations.　Note: Total acres treated may not add due to rounding.
FSA, Conservation and Environmental Programs Division, (530) 792-5594

Table 12-4.—Conservation Reserve Program (CRP): Enrollment by practice, under contract, January 2010

(CP 8, CP 9, and CP 10)

State	CP 8 Grass waterways			CP 9 Shallow water areas for wildlife			CP 10 Vegetative-cover-grass-already established		
	Total acres treated	Total cost share	Avg cost share per acre treated [1]	Total acres treated	Total cost share	Avg cost share per acre treated [1]	Total acres treated	Total cost share	Avg cost share per acre treated [1]
AL	20.7	2,733	198.04	152.6	121,040	830.18	80,531.2	23,348	15.24
AK	0.0	0	....	4.6	56,864	12,361.74	20,327.6	0	....
AR	21.6	1,322	76.42	766.8	245,318	404.15	17,253.7	5,093	49.54
CA	0.0	0	....	181.8	117,865	648.32	99,370.6	0	....
CO	729.4	192,661	469.45	30.0	8,309	393.79	1,006,114.0	27,083	48.27
CT	0.0	0	....	0.0	0	....	11.1	0	....
DE	7.3	6,919	2,661.5	359.3	812,400	2,454.38	25.0	0	....
FL	0.0	0	....	0.0	0	....	1,348.7	500	45.05
GA	35.5	29,830	1,142.91	16.3	15,387	2,442.38	3,542.0	0	....
HI	*	*	*	*	*	*	*	*	*
ID	9.4	11,238	3,305.29	43.4	43,712	1,539.15	450,511.8	1,844,537	356.73
IL	33,068.4	54,829,436	2,047.64	5,721.6	2,649,450	572.64	208,939.7	986,050	201.46
IN	19,282.6	79,463,186	4,691.50	1,407.3	970,784	839.20	59,293.3	1,800	56.78
IA	37,838.7	51,164,724	1,633.68	13,509.9	3,023,624	281.19	395,737.9	8,555,050	312.59
KS	9,868.7	3,662,050	435.82	1,152.4	267,727	384.28	1,337,834.1	1,217,140	110.17
KY	4,213.0	7,424,573	1,901.25	2,952.6	1,432,029	552.42	91,966.0	19,123	20.68
LA	13.4	9,610	1,079.78	814.2	182,067	359.11	11,343.0	638	11.71
ME	47.6	296,912	6,371.50	0.0	0	....	15,325.3	0	....
MD	259.7	715,941	4,694.70	1,268.2	1,819,218	1,911.55	2,029.5	15,040	85.31
MA	0.0	0	....	0.0	0	....	0.0	0	....
MI	890.9	3,054,524	4,132.20	2,497.8	1,573,812	715.50	66,884.2	112,345	48.84
MN	5,181.2	5,997,304	1,305.04	588.5	82,896	176.49	229,566.0	5,736	10.37
MS	63.9	300	93.75	673.7	120,333	297.19	88,905.4	768	19.59
MO	2,238.7	2,183,908	1,115.38	2,678.4	687,857	342.00	679,844.8	34,560	46.09
MT	105.3	8,273	99.92	85.0	11,080	130.35	1,361,920.8	244,432	4.27
NE	1,954.8	728,966	415.86	255.8	62,820	285.16	404,098.7	3,073,434	372.01
NV	*	*	*	*	*	*	*	*	*
NH	0.0	0	....	0.0	0	....	0.0	0	....
NJ	118.3	1,204,548	10,182.15	2.8	7,181	2,564.64	195.7	0	....
NM	0.0	0	....	0.0	0	....	371,744.5	0	....
NY	90.1	233,112	3,468.93	16.7	5,676	1,669.41	25,623.8	405,290	88.13
NC	299.3	815,244	3,122.34	2,113.6	1,809,421	911.09	10,462.2	7,081	94.41
ND	115.3	52,315	508.41	1.1	0	....	1,051,268.1	51,984	8.00
OH	11,005.6	39,824,655	4,159.93	810.2	602,530	995.42	69,864.8	4,519,496	4,934.49
OK	271.1	72,750	298.40	79.3	33,699	424.96	443,417.0	301,465	65.52
OR	31.9	13,507	608.42	0.0	0	....	278,369.4	0	....
PA	605.6	1,684,383	2,958.69	60.4	95,252	1,849.55	25,428.6	395,587	118.32
PR	0.0	0	....	0.0	0	....	188.0	0	....
RI	*	*	*	*	*	*	*	*	*
SC	67.8	136,948	2,019.88	1,601.3	2,371,144	1,499.49	5,721.0	774	73.02
SD	1,364.7	760,069	711.34	189.6	198,257	1,235.25	265,849.9	20,963	10.72
TN	210.7	231,858	1,171.59	141.0	56,209	477.97	78,467.2	3,313	32.80
TX	2,455.5	2,030,597	873.49	162.2	99,343	612.47	1,626,469.2	2,370,237	1,236.62
UT	14.0	347	43.38	0.0	0	....	80,019.4	0	....
VT	16.2	13,669	1,051.46	0.0	0	....	94.2	0	....
VA	55.3	55,394	1,105.67	68.2	93,978	1,702.50	8,692.7	54,445	71.27
WA	428.0	122,291	340.83	58.2	31,240	673.28	327,439.2	116,468	25.16
WV	0.0	0	....	0.0	0	....	482.1	0	....
WI	2,013.3	4,124,494	2,124.71	3,218.5	9,738,041	3,410.16	191,092.1	466,237	64.84
WY	4.1	4,671	1,139.27	0.0	0	....	131,512.8	0	....
US	135,018	261,165,262	2,287.98	43,683	29,446,586	824.64	11,625,126	24,880,017	157.91

[1] Not including acres which receive no cost share.   * Data withheld to avoid disclosure of individual operations.   Note: Total acres treated may not add due to rounding.

FSA, Conservation and Environmental Programs Division, (530) 792-5594.

Table 12-5.—Conservation Reserve Program (CRP): Enrollment by practice, under contract, January 2010
(CP 11 and CP 12)

State	CP 11 Vegetative-cover-trees-already established			CP 12 Wildlife food plots		
	Total acres treated	Total cost share	Avg cost share per acre treated [1]	Total acres treated	Total cost share	Avg cost share per acre treated [1]
AL	138,475.7	515,573	41.67	1,524.1	0	
AK	0.0	0		0.0	0	
AR	51,771.6	330,943	43.23	598.7	0	
CA	335.0	13,315	43.23	96.0	0	
CO	183.2	0		921.5	0	
CT	0.0	0		0.0	0	
DE	21.6	0		14.0	0	
FL	40,482.5	217,821	43.14	160.8	0	
GA	87,881.1	476,813	43.16	1,783.8	0	
HI	*	*	*	*	*	*
ID	2,013.6	25,795	43.23	952.8	0	
IL	16,696.9	270,773	43.69	6,340.2	0	
IN	8,998.4	169,535	43.23	1,140.7	0	
IA	9,202.8	181,097	44.36	5,234.5	0	
KS	988.9	5,111	39.90	5,671.5	0	
KY	1,634.9	19,943	42.43	1,375.5	0	
LA	38,804.8	294,214	38.77	1,720.0	0	
ME	610.0	0		1.6	0	
MD	442.9	5,899	42.44	82.9	0	
MA	0.0	0		0.0	0	
MI	5,454.5	60,147	42.93	1,701.7	0	
MN	17,445.4	140,174	44.84	4,809.1	0	
MS	344,006.3	2,106,052	43.02	4,230.9	0	
MO	9,834.1	250,423	43.23	4,003.3	0	
MT	816.2	813	43.23	3,395.1	0	
NE	2,372.1	30,660	45.44	2,689.5	0	
NV	*	*	*	*	*	*
NH	0.0	0		0.0	0	
NJ	22.4	968	43.23	7.0	0	
NM	30.0	0		24.0	0	
NY	943.0	23,180	73.15	62.2	0	
NC	31,471.4	167,141	42.79	51.2	0	
ND	1,236.5	4,384	43.23	4,731.8	0	
OH	4,896.9	53,815	42.59	927.9	0	
OK	295.3	1,749	34.03	1,542.6	0	
OR	1,139.0	695	24.30	188.9	0	
PA	376.6	7,412	63.19	1,753.9	0	
PR	121.0	0		0.0	0	
RI	*	*	*	*	*	*
SC	72,457.3	139,165	28.42	490.1	0	
SD	1,199.2	13,453	43.23	9,118.8	0	
TN	16,513.4	90,774	43.23	446.6	0	
TX	5,328.9	23,617	43.23	5,696.2	0	
UT	0.0	0		38.6	0	
VT	0.0	0		0.0	0	
VA	10,810.4	46,697	43.23	34.7	0	
WA	1,139.6	14,058	44.06	1,049.3	0	
WV	4.0	0		0.3	0	
WI	25,269.4	400,418	43.23	2,958.3	0	
WY	56.3	0		130.8	0	
US	951,783	6,105,628	42.42	77,701	0	

[1] Not including acres which receive no cost share. * Data withheld to avoid disclosure of individual operations. Note: Total acres treated may not add due to rounding.
FSA, Conservation and Environmental Programs Division, (530) 792-5594.

Table 12-6.—Conservation Reserve Program (CRP): Enrollment by practice, under contract, January 2010
(CP 15, CP 16, and CP 17)

State	CP 15 Contour grass strips			CP 16 Shelter belts			CP 17 Living snow fences		
	Total acres treated	Total cost share	Avg cost share per acre treated[1]	Total acres treated	Total cost share	Avg cost share per acre treated[1]	Total acres treated	Total cost share	Avg cost share per acre treated[1]
AL	51.8	4,390	84.75	0.0	0	....	0.0	0	....
AK	0.0	0	....	0.0	0	....	0.0	0	....
AR	0.0	0	....	0.0	0	....	0.0	0	....
CA	0.0	0	....	0.0	0	....	0.0	0	....
CO	0.0	0	....	4,583.8	4,357,886	1,060.44	35.4	18,880	1,026.09
CT	0.0	0	....	0.0	0	....	0.0	0	....
DE	4.3	1,290	300.00	0.0	0	....	0.0	0	....
FL	0.0	0	....	0.0	0	....	0.0	0	....
GA	8.9	1,544	173.48	0.0	0	....	0.0	0	....
HI	*	*	*	*	*	*	*	*	*
ID	52.2	6,758	129.46	211.5	554,659	2,671.77	63.4	51,844	817.73
IL	1,697.7	88,914	71.61	158.0	36,560	260.03	58.2	17,803	315.10
IN	135.4	11,022	87.48	26.3	6,645	252.66	1.9	171	244.29
IA	19,102.8	632,190	55.23	2,337.4	2,127,889	1,002.26	590.4	135,644	258.96
KS	5,146.3	171,622	49.80	844.5	402,453	558.11	73.8	37,943	555.53
KY	72.3	6,223	98.47	0.0	0	....	0.0	0	....
LA	0.0	0	....	0.0	0	....	0.0	0	....
ME	0.0	0	....	0.0	0	....	0.0	0	....
MD	0.0	0	....	0.0	0	....	0.0	0	....
MA	0.0	0	....	0.0	0	....	0.0	0	....
MI	12.3	2,597	218.24	78.4	14,489	249.38	2.5	900	360.00
MN	1,396.0	94,780	74.15	4,252.0	1,548,679	388.55	3,951.7	778,966	220.65
MS	27.7	60	60.00	0.0	0	....	0.0	0	....
MO	1,579.5	54,866	69.28	58.0	25,237	503.73	0.0	0	....
MT	0.0	0	....	233.4	143,637	670.89	51.7	32,858	635.55
NE	567.6	27,238	57.51	2,432.4	1,238,366	536.65	118.8	42,381	475.12
NV	*	*	*	*	*	*	*	*	*
NH	0.0	0	....	0.0	0	....	0.0	0	....
NJ	0.0	0	....	0.3	175	583.33	0.0	0	....
NM	0.0	0	....	0.0	0	....	0.0	0	....
NY	7.1	1,038	146.20	0.2	422	2,110.00	0.0	0	....
NC	0.0	0	....	13.4	644	67.08	0.0	0	....
ND	0.0	0	....	5,397.0	3,553,520	690.19	582.5	270,088	572.95
OH	15.2	488	44.36	105.0	27,165	275.79	2.8	400	142.86
OK	1.8	0	....	37.1	7,118	191.86	0.0	0	....
OR	0.0	0	....	1.6	710	887.50	0.0	0	....
PA	123.4	19,367	181.68	0.0	0	....	0.0	0	....
PR	0.0	0	....	0.0	0	....	0.0	0	....
RI	*	*	*	*	*	*	*	*	*
SC	0.0	0	....	0.0	0	....	0.0	0	....
SD	73.0	5,211	172.55	16,225.3	12,753,520	830.15	517.9	354,930	784.72
TN	61.3	7,597	123.93	0.0	0	....	0.0	0	....
TX	171.4	9,421	65.29	11.6	5,545	513.43	0.0	0	....
UT	0.0	0	....	0.0	0	....	0.0	0	....
VT	0.0	0	....	0.0	0	....	0.0	0	....
VA	1.3	0	....	0.0	0	....	0.0	0	....
WA	38,447.6	2,266,332	84.32	8.2	24,954	5,545.33	0.0	0	....
WV	0.0	0	....	0.0	0	....	0.0	0	....
WI	1,094.9	102,428	120.83	26.5	8,790	331.70	34.0	8,573	252.15
WY	0.0	0	....	72.4	93,675	1,293.85	3.4	729	214.41
US	69,852	3,515,366	74.66	37,114	26,932,738	775.80	6,088	1,752,110	326.36

[1] Not including acres which receive no cost share.   * Data withheld to avoid disclosure of individual operations.   Note: Total acres treated may not add due to rounding.
FSA, Conservation and Environmental Programs Division, (530) 792-5594.

Table 12-7.—Conservation Reserve Program (CRP): Enrollment by practice, under contract, January 2010
(CP 18 and CP 21)

State	CP 18 Salt tolerant grasses			CP 21 Alternative perennials		
	Total acres treated	Total cost share	Avg cost share per acre treated [1]	Total acres treated	Total cost share	Avg cost share per acre treated [1]
AL	0.0	0	....	706.8	71,414	153.35
AK	0.0	0	....	0.0	0	....
AR	0.0	0	....	5,998.2	436,911	91.34
CA	0.0	0	....	0.0	0	....
CO	77.2	6,621	85.76	354.7	12,330	142.05
CT	0.0	0	....	10.0	1,500	150.00
DE	0.0	0	....	1,282.1	394,545	338.87
FL	0.0	0	....	0.0	0	....
GA	0.0	0	....	390.1	9,453	29.32
HI	*	*	*	*	*	*
ID	0.0	0	....	1,161.8	97,082	112.34
IL	5.7	1,000	714.29	142,031.2	7,567,067	65.14
IN	0.5	85	170.00	60,399.2	7,161,879	143.38
IA	4.4	180	40.91	236,820.2	12,978,482	69.77
KS	870.9	2,409	63.39	31,389.6	1,571,950	59.76
KY	0.0	0	....	32,079.0	3,464,953	137.68
LA	0.0	0	....	628.0	24,948	48.93
ME	0.0	0	....	106.9	31,846	481.79
MD	0.0	0	....	36,644.5	5,072,836	164.83
MA	0.0	0	....	9.9	7,074	714.55
MI	0.0	0	....	46,651.0	6,430,733	149.94
MN	7,025.5	430,225	64.63	158,170.8	10,164,117	76.57
MS	0.0	0	....	8,172.5	589,359	89.29
MO	0.0	0	....	40,933.1	2,378,889	79.41
MT	100,520.1	351,681	13.52	225.4	5,073	25.52
NE	1,034.0	42,485	41.09	21,181.8	1,193,573	62.48
NV	*	*	*	*	*	*
NH	0.0	0	....	45.3	1	0.02
NJ	0.0	0	....	335.5	112,838	358.10
NM	0.0	0	....	0.0	0	....
NY	0.0	0	....	451.5	193,436	453.76
NC	0.0	0	....	5,888.8	1,227,813	222.73
ND	112,340.9	2,482,705	41.84	9,586.2	366,659	49.08
OH	0.0	0	....	75,079.5	5,045,645	79.47
OK	2,316.1	25,160	42.73	700.9	36,059	66.99
OR	0.0	0	....	2,382.3	178,160	127.80
PA	0.0	0	....	1,855.8	410,489	230.87
PR	0.0	0	....	0.0	0	....
RI	*	*	*	*	*	*
SC	0.0	0	....	4,317.8	105,134	63.74
SD	13,635.0	571,614	51.75	9,942.6	457,909	55.85
TN	0.0	0	....	9,480.0	1,086,772	139.86
TX	500.8	20,917	55.65	1,541.3	299,698	253.44
UT	0.0	0	....	38.6	4,465	115.67
VT	0.0	0	....	225.0	43,951	198.78
VA	0.0	0	....	4,891.0	498,247	105.02
WA	24.1	0	....	48,056.2	3,844,048	95.26
WV	0.0	0	....	372.1	49,157	633.47
WI	0.0	0	....	26,328.5	2,736,720	124.48
WY	0.0	0	....	9.4	1,382	147.02
US	238,355	3,935,082	37.42	1,026,875	76,364,597	90.76

[1] Not including acres which receive no cost share.   * Data withheld to avoid disclosure of individual operations.   Note: Total acres treated may not add due to rounding.
FSA, Conservation and Environmental Programs Division, (530) 792-5594.

Table 12-8.—Conservation Reserve Program (CRP): Enrollment by practice, under contract, January 2010

(CP 22, CP 23, and CP 24)

State	CP 22 Riparian buffer			CP 23 Wetland restoration			CP 24 Cross wind trap strips		
	Total acres treated	Total cost share	Avg cost share per acre treated [1]	Total acres treated	Total cost share	Avg cost share per acre treated [1]	Total acres treated	Total cost share	Avg cost share per acre treated [1]
AL	33,725.4	4,726,618	152.62	65.4	4,279	81.97	0.0	0	
AK	184.6	33,371	180.77	0.0	0		0.0	0	
AR	60,724.5	5,516,480	113.80	31,907.5	2,182,069	98.63	0.0	0	
CA	12,960.1	1,998,608	173.27	5,108.3	103,304	22.76	0.0	0	
CO	797.7	836,414	1,090.79	1,155.4	92,791	133.13	31.8	23,065	725.31
CT	35.9	22,887	637.52	0.0	0		0.0	0	
DE	113.6	40,817	359.30	338.1	276,211	889.28	0.0	0	
FL	64.0	0		0.0	0		0.0	0	
GA	1,376.5	749,536	671.87	511.3	11,850	91.79	0.0	0	
HI	*	*	*	*	*	*	*	*	*
ID	7,271.5	3,380,442	507.51	1,259.1	39,202	36.44	0.0	0	
IL	110,824.8	19,721,293	194.19	49,896.3	6,272,316	157.19	0.0	0	
IN	5,853.3	1,282,988	259.64	8,297.2	1,286,528	259.62	0.0	0	
IA	65,669.4	19,703,121	322.28	75,728.8	9,541,875	171.94	15.8	1,570	99.37
KS	3,922.9	286,447	91.07	9,055.5	319,635	51.40	175.4	6,819	44.71
KY	25,061.5	10,044,171	418.21	130.9	22,455	207.53	0.0	0	
LA	5,432.8	533,171	112.96	60,022.2	4,464,168	98.01	0.0	0	
ME	160.1	415,689	2,596.43	0.0	0		0.0	0	
MD	16,621.4	6,006,886	449.39	2,573.7	2,047,334	899.49	0.0	0	
MA	5.0	750	150.00	0.0	0		0.0	0	
MI	3,486.5	1,172,871	355.59	20,084.1	4,927,465	255.16	0.0	0	
MN	47,404.3	8,453,114	195.16	324,017.2	20,258,305	80.11	5.4	485	89.81
MS	165,273.8	10,365,002	71.66	13,992.5	531,180	74.69	0.0	0	
MO	29,039.0	5,830,691	250.54	11,326.1	881,731	133.91	0.0	0	
MT	2,326.3	406,567	180.46	4,301.3	261,736	77.61	0.0	0	
NE	3,245.9	803,028	256.39	13,364.8	329,045	43.44	0.0	0	
NV	*	*	*	*	*	*	*	*	*
NH	14.5	21,642	2,061.14	0.0	0		0.0	0	
NJ	230.2	182,847	794.30	1.0	1,500	1,500.00	0.0	0	
NM	5,280.3	2,022,232	399.43	0.0	0		0.0	0	
NY	13,440.8	10,717,403	890.73	88.6	12,834	175.81	0.0	0	
NC	31,487.8	2,736,842	96.18	2,209.4	335,342	161.47	0.0	0	
ND	527.4	175,263	357.75	618,879.0	8,538,946	27.60	9.5	220	23.16
OH	7,027.7	2,086,343	330.67	9,081.0	4,161,366	585.45	3.5	1,656	473.14
OK	1,518.6	363,575	265.44	2,313.9	127,170	65.65	0.0	0	
OR	35,537.7	17,603,606	617.61	365.3	71,296	219.91	0.0	0	
PA	24,699.1	30,896,282	1,286.12	1,013.0	1,176,066	1,192.89	0.0	0	
PR	716.5	37,171	59.71	0.0	0		0.0	0	
RI	*	*	*	*	*	*	*	*	*
SC	26,909.3	1,434,401	72.75	254.0	4,391	18.86	0.0	0	
SD	5,476.5	3,375,616	682.27	281,767.4	7,148,529	36.94	11.4	969	85.00
TN	6,159.6	1,071,990	190.69	703.2	13,174	45.21	0.0	0	
TX	33,760.1	4,358,584	152.50	9,341.5	248,832	64.78	167.0	5,716	36.18
UT	209.4	83,915	409.74	0.0	0		0.0	0	
VT	2,363.1	2,729,438	1,215.79	0.0	0		0.0	0	
VA	23,783.2	29,195,917	1,317.18	236.3	127,376	540.87	37.7	16,362	434.01
WA	23,031.7	24,125,419	1,131.15	3,379.5	305,656	93.70	0.0	0	
WV	4,426.7	3,218,149	783.02	0.0	0		0.0	0	
WI	16,620.0	5,489,115	339.49	9,821.0	1,167,300	199.56	0.0	0	
WY	5,827.4	1,465,588	252.19	0.0	0		0.0	0	
US	870,700	245,911,228	319.00	1,572,590	77,293,257	76.58	458	56,862	133.60

[1] Not including acres which receive no cost share.   * Data withheld to avoid disclosure of individual operations.   Note: Total acres treated may not add due to rounding.

FSA, Conservation and Environmental Programs Division, (530) 792-5594.

Table 12-9.—Conservation Reserve Program (CRP): Enrollment by practice, under contract, January 2010
(CP 25, CP 26, and CP 27)

State	CP 25 Rare and declining habitat			CP 26 Sediment retention			CP 27 Farmable wetland pilot (wetland)		
	Total acres treated	Total cost share	Avg cost share per acre treated[1]	Total acres treated	Total cost share	Avg cost share per acre treated[1]	Total acres treated	Total cost share	Avg cost share per acre treated[1]
AL	474.1	14,525	101.64	0.0	0	....	0.0	0	....
AK	0.0	0	....	0.0	0	....	0.0	0	....
AR	0.0	0	....	0.0	0	....	0.0	0	....
CA	0.0	0	....	0.0	0	....	0.0	0	....
CO	1,602.5	179,899	115.50	0.0	0	....	2.0	0	....
CT	0.0	0	....	0.0	0	....	0.0	0	....
DE	0.0	0	....	0.0	0	....	0.0	0	....
FL	0.0	0	....	0.0	0	....	0.0	0	....
GA	0.0	0	....	0.0	0	....	0.0	0	....
HI	*	*	*	*	*	*	*	*	*
ID	37.3	4,280	114.74	0.0	0	....	3.5	606	173.14
IL	2,173.0	244,059	112.89	0.0	0	....	137.8	45,174	394.88
IN	1,908.4	218,465	114.74	0.0	0	....	322.3	503,068	1,886.27
IA	79,603.7	9,073,292	121.33	0.0	0	....	22,000.5	5,222,728	269.59
KS	503,151.9	50,517,927	103.04	0.0	0	....	552.0	17,693	120.52
KY	16,937.0	2,173,332	132.19	0.0	0	....	0.0	0	....
LA	0.0	0	....	0.0	0	....	0.0	0	....
ME	0.0	0	....	0.0	0	....	0.0	0	....
MD	264.9	82,959	313.17	0.0	0	....	1.2	1,062	885.00
MA	0.0	0	....	0.0	0	....	0.0	0	....
MI	216.4	24,830	114.74	36.4	112,660	3,095.05	22.5	3,047	507.83
MN	135,481.3	15,396,911	115.29	0.0	0	....	12,764.9	1,730,863	168.22
MS	0.0	0	....	0.0	0	....	0.0	0	....
MO	73,303.9	7,395,060	106.05	0.0	0	....	4.3	2,258	525.12
MT	188,394.8	11,778,461	68.97	0.0	0	....	50.1	103	16.89
NE	130,906.7	12,333,892	99.16	0.0	0	*	1,564.3	68,383	167.77
NV	*	*	*	*	*	*	*	*	*
NH	0.0	0	....	0.0	0	....	0.0	0	....
NJ	0.0	0	....	0.0	0	....	0.0	0	....
NM	0.0	0	....	0.0	0	....	0.0	0	....
NY	0.0	0	....	0.0	0	....	0.0	0	....
NC	0.0	0	....	0.0	0	....	0.0	0	....
ND	6,250.7	716,568	114.64	0.0	0	....	7,513.4	467,987	74.69
OH	6,516.8	887,291	136.15	0.0	0	....	80.1	322,276	4,366.88
OK	25,489.8	2,924,700	114.74	0.0	0	....	9.3	7,732	831.40
OR	7.3	838	114.74	0.0	0	....	0.0	0	....
PA	0.0	0	....	0.0	0	....	0.0	0	....
PR	0.0	0	....	0.0	0	....	0.0	0	....
RI	*	*	*	*	*	*	*	*	*
SC	0.0	0	....	0.0	0	....	0.0	0	....
SD	13,927.4	1,558,462	112.70	0.0	0	....	18,372.2	704,827	71.09
TN	0.0	0	....	0.0	0	....	0.0	0	....
TX	0.0	0	....	0.0	0	....	0.0	0	....
UT	0.0	0	....	0.0	0	....	0.0	0	....
VT	0.0	0	....	0.0	0	....	0.0	0	....
VA	0.0	0	....	0.0	0	....	0.0	0	....
WA	107.6	12,346	114.74	0.0	0	....	0.0	0	....
WV	0.0	0	....	0.0	0	....	0.0	0	....
WI	15,091.8	3,101,473	211.46	0.0	0	....	21.6	3,985	184.49
WY	20.0	617	30.85	0.0	0.0	....	0.0	0	....
US	1,201,867	118,640,186	102.89	36	112,660	3,095.05	63,422	9,101,792	194.05

[1] Not including acres which receive no cost share.   * Data withheld to avoid disclosure of individual operations.   Note: Total acres treated may not add due to rounding.
FSA, Conservation and Environmental Programs Division, (530) 792-5594.

Table 12-10.—Conservation Reserve Program (CRP): Enrollment by practice, under contract, January 2010
(CP 28, CP 29, and CP 30)

State	CP 28 Farmable wetland pilot (buffer)			CP 29 Wildlife habitat buffer (marginal pastureland)			CP 30 Wetland buffer (marginal pastureland)		
	Total acres treated	Total cost share	Avg cost share per acre treat-ed[1]	Total acres treated	Total cost share	Avg cost share per acre treat-ed[1]	Total acres treated	Total cost share	Avg cost share per acre treat-ed[1]
AL	0.0	0		46.0	0		0.0	0	
AK	0.0	0		0.0	0		433.2	147,782	341.14
AR	0.0	0		391.7	158,615	405.46	1.0	6,502	6,502.00
CA	0.0	0		577.7	103,019	216.24	0.0	0	
CO	4.0	0		189.4	44,713	461.43	18.8	8,221	437.29
CT	0.0	0		0.0	0		0.0	0	
DE	0.0	0		0.0	0		0.0	0	
FL	0.0	0		0.0	0		0.0	0	
GA	0.0	0		2.8	9,463	4,301.36	0.0	0	
HI	*	*	*	*	*	*	*	*	*
ID	2.0	62	31.00	219.9	59,700	303.82	171.0	21,145	128.15
IL	256.8	21,014	98.98	232.9	45,584	224.77	23.6	2,368	100.34
IN	610.7	105,117	184.77	78.0	93,668	1,252.25	50.8	9,398	185.00
IA	54,052.2	4,153,820	85.76	9,735.8	3,715,899	442.22	2,315.1	286,289	179.06
KS	1,099.4	42,587	47.50	19.4	4,056	209.07	0.0	0	
KY	0.0	0		73,249.1	11,445,435	159.38	5.1	1,508	295.69
LA	0.0	0		0.0	0		0.0	0	
ME	0.0	0		0.1	7,687	76,870.00	12.2	37,630	3,084.43
MD	3.8	0		687.1	125,236	532.01	5.5	2,592	471.27
MA	0.0	0		0.0	0		0.0	0	
MI	48.2	3,4444	186.16	0.0	0		230.5	106,936	486.29
MN	29,832.9	2,522,493	100.77	872.7	49,032	84.26	4,069.4	299,515	120.83
MS	0.0	0		37.2	4,164	277.60	23.6	2,466	104.49
MO	4.6	2,416	525.22	1,089.7	373,333	388.73	1,988.7	1,286,121	754.37
MT	89.7	2,037	35.43	96.3	5,876	63.80	0.0	0	
NE	2,591.6	166,905	71.91	1,109.1	232,467	234.11	208.8	35,471	197.17
NV	*	*	*	*	*	*	*	*	*
NH	0.0	0		0.0	0		0.0	0	
NJ	0.0	0		0.0	0		0.0	0	
NM	0.0	0		0.0	0		0.0	0	
NY	0.0	0		2,277.9	1,521,768	751.42	920.7	505,913	592.82
NC	0.0	0		50.8	75,058	1,698.14	0.0	0	
ND	25,545.1	1,184,175	55.10	0.0	0		0.0	0	
OH	181.9	16,490	96.72	2,652.7	387,944	150.48	74.3	180,837	2,624.63
OK	20.8	1,947	93.61	6.2	4,324	697.42	8.5	850	100.00
OR	0.0	0		10,366.7	2,249,600	276.27	105.1	115,242	1,096.50
PA	0.0	0		1,167.7	563,297	617.45	441.5	125,552	396.94
PR	0.0	0		961.4	37,170	60.00	0.0	0	
RI	*	*	*	*	*	*	*	*	*
SC	0.0	0		49.9	135,107	2,757.29	86.2	186,212	2,160.23
SD	42,511.2	2,175,565	59.93	4,129.7	308,279	110.43	19,578.8	526,838	52.39
TN	0.0	0		8.5	400	47.06	0.0	0	
TX	0.0	0		2,085.9	356,380	177.00	2.2	2,973	1,351.36
UT	0.0	0		34.5	6,563	321.72	0.0	0	
VT	0.0	0		0.0	0		3.3	9,394	2,846.67
VA	0.0	0		475.4	533,780	1,143.00	20.1	0	0
WA	0.0	0		843.5	281,355	696.42	0.0	0	
WV	0.0	0		0.0	0		0.0	0	
WI	35.3	3,390	125.09	1,175.6	306,531	277.13	30.4	18,332	603.03
WY	0.0	0		745.5	141,497	223.46	0.0	0	
US	156,890	10,401,462	76.73	115,667	23,387,000	219.86	30,828	3,926,087	212.79

[1] Not including acres which receive no cost share.   * Data withheld to avoid disclosure of individual operations.   Note: Total acres treated may not add due to rounding.
    FSA, Conservation and Environmental Programs Division, (530) 792-5594.

Table 12-11.—Conservation Reserve Program (CRP): Enrollment by practice, under contract, January 2010

(CP 31, CP 32, and CP 33)

State	CP 31 Bottomland hardwood			CP 32 Hardwood trees			CP 33 Upland bird habitat buffers		
	Total acres treated	Total cost share	Avg cost share per acre treated[1]	Total acres treated	Total cost share	Avg cost share per acre treated[1]	Total acres treated	Total cost share	Avg cost share per acre treated[1]
AL	800.5	88,520	114.62	0.0	0	....	1,099.4	69,133	112.52
AK	0.0	0	....	0.0	0	....	0.0	0	....
AR	9,857.8	913,330	113.94	392.8	6,000	594.06	5,387.9	471,196	131.19
CA	0.0	0	....	0.0	0	....	0.0	0	....
CO	0.0	0	....	0.0	0	....	0.0	0	....
CT	0.0	0	....	0.0	0	....	0.0	0	....
DE	0.0	0	....	0.0	0	....	0.0	0	....
FL	0.0	0	....	0.0	0	....	0.0	0	....
GA	24.9	2,000	103.63	0.0	0	....	2,197.4	139,467	87.43
HI	*	*	*	*	*	*	*	*	*
ID	0.0	0	....	0.0	0	....	0.0	0	....
IL	2,716.2	528,836	195.60	637.3	0	....	50,626.4	5,695,558	122.23
IN	3,190.3	684,986	260.75	575.4	9,876	50.18	12,413.7	1,727,990	148.93
IA	2,032.2	655,324	359.36	1,550.9	54,415	275.38	23,948.6	3,579,095	170.35
KS	160.3	28,515	271.57	0.0	0	....	37,846.5	1,437,481	44.86
KY	280.0	104,379	372.78	234.0	9	0.50	7,629.9	1,119,775	153.73
LA	24,170.0	2,249,002	102.88	921.7	0	....	439.2	24,290	96.43
ME	0.0	0	....	0.0	0	....	0.0	0	....
MD	3.7	0	....	0.0	0	....	682.5	10,421	149.99
MA	0.0	0	....	0.0	0	....	0.0	0	....
MI	10.8	7,300	675.93	6.1	610	100.00	773.2	93,193	164.94
MN	228.1	12,846	332.80	1,862.0	12,859	82.32	397.6.1	37,478	94.81
MS	9,742.0	687,326	85.04	775.0	0	....	2,230.9	157,144	89.81
MO	935.7	94,705	161.28	545.6	0	....	31,420.6	2,653,595	98.73
MT	0.0	0	....	0.0	0	....	0.0	0	....
NE	8.8	4,778	542.95	0.0	0	....	5,812.6	416,765	77.46
NV	*	*	*	*	*	*	*	*	*
NH	0.0	0	....	0.0	0	....	0.0	0	....
NJ	0.0	0	....	0.0	0	....	0.0	0	....
NM	0.0	0	....	0.0	0	....	0.0	0	....
NY	2.3	2,000	869.57	0.0	0	....	0.0	0	....
NC	14.2	1,978	139.30	0.0	0	....	7,454.7	470,623	83.24
ND	0.0	0	....	0.0	0	....	0.0	0	....
OH	76.1	24,379	320.35	39.2	473	105.11	15,057.3	1,525,588	107.16
OK	415.5	63,174	152.04	79.7	0	....	1,053.8	45,862	58.78
OR	0.0	0	....	0.0	0	....	0.0	0	....
PA	2.0	1,500	750.00	0.0	0	....	0.0	0	....
PR	0.0	0	....	0.0	0	....	0.0	0	....
RI	*	*	*	*	*	*	*	*	*
SC	0.0	0	....	0.0	0	....	5670.0	372,187	70.96
SD	0.0	0	....	0.0	0	....	1,291.9	86,175	72.59
TN	2,828.3	300,723	110.70	0.7	0	....	4,982.9	380,548	85.00
TX	381.1	53,558	140.54	0.0	0	....	4,748.3	584,945	130.63
UT	0.0	0	....	0.0	0	....	0.0	0	....
VT	0.0	0	....	0.0	0	....	0.0	0	....
VA	0.0	0	....	0.0	0	....	1,540.0	104,846	82.13
WA	0.0	0	....	0.0	0	....	0.0	0	....
WV	0.0	0	....	0.0	0	....	0.0	0	....
WI	0.0	0	....	948.3	71	14.49	303.7	53,607	176.51
WY	0.0	0	....	0.0	0	....	0.0	0	....
US	57,881	6,509,159	128.78	8,569	84,313	141.89	225,009	21,347,962	107.89

[1] Not including acres which receive no cost share.   * Data withheld to avoid disclosure of individual operations.   Note: Total acres treated may not add due to rounding.

FSA, Conservation and Environmental Programs Division, (530) 792-5594.

Table 12-12.—Conservation Reserve Program (CRP): Enrollment by practice, under contract, January 2010
(CP 34, CP 36, and CP 37)

State	CP 34 Flood control structure			CP 36 Longleaf pine			CP 37 Duck nesting habitat		
	Total acres treated	Total cost share	Avg cost share per acre treat-ed[1]	Total acres treated	Total cost share	Avg cost share per acre treat-ed[1]	Total acres treated	Total cost share	Avg cost share per acre treat-ed[1]
AL	0.0	0		7,929.7	1,327,671	192.37	0.0	0	
AK	0.0	0		0.0	0		0.0	0	
AR	0.0	0		0.0	0		0.0	0	
CA	0.0	0		0.0	0		0.0	0	
CO	0.0	0		0.0	0		0.0	0	
CT	0.0	0		0.0	0		0.0	0	
DE	0.0	0		0.0	0		0.0	0	
FL	0.0	0		740.7	130,366	185.26	0.0	0	
GA	0.0	0		66,807.6	12,437,572	288.16	0.0	0	
HI	*	*		*	*	*	*	*	*
ID	0.0	0		0.0	0		0.0	0	
IL	0.0	0		0.0	0		0.0	0	
IN	0.0	0		0.0	0		0.0	0	
IA	0.0	0		0.0	0		0.0	0	
KS	0.0	0		0.0	0		0.0	0	
KY	0.0	0		0.0	0		0.0	0	
LA	0.0	0		58.6	7,030	119.97	0.0	0	
ME	0.0	0		0.0	0		0.0	0	
MD	0.0	0		0.0	0		0.0	0	
MA	0.0	0		0.0	0		0.0	0	
MI	0.0	0		0.0	0		77.0	23,331	303.00
MN	71.0	5,825	113.11	0.0	0		5,198.2	406,811	92.44
MS	0.0	0		397.1	37,880	95.39	0.0	0	
MO	0.0	0		0.0	0		0.0	0	
MT	0.0	0		0.0	0		28.2	113	4.01
NE	0.0	0		0.0	0		0.0	0	
NV	*	*		*	*	*	*	*	*
NH	0.0	0		0.0	0		0.0	0	
NJ	0.0	0		0.0	0		0.0	0	
NM	0.0	0		0.0	0		0.0	0	
NY	0.0	0		0.0	0		0.0	0	
NC	0.0	0		4,205.3	604,424	192.87	0.0	0	
ND	0.0	0		0.0	0		46,929.6	1,201,190	40.05
OH	0.0	0		0.0	0		0.0	0	
OK	0.0	0		0.0	0		0.0	0	
OR	0.0	0		0.0	0		0.0	0	
PA	0.0	0		0.0	0		0.0	0	
PR	0.0	0		0.0	0		0.0	0	
RI	*	*	*	*	*	*	*	*	*
SC	0.0	0		3,887.4	584,104	154.90	0.0	0	
SD	0.0	0		0.0	0		46,955.2	1,233,937	32.09
TN	0.0	0		0.0	0		0.0	0	
TX	0.0	0		0.0	0		0.0	0	
UT	0.0	0		0.0	0		0.0	0	
VT	0.0	0		0.0	0		0.0	0	
VA	0.0	0		384.9	53,594	139.24	0.0	0	
WA	0.0	0		0.0	0		0.0	0	
WV	0.0	0		0.0	0		0.0	0	
WI	0.0	0		0.0	0		0.0	0	
WY	0.0	0		0.0	0		0.0	0	
US	71	5,825	113.11	84,411.3	15,182,641	259.47	99,188	2,865,382	39.28

[1] Not including acres which receive no cost share.   * Data withheld to avoid disclosure of individual operations.   Note: Total acres treated may not add due to rounding.
FSA, Conservation and Environmental Programs Division, (530) 792-5594.

Table 12-13.—Conservation Reserve Program (CRP): Enrollment by practice, under contract, January 2010

(CP 38 and CP 39)

State	CP 38 State acres for wildlife enhancement			CP 39 Constructed wetlands		
	Total acres treated	Total cost share	Avg cost share per acre treat-ed [1]	Total acres treated	Total cost share	Avg cost share per acre treat-ed [1]
Alabama	2,050.5	239,997	151.61	0.0	0	
Alaska	0.0	0		0.0	0	
Arkansas	10,326.6	484,469	156.94	0.0	0	
California	0.0	0		0.0	0	
Colorado	10,915.1	576,802	75.44	0.0	0	
Connecticut	0.0	0		0.0	0	
Delaware	0.0	0		0.0	0	
Florida	0.0	0		0.0	0	
Georgia	3,020.9	2,022	26.46	0.0	0	
Hawaii	*	*	*	*	*	*
Idaho	11,002.3	415,494	43.67	0.0	0	
Illinois	7,160.4	1,217,457	222.96	0.0	0	
Indiana	12,220.3	1732,431	172.32	0.0	0	
Iowa	29,763.5	4,717,524	206.82	0.0	0	
Kansas	6,840.3	260,003	56.98	0.0	0	
Kentucky	7,671.9	722,193	264.97	0.0	0	
Louisiana	215.4	21,000	150.00	0.0	0	
Maine	2,136.7	384,708	418.57	0.0	0	
Maryland	0.0	0		0.0	0	
Massachusetts	0.0	0		0.0	0	
Michigan	2,796.9	566,325	357.53	0.0	0	
Minnesota	25,450.4	2,543,821	119.79	13.6	6,172	453.82
Mississippi	4,315.2	766,424	347.38	0.0	0	
Missouri	16,521.8	2,904,149	272.78	0.0	0	
Montana	17,476.9	2,831,606	241.00	0.0	0	
Nebraska	24,283.4	1,615,674	84.82	0.0	0	
Nevada	*	*	*	*	*	*
New Hampshire	0.0	0		0.0		
New Jersey	180.5	46,269	289.72	0.0	0	
New Mexico	0.0	0		0.0	0	
New York	389.7	19,980	162.70	0.0	0	
North Carolina	347.8	51,367	201.04	0.0	0 D	
North Dakota	27,336.5	367,926	41.28	0.0	0	
Ohio	2,459.0	325,808	193.75	0.0	0	
Oklahoma	1,437.8	88,564	61.60	0.0	0	
Oregon	0.0	0		0.0	0	
Pennsylvania	0.0	0		0.0	0	
Puerto Rico	0.0	0		0.0	0	
Rhode Island	*	*	*	*	*	*
South Carolina	301.8	39,041	228.04	0.0	0	
South Dakota	54,202.4	1,237,989	39.55	0.0	0	
Tennessee	3,107.5	432,728	149.91	0.0	0	
Texas	6,196.8	331,202	56.25	0.0	0	
Utah	0.0	0		0.0	0	
Vermont	0.0	0		0.0	0	
Virginia	126.2	0			0	
Washington	1,736.2	252,797	192.17	0.0	0	
West Virginia	0.0	0				
Wisconsin	4,330.3	762,877	198.27	0.0	0	
Wyoming	0.0	0		0.0	0	
United States, total	296,321	26,012,647	134.51	17	6,172	453.82

[1] Not including acres which receive no cost share.   * Data withheld to avoid disclosure of individual operations.   Note: Total acres treated may not add due to rounding.
FSA, Conservation and Environmental Programs Division, (530) 792-5594

Table 12-14.—Conservation Reserve Program (CRP): Enrollment by practice, under contract, January 2010
(CP 40 and CP 41)

State	CP 40 Aquaculture wetlands			CP 41 Flooded praire wetlands		
	Total acres treated	Total cost share	Avg cost share per acre treated[1]	Total acres treated	Total cost share	Avg cost share per acre treated[1]
Alabama	0.0	0		0.0	0.0	
Alaska	0.0	0		0.0	0	
Arkansas	0.0	0		0.0	0	
California	0.0	0		0.0	0	
Colorado	0.0	0		0.0	0	
Connecticut	0.0	0		0.0	0	
Delaware	0.0	0		0.0	0	
Florida	0.0	0		0.0	0	
Georgia	0.0	0		0.0	0	
Hawaii	*	*	*	*	*	*
Idaho	0.0	0		0.0	0	
Illinois	0.0	0		0.0	0	
Indiana	0.0	0		0.0	0	
Iowa	0.0	0		33.5	9,477	282.90
Kansas	0.0	0		0.0	0	
Kentucky	0.0	0		0.0	0	
Louisiana	1,123.0	0		0.0	0	
Maine	0.0	0		0.0	0	
Maryland	0.0	0		0.0	0	
Massachusetts	0.0	0		0.0	0	
Michigan	0.0	0		0.0	0	
Minnesota	0.0	0		82.4	6,298	81.58
Mississippi	2,719.9	0		0.0	0	
Missouri	206.5	0		0.0	0	
Montana	0.0	0		0.0	0	
Nebraska	0.0	0		0.0	0	
Nevada	*	*	*	*	*	*
New Hampshire	0.0	0		0.0		
New Jersey	0.0	0		0.0	0	
New Mexico	0.0	0		0.0	0	
New York	0.0	0		0.0	0	
North Carolina	0.0	0		0.0	0	
North Dakota	0.0	0		2,956.8	35,032	51.77
Ohio	0.0	0		0.0	0	
Oklahoma	0.0	0		0.0	0	
Oregon	0.0	0		0.0	0	
Pennsylvania	0.0	0		0.0	0	
Puerto Rico	0.0	0		0.0	0	
Rhode Island	*	*	*	*	*	*
South Carolina	0.0	0		0.0	0	
South Dakota	0.0	0		673.1	28,198	66.60
Tennessee	0.0	0		0.0	0	
Texas	0.0	0		0.0	0	
Utah	0.0	0		0.0	0	
Vermont	0.0	0		0.0	0	
Virginia	0.0	0		0.0	0	
Washington	0.0	0		0.0	0	
West Virginia	0.0	0		0.0	0	
Wisconsin	0.0	0		0.0	0	
Wyoming	0.0	0		0.0	0	
United States, total	4,046.0	0		3,746.0	79,005	65.25

[1] Not including acres which receive no cost share.   * Data withheld to avoid disclosure of individual operations.   Note: Total acres treated may not add due to rounding.
  FSA, Conservation and Environmental Programs Division, (530) 792-5594

Table 12-15.—Emergency Conservation Program: Assistance, fiscal years 1999–2009 [1]

Year	Emergency Conservation Program
1999	40,226
2000	97,970
2001	55,246
2002	32,601
2003	37,548
2004	22,480
2005	56,376
2006	58,973
2007	30,754
2008	27,845
2009	28,483

[1] Totals are from unrounded data.
  FSA, Conservation and Environmental Protection Division, (202) 720-0048.

Table 12-16.—Conservation Reserve Program (CRP): Enrollment by State, January 2010

State [1]	Number of contracts	Number of farms	Acres	Annual rent ($1,000)	Avg Payments [2] ($1 acre)
AL	9,459	6,772	418,528.2	19,196	45.86
AK	61	40	25,823.3	900	34.83
AZ	*	*	*	*	*
AR	5,887	3,272	247,863.3	13,931	56.20
CA	508	398	125,705.0	4,435	35.28
CO	11,709	5,776	2,022,708.0	65,993	32.63
CT	19	17	175.8	14	80.32
DE	672	355	6,784.9	745	109.76
FL	1,444	1,172	62,523.4	2,455	39.26
GA	9,001	6,513	321,442.3	14,369	44.70
HI	*	*	*	*	*
ID	5,501	3,116	718,125.1	30,174	42.02
IL	79,796	44,045	1,020,912.7	112,001	109.71
IN	37,612	21,199	287,496.3	30,122	104.77
IA	103,801	52,724	1,637,313.4	196,327	119.91
KS	46,467	26,760	2,781,280.3	110,957	39.89
KY	17,999	9,775	382,556.1	40,437	105.70
LA	4,860	3,116	318,997.7	18,042	56.56
ME	761	508	20,601.9	1,084	52.64
MD	6,327	3,493	78,534.6	10,454	133.12
MA	4	4	14.9	3	172.55
MI	15,297	8,826	232,962.4	19,929	85.54
MN	61,977	32,892	1,645,162.0	107,892	65.58
MS	19,937	12,662	861,096.6	39,260	45.59
MO	36,058	21,147	1,391,755.5	98,227	70.58
MT	16,141	6,261	3,080,767.4	99,774	32.39
NE	28,347	15,982	1,093,129.4	65,439	59.86
NV	*	*	*	*	*
NH	6	6	59.8	3	54.82
NJ	254	180	2,497.0	169	67.75
NM	2,369	1,493	543,127.7	17,839	32.84
NY	2,872	2,047	54,210.7	3,755	69.26
NC	8,428	5,437	122,306.2	8,123	66.42
ND	34,147	16,880	2,717,638.9	94,898	34.92
OH	37,256	20,908	343,304.4	39,673	115.56
OK	7,428	5,071	861,313.3	28,429	33.01
OR	4,199	2,209	547,644.1	27,335	49.91
PA	11,988	7,564	221,163.0	22,638	102.36
PR	22	21	2,051.9	132	64.40
RI	*	*	*	*	*
SC	7,992	4,586	172,794.4	6,481	37.51
SD	29,697	14,195	1,106,182.4	55,442	50.12
TN	7,564	5,058	217,649.3	13,697	62.93
TX	21,157	15,669	3,301,885.1	117,017	35.44
UT	829	515	145,422.8	4,468	30.73
VT	351	250	2,702.6	267	98.86
VA	5,640	4,321	62,784.2	3,698	58.90
WA	12,198	5,033	1,443,100.9	79,492	55.08
WV	403	332	5,389.1	391	72.62
WI	25,562	15,791	428,837.4	32,909	76.74
WY	901	607	209,142.7	5,800	27.73
National	740,914	415,004	31,293,691	1,664,823	53.20

[1] State in which land is located. [2] Payments scheduled to be made October 2010. * Data withheld to avoid disclosure of individual operations.

FSA, Conservation and Environmental Programs Division, (530) 792-5594.

Table 12-17.—Small watershed protection and flood prevention projects: Accomplishments for years ending Sept. 30, 1994–98

Item	Unit of measure	1994	1995	1996	1997	1998
Small watershed protection: [1]						
Land treatment: [2]						
Forest land	Acres	38,322	16,806	1,905	2,193	8,402
Cropland	......do	501	626	0	1,160	741
Pastureland	......do	170	28	7,284	45	88
Total land treatment	......do	38,993	17,460	9,189	3,398	9,233
Land owners assisted	Number	3,534	1,483	1,465	1,348	1,186
Flood prevention: [3]						
Land treatment: [2]						
Forest land	Acres	2,196	6,335	63,028	8,682	6,541
Cropland	......do			575	1,668	20
Pastureland	......do		40	83	92	78
Total land treatment	......do	2,196	6,375	63,686	10,442	6,639
Land owners assisted	Number	1,452	1,528	2,461	2,265	1,183

[1] As authorized by the Watershed Protection and Flood Prevention Act of 1954 (Public Law 83–566), as amended. Accomplishments are limited to activities accomplished solely by small watershed protection program funds.   [2] Reported in land use categories consistent with those reported by the National Resources Conservation Service.   [3] As authorized by the Navigation and Flood Control Act of 1944 (Public Law 78–534), as amended. Accomplishments are limited to activities accomplished solely by small watershed protection program funds.

FS, Timber Demand and Technology Assessment, RWU-4851, (608) 231–9376.

Table 12-18.—Tree planting: Acres seeded and acres of tree planting, in States and Territories, fiscal year 2002

State or other area	Total	Federal lands			Non-federal public lands [1]	Private [2] lands
		Total	National Forest System	Other [3]		
	Acres	Acres	Acres	Acres	Acres	Acres
AL	69,725	1,691	986	705	30	68,004
AK	2,086	333	329	4	534	1,219
AZ	342	56	56	0	0	286
AR	25,768	1,919	1,919	0	5,696	18,153
CA	17,396	15,667	15,649	18	0	1,729
CO	4,493	774	773	1	0	3,719
CT	88	4	0	4	8	76
DE	1,772	0	0	0	45	1,727
FL	88,665	7,895	4,374	3,521	5,791	74,979
GA	193,905	2,371	266	2,105	446	191,088
HI	1,379	0	0	0	14	1,365
ID	18,224	11,464	11,464	0	3,496	3,264
IL	69,625	1,525	1,525	0	100	68,000
IN	8,096	97	42	55	143	7,856
IA	13,387	0	0	0	127	13,260
KS	1,863	7	0	7	0	1,856
KY	5,406	39	36	3	50	5,317
LA	117,608	953	908	45	11,224	105,431
ME	236	0	0	0	126	110
MD	20,849	3	0	3	136	20,710
MA	20	0	0	0	0	20
MI	6,772	3,844	3,812	32	2,499	429
MN	24,704	3,472	3,472	0	9,750	11,482
MS	222,401	3,179	3,036	143	1,896	217,326
MO	15,357	267	231	36	1,052	14,038
MT	9,386	8,651	8,651	0	735	0
NE	584	0	0	0	0	584
NV	346	40	40	0	104	202
NH	74	0	0	0	15	59
NJ	1,086	1	0	1	25	1,060
NM	1,262	135	135	0	0	1,127
NY	4,136	0	0	0	1,848	2,288
NC	85,049	1,293	467	826	440	83,316
ND	16,719	13	0	13	13	16,693
OH	1,962	138	138	0	73	1,751
OK	7,875	25	0	25	120	7,730
OR	38,638	13,914	13,914	0	4,694	20,030
PA	2,214	153	153	0	1,279	782
RI	0	0	0	0	0	0
SC	77,056	1,116	83	1,033	2,455	73,485
SD	10,301	0	0	0	68	10,233
TN	5,920	543	444	99	613	4,764
TX	40,474	493	183	310	840	39,141
UT	2,951	1,871	1,277	594	0	1,080
VT	263	43	40	3	0	220
VA	67,518	193	54	139	246	67,079
WA	45,771	7,974	7,637	337	13,227	24,570
WV	1,755	0	0	0	15	1,740
WI	17,529	1,435	1,410	25	1,208	14,886
WY	1,308	457	457	0	0	851
State totals	1,370,344	94,048	83,961	10,087	71,181	1,205,115
PR	5,283	0	0	0	0	5,283
Other [4]	158	0	0	0	53	105
Total	1,375,785	94,048	83,961	10,087	71,234	1,210,503

[1] State forest, other State, and other public agencies lands. [2] Forest industry, other industry, and nonindustrial lands. [3] U.S. Department of Interior and Indian Reservations, and other federal lands. [4] Guam and the Trust Territories of the Pacific Islands.

FS, Timber Demand and Technology Assessment, RWU-4851, (608) 231-9376.

Table 12-19.—Forest land: Total forest land and area and ownership of timberland, by regions, Jan. 1, 2007 [1]

Region	Total forest land [2]	Timberland [3]							
		All owner-ships	Federal			State, county, and mu-nicipal	Private		
			Total	National forest	Other		Total	Forest industry	Farmer and other private [4]
	1,000 acres	1,000 acres	1,000 acres	1,000 acres	1,000 acres	1,000 acres	1,000 acres	1,000 acres	1,000 acres
Northeast	85,796	79,803	2,971	2,401	570	9,308	67,523	20,860	46,663
North Central	87,243	84,215	8,926	7,725	1,201	15,944	59,345	7,297	52,048
North	172,039	164,018	11,897	10,126	1,771	25,252	126,868	28,157	98,711
Southeast	87,889	85,665	7,559	4,969	2,590	4,689	73,417	24,711	48,706
South Central	126,756	118,365	9,606	7,256	2,350	3,191	105,569	32,291	73,278
South	214,645	204,030	17,165	12,225	4,940	7,880	178,986	57,002	121,984
Great Plains	5,757	5,287	1,294	1,056	238	198	3,795	79	3,716
Intermountain	144,905	65,681	47,318	44,330	2,988	2,987	15,375	3,870	11,505
Rocky Mountains	150,662	70,968	48,612	45,386	3,226	3,185	19,170	3,949	15,221
Alaska	126,869	11,865	4,750	3,772	978	4,344	2,771	2,022	749
Pacific Northwest	52,449	43,489	20,403	17,938	2,465	3,704	19,383	10,681	8,702
Pacific Southwest [5]	34,565	19,843	9,907	9,275	632	629	9,308	4,320	4,988
Pacific Coast	213,883	75,197	35,060	30,985	4,075	8,677	31,462	17,023	14,439
All regions	751,229	514,213	112,734	98,722	14,012	44,994	356,486	106,131	250,355

[1] Data may not add to totals because of rounding.   [2] Forest land is land at least 10 percent stocked by forest trees of any size, including land that formerly had such tree cover and that will be naturally or artificially regenerated. Forest land includes transition zones, such as areas between heavily forested and nonforested lands that are at least 10 percent stocked with forest trees, and forest areas adjacent to urban and built-up lands. Also included are pinyon-juniper and chaparral areas in the West and afforested areas. The minimum area for classification of forest land is 1 acre. Roadside, streamside, and shelterbelt strips of timber must have a crown width at least 120 feet wide to qualify as forest land. Unimproved roads and trails, streams, and clearings in forest areas are classified as forest if less than 120 feet in width.   [3] Timberland is forest land that is producing or is capable of producing crops of industrial wood and that is not withdrawn from timber utilization by statute or administrative regulation. Areas qualifying as timberland have the capability of producing more than 20 cubic feet per acre per year of industrial wood in natural stands. Currently inaccessible and inoperable areas are included.   [4] Includes Indian lands.   [5] Includes Hawaii.

FS, Timber Demand and Technology Assessment, RWU-4851, (608) 231–9376.

Table 12-20.—Timber volume: Net volume of growing stock and sawtimber on timberland, by softwoods and hardwoods, and regions, 2007 [1]

Region	Growing stock [2]			Sawtimber [3]		
	All species	Softwoods	Hardwoods	All species	Softwoods	Hardwoods
	Million cubic feet	*Million cubic feet*	*Million cubic feet*	*Million board feet*	*Million board feet*	*Million board feet*
Northeast	137,585	34,252	103,333	145,976	36,805	109,171
North Central	110,422	21,614	88,808	122,328	22,896	99,432
North	248,007	55,866	192,141	268,304	59,701	208,603
Southeast	126,747	56,722	70,025	142,582	58,462	84,120
South Central	161,775	61,749	100,026	182,486	64,226	118,260
South	288,522	118,471	170,051	325,068	122,688	202,380
Great Plains	4,539	1,641	2,898	6,591	1,812	4,779
Intermountain	137,724	123,168	9,556	152,549	141,869	10,680
Rocky Mountains	137,263	124,809	12,454	159,140	143,681	15,459
Alaska	31,998	29,125	2,873	34,267	31,191	3,076
Pacific Northwest	158,896	146,006	12,890	159,047	146,048	12,999
Pacific Southwest [4]	67,410	54,926	12,484	67,580	54,983	12,597
Pacific Coast	258,304	230,057	28,247	260,894	232,222	28,672
All regions	932,096	529,203	402,893	1,013,406	558,292	455,114

[1] Data may not add to totals because of rounding.　　[2] Live trees of commercial species meeting specified standards of quality or vigor. Cull trees are excluded. Includes only trees 5.0-inches diameter or larger at 4½ feet above ground.　　[3] Live trees of commercial species containing at least one 12-foot sawlog or two noncontiguous 8-foot logs, and meeting regional specifications for freedom from defect. Softwood trees must be at least 9.0-inches diameter and hardwood trees must be at least 11.0-inches diameter at 4½ feet above ground.　　[4] Includes Hawaii.

FS, Timber Demand and Technology Assessment, RWU-4851, (608) 231–9376.

Table 12-21.—Timber removals: Roundwood product output, logging residues and other removals from growing stock and other sources, by softwoods and hardwoods, 2006 [1]

Roundwood products, logging residues, and other removals	All sources			Growing stock [2]			Other sources [3]		
	All species	Soft-woods	Hard-woods	All species	Soft-woods	Hard-woods	All species	Soft-woods	Hard-woods
	Million cubic feet	Million cubic feet	Million cubic feet	Million cubic feet	Million cubic feet	Million cubic feet	Million cubic feet	Million cubic feet	Million cubic feet
Roundwood products:									
Sawlogs	7,179	5,289	1,890	6,781	5,030	1,752	398	260	138
Pulpwood	4,394	2,634	1,760	3,872	2,345	1,527	522	289	233
Veneer logs	1,211	1,068	143	1,156	1,020	136	55	48	7
Other products [4]	255	215	40	217	183	35	37	32	5
Fuelwood [5]	1,408	477	931	490	86	404	918	391	526
Total	14,447	9,684	4,763	12,517	8,663	3,854	1,930	1,021	909
Logging residues [6]	4,543	2,253	2,290	1,253	552	701	3,290	1,700	1,589
Other removals [7]	1,658	489	1,170	1,278	409	869	380	80	301
Total	6,201	2,741	3,460	2,531	962	1,569	3,670	1,780	1,890

[1] Data may not add to totals because of rounding. [2] Includes live trees of commercial species meeting specified standards of quality or vigor. Cull trees are excluded. Includes only trees 5.0-inches diameter or larger at 4½ feet above ground. [3] Includes salvable dead trees, rough and rotten trees, trees of noncommercial species, trees less than 5.0-inches diameter at 4½ feet above ground, tops, and roundwood harvested from nonforest land (for example, fence rows). [4] Includes such items as cooperage, pilings, poles, posts, shakes, shingles, board mills, charcoal and export logs. [5] Downed and dead wood volume left on the ground after trees have been cut on timberland. [6] Net of wet rot or advanced dry rot, and excludes old punky logs; consists of material sound enough to chip; excludes stumps and limbs. [7] Unutilized wood volume from cut or otherwise killed growing stock, from nongrowing stock sources on timberland (for example, precommercial thinnings), or from timberland clearing. Does not include volume removed from inventory through reclassification of timberland to reserved timberland.
FS, Timber Demand and Technology Assessment, RWU-4851, (608) 231–9376.

Table 12-22.—Timber growth, removals and mortality: Net annual growth, removals, and mortality of growing stock on timberland by softwoods and hardwoods and regions, 2006 [1]

Region	Growth [2]			Removals [3]			Mortality [4]		
	All species	Soft-woods	Hard-woods	All species	Soft-woods	Hard-woods	All species	Soft-woods	Hard-woods
	Million cubic feet	Million cubic feet	Million cubic feet	Million cubic feet	Million cubic feet	Million cubic feet	Million cubic feet	Million cubic feet	Million cubic feet
Northeast	3,249	836	2,412	1,169	353	815	935	300	636
North Central	3,327	652	2,675	1,651	324	1,328	1,098	247	851
North	6,576	1,489	5,087	2,820	677	2,034	1,683	547	1,487
Southeast	6,115	3,876	2,239	4,306	2,961	1,345	1,192	611	581
South Central	7,157	3,756	3,401	5,391	3,357	2,034	1,668	754	913
South	13,272	7,632	5,640	9,696	6,317	3,379	2,860	1,366	1,494
Great Plains	72	27	45	41	25	16	54	11	43
Intermountain	1,689	1,550	139	502	496	6	1,310	1,227	83
Rocky Mountains	1,761	1,577	184	543	521	22	1,364	1,238	126
Alaska	248	130	118	66	59	7	256	236	20
Pacific Northwest	3,340	3,039	301	1,939	1,818	121	950	836	114
Pacific Southwest [5]	1,548	1,374	174	469	466	3	363	288	75
Pacific Coast	5,135	4,543	593	2,474	2,344	131	1,569	1,360	209
All regions	26,744	15,241	11,503	15,533	9,859	5,675	7,826	4,511	3,316

[1] Data may not add to totals because of rounding. [2] The net increase in the volume of trees during a specified year. Components include the increment in net volume of trees at the beginning of the specific year surviving to its end, plus the net volume of trees reaching the minimum size class during the year, minus the volume of trees that died during the year, and minus the net volume of trees that became cull trees during the year. [3] The net volume of trees removed from the inventory during a specified year by harvesting, cultural operations such as timber stand improvement, or land clearing. [4] The volume of sound wood in trees that died from natural causes during a specified year. [5] Includes Hawaii.
FS, Timber Demand and Technology Assessment, RWU-4851, (608) 231–9376.

Table 12-23.—Timber volume: Net volume of sawtimber on timberland in the West, by regions and species, Jan. 1, 2007[1]

Species	Total West	Inter-mountain	Alaska	Pacific Northwest	Pacific South-west[2]	Great Plains
	Million board feet	Million board feet	Million board feet	Million board feet	Million board feet	Million board feet
Softwoods:						
Douglas-fir	124,628	30,504	....	75,516	18,608	....
Ponderosa and Jeffrey pines	41,589	17,383	....	12,420	10,379	1,407
True fir	53,046	23,024	6	17,213	12,803	....
Western hemlock	33,940	941	11,224	21,697	78	....
Sugar pine	3,394	....	....	677	2,717	....
Western white pine	1,162	443	....	436	283	....
Redwood	4,711	....	....	1	4,710	....
Sitka spruce	10,233	....	8,641	1,486	106	....
Engelmann and other spruces	25,128	18,934	4,287	1,889	18	....
Western larch	6,099	3,961	3	2,135	....	....
Incense cedar	4,031	....	....	695	3,336	....
Lodgepole pine	26,537	21,855	81	3,678	923	....
Western redcedar	....	....	....	....	....	....
Other	20,369	6,123	4,884	8,164	964	234
Total	354,867	123,168	29,126	146,007	54,925	1,641
Hardwoods:						
Cottonwood and aspen	12,163	9,198	843	969	124	1,029
Red alder	6,791	68	73	6,317	333	....
Oak	7,427	18	....	777	6,068	564
Other	14,317	272	1,957	4,826	5,957	1,305
Total	40,698	9,556	2,873	12,889	12,482	2,898
All species	395,565	132,724	31,999	158,896	67,407	4,539

[1] International ¼-inch rule. Data may not add to totals because of rounding.　[2] Includes Hawaii.
FS, Timber Demand and Technology Assessment, RWU-4851, (608) 231–9376.

Table 12-24.—Timber volume: Net volume of sawtimber on timberland in the East, by regions and species, Jan. 1, 2007[1]

Species	Total East	North			South		
		Total	Northeast	North Central	Total	Southeast	South Central
	Million board feet	Million board feet	Million board feet	Million board feet	Million board feet	Million board feet	Million board feet
Softwoods:							
Longleaf and slash pines	16,830	....	....	....	16,830	12,212	4,618
Loblolly and shortleaf pines	84,313	1,584	658	926	82,729	32,873	49,856
Other yellow pines	8,979	1,984	1,605	379	6,995	4,907	2,088
White and red pines	21,456	18,759	11,093	7,666	2,697	2,180	517
Jack pine	1,172	1,172	3	1,169	....	....	....
Spruce and balsam fir	13,599	13,554	9,413	4,141	45	45	....
Eastern hemlock	10,509	9,558	8,281	1,277	951	502	449
Cypress	6,543	13	6	7	6,530	3,529	3,001
Other	10,928	9,240	3,193	6,047	1,688	474	1,214
Total	174,329	55,864	34,252	21,612	118,465	56,722	61,743
Hardwoods:							
Select white oaks	34,050	15,375	5,395	9,980	18,675	7,056	11,619
Select red oaks	25,241	16,236	9,775	6,461	9,005	3,190	5,815
Other white oaks	21,759	7,075	4,785	2,290	14,684	5,300	9,384
Other red oaks	44,029	13,149	5,141	8,008	30,880	11,338	19,542
Hickory	21,023	8,334	3,499	4,835	12,689	3,591	9,098
Yellow birch	4,231	4,162	3,355	807	69	58	11
Hard maple	24,409	22,100	12,696	9,404	2,309	470	1,839
Soft maple	38,480	30,239	20,418	9,821	8,241	5,149	3,092
Beech	8,350	6,075	4,922	1,153	2,275	770	1,505
Sweetgum	19,361	877	658	219	18,484	7,637	10,847
Tupelo and black gum	11,583	985	697	288	10,598	6,005	4,593
Ash	17,258	12,086	5,881	6,205	5,172	1,545	3,627
Basswood	5,741	5,031	1,846	3,185	710	316	394
Yellow-poplar	27,826	8,218	5,780	2,438	19,608	12,009	7,599
Cottonwood and aspen	15,451	14,744	3,740	11,004	707	99	608
Black walnut	2,563	1,912	358	1,554	651	196	455
Black cherry	8,964	7,881	5,688	2,193	1,083	427	656
Other	31,871	17,660	8,696	8,964	14,211	4,866	9,345
Total	362,190	192,139	103,330	88,809	170,051	70,022	100,029
All species	536,519	248,003	137,582	110,421	288,516	126,744	161,772

[1] International ¼-inch rule. Data may not add to totals because of rounding.
FS, Timber Demand and Technology Assessment, RWU-4851, (608) 231–9376.

Table 12-25.—National Forest System: National Forest System lands and other lands in States and Territories, 2009

State or other area	Gross acreage	National Forest System acreage [1]	Other acreage [2]
	1,000 acres	*1,000 acres*	*1,000 acres*
AL	1,289	669	620
AK	24,359	21,966	2,393
AZ	11,892	11,265	627
AR	3,553	2,599	954
CA	24,444	20,819	3,625
CO	16,022	14,520	1,502
CT	24	24	
FL	1,435	1,176	259
GA	1,777	865	911
HI	1	1	
ID	21,659	20,465	1,194
IL	923	298	626
IN	644	202	442
KS	116	108	8
KY	2,208	814	1,394
LA	1,025	604	420
ME	93	54	40
MI	4,894	2,875	2,019
MN	5,467	2,841	2,626
MS	2,318	1,174	1,144
MO	3,060	1,492	1,568
MT	19,136	16,969	2,167
NE	443	352	90
NV	6,274	5,764	510
NH	828	736	93
NM	10,455	9,416	1,039
NY	16	16	
NC	3,165	1,255	1,910
ND	1,106	1,106	
OH	834	241	593
OK	755	401	354
OR	17,582	15,685	1,897
PA	743	513	230
SC	1,379	631	748
SD	2,370	2,018	352
TN	1,276	716	560
TX	1,994	755	1,239
UT	9,213	8,206	1,008
VT	823	399	424
VA	3,223	1,664	1,559
WA	10,113	9,281	832
WV	1,896	1,043	853
WI	2,023	1,533	490
WY	9,706	9,241	465
PR	56	28	28
VI	0	0	
Total	232,611	192,803	39,809

[1] *National Forest System acreage.*—A nationally significant system of Federally owned units of forest, range, and related land consisting of national forests, purchase units, national grasslands, land utilization project areas, experimental forest areas, experimental range areas, designated experimental areas, other land areas; water areas, and interests in lands that are administered by USDA Forest Service or designated for administration through the Forest Service.

National forests.—Units formally established and permanently set aside and reserved for national forest purposes.

Purchase units.—Units designated by the Secretary of Agriculture or previously approved by the National Forest Reservation Commission for purposes of Weeks Law Acquisition.

National grasslands.—Units designated by the Secretary of Agriculture and permanently held by the Department of Agriculture under Title III of the Bankhead-Jones Farm Tenant Act.

Land utilization projects.—Units designated by the Secretary of Agriculture for conservation and utilization under Title III of the Bankhead-Jones Farm Tenant Act.

Research and experimental areas.—Units reserved and dedicated by the Secretary of Agriculture for forest or range research and experimentation.

Other areas.—Units administered by the Forest Service that are not included in the above groups.   [2] *Other acreage.*— Lands within the unit boundaries in private, State, county, and municipal ownership and Federal lands over which the Forest Service has no jurisdiction. Areas of such lands which have been offered to the United States and have been approved for acquisition and subsequent Forest Service administration, but to which title had not yet been accepted by the United States.

FS, Timber, Demand and Technology Assessment, RWU-4851, (608) 231–9376.

Table 12-26.—Forest products cut on National Forest System lands: Volume and value of timber cut and value of all products, United States, fiscal years 2000–2009

Year [1]	Timber cut [2]		Value of miscellaneous forest products [4]	Total value including free-use timber [5]
	Volume	Value [3]		
	Million bd. ft.	1,000 dollars	1,000 dollars	1,000 dollars
2000	2,542	302,934	3,262	305,921
2001	1,938	177,634	3,262	180,708
2002	1,728	164,051	3,262	167,313
2003	1,818	157,323	3,262	160,585
2004	2,032	217,534	3,262	220,796
2005	2,098	224,143	3,262	227,405
2006	2,296	218,520	3,262	221,512
2007	1,960	173,774	3,262	169,992
2008	2,049	131,261	3,262	141,231
2009	1,954	98,088	3,262	78,050

[1] Fiscal years Oct. 1–Sept. 30. [2] Commercial and cost sales and land exchanges. [3] Includes collections for forest restoration or improvement under the Knutson-Vandenberg Act, 1930. [4] Includes materials not measurable in board feet, such as Christmas trees, tanbark, turpentine, seedlings, Spanish moss, etc. [5] Total value including free-use timber from 1996-2002 has been estimated.
FS, Timber Demand and Technology Assessment, RWU-4851, (608) 231–9376.

Table 12-27.—National Forest System lands: Receipts, United States and Puerto Rico, fiscal years 1998–2007

Year [1]	From the use of timber [2]	From the use of grazing	From special land uses, water power, etc.	Total [2]
	1,000 dollars	1,000 dollars	1,000 dollars	1,000 dollars
1998	207,938	6,992	78,869	293,799
1999	NA	NA	NA	NA
2000	NA	NA	NA	NA
2001	NA	NA	NA	NA
2002	NA	NA	NA	NA
2003	NA	NA	NA	NA
2004	NA	NA	NA	NA
2005	NA	NA	NA	NA
2006	NA	NA	NA	NA
2007	NA	NA	NA	NA

[1] Fiscal years Oct. 1–Sept. 30. [2] Includes receipts from Oregon and California Railroad Grant Lands.
FS, Timber Demand and Technology Assessment, RUW-4851, (608) 231–9376.

Table 12-28.—National forests: Payments to States and Puerto Rico from receipts from timber sales, grazing fees, and miscellaneous uses, fiscal years 2000–2002 [1] [2]

State or other areas	2000	2001	2002
	1,000 dollars	*1,000 dollars*	*1,000 dollars*
AL	617	2,032	2,015
AK	2,304	8,796	8,875
AZ	1,781	7,002	7,057
AR	6,707	6,410	5,988
CA	26,418	61,909	60,937
CO	4,530	5,595	5,434
FL	945	2,381	2,366
GA	53	1,221	1,231
ID	7,584	20,202	20,022
IL	167	285	287
IN	5	122	123
KY	72	418	391
LA	1,839	3,644	3,518
ME	27	39	39
MI	3,856	3,036	2,456
MN	4,072	3,908	3,852
MS	6,504	7,619	7,311
MO	1,168	2,387	2,499
MT	7,051	13,446	12,464
NE	34	40	40
NV	295	422	428
NH	397	445	220
NM	681	1,894	2,022
NY	8	8	8
NC	455	956	964
ND	3	3	3
OH	([3])	40	61
OK	1,250	1,303	1,214
OR	76,323	141,075	140,987
PA	2,982	4,831	3,665
SC	577	3,080	3,104
SD	3,070	3,669	3,699
TN	374	525	529
TX	666	4,447	4,435
UT	1,900	1,865	1,913
VT	328	336	283
VA	487	790	718
WA	24,658	41,229	40,191
WV	1,285	1,861	1,869
WI	1,788	2,230	1,596
WY	1,592	2,184	2,193
PR	21	21	8
Total	194,869	363,702	357,009

[1] Fiscal years Oct. 1–Sept. 30.   [2] Payments under the acts of May 23, 1908 (as amended), July 24, 1956, and Oct. 22, 1976, are 25 percent of total receipts remaining after deducting (a) payments to Arizona and New Mexico on account school section lands administered by Forest Service, (b) appropriations of receipts under laws authorizing such appropriations for acquisition of lands in specified national forests or portions thereof, and (c) receipts from an area of the Superior National Forest, Minnesota, on account of which the State (for the counties) is paid 0.75 percent of the appraised valuation in lieu of 25 percent of the receipts. Payments made in the following year.   [3] Less than $500.
FS, Timber Demand and Technology Assessment, RWU-4851, (608) 231–9376.

Table 12-29.—Livestock on National Forest System lands: Number grazed and grazing receipts, United States, 1993–2002

Year	Number grazed [1]		Receipts from grazing [2]
	Cattle, horses, and burros	Sheep and goats	
	Thousands	*Thousands*	*1,000 dollars*
1993	1,318	1,111	10,518
1994	1,229	941	11,056
1995	1,227	940	8,756
1996	1,174	868	7,352
1997	1,225	932	6,972
1998	1,208	909	6,992
1999	NA	NA	NA
2000	1,246	954	NA
2001	1,233	960	NA
2002	1,079	916	NA

[1] Calendar year data for number actually grazed.   [2] Fiscal years Oct. 1–Sept. 30.
FS, Timber Demand and Technology Assessment, RWU-4851, (608) 231–9376.

Table 12-30.—Timber prices: Average stumpage prices for sawtimber sold from national forests, by selected species, 2000–2009

Year	Douglas-fir[1]	Southern pine[2]	Ponderosa pine[3]	Western hemlock[4]	All eastern hardwoods[5]	Oak, white, red, and black[5]	Maple, sugar[6]
	Dollars per 1,000 bd. ft.	Dollars per 1,000 bd. ft.	Dollars per 1,000 bd. ft.	Dollars per 1,000 bd. ft.	Dollars per 1,000 bd. ft.	Dollars per 1,000 bd. ft.	Dollars per 1,000 bd. ft.
2000	433.40	258.10	154.60	46.12	368.61	265.63	445.80
2001	255.38	153.49	115.47	33.98	530.45	326.38	587.22
2002	184.83	166.40	117.75	73.19	382.04	273.73	484.97
2003	279.00	148.00	32.00	95.00	279.00	236.00	586.00
2004	114.00	84.00	60.00	32.00	351.00	291.00	618.00
2005	320.50	192.80	103.30	70.10	415.10	329.20	648.00
2006	NA	112.50	39.20	101.10	275.30	180.30	533.30
2007	NA	176.40	60.90	26.30	276.60	220.40	361.60
2008	NA	152.65	33.52	19.67	198.25	156.30	479.60
2009	NA	104.46	18.16	23.95	152.30	119.53	274.98

[1] Western Washington and western Oregon. [2] Southern region. [3] Pacific Southwest region. Includes Jeffrey pine. [4] Pacific Northwest region. [5] Eastern and Southern regions. [6] Eastern region.

Forest Service National Forest prices in this table are for timber sold on a Scribner Decimal C log rule basis, except in the Northeastern States where International ¼-inch log rule is used. Prices include KV payments; exclude timber sold by land exchanges and from land utilization project lands. Data for 1983 are statistical high bid prices; beginning in 1984, data are high bid prices which include specified road costs.

FS, Timber Demand and Technology Assessment, RWU-4851, (608) 231–9376.

Table 12-31.—Timber products: Production, imports, exports, and consumption, United States, 2000–2009 [1]

Year	Industrial roundwood used for—											
	Lumber				Plywood and veneer				Pulp products			
	Production	Imports	Exports	Consumption	Production	Imports	Exports	Consumption	Production	Imports [2]	Exports [2]	Consumption
	Million cu. ft.[3]	Million cu. ft.[3]	Million cu. ft.[3]	Million cu. ft.[3]	Million cu. ft.[3]	Million cu. ft.[3]	Million cu. ft.[3]	Million cu. ft.[3]	Million cu. ft.[3]	Million cu. ft.[3]	Million cu. ft.[3]	Million cu. ft.[3]
2000	7,345	2,924	434	9,835	1,187	155	42	1,300	6,021	1,493	865	6,649
2001	7,110	3,071	362	9,819	1,067	173	32	1,208	5,853	1,499	827	6,524
2002	7,293	3,170	359	10,103	1,074	206	31	1,249	5,708	1,472	785	6,395
2003	7,131	3,193	347	9,977	1,054	240	35	1,259	5,557	1,579	643	6,493
2004	7,510	3,704	348	10,866	1,086	354	43	1,397	5,692	1,669	680	6,680
2005	7,889	3,737	389	11,237	1,068	373	37	1,403	5,822	1,605	727	6,699
2006	7,552	3,415	390	10,577	1,003	339	35	1,308	5,470	1,440	681	6,229
2007	6,964	2,743	359	9,347	912	264	40	1,135	5,176	1,071	526	5,721
2008	5,928	1,922	345	7,506	743	185	45	882	4,918	897	570	5,246
2009	5,020	1,336	272	6,084	617	177	37	757	4,818	434	423	4,829

Year	Industrial roundwood used for—Continued									Fuelwood production and consumption	Production, all products	Consumption, all products
	Other industrial products, production and consumption [4]	Logs [5]		Pulpwood chip imports	Pulpwood chip exports	Total						
		Imports	Exports			Production	Imports	Exports	Consumption			
	Million cu. ft.[3]	Million cu. ft.[3]	Million cu. ft.[3]	Million cu. ft.[3]	Million cu. ft.[3]	Million cu. ft.[3]	Million cu. ft.[3]	Million cu. ft.[3]	Million cu. ft.[3]	Million cu. ft.[3]	Million cu. ft.[3]	Million cu. ft.[3]
2000 ..	300	72	422	2	354	15,630	4,310	2,117	18,158	1,622	17,252	19,780
2001 ..	270	73	403	1	264	14,966	4,700	1,888	17,896	1,640	16,606	19,536
2002 ..	263	86	388	2	189	14,915	4,877	1,738	18,101	1,618	16,533	19,717
2003 ..	318	80	356	4	155	14,571	5,096	1,535	18,132	1,515	16,086	19,647
2004 ..	318	73	366	5	168	15,139	5,805	1,604	19,339	1,540	16,679	20,879
2005 ..	318	114	345	9	166	15,608	5,837	1,665	19,780	1,550	17,158	21,330
2006 ..	320	94	339	4	151	14,836	5,292	1,596	18,532	1,555	16,391	20,087
2007 ..	325	67	350	3	205	13,932	4,147	1,481	16,598	1,605	15,537	18,203
2008 ..	290	35	313	5	257	12,450	3,045	1,531	13,964	1,510	13,960	15,474
2009 ..	294	29	321	9	195	11,264	1,986	1,248	12,002	1,400	12,664	13,402

[1] Data may not add to totals because of rounding.   [2] Includes both pulpwood and the pulpwood equivalent of woodpulp, paper, and board.   [3] Roundwood equivalent.   [4] Includes cooperage logs, poles and piling, fence posts, hewn ties, round mine timbers, box bolts, excelsior bolts, chemical wood, shingle bolts, and miscellaneous items.   [5] Prior to 2000, Pulpwood Logs are not included in logs.
FS, Timber Demand and Technology Assessment, RWU-4851, (608) 231–9376.

Table 12-32.—Timber products: Pulpwood consumption, woodpulp production, and paper and board production and consumption, United States, 2000–2009[1]

Year	Pulpwood consumption[2]	Woodpulp production[3]	Paper and board[4]		
			Production	Consumption or new supply[5]	Per capita consumption
	1,000 cords[6]	*1,000 tons*	*1,000 tons*	*1,000 tons*	*Pounds*
2000	95,904	62,758	94,491	103,147	731
2001	92,181	58,198	88,913	97,303	683
2002	90,500	58,069	89,636	97,227	676
2003	85,436	53,197	80,712	94,422	629
2004	87,110	54,301	83,612	95,068	627
2005	88,595	60,267	91,031	101,864	687
2006	84,561	60,568	91,800	102,439	685
2007	80,696	60,568	91,570	99,825	661
2008	74,039	60,568	87,619	93,640	613
2009	72,231	46,990	78,521	79,141	515

[1] Revised to match data from American Forest and Paper Association and American Pulpwood Association. [2] Includes changes in stocks. [3] Excludes defibrated and exploded woodpulp used for hard pressed board. [4] Excludes hardboard. [5] Production plus imports and minus exports (excludes products); changes in inventories not taken into account. [6] One cord equals 128 cubic feet.

FS, Timber Demand and Technology Assessment, RWU-4851, (608) 231–9376. Compiled from U.S. Department of Commerce and American Forest and Paper Association.

Table 12-33.—Timber products: Producer price indexes, selected products, United States, 2000–2009

[1982=100]

Year	Lumber	Softwood plywood	Woodpulp	Paper	Paperboard
2000	90.6	98.9	113.1	104.2	122.1
2001	87.0	95.7	98.0	104.8	118.9
2002	86.5	93.6	90.6	100.8	113.7
2003	88.3	111.8	94.5	101.6	112.4
2004	103.2	143.1	102.9	103.9	117.6
2005	100.6	127.5	107.5	111.0	121.3
2006	95.6	108.7	112.2	116.4	132.7
2007	88.5	112.8	125.8	117.8	139.4
2008	82.9	110.2	133.5	128.2	150.5
2009	75.7	98.2	131.5	127.5	147.9

FS, Timber Demand and Technology Assessment, RWU-4851, (608) 231–9376. Compiled from reports of the U.S. Department of Labor, Bureau of Labor Statistics.

Table 12-34.—Timber products: Structual panels, LVL, and lumber production, United States, 2000–2009

Year	Laminated ve-neer lumber [1]	Oriented strand board	Plywood	Medium-den-sity fiberboard	Lumber	
					Hardwood	Softwood [2]
	Million cubic meters	*Million cubic meters*	*Million cubic meters*	*Million cubic meters*	*Million cubic meters*	*Million cubic meters*
2000	1.35	10.54	15.47	2.63	29.74	61.20
2001	1.51	11.09	13.38	2.45	27.93	58.78
2002	1.58	11.88	13.45	2.87	27.73	60.86
2003	1.63	12.05	13.01	2.88	25.02	61.71
2004	1.69	12.63	12.98	2.91	25.72	65.28
2005	2.57	13.26	12.68	3.26	27.36	69.19
2006	2.50	13.24	11.88	3.29	26.96	65.55
2007	2.13	13.07	10.84	3.34	25.60	59.80
2008	1.47	11.51	9.06	3.02	23.50	49.40
2009	0.93	8.49	7.62	2.96	22.40	39.60

[1] Prior to 1994, data are estimates from various articles and reports. [2] Revised due to softwood conversion factor of 1.7 (2.36 was previously used).
FS, Timber Demand and Technology Assessment, RWU-4851, (608) 231–9376.

Table 12-35.—Lumber: Production, United States, 2000–2009

Year	Total	Softwoods	Hardwoods
	Million bd. ft.	*Million bd. ft.*	*Million bd. ft.*
2000	48,565	35,967	12,598
2001	46,411	34,577	11,834
2002	47,580	35,830	11,750
2003	46,784	36,290	10,494
2004	49,314	38,360	10,954
2005	52,289	40,698	11,591
2006	49,980	38,558	11,422
2007	46,009	35,158	10,851
2008	39,006	29,068	9,938
2009	32,781	23,280	9,501

FS, Timber Demand and Technology Assessment, RWU-4851, (608) 231–9376. From data published by the American Forest and Paper Association.

CHAPTER XIII

CONSUMPTION AND FAMILY LIVING

The statistics in this chapter deal with the consumption of food by both rural and urban people, retail price levels, and other aspects of family living of farm people. Data presented here on quantities of food available for consumption are based on material presented in the earlier commodity chapters, but they are shown here at the retail level, a form that is more useful for an analysis of the demand situation faced by the producer. Data on quantities of farm-produced food consumed directly by farm households are presented in the commodity chapters. Its value and the rental value of the farm home are given in the section on farm income.

Table 13-1.—Population: Number of people eating from civilian food supplies, United States, Jan. 1 and July 1, 2000-2010

Year	Jan. 1	July 1
	Millions	*Millions*
2000	279.5	280.9
2001	282.5	283.8
2002	285.3	286.5
2003	288.0	289.1
2004	290.5	291.8
2005	293.2	294.6
2006	296.0	297.4
2007	299.0	300.4
2008	301.9	303.2
2009	304.6	305.8
2010	307.2	308.4

ERS, Farm and Rural Household Well-Being Branch (202) 694–5435. Compiled from reports of the U.S. Department of Commerce, Census Bureau.

Table 13-2.—Macronutrients: Quantities available for consumption per capita per day, United States, 1971–2005

Year	Food energy	Protein	Fat				Cholesterol	Carbohydrate	Dietary fiber
			Total fat	Monounsaturated	Saturated	Polyunsaturated			
	Kilocalories	*Grams*	*Grams*	*Grams*	*Grams*	*Grams*	*Milligrams*	*Grams*	*Grams*
1971	3,200	99	146	59	51	26	470	394	19
1972	3,200	98	147	59	51	26	460	389	19
1973	3,200	97	143	57	49	27	430	394	20
1974	3,200	97	144	57	50	27	440	389	19
1975	3,100	95	140	55	47	27	420	389	20
1976	3,300	98	146	59	49	29	430	402	20
1977	3,200	98	143	57	48	28	430	401	20
1978	3,200	97	145	58	49	29	430	398	20
1979	3,200	97	145	58	49	29	430	401	20
1980	3,300	97	146	59	49	29	430	402	20
1981	3,200	97	147	59	49	30	420	400	20
1982	3,200	96	147	59	49	30	420	397	20
1983	3,300	98	150	60	50	31	420	402	21
1984	3,300	99	153	62	51	31	420	409	21
1985	3,500	102	158	64	53	32	430	426	22
1986	3,500	104	156	64	52	31	420	431	22
1987	3,500	104	155	63	51	31	420	441	22
1988	3,500	106	156	63	51	32	420	448	23
1989	3,500	105	151	62	49	31	410	445	23
1990	3,500	106	150	62	49	31	400	457	24
1991	3,500	107	148	63	48	31	400	460	24
1992	3,600	109	153	65	49	32	400	468	25
1993	3,700	109	154	66	49	32	400	478	24
1994	3,700	110	151	65	48	31	400	483	25
1995	3,600	109	148	63	48	31	400	482	24
1996	3,600	110	147	63	47	30	400	491	25
1997	3,700	109	146	62	46	31	400	494	25
1998	3,700	110	148	63	48	30	410	495	25
1999	3,700	112	153	65	49	32	420	499	25
2000	3,900	112	173	76	54	36	420	498	25
2001	3,900	111	172	76	53	36	410	492	25
2002	4,000	110	184	81	56	39	420	486	24
2003	3,900	111	183	81	56	39	420	483	25
2004	3,900	112	179	79	55	39	420	483	25
2005	4,000	115	190	85	59	37	430	479	25

Center for Nutrition Policy and Promotion (CNPP), (703) 305–7600.

CONSUMPTION AND FAMILY LIVING

Table 13-3.—Vitamins: Quantities available for consumption per capita per day, United States, 1971–2005 [1]

Year	Vita- min A	Caro- tenes	Vita- min E	Vita- min C	Thia- min	Ribo- flavin	Niacin	Vita- min B$_6$	Total Folate	Folate DFE	Vita- min B$_{12}$
	Micro- grams retinol activity equiv- alent	Micro- grams retinol equiv- alent	Milli- grams alpha-to- copherol	Milli- grams	Milli- grams	Milli- grams	Milli- grams	Milli- grams	Micro- grams	Micro- grams	Micro- grams
1971	1,280	520	13.1	108	2.1	2.4	22	2.0	301	303	9.5
1972	1,240	560	13.5	108	2.0	2.3	23	2.0	300	302	9.4
1973	1,220	590	14.0	107	2.0	2.3	22	1.9	306	308	8.9
1974	1,280	610	13.9	112	2.4	2.7	26	2.1	332	358	9.1
1975	1,270	630	14.1	117	2.4	2.7	26	2.0	343	370	8.6
1976	1,300	630	14.5	118	2.5	2.7	27	2.1	348	376	8.9
1977	1,260	590	14.1	117	2.5	2.7	27	2.1	349	377	8.8
1978	1,240	580	14.4	113	2.4	2.7	27	2.1	337	365	8.5
1979	1,250	620	14.5	114	2.5	2.7	28	2.1	349	377	8.2
1980	1,240	600	14.4	117	2.5	2.7	27	2.1	344	373	8.2
1981	1,240	610	14.6	115	2.5	2.7	28	2.1	342	371	8.2
1982	1,220	630	14.8	116	2.5	2.6	27	2.1	348	377	7.9
1983	1,220	600	15.2	121	2.5	2.7	28	2.2	352	382	8.1
1984	1,240	640	15.6	118	2.5	2.7	28	2.2	347	376	8.2
1985	1,230	630	16.1	119	2.6	2.8	29	2.2	362	393	8.3
1986	1,230	610	16.0	123	2.7	2.8	29	2.3	367	398	8.2
1987	1,240	640	16.0	120	2.7	2.9	30	2.3	357	390	8.2
1988	1,200	610	16.6	121	2.8	2.9	30	2.3	372	406	8.0
1989	1,230	650	16.2	122	2.8	2.9	30	2.3	366	400	8.0
1990	1,240	670	16.4	118	2.9	3.0	31	2.4	374	410	8.0
1991	1,220	640	16.9	122	2.9	2.9	31	2.4	385	421	7.9
1992	1,250	680	17.1	125	3.0	3.0	32	2.5	396	432	7.9
1993	1,280	750	17.6	129	3.0	2.9	32	2.5	393	422	7.7
1994	1,320	830	16.8	129	3.0	3.0	32	2.5	392	421	7.9
1995	1,270	750	16.2	125	2.9	2.9	31	2.4	382	411	8.0
1996	1,290	800	16.1	130	3.0	2.9	32	2.4	384	414	8.0
1997	1,330	850	16.3	130	3.0	2.9	32	2.4	382	412	7.8
1998	1,240	710	16.2	131	3.0	2.9	32	2.4	697	913	8.0
1999	1,250	700	17.0	130	3.0	2.9	33	2.5	704	919	8.0
2000	1,260	710	19.8	131	3.0	2.9	33	2.5	707	927	7.9
2001	1,080	680	20.2	120	3.0	2.9	32	2.4	693	908	7.9
2002	1,070	660	21.3	115	2.9	2.8	32	2.4	681	891	7.9
2003	1,070	680	21.3	119	2.9	2.8	32	2.4	688	900	7.9
2004	1,080	680	21.1	118	2.9	2.8	32	2.4	686	898	8.0
2005	1,030	660	21.4	115	2.9	2.8	33	2.5	682	893	8.5

[1] Computed by Center for Nutrition Policy and Promotion (CNPP), USDA. Based on Economic Research Service estimates of per capita quantities of food available for consumption (retail weight) and on CNPP estimates of quantities of produce from home gardens and certain other foods. No deduction is made in food supply estimates for loss of food or nutrients in further processing, in marketing, or in the home. Data include iron, thiamin, riboflavin, niacin, vitamin A, vitamin B$_6$, vitamin B$_{12}$, ascorbic acid, and zinc added by enrichment and fortification.

Center for Nutrition Policy and Promotion (CNPP), (703) 305–7600.

Table 13-4.—Minerals: Quantities available for consumption per capita per day, United States, 1971–2005 [1]

Year	Calcium	Phos-phorus	Magne-sium	Iron	Zinc	Copper	Potas-sium	So-dium [2]	Sele-nium
	Milli-grams	*Milli-grams*	*Milli-grams*	*Milli-grams*	*Milli-grams*	*Milli-grams*	*Micro-grams*	*Milli-grams*	*Milli-grams*
1971	970	1,560	340	16.1	12.8	1.7	3,670	1,280	125.4
1972	960	1,560	350	16.2	12.7	1.7	3,660	1,280	126.3
1973	970	1,540	350	16.4	12.4	1.7	3,650	1,260	122.8
1974	940	1,540	340	16.7	13.8	1.7	3,590	1,260	117.4
1975	920	1,490	340	16.9	13.6	1.7	3,580	1,240	136.2
1976	930	1,540	350	17.4	14.0	1.8	3,650	1,290	139.5
1977	930	1,530	350	17.3	14.0	1.8	3,590	1,280	133.5
1978	920	1,510	340	16.8	13.7	1.7	3,510	1,270	135
1979	920	1,530	350	17.3	13.8	1.8	3,590	1,270	134
1980	910	1,510	340	17.2	13.7	1.7	3,550	1,240	131.9
1981	900	1,510	340	17.3	13.8	1.8	3,510	1,220	132
1982	910	1,510	350	17.5	13.8	1.8	3,520	1,230	134.5
1983	920	1,530	350	19.9	14.0	1.8	3,590	1,240	137.1
1984	930	1,560	360	20.0	14.2	1.8	3,610	1,270	137.3
1985	960	1,600	370	20.9	14.5	1.9	3,700	1,290	140.7
1986	970	1,620	380	21.1	14.8	1.9	3,760	1,300	143
1987	960	1,630	380	21.4	14.6	1.9	3,700	1,290	143.6
1988	960	1,650	380	21.9	14.9	1.9	3,740	1,260	145
1989	950	1,640	380	22.0	14.9	1.9	3,730	1,270	146
1990	980	1,670	390	22.7	15.3	2.0	3,760	1,300	147.9
1991	970	1,670	400	23.0	15.4	2.0	3,810	1,300	156.9
1992	990	1,700	400	23.4	15.8	2.0	3,860	1,320	160.7
1993	970	1,690	400	23.3	15.5	2.0	3,850	1,310	161.1
1994	1,000	1,700	400	23.2	15.4	2.0	3,890	1,310	161.6
1995	980	1,680	390	22.9	15.2	2.0	3,800	1,290	158.8
1996	990	1,690	400	23.2	15.1	2.0	3,870	1,280	162.7
1997	980	1,680	390	23.1	14.8	2.0	3,840	1,280	163.0
1998	980	1,690	390	23.4	15.1	2.0	3,860	1,260	176.5
1999	990	1,710	400	23.8	15.5	2.1	3,910	1,270	177.3
2000	990	1,720	400	23.8	15.4	2.1	3,910	1,280	178.5
2001	980	1,690	400	23.5	15.2	2.0	3,820	1,240	180.1
2002	970	1,680	390	23.4	15.2	2.0	3,760	1,320	181.7
2003	980	1,690	400	23.6	15.2	2.0	3,810	1,300	185.1
2004	990	1,710	400	23.7	15.4	2.1	3,840	1,250	185.1
2005	950	1,720	400	24.1	16.2	2.1	3,820	1,270	184.4

[1] Computed by Center for Nutrition Policy and Promotion (CNPP), USDA. Based on Economic Research Service estimates of per capita quantities of food available for consumption (retail weight) and on CNPP estimates of quantities of produce from home gardens and certain other foods. No deduction is made in food supply estimates for loss of food or nutrients in further processing, in marketing, or in the home. Data include iron, thiamin, riboflavin, niacin, vitamin A, vitamin B_6, vitamin B_{12}, ascorbic acid, and zinc added by enrichment and fortification.   [2] Sodium levels do not reflect sodium from most processed foods and therefore underestimate total sodium available in the U.S. food supply.

Center for Nutrition Policy and Promotion (CNPP), (703) 305–7600.

Table 13-5.—Food nutrients: Percentage of total contributed by major food groups, 1971[1]

Nutrient	Meat, poultry, fish	Dairy[2] products	Eggs	Fats,[3] oils	Fruits Citrus	Fruits Non-citrus	Fruits Total[5]
	Percent	Percent	Percent	Percent	Percent	Percent	Percent
Food energy	18.0	11.4	1.9	18.2	1.0	2.0	3.0
Carbohydrate	0.1	6.8	0.1	0.0	1.9	4.1	6.1
Fiber	0.0	0.4	0.0	0.0	2.9	9.7	12.6
Protein	39.7	22.3	5.3	0.2	0.5	0.7	1.2
Total fat	31.3	13.5	2.9	45.4	0.1	0.3	0.4
Saturated fat	34.6	24.2	2.6	34.5	0.0	0.2	0.2
Monounsaturated fat	34.9	9.7	2.7	46.6	0.0	0.4	0.4
Polyunsaturated fat	15.9	2.6	2.2	67.3	0.1	0.4	0.5
Cholesterol	39.4	15.8	39.3	5.5	0.0	0.0	0.0
Vitamin A (retinol activity equivalents)	34.1	21.6	6.4	11.1	0.3	1.4	1.7
Carotene (retinol equivalents)	0.0	3.0	0.0	4.2	1.5	6.5	8.0
Vitamin E	5.0	3.8	3.3	66.4	1.0	3.0	4.0
Vitamin C	2.3	4.2	0.0	0.0	26.6	13.7	40.3
Thiamin	24.6	8.4	1.2	0.1	2.3	1.7	4.0
Riboflavin	20.7	37.5	8.7	0.0	0.6	1.6	2.2
Niacin	41.3	2.1	0.1	0.0	0.7	1.7	2.5
Vitamin B[6]	38.6	12.3	2.7	0.1	1.7	7.0	8.7
Folate	9.2	9.3	6.4	0.1	7.1	2.5	9.6
Folate DFE	9.0	8.7	6.3	0.1	7.0	2.5	9.4
Vitamin B[12]	72.9	21.1	4.5	0.2	0.0	0.0	0.0
Calcium	2.7	76.2	2.1	0.5	1.2	1.0	2.3
Phosphorus	25.0	37.1	4.8	0.3	0.7	1.0	1.6
Magnesium	12.6	21.1	1.2	0.1	2.1	3.7	5.8
Iron	21.7	2.5	3.7	0.1	0.6	2.3	4.5
Zinc	45.1	19.3	3.6	0.1	0.3	0.9	1.2
Copper	17.4	3.3	0.2	0.0	1.7	4.4	6.1
Potassium	16.2	23.2	1.4	0.2	3.5	5.6	9.1
Sodium	17.0	29.4	4.2	14.0	0.1	1.4	1.5
Selenium	16.3	17.3	10.5	0.1	0.2	0.4	0.6

Nutrient	Vegetables White potatoes	Vegetables Dark green, deep yellow	Vegetables Tomatoes	Vegetables Other	Vegetables Total[5]	Legumes, nuts, soy	Grain products	Sugars, sweeteners	Miscellaneous[4]
	Percent	Percent	Percent	Percent	Percent	Percent	Percent	Percent	Percent
Food energy	2.7	0.4	0.6	1.8	5.5	3.1	19.5	18.6	0.9
Carbohydrate	5.1	0.8	1.2	3.3	10.3	2.2	33.4	39.6	1.3
Fiber	8.8	3.8	4.3	15.2	32.2	14.5	29.6	0.0	10.7
Protein	2.4	0.4	0.6	2.5	5.9	5.6	17.9	0.0	1.7
Total fat	0.1	0.0	0.1	0.3	0.5	3.7	1.5	0.0	0.8
Saturated fat	0.1	0.0	0.0	0.1	0.2	2.1	0.7	0.0	0.9
Monounsaturated fat	0.0	0.0	0.0	0.1	0.2	4.1	0.6	0.0	0.8
Polyunsaturated fat	0.2	0.1	0.2	0.7	1.2	6.5	3.2	0.0	0.7
Cholesterol	0.0	0.0	0.0	0.0	0.0	0.0	0.0	0.0	0.0
Vitamin A (retinol activity equivalents)	0.0	15.5	1.5	2.5	19.4	0.0	0.7	0.0	4.8
Carotene (retinol equivalents)	0.0	68.0	3.7	11.0	82.7	0.1	0.3	0.0	1.7
Vitamin E	0.5	1.2	4.0	2.7	8.4	6.3	2.6	0.0	0.4
Vitamin C	17.1	6.8	8.9	15.3	48.0	0.0	0.9	0.0	4.2
Thiamin	5.1	0.8	1.4	3.9	11.2	5.2	44.6	0.1	0.6
Riboflavin	1.1	0.9	1.0	3.0	6.1	1.5	21.1	0.7	1.3
Niacin	6.8	0.8	2.2	3.8	13.5	4.8	31.2	0.0	4.5
Vitamin B[6]	12.3	2.3	3.1	5.5	23.2	3.6	9.9	0.2	0.8
Folate	4.3	2.9	2.9	17.4	27.5	18.9	16.7	0.0	2.3
Folate DFE	4.9	2.9	2.9	17.1	27.7	18.5	17.9	0.0	2.3
Vitamin B[12]	0.0	0.0	0.0	0.0	0.0	0.0	1.3	0.0	0.0
Calcium	0.8	1.0	0.9	3.5	6.2	3.6	3.3	0.6	2.4
Phosphorus	2.9	0.6	1.0	3.6	8.0	5.3	14.0	0.3	3.5
Magnesium	5.7	1.4	2.2	6.9	16.2	12.2	16.1	0.7	14.1
Iron	4.5	1.3	2.1	5.7	13.5	9.2	37.2	1.1	7.9
Zinc	2.8	0.6	0.7	3.3	7.4	5.8	13.4	0.5	3.7
Copper	10.7	1.4	4.0	5.2	21.2	17.4	16.2	3.8	14.2
Potassium	13.2	1.9	4.3	7.4	26.8	7.8	6.2	0.5	8.6
Sodium	3.0	1.3	9.3	16.6	30.2	0.2	0.6	2.6	0.4
Selenium	1.8	0.2	0.3	0.8	3.0	9.6	40.0	0.9	1.7

[1] Percentages of food groups are based on aggregate data.　[2] Excludes butter.　[3] Includes butter.　[4] Coffee, tea, spices, chocolate liquor equivalent of cocoa beans, and fortification not assigned to a specific group.　[5] Components may not add to total due to rounding.
Center for Nutrition Policy and Promotion, (703) 305–7600.

Table 13-6.—Food nutrients: Percentage of total contributed by major food groups, 2005 [1]

Nutrient	Meat, poultry, fish	Dairy products [2]	Eggs	Fats, oils [3]	Fruits		
					Citrus	Non-citrus	Total [5]
	Percent	Percent	Percent	Percent	Percent	Percent	Percent
Food energy	15.2	7.6	1.3	24.7	0.8	2.1	2.9
Carbohydrate	0.1	4.3	0.1	0.0	1.7	4.2	5.9
Fiber	0.0	0.4	0.0	0.0	2.2	8.8	10.9
Protein	42.3	18.1	3.9	0.1	0.5	0.7	1.2
Total fat	23.3	8.3	1.9	58.8	0.0	0.5	0.5
Saturated fat	26.4	16.6	1.9	49.6	0.0	0.3	0.3
Monounsaturated fat	23.8	5.2	1.6	63.2	0.0	0.6	0.6
Polyunsaturated fat	14.3	1.3	1.3	71.1	0.0	0.5	0.5
Cholesterol	47.2	12.5	35.2	5.0	0.0	0.0	0.0
Vitamin A (retinol activity equivalents)	33.2	15.7	6.6	7.4	0.4	2.1	2.5
Carotene (retinol equivalents)	0.0	1.7	0.0	1.7	1.2	6.7	7.8
Vitamin E	3.8	1.7	1.8	74.6	0.7	2.2	2.9
Vitamin C	2.3	2.5	0.0	0.0	26.5	14.0	40.5
Thiamin	18.0	4.3	0.7	0.0	1.8	1.8	3.6
Riboflavin	17.9	25.0	6.3	0.1	0.4	1.9	2.3
Niacin	37.7	1.1	0.1	0.0	0.5	1.5	2.0
Vitamin B [6]	38.1	6.8	1.9	0.0	1.3	7.8	9.1
Folate	3.8	3.3	2.5	0.0	4.1	1.8	5.9
Folate DFE	2.9	2.5	1.9	0.0	3.2	1.4	4.5
Vitamin B [12]	77.4	18.2	4.2	0.1	0.0	0.0	0.0
Calcium	3.5	70.3	1.8	0.3	1.2	1.4	2.6
Phosphorus	26.6	30.1	3.7	0.1	0.7	1.1	1.8
Magnesium	13.2	13.9	0.1	0.1	1.8	4.1	5.9
Iron	16.5	1.8	2.1	0.1	0.4	2.0	2.4
Zinc	40.5	15.0	2.4	0.0	0.3	0.8	1.1
Copper	13.6	2.5	0.2	0.0	1.4	4.5	5.9
Potassium	18.4	16.0	1.1	0.1	3.5	7.3	10.9
Sodium	21.3	33.2	3.5	7.5	0.1	2.1	2.2
Selenium	28.9	9.8	6.0	0.0	0.1	0.3	0.4

Nutrient	Vegetables					Legumes, nuts, soy	Grain products	Sugars, sweeteners	Miscellaneous [4]
	White potatoes	Dark-green, deep-yellow	Tomatoes	Other	Total [5]				
	Percent	Percent	Percent	Percent	Percent	Percent	Percent	Percent	Percent
Food energy	2.2	0.4	0.6	1.4	4.4	2.9	22.9	17.0	1.1
Carbohydrate	4.1	0.7	1.1	2.5	8.4	1.9	40.0	37.6	1.7
Fiber	6.4	3.7	3.9	10.6	24.7	12.9	35.5	0.0	15.6
Protein	2.0	0.5	0.6	2.0	5.2	5.8	21.3	0.0	2.3
Total fat	0.1	0.0	0.1	0.2	0.4	3.5	2.2	0.0	1.1
Saturated fat	0.0	0.0	0.0	0.1	0.2	2.1	1.4	0.0	1.4
Monounsaturated fat	0.0	0.0	0.0	0.1	0.1	3.5	1.1	0.0	0.9
Polyunsaturated fat	0.1	0.1	0.2	0.5	0.9	5.5	4.1	0.0	0.9
Cholesterol	0.0	0.0	0.0	0.0	0.0	0.0	0.0	0.0	0.0
Vitamin A (retinol activity equivalents)	0.0	22.1	2.1	2.9	27.1	0.0	5.3	0.0	2.2
Carotene (retinol equivalents)	0.0	68.3	4.1	9.0	81.5	0.1	0.8	0.0	6.4
Vitamin E	0.3	1.2	3.0	1.4	5.9	4.8	3.9	0.0	0.6
Vitamin C	15.0	12.6	9.0	11.6	48.3	0.1	4.7	0.0	1.7
Thiamin	4.0	0.8	1.2	2.6	8.6	4.5	59.3	0.1	0.9
Riboflavin	0.9	1.0	1.1	2.7	5.7	1.6	38.5	0.7	1.9
Niacin	4.7	0.7	1.7	2.6	9.8	4.0	42.3	0.0	3.2
Vitamin B [6]	9.7	2.5	2.8	5.1	20.1	3.6	18.3	0.2	1.9
Folate	1.9	2.2	1.4	6.8	12.3	9.3	61.2	0.0	1.7
Folate DFE	1.6	1.7	1.0	5.2	9.5	7.1	70.0	0.0	1.3
Vitamin B [12]	0.0	0.0	0.0	0.0	0.0	0.0	0.1	0.0	0.0
Calcium	0.8	1.2	0.9	3.8	6.7	4.2	5.0	0.6	5.0
Phosphorus	2.4	0.8	1.0	3.1	7.4	5.9	19.2	0.3	4.9
Magnesium	4.4	1.6	2.2	5.1	13.3	12.4	23.3	0.6	16.5
Iron	3.2	1.1	1.5	3.6	9.4	7.0	49.6	0.8	10.3
Zinc	2.1	0.6	0.6	2.5	5.8	5.2	24.5	0.4	4.9
Copper	7.4	1.2	3.8	4.2	16.6	18.9	21.1	3.2	18.0
Potassium	11.6	2.5	4.7	7.0	25.8	8.7	9.4	0.4	9.2
Sodium	3.0	1.1	12.4	10.5	27.1	0.3	1.0	3.3	0.6
Selenium	1.1	0.2	0.2	0.7	2.2	10.0	40.3	0.8	1.5

[1] Percentages of food groups are based on aggregate nutrient data　[2] Excludes butter.　[3] Includes butter.　[4] Coffee, tea, spices, chocolate liquor equivalent of cocoa beans, and fortification not assigned to a specific food group.　[5] Components may not add to total due to rounding.
Center for Nutrition Policy and Promotion, (703) 305–7600.

Table 13-7.—Consumption: Per capita consumption of major food commodities, United States, 2001–2008 [1]

Commodity	2001	2002	2003	2004	2005	2006	2007	2008 [2]
	Pounds	Pounds	Pounds	Pounds	Pounds	Pounds	Pounds	Pounds
Red meats [3][4]	111.4	114.1	111.8	112.2	110.3	110.0	110.7	108.3
Beef	63.1	64.5	62.0	63.0	62.5	62.8	62.2	61.2
Veal	0.5	0.5	0.5	0.4	0.4	0.4	0.3	0.3
Lamb and mutton	0.8	0.9	0.8	0.8	0.8	0.8	0.8	0.7
Pork	47.0	48.2	48.5	47.9	46.6	46.0	47.3	46.0
Fish [3]	14.7	15.6	16.3	16.5	16.2	16.5	16.3	16.0
Canned	4.2	4.3	4.7	4.5	4.3	3.9	3.9	3.9
Fresh and frozen	10.2	11.0	11.4	11.8	11.6	12.3	12.1	11.7
Cured	0.3	0.3	0.3	0.3	0.3	0.3	0.3	0.3
Poultry [3][4]	67.8	70.8	71.3	72.8	73.7	74.3	73.7	72.6
Chicken	54.0	56.8	57.6	59.3	60.6	60.9	59.9	58.8
Turkey	13.8	14.0	13.8	13.5	13.2	13.3	13.8	13.9
Eggs	32.5	32.8	32.8	33.1	32.8	33.1	32.1	31.9
Dairy products [5]								
Total dairy products	585.2	585.9	594.2	591.5	597.9	606.6	603.8	600.5
Fluid milk and cream	207.7	206.9	206.4	205.4	204.5	204.7	204.7	203.7
Plain and flavored whole milk	67.2	66.6	65.6	62.8	59.8	58.0	54.9	52.7
Plain reduced fat and light milk (2%, 1%, and 0.5%)	82.9	82.1	81.2	80.6	80.8	81.3	82.0	85.3
Plain fat free milk (skim)	28.9	27.9	26.8	26.6	27.0	27.2	27.2	27.1
Flavored lower fat free milk	9.0	10.5	10.8	11.7	12.0	12.5	12.3	12.3
Buttermilk	2.1	2.0	1.9	1.8	1.7	1.7	1.7	1.8
Eggnog	0.4	0.4	0.5	0.4	0.4	0.4	0.4	0.4
Yogurt (excl. frozen)	7.0	7.4	8.2	9.2	10.3	11.1	11.5	11.8
Heavy cream, light cream and half and half	6.8	6.5	7.4	7.9	8.0	8.2	8.4	8.2
Sour cream and dip	3.5	3.6	4.0	4.2	4.4	4.2	4.4	4.2
Cheese (excluding cottage) [6]	30.0	30.5	30.5	31.3	31.6	32.5	33.1	32.4
American	12.8	12.8	12.5	12.9	12.7	13.1	12.8	13.0
Cheddar	9.9	9.6	9.2	10.3	10.3	10.4	10.0	9.9
Italian	12.4	12.5	12.6	12.9	13.3	13.7	14.5	14.0
Mozzarella	9.7	9.7	9.7	9.9	10.2	10.5	11.0	10.6
Cottage cheese	2.6	2.6	2.6	2.7	2.7	2.6	2.6	2.3
Condensed and evaporated milk	5.4	6.0	5.9	5.4	5.9	6.4	7.6	7.5
Ice cream	16.3	16.7	16.4	13.8	14.6	14.7	14.2	13.9
Fats and oils [7]	82.7	87.3	86.9	86.5	85.5	84.6	84.9	85.2
Butter	4.4	4.4	4.4	4.5	4.5	4.7	4.7	5.0
Margarine	7.0	6.5	5.3	5.2	4.0	4.6	4.5	4.2
Shortening	32.5	32.8	32.5	32.5	29.0	24.9	21.0	18.0
Lard (direct use)	1.1	1.3	1.3	0.8	1.6	1.7	1.6	1.0
Edible tallow (direct use)	3.0	3.4	3.8	4.0	3.8	3.9	2.9	2.9
Salad and cooking oils	35.6	39.7	40.2	40.0	42.8	44.6	50.2	54.3
Fruits and vegetables [4][8]	695.1	689.5	702.2	703.1	685.4	672.5	670.1	643.6
Fruits	280.0	274.9	279.6	278.3	270.1	268.8	261.7	250.9
Fresh	125.8	126.8	128.1	127.7	125.3	127.9	123.6	126.8
Citrus	23.9	23.4	23.8	22.7	21.6	21.7	17.9	20.6
Noncitrus	101.9	103.5	104.3	105.0	103.7	106.3	105.6	106.1
Processing	154.2	148.0	151.5	150.6	144.7	140.8	138.1	124.1
Citrus	90.2	84.2	84.0	83.7	78.2	71.4	67.2	55.4
Noncitrus	64.0	63.8	67.5	66.9	66.5	69.4	70.9	68.7
Vegetables	415.1	414.6	422.6	424.8	415.4	403.7	408.4	392.7
Fresh	198.1	197.5	201.0	204.8	196.8	194.1	194.7	187.7
Processing	217.0	217.1	221.6	220.0	218.6	209.6	213.8	205.0
Flour and cereal products [4]	195.1	192.3	193.8	192.1	192.3	194.2	197.3	196.5
Wheat flour [9]	141.1	136.8	136.8	134.6	134.4	135.8	138.3	136.6
Rice (milled basis)	19.3	20.1	20.9	20.8	20.8	20.7	20.8	21.0
Corn products	29.0	29.7	30.3	30.9	31.4	31.9	32.4	33.0
Oat products	4.5	4.5	4.6	4.6	4.6	4.6	4.7	4.8

See footnotes at end of table.

Table 13-7.—Consumption: Per capita consumption of major food commodities, United States, 2001–2008 [1]—Continued

Commodity	2001	2002	2003	2004	2005	2006	2007	2008[2]
	Pounds	Pounds	Pounds	Pounds	Pounds	Pounds	Pounds	Pounds
Barley and rye products	1.2	1.2	1.2	1.2	1.2	1.2	1.2	1.2
Caloric sweeteners (dry weight basis)[4]	147.1	146.3	141.6	141.8	142.3	139.1	136.2	136.3
Sugar (refined)	64.5	63.2	61.0	61.7	63.2	62.4	62.0	65.7
Corn sweeteners [10]	81.3	81.5	79.2	78.8	77.7	75.1	72.9	69.2
Honey and edible syrups	1.4	1.5	1.4	1.3	1.5	1.6	1.4	1.4
Other:.								
Coffee (green bean equivalent)	9.5	9.2	9.5	9.6	9.5	9.5	9.6	9.5
Cocoa(chocolate liquor equiva- lent) [11]	4.5	3.9	4.2	4.8	5.2	5.2	4.8	4.5
Tea (dry leaf equivalent)	0.9	0.8	0.8	0.8	0.8	0.9	0.9	0.9
Peanuts (shelled)	5.9	5.8	6.4	6.7	6.7	6.5	6.3	6.4
Tree nuts (shelled)	2.9	3.3	3.5	3.5	2.6	3.3	3.5	3.5

[1] Quantity in pounds, retail weight unless otherwise shown.　[2] Preliminary.　[3] Boneless, trimmed weight equivalent.　[4] Total may not add due to rounding.　[5] Total dairy products reported on a milk-equivalent, milkfat basis. All other dairy categories reported on a product weight basis.　[6] Natural equivalent of cheese and cheese products.　[7] Total fats and oils reported on a fat content basis. All other fats and oils categories reported on a product weight basis.　[8] Farm weight.　[9] White, whole wheat, semolina, and durum flour.　[10] High fructose, glucose, and dextrose.　[11] Chocolate liquor is what remains after cocoa beans have been roasted and hulled; it is sometimes called ground or bitter chocolate.

ERS, Food Economics Division, (202) 694-5400. Historical consumption and supply-disappearance data for food may be found at,www.ers.USDA.gov/data/foodconsumption/, ERS, USDA, 2010.

Table 13-8.—Food plans: Food cost at home, at four cost levels, for families and individuals in the United States, for week and month, November 2009 [1]

Age-gender groups	Weekly cost[2]				Monthy cost[2]			
	Thrifty plan	Low-cost plan	Mod-erate-cost plan	Liberal plan	Thrifty plan	Low-cost plan	Mod-erate-cost plan	Liberal plan
	Dollars	Dollars	Dollars	Dollars	Dollars	Dollars	Dollars	Dollars
Individuals: [3].								
Child:.								
1 year	20.10	26.40	30.20	36.50	87.10	114.30	131.00	158.10
2-3 year	21.70	27.10	32.90	39.90	93.80	117.60	142.40	173.10
4-5 years	22.40	28.40	35.00	42.70	97.20	122.80	151.70	185.00
6-8 years	28.60	38.70	47.70	56.20	123.80	167.70	206.60	243.30
9-11 years	32.70	43.10	55.20	64.60	141.60	186.60	239.30	280.00
Male:.								
12-13 years	35.00	49.20	61.10	72.00	151.50	213.10	264.60	311.80
14-18 years	36.30	50.80	63.70	72.70	157.30	220.30	275.90	314.80
19-50 years	38.90	50.00	62.70	76.80	168.40	216.50	271.80	332.90
51-70 years	35.60	47.30	58.10	70.30	154.00	205.20	251.90	304.50
71+ years	35.80	46.80	58.00	71.70	155.20	202.70	251.50	310.50
Female:.								
12-13 years	35.20	42.80	51.20	62.40	152.70	185.40	221.80	270.30
14-18 years	34.70	42.90	52.00	63.80	150.40	185.90	225.40	276.50
19-50 years	34.50	43.40	53.50	68.60	149.60	187.90	231.80	297.10
51-70 years	34.20	42.40	52.70	62.70	148.30	183.60	228.40	271.50
71+ years	33.40	41.90	52.10	62.70	144.70	181.50	225.70	271.70
Families:.								
Family of 2: [4].								
19-50 years	80.70	102.70	127.60	159.90	349.70	448.80	554.00	692.90
51-70 years	76.70	98.70	121.90	146.20	332.60	427.60	528.30	633.50
Family of 4:.								
Couple, 19-50 years and children.								
2-3 and 4-5 years	117.50	148.80	184.10	228.00	509.00	644.80	797.80	988.10
6-8 and 9-11 years	134.60	175.10	219.10	266.20	583.40	758.70	949.50	1,153.00

[1] The Food Plans represent a nutritious diet at four different cost levels. The nutritional bases of the Food Plans are the 1997-2005 Dietary References Intakes, 2005 Dietary Guidelines for Americans, and 2005 MyPyramid food intake recommendations. In addition to cost, differences among plans are in specific foods and quantities of foods. Another basis of the Food Plans is that all meals and snacks are prepared at home. All four Food Plans are based on 2001-02 data and are updated to current dollars by using the Consumer Price Index for specific food items.　[2] All costs are rounded to nearest 10 cents.　[3] The costs given are for individuals in 4–person families. For individuals in other size families, the following adjustments are suggested: 1 person-add 20 percent; 2 person-add 10 percent; 3 person-add 5 percent; 4 person-no adjustment; 5 or 6 person-subtract 5 percent; 7 (or more) person-subtract 10 percent. To calculate overall household food costs, (1) adjust food costs for each person in household and then (2) sum these adjusted food costs.　[4] Ten percent added for family size adjustment.

Center for Nutrition Policy and Promotion, (703) 305–7600.

Table 13-9.—Special Nutrition Assistance Program: Participation and federal costs, fiscal years 2000–2009

Fiscal year [1]	Average monthly participation [2]		Recipient benefits	Total cost [3]	Average monthly benefit [4]	
	Persons	Housholds			Per person	Per household
	1,000	*1,000*	*1,000 dollars*	*1,000 dollars*	*Dollars*	*Dollars*
2000	17,194	7,351	14,983,319	17,054,017	72.62	169.85
2001	17,318	7,449	15,547,390	17,789,386	74.81	173.93
2002	19,096	8,195	18,256,204	20,637,025	79.67	185.65
2003	21,259	9,154	21,404,276	23,816,283	83.90	194.86
2004	23,858	10,279	24,618,890	27,099,055	85.99	199.60
2005	25,718	11,197	28,567,876	31,072,135	92.57	212.61
2006	26,672	11,734	30,187,347	32,911,987	94.32	214.38
2007	26,469	11,790	30,373,271	33,193,148	95.63	214.69
2008	28,410	12,729	34,608,397	37,659,138	101.52	226.57
2009 [5]	33,722	15,232	50,360,147	53,637,582	124.45	275.52

* Note: SNAP is the Special Nutrition Assistance Program, formerly known as the Food Stamp Program.   [1] October 1 to September 30.   [2] Participation data are 12-month averages.   [3] Total cost includes matching funds for state administrative expenses (e.g., certification of households, quality control, anti-fraud activities; employment and training); and for other Federal costs (e.g., benefit redemption processing; computer support; electronic benefit transfer systems; retailer redemption and monitoring; certification of SSI recipients; nutrition education and program information).   [4] The sharp rise in FY 2009 reflects April 2009 implementation of higher benefits mandated by the American Recovery Reinvestment Act.   [5] Preliminary.
FNS, Budget Division/Program Reports, Analysis and Monitoring Branch, (703) 305–2165.

Table 13-10.—Food and Nutrition Service Programs: Federal costs of the National School Lunch, School Breakfast, Child Care Food, Summer Food Service, WIC, Special Milk, and Food Distribution Programs, fiscal years 2000–2009 [1]

Fiscal year [2]	Child Nutrition					WIC [6]	Special Milk	Food Distribution Programs [7]
	Cash payments				Cost of food distributed [5]			
	School Lunch	School Breakfast	Child & Adult Care [3]	Summer Food [4]				
	1,000 dollars	*1,000 dollars*	*1,000 dollars*	*1,000 dollars*	*1,000 dollars*	*1,000 dollars*	*1,000 dollars*	*1,000 dollars*
2000	5,492,909	1,393,282	1,635,294	265,595	704,159	3,982,050	15,439	538,217
2001	5,612,344	1,450,113	1,685,143	268,339	917,015	4,149,431	15,547	716,419
2002	6,049,563	1,566,681	1,795,902	260,465	862,262	4,339,797	16,056	802,720
2003	6,340,568	1,651,789	1,866,989	255,100	908,813	4,524,369	14,298	662,886
2004	6,663,108	1,775,769	1,954,795	260,188	1,030,518	4,887,275	14,208	676,888
2005	7,055,436	1,927,223	2,040,557	264,982	1,047,236	4,992,568	16,437	626,758
2006	7,389,104	2,042,625	2,079,416	274,284	876,548	5,072,718	14,581	529,093
2007	7,707,307	2,164,383	2,160,125	288,645	1,111,903	5,410,417	13,619	488,051
2008	8,265,152	2,365,978	2,315,170	325,089	1,141,091	6,190,843	14,853	543,150
2009 [8]	8,873,087	2,582,469	2,439,339	345,942	1,216,544	6,475,912	13,969	873,982

[1] See table 13-7 for Food Stamp Program costs.   [2] October 1–September 30.   [3] Includes sponsor administrative, audit, and startup costs.   [4] Includes State administrative and health clinic expenses.   [5] Includes entitlement commodities, bonus commodities, and cash-in-lieu for the National School Lunch, School Breakfast, Child and Adult Care Food, and Summer Food Service Programs.   [6] Includes food costs, administrative costs, program evaluation funds, special grants, and Farmers Market projects for the Special Supplemental Food Program for Women, Infants and Children.   [7] Includes entitlement and bonus commodities, cash-in-lieu of commodities, and administrative costs of the following programs: Food Distribution to Indian Reservations, Nutrition Services Incentive Program (formerly Nutrition Program for the Elderly), Commodity Supplemental Food, Charitable Institutions, Summer Camps, Emergency Food Assistance Program (TEFAP), Disaster Feeding, Bureau of Federal Prisons, Veteran Affairs Administration, and the Food Stamp Program Elderly Pilot Project.   [8] Preliminary.
FNS, Budget Division/Program Reports, Analysis and Monitoring Branch, (703) 305–2163.

Table 13-11.—Food and Nutrition Service program benefits: Cash payments made under the National School Lunch, School Breakfast, Child and Adult Care, Summer Food and Special Milk Programs and the value of food benefits provided under the Food Stamp, WIC, Commodity Distribution and the Emergency Feeding Food Programs, fiscal year 2009 [1]

State/Territory	Child Nutrition Program (cash payments only) [2]					Special Supplemental Food (WIC) [3]	Food Stamp Program	Commodity distribution [4]	Emergency food assistance (TEFAP)	Total [5]
	Child and Adult Care Food	Summer Food	Special Milk	National School Lunch	Breakfast					
	1,000 dollars	1,000 dollars	1,000 dollars	1,000 dollars	1,000 dollars	1,000 dollars	1,000 dollars	1,000 dollars	1,000 dollars	1,000 dollars
Alabama	34,711	3,215	40	167,139	50,822	76,413	970,949	9,466	23,048	1,335,803
Alaska	7,149	834	11	25,873	6,108	16,786	129,624	1,849	2,363	190,598
American Samoa [5]	0	0	0	0	0	5,975	0	0	0	5,975
Arizona	43,473	3,003	65	204,597	54,990	105,372	1,223,846	11,194	41,731	1,688,272
Arkansas	35,170	2,980	9	106,886	36,680	48,132	569,987	7,619	16,327	823,791
California	241,113	14,026	518	1,205,291	320,988	759,816	4,382,008	68,507	149,301	7,141,567
Colorado	19,818	1,907	166	100,484	23,844	46,114	502,657	7,981	20,547	723,518
Connecticut	11,874	1,592	326	69,634	17,139	35,818	417,159	4,517	11,927	569,985
Delaware	11,040	2,336	41	21,089	6,688	10,815	129,098	1,267	4,003	186,377
District of Columbia	4,204	2,967	11	16,848	4,768	9,498	159,507	1,466	2,787	202,057
Florida	137,218	22,074	42	499,266	140,380	272,455	2,968,375	29,529	74,679	4,144,019
Georgia	93,819	9,034	33	377,924	131,418	187,897	1,943,840	20,699	55,045	2,819,708
Guam	312	0	0	5,608	1,932	6,620	78,829	351	84	93,736
Hawaii	5,131	899	2	31,733	8,550	23,613	273,684	1,746	3,733	349,092
Idaho	5,750	3,863	209	40,979	14,153	20,355	200,937	2,176	7,587	296,008
Illinois	104,440	8,149	3,357	334,927	74,301	170,092	2,322,771	22,862	43,825	3,084,725
Indiana	38,169	6,163	243	186,437	48,018	76,644	1,071,249	11,379	30,657	1,468,959
Iowa	22,473	1,554	75	75,687	16,836	34,995	419,857	4,614	17,502	593,594
Kansas	28,630	1,909	107	78,642	20,436	32,401	301,564	4,711	26,421	494,820
Kentucky	26,406	6,426	105	147,846	52,752	69,053	1,002,094	7,559	26,633	1,338,874
Louisiana	54,799	6,993	36	173,837	56,709	93,007	1,119,137	9,843	38,794	1,553,155
Maine	8,762	955	34	26,989	8,011	12,828	292,705	2,936	4,293	357,513
Maryland	34,239	5,990	426	108,386	31,318	72,979	668,683	8,014	19,531	949,566
Massachusetts	46,717	5,583	394	122,969	32,593	62,989	925,604	10,098	25,880	1,232,825
Michigan	54,894	6,473	599	230,654	63,159	115,584	2,106,871	22,432	34,500	2,635,166
Minnesota	52,687	4,230	797	111,857	27,341	68,989	472,664	8,290	27,374	774,228
Mississippi	31,742	3,853	3	138,313	50,800	73,775	691,068	6,733	20,283	1,016,569
Missouri	40,460	9,830	575	159,393	51,512	55,006	1,135,613	11,126	28,072	1,491,586
Montana	8,806	964	21	20,411	5,491	9,231	134,564	1,714	6,585	187,788
Nebraska	25,356	1,724	48	49,704	11,305	20,972	179,068	2,676	14,847	305,700
Nevada	3,990	1,200	80	60,093	14,271	27,028	285,774	3,037	9,480	404,953
New Hampshire	3,230	604	190	18,336	3,748	8,777	115,949	1,652	5,989	158,475
New Jersey	57,321	6,727	748	173,366	41,431	94,549	750,159	11,655	28,973	1,164,930
New Mexico	30,412	5,479	7	76,722	30,041	32,408	410,845	4,019	15,911	605,843
New York	168,019	41,805	928	511,643	134,787	314,990	3,955,033	33,209	86,779	5,247,192
North Carolina	75,566	4,757	196	283,317	86,765	142,030	1,625,497	17,704	46,795	2,282,628
Northern Mariana	8,729	423	39	13,727	3,374	7,590	79,565	990	6,775	121,213
North Dakota	0	0	0	0	0	0	3,191	0	0	3,191
Ohio	76,583	8,681	571	273,827	79,102	134,074	2,167,118	23,461	53,336	2,816,754
Oklahoma	49,873	2,992	42	127,140	46,210	61,879	666,447	7,857	38,150	1,000,590
Oregon	25,519	4,020	139	84,594	28,391	50,634	831,409	7,285	11,805	1,043,797
Pennsylvania	73,337	11,384	577	261,468	63,924	136,480	1,900,788	18,658	58,125	2,524,741
Puerto Rico [5]	24,887	12,353	0	122,555	32,035	206,766	0	12,018	21,269	431,883
Rhode Island	6,066	666	83	23,215	5,759	13,461	170,464	1,704	3,449	224,867
South Carolina	24,815	6,004	16	155,786	58,338	70,849	1,001,692	10,420	22,398	1,350,318
South Dakota	7,256	716	49	22,173	5,676	9,796	111,278	972	11,282	169,198

See footnotes at end of table.

Table 13-11.—Food and Nutrition Service program benefits: Cash payments made under the National School Lunch, School Breakfast, Child and Adult Care, Summer Food and Special Milk Programs and the value of food benefits provided under the Food Stamp, WIC, Commodity Distribution and the Emergency Feeding Food Programs, fiscal year 2009 [1]—Continued

State/Territory	Child Nutrition Program (cash payments only) [2]					Special Supplemental Food (WIC) [3]	Food Stamp Program	Commodity distribution [4]	Emergency Food Assistance (TEFAP)	Total [5]
	Child and Adult Care Food	Summer Food	Special Milk	National School Lunch	Break-fast					
	1,000 dollars	1,000 dollars	1,000 dollars	1,000 dollars	1,000 dollars	1,000 dollars	1,000 dollars	1,000 dollars	1,000 dollars	1,000 dollars
Tennessee	43,857	5,935	29	194,876	63,903	78,337	1,603,676	9,683	34,676	2,034,972
Texas	232,499	31,091	62	1,038,988	359,519	375,588	4,399,125	42,939	140,093	6,619,903
Utah	19,011	1,816	82	70,486	14,174	26,628	263,258	3,312	14,152	412,918
Vermont	3,829	388	79	11,534	4,080	9,626	99,238	1,438	2,845	133,057
Virginia	31,626	8,146	292	166,626	48,013	62,652	922,880	11,650	27,833	1,279,717
Virgin Islands	799	678	2	4,706	984	5,880	33,700	59	592	47,400
Washington	39,060	3,452	246	142,527	38,549	100,650	1,046,741	11,340	22,600	1,405,164
West Virginia	12,720	1,776	35	52,045	18,508	27,770	408,456	4,691	7,230	533,230
Wisconsin	36,275	4,529	1,168	124,253	29,128	61,053	679,971	9,409	24,511	970,297
Wyoming	4,381	367	16	10,955	2,700	4,815	37,075	749	2,579	63,637
Outly Areas	0	0	0	0	0	3,191	0	0	0	3,191
Dpt. of Defense [6]	0	0	0	8,725	18	0	0	0	0	8,743
Total	2,289,022	305,513	13,969	8,873,087	2,582,469	4,661,723	50,360,147	553,266	1,496,692	71,135,888

[1] Preliminary. Excludes all administrative and program evaluation costs. [2] Excludes $2.1 million for Food Safety Education and $14.7 million for Team Nutrition. [3] Includes $19.7 million for WIC Farmers Market Nutrition Program benefits. [4] Includes distribution of bonus and entitlement commodities to the National School Lunch, Child and Adult Care, Summer Food Service, Charitable Institutions, Summer Camps, Food Distribution on Indian Reservations, Nutrition Services Incentive Program (NSIP, formerly Nutrition Program for the Elderly), Commodity Supplemental Food, and Disaster Feeding Programs. Also includes cash-in-lieu of commodities for the National School Lunch and the Child and Adult Care Food programs (NSIP cash grants were transferred to the Agency on Aging, DHHS, in FY 2003). [5] Excludes Nutrition Assistance grants of $2,001 million for Puerto Rico, $19.9 million for the Northern Marianas, $25.7 million for American Samoa. [6] Dept. of Defense represents food service to children of armed forces personnel in overseas schools.

FNS, Budget Division/Program Reports, Analysis and Monitoring Branch (703) 305–2165.

Table 13-12.—Food and Nutrition Service Programs: Persons participating, fiscal years 2000–2009

Fiscal year	National School Lunch Program [1]	School Breakfast Program [1]	Child and Adult Care Program [2]	Summer Food Service [3]	WIC Program [4]
	Thousands	Thousands	Thousands	Thousands	Thousands
2000	27,305	7,553	2,707	2,103	7,192
2001	27,514	7,794	2,726	2,090	7,306
2002	28,002	8,148	2,850	1,923	7,491
2003	28,392	8,430	2,917	2,070	7,631
2004	28,962	8,905	3,009	1,997	7,904
2005	29,646	9,357	3,108	1,946	8,023
2006	30,133	9,763	3,112	1,912	8,088
2007	30,513	10,122	3,207	1,977	8,285
2008	30,015	10,609	3,252	2,130	8,705
2009 [5]	31,313	11,075	3,329	2,230	9,122

[1] Average monthly participation (excluding summer months). [2] Average daily attendance (data reported quarterly). [3] Average daily attendance for peak month (July). [4] Average monthly participation. WIC is an abbreviation for the Special Supplemental Food Program for Women, Infants and Children. [5] Preliminary.

FNS, Budget Division/Program Reports, Analysis and Monitoring Branch (703) 305–2165.

Table 13-13.—Consumers' prices: Index number of prices paid for goods and services, United States, 2000–2009 [1]

[1982–84=100]

Year	Food	Nonfood items					All items
		Apparel and upkeep	Housing		Transportation	Medical care	
			Total	Rent			
2000	167.8	129.6	169.6	201.3	153.3	260.8	172.2
2001	173.1	127.3	176.4	208.9	154.3	272.8	177.1
2002	176.2	124.0	180.3	216.7	152.9	285.6	179.9
2003	180.0	120.9	184.8	221.9	157.6	297.1	184.0
2004	186.2	120.4	189.5	227.9	163.1	310.1	188.9
2005	190.7	119.5	195.7	233.7	173.9	323.2	195.3
2006	195.2	119.5	203.2	241.9	180.9	336.2	201.6
2007	202.9	119.0	209.6	250.8	184.7	351.1	207.3
2008	214.1	118.9	216.3	257.2	195.5	364.1	215.3
2009 [1]	218.0	120.1	217.1	259.9	179.3	375.6	214.5

[1] Reflects retail prices of goods and services usually bought by average families in urban areas of the United States. This index is the official index released monthly by the U.S. Department of Labor. Beginning 1978 data are for all urban consumers; earlier data are for urban wage earners and clerical workers.

ERS, Food Markets Branch, (202) 694–5349. Compiled from data of the U.S. Department of Labor.

CHAPTER XIV

STATISTICS OF FERTILIZERS AND PESTICIDES

This chapter contains statistics on percentages of crop acres treated by various types of fertilizers and pesticides. Nitrogen, phosphate, potash, and sulfur are the most common fertilizers; herbicides, insecticides, fungicides, and other chemicals are the main categories of pesticides. Other chemicals include soil fumigants, vine killers, and dessicants. The tables show data for field crops for 2001–2009, fruits for 2007, and vegetables for 2006. NASS collects data for field crops on an annual basis and data for fruits and vegetables on a bi-yearly alternating basis. The surveyed States are generally the major producing States for each crop shown in the tables and represent 65–95 percent of the U.S. planted acres, depending on the selected crop. Application data for specific pesticide active ingredients and additional fertilizer data are available in the series of NASS "Agricultural Chemical Usage" reports and data sets.

Table 14-1.—Field crops: Fertilizer, and percent of area receiving applications, all States surveyed, 2004–2009 [1]

Crop	Nitrogen	Phosphate	Potash	Sulfur [2]
	Percent	*Percent*	*Percent*	*Percent*
2004:				
Peanuts	60	66	63	
Soybeans	21	26	23	
Wheat, Durum	95	73	7	
Wheat, Other Spring	93	79	25	
Wheat, Winter	84	55	16	
2005:				
Corn	96	81	65	13
Fall Potatoes	99	98	92	72
Oats	56	40	28	9
Upland Cotton	88	65	55	38
2006:				
Rice	97	67	54	18
Soybeans	18	23	25	3
Wheat, Durum	92	74	7	4
Wheat, Other Spring	95	85	27	13
Wheat, Winter	80	57	17	14
2007:				
All Cotton	92	67	52	42
2009:				
Wheat, Durum	99	85	11	9
Wheat, Other Spring	94	84	21	14
Wheat, Winter	83	54	16	16

[1] Refers to percent of planted acres receiving one or more applications of a specific fertilizer ingredient. [2] Estimates began in 2005. See tables 14-2 through 14-21 for surveyed States.   Note: See planted acreage estimates in tables 1-37 for corn, 2-2 for upland and all cotton, 1-45 for oats, 3-20 for peanuts, 1-27 for rice, 3-36 for soybeans, and 1-8 for wheat.
NASS, Environmental, Economics, and Demographics Branch, (202) 720–6146.

STATISTICS OF FERTILIZERS AND PESTICIDES

Table 14-2.—Barley: Pesticide usage, 2003 [1]

State and Year	Percent treated and amount applied							
	Herbicide		Insecticide		Fungicide		Other Chemicals	
	Area applied	Pounds applied	Area applied	Pounds applied	Area applied	Pounds applied	Area applied	Pounds applied
	Percent	*Thousands*	*Percent*	*Thousands*	*Percent*	*Thousands*	*Percent*	*Thousands*
CA:								
2003	67	32	*	*	*	*	*	*
ID:								
2003	94	573	3	16	*	*	5	9
MN:								
2003	89	88	8	3	39	9	....	....
MT:								
2003	93	1,005	2	5	*	*	*	*
ND:								
2003	98	1,067	4	12	11	20	....	....
PA:								
2003	32	8	*	*	*	*	....	....
SD:								
2003	86	34	....	....	*	*	....	....
UT:								
2003	75	17	*	*	....	....	....	....
WA:								
2003	94	358	....	....	....	....	*	*
WI:								
2003	21	5	....	....	....	....	....	....
WY:								
2003	83	57	10	**	....	....	....	....

[1] Data not available for all States for all years.   * Insufficient number of reports to publish data.   ** Amount applied is less than 500 lbs.   Note: Planted acres are in table 1-56.
NASS, Environmental, Economics, and Demographics Branch, (202) 720–6146.

Table 14-3.—Barley: Fertilizer usage, 2003 [1]

State and Year	Percent treated and amount applied					
	Nitrogen		Phosphate		Potash	
	Area applied	Pounds applied	Area applied	Pounds applied	Area applied	Pounds applied
	Percent	*Millions*	*Percent*	*Millions*	*Percent*	*Millions*
CA:						
2003	72	5.2	32	0.6	2	0
ID:						
2003	91	56.2	58	15.4	25	5.7
MN:						
2003	91	11.4	87	5.6	66	4.0
MT:						
2003	92	44.2	88	30.2	52	9.7
ND:						
2003	98	116.5	91	50.7	20	4.2
PA:						
2003	69	2.2	39	1.1	40	1.2
SD:						
2003	82	2.6	78	1.9	13	0.2
UT:						
2003	58	2.1	14	0.3	0	0
WA:						
2003	99	22.5	58	2.5	8	0.5
WI:						
2003	37	0.5	36	0.7	44	1.8
WY:						
2003	78	7.3	60	2.4	22	0.7

[1] Data not available for all States for all years.   Note: Planted acres are in table 1-56.
NASS, Environmental, Economics, and Demographics Branch, (202) 720–6146.

Table 14-4.—Corn: Pesticide usage, 2001–2005 [1]/[2]/

State and Year	Percent treated and amount applied			
	Herbicide		Insecticide [3]	
	Area applied	Pounds applied	Area applied	Pounds applied
	Percent	*Thousands*	*Percent*	*Thousands*
CO:				
2001	92	1,506	51	431
2003	77	1,099	39	278
2005	90	1,494	24	252
GA:				
2001	95	398	34	57
2005	91	495	14	25
IL:				
2001	100	31,868	42	1,787
2002	90	25,157	36	1,088
2003	98	28,926	58	1,640
2005	99	30,967	52	1,426
IN:				
2001	99	16,007	47	1,103
2002	90	11,535	39	729
2003	93	13,064	52	1,323
2005	97	14,136	41	722
IA:				
2001	99	20,627	7	864
2002	91	22,485	12	432
2003	96	25,328	14	623
2005	96	24,726	11	187
KS:				
2001	95	9,958	24	657
2003	97	6,041	29	337
2005	87	7,436	11	89
KY:				
2001	97	2,834	18	43
2003	97	2,716	16	52
2005	100	3,187	18	26
MI:				
2001	88	4,944	22	288
2003	98	4,934	14	206
2005	99	5,145	14	153
MN:				
2001	99	13,446	*	*
2002	96	10,002	6	212
2003	95	10,927	13	454
2005	100	10,361	12	214
MO:				
2001	97	7,232	37	167
2003	98	7,733	33	139
2005	96	7,707	11	41
NE:				
2001	99	15,159	48	1,104
2002	83	12,869	38	986
2003	93	15,209	36	742
2005	98	18,416	20	456
NY:				
2001	96	2,610	19	69
2003	96	2,107	28	141
2005	96	2,325	21	146
NC:				
2001	96	1,558	37	181
2003	97	1,854	28	213
2005	98	1,669	17	130
ND:				
2001	90	745	*	*
2003	96	1,564	*	*
2005	99	1,094	....	....
OH:				
2001	99	9,986	26	647
2002	91	8,424	14	125
2003	96	9,198	11	110
2005	99	9,322	9	215
PA:				
2001	99	4,484	60	550
2003	92	3,620	31	179
2005	97	3,346	21	154
SD:				
2001	96	5,622	8	87
2003	96	6,003	*	*
2005	100	6,036	12	239
TX:				
2001	90	1,990	76	664
2003	87	2,273	53	594
2005	94	3,344	24	236
WI:				
2001	98	6,265	16	155
2002	81	5,304	20	356
2003	98	6,533	22	273
2005	97	6,369	22	134

[1] Data not available for all States for all years.　[2] Insufficient number of reports to publish data for fungicides and other chemicals.　[3] Amount applied excludes Bt (bacillus thuringiensis) and other biologicals.　* Insufficient number of reports to publish data.　Note: Planted acres are in table 1-37.
NASS, Environmental, Economics, and Demographics Branch, (202) 720–6146.

Table 14-5.—Corn: Fertilizer usage, 2001–2005 [1]

State and Year	Percent treated and amount applied							
	Nitrogen		Phosphate		Potash		Sulfur [2]	
	Area applied	Pounds applied	Area applied	Pounds applied	Area applied	Pounds applied	Area applied	Pounds applied
	Percent	*Millions*	*Percent*	*Millions*	*Percent*	*Millions*	*Percent*	*Millions*
CO:								
2001	93	141.5	65	32.1	24	10.8		
2003	89	138.2	59	30.0	31	8.3		
2005	89	126.2	63	24.4	21	4.2	33	3.3
GA:								
2001	97	28.6	91	12.6	87	20.8		
2005	98	38.7	86	16.1	87	24.5	53	2.5
IL:								
2001	99	1,682.8	81	720.6	85	1,092.2		
2002	94	1,698.3	77	754.1	77	1,028.7		
2003	98	1,758.5	83	751.4	78	963.9		
2005	98	1,728.3	84	780.4	84	1,160.5	4	14.9
IN:								
2001	98	837.4	85	331.7	86	660.0		
2002	99	786.7	92	350.4	84	567.1		
2003	99	854.4	85	376.4	83	640.0		
2005	100	869.3	93	420.2	88	648.2	14	8.1
IA:								
2001	87	1,272.8	62	415.8	60	482.4		
2002	94	1,408.0	72	515.8	69	607.4		
2003	93	1,544.3	59	468.6	65	670.6		
2005	92	1,653.2	70	579.0	71	762.3	5	4.5
KS:								
2001	97	444.4	71	93.5	19	24.8		
2003	99	453.9	81	92.7	30	33.5		
2005	97	482.1	81	112.7	26	34.9	17	5.3
KY:								
2001	91	173.4	87	92.5	82	99.9		
2003	98	189.0	83	81.0	78	76.1		
2005	98	210.5	78	75.5	77	86.9	....	....
MI:								
2001	91	251.3	78	85.9	78	175.2		
2003	99	281.8	86	95.3	88	201.6		
2005	97	277.8	88	89.6	81	148.4	21	3.7
MN:								
2001	97	750.2	90	283.4	81	340.5		
2002	95	839.9	86	330.1	78	344.8		
2003	95	835.9	89	309.2	73	349.2		
2005	94	953.9	86	378.1	77	400.3	9	8.2
MO:								
2001	99	411.6	82	129.6	83	161.2		
2003	99	482.2	91	162.0	88	210.7		
2005	99	489.5	79	149.5	78	180.1	19	10.0
NE:								
2001	100	1,067.0	77	219.4	25	42.8		
2002	97	1,195.5	70	220.3	21	32.3		
2003	95	1,005.1	76	232.1	25	39.3		
2005	99	1,162.5	75	237.3	22	38.8	30	35.0
NY:								
2001	100	76.8	98	49.4	90	45.6		
2003	98	81.7	81	43.3	75	50.9		
2005	94	62.2	88	33.2	79	34.9	....	....
NC:								
2001	98	81.8	85	41.6	84	56.6		
2003	99	95.9	89	37.9	86	61.8		
2005	97	90.5	74	25.5	86	53.1	18	1.1
ND:								
2001	94	89.9	83	33.8	38	10.1		
2003	98	157.2	87	62.8	37	20.0		
2005	99	169.3	94	58.8	38	13.3	8	0.9
OH:								
2001	100	572.1	92	210.8	89	338.9		
2002	99	500.1	85	183.2	78	283.1		
2003	100	538.6	91	225.7	85	284.6		
2005	99	551.7	87	224.9	76	264.5	12	3.2
PA:								
2001	98	130.2	79	55.8	76	43.4		
2003	91	98.6	72	52.2	66	33.5		
2005	88	108.4	64	40.7	58	37.4	6	3.0
SD:								
2001	95	393.8	69	119.4	32	38.9		
2003	92	396.5	78	159.8	25	27.9		
2005	95	477.7	79	154.2	37	41.9	13	5.5
TX:								
2001	100	245.6	83	66.3	40	18.4		
2003	98	261.4	85	70.9	37	17.1		
2005	94	282.0	81	73.9	28	10.6	29	6.9
WI:								
2001	98	355.3	95	120.9	89	169.5		
2002	98	325.0	87	102.2	88	202.2		
2003	99	380.1	90	138.6	89	233.6		
2005	93	380.9	84	118.8	84	191.7	22	9.1

[1] Data not available for all States for all years.   [2] Estimates began in 2005.   Note: Planted acres are in table 1-37.
NASS, Environmental, Economics, and Demographics Branch, (202) 720–6146.

Table 14-6.—Upland Cotton: Pesticide usage, 2001–2007 [1]/[2]/

State and Year	Percent treated and amount applied							
	Herbicide		Insecticide [3]		Fungicide		Other Chemicals	
	Area applied	Pounds applied	Area applied	Pounds applied	Area applied	Pounds applied	Area applied	Pounds applied
	Percent	Thousands	Percent	Thousands	Percent	Thousands	Percent	Thousands
AL:								
2003	99	1,336	84	260	15	44	93	930
2005	98	1,186	74	192	2	3	89	697
2007	98	941	55	88	*	*	75	423
AZ:								
2003	94	382	74	374	*	*	80	323
AR:								
2001	96	2,312	53	2,038	8	9	78	1,395
2003	96	2,703	89	3,575	17	64	92	1,947
2005	95	2,997	84	2,669	6	18	87	1,910
2007	97	2,399	92	1,092	2	16	96	1,780
CA:								
2001	*	*	*	*	*	*	*	*
2003	97	1,005	95	899	7	13	96	2,091
2005	92	551	96	574	4	2	96	1,570
2007	90	565	90	506	2	1	93	1,414
GA:								
2001	93	2,958	59	366	*	*	65	1,902
2003	96	2,994	73	746	4	43	91	2,709
2005	99	2,958	88	1,145	*	1	95	2,539
2007	100	3,163	85	956	*	*	96	3,955
LA:								
2001	95	2,552	93	2,217	16	70	88	931
2003	100	1,448	97	2,007	17	11	99	690
2005	98	1,897	94	1,358	3	7	99	888
2007	98	992	99	562	*	*	100	567
MS:								
2001	99	3,913	92	3,306	5	22	95	2,461
2003	100	3,475	94	1,534	17	63	99	1,590
2005	100	3,947	92	1,917	6	28	98	1,880
2007	100	2,132	97	1,231	2	3	99	1,146
MO:								
2003	96	636	74	146	*	*	95	822
2007	100	995	83	270	*	*	100	867
NC:								
2001	*	*	*	*	*	*	*	*
2003	97	2,118	88	420	7	41	90	2,041
2005	99	2,181	82	597	7	41	92	1,642
2007	100	1,479	79	300	3	15	99	896
SC:								
2003	92	470	97	141	3	4	79	307
2007	100	535	92	85	13	13	86	291
TN:								
2003	98	1,270	88	422	20	33	90	863
2005	99	1,339	87	253	11	23	94	1,030
2007	100	1,482	94	228	*	*	99	985
TX:								
2001	85	5,921	58	14,587	1	19	20	1,330
2003	99	7,701	36	3,102	2	22	31	1,400
2005	93	8,677	53	5,946	....	....	47	3,075
2007	96	11,532	43	2,624	*	*	74	5,702

[1] Data not available for all States for all years.　[2] 2007 data are for all cotton　[3] Amount applied excludes Bt (bacillus thuringiensis).　* Insufficient number of reports to publish data.　Note: Planted acres are in table 2-2.
NASS, Environmental, Economics, and Demographics Branch, (202) 720–6146.

Table 14-7.—Upland Cotton: Fertilizer usage, 2001–2007 [1]

State and Year	Percent treated and amount applied							
	Nitrogen		Phosphate		Potash		Sulfur [2]	
	Area applied	Pounds applied	Area applied	Pounds applied	Area applied	Pounds applied	Area applied	Pounds applied
	Percent	*Millions*	*Percent*	*Millions*	*Percent*	*Millions*	*Percent*	*Millions*
AL:								
2003	97	51.9	84	31.2	83	33.4		
2005	98	51.4	87	27.0	90	37.0	39	3.4
2007	97	34.2	87	17.0	90	23.3	46	2.1
AZ:								
2003	93	35.3	35	4.6	11	0.8		
AR:								
2001	93	80.3	63	24.6	68	54.0		
2003	97	89.7	84	33.5	90	79.9		
2005	96	112.8	73	33.3	82	71.2	33	8.5
2007	98	94.1	83	29.4	85	63.9	46	5.5
CA:								
2001	*	*	*	*	*	*		
2003	94	72.9	47	14.3	25	11.6		
2005	96	79.8	32	10.2	22	8.3	4	0.2
2007	96	53.6	39	13.2	20	4.3	*	*
GA:								
2001	99	116.2	92	71.9	93	119.3		
2003	100	124.5	90	65.8	91	105.8		
2005	97	112.6	88	63.8	90	103.7	56	11.7
2007	98	90.9	91	56.3	91	81.3	67	10.5
LA:								
2001	95	70.8	50	18.4	52	35.1		
2003	99	45.1	45	8.8	59	16.1		
2005	99	47.5	47	12.3	49	23.3	35	1.3
2007	100	29.3	70	8.1	63	16.1	*	*
MS:								
2001	99	179.9	31	25.8	46	72.5		
2003	99	119.8	45	23.0	70	82.2		
2005	99	144.5	35	22.6	58	82.7	17	2.8
2007	100	77.3	33	12.4	54	37.7	28	2.0
MO:								
2003	100	35.5	73	11.6	81	26.2		
2007	98	36.1	88	10.3	95	24.5	64	2.6
NC:								
2001	*	*	*	*	*	*		
2003	97	59.9	74	24.4	93	79.7		
2005	95	57.9	74	25.7	95	79.0	40	7.1
2007	92	31.3	71	11.0	89	44.2	25	2.9
SC:								
2003	95	16.0	78	7.9	90	21.6		
2007	99	16.2	79	6.5	94	16.0	33	0.8
TN:								
2003	97	50.0	92	27.3	96	46.4		
2005	100	60.6	90	31.1	99	58.3	42	2.1
2007	100	52.3	95	25.2	100	45.1	60	2.8
TX:								
2001	52	195.9	37	85.2	14	16.4		
2003	61	258.0	50	141.7	20	28.6		
2005	77	310.9	64	144.9	32	35.4	40	32.3
2007	86	347.7	60	109.8	24	19.8	42	26.2

[1] Data not available for all States for all years.   [2] Estimates began in 2005. *Insufficient number of reports to publish data.   Note: Planted acres are in table 2-2.
NASS, Environmental, Economics, and Demographics Branch, (202) 720–6146.

Table 14-8.—Peanuts: Pesticide usage, 2004

State and Year	Percent treated and amount applied							
	Herbicide		Insecticide		Fungicide		Other Chemicals	
	Area applied	Pounds applied	Area applied	Pounds applied	Area applied	Pounds applied	Area applied	Pounds applied
	Percent	Thousands	Percent	Thousands	Percent	Thousands	Percent	Thousands
AL:								
2004	100	277	81	200	100	896	....	....
FL:								
2004	100	298	88	199	100	835	....	....
GA:								
2004	99	878	77	569	99	2,275	....	....
NC:								
2004	100	221	92	161	96	164	43	1,404
TX:								
2004	94	258	3	2	67	154	....	....

Note: Planted acres are in table 3-26.
NASS, Environmental, Economics, and Demographics Branch, (202) 720–6146.

Table 14-9.—Peanuts: Fertilizer usage, 2004

State and Year	Percent treated and amount applied					
	Nitrogen		Phosphate		Potash	
	Area applied	Pounds applied	Area applied	Pounds applied	Area applied	Pounds applied
	Percent	Millions	Percent	Millions	Percent	Millions
AL:						
2004	70	4.3	79	8.6	75	12.4
FL:						
2004	71	3.3	80	5.4	94	12.7
GA:						
2004	48	5.3	59	17.5	51	23.7
NC:						
2004	37	1.0	35	1.2	64	6.7
TX:						
2004	86	14.4	77	10.6	62	9.3

Note: Planted acres are in table 3-20.
NASS, Environmental, Economics, and Demographics Branch, (202) 720–6146.

Table 14-10.—Oats: Pesticide usage, 2005[1]

State and Year	Percent treated and amount applied			
	Herbicide		Insecticide[2]	
	Area applied	Pounds applied	Area applied	Pounds applied
	Percent	*Thousands*	*Percent*	*Thousands*
CA:				
2005	36	59	....	....
ID:				
2005	26	17	....	....
IL:				
2005	7	1	....	....
IA:				
2005	3	2	....	....
KS:				
2005	27	13	....	....
MI:				
2005	61	26	....	....
MN:				
2005	21	26	....	....
MT:				
2005	34	18	....	....
NE:				
2005	7	4	....	....
NY:				
2005	51	23	....	....
ND:				
2005	54	167	....	....
PA:				
2005	58	46	....	....
SD:				
2005	37	52	....	....
TX:				
2005	26	80	18	35
WI:				
2005	18	25	....	....

[1] Insufficient number of reports to publish data for fungicides and other chemicals.   Note: Planted acres are in table 1-49.
NASS, Environmental, Economics, and Demographics Branch, (202) 720–6146.

Table 14-11.—Oats: Fertilizer usage, 2005[1]

State and Year	Percent treated and amount applied							
	Nitrogen		Phosphate		Potash		Sulfur[2]	
	Area applied	Pounds applied	Area applied	Pounds applied	Area applied	Pounds applied	Area applied	Pounds applied
	Percent	*Millions*	*Percent*	*Millions*	*Percent*	*Millions*	*Percent*	*Millions*
CA:								
2005	26	4.4	....	....	....	....	....	....
ID:								
2005	42	1.6	22	1.4	5	0.1	12	0.2
IL:								
2005	15	0.4	12	0.4	26	1.7	....	....
IA:								
2005	31	1.8	30	2.5	40	6.9	....	....
KS:								
2005	84	4.4	39	1.4	17	0.8	....	....
MI:								
2005	82	2.6	72	2.8	77	3.4	....	....
MN:								
2005	28	4.2	22	2.4	28	5.9	5	0.2
MT:								
2005	53	2.0	35	1.0	14	0.4	9	0.1
NE:								
2005	68	4.5	24	1.3	7	0.1	5	0.0
NY:								
2005	75	1.9	72	2.7	72	2.8	....	....
ND:								
2005	71	15.8	49	5.7	9	0.7	5	0.1
PA:								
2005	90	4.5	81	4.9	82	5.1	2	0.1
SD:								
2005	. 64	11.8	46	5.6	17	1.7	....	....
TX:								
2005	79	45.4	56	12.7	39	4.9	25	1.7
WI:								
2005	23	2.1	24	3.9	35	15.1	8	0.4

[1] Data not available for all States for all years.   [2] Estimates began in 2005.   Note: Planted acres are in table 1-49.
NASS, Environmental, Economics, and Demographics Branch, (202) 720–6146.

Table 14-12.—Fall potatoes: Pesticide usage, 2001–2005 [1]

State and Year	Percent treated and amount applied							
	Herbicide		Insecticide [2]		Fungicide		Other Chemicals	
	Area applied	Pounds applied	Area applied	Pounds applied	Area applied	Pounds applied	Area applied	Pounds applied
	Percent	*Thousands*	*Percent*	*Thousands*	*Percent*	*Thousands*	*Percent*	*Thousands*
CO:								
2003	84	168	71	40	90	122	57	14,815
2005	78	101	57	10	78	87	34	9,678
ID:								
2001	75	714	93	853	70	691	59	46,698
2003	89	693	78	458	78	606	57	31,892
2005	90	694	65	331	81	813	49	37,732
ME:								
2001	92	28	88	13	98	530	97	405
2003	100	34	88	18	100	576	21	52
2005	100	35	91	18	100	607	12	46
MI:								
2003	94	68	99	19	96	382	48	696
2005	98	68	97	20	98	391	2	55
MN:								
2001	78	53	95	18	97	431	56	456
2003	94	42	69	6	98	461	4	1,294
2005	97	33	97	10	98	578	8	7
ND:								
2001	*	*	*	*	*	*	*	*
2003	82	57	80	29	99	1,350	3	311
2005	89	57	76	11	96	854	7	15
OR:								
2001	*	*	*	*	*	*	*	*
2003	95	71	83	140	94	169	70	3,626
PA:								
2003	91	28	99	23	96	126	6	3
WA:								
2001	92	290	95	647	91	1,108	78	14,470
2003	94	339	97	701	99	1,704	77	20,847
2005	96	328	97	517	99	1,394	70	17,171
WI:								
2001	88	73	100	110	97	1,193	86	2,644
2003	94	72	99	133	99	1,038	38	1,846
2005	99	78	97	62	99	810	49	3,327

[1] Data not available for all States for all years.   [2] Amount applied excludes Bt (bacillus thuringiensis).   *Insufficient number of reports to publish data.

NASS, Environmental, Economics, and Demographics Branch, (202) 720–6146.

STATISTICS OF FERTILIZERS AND PESTICIDES

Table 14-13.—Fall potatoes: Fertilizer usage, 2001–2005 [1]

State and Year	Percent treated and amount applied							
	Nitrogen		Phosphate		Potash		Sulfur [2]	
	Area applied	Pounds applied	Area applied	Pounds applied	Area applied	Pounds applied	Area applied	Pounds applied
	Percent	Millions	Percent	Millions	Percent	Millions	Percent	Millions
CO:								
2003	98	15.9	96	9.7	90	7.0		
2005	92	9.4	86	7.9	64	3.2	89	2.6
ID:								
2001	99	79.6	97	63.2	77	35.1		
2003	100	81.4	95	63.2	86	37.3		
2005	100	72.9	99	56.9	92	40.0	82	21.7
ME:								
2001	98	11.0	98	11.4	98	11.8		
2003	100	12.0	100	12.3	100	13.8		
2005	100	10.2	100	10.1	100	11.9	....	....
MI:								
2003	100	8.5	98	4.0	98	9.1		
2005	99	9.2	94	4.9	100	10.2	58	1.4
MN:								
2001	93	6.4	89	4.5	89	7.6		
2003	100	8.6	94	4.9	92	8.5		
2005	100	8.2	100	5.0	81	7.7	55	0.7
ND:								
2001	*	*	*	*	*	*		
2003	97	16.5	92	10.0	84	13.7		
2005	100	14.7	100	8.4	96	13.7	54	1.3
OR:								
2001	*	*	*	*	*	*		
2003	100	10.7	96	7.4	84	8.8		
PA:								
2003	100	1.9	99	1.3	99	1.4		
WA:								
2001	97	37.6	92	33.0	92	37.4		
2003	100	43.1	85	33.2	82	30.7		
2005	100	37.8	98	30.2	92	38.2	89	9.5
WI:								
2001	100	22.0	98	13.7	100	24.3		
2003	100	19.9	99	12.2	100	25.5		
2005	100	17.9	99	9.1	99	20.5	72	4.1

[1] Data not available for all States for all years. [2] Estimates began in 2005. *Insufficient number of reports to publish data. NASS, Environmental, Economics, and Demographics Branch, (202) 720–6146.

Table 14-14.—Rice: Pesticide usage, 2006 [1]

State and Year	Percent treated and amount applied							
	Herbicide		Insecticide [2]		Fungicide		Other Chemicals	
	Area applied	Pounds applied	Area applied	Pounds applied	Area applied	Pounds applied	Area applied	Pounds applied
	Percent	Thousands	Percent	Thousands	Percent	Thousands	Percent	Thousands
AR: 2006	95	3,054	10	14	37	109	5	269
CA: 2006	93	2,500	14	2	50	738	....	....
LA: 2006	96	475	42	49	46	30	*	*
MS: 2006	100	502	55	14	46	16	3	36
MO: 2006	100	454	....	....	25	12	*	*
TX: 2006	97	496	77	83	55	21	....	....

[1] Data not available for all States for all years.　[2] Amount applied excludes Bt (bacillus thuringiensis).　*Insufficient number of reports to publish data.　Note: Planted acres are in table 1-27.
NASS, Environmental, Economics, and Demographics Branch, (202) 720–6146.

Table 14-15.—Rice: Fertilizer usage, 2006 [1]

State and Year	Percent treated and amount applied							
	Nitrogen		Phosphate		Potash		Sulfur	
	Area applied	Pounds applied	Area applied	Pounds applied	Area applied	Pounds applied	Area applied	Pounds applied
	Percent	Millions	Percent	Millions	Percent	Millions	Percent	Millions
AR: 2006	97	281.2	68	54.7	60	64.9	9	6.0
CA: 2006	94	61.4	75	18.2	40	7.2	31	4.0
LA: 2006	99	52.8	78	14.6	75	16.2	4	0.3
MS: 2006	99	35.8	29	2.5	4	0.5	42	1.5
MO: 2006	100	45.2	47	5.5	42	5.7	29	0.7
TX: 2006	97	29.2	92	5.8	89	6.0	30	0.6

[1] Data not available for all States for all years.　Note: Planted acres are in table 1-27.
NASS, Environmental, Economics, and Demographics Branch, (202) 720–6146.

Table 14-16.—Sorghum: Pesticide usage, 2003 [1]/[2]/

State and Year	Percent treated and amount applied			
	Herbicide		Insecticide	
	Area applied	Pounds applied	Area applied	Pounds applied
	Percent	Thousands	Percent	Thousands
CO: 2003	52	132	*	*
KS: 2003	90	9,014	....	....
MO: 2003	98	571	6	4
NE: 2003	98	2,030	4	29
OK: 2003	84	329	*	*
SD: 2003	87	430	*	*
TX: 2003	78	2,881	20	208

[1] Data not available for all States for all years.　[2] Insufficient number of reports to publish data for fungicides and other chemicals.　[3] Amount applied excludes Bt (bacillus thuringiensis).　* Insufficient number of reports to publish data.　Note: Planted acres are in table 1-65.
NASS, Environmental, Economics, and Demographics Branch, (202) 720–6146.

Table 14-17.—Sorghum: Fertilizer usage, 2003[1]

State and Year	Nitrogen		Phosphate		Potash	
	Area applied	Pounds applied	Area applied	Pounds applied	Area applied	Pounds applied
	Percent	*Millions*	*Percent*	*Millions*	*Percent*	*Millions*
CO:						
2003	61	7.8	39	5.5	0	0
KS:						
2003	97	261.8	55	57.5	4	4.7
MO:						
2003	100	25.0	75	9.1	72	10.8
NE:						
2003	99	56.7	40	6.1	1	0.1
OK:						
2003	69	15.5	36	3.6	11	0.8
SD:						
2003	84	13.0	54	4.4	3	0.1
TX:						
2003	63	182.8	43	45.5	14	5.5

[1] Data not available for all States for all years.   Note: Planted acres are in table 1-65.
NASS, Environmental, Economics, and Demographics Branch, (202) 720–6146.

Table 14-18.—Soybeans: Pesticide usage, 2002–2006 [1][2]/

State and Year	Herbicide		Insecticide [3]		Fungicide		Other Chemicals	
	Area applied	Pounds applied	Area applied	Pounds applied	Area applied	Pounds applied	Area applied	Pounds applied
	Percent	*Thousands*	*Percent*	*Thousands*	*Percent*	*Thousands*	*Percent*	*Thousands*
AR:								
2002	90	2,945	14	112				
2004	92	3,642	7	57	6	23		
2005	95	4,152	14	344	8	21		
2006	88	4,317	12	96	9	26	*	*
IL:								
2002	100	12,939	*	*				
2004	98	10,832	1	15				
2005	99	11,767	9	384				
2006	99	13,794	5	141	2	12		
IN:								
2002	100	7,853	*	*				
2004	99	7,037						
2005	99	6,511	18	209				
2006	100	8,910		(¹)	6	44		
IA:								
2002	99	13,143	9	58				
2004	98	11,964	1	5				
2005	96	11,281	16	509				
2006	99	13,946	9	127	*	*		
KS:								
2002	98	2,931	*	*				
2004	97	3,225						
2005	100	3,549						
2006	100	4,386	6	7		*		*
KY:								
2002	100	1,479	*	*				
2005	89	1,385	2	9				
2006	97	1,978	7	1	8	40	*	
LA:								
2002	98	1,257	72	470	14	8		
2005	97	1,285	44	277	13	15		
2006	97	1,664	75	499	37	66	*	
MD:								
2002	98	753	3	*				
MI:								
2002	98	2,496	*	*				
2005	92	2,061	42	172				
2006	98	2,390	*	*	*			
MN:								
2002	99	7,073	*	*				
2004	98	8,289						
2005	99	7,310	30	125				
2006	99	9,715	56	896	*	*	*	*
MS:								
2002	98	2,392	24	24				
2005	100	2,860	10	9				
2006	100	3,770	26	65	12	30	*	*
MO:								
2002	99	5,924	*	*				
2004	98	5,394						
2005	99	5,382						
2006	95	6,577	8	28	6	70		

See footnotes at end of table.

Table 14-19.—Soybeans: Fertilizer usage, 2002–2006 [1]

State and Year	Percent treated and amount applied							
	Nitrogen		Phosphate		Potash		Sulfur [2]	
	Area applied	Pounds applied	Area applied	Pounds applied	Area applied	Pounds applied	Area applied	Pounds applied
	Percent	*Millions*	*Percent*	*Millions*	*Percent*	*Millions*	*Percent*	*Millions*
AR:								
2002	7	5.2	36	57.8	35	66.1		
2004	10	9.3	38	67.2	38	98.4		
2006	3	0.9	34	60.4	36	94.4	*	*
IL:								
2002	18	37.5	25	143.1	38	422.6		
2004	14	49.5	18	185.1	32	525.2		
2006	11	18.1	16	96.0	31	290.2	*	*
IN:								
2002	18	17.4	24	67.9	46	276.0		
2004	15	30.7	25	121.4	40	331.5		
2006	16	15.2	20	54.6	32	177.4	*	*
IA:								
2002	3	9.3	7	48.3	12	163.7		
2004	10	38.4	11	99.8	15	157.2		
2006	7	10.8	12	64.4	20	172.6	1	0.9
KS:								
2002	24	12.2	25	28.7	8	5.9		
2004	22	22.0	25	34.2	5	7.1		
2006	21	10.5	25	32.0	8	8.8	*	*
KY:								
2002	21	9.6	37	30.3	38	46.6		
2006	28	14.6	40	35.3	41	44.5	*	*
LA:								
2002	2	0.1	18	5.5	18	7.5		
2006	4	0.4	13	4.9	16	9.3	*	*
MD:								
2002	23	2.7	17	2.9	26	7.0	....	....
MI:								
2002	44	24.4	34	32.0	67	119.1		
2006	28	5.9	28	19.5	56	96.7	3	0.2
MN:								
2002	11	16.1	12	34.2	10	39.1		
2004	19	41.3	18	81.2	16	85.6		
2006	16	15.3	18	53.2	16	57.4	*	*
MS:								
2002	12	3.7	20	15.8	20	25.7		
2006	6	1.0	14	9.0	19	25.6	*	*
MO:								
2002	13	11.8	29	62.9	36	158.1		
2004	20	23.4	35	128.1	38	206.3		
2006	12	10.9	19	45.7	22	76.2	*	*

See footnotes at end of table.

Table 14-19.—Soybeans: Fertilizer usage, 2002–2006 [1]—Continued

State and Year	Percent treated and amount applied							
	Nitrogen		Phosphate		Potash		Sulfur [2]	
	Area applied	Pounds applied	Area applied	Pounds applied	Area applied	Pounds applied	Area applied	Pounds applied
	Percent	Millions	Percent	Millions	Percent	Millions	Percent	Millions
NE:								
2002	31	23.1	36	79.9	11	14.6		
2004	25	24.6	28	76.8	7	12.4		
2006	32	20.2	32	70.4	12	15.8	12	8.0
NC:								
2002	36	14.4	36	25.0	41	51.3		
2006	39	11.0	42	25.6	44	50.8	*	*
ND:								
2002	64	44.1	59	50.5	11	3.3		
2004	64	61.3	63	113.1	11	15.7		
2006	43	22.6	42	58.3	3	1.9	*	*
OH:								
2002	20	14.1	27	62.6	56	276.4		
2004	20	19.0	24	73.0	43	282.0		
2006	19	11.9	20	40.5	40	171.4	2	1.7
SD:								
2002	37	32.5	41	102.0	15	24.4		
2004	42	38.6	45	116.0	8	12.5		
2006	29	19.7	31	49.4	8	8.6	*	*
TN:								
2002	42	14.5	47	31.1	57	48.6		
2006	42	12.0	48	28.1	63	63.8	8	0.7
VA:								
2002	25	3.6	33	7.3	46	18.4		
2006	32	3.9	34	7.4	39	15.1	4	0.3
WI:								
2002	40	9.2	35	18.9	48	54.7		
2006	31	7.4	33	18.0	55	74.2	10	2.3

[1] Data not available for all States for all years.   [2] Estimates began in 2005.   * Insufficient number of reports to publish data.   Note: Planted acres are in table 3-36.

NASS, Environmental, Economics, and Demographics Branch, (202) 720–6146.

Table 14-20.—Wheat: Pesticide usage, 2004–2009 [1][2]

State and Year	Percent treated and amount applied							
	Herbicide		Insecticide [3]		Fungicide		Other Chemicals	
	Area applied	Pounds applied	Area applied	Pounds applied	Area applied	Pounds applied	Area applied	Pounds applied
	Percent	*Thousands*	*Percent*	*Thousands*	*Percent*	*Thousands*	*Percent*	*Thousands*
Winter								
CO:								
2004	54	908						
2006	54	1,018	*	*				
2009	75	2,535	17	160	*	*		
ID:								
2004	94	380	1	2				
2006	84	349	*	*	5	3		
2009	100	478	*	*	16	12	*	*
IL:								
2004	35	41			9	11		
2006	46	62	*	*	6	7		
2009	40	31	*	*	11	10		
KS:								
2004	38	1,138						
2006	53	2,600		*	*			
2009	51	4,789	*	*	*	*		
MI:								
2004	50	94	11	3	11	11		
2006	71	148	3	**	23	17		
2009	52	113	*	*	17	25		
MN:								
2009	83	51	*	*	30	3		
MO:								
2004	35	109	8	9				
2006	28	49	12	12	6	10		
2009	42	57	16	3	18	16		
MT:								
2004	95	2,533						
2006	92	2,315	*	*	*	*		
2009	100	3,746	*	*	5	8		
NE:								
2004	51	537						
2006	56	399			4	8		
2009	61	787			12	25		
ND:								
2009	99	652	*	*	66	41		
OH:								
2004	29	96						
2006	44	93	*	*	*	*		
2009	31	78	*	*	*	*		
OK:								
2004	34	267	24	511				
2006	20	495	7	138				
2009	53	2,359	12	159	*	*		
OR:								
2004	98	694	3	7	3	5		
2006	87	366	*	*	3	3		
2009	94	690	*	*	4	5		
SD:								
2004	66	646			13	21		
2006	74	749	*	*	21	27		
2009	91	1,183	*	*	37	54		
TX:								
2004	19	810	7	189				
2006	22	1,299	4	92				
2009	36	2,323	9	228	*	*		
WA:								
2004	88	1,007			4	17		
2006	94	1,077	*	*	2	5		
2009	99	1,723			4	8	*	*

See footnotes at end of table.

Table 14-20.—Wheat: Pesticide usage, 2004–2009 [1][2]—Continued

State and Year	Percent treated and amount applied							
	Herbicide		Insecticide [3]		Fungicide		Other Chemicals	
	Area applied	Pounds applied	Area applied	Pounds applied	Area applied	Pounds applied	Area applied	Pounds applied
	Percent	*Thousands*	*Percent*	*Thousands*	*Percent*	*Thousands*	*Percent*	*Thousands*
Durum								
ID:								
2009	*	*	....	....	*	*	*	*
MT:								
2004	99	508	....	....	....	....	....	....
2006	89	250	....	....	*	*	....	....
2009	100	522	*	*	*	*	....	....
ND:								
2004	99	1,216	....	....	*	*	....	....
2006	97	862	*	*	*	*	....	....
2009	100	1,618	*	*	30	47	....	....
SD:								
2009	*	*	....	....	*	*	....	....
Other Spring								
CO:								
2009	50	5	*	*	....	....	....	....
ID:								
2004	92	288	4	6	....	....	....	....
2006	95	272	8	9	12	6	*	*
2009	96	298	4	4	15	7	*	*
MN:								
2004	99	1,054	10	28	46	84	....	....
2006	96	952	5	12	40	45	....	....
2009	97	786	23	118	59	136	....	....
MT:								
2004	95	1,652	....	....	....	....	....	....
2006	91	2,172	....	....	*	*	....	....
2009	96	2,306	*	*	*	*	....	....
ND:								
2004	97	3,452	....	....	28	190	....	....
2006	95	4,723	....	....	14	88	....	....
2009	98	4,824	*	*	47	309	....	....
OR:								
2004	95	133	4	1	9	2	....	....
2009	88	114	10	1	6	1	*	*
SD:								
2004	89	702	....	....	14	26	....	....
2006	84	943	*	*	24	31	....	....
2009	96	864	13	29	43	62	....	....
WA:								
2004	99	364	4	8	3	2	....	....
2006	96	261	11	19	12	5	....	....
2009	99	505	*	*	*	*	....	....

[1] Data not available for all States for all years.   [2] Insufficient number of reports to publish data for fungicides and other chemicals.   [3] Amount applied excludes Bt (bacillus thuringiensis).   *Withheld to avoid disclosing data for individual farms/insufficient reports.   **Total applied is less than 500 pounds.   Note: Planted acres are in tables 1-6 and 1-8.
NASS, Environmental, Economics, and Demographics Branch, (202) 720–6146.

Table 14-21.—Wheat: Fertilizer usage, 2004–2009 [1]

State and Year	Percent treated and amount applied							
	Nitrogen		Phosphate		Potash		Sulfur [2]	
	Area applied	Pounds applied	Area applied	Pounds applied	Area applied	Pounds applied	Area applied	Pounds applied
	Percent	*Millions*	*Percent*	*Millions*	*Percent*	*Millions*	*Percent*	*Millions*
Winter								
CO:								
2004	59	51.2	31	15.8	5	2.7		
2006	54	36.8	36	13.5	*	*	4	0.7
2009	54	49.6	33	15.1	*	*	6	1.1
ID:								
2004	89	89.2	62	18.5	31	6.1		
2006	93	80.9	66	13.7	16	2.2	63	9.6
2009	99	85.4	54	11.2	24	5.5	64	8.5
IL:								
2004	98	103.2	85	74.2	77	92.3		
2006	93	82.1	76	49.8	76	68.4	3	0.5
2009	99	84.2	80	47.3	73	56.8	4	1.0
KS:								
2004	90	788.6	62	281.8	6	23.4		
2006	88	493.0	66	197.5	8	29.0	5	5.3
2009	94	466.8	62	170.6	7	18.0	10	6.3
MI:								
2004	97	73.5	71	27.5	77	38.4		
2006	98	57.6	74	22.2	85	33.9	37	3.0
2009	96	48.4	58	17.2	67	31.1	26	1.9
MN:								
2009	92	3.3	67	1.3	46	0.6	15	0.1
MO:								
2004	97	125.9	84	52.9	86	70.0		
2006	97	90.7	73	35.5	74	44.8	12	1.8
2009	91	67.7	75	30.2	80	39.8	15	1.6
MT:								
2004	92	83.0	83	47.3	21	3.9		
2006	87	96.8	84	46.2	31	9.9	12	2.0
2009	96	130.4	87	54.8	22	4.9	16	2.7
NE:								
2004	73	76.4	42	24.3	3	1.2		
2006	75	73.3	57	34.0	4	1.4	13	1.9
2009	84	80.9	65	33.4	6	1.8	19	3.3
ND:								
2009	100	53.9	97	19.2	26	2.0	24	0.7
OH:								
2004	100	91.6	95	65.8	90	69.5		
2006	98	86.2	84	53.0	82	57.5	23	7.2
2009	98	93.5	72	39.2	73	46.9	32	6.7
OK:								
2004	92	571.0	62	147.8	13	22.0		
2006	89	283.4	65	130.9	8	9.8	*	*
2009	95	295.7	55	95.5	13	7.7	*	*
OR:								
2004	96	64.7	11	5.3	6	2.5		
2006	95	46.2	12	2.8	10	1.4	48	4.9
2009	97	39.5	23	3.6	*	*	41	3.6
SD:								
2004	77	105.8	58	44.6	7	5.1		
2006	82	78.7	57	28.1	15	4.7	12	1.1
2009	83	99.9	55	25.9	15	2.9	*	*
TX:								
2004	64	347.7	35	116.6	9	9.6		
2006	44	152.1	29	47.3	8	20.8	11	5.3
2009	47	141.4	29	46.2	11	6.3	19	10.2
WA:								
2004	97	161.2	24	11.6	3	1.4		
2006	99	140.8	36	12.0	10	3.5	71	18.0
2009	100	130.0	27	8.8	7	3.3	60	13.9

See footnotes at end of table.

Table 14-21.—Wheat: Fertilizer usage, 2004–2009 [1]—Continued

State and Year	Percent treated and amount applied							
	Nitrogen		Phosphate		Potash		Sulfur [2]	
	Area applied	Pounds applied	Area applied	Pounds applied	Area applied	Pounds applied	Area applied	Pounds applied
	Percent	Millions	Percent	Millions	Percent	Millions	Percent	Millions
Durum								
ID:								
2009	*	*	*	*	*	*	*	*
MT:								
2004	96	32.5	84	11.8	10	0.6		
2006	93	20.6	82	7.3	8	0.3	4	0.1
2009	97	29.8	82	11.2	9	1.1	8	0.2
ND:								
2004	95	115.3	70	35.1	6	1.1		
2006	92	77.4	71	21.3	7	0.8	4	0.1
2009	99	101.1	87	35.3	11	1.4	9	0.8
SD:								
2009	*	*	*	*	*	*	*	*
Other Spring:								
CO:								
2009	*	*	*	*	....	....	....	....
ID:								
2004	93	56.1	63	12.7	23	4.4		
2006	96	60.7	56	9.5	25	3.5	59	8.5
2009	98	59.1	78	12.5	18	3.0	74	7.7
MN:								
2004	98	180.1	91	75.5	54	34.8		
2006	99	148.5	97	64.0	72	31.6	2	0.4
2009	94	148.2	88	56.5	56	31.9	9	0.9
MT:								
2004	79	134.6	69	72.6	13	9.0		
2006	86	129.5	81	57.7	21	9.0	10	2.5
2009	85	115.6	80	45.8	15	3.6	10	2.0
ND:								
2004	98	691.9	86	269.0	27	39.9		
2006	99	504.6	88	202.2	21	13.0	11	4.3
2009	96	461.2	89	172.0	18	18.8	6	3.7
OR:								
2004	91	9.7	28	1.7	9	0.5		
2009	97	8.4	*	*	11	0.4	68	1.1
SD:								
2004	92	132.5	68	53.2	19	8.5		
2006	90	119.4	80	55.6	22	11.9	10	3.5
2009	*	*	86	48.7	19	5.9	9	0.8
WA:								
2004	100	45.4	67	7.4	9	2.1		
2006	100	43.6	60	4.7	9	1.6	89	6.4
2009	99	45.8	62	6.0	6	0.7	87	7.8

[1] Data not available for all States for all years. [2] Estimates began in 2005. *Withheld to avoid disclosing data for individual farms/or insufficient reports. Note: Planted acres are in tables 1-6 and 1-8.
NASS, Environmental, Economics, and Demographics Branch, (202) 720–6146.

Table 14-22.—Fruits, Pesticides: Percent of acres receiving applications, for surveyed States, 2007 [1]

Crop	Herbicide	Insecticide	Fungicide	Other
	Percent			
Apples	61	97	91	65
Apples Organic		79	75	51

[1] Refers to acres receiving one or more applications of a specific agricultural chemical.
NASS, Environmental, Economics, and Demographics Branch, (202) 720–6146.

Table 14-23.—Fruit, Fertilizers: Percent of acres receiving applications, for surveyed States, 2007[1]

Crop	Nitrogen	Phosphate	Potash	Sulfur
		Percent		
Apples	71	24	34	12
Apples Organic	53	25	26	12

[1] Refers to acres receiving one or more applications of a specific agricultural chemical.
NASS, Environmental, Economics, and Demographics Branch, (202) 720–6146.

Table 14-24.—Vegetables: Percent of acres receiving applications, for surveyed States, 2006[1]

Crop	Herbicide	Insecticide	Fungicide	Other
		Percent		
Asparagus	78	68	34	*
Beans, Snap, Fresh	48	72	67	7
Beans, Snap, Proc	95	73	53	1
Broccoli	48	84	19	*
Cabbage, Fresh	56	94	60	8
Cantatoupes	43	79	62	26
Carrots, Fresh	67	23	62	*
Carrots, Proc	96	75	86	*
Cauliflower	42	77	3	*
Celery	55	92	74	*
Sweet Corn, Fresh	83	88	20	2
Sweet Corn, Proc	86	72	8	*
Cucumbers, Fresh	40	75	75	25
Cucumbers, Pickles	83	34	68	*
Eggplant	28	85	62	*
Garlic	49	31	55	*
Honeydews	*	*	*	*
Head Lettuce	63	98	87	1
Other Lettuce	62	93	74	*
Onions	79	78	76	18
Green Peas, Proc	92	19	3	*
Bell Peppers	57	91	83	26
Pumpkins	75	79	75	*
Spinach	54	74	61	24
Squash	34	72	71	13
Strawberries	22	80	89	53
Tomatoes, Fresh	41	82	81	27
Tomatoes, Proc	65	71	76	23
Watermelons	38	56	80	13

[1] Refers to acres receiving one or more applications of a specific agricultural chemical.   * Insufficient number of reports to publish data.
NASS, Environmental, Economics, and Demographics Branch, (202) 720–6146.

Table 14-25.—Vegetables, fertilizers: Percent of acres receiving applications, for surveyed States, 2006[1]

Crop	Nitrogen	Phosphate	Potash	Sulfur
		Percent		
Asparagus	79	43	57	28
Beans, Snap, Fresh	87	71	84	14
Beans, Snap, Proc	96	81	81	30
Broccoli	96	63	40	27
Cabbage, Fresh	95	89	77	27
Cantatoupes	98	92	28	33
Carrots, Fresh	83	67	41	19
Carrots, Proc	86	65	61	30
Cauliflower	95	72	41	38
Celery	94	83	86	*
Sweet Corn, Fres	95	91	85	27
Sweet Corn, Proc	94	78	64	33
Cucumbers, Fresh	96	72	95	23
Cucumbers, Pickles	96	84	75	15
Eggplant	99	92	99	*
Garlic	98	88	42	*
Honeydews	95	88	*	*
Head Lettuce	98	84	32	38
Other Lettuce	93	79	*	*
Onions	95	77	53	38
Green Peas, Proc	74	43	45	21
Bell Peppers	99	89	92	23
Pumpkins	90	63	85	13
Spinach	88	83	53	40
Squash	90	64	82	15
Strawberries	97	91	94	22
Tomatoes, Fresh	98	97	96	55
Tomatoes, Proc	98	77	50	20
Watermelons	98	87	87	26

[1] Refers to acres receiving one or more applications of a specific agricultural chemical. * Insufficient number of reports to publish data.

NASS, Environmental, Economics, and Demographics Branch, (202) 720–6146.

CHAPTER XV

MISCELLANEOUS AGRICULTURAL STATISTICS

This chapter contains miscellaneous data which do not fit into the preceding chapters. Included here are summary tables on foreign trade in agricultural products; statistics on fishery products; tables on refrigerated warehouses; and statistics on crops in Alaska.

Foreign Agricultural Trade Statistics

Agricultural products, sometimes referred to as food and fiber products, cover a broad range of goods from unprocessed bulk commodities like soybeans, feed corn and wheat to highly-processed, high-value foods and beverages like sausages, bakery goods, ice cream, or beer sold in retail stores and restaurants. All of the products found in Chapters 1-24 (except for fishery products in Chapter 3) of the U.S. Harmonized Tariff Schedule are considered agricultural products. These products generally fall into the following categories: grains, animal feeds, and grain products (like bread and pasta); oilseeds and oilseed products (like canola oil); livestock, poultry and dairy products including live animals, meats, eggs, and feathers; horticultural products including all fresh and processed fruits, vegetables, tree nuts, as well as nursery products and beer and wine; unmanufactured tobacco; and tropical products like sugar, cocoa, and coffee. Certain other products are considered "agricultural," the most significant of which are essential oils (Chapter 33), raw rubber (Chapter 40), raw animal hides and skins (Chapter 41), and wool and cotton (Chapters 51-52). Manufactured products derived from plants or animals, but which are not considered "agricultural" are cotton yarn, textiles and clothing; leather and leather articles of apparel; and cigarettes and spirits.

U.S. foreign agricultural trade statistics are based on documents filed by exporters and importers and compiled by the Bureau of the Census. Puerto Rico is a Customs district within the U.S. Customs territory, and its trade with foreign countries is included in U.S. export and import statistics. U.S. export and import statistics include merchandise trade between the U.S. Virgin Islands and foreign countries even though the Virgin Islands of the United States are not officially a part of the U.S. Customs territory.

Data on trade of other U.S. outlying possessions with foreign countries is not compiled by the United States. Export statistics are fully compiled on shipments to all countries, except Canada, where the value of commodities classified under each individual Schedule B number is over $2,500. Value data for such commodities valued under $2,501 are estimated for individual countries using factors based on the ratios of low-valued shipments to individual country totals for past periods. The estimates for low-valued shipments are shown under a single Schedule B number and are omitted from the statistics for the detailed commodity classifications. Shipments valued under $2,501 to all counties, except Canada, represent slightly less that 2.5 percent of the monthly value of U.S. exports to those countries. As a result of the data exchange between the United States and Canada, the United States has adopted the Canadian import exemption level for its export statistics on shipments to Canada. The Canadian import exemption level is based on total value per shipment rather than value per commodity classification line item.

The export value, the value at the port of exportation, is based on the selling price and includes inland freight, insurance, and other charges to the port. The country of destination is the country of ultimate destination or where the commodities are consumed or further processed. When the shipper does not know the ultimate destination, the shipments are credited to the last country, as known at the time of shipment from the United States.

Agricultural products, like manufactured goods, are often transhipped from the one country to another. Shippers are asked to identify the ultimate destination of a shipment. However, transhipment points are often recorded as the ultimate destination even though the actual point of consumption may be in a neighboring state. Thus, exports to countries which act as transhipment points are generally overstated, while exports to neighboring countries are often understated. Major world transhipment points include the Netherlands, Hong Kong, and Singapore. In such cases, exports are over reported for the Netherlands, but under reported for Germany, Belgium and the United Kingdom. They are overstated to Hong Kong, but under reported to China, and they overstated to Singapore, but understated to Malaysia and Indonesia. After the collapse of communism in Eastern Europe and Russia, Germany and the Baltic countries became important transhipment points to those countries further east.

Imports for consumption are a combination of entries for immediate consumption and withdrawals from warehouses for consumption. The import value, defined generally as the market value in the foreign country, excludes import duties, ocean freight, and marine insurance. The country of origin is defined as the country where the commodities were grown or processed. Where the country of origin is not known, the imports are credited to the country of shipment.

Import statistics are fully compiled on shipments valued over $1,250. Value data for shipments valued under $1,251 are not required to be reported on formal entries. They are estimated for individual countries using factors based on the ratios of low-valued shipments to individual country totals for past periods. The estimates for low-valued shipments are shown under a single HTS number. The total value excluded represents slightly less than 1 percent of the monthly import value.

Table 15-1.—Foreign trade: Value of total agricultural exports and imports, United States, fiscal years 2000–2009

Fiscal year ending Sep. 30 [1]	U.S. total domestic exports			U.S. total imports for consumption, customs value			Surplus agricultural exports over agricultural imports
	Total merchandise exports	Agricultural exports [2]	Agricultural exports share of total exports	Total merchandise imports	Agricultural imports	Agricultural imports share of total imports	
	Million dollars	Million dollars	Percent	Million dollars	Million dollars	Percent	Million dollars
2000	701,651	50,744	7	1,167,768	38,857	3	11,887
2001	690,634	52,698	8	1,152,642	39,027	3	13,671
2002	628,263	53,319	8	1,120,323	40,960	4	12,360
2003	637,160	56,014	9	1,222,580	45,692	4	10,322
2004	712,326	62,409	9	1,397,129	52,668	4	9,741
2005	783,806	62,516	8	1,610,655	57,711	4	4,805
2006	895,629	68,593	8	1,824,308	64,026	4	4,566
2007	1,016,871	82,217	8	1,899,097	70,063	4	12,154
2008 [1]	1,183,448	115,305	10	2,147,849	79,320	4	35,985
2009 [3]	943,803	96,632	10	1,595,068	73,418	5	23,215

[1] Fiscal years Oct. 1–Sept. 30 revised.   [2] Included food exported for relief or charity by individuals and private agencies.   [3] Fiscal 2009 is nonrevised data.
ERS, Market and Trade Economics Division, (202) 694–5211.

Table 15-2.—Foreign trade: Value and quantity of bulk commodity exports, United States, fiscal years 2004–2009 [1]

Fiscal year	Wheat, unmilled	Rice, milled	Feed grains [2]	Oilseeds [3]	Tobacco unmanufactured	Cotton and linters	Bulk commodities
	Value						
	Million dollars	Million dollars	Million dollars	Million dollars	Million dollars	Million dollars	Million dollars
2004	5,095	1,198	6,611	8,178	1,050	4,534	26,666
2005	4,252	1,235	5,316	7,685	988	3,880	23,356
2006	4,289	1,291	6,808	7,161	1,058	4,678	25,286
2007	6,579	1,273	9,783	9,339	1,143	4,305	32,423
2008	12,332	2,010	15,750	15,580	1,280	4,762	51,714
2009	5,997	2,249	10,018	14,872	1,199	3,628	37,963
	Quantity						
	1,000 metric tons	1,000 metric tons	1,000 metric tons	1,000 metric tons	1,000 metric tons	1,000 metric tons	1,000 metric tons
2004	31,179	3,690	53,770	25,482	163	3,021	117,306
2005	26,505	4,248	50,538	30,385	152	3,375	115,204
2006	25,005	4,014	61,363	27,593	169	3,707	121,851
2007	29,636	3,306	59,051	31,592	180	3,128	126,893
2008	32,847	3,889	68,205	32,148	184	2,970	140,253
2009	22,522	3,401	51,604	35,915	167	2,814	116,453

[1] Fiscal years, Oct. 1–Sept. 30.   [2] Corn, barley, sorghum, rye, and oats.   [3] Soybeans, peanuts, rapeseed, cottonseed, sunflowerseed, safflowerseed, and others.
ERS, Market and Trade Economics Division, (202) 694–5211.

Table 15-3.—Agricultural exports: Value to top 50 countries of destination, United States, fiscal years 2007–2009 [1]

Country	2007	2008	2009
	Million dollars	Million dollars	Million dollars
Canada	13,260.9	16,257.4	15,518.3
Mexico	12,331.0	15,581.7	13,459.7
Japan	9,738.8	13,061.0	11,221.1
China	7,051.0	11,169.6	11,157.3
European Union-27	8,039.6	10,659.9	7,620.0
South Korea	3,189.6	5,552.3	3,825.6
Taiwan	2,917.1	3,509.4	2,891.6
Hong Kong	1,082.1	1,597.5	1,789.0
Indonesia	1,374.5	2,205.3	1,667.1
Russian Federation	1,124.4	1,887.5	1,429.2
Egypt	1,685.2	2,198.7	1,420.8
Turkey	1,362.9	1,731.9	1,387.5
Philippines	962.9	1,730.4	1,246.8
Venezuela	517.8	1,451.0	1,051.3
Colombia	1,123.6	1,756.9	954.3
Thailand	802.3	1,144.2	910.0
Dominican Republic	723.1	1,038.4	893.1
Vietnam	367.5	896.1	853.6
Nigeria	581.9	1,034.1	839.8
Australia	611.1	810.2	809.4
Guatemala	644.9	824.9	729.5
United Arab Emirates	449.6	558.7	708.4
Saudi Arabia	536.5	978.1	686.1
India	460.1	505.9	600.8
Cuba	373.2	657.7	575.1
Malaysia	508.4	630.9	562.3
Israel	505.0	898.4	502.1
Costa Rica	385.7	624.3	468.6
Honduras	387.9	493.8	446.1
Singapore	339.3	493.4	439.0
Peru	348.7	502.3	432.7
Switzerland	217.8	334.5	412.7
Morocco	468.2	722.2	390.6
Panama	264.0	405.3	385.5
Brazil	386.0	663.5	382.1
El Salvador	314.9	462.0	378.5
Iran	17.9	365.5	377.6
Jamaica	275.0	402.7	335.2
Haiti	221.4	398.2	313.1
Pakistan	282.7	376.5	283.4
Syria	317.2	449.4	278.9
Trinidad and Tobago	199.1	287.2	272.9
Chile	335.7	504.3	255.5
Republic of South Africa	261.6	342.2	229.0
New Zealand	204.4	212.0	225.5
Ecuador	181.9	265.5	216.3
Bahamas	182.8	194.1	214.7
Jordan	133.4	149.0	213.5
Nicaragua	155.9	214.6	213.3
Bangladesh	170.8	169.6	189.9
Other	3,839.6	5,945.4	3,967.6
Total World agricultural exports [2]	82,216.8	115,305.4	96,632.2

[1] Fiscal years Oct. 1–Sept. 30.　　[2] Totals may not add due to rounding.
ERS, Market and Trade Economics Divison, (202) 694–5211.

Table 15-4.—Foreign trade in agricultural products: Value of exports by principal commodity groups, United States, fiscal years 2006–2009 [1]

Commodity	2006	2007	2008	2009
	1,000 dollars	1,000 dollars	1,000 dollars	1,000 dollars
Total Merchandise Exports	895,629,382	1,016,870,752	1,183,447,973	943,802,506
Nonagricultural U. S. Exports	827,036,723	925,288,615	1,068,142,534	847,170,313
Total Agricultural exports	68,592,956	82,216,762	115,305,439	96,632,193
Animals and animal products	13,074,386	16,007,989	21,671,168	18,284,185
Animals, live excluding poultry	495,675	574,101	592,693	526,475
Cattle and calves-live	16,562	31,462	98,904	74,398
Horses,mules,burros-live	444,349	513,702	452,983	428,905
Swine-live	26,023	18,830	28,514	11,653
Sheep-live	7,776	8,976	10,857	10,378
Other live Animals	966	1,131	1,435	1,141
Red meat and products	4,953,740	5,808,287	8,365,990	7,853,666
Beef and Veal	1,398,840	1,899,374	2,663,510	2,584,985
Beef or veal-fr or frozen	1,302,443	1,789,629	2,530,734	2,462,426
Beef prep or pres	96,397	109,745	132,776	122,559
Horsemeat fr chill. Froz	64,841	34,553	514	50
Lamb, mut or goat-fr. ch, frz	22,001	13,168	18,148	23,364
Pork	2,399,029	2,624,671	3,932,493	3,654,395
Pork-fr or froz	2,122,238	2,356,011	3,632,213	3,262,799
Pork prep or pres	276,791	268,660	300,280	391,596
Variety meats, ed. offals	795,909	862,461	1,301,622	1,143,551
Beef variety meats	511,916	571,893	802,191	555,016
Pork variety meats	249,247	258,562	469,183	560,717
Other variety meats	34,746	32,006	30,248	27,819
Other meats- fr. or prep	273,121	374,061	449,703	447,321
Poultry and poultry products	2,966,888	3,777,088	4,928,897	4,849,922
Poultry - live	118,346	135,454	170,279	176,967
Baby chicks	106,466	122,098	158,728	166,558
Other live poultry	11,880	13,356	11,550	10,409
Poultry meats	2,395,004	3,032,858	4,050,726	3,948,921
Chickens - fresh or frozen	1,773,849	2,366,361	3,281,030	3,210,141
Turkeys - fresh or frozen	314,161	342,969	409,425	360,801
Other poultry - fresh or frozen	16,889	16,594	17,049	14,582
Poultry meat-prep or pres	290,105	306,934	343,222	363,397
Poultry, misc	203,375	297,559	388,910	381,842
Eggs	250,164	311,216	318,983	342,191
Dairy products	1,820,152	2,522,071	4,097,482	2,335,252
Evaporated and condensed milk	12,517	37,242	51,650	19,752
Nonfat dry milk	584,869	674,312	1,582,894	606,801
Butter and anhydrous milkfat	15,348	50,728	261,372	58,780
Cheese	232,609	338,011	567,170	425,695
Whey,fluid or dried	303,748	544,536	629,182	388,717
Other dairy products	671,061	877,242	1,005,213	835,507
Fats, oils and greases	477,843	748,353	1,061,212	682,690
Lard	29,590	77,103	74,830	56,452
Tallow, inedible	296,067	439,084	638,036	390,017
Other animal fats and oils	152,186	232,165	348,346	236,221
Hides and skins, including furskins	1,978,246	2,160,521	2,130,935	1,509,774
Bovine hides, whole	1,207,363	1,177,368	1,077,525	585,069
Other cattle hides-pieces	26,229	28,174	14,271	15,136
Calf skins, whole	173,926	329,994	331,513	209,881
Horse hides, whole	54,485	54,657	64,674	94,058
Sheep and lamb skins	17,863	20,415	20,959	20,216
Other hides and Skin, Ex furs	293,598	300,620	342,895	419,422
Furskins	204,782	249,292	279,098	165,991
Mink pelts	164,471	203,048	229,812	143,267
Other furskins	40,311	46,244	49,286	22,724
Wool and mohair	33,204	34,283	27,814	17,691
Sausage casings	113,771	141,259	199,240	251,894
Bull semen	67,833	83,018	103,126	99,076
Misc animal prods - Other	167,034	159,008	163,780	157,746
Grains and feeds	18,302,021	24,433,300	38,481,170	26,464,317
Wheat,unmilled	4,289,500	6,579,492	12,332,402	5,996,595
Wheat flour	57,631	95,867	165,430	121,245
Other wheat products	107,759	118,668	151,698	169,020
Rice-paddy, milled, parb	1,290,832	1,273,429	2,009,662	2,248,871
Feed grains and products	7,498,821	10,714,344	16,368,524	10,625,031
Feed grain	6,808,162	9,782,680	15,749,639	10,017,823
Barley	47,077	94,538	264,008	51,088
Corn	6,186,650	8,932,911	13,999,314	9,311,737
Grain sorghum	568,489	750,118	1,476,894	648,377
Oats	5,578	4,865	8,657	6,142
Rye	369	247	766	480

See footnotes at end of table.

Table 15-4.—Foreign trade in agricultural products: Value of exports by principal commodity groups, United States, fiscal years 2006–2009 [1]—Continued

Commodity	2006	2007	2008	2009
	1,000 dollars	*1,000 dollars*	*1,000 dollars*	*1,000 dollars*
Feed grains and products--Continued				
Feed grain products	690,660	931,664	618,885	607,208
Popcorn	39,002	47,979	80,404	84,878
Blended food prods	71,923	47,165	71,937	99,039
Other grain prods	1,855,261	2,099,043	2,483,463	2,606,879
Feed and fodders, ex oilcake	3,091,291	3,457,313	4,817,650	4,512,757
Corn by-products	540,722	504,828	737,297	510,440
Alfalfa meal and cubes	25,176	27,513	35,479	33,229
Beef pulp	72,113	80,367	77,048	83,262
Citrus pulp pellets	26,358	41,239	91,524	25,508
Other feeds and fodders	2,426,922	2,803,366	3,876,302	3,860,318
Fruit and prep, ex juice	3,687,476	4,001,176	4,770,291	4,679,383
Fruits-fresh	2,841,751	3,010,394	3,571,948	3,519,694
Fruits-fresh-citrus	672,677	668,475	856,030	727,159
Grapefruit-fresh	163,383	232,836	199,705	184,761
Lemons and limes-fresh	79,719	121,938	158,302	106,402
Oranges and tanger-fresh	428,205	310,360	496,429	432,001
Other citrus-fresh	1,370	3,340	1,594	3,995
Fruit fresh-noncitrus	2,169,074	2,341,919	2,715,919	2,792,536
Apples-fresh	527,367	620,233	710,351	764,752
Berries-fresh	370,657	420,836	545,635	527,275
Cherries-fresh	210,339	255,595	271,098	285,700
Grapes-fresh	542,191	512,997	558,925	617,942
Melons-fresh	118,775	122,506	135,361	134,022
Peaches-fresh	116,116	133,371	147,852	144,935
Pears-fresh	118,215	130,464	171,459	148,480
Plums-fresh	60,922	58,196	65,460	65,530
Other noncitrus-fresh	104,492	87,721	109,778	103,900
Fruits dried	402,089	474,106	576,751	536,303
Raisin dried	202,852	203,263	296,515	283,764
Prunes-dried	122,791	179,852	172,406	159,351
Other dried fruits	76,446	90,990	107,830	93,188
Fruits-canned ex juice	230,006	267,876	346,291	364,029
Fruits-frozen ex juice	60,603	73,900	88,082	68,104
Other fruits-prep or pres	153,026	174,901	187,219	191,254
Fruits juices incl frozen	893,467	1,021,740	1,156,367	1,108,127
Apple juice	22,800	25,571	31,239	30,513
Grape juice	55,936	64,639	88,066	85,191
Grapefruit juice	52,440	63,895	59,569	53,524
Orange juice	315,254	360,449	408,090	380,339
Other fruit juices	447,037	502,187	569,404	558,560
Wine	754,979	867,580	926,934	798,537
Nuts and prep	3,206,596	3,254,059	3,777,077	3,774,401
Almonds	1,970,989	1,858,023	1,952,846	1,861,954
Filberts	43,604	54,454	75,512	62,467
Peanuts, shelled of prep	198,013	229,130	290,438	279,199
Pistachios	289,190	348,943	510,987	616,392
Walnuts Shelled/unshelled	373,766	391,282	495,609	566,175
Pecans shelled or unshelled	141,297	172,813	246,346	165,544
Other nuts shelled or prepared	189,736	199,413	205,339	222,671
Vegetables and preparations	3,813,788	4,161,925	4,950,280	5,004,931
Vegetables fresh	1,628,947	1,775,412	1,936,076	1,892,473
Aspargus-fresh	25,617	25,367	30,710	25,802
Broccoli-fresh	122,213	122,802	133,312	123,743
Carrots-fresh	109,844	125,263	139,120	130,053
Cabbage-fresh	25,397	21,836	20,082	19,622
Celery-fresh	51,327	69,389	62,921	65,232
Cauliflower-fresh	63,975	63,861	77,992	80,667
Corn sweet-fresh	41,346	36,371	47,958	41,070
Cucumbers-fresh	16,067	21,548	19,498	15,996
Garlic-fresh	9,226	8,087	9,384	9,237
Lettuce-fresh	365,026	393,566	426,180	418,352
Mushrooms-fresh	20,112	22,152	37,913	33,985
Onions and shallots-fresh	116,933	179,032	129,827	137,523
Peppers-fresh	81,468	94,051	106,056	89,714
Potatoes-fresh	131,234	122,706	156,270	155,982
Tomatoes-fresh	166,931	179,350	214,860	205,674
Other fresh vegetables	282,232	290,032	323,992	339,824
Vegetables-frozen	656,755	771,128	951,877	929,397
Corn, sweet, frozen	62,665	63,844	66,467	68,432
Potatoes frozen	486,638	570,784	693,818	698,065
Other frozen vegetables	107,452	136,499	191,592	162,900
Vegetables-canned	300,044	291,559	472,448	494,801
Pulses	352,700	367,073	543,765	609,496
Dried Beans	198,627	182,869	252,365	329,266
Dried Peas	85,172	119,220	198,300	170,503
Dried Lentils	57,541	50,763	75,567	93,667
Dried chickpeas	11,361	14,222	17,532	16,060
Other vegetables-prep or pres	875,342	956,753	1,046,114	1,078,763

See footnotes at end of table.

Table 15-4.—Foreign trade in agricultural products: Value of exports by principal commodity groups, United States, fiscal years 2006–2009 [1]—Continued

Commodity	2006	2007	2008	2009
	1,000 dollars	*1,000 dollars*	*1,000 dollars*	*1,000 dollars*
Oilseeds and products	10,563,511	13,579,777	22,756,360	20,936,360
Oilcake and meal	1,666,143	1,980,137	3,301,648	3,019,851
Bran and residues, legum. Veg.	14,512	14,353	20,480	18,761
Corn oilcake and meal	1,109	1,709	6,922	15,448
Soybean meal	1,584,445	1,923,058	3,210,663	2,931,075
Other oilcake and meal	66,078	41,017	63,584	54,567
Oilseeds	7,160,697	9,338,961	15,579,843	14,872,333
Rapeseed	46,295	76,643	111,355	76,449
Safflower seeds	1,212	430	0	0
Soybeans	6,333,987	8,482,670	14,515,719	13,904,168
Sunflowerseeds	110,290	121,896	143,333	162,052
Peanuts including oilstock	21,753	14,760	40,241	44,367
Other oilseeds	129,492	148,000	173,384	89,588
Protein substances	517,669	494,561	595,811	595,709
Vegetable oils	1,736,671	2,260,679	3,874,869	3,044,176
Soybean oil	311,385	608,124	1,531,516	854,772
Cottonseed oil	16,818	37,584	53,682	52,540
Sunflower oil	81,218	72,490	81,705	92,922
Corn oil	254,442	282,945	538,809	347,048
Peanut oil	3,678	5,144	5,883	3,792
Rapeseed oil	132,632	210,067	191,241	203,966
Safflower oil	18,119	19,091	24,124	30,131
Other vegetable oils and waxes	918,378	1,025,235	1,447,908	1,459,005
Tobacco-unmfg	1,058,440	1,143,483	1,279,734	1,199,467
Tobacco-light air cured	391,064	481,190	403,634	314,235
Tobacco-flue-cured	516,761	500,971	692,450	719,066
Other-tobacco-unmfg	150,615	161,322	183,650	166,166
Cotton, ex linters	4,665,668	4,293,770	4,754,444	3,581,250
Cotton linters	12,223	11,087	8,341	46,729
Essential oils	1,041,163	1,142,332	1,278,773	1,232,752
Seeds, field and garden	870,703	933,118	1,154,923	1,238,255
Sugar and tropical products	2,425,807	2,837,950	3,294,527	3,169,251
Sugar and related products	888,413	1,110,105	1,209,411	1,101,648
Sugar-cane or beet	76,969	144,166	126,248	86,159
Related sugar products	811,444	965,939	1,083,163	1,015,489
Coffee	433,866	496,905	598,678	645,671
Cocoa	136,157	157,090	173,478	136,460
Chocolate and prep	632,363	696,569	874,747	868,966
Tea and mate	210,990	240,651	277,551	253,236
Spices	86,943	90,234	110,832	118,635
Rubber-crude natural	32,775	43,044	45,074	41,002
Fibers ex cotton	4,299	3,352	4,757	3,634
Other hort products	3,176,056	3,361,302	3,732,681	3,792,472
Hops, including extract	94,783	136,575	204,461	274,144
Starches, not wheat/corn	100,574	101,941	110,080	108,341
Yeasts	59,586	71,030	80,982	74,977
Misc hort products	2,921,113	3,051,756	3,337,157	3,335,011
Nursery & greenhouse	319,106	366,570	385,624	366,659
Beverages ex juice	727,565	799,605	926,745	955,116

[1] Fiscal years, Oct. 1–Sept. 30. Totals may not add due to rounding.
ERS, Market and Trade Economics Division, (202) 694–5211. Compiled from reports of the U.S. Department of Commerce.

Table 15-5.—Foreign trade in agricultural products: Value of imports by principal groups, United States, fiscal years 2006–2009 [1]

Product	2006	2007	2008	2009
	1,000 dollars	*1,000 dollars*	*1,000 dollars*	*1,000 dollars*
Total merchandise imports	1,824,307,511	1,899,097,214	2,147,848,736	1,595,067,672
Non-agricultural U.S. imports	1,760,281,117	1,826,318,891	2,051,506,665	1,518,255,364
Total agricultural imports	64,026,394	70,062,946	79,319,971	73,417,626
Animals & prods.	11,627,195	12,101,935	12,235,060	10,679,766
Animals - live ex. poultry	2,485,482	2,668,098	2,903,208	1,951,178
Cattle and calves	1,592,101	1,681,213	1,996,837	1,359,810
Horses, mules, burros	302,100	323,007	336,808	231,408
Swine	579,888	645,865	543,550	332,499
Sheep, Live	395	55	30	8
Other live animals	10,997	17,958	25,982	27,452
Red meat & products	5,130,738	5,247,467	4,760,523	4,602,912
Beef & veal	3,250,626	3,384,119	2,963,195	2,931,288
Beef & veal - fr. or froz.	2,813,841	2,935,487	2,572,024	2,525,122
Beef & veal - prep. or pres.	436,785	448,633	391,171	406,166
Pork	1,228,126	1,182,349	1,074,781	988,054
Pork - fr. or froz.	896,248	849,351	740,699	706,074
Pork - prep. or pres.	331,878	332,997	334,082	281,980
Mutton, goat & lamb	459,576	469,719	492,990	467,655
Horsemeat - fr. or froz.	17	236	1,032	943
Variety meats - fr. or froz.	98,330	126,428	132,861	113,647
Other meats - fr. or froz.	30,071	26,673	34,749	32,290
Other meats & prods.	63,993	57,943	60,915	69,037
Poultry and prods.	404,601	459,135	461,767	393,160
Poultry - live	35,119	33,277	33,035	30,396
Poultry meat	154,023	207,609	226,891	230,134
Eggs	28,112	40,031	50,032	30,505
Poultry, misc.	187,348	178,217	151,809	102,125
Dairy products.	2,714,395	2,747,246	3,104,442	2,747,860
Milk & cream, fr. or dried	77,160	71,771	73,134	79,410
Butter & butterfat mixtures	67,335	65,884	32,908	65,730
Cheese	993,467	1,076,454	1,172,143	1,044,631
Casein & mixtures	575,873	553,438	784,997	618,814
Other dairy prods.	1,000,560	979,698	1,041,259	939,275
Fats, oils, & greases	85,455	105,101	119,047	97,654
Hides & skins	147,581	168,274	152,882	123,191
Sheep & lamb skins	1,360	1,067	1,324	729
Other hides & skins	66,426	56,636	50,680	29,779
Furskins	79,795	110,571	100,878	92,683
Wool - unmfg.	32,470	31,631	32,955	17,849
Apparel grade wool	17,807	17,002	17,366	7,824
Carpet grade wool	14,663	14,629	15,590	10,024
Sausage casings	83,445	105,513	131,036	158,001
Bull semen	34,607	34,715	35,711	25,504
Misc. animal prods	508,251	534,600	533,322	562,266
Silk, raw	171	156	168	192
Grains & feeds	5,069,009	6,150,862	8,038,538	7,549,004
Wheat, ex. seed	253,024	482,898	994,499	779,682
Corn, unmilled	17,160	44,555	86,416	56,766
Oats, unmilled	251,081	319,439	513,382	350,856
Barley, unmilled	15,785	68,106	197,334	170,630
Rice	289,875	373,549	537,294	540,754
Biscuits & wafers	1,828,673	2,023,651	2,233,123	2,167,568
Pasta & noodles	343,846	391,618	455,923	434,877
Other grains & preps.	1,392,119	1,642,868	2,062,138	2,034,940
Feeds & fodders, ex. oilcake	677,445	804,178	958,430	1,012,932
Fruits & preps.	6,341,896	7,317,536	7,721,478	8,162,541
Fruits - fr. or froz.	5,017,009	5,826,482	5,988,803	6,436,395
Apples, fresh	133,337	168,508	145,870	142,108
Avocados	262,556	483,648	585,324	705,323
Berries, excl. strawberries	447,824	600,206	672,905	657,359
Bananas & plantains - fr. or froz.	1,201,037	1,206,503	1,322,345	1,527,338
Citrus, fresh	398,319	498,872	417,034	441,962
Grapes, fresh	872,381	1,038,107	942,405	1,030,401
Kiwifruit, fresh	37,731	56,907	58,617	60,544
Mangoes	226,306	222,842	246,173	239,886
Melons	343,942	394,087	422,662	441,663
Peaches	60,277	72,261	73,188	59,986
Pears	98,253	124,066	100,695	97,768
Pineapples - fr. or froz.	452,535	425,938	472,484	470,295
Plums	42,061	47,232	40,989	38,100
Strawberries - fr. or froz.	210,282	235,610	217,890	246,035
Other fruits - fr. or froz.	230,167	251,695	270,224	277,626
Fruits - prep. or pres.	1,324,887	1,491,053	1,732,675	1,726,146
Bananas & plantains - prep. or pres.	37,252	51,465	60,121	61,647
Pineapples - canned or prep.	251,575	234,285	284,142	302,792
Other fruits - prep. or pres.	1,036,060	1,205,303	1,388,412	1,361,707
Fruit juices	1,056,273	1,616,388	1,932,292	1,414,578
Apple juice	307,745	496,540	815,770	489,995
Grape juice	80,632	100,980	125,693	116,849
Grapefruit juice	9,815	1,510	695	676
Lemon juice	14,299	15,293	33,496	48,856
Lime juice	9,349	10,839	13,864	14,673
Orange juice	279,077	580,473	498,763	330,417
Pineapple juice	88,042	71,538	91,847	117,801
Other fruit juice	267,314	339,215	352,164	295,312

See footnotes at end of table.

Table 15-5.—Foreign trade in agricultural products: Value of imports by principal groups, United States, fiscal years 2006–2009 [1]—Continued

Product	2006	2007	2008	2009
	1,000 dollars	*1,000 dollars*	*1,000 dollars*	*1,000 dollars*
Nuts & preps	1,114,408	1,128,863	1,352,658	1,215,964
Brazil nuts	45,633	45,879	43,965	37,687
Cashew nuts	544,071	592,841	707,979	631,951
Chestnuts	10,870	11,831	10,475	10,446
Coconut meat	65,134	79,074	82,598	99,156
Filberts	47,034	30,699	42,309	23,885
Macadamia nuts	75,183	45,844	48,265	51,087
Pecans	169,936	137,404	180,775	145,640
Pistachio nuts	4,500	7,103	4,445	5,024
Other nuts	152,048	178,187	231,847	211,088
Vegetables & preps.	6,944,659	7,545,119	8,236,742	7,994,608
Vegetables, fresh	3,982,073	4,171,990	4,448,886	4,237,201
Tomatoes, fresh	1,256,262	1,157,473	1,439,579	1,346,238
Asparagus fresh	246,824	272,681	310,917	309,041
Beans, fresh	53,928	43,871	55,988	58,607
Cabbage, fresh	17,689	19,555	18,170	14,030
Carrots, fresh	39,563	49,227	38,999	46,281
Cauliflower & broccoli, fresh	48,556	49,099	70,792	77,594
Celery, fresh	10,706	14,128	12,280	10,040
Cucumbers, fresh	394,830	526,972	343,436	335,572
Eggplant, fresh	40,646	45,174	46,591	52,667
Endive, fresh	5,154	5,175	4,799	4,099
Garlic, fresh	86,924	104,716	109,332	58,290
Lettuce, fresh	53,448	66,867	97,400	106,523
Okra, fresh	18,875	20,260	19,159	18,528
Onions, fresh	191,193	262,022	264,354	239,037
Peas, fresh	41,519	44,324	47,737	53,265
Peppers, fresh	784,706	765,761	818,521	749,934
Potatoes, fresh	117,903	124,665	152,073	158,934
Radishes, fresh	15,825	17,690	13,862	14,054
Squash, fresh	209,279	223,569	199,029	189,260
Other vegs., fresh	348,245	358,759	385,868	395,205
Vegetables - prep. or pres.	1,779,276	2,043,466	2,211,672	2,174,939
Bamboo shoots, preserved	17,032	26,248	23,958	25,140
Cucumbers, preserved	38,314	33,176	38,231	62,267
Garlic, dried	36,108	63,298	40,047	31,027
Olives - prep. or pres.	299,391	388,073	432,743	376,715
Mushrooms, canned	92,851	141,951	167,535	103,476
Mushrooms, dried	20,203	29,383	26,710	18,157
Onions, preserved	20,960	17,637	24,303	29,291
Artichokes - prep.	112,593	124,395	127,458	121,762
Asparagus- prep.	26,055	26,115	31,810	37,426
Tomatoes, incl. paste & sauce	135,160	170,014	153,374	163,478
Waterchestnuts	18,544	19,019	25,477	30,251
Peppers & pimentos, prep.	54,344	63,909	77,835	79,612
Veg Starches, excl. wheat & corn	55,041	70,003	86,387	79,721
Soups & sauces	180,328	190,825	227,018	216,538
Other vegetables - prep. or pres.	672,355	679,420	728,785	800,077
Vegetables, frozen	1,071,536	1,201,993	1,392,442	1,396,261
Tomatoes, frozen	510	1,258	3,934	4,596
Asparagus, frozen	8,434	11,496	13,662	10,632
Beans, frozen	44,536	53,870	61,096	58,971
Carrots, frozen	4,932	4,657	4,827	4,077
Cauliflower & broccoli, frozen	192,264	216,441	283,170	265,173
Okra, frozen	6,639	6,116	11,158	17,548
Peas, frozen	26,147	27,606	36,576	41,934
Potatoes, frozen	551,429	596,221	648,696	659,771
Other vegetables, frozen	236,644	284,328	329,321	333,559
Pulses	111,773	127,669	183,742	186,208
Dried peas	18,461	17,333	22,179	28,119
Dried beans	70,991	87,546	127,371	119,879
Dried lentils	12,246	10,337	16,567	18,486
Dried chickpeas	10,075	12,453	17,626	19,724
Sugar & related prods.	3,083,047	2,588,893	2,831,717	3,118,533
Sugar - cane & beet	1,405,812	817,754	949,423	1,249,977
Molasses	133,238	129,275	109,415	161,959
Confectionery prods.	1,168,088	1,218,572	1,253,353	1,170,532
Other sugar & related prods.	375,909	423,292	519,525	536,065
Cocoa & products	2,669,596	2,628,869	3,094,670	3,342,692
Coffee & products	3,205,768	3,653,599	4,348,886	4,147,706
Tea	419,972	446,587	487,192	498,833
Spices & herbs	617,617	756,056	929,884	879,204
Pepper	273,498	372,842	426,056	398,840
Other spices & herbs	344,119	383,214	503,829	480,363

See footnotes at end of table.

Table 15-5.—Foreign trade in agricultural products: Value of imports by principal groups, United States, fiscal years 2006–2009 [1]—Continued

Product	2006	2007	2008	2009
	1,000 dollars	*1,000 dollars*	*1,000 dollars*	*1,000 dollars*
Drugs, crude & natural	762,891	854,243	1,062,328	1,094,376
Essential oils	2,237,642	2,161,376	2,310,362	2,002,964
Fibers, excl. cotton	21,955	80,489	105,126	100,541
Rubber & gums	1,950,361	2,086,899	2,711,004	1,568,910
Tobacco - unmfg.	768,333	803,772	842,709	869,679
Tobacco - filler	707,507	747,960	794,410	830,180
Tobacco - scrap	17,862	15,462	15,415	18,791
Other tobacco	42,964	40,350	32,885	20,708
Beverages, ex. fruit juice	8,985,896	10,218,465	10,370,454	9,399,020
Wine	4,014,523	4,524,452	4,732,878	4,067,957
Malt beverages	3,394,143	3,709,183	3,682,245	3,442,225
Other beverages	1,577,230	1,984,830	1,955,330	1,888,838
Oilseeds & prods.	3,377,624	3,917,047	6,446,659	5,223,699
Oilseeds & oilnuts	306,707	449,383	883,111	629,565
Flaxseed	41,946	60,180	115,372	68,280
Rapeseed	176,602	201,299	485,009	250,042
Soybeans	37,735	79,571	141,798	176,634
Sunflower seeds	17,599	45,236	48,505	48,615
Other oilseeds & oilnuts	32,824	63,098	92,427	85,994
Oils & waxes - vegetable	2,864,057	3,241,515	5,149,274	4,240,866
Castor oil	29,098	34,547	67,767	47,944
Coconut oil	291,186	295,978	637,155	418,808
Cottonseed oil	250	315	8	45
Olive oil	994,962	956,944	1,053,580	927,881
Palm oil	259,747	421,071	919,598	761,684
Palm kernel oil	158,816	202,232	271,969	288,082
Peanut oil	28,987	39,611	59,816	30,050
Rapeseed oil	456,303	586,153	1,288,964	1,036,839
Soybean oil	9,746	12,504	33,231	35,184
Sesame oil	35,839	36,138	46,613	48,139
Other vegetable oils	599,123	656,023	770,274	646,210
Oilcake & meal	206,859	226,148	414,274	353,269
Cotton, excl. linters	13,128	8,894	4,961	74
Cotton, linters	2,609	5,101	7,939	1,284
Seeds - field & garden	594,480	683,338	763,959	780,206
Cut flowers	754,840	825,612	818,157	766,598
Nursery stock, bulbs, etc.	674,935	709,809	698,556	594,369
Other hort products	1,732,262	1,773,194	1,968,641	2,012,478
Hops, including extract	35,066	39,511	47,601	53,988
Starches, ex wheat/corn	60,216	73,568	89,245	81,438
Yeasts	157,311	171,148	194,324	204,382
Misc hort products	1,479,670	1,488,966	1,637,470	1,672,670

[1] Fiscal years, Oct. 1–Sept. 30.
ERS, Market and Trade Economics Division, (202) 694–5211. Compiled from reports of the U.S. Depart. of Commerce.

Table 15-6.—Agricultural exports: Value of U.S. exports to the top market, Canada, by commodity, fiscal years 2007–2009 [1]

Commodity	Value		
	2007	2008	2009
	1,000 dollars	*1,000 dollars*	*1,000 dollars*
Total agricultural exports	13,260,945	16,257,417	15,518,313
Animals and animal products	2,192,527	2,646,586	2,391,327
Animals Live-Ex Poultry	60,617	73,883	60,767
Cattle and calves-live	14,341	9,689	11,304
Horses, Mules, Burros-live	41,831	57,654	43,431
Swine-Live	655	912	1,051
Sheep-Live	3,314	5,107	4,463
Other live animals	477	522	518
Red meat and Products	1,103,289	1,413,418	1,255,997
Beef and Veal	531,290	733,321	610,129
Beef and Veal-fresh or frozen	433,186	614,579	501,664
Beef-prep or pres	98,104	118,742	108,464
Lamb-mutton or goat-fr-ch-froz	2,498	2,350	1,205
Pork	447,309	540,715	499,577
Pork-fresh or frozen	337,763	417,330	360,541
Pork-prep or pres	109,546	123,385	139,037
Variety meats, Ed Offals	32,280	33,995	35,375
Beef variety meats	11,248	17,780	17,555
Pork variety meats	14,249	9,057	11,378
Other variety meats	6,783	7,157	6,442
Other meats-fr or froz	89,912	103,038	109,711
Poultry and poultry products	488,146	532,163	534,650
Poultry-Live	24,197	29,113	29,995
Baby chicks	19,423	23,594	23,913
Other live poultry	4,775	5,519	6,082
Poultry meats	403,539	426,869	432,877
Chickens-fresh or frozen	231,038	238,700	247,028
Turkeys-fresh or frozen	6,329	4,674	4,930
Other poultry-fresh or frozen	6,451	6,463	4,077
Poultry meats-prep or pres.	159,721	177,032	176,842
Poultry misc.	4,498	2,406	1,834
Eggs	55,911	73,776	69,945
Dairy prods	341,537	435,056	393,401
Evap and condensed milk	2,586	4,937	1,708
Nonfat dry milk	10,171	14,858	10,995
Butter and Anhydrous Milkfat	3,277	12,311	3,590
Cheese	34,115	47,527	48,599
Whey, fluid or dried	63,776	62,622	47,433
Other dairy products	227,612	292,802	281,076
Fats, oils and greases	43,095	52,297	34,908
Lard	4,657	3,994	3,483
Tallow-inedible	9,395	18,278	12,704
Other animal fats and oils	29,043	30,026	18,722
Hides and skins include furs	97,371	78,794	50,498
Bovine hides, whole	1,978	8,673	8,260
Other cattle hides-pieces	4	0	0
Calf skins, whole	55	85	0
Horse hides whole	206	29	94
Sheep and lamb skins	417	616	240
Other hides and skins, ex.furs	1,183	171	386
Furskins	93,529	69,220	41,518
Mink pelts	72,391	51,370	29,064
Other furskins	21,138	17,850	12,454
Wool and Mohair	637	338	48
Sausage casings	8,484	9,792	11,848
Bull semen	4,220	4,639	5,498
Misc animal products-other	45,130	46,206	43,712
Grains and feeds	2,464,629	3,284,402	3,092,172
Wheat, unmilled	4,663	10,958	11,607
Wheat flour	34,242	79,553	45,693
Other wheat products	62,579	76,334	92,725
Rice-paddy,milled parb	115,435	150,765	176,883
Feed grains and products	403,131	678,341	396,188
Feed grains	309,781	575,128	312,526
Barley	6,769	13,481	9,654
Corn	300,072	557,633	299,746
Grain sorghums	722	1,298	1,170
Oats	2,151	2,579	1,865
Rye	67	137	90
Feed grain products	93,350	103,213	83,662
Popcorn	4	233	0
Other grain prods	1,203,072	1,406,006	1,511,676
Feeds and fodders, ex.oilcakes	641,502	882,213	857,400
Corn by-products	30,449	32,681	35,577
Alfalfa meal and cubes	41	136	37
Beet pulp	4,347	4,251	5,702
Other feeds and fodders	606,665	845,146	816,084

See footnotes at end of table.

Table 15-6.—Agricultural exports: Value of U.S. exports to the top market, Canada, by commodity, fiscal years 2007–2009 [1]—Continued

Commodity	Value		
	2007	2008	2009
	1,000 dollars	*1,000 dollars*	*1,000 dollars*
Fruits and prep. ex.juice	1,455,241	1,764,338	1,680,310
Fruits-fresh	1,189,991	1,446,462	1,352,407
Fruits-fresh-citrus	156,491	220,821	200,343
Grapefruit-fresh	28,449	27,059	25,529
Lemons and limes-fresh	30,339	50,541	33,202
Oranges and tangerines fresh	97,223	142,907	141,441
Other citrus-fresh	481	314	171
Fruits-fresh-noncitrus	1,033,500	1,225,641	1,152,063
Apple-fresh	142,050	159,689	130,617
Berries-fresh	323,662	432,148	423,716
Cherries-fresh	102,423	111,893	105,491
Grapes, fresh	160,880	180,130	176,386
Melon-fresh	109,021	119,202	114,687
Peaches-fresh	71,191	75,483	72,738
Pears-fresh	47,132	49,893	41,823
Plums-fresh	28,227	28,868	29,790
Other noncitrus-fresh	48,914	68,335	56,815
Fruits, dried	69,407	76,658	73,729
Raisins, dried	25,896	28,455	31,492
Prunes,dried	11,428	13,615	13,443
Other dried-fruits	32,082	34,588	28,794
Fruits-canned excl. juice	89,662	117,437	135,605
Fruits-froz. excl. juice	36,399	43,148	34,038
Other fruits-prep. or pres	69,782	80,633	84,531
Fruit juices incl. frozen	455,961	545,876	514,247
Apple juice	14,104	20,466	17,505
Grape juice	36,443	40,248	41,342
Grapefruit juice	9,429	10,296	8,541
Orange juice	242,051	285,155	251,356
Other fruit juices	153,933	189,710	195,503
Wine	188,424	235,968	206,964
Nuts and prep	394,704	482,699	421,602
Almonds	115,832	119,877	111,211
Filbert	5,649	14,033	4,435
Peanuts,shelled or prep	77,871	117,465	109,240
Pistachios	29,004	38,732	28,370
Walnuts	37,055	49,858	35,829
Pecan	37,197	35,399	28,124
Other nuts	92,096	107,335	104,394
Vegetables and prep	2,016,348	2,241,702	2,229,144
Vegetables-fresh	1,379,296	1,511,286	1,475,573
Asparagus-fresh	14,010	18,589	13,356
Broccoli-fresh	69,684	75,042	70,749
Carrots-fresh	114,543	127,678	118,109
Cabbage-fresh	19,997	17,774	17,428
Celery-fresh	58,042	49,890	51,710
Cauliflower-fresh	50,111	62,498	65,488
Corn, sweet-fresh	26,207	31,721	31,563
Cucumber-fresh	21,342	18,895	15,170
Garlic-fresh	3,175	4,332	4,253
Lettuce-fresh	347,408	375,869	380,487
Mushroom-fresh	15,596	22,486	20,948
Onion and Shallots-fresh	96,469	69,252	85,738
Peppers-fresh	89,716	98,718	84,981
Potatoes-fresh	82,329	97,230	98,283
Tomatoes-fresh	135,680	168,307	136,694
Other fresh vegetables	234,987	273,006	280,617
Vegetables-frozen	125,786	143,410	142,745
Corn, sweet-frozen	2,500	2,608	3,490
Potatoes-frozen	68,981	79,485	71,239
Other frozen vegetables	54,305	61,317	68,016
Vegetables-canned	104,547	118,956	127,302
Pulses	23,125	42,693	40,668
Dried beans	12,281	30,361	29,990
Dried peas	7,980	7,002	7,480
Dried lentils	1,133	1,330	1,550
Dried chick peas	1,731	4,000	1,648
Other veg-prep or pres	383,594	425,357	442,856

See footnotes at end of table.

Table 15-6.—Agricultural exports: Value of U.S. exports to the top market, Canada, by commodity, fiscal years 2007–2009 [1]—Continued

Commodity	Value		
	2007	2008	2009
	1,000 dollars	*1,000 dollars*	*1,000 dollars*
Oilseeds and prods	1,039,159	1,501,655	1,421,959
Oilcake and meal	317,248	507,262	432,313
Bran and residues, legum.veg.	1,134	2,316	5,042
Corn oilcake and meal	251	185	56
Soybean meal	313,080	494,733	419,900
Other oilcake and meal	2,782	10,028	7,314
Oilseeds	203,401	350,045	270,084
Rapeseed	69,558	104,041	70,897
Safflower seeds	161	0	0
Soybeans	57,926	140,658	121,275
Sunflowerseeds	7,096	14,512	18,305
Peanuts, including oilstock	2,800	5,604	3,462
Other oilseeds	5,002	8,478	4,919
Protein substances	60,858	76,752	51,226
Vegetable oils	518,509	644,347	719,562
Soybean oil	58,132	100,477	51,322
Cottonseed oil	15,088	27,668	24,729
Sunflower oil	50,133	67,448	77,396
Corn oil	18,881	29,186	27,400
Peanut oil	2,140	2,573	1,969
Rapeseed oil	99,734	46,730	118,346
Safflower oil	1,482	1,216	1,530
Other Vegetable oils & Waxes	272,918	369,049	416,870
Tobacco-unmfg	828	4,058	7,258
Tobacco-light air cured	0	0	14
Tobacco-flue cured	275	2,284	6,337
Other tobacco-unmfg	553	1,774	907
Cotton, ex. linters	31,584	7,892	4,938
Cotton linters	184	96	111
Essential oils	289,906	330,998	294,939
Seeds-field and garden	151,090	202,534	208,639
Sugar and tropical prods	1,280,596	1,470,493	1,513,946
Sugar and related products	360,048	410,363	418,799
Sugar cane or beet	13,038	13,782	5,011
Related sugar product	347,010	396,581	413,788
Coffee	375,838	443,489	492,264
Cocoa	112,361	110,965	79,895
Chocolate and prep	321,066	377,760	394,338
Tea and Mate	68,811	75,763	77,591
Spices	37,243	46,095	46,397
Ruber-crude-natural	4,289	4,576	3,793
Fibers ex cotton	940	1,481	869
Other hort products	826,352	953,210	939,684
Hops, including extract	8,055	15,775	13,519
Starches, not wheat/corn	70,744	71,000	69,784
Yeasts	22,870	24,179	25,052
Misc hort products	724,683	842,255	831,330
Nursery & greenhouse	189,006	208,932	194,913
Beverages ex juice	284,408	375,979	396,159

[1] Fiscal years Oct. 1–Sept. 30.
ERS, Market and Trade Economics Division, (202) 694–5211.

Table 15-7.—Agricultural imports for consumption: Value of Top 50 countries of origin, United States, fiscal years 2007–2009 [1]

Country	2007	2008	2009
	Million dollars	Million dollars	Million dollars
Canada	14,703.2	17,936.2	15,351.2
European Union-27	14,989.6	15,781.5	13,646.1
Mexico	9,916.0	10,760.9	11,255.1
China	2,798.1	3,426.3	2,914.5
Brazil	2,538.6	2,597.5	2,550.7
Australia	2,609.6	2,403.5	2,445.1
Chile	1,919.1	1,960.6	2,135.0
Indonesia	1,938.3	2,669.0	1,998.8
New Zealand	1,700.8	1,739.6	1,784.1
Colombia	1,518.2	1,716.1	1,771.2
Thailand	1,497.1	1,831.0	1,593.7
Malaysia	1,028.1	1,710.3	1,479.5
India	1,098.7	1,533.2	1,319.0
Guatemala	1,025.4	1,259.3	1,291.5
Argentina	1,102.2	1,176.4	1,186.7
Costa Rica	1,214.0	1,201.6	1,121.0
Ecuador	684.7	722.4	875.0
Peru	660.7	778.2	802.5
Philippines	625.1	895.6	731.7
Vietnam	622.5	765.0	705.1
Switzerland	381.6	603.3	668.1
Ivory Coast	481.8	632.6	603.4
Japan	443.1	497.6	498.2
Turkey	477.8	489.5	484.3
Dominican Republic	325.3	293.2	380.1
Honduras	350.5	416.7	365.7
Nicaragua	238.8	350.3	302.3
South Korea	229.2	249.5	240.7
Israel	222.2	222.2	239.4
Taiwan	220.3	228.2	224.7
El Salvador	180.2	229.3	203.3
South Africa	184.1	181.2	168.6
Uruguay	362.2	116.7	145.8
Singapore	117.2	131.4	114.5
Tunisia	68.5	107.2	113.2
Jamaica	88.5	91.8	102.6
Morocco	87.6	96.2	98.0
Egypt	56.7	52.8	85.8
Liberia	118.8	137.9	84.1
Ethiopia	68.2	115.0	83.5
Papua New Guinea	69.7	73.9	72.1
Malawi	49.9	29.6	70.1
Pakistan	52.6	70.2	69.1
Hong Kong	69.5	74.9	68.7
Nigeria	17.0	67.9	67.8
Bolivia	34.3	46.7	65.0
Ghana	55.8	31.5	61.3
Kenya	45.1	61.9	58.4
Norway	54.4	55.9	53.5
Sri Lanka	56.2	62.0	49.4
Other	665.9	638.9	618.4
Total U. S. Agricultural Imports [2]	70,062.9	79,320.0	73,417.6

[1] Fiscal years Oct. 1–Sept. 30. [2] Totals may not add due to rounding.
ERS, Market and Trade Economics Division, (202) 694–5211. Compiled from reports of the U.S. Department of Commerce.

Table 15-8.—European Union: Value of agricultural imports by origin, 1998–2007 [1]

Year [2]	United States	EU countries	Other countries	Total
	Million dollars	Million dollars	Million dollars	Million dollars
1998	7,961	133,739	52,482	194,182
1999	6,603	132,666	49,032	188,301
2000	6,312	117,228	48,673	172,213
2001	6,429	117,910	48,004	172,343
2002	6,290	133,948	59,540	191,778
2003	6,450	167,970	61,746	236,168
2004	6,521	196,459	70,216	273,196
2005	6,850	216,394	74,096	297,342
2006	7,302	220,542	93,301	321,145
2007 [3]	8,664	245,574	106,694	360,872

[1] EU-15 (1996-2004). Based on bilateral import data from the United Nations.   [2] Data on calendar year basis.   [3] EU-25 included in 2005.
ERS, Market and Trade Economics Division, (202) 694–5232.

Table 15-9.—Fisheries: Landings and value of principal species: 2001–2008 [1]

Species	Landings							
	2001	2002	2003	2004	2005	2006	2007	2008
	Million pounds	Million pounds	Million pounds	Million pounds	Million pounds	Million pounds	Million pounds	Million pounds
Fish:								
Cod, Atlantic	33	29	24	16	14	13	17	19
Flounder	352	373	365	360	419	446	483	663
Haddock	13	17	15	18	17	7	8	14
Halibut	78	82	80	80	76	71	70	67
Herring, sea	300	214	287	265	303	290	232	259
Jack mackerel	8	2	1	3	1	3	1	1
Menhaden	1,741	1,751	1,599	1,498	1,244	1,307	1,482	1,341
Ocean perch, Atlantic	1	1	1	1	1	1	2	3
Pollock	3,188	3,349	3,372	3,365	3,426	3,414	3,085	2,298
Salmon, Pacific	723	567	674	739	899	664	885	658
Tuna	52	49	62	57	44	50	51	48
Whiting	28	18	19	19	17	12	14	14
Shellfish:								
Clams (meats)	123	130	128	119	106	111	116	108
Crabs	272	308	332	316	299	340	294	325
Lobsters, American	74	82	74	88	88	93	81	82
Oysters (meats)	33	34	37	39	34	34	38	30
Scallops (meats)	47	53	56	65	57	59	59	54
Shrimp	324	317	315	309	261	320	281	257
	Value							
	Million dollars	Million dollars	Million dollars	Million dollars	Million dollars	Million dollars	Million dollars	Million dollars
Fish:								
Cod, Atlantic	32	31	28	22	21	20	27	31
Flounder	105	102	94	124	135	151	154	184
Haddock	15	19	17	18	19	11	12	16
Halibut	115	136	172	176	178	202	227	218
Herring, sea	26	21	26	30	34	30	35	45
Jack mackerel	(2)	(2)	(2)	(2)	(2)	(2)	(2)	(2)
Menhaden	103	105	96	72	62	66	93	91
Ocean perch, Atlantic	(2)	(2)	(2)	(2)	1	1	1	1
Pollock	237	210	208	277	315	337	306	334
Salmon, Pacific	209	155	201	303	331	311	381	395
Tuna	93	84	87	91	86	87	94	107
Whiting	13	7	9	10	8	7	8	8
Shellfish:								
Clams (meats)	162	167	162	166	174	189	194	187
Crabs	382	398	481	450	415	414	472	562
Lobsters, American	254	293	292	366	417	395	376	306
Oysters (meats)	81	89	104	112	111	121	140	132
Scallops (meats)	175	204	229	322	434	386	387	372
Shrimp	569	461	421	428	406	441	433	442

[1] Data exclude landings by U.S. flag vessels at Puerto Rico and other ports outside the 50 States, and production of artificially cultivated fish and shellfish.   [2] Less than $500,000.
U.S. Department of Commerce, NOAA, NMFS, Fisheries Statistics Division. (301) 713–2328.

Table 15-10.—Fresh and frozen fishery products: Production and value, 2001–2008 [1]

Product	Production							
	2001	2002	2003	2004	2005	2006	2007 [4]	2008
	Million pounds	*Million pounds*	*Million pounds*	*Million pounds*	*Million pounds*	*Million pounds*	*Million pounds*	*Million pounds*
Fish fillets and steaks [2]	480	517	612	567	615	631	632	656
Cod	40	50	56	15	47	42	32	39
Flounder	30	25	21	20	20	18	21	21
Haddock	6	8	8	10	24	16	11	9
Ocean perch, Atlantic	([3])	([3])	1	1	1	1	1	1
Rockfish	7	7	5	4	3	2	2	2
Pollock, Atlantic	2	4	7	3	3	2	2	3
Pollock, Alaska	271	308	367	384	383	398	401	364
Other	124	115	147	130	134	152	162	217
	Value							
	Million dollars	*Million dollars*	*Million dollars*	*Million dollars*	*Million dollatrs*	*Million dollars*	*Million dollars*	*Million dollars*
Fish fillets and steaks [2]	914	981	1,133	933	1,136	1,300	1,304	1,392
Cod	123	155	171	54	116	123	102	112
Flounder	74	73	62	66	65	73	69	69
Haddock	27	32	35	42	89	70	59	44
Ocean perch, Atlantic	1	1	3	3	4	3	3	3
Rockfish	17	15	12	9	8	5	6	4
Pollock, Atlantic	8	11	10	6	6	4	5	8
Pollock, Alaska	296	330	395	366	404	488	494	450
Other	368	364	445	387	444	564	566	702

[1] Excludes Alaska and Hawaii, except frozen products includes Alaska and Hawaii. [2] Fresh and frozen. [3] Less than 500,000 lb. [4] 2007 Revised.
U.S. Department of Commerce, NOAA, NMFS, Fisheries Statistics Division (301) 713–2328.

Table 15-11.—Canned fishery products: Production and value, 2001–2008 [1]

Product	Production							
	2001	2002	2003	2004	2005	2006	2007 [2]	2008
	Million pounds	*Million pounds*	*Million pounds*	*Million pounds*	*Million pounds*	*Million pounds*	*Million pounds*	*Million pounds*
Total [3]	1,664	1,317	1,295	1,106	1,082	1,081	1,070	1,316
Tuna	507	547	529	434	446	445	436	474
Salmon	185	224	188	199	219	152	142	124
Clam products	126	140	123	108	123	112	110	105
Sardines, Maine	([4])	([4])	([4])	([4])	([4])	([5])	([5])	([4])
Shrimp	2	2	1	1	1	([5])	([5])	([4])
Crabs	([5])	([5])	([5])	([5])	([5])	([5])	([5])	([5])
Oysters [6]	1	([5])	([5])	([5])	([5])	([5])	([5])	([5])
	Value							
	Million dollars	*Million dollars*	*Million dollars*	*Million dollars*	*Million dollatrs*	*Million dollars*	*Million dollars*	*Million dollars*
Total [3]	1,400	1,290	1,239	1,100	1,211	1,330	1,324	1,422
Tuna	658	675	669	569	628	705	702	845
Salmon	259	296	242	251	301	250	274	225
Clam products	125	118	132	113	127	123	89	95
Sardines, Maine	([4])	([4])	([4])	([4])	([4])	([4])	([4])	([4])
Shrimp	10	9	5	5	3	1	1	([4])
Crabs	([5])	([5])	([5])	([5])	([5])	([5])	([5])	([5])
Oysters [6]	1	([5])	([5])	1	1	([5])	([5])	([5])

[1] Natural pack only. [2] Revised. [3] Includes other products not shown separately. [4] Confidential data. [5] Less than 500,000 pounds or $500,000. [6] Includes oyster specialties.
U.S. Dept. of Commerce, NOAA, NMFS, Fisheries Statistics Division (301) 713–2328.

Table 15-12.—Fisheries: Fishermen and craft, 1977, and catch, 2003–2008 by area

Area	1977 [1]			2003		2004	
	Fisher-men	Fishing vessels	Fishing boats [2]	Total catch	Value	Total catch	Value
	1,000	*Number*	*1,000*	*Million ponds*	*Million dollars*	*Million pounds*	*Million dollars*
United States	182.1	17,545	89.2	9,507	3,347	9,683	3,756
New England States	31.7	929	15.4	661	691	717	813
Middle Atlantic States	17.3	573	11.3	215	177	227	199
Chesapeake Bay States	27.9	2,086	19.0	496	180	531	210
South Atlantic States	11.6	1,463	6.7	197	153	198	155
Gulf States	29.3	5,328	11.0	1,600	683	1,477	669
Pacific Coast States	54.0	7,643	15.4	6,291	1,382	6,485	1,623
Great Lakes States	1.2	217	0.5	17	13	17	12
Hawaii	2.7	101	1.3	24	52	24	57
Utah	....	....	....	6	16	7	18

	2005		2006		2007		2008 [3]	
	Total catch	Value	Total catch	Value	Total catch	Value	Total catch	Value
	Million pounds	*Million dollars*	*Million pounds*	*Million dollars*	*Million pounds*	*Million dollars*	*Million pounds*	*Million dollars*
United States	9,707	3,942	9,483	4,024	9,309	4,192	8,326	4,384
New England States	684	971	701	953	573	875	590	792
Middle Atlantic States	200	222	190	199	195	219	201	233
Chesapeake Bay States	509	219	477	163	532	182	477	219
South Atlantic States	122	125	116	141	105	142	116	167
Gulf States	1,196	621	1,346	674	1,353	654	1,273	698
Pacific Coast States	6,951	1,701	6,609	1,814	6,426	1,927	5,619	2,174
Great Lakes States	17	12	18	13	19	14	18	17
Hawaii	28	71	26	67	29	76	31	85
Utah	-	-	-	-	-	-	-	-

[1] Exclusive of duplication among regions. Computation of area amounts will not equal U.S. total. Mississippi River data included with total.   [2] Refers to craft having capacity of less than 5 net tons.   [3] Preliminary.   Note: Table may not add due to rounding.
U.S. Department of Commerce, NOAA, NMFS, Fisheries Statistics Division (301) 713–2328.

Table 15-13.—Fisheries: Quantity and value of domestic catch, 1999–2008

Year	Quantity [1]			Ex-vessel value	Average price per lb.
	Total	For human food	For industrial products [2]		
	Million pounds	*Million pounds*	*Milion pounds*	*Million dollars*	*Cents*
1999	9,339	6,832	2,507	3,467	37.1
2000	9,069	6,912	2,157	3,550	39.1
2001	9,489	7,314	2,178	3,228	34.0
2002	9,397	7,205	2,192	3,092	32.9
2003	9,507	7,521	1,986	3,347	35.2
2004	9,683	7,794	1,889	3,756	38.8
2005	9,707	7,997	1,710	3,942	40.6
2006	9,483	7,842	1,641	4,024	42.4
2007	9,309	7,490	1,819	4,192	45.0
2008 [3]	8,326	6,633	1,692	4,383	52.6

[1] Live weight.   [2] Meals, oil, fish solubles, homogenized condensed fish, shell products, bait, and animal food.
[3] Preliminary.
U.S. Department of Commerce, NOAA, NMFS Fisheries Statistics Division (301) 723–2328.

Table 15-14.—Fishery products: Supply, 1999–2008 [1]

Item	1999	2000	2001	2002	2003
	Million pounds	Million pounds	Milion pounds	Million pounds	Million pounds
Total	17,378	17,339	18,119	19,028	19,850
For human food	14,462	14,740	15,306	16,007	17,187
Finfish	10,831	11,006	11,330	11,770	12,617
Shellfish [2]	3,630	3,734	3,977	4,237	4,570
For industrial use	2,916	2,599	2,812	3,022	2,663
Domestic catch	9,339	9,068	9,492	9,397	9,507
Percent of total	53.7	52.3	52.4	49.4	47.9
For human food	6,832	6,912	7,314	7,205	7,521
Finfish	5,490	5,637	6,162	6,013	6,388
Shellfish [2]	1,341	1,275	1,152	1,192	1,133
For industrial use	2,507	2,157	2,178	2,193	1,986
Imports [3]	8,039	8,271	8,627	9,631	10,343
Percent of total	46.3	47.7	47.6	50.6	52.1
For human food	7,630	7,828	7,992	8,802	9,666
Finfish	5,341	5,369	5,168	5,757	6,229
Shellfish [2]	2,289	2,459	2,825	3,045	3,437
For industrial use [4]	409	442	634	829	677

Item	2004	2005	2006	2007	2008 [5]
	Million pounds	Million pounds	Million pounds	Million pounds	Million pounds
Total	20,413	20,612	20,960	20,561	19,252
For human food	17,648	18,155	18,594	18,253	17,089
Finfish	12,959	13,567	13,484	13,339	12,347
Shellfish [2]	4,689	4,588	5,110	4,914	4,742
For industrial use	2,765	2,457	2,366	2,308	2,163
Domestic catch	9,683	9,707	9,483	9,309	8,326
Percent of total	47.4	47.1	45.2	45.3	43.2
For human food	7,794	7,997	7,842	7,490	6,633
Finfish	6,641	6,914	6,671	6,415	5,590
Shellfish [2]	1,153	1,084	1,171	1,075	1,043
For industrial use	1,889	1,710	1,641	1,819	1,692
Imports [3]	10,730	10,905	11,477	11,252	10,927
Percent of total	52.6	52.9	54.8	54.7	56.8
For human food	9,854	10,158	10,752	10,763	10,456
Finfish	6,318	6,653	6,813	6,925	6,757
Shellfish [2]	3,536	3,505	3,939	3,838	3,699
For industrial use [4]	876	747	725	489	471

[1] Live weight, except percent. May not add due to rounding. [2] For univalve and bivalve mollusks (conchs, clams, oysters, scallops, etc.), the weight of meats, excluding the shell is reported. [3] Excluding imports of edible fishery products consumed in Puerto Rico; includes landings of tuna caught by foreign vessels in American Samoa. [4] Fish meal and sea herring. [5] Preliminary.
U.S. Department of Commerce, NOAA, NMFS Fisheries Statistics Division (301) 713–2328.

Table 15-15.—Fisheries: Disposition of domestic catch, 2000–2008 [1]

Disposition	2000	2001	2002	2003	2004	2005	2006	2007	2008 [2]
	Million pounds	Million pounds	Million pounds	Million pound	Million pounds	Million	Million pounds	Million pounds	Million pounds
Fresh and frozen	6,657	7,082	6,826	7,266	7,488	7,776	7,627	7,450	6,538
Canned	530	536	652	498	552	563	573	514	336
Cured	119	123	117	119	137	160	117	121	138
Reduced to meal, oil, etc	1,763	1,748	1,802	1,624	1,506	1,208	1,166	1,224	1,313
Total	9,069	9,489	9,397	9,507	9,683	9,707	9,483	9,309	8,325

[1] Live weight catch. In addition to whole fish, a large portion of waste (400–500 mil. lb.) derived from canning, filleting, and dressing fish and shellfish is utilized in production of fish meal and oil in each year shown. [2] Preliminary.
U.S. Department of Commerce, NOAA, NMFS Fisheries Statistics Division (301) 713–2328.

Table 15-16.—Processed fishery products: Production and value, 2001–2008 [1]

Item	Production							
	2001	2002	2003	2004	2005	2006	2007	2008 [2]
	Million pounds	*Million pounds*	*Million pounds*	*Million pounds*	*Million pounds*	*Million pounds*	*Million pounds*	*Million pounds*
Fresh and frozen:.								
Fillets	450	495	588	551	601	617	617	643
Steaks	30	22	25	16	14	14	15	13
Fish sticks	43	48	31	60	62	59	74	82
Fish portions	189	187	162	138	181	179	194	204
Breaded shrimp	152	147	152	110	120	140	86	74
Canned products [3]	1,664	1,317	1,295	1,106	1,082	1,081	1,070	1,316
Fish and shellfish	885	953	858	762	802	721	699	714
Animal feed	779	365	437	344	280	360	371	602
Industrial products	NA	NA	NA	NA	NA	NA	NA	NA
Meal and scrap	644	638	603	571	565	583	563	493
Oil (body and liver)	279	211	196	179	158	143	152	190
Other	NA	NA	NA	NA	NA	NA	NA	NA

Item	Value							
	2001	2002	2003	2004	2005	2006	2007	2008 [2]
	Million dollars	*Million dollars*	*Million dollars*	*Million dollars*	*Million dollars*	*Million dollars*	*Million dollars*	*Million dollars*
Fresh and frozen:.								
Fillets	845	920	1,064	881	1,090	1,246	1,249	1,340
Steaks	70	62	69	51	46	54	55	52
Fish sticks	42	51	35	71	76	62	105	121
Fish portions	235	237	227	209	323	303	300	310
Breaded shrimp	540	464	465	306	278	347	200	159
Canned products [3]	1,400	1,290	1,239	1,099	1,210	1,330	1,324	1,422
Fish and shellfish	1,110	1,150	1,076	966	1,081	1,101	1,090	1,191
Animal feed	290	140	163	133	129	229	234	231
Industrial products	237	233	222	202	207	242	340	310
Meal and scrap	126	140	134	153	123	152	218	182
Oil (body and liver)	48	41	34	35	31	34	60	63
Other	83	52	54	14	53	57	62	65

[1] Includes cured fish.   [2] Preliminary.   [3] Includes salmon eggs for baits.   NA-not applicable.
U.S. Department of Commerce, NOAA, NMFS, Fisheries Statistics Division (301) 713–2328.

Table 15-17.—Selected fishery products: Imports and exports, 2001–2008 [1]

Product	Quantity							
	2001	2002	2003	2004	2005	2006	2007	2008
	Million pounds	Million pounds	Million pounds	Million pounds	Million pounds	Million pounds	Million pounds	Million pounds
Imports								
Edible	4,102	4,427	4,907	4,951	5,115	5,400	5,346	5,226
Fresh or frozen	3,449	3,670	4,032	4,075	4,219	4,529	4,497	4,363
Salmon [2]	159	182	163	153	171	200	204	201
Tuna	405	358	462	407	394	429	417	372
Groundfish fillets, blocks [3]	310	347	332	361	372	350	341	298
Other fillets and steaks ...	601	691	760	813	875	944	1,040	1,027
Scallops (meats)	40	48	52	45	51	59	55	56
Lobster, American and spiny	92	100	99	97	93	94	65	96
Shrimp and prawn	878	942	1,108	1,138	1,163	1,297	1,224	1,241
Canned	539	632	748	745	748	724	702	707
Sardines, in oil	19	15	16	18	18	17	19	21
Sardines and herring, not in oil	42	42	45	43	37	39	38	40
Tuna	292	378	459	443	452	420	378	378
Oysters	12	13	15	15	13	13	14	12
Pickled or salted	43	46	49	49	49	52	49	54
Cod, haddock, hake, pollock, cusk	8	8	8	8	8	9	7	7
Nonedible scrap and metal	113	148	121	156	133	129	87	84
Exports								
Canned salmon	110	99	96	118	115	116	114	117
Fish oil, nonedible	249	213	147	110	124	148	123	126

Product	Value							
	2001	2002	2003	2004	2005	2006	2007	2008
	Million dollars	Million dollars	Million dollars	Million dollars	Million dollars	Million dollars	Million dollars	Million dollars
Imports								
Edible	9,864	10,121	11,095	11,331	12,099	13,355	13,696	14,171
Fresh or frozen	8,832	8,948	9,815	9,916	10,506	11,738	11,954	12,138
Salmon [2]	323	344	324	307	366	494	523	516
Tuna	515	417	543	551	589	611	584	601
Groundfish fillets, blocks [3]	479	544	505	537	581	602	614	570
Other fillets and steaks ...	1,263	1,383	1,580	1,726	1,949	2,333	2,642	2,793
Scallops (meats)	128	144	157	146	226	238	231	239
Lobster, American and spiny	727	825	883	876	914	928	935	914
Shrimp and prawn	3,617	3,414	3,753	3,675	3,633	4,104	3,896	4,084
Canned	774	907	1,010	1,123	1,232	1,259	1,367	1,625
Sardines, in oil	30	23	28	30	31	32	35	41
Sardines and herring, not in oil	39	38	41	40	36	36	37	45
Tuna	314	399	455	483	533	526	524	661
Oysters	24	24	28	32	28	28	31	28
Pickled or salted	61	68	72	72	75	85	85	95
Cod, haddock, hake, pollock, cusk	16	18	16	16	18	20	16	15
Nonedible scrap and metal	27	39	32	43	40	41	33	33
Exports								
Canned salmon	168	141	148	177	179	182	203	218
Fish oil, nonedible	42	49	38	32	44	57	64	101

[1] Includes Puerto Rico. [2] Excludes fillets. [3] Includes cod, cusk, haddock, hake, pollock, ocean perch, and whiting.
U.S. Dept. of Commerce, NOAA, NMFS, Fisheries Statistics Division (301) 713–2328.

Table 15-18.—Fishery products: Imports and exports, 1999–2008[1]

Year	Imports[2]				Exports			
	Total value	Edible products		Non-edi-ble, value	Total value	Edible products		Non-edi-ble, value
		Quantity	Value			Quantity	Value	
	Million pounds	*Millions pounds*	*Millions pounds*	*Million pounds*	*Million pounds*	*Million pounds*	*Million pounds*	*Million pounds*
1999	17,040	3,888	9,014	8,026	10,007	1,961	2,849	7,158
2000	19,013	3,978	10,054	8,959	10,782	2,165	2,952	7,830
2001	18,547	4,102	9,864	8,683	11,834	2,565	3,195	8,639
2002	19,691	4,427	10,121	9,570	11,713	2,398	3,120	8,593
2003	21,283	4,907	11,095	10,187	11,999	2,396	3,268	8,731
2004	22,949	4,951	11,331	11,618	13,592	2,888	3,708	9,884
2005	25,120	5,115	12,099	13,021	15,431	2,929	4,074	11,357
2006	27,712	5,401	13,355	14,357	17,760	2,967	4,238	13,522
2007	28,777	5,346	13,696	15,081	20,054	2,869	4,269	15,785
2008	28,457	5,226	14,171	14,286	23,367	2,650	4,257	19,110

[1] Includes Puerto Rico.   [2] Includes landings of tuna by foreign vessels in American Samoa.
U.S. Department of Commerce, NMFS, Fisheries Statistics Division (301) 713–2328.

Table 15-19.—Fish trips: Estimated number of fishing trips taken by marine recreational fishermen by subregion and year, Atlantic, Gulf, and Pacific Coasts, 2005–2008

Subregion	2005	2006	2007	2008
	Thousands	*Thousands*	*Thousands*	*Thousands*
Atlantic and Gulf:				
North Atlantic	9,254	9,656	9,699	9,185
Mid-Atlantic	20,817	21,366	22,718	20,599
South Atlantic[1]	21,809	23,860	25,652	22,254
Gulf[1]	21,871	23,863	24,267	24,109
Total	73,751	78,745	82,336	76,147

Subregion	2005	2006	2007	2008
	Thousands	*Thousands*	*Thousands*	*Thousands*
Pacific:[2]				
Southern California	2,780	3,755	2,833	2,753
Northern California	1,561	1,890	1,344	1,284
Oregon	172	162	187	128
Washington	135	144	143	106
Total	4,647	5,951	4,507	4,271

[1] Does not include trips from headboats (party boats) in the South Atlantic or Gulf of Mexico.   [2] Data do not include recreational trips in Hawaii or Alaska. Pacific state estimates do not include salmon data collected by recreational surveys.
U.S. Department of Commerce, NOAA, NMFS, Fisheries Statistics Division (301) 713–2328.

Table 15-20.—Fish harvested: Estimated number of fish harvested by marine recreational anglers by subregion and year, Atlantic, Gulf Coasts, and Pacific Coasts, 2005–2008

Subregion	2005	2006	2007	2008
	Thousands	Thousands	Thousands	Thousands
Atlantic and Gulf:				
North Atlantic	9,865	10,601	16,179	10,951
Mid-Atlantic	34,670	35,075	40,350	33,505
South Atlantic [1]	40,502	45,539	48,705	42,468
Gulf [1]	68,637	100,658	79,214	92,472
Total	153,675	191,872	184,448	179,397

Subregion	2005	2006	2007	2008
	Thousands	Thousands	Thousands	Thousands
Pacific: [2]				
Southern California	7,835	7,970	4,451	5,084
Northern California	2,929	5,321	2,311	1,984
Oregon	559	491	573	440
Washington	478	489	578	374
Total	11,801	14,270	7,913	7,882

[1] Does not include trips from headboats (party boats) in the South Atlantic or Gulf of Mexico.   [2] Data do not include recreational trips in Hawaii or Alaska. Pacific state estimates do not include salmon data collected by recreational surveys. NOTE: "Harvested" includes dead discards and fish used for bait but does not include fish released alive; totals may not match due to rounding.
U.S. Department of Commerce, NOAA, NMFS, Fisheries Statistics Division (301) 713–2328.

Table 15-21.—Fish harvested: Estimated number of fish harvested by marine recreational anglers by mode and year, Atlantic, Gulf Coasts, and Pacific Coasts, 2005–2008

Mode	2005	2006	2007	2008
	Thousands	Thousands	Thousands	Thousands
Atlantic and Gulf: [1]				
Shore	40,756	50,514	54,360	50,344
Party/charter [2]	11,746	11,474	11,855	11,028
Private/rental	101,174	129,885	118,234	118,025
Total	153,675	191,872	184,448	179,397

Mode	2005	2006	2007	2008
	Thousands	Thousands	Thousands	Thousands
Pacific: [2]				
Shore	7,348	9,970	3,735	4,707
Party/charter	1,902	2,054	2,089	1,789
Private/rental	2,552	2,247	2,090	1,385
Total	11,801	14,270	7,913	7,882

[1] Does not include trips from headboats (party boats) in the South Atlantic or Gulf of Mexico.   [2] Data do not include recreational trips in Hawaii or Alaska. Pacific state estimates do not include salmon data collected by recreational surveys. NOTE: "Harvested" includes dead discards and fish used for bait but does not include fish released alive; totals may not match due to rounding.
U.S. Department of Commerce, NOAA, NMFS, Fisheries Statistics Division (301) 713–2328.

Table 15-22.—Fish harvested: Estimated number of fish harvested by marine recreational anglers by species group and year, Atlantic and Gulf coasts, 2005–2008 [1]

Species group	2005	2006	2007	2008
	Thousands	*Thousands*	*Thousands*	*Thousands*
Barracudas	69	81	139	114
Bluefish	8,902	7,833	8,659	7,120
Dogfish sharks	76	57	166	140
Other sharks	219	171	241	131
Skates/rays	109	197	120	71
Freshwater catfishes	266	162	470	448
Saltwater catfishes	604	713	673	418
Atlantic cod	732	267	314	502
Other cods/hakes	505	379	526	479
Pollock	157	175	161	242
Red hake	118	113	58	187
Dolphins	1,450	1,500	1,511	1,355
Other croaker	0	0	0	0
Atlantic croaker	12,343	11,662	11,979	10,603
Black drum	686	1,066	1,135	1,491
Kingfishes	6,602	5,571	5,832	6,076
Other drum	663	584	400	323
Red drum	2,674	3,195	3,500	3,689
Sand seatrout	1,753	2,677	2,909	3,562
Silver perch	371	309	199	212
Spot	8,894	11,431	15,929	12,505
Spotted seatrout	11,362	17,256	13,171	14,774
Weakfish	1,504	743	585	543
Eels	13	23	64	7
Gulf flounder	163	163	244	211
Other flounders	57	99	86	78
Southern flounder	950	1,046	1,156	1,007
Summer flounder	4,110	4,214	3,397	2,312
Winter flounder	246	309	263	244
Other grunts	737	537	999	772
Pigfish	726	533	773	888
White grunt	1,998	1,051	1,123	2,032
Herrings	36,920	61,764	44,341	51,248
Blue runner	964	3,243	3,169	1,907
Crevalle Jack	531	527	453	396
Florida pompano	676	573	515	535
Greater amberjack	100	87	92	128
Other jacks	1,156	1,387	1,975	1,368
Mullets	7,190	9,169	8,499	8,311
Other fishes	3,255	4,021	10,068	4,109
Other porgies	202	186	197	186
Pinfishes	7,487	7,553	7,916	9,464
Red porgy	117	126	170	192
Scup	2,393	2,796	3,592	3,674
Sheepshead	2,613	1,951	2,077	2,643
Puffers	242	92	50	291
Sculpins	(2)	3	3	1
Black sea bass	2,282	2,422	2,650	1,780
Epinephelus groupers	286	206	242	244
Mycteroperca groupers	587	375	425	527
Other sea basses	467	510	308	191
Searobins	193	123	201	276
Gray snapper	1,433	1,465	1,875	1,958
Lane snapper	383	212	290	256
Other snappers	158	177	239	343
Red snapper	865	996	1,159	820
Vermilion snapper	600	663	566	596
Yellowtail snapper	453	514	666	586
Other temperate basses	0	3	0	0
Striped bass	2,338	2,709	2,203	2,056
White perch	2,672	3,037	3,517	3,029
Toadfishes	20	7	47	38
Triggerfishes/filefishes	484	314	458	357
Atlantic mackerel	3,131	4,851	3,079	3,478
King mackerel	665	967	1,124	718
Little tunny/Atlantic bonito	174	255	294	198
Other tunas/mackerels	511	598	513	321
Spanish mackerel	2,266	2,753	2,932	3,327
Cunner	112	14	341	223
Other wrasses	74	57	150	153
Tautog	613	1,049	1,274	931
Total [2]	153,675	191,872	184,448	179,395

[1] Data do not include headboats (party boats) in the South Atlantic and the Gulf of Mexico. [2] Totals may not add due to rounding. NOTE: "Harvested" includes dead discards and fish used for bait but does not include fish released alive.
U.S. Department of Commerce, NOAA, NMFS, Fisheries Statistics Division (301) 713–2328.

Table 15-23.—Fish harvested: Estimated number of fish harvested by marine recreational anglers by species group and year, Pacific coast, 2005–2008 [1]

Species group	2005	2006	2007	2008
	Thousands	Thousands	Thousands	Thousands
Northern anchovy	1,958	1,266	235	194
Other anchovies	(2)	1	13	5
California scorpionfish	0	0	0	0
Dogfish sharks	2	2	1	1
Other sharks	41	31	13	12
Skates/rays	17	11	12	15
Other cods/hakes	(2)	1	0	0
Pacific cod	4	1	0	0
Pacific hake	(2)	(2)	0	0
Pacific tomcod	(2)	(2)	0	0
California corbina	40	64	26	6
Other croakers	0	0	0	0
Queenfish	332	287	162	144
White croaker	313	205	334	83
Dolphins	0	0	0	0
Other drum	137	235	236	121
California halibut	80	48	35	73
Other flounders	32	42	31	27
Rock sole	1	1	1	1
Sanddabs	472	194	166	203
Starry flounder	9	2	1	1
Kelp greenling	25	27	23	24
Lingcod	145	154	98	76
Other greenlings	1	2	2	2
Herrings	475	1,109	241	598
Other jacks	80	48	33	87
Yellowtail	15	74	10	6
Mullets	0	0	0	0
Other fishes	610	392	654	318
Pacific barracuda	60	50	111	43
Black rockfish	737	745	628	582
Blue rockfish	381	651	284	203
Bocaccio	38	38	50	34
Brown rockfish	92	139	92	76
Canary rockfish	12	19	13	7
Chilipepper rockfish	7	4	15	9
Copper rockfish	64	66	79	66
Greenspotted rockfish	41	18	34	27
Olive rockfish	86	85	69	57
Other rockfishes	582	493	546	422
Quillback rockfish	9	21	20	11
Gopher rockfish	106	134	84	96
Widow rockfish	5	6	12	8
Yellowtail rockfish	51	71	135	69
Sablefishes	(2)	(2)	1	0
Cabezon	34	26	20	21
Sculpins	0	0	0	0
Barred sand bass	369	175	123	136
Kelp bass	190	194	157	133
Other sea basses	2	1	0	0
Spotted sand bass	40	22	12	14
Halfmoon	27	37	30	28
Opaleye	70	57	25	27
Jacksmelt	564	1,104	346	581
Other silversides	376	567	290	305
Other smelts	1	(2)	0	0
Surf smelt	7	30	61	9
Sturgeons	3	2	1	1
Barred surfperch	307	611	220	310
Black perch	145	130	51	35
Other surfperches	67	83	58	90
Pile perch	5	8	5	9
Redtail surfperch	27	35	26	34
Shiner perch	257	186	111	60
Silver surfperch	12	24	4	15
Striped seaperch	44	14	37	34
Walleye surfperch	202	157	96	87
White seaperch	22	32	15	14
Striped bass	126	36	21	20
Other tunas/mackerels	1,666	3,564	1,432	1,958
Pacific bonito	71	297	102	76
California sheephead	19	23	22	26
Other wrasses	28	12	2	5
Total[2]	11,741	14,164	7,913	7,885

[1] Data do not include recreational harvest in Hawaii or Alaska. Pacific estimates do not include salmon data collected by state recreational surveys. [2] Totals may not add exactly due to rounding. [3] Harvest less than 500 fish. NOTE: "Harvested" includes dead discards and fish used for bait but does not include fish released alive.

U.S. Department of Commerce, NOAA, NMFS, Fisheries Statistics Division. (301) 713–2328.

Table 15-24.—Fish harvested: Estimated number of fish harvested by marine recreational anglers, by area of fishing and year, Atlantic and Gulf and Pacific Coast, 2005–2008

Area	2005	2006	2007	2008
	Thousands	Thousands	Thousands	Thousands
Atlantic and Gulf: [1]				
Inland	94,112	116,891	118,865	119,455
State Territorial Sea [2]	43,878	56,050	48,739	45,320
Federal Exclusive Ecomomic Zone [3]	17,691	18,932	16,844	14,622
Total [3]	155,681	191,872	184,448	179,397

Area	2005	2006	2007	2008
	Thousands	Thousands	Thousands	Thousands
Pacific: [4]				
Inland	2,978	2,559	963	1,179
State Territorial Sea [2]	8,355	11,499	6,306	6,214
Federal Exclusive Ecomomic Zone [3]	468	211	645	489
Total [3]	11,801	14,270	7,913	7,882

[1] Data do not include headboats (party boats) in the South Atlantic and the Gulf of Mexico.   [2] Open Ocean extending 0 to 3 miles from shore, except West Florida (10 miles).   [3] Open ocean extending to 200 miles offshore from the outer edge of the State Territorial Sea.   [4] Data do not include recreational catch in Hawaii or Alaska. Pacific state estimates do not include salmon data collected by recreational surveys. Note: "Harvested" includes dead discards and fish used for bait but does not include fish released alive.
U.S. Department of Commerce, NOAA, NMFS, Fisheries Statistics Division. (301) 713–2328.

Table 15-25.—Farm-raised catfish: Processed, prices received by producers, sales, inventory, and imports, 2000–2009

Year	Round weight processed	Prices per pound [1]	Fresh sales	Frozen sales	Total sales	Inventory end of year	Imports [2]
	(000) pounds	Cents	(000) pounds	(000) pounds	(000) pounds	(000) pounds	(000) pounds
2000	593,603	75.1	116,734	180,422	297,156	13,598	8,236
2001	597,108	64.7	120,775	175,592	296,367	14,997	18,079
2002	630,601	56.8	123,451	194,198	317,649	12,283	10,201
2003	661,504	58.1	126,841	192,486	319,327	13,592	5,430
2004	630,450	69.7	117,599	189,180	306,779	15,172	9,224
2005	600,670	72.5	107,984	191,984	299,968	13,707	30,105
2006	566,131	79.5	100,286	183,722	284,008	18,174	74,964
2007	496,246	76.7	90,741	161,709	252,450	15,064	84,605
2008	509,597	77.6	90,479	160,728	251,207	15,520	102,428
2009	466,100	77.1	82,850	146,335	229,185	14,456	129,380

[1] Quantity processed by major processors and the prices received for fish delivered to the processing plant's door. Price includes charges for any services provided by the processing plant, such as seining and hauling. Price also includes any discounts or premiums for size or quality, but does not include adjustments based on year-end settlements.   [2] Data furnished by U.S. Bureau of Census. Includes freshwater imports for consumption of "Ictalurus" spp., "Pangasius" spp., and other catfish of the order Siluriformes.
NASS, Livestock Branch, (202) 720–3570.

Table 15-26.—Farm-raised catfish: Prices received by processors, 2000–2009 [1]

Year	Fresh			Frozen		
	Whole fish [2]	Fillets [3]	Other [4]	Whole fish [2]	Fillets [3]	Other [4]
	Dollars per/lb	Dollars per/lb	Dollars per/lb	Dollars per/lb	Dollars per/lb	Dollars per/lb
2000	1.66	2.86	1.68	2.03	2.83	1.65
2001	1.57	2.74	1.60	1.98	2.61	1.63
2002	1.32	2.52	1.51	1.84	2.39	1.54
2003	1.35	2.48	1.52	1.84	2.41	1.44
2004	1.56	2.71	1.71	1.95	2.62	1.46
2005	1.59	2.83	1.69	2.00	2.67	1.50
2006	1.68	3.07	1.75	2.15	2.91	1.59
2007	1.69	3.15	1.68	2.17	2.92	1.39
2008	1.63	3.13	1.65	2.16	2.89	1.52
2009	1.65	3.22	1.64	2.21	2.96	1.70

[1] Prices are gross value f.o.b. plant.   [2] Includes round and gutted (viscera only removed) and whole dressed (head, viscera and skin removed).   [3] Includes regular, shank, and strip fillets; excludes any breaded product.   [4] Includes nuggets, steaks, and all other products not already reported, includes weight of breading and added ingredients.
NASS, Livestock Branch, (202) 720–3570.

Table 15-27.—Catfish: Number of operations and water surface acres used for production, 2009–2010, and total sales, 2008–2009, by State and United States

State	Number of operations on Jan. 1[1]		Water surface accres used for production during Jan 1 - Jun 30		Total sales	
	2009	2010	2009	2010	2008	2009
	Number	*Number*	*Acres*	*Acres*	*1,000 dollars*	*1,000 dollars*
AL	....	....	22,100	19,800	93,254	90,688
AR	....	....	25,000	19,200	64,263	44,914
CA	....	....	2,400	1,500	7,913	8,074
LA	....	....	6,300	1,800	11,883	8,395
MS	....	....	80,200	64,000	206,288	196,787
NC	....	....	2,200	1,900	7,221	5,495
TX	....	....	3,800	2,900	13,212	12,644
Oth Sts[2]	....	....	4,900	3,700	5,964	5,570
US	1,306	994	146,900	114,800	409,998	372,567

[1] State level number of operations will only be published every five years in conjunction with the Census of Agriculture. [2] Other States include State estimates not shown and States supressed due to disclosure.
NASS, Livestock Branch, (202) 720–3570.

Table 15-28.—Catfish production: Water surface acre usage by State and United States, 2009–2010

State	Acres intended for utilization during Jan 1-Jun 30					Acres taken out of production during Jul 1-Dec 31 prev. year
	Foodsize	Fingerlings	Broodfish	Currently under or scheduled for:		
				Renovation	New construction	
	Acres	*Acres*	*Acres*	*Acres*	*Acres*	*Acres*
2009[1]						
AL	21,100	480	320	180	85	1,000
AR	21,300	2,800	250	700	....	2,500
CA	1,700	290	100	140	320	*
LA	4,900	480	....	640	....	1,300
MS	64,000	12,100	1,400	2,800	....	6,700
NC	2,000	160	70	150	*	55
TX	3,100	420	220	205	75	550
Oth Sts[1]	2,500	1,700	340	200	40	660
US	120,600	18,430	2,700	5,015	520	12,765
2010						
AL	19,200	380	120	120	30	370
AR	16,600	2,200	250	570	....	2,200
CA	1,100	190	80	70	*	*
LA	1,700	50	....	10	....	2,800
MS	52,000	9,700	1,300	2,100	50	3,500
NC	1,600	200	50	90	*	40
TX	2,600	190	70	85	50	135
Oth Sts[1]	1,900	1,300	370	65	10	840
US	96,700	14,210	2,240	3,110	140	9,885

[1] Other States include State estimates not shown and States supressed due to disclosure. * Not published to avoid disclosure of individual operations.
NASS, Livestock Branch, (202) 720–3570.

Table 15-29.—Catfish: Sales by size category, by State and United States, 2008–2009

Size category and State	Number of fish		Live weight		Sales			
	2008	2009	2008	2009	Total		Average price per pound	
					2008	2009	2008	2009
	1,000	*1,000*	*1,000 pounds*	*1,000 pounds*	*1,000 dollars*	*1,000 dollars*	*Dollars*	*Dollars*
Foodsize:								
AL	82,600	66,600	131,600	128,900	92,120	90,230	0.70	0.70
AR	51,100	31,200	83,700	58,100	62,775	42,994	0.75	0.74
CA	2,220	2,550	3,150	3,400	7,592	7,820	2.41	2.30
LA	7,420	4,660	15,400	11,500	11,827	8.395	0.77	0.73
MS	143,000	146,000	252,370	249,000	191,801	181,770	0.76	0.73
NC	4,040	3,120	8,050	6,150	6,843	5,166	0.85	0.84
TX	11,100	10,500	16,900	16,100	12,844	12,558	0.76	0.78
Oth Sts [1]	2,530	1,680	3,750	2,800	3,488	3,080	0.93	1.10
US	304,010	266,310	514,920	475,950	389,290	352,013	0.76	0.74
Broodfish:								
AL	7	*	36	*	80	*	2.21	*
AR	*	*	*	*	*	*	*	*
CA	*	2	*	13	*	32	*	2.46
LA	....	*	....	*	....	*	....	*
MS	39	25	112	100	75	87	0.67	0.87
NC	*	1	*	*	*	*	*	*
TX	*	1	*	2	*	2	*	1.17
Oth Sts [1]	29	15	138	69	139	53	1.01	0.77
US	75	43	286	184	294	174	1.03	0.95

See footnotes at end of table.

Table 15-29.—Catfish: Sales by size category, by State and United States, 2008–2009—Continued

Size category and State	Number of fish		Live weight		Sales			
					Total		Average price per pound	
	2008	2009	2008	2009	2008	2009	2008	2009
	1,000	*1,000*	*1,000 pounds*	*1,000 pounds*	*1,000 dollars*	*1,000 dollars*	*Dollars*	*Dollars*
Stockers:								
AL	3,700	500	262	280	472	202	1.80	0.72
AR	*	*	*	*	*	*	*	*
CA	135	....	76	....	219	....	2.88	....
LA	....	....	....	....	....	....	....	....
MS	60,200	40,000	6,000	4,500	6,060	4,635	1.01	1.03
NC	*	*	*	*	*	*	*	*
TX	*	130	*	20	*	....	*	1.08
Oth Sts [1]	10,440	17,790	1,574	2,657	1,587	2,626	1.01	0.99
US	74,475	58,420	7,912	7,457	8,338	7,485	1.05	1.00
Fingerlings and fry:								
AL	11,100	3,800	210	95	582	253	2.77	2.66
AR	52,300	29,100	453	272	761	541	1.68	1.99
CA	370	1,300	16	73	65	222	4.08	3.04
LA	1,540	....	31	....	56	....	1.80	....
MS	207,000	202,000	5,800	7,100	8,352	10,295	1.44	1.45
NC	3,540	3,200	200	190	356	329	1.78	1.73
TX	3,170	1,270	167	65	212	62	1.27	0.96
Oth Sts [1]	20,900	14,700	544	481	1,692	1,193	3.11	2.48
US	299,920	255,370	7,421	8,276	12,076	12,895	1.63	1.56

[1] Other States include State estimates not shown and States supressed due to disclosure.   *Not published to avoid disclosure of individual operations.

NASS, Livestock Branch, (202) 720–0585.

Table 15-30.—Trout: Number of operations selling and/or distributing fish and/or eggs, United States, 2008–2009 [1]

US	Total	Selling trout	Distributing trout [2]
	Number	*Number*	*Number*
2008 [3]	1,017	463	584
2009	888	348	528

[1] State level number of operations will only be published every five years in conjunction with the Census of Agriculture.   [2] Trout distributed for restoration, conservation, or recreational purposes.   [3] Revised.
NASS, Livestock Branch, (202) 720-3570.

Table 15-31.—Trout: Value of fish sold and distributed, by State (excluding eggs), and United States (including and excluding eggs), 2008–2009

State	Total value of fish sold		Total value of distributed fish	
	2008 [1]	2009	2008 [1]	2009
	1,000 dollars	*1,000 dollars*	*1,000 dollars*	*1,000 dollars*
AR	....	....	*	*
CA	8,318	5,270	15,268	12,046
CO	621	1,685	6,733	5,966
GA	547	698	951	1,119
ID	35,583	36,313	7,271	5,582
MI	1,027	933	1,078	1,607
MO	2,245	4,675	1,823	2,707
NY	841	386	*	*
NC	7,135	7,180	1,279	612
OR	952	829	4,252	3,471
PA	5,427	5,149	14,691	12,071
UT	535	529	*	*
VA	1,605	1,619	1,876	1,207
WA	5,805	2,537	9,154	10,053
WV	777	1,562	*	*
WI	1,421	1,791	2,650	2,001
Oth Sts [2]	7,132	5,757	35,567	41,888
US [3]	79,971	76,913	102,593	100,330
US [4]	86,618	84,364	N/A	N/A

[1] Revised.   [2] Other States include State estimates not listed and States supressed due to disclosure.   [3] Excludes value of eggs.   [4] Includes value of eggs.   * Not published to avoid disclosure of individual operations. N/A Data not available.
NASS, Livestock Branch, (202) 720-3570.

Table 15-32.—Trout: Sales by size category, by State and United States, 2008–2009

Size category and State	Number of fish		Live weight		Sales			
					Total		Average price per pound	
	2008 [1]	2009	2008 [1]	2009	2008 [1]	2009	2008 [1]	2009
	1,000	1,000	1,000 pounds	1,000 pounds	1,000 dollars	1,000 dollars	Dollars	Dollars
12 inch or longer:								
AR	....	....	....	....	....	....	....	....
CA	2,290	1,400	2,950	1,660	7,877	4,864	2.67	2.93
CO	*	440	*	420	*	1,357	*	3.23
GA	174	155	206	197	441	593	2.14	3.01
ID	27,600	29,800	35,400	35,600	35,046	35,956	0.99	1.01
MI	300	300	296	340	864	751	2.92	2.21
MO	*	*	*	*	*	*	*	*
NY	103	43	98	43	451	210	4.60	4.89
NC	3,820	3,400	3,550	3,750	6,390	6,488	1.80	1.73
OR	*	140	*	165	*	530	*	3.21
PA	1,340	1,240	1,460	1,320	4,336	3,788	2.97	2.87
UT	109	99	124	106	433	333	3.49	3.14
VA	640	600	634	600	1,433	1,380	2.26	2.30
WA	1,040	400	3,700	1,220	4,921	1,269	1.33	1.04
WV	305	630	450	812	775	1,437	1.72	1.77
WI	480	480	446	459	1,244	1,519	2.79	3.31
Oth Sts [2]	2,200	1,990	3,096	2,400	8,221	8,173	2.66	3.41
US	40,401	41,117	52,410	49,092	72,432	68,648	1.38	1.40
6 inch-12 inch:								
AR	....	....	....	....	....	....	....	....
CA	*	*	*	*	*	*	*	*
CO	*	*	*	*	*	*	*	*
GA	*	*	*	*	*	*	*	*
ID	*	*	*	*	*	*	*	*
MI	*	*	*	*	*	*	*	*
MO	*	*	*	*	*	*	*	*
NY	136	71	58	26	351	119	6.05	4.59
NC	420	390	204	190	396	353	1.94	1.86
OR	250	180	78	71	253	279	3.24	3.93
PA	540	630	202	289	891	1,196	4.41	4.14
UT	*	*	*	*	*	*	*	*
VA	*	*	*	*	*	*	*	*
WA	670	810	248	270	630	751	2.54	2.78
WV	*	*	*	*	*	*	*	*
WI	*	*	*	*	*	*	*	*
Oth Sts [2]	3,092	3,463	1,260	1,439	3,585	3,872	2.85	2.69
US	5,108	5,544	2,050	2,285	6,106	6,570	2.98	2.88

See footnotes at end of table.

Table 15-32.—Trout: Sales by size category, by State and United States, 2008–2009—Continued

Size category and State	Number of fish		Live weight		Sales			
					Total		Average value per 1,000 fish	
	2008 [1]	2009	2008 [1]	2009	2008 [1]	2009	2008 [1]	2009
	dollars	pounds	pounds	1,000 dollars	1,000 dollars	1,000 dollars	Dollars	Dollars
1 inch-6 inch:								
AR	....	....	....	....	....	....	....	....
CA	*	*	*	*	*	*	*	*
CO	*	*	*	*	*	*	*	*
GA	*	*	*	*	*	*	*	*
ID	*	*	*	*	*	*	*	*
MI	*	*	*	*	*	*	*	*
MO	*	*	*	*	*	*	*	*
NY	88	125	4	3	39	57	448	453
NC	3,360	3,000	48	45	349	339	104	113
OR	*	80	*	1	*	20	*	256
PA	540	450	6	5	200	165	370	367
UT	*	*	*	*	*	*	*	*
VA	*	*	*	*	*	*	*	*
WA	980	1,690	36	83	254	517	259	306
WV	*	*	*	*	*	*	*	*
WI	*	*	*	*	*	*	*	*
Oth Sts [2]	4,057	2,581	89	66	591	597	146	231
US	9,025	7,926	183	203	1,433	1,695	159	214

[1] Revised.   [2] Other States include State estimates not listed and States supressed due to disclosure.   * Not published to avoid disclosure of individual operations.

NASS, Livestock Branch, (202) 720-3570.

Table 15-33.—Trout: Egg Sales, United States, 2008–2009 [1] [2]

US	Number of Eggs	Average Price e per 1,000 Eggs	Total Sales
	1,000	Dollars	1,000 dollars
2008	364,982	18.20	6,647
2009	358,750	20.80	7,451

[1] Regional numbers for eggs sold have been discontinued.   [2] Total sales may not calculate due to rounding.
NASS, Livestock Branch, (202) 720-3570.

Table 15-34.—Refrigerated warehouses: Gross refrigerated space by type of warehouse, United States, biennially, October 1991–2009 [1] [2]

Type	1991	1993	1995	1997	1999
	1,000 Cubic Feet				
General:					
Public	1,572,879	1,678,461	1,741,585	2,043,908	2,146,643
Private and Semiprivate	624,005	658,893	674,649	683,372	756,505
Total	2,196,884	2,337,354	2,416,234	2,727,280	2,903,152
Apple:					
Public	27,227	21,645	23,419	23,907	21,690
Private and Semiprivate	584,296	613,093	647,993	675,838	680,736
Total	611,523	634,737	671,412	699,745	702,426
Total, all	2,808,407	2,972,092	3,087,646	3,427,025	3,605,578

Type	2001	2003	2005	2007	2009
	1,000 Cubic Feet				
General:					
Public	2,251,943	2,357,080	2,435,773	2,498,198	2,900,511
Private and Semiprivate	788,853	802,454	771,725	821,998	894,463
Total	3,040,796	3,159,535	3,207,497	3,320,194	3,794,974
Apple:					
Public	14,183	12,517	9,270	8,170	23,474
Private and Semiprivate	712,412	723,499	711,951	683,798	613,118
Total	726,595	736,016	721,221	691,968	636,593
Total, all	3,767,394	3,895,551	3,928,718	4,012,162	4,431,567

[1] Warehouse space is defined as all space artificially cooled to temperatures of 50 degrees F. or less, in which food commodities are normally held for 30 days or longer. [2] Totals may not add due to rounding.
NASS, Livestock Branch, (202) 720–8784.

Table 15-35.—Apple and pear storages: Number of refrigerated warehouses, gross and usable refrigerated space, regular and CA capacity, by State and United States, October 1, 2009 [1] [2]

State	Number of warehouses	Refrigerated space		Apple & pear storage capacity		
		Gross	Usable	Regular	Controlled atmosphere	Total
		1,000 Cubic feet	1,000 Cubic feet	1,000 Bushels	1,000 Bushels	1,000 Bushels
CA	266	19,261	14,761	3,002	2,125	5,127
CT	31	1,174	994	217	142	359
ID	24	3,467	3,018	764	586	1,350
IL	52	893	773	177	8	185
IN	53	1,907	1,481	273	203	477
KY	10	133	100	33	....	33
ME	37	2,505	2,220	418	526	944
MD	17	737	623	163	54	217
MA	106	2,775	2,332	622	345	967
MI	195	31,060	27,647	3,696	6,360	10,057
MN	54	1,144	945	321	33	354
NH	25	1,684	1,488	300	373	672
NJ	57	1,637	1,410	393	35	428
NY	195	34,448	31,029	4,273	8,647	12,920
NC	33	3,784	3,097	955	90	1,045
OH	74	3,233	2,734	711	330	1,041
OR	109	50,273	41,496	7,379	3,922	11,301
PA	182	22,325	17,468	3,193	2,290	5,482
UT	32	1,567	1,197	268	176	444
VT	16	2,401	1,715	208	320	528
VA	58	11,281	9,802	1,499	2,002	3,501
WA	257	428,434	361,133	46,357	124,054	170,411
WV	17	6,698	5,550	1,401	376	1,777
WI	132	1,102	910	248	112	360
Oth Sts	540	2,670	2,069	500	79	578
US	2,572	636,593	535,992	77,371	153,188	230,558

[1] Totals may not add due to rounding. [2] Firms in this table store only apples or pears. Nearly all the storages are private and nearly all the space is cooler, thus public use and freezer space breakouts are not presented at the State level.
NASS, Livestock Branch, (202) 720–8784.

Table 15-36.—General storages: Gross and usable cooler and freezer space, by State and United States, October 1, 2009[1][2]

State	Cooler		Freezer		Total	
	Gross	Usable	Gross	Usable	Gross	Usable
	1,000 Cubic Feet					
AL	3,018	2,574	30,999	26,522	34,017	29,096
AK	*	*	*	*	2,189	1,915
AZ	3,916	2,860	12,449	9,708	16,365	12,567
AR	*	*	*	*	95,200	80,302
CA	186,498	153,131	308,456	256,083	494,953	409,214
CO	2,521	1,923	23,457	19,248	25,978	21,172
CT	2,396	1,773	6,066	4,782	8,462	6,555
DE	*	*	*	*	29,131	23,452
FL	101,076	85,822	172,732	143,099	273,808	228,921
GA	60,131	51,250	158,290	132,241	218,421	183,490
HI	*	*	*	*	*	*
ID	*	3,540	*	46,949	60,168	50,489
IL	29,176	24,036	168,435	131,814	197,610	155,850
IN	14,106	*	78,954	*	93,061	77,172
IA	12,660	9,136	79,652	66,921	92,312	76,057
KS	8,148	5,290	40,441	29,795	48,589	35,084
KY	5,017	3,861	17,890	14,492	22,907	18,353
LA	1,547	1,391	14,134	11,963	15,681	13,354
ME	*	*	*	*	10,341	7,352
MD	4,249	3,809	37,791	29,792	42,040	33,601
MA	15,023	12,453	86,474	72,574	101,497	85,027
MI	14,392	11,715	88,895	72,647	103,287	84,362
MN	24,478	16,447	66,358	51,648	90,836	68,096
MS	2,760	2,103	20,982	17,379	23,742	19,482
MO	32,885	26,303	83,642	65,966	116,528	92,268
MT	441	339	580	430	1,020	769
NE	5,739	3,830	36,769	29,530	42,509	33,359
NV	*	*	*	*	*	*
NH	*	*	*	*	4,497	3,568
NJ	27,790	26,381	87,446	68,708	115,236	95,089
NM	*	*	*	*	2,442	2,078
NY	25,133	20,113	67,528	55,688	92,661	75,800
NC	7,874	7,521	60,859	51,182	68,733	58,702
ND	*	*	*	*	9,063	6,866
OH	8,501	6,950	55,252	47,470	63,752	54,420
OK	6,375	5,501	10,471	8,322	16,847	13,823
OR	19,278	16,004	107,750	87,397	127,028	103,401
PA	39,470	32,184	187,922	161,681	227,391	193,865
RI	*	*	*	*	*	*
SC	1,617	1,374	21,648	16,524	23,265	17,898
SD	*	*	*	*	13,740	8,158
TN	3,094	2,249	47,671	43,005	50,765	45,254
TX	41,998	32,265	155,740	119,654	197,738	151,919
UT	7,605	6,185	30,152	25,379	37,757	31,564
VT	*	*	*	*	*	*
VA	23,594	18,864	49,777	44,274	73,371	63,138
WA	16,171	11,402	178,368	142,112	194,539	153,514
WV	*	*	*	*	*	*
WI	82,471	63,142	111,932	95,973	194,403	159,115
WY	*	*	*	*	*	*
Oth Sts	27,037	28,365	220,828	197,405	21,094	14,909
US	868,185	702,086	2,926,790	2,398,357	3,794,974	3,100,440

[1] Totals may not add due to rounding.   [2] Excludes storages used exclusively for storing apples and pears. Includes frozen juice tank storage capacity.   *Not published to avoid disclosure of individual operations. Included in "Other States" and U.S. totals.

NASS, Livestock Branch, (202) 720–8784.

Table 15-37.—Alaska crops: Acreage harvested, volume harvested, and value of production, 2000–2009

Year	Oats for grain	Barley for grain	All hay	Potatoes	All vegetables [1]
			Acreage harvested		
	Acres	Acres	Acres	Acres	Acres
2000	300	3,300	18,000	840	370
2001	1,200	5,100	23,000	910	361
2002	1,200	3,800	23,000	850	368
2003	1,200	3,500	22,000	800	359
2004	1,300	4,200	21,000	810	328
2005	900	4,300	21,000	780	351
2006	800	4,200	20,000	840	341
2007	1,000	3,900	23,000	870	326
2008	500	3,400	18,000	780	347
2009 [2]	900	4,400	20,000	740	336

Year	Oats for grain	Barley for grain	All hay	Potatoes	All vegetables [1]
			Volume harvested		
	Bushels	Bushels	Tons	Cwt.	Cwt.
2000	7,000	102,500	17,000	129,000	58,042
2001	61,000	208,000	30,000	230,000	49,989
2002	48,000	149,000	26,000	154,000	51,762
2003	34,000	135,000	29,000	168,000	52,690
2004	41,000	145,000	28,000	177,000	47,762
2005	58,000	208,000	30,000	166,000	57,833
2006	28,000	157,000	22,000	186,000	55,573
2007	47,000	158,000	31,000	176,000	47,340
2008	13,000	99,000	20,000	135,000	40,197
2009 [2]	37,000	183,000	23,000	137,000	43,418

Year	Oats for grain	Barley for grain	All hay	Potatoes	All vegetables [1]
			Value of production		
	Dollars	Dollars	Dollars	Dollars	Dollars
2000	22,000	369,000	3,740,000	2,670,000	2,080,000
2001	153,000	707,000	6,300,000	4,669,000	2,169,000
2002	125,000	529,000	5,590,000	3,080,000	2,318,000
2003	87,000	479,000	6,525,000	3,310,000	2,619,000
2004	100,000	500,000	6,440,000	3,469,000	2,439,000
2005	148,000	759,000	7,200,000	3,403,000	3,507,000
2006	69,000	557,000	5,500,000	3,757,000	3,302,000
2007	132,000	577,000	8,370,000	3,538,000	3,072,000
2008	39,000	446,000	6,300,000	3,348,000	2,954,000
2009 [2]	113,000	814,000	7,130,000	3,425,000	3,155,000

[1] Excludes greenhouse-grown vegetables. [2] Preliminary. NA-not available.
NASS, Crops Branch, (202) 720–2127.

Table 15-38.—Crop ranking: Major field crops, rank by production, major States, 2009

Rank	State	Corn, grain	State	Soybeans	State	All wheat
		1,000 Bushels		*1,000 Bushels*		*1,000 Bushels*
1	IA	2,438,800	IA	486,030	ND	377,190
2	IL	2,065,000	IL	430,100	KS	369,600
3	NE	1,575,300	MN	284,800	MT	176,625
4	MN	1,251,250	IN	266,560	SD	129,147
5	IN	933,660	NE	259,420	WA	123,085
6	SD	719,100	MO	230,550	CO	100,610
7	KS	598,300	OH	221,970	ID	99,130
8	OH	546,360	SD	175,980	MN	84,175
9	WI	448,290	KS	160,600	OK	77,000
10	MO	446,760	AR	122,625	NE	76,800
	US	13,151,062	US	3,361,028	US	2,216,171

Rank	State	Winter wheat	State	Durum wheat	State	Other spring wheat
		1,000 Bushels		*1,000 Bushels*		*1,000 Bushels*
1	KS	369,600	ND	61,230	ND	289,800
2	CO	98,000	CA	17,000	MN	82,150
3	WA	96,760	MT	16,585	MT	70,500
4	MT	89,540	AZ	12,400	SD	64,680
5	OK	77,000	ID	1,620	ID	40,810
6	NE	76,800	SD	207	WA	26,325
7	OH	70,560		-	OR	6,858
8	SD	64,260		-	CO	2,610
9	TX	61,250		-	UT	528
10	ID	56,700		-	NV	130
	US	1,522,718	US	109,042	US	584,411

Rank	State	Sorghum, grain	State	Barley	State	Oats
		1,000 bushels		*1,000 bushels*		*1,000 bushels*
1	KS	224,400	ND	79,100	WI	13,260
2	TX	98,400	ID	48,450	MN	12,070
3	NE	13,020	MT	41,040	ND	11,220
4	OK	12,320	CO	10,395	SD	6,570
5	SD	7,320	WY	6,720	IA	6,175
6	CO	6,750	WA	6,208	PA	4,880
7	LA	5,330	AZ	5,175	NY	4,620
8	MO	3,698	MN	4,880	MI	3,465
9	IL	2,952	PA	3,375	OH	3,375
10	AR	2,923	MD	3,360	CA	3,150
	US	382,983	US	227,323	US	93,081

Rank	State	All cotton	State	Peanuts	State	Rice
		1,000 bales		*1,000 pounds*		*1,000 cwt.*
1	TX	4,932	GA	1,782,650	AR	99,924
2	GA	1,820	TX	542,500	CA	47,804
3	AR	830	AL	471,200	LA	29,217
4	NC	760	FL	336,000	MS	16,281
5	CA	600	NC	244,200	MO	13,423
6	MO	520	SC	148,800	TX	13,201
7	TN	500	MS	54,000		-
8	AZ	444	VA	44,400		-
9	MS	425	OK	42,900		-
10	AL	360	NM	21,700		-
	US	12,401.3	US	3,688,350	US	219,850

Rank	State	All hay, baled	State	Alfalfa hay, baled	State	Other hay, baled
		1,000 tons		*1,000 tons*		*1,000 tons*
1	CA	8,632	CA	6,958	TX	7,650
2	TX	8,250	SD	5,750	MO	7,200
3	MO	8,040	ID	4,788	KY	5,520
4	SD	7,830	MN	3,900	OK	4,350
5	KS	7,225	WI	3,875	TN	4,180
6	KY	6,290	KS	3,655	KS	3,570
7	NE	6,235	NE	3,610	AR	3,080
8	ID	5,528	MT	3,570	NE	2,625
9	OK	5,278	CO	3,315	VA	2,398
10	MN	5,250	IA	3,312	PA	2,205
	US	147,442	US	71,030	US	76,412

Rank	State	All tobacco	State	Dry edible beans	State	Potatoes
		1,000 pounds		*1,000 cwt.*		*1,000 cwt.*
1	NC	423,856	ND	8,526	ID	131,000
2	KY	206,900	MI	3,510	WA	88,450
3	TN	49,960	MN	2,520	WI	28,980
4	VA	47,435	NE	2,461	CO	23,640
5	SC	38,850	ID	1,980	OR	21,460
6	GA	28,000	CA	1,508	MN	20,700
7	PA	18,660	WA	1,140	ND	19,125
8	OH	6,800	CO	848	MI	15,660
9	CT	2,310	WY	680	ME	15,263
10	MA	519	TX	425	CA	14,833
	US	823,290	US	25,360	US	431,425

NASS, Crops Branch, (202) 720–2127.

Table 15-39.—U.S. crop progress: 2009 crop and 5-year average

[In percent]

Week-end-ing date	Winter wheat							
	Planted		Emerged		Headed		Harvested	
	2009	Avg	2009	Avg	2009	Avg	2009	Avg
2008: [1]								
Sep 14	11	16						
Sep 21	22	30						
Sep 28	42	45	14	18				
Oct 5	59	60	28	30				
Oct 12	73	73	46	44				
Oct 19	79	81	60	58				
Oct 26	84	88	69	69				
Nov 2	90	92	76	78				
Nov 9	94	94	83	84				
Nov 16	96	96	88	88				
Nov 23			92	91				
2009:								
Apr 12					9	6		
Apr 19					14	12		
Apr 26					21	23		
May 3					27	35		
May 10 ...					40	48		
May 17 ...					56	60		
May 24 ...					68	71		
May 31 ...					77	81		
Jun 7					84	88	5	10
Jun 14					90	93	9	19
Jun 21					95	97	20	31
Jun 28							40	46
Jul 5							56	59
Jul 12							66	69
Jul 19							72	77
Jul 26							79	84
Aug 2							85	90
Aug 9							91	94
Aug 16							94	97
Aug 23							97	98

Week-end-ing date	Spring wheat							
	Planted		Emerged		Headed		Harvested	
	2009	Avg	2009	Avg	2009	Avg	2009	Avg
2009:								
Apr 12	2	11						
Apr 19	6	21						
Apr 26	15	36	2	9				
May 3	23	59	7	20				
May 10 ...	35	78	13	38				
May 17 ...	50	90	21	59				
May 24 ...	79	95	45	77				
May 31 ...	89	98	67	90				
Jun 7	96	100	84	97				
Jun 14			93	99				
Jun 21			99	100				
Jun 28					15	40		
Jul 5					30	65		
Jul 12					57	83		
Jul 19					84	93		
Jul 26					93	98		
Aug 2					98	100	3	15
Aug 9							8	31
Aug 16							13	48
Aug 23							22	66
Aug 30							38	79
Sep 6							58	88
Sep 13							69	92
Sep 20							85	96
Sep 27							94	98
Oct 4							97	99

See footnote at end of table.

Table 15-39.—U.S. crop progress: 2009 crop and 5-year average—Continued

[In percent]

Week-ending date	Rice								Sorghum									
	Planted		Emerged		Headed		Harvested		Planted		Headed		Coloring		Mature		Harvested	
	2009	Avg	2009	Avg	2009	Avg	2009	Avg	2009	Avg	2009	Avg	2009	Avg	2009	Avg	2009	Avg
2009:																		
Apr 5	11	15							19	20								
Apr 12	22	25	7	10					23	24								
Apr 19	29	39	11	18					26	26								
Apr 26	47	56	21	30					28	28								
May 3	64	69	35	46					30	30								
May 10	69	80	51	60					31	33								
May 17	76	87	58	72					38	39								
May 24	86	93	69	82					47	48								
May 31	94	97	81	90					57	58								
Jun 7	97	99	90	94					74	68								
Jun 14			94	96					81	78								
Jun 21			98	98					87	87	20	21						
Jun 28					6	6			93	93	23	23						
Jul 5					9	11			97	96	26	27	20	20				
Jul 12					14	16					29	32	24	23				
Jul 19					21	24					31	38	25	25				
Jul 26					28	37					38	46	29	28				
Aug 2					41	53					49	58	31	30				
Aug 9					56	71					61	70	33	34	27	25		
Aug 16					71	83	8	10			74	79	36	41	29	27	26	22
Aug 23					81	91	10	14			84	86	40	48	30	28	27	24
Aug 30					90	96	16	19			92	91	49	57	31	31	28	26
Sep 6					95	98	25	27			96	95	60	66	33	36	29	30
Sep 13							32	38					70	75	35	42	30	32
Sep 20							37	52					81	83	40	49	31	35
Sep 27							45	66					87	89	45	59	33	39
Oct 4							62	78					91	93	55	68	35	44
Oct 11							69	87					93	96	64	76	37	49
Oct 18							76	93					94	98	72	82	39	54
Oct 25							85	96					95	99	79	88	42	60
Nov 1							89	98							83	93	45	68
Nov 8							96	98							94	97	56	77
Nov 15															97	99	68	84
Nov 22																	75	90
Nov 29																	87	93
Dec 6																	94	

Week-ending date	Corn													
	Planted		Emerged		Silking		Dough		Dented		Mature		Harvested	
	2009	Avg	2009	Avg	2009	Avg	2009	Avg	2009	Avg	2009	Avg	2009	Avg
2009:														
Apr 12	2	6												
Apr 19	5	14												
Apr 26	22	28												
May 3	33	50	5	14										
May 10	48	71	14	28										
May 17	62	85	30	49										
May 24	82	93	52	71										
May 31	93	97	73	86										
Jun 7	97	99	87	94										
Jun 14			95	98										
Jun 21														
Jun 28					4	8								
Jul 5					8	16								
Jul 12					16	32								
Jul 19					31	54								
Jul 26					55	76	7	17						
Aug 2					76	89	14	29						
Aug 9					89	96	24	46	5	14				
Aug 16					96	98	40	64	9	26				
Aug 23							57	79	18	43				
Aug 30							75	88	32	60	5	13		
Sep 6							86	94	50	75	8	23		
Sep 13							93	98	66	86	12	37		
Sep 20							97	99	80	93	21	55		
Sep 27									90	97	37	72	6	18
Oct 4									95	99	57	84	10	25
Oct 11											74	92	13	35
Oct 18											83	97	17	46
Oct 25											90	99	20	58
Nov 1											94	99	25	71
Nov 8											97	100	37	82
Nov 15													54	89
Nov 22													68	94
Nov 29													79	97
Dec 6													88	
Dec 13													92	
Dec 20													95	

See footnote at end of table.

Table 15-39.—U.S. crop progress: 2009 crop and 5-year average—Continued

[In percent]

Week-ending date	Soybeans											
	Planted		Emerged		Blooming		Setting Pods		Dropping Leaves		Harvested	
	2009	Avg	2009	Avg	2009	Avg	2009	Avg	2009	Avg	2009	Avg
2009:												
Apr 26	3	5										
May 3	6	11										
May 10 ...	14	25										
May 17 ...	25	44										
May 24 ...	48	65	17	31								
May 31 ...	66	79	36	51								
Jun 7	78	87	55	70								
Jun 14	87	92	72	83								
Jun 21	91	95	84	90								
Jun 28	96	98	91	95	5	10						
Jul 5			96	98	14	24						
Jul 12					24	43						
Jul 19					44	62						
Jul 26					63	76	20	36				
Aug 2					76	86	36	54				
Aug 9					86	93	55	72				
Aug 16					93	96	72	85				
Aug 23					97	99	85	92				
Aug 30							93	96	3	8		
Sep 6							97	99	7	18		
Sep 13									17	36		
Sep 20									40	58		
Sep 27									63	77	5	18
Oct 4									79	88	15	36
Oct 11									89	95	23	57
Oct 18									95	97	30	72
Oct 25											44	80
Nov 1											51	87
Nov 8											75	92
Nov 15											89	96
Nov 22											94	97
Nov 29											96	98

Week-ending date	Cotton									
	Planted		Squaring		Setting Bolls		Bolls Opening		Harvested	
	2009	Avg	2009	Avg	2009	Avg	2009	Avg	2009	Avg
2009:										
Apr 5	4	8								
Apr 12	8	11								
Apr 19	11	14								
Apr 26	16	20								
May 3	24	28								
May 10 ...	32	39								
May 17 ...	42	53								
May 24 ...	61	69								
May 31 ...	77	81								
Jun 7	89	90								
Jun 14	95	95	10	21						
Jun 21			20	33	5	7				
Jun 28			32	46	8	11				
Jul 5			61	60	14	18				
Jul 12			77	72	22	28				
Jul 19			84	81	32	40				
Jul 26			88	88	48	54				
Aug 2			94	93	65	68				
Aug 9			97	96	75	78	8	9		
Aug 16					84	86	9	13		
Aug 23					90	91	13	18		
Aug 30					93	96	19	25		
Sep 6					94	98	25	35		
Sep 13					97	99	35	45		
Sep 20							46	57	7	11
Sep 27							57	68	8	15
Oct 4							68	77	10	21
Oct 11							79	83	12	29
Oct 18							86	88	15	35
Oct 25							91	91	19	43
Nov 1							92	95	28	50
Nov 8							97	97	44	59
Nov 15									60	68
Nov 22									72	75
Nov 29									83	82
Dec 6									88	
Dec 13									91	
Dec 20									94	

See footnote at end of table.

Table 15-39.—U.S. crop progress: 2009 crop and 5-year average—Continued

[In percent]

Week-ending date	Oats								Barley							
	Planted		Emerged		Headed		Harvested		Planted		Emerged		Headed		Harvested	
	2009	Avg	2009	Avg	2009	Avg	2009	Avg	2009	Avg	2009	Avg	2009	Avg	2009	Avg
2009:																
Apr 5	32	34														
Apr 12	37	41	29	30					3	13						
Apr 19	48	52	31	33					9	21						
Apr 26	61	65	37	40					17	34						
May 3	69	78	47	51					22	54	6	18				
May 10	80	89	60	65					33	74	12	34				
May 17	88	95	71	79					50	87	20	54				
May 24	95	98	82	89	29	28			77	94	40	73				
May 31			92	96	31	31			87	98	60	88				
Jun 7			96	98	33	35			96	99	79	95				
Jun 14					40	45					93	98				
Jun 21					52	59					99	100				
Jun 28					68	74							12	37		
Jul 5					77	87							27	61		
Jul 12					90	94	11	13					55	80		
Jul 19					97	98	14	20					84	92		
Jul 26							19	33					95	97		
Aug 2							31	51								
Aug 9							48	69							5	33
Aug 16							62	82							11	52
Aug 23							72	91							27	70
Aug 30							85	96							46	81
Sep 6							93	99							71	90
Sep 13							96	99							83	95
Sep 20															91	97
Sep 27															95	98

Week-ending date	Peanuts						Sunflower				Sugarbeets			
	Planted		Pegging		Harvested		Planted		Harvested		Planted		Harvested	
	2009	Avg	2009	Avg	2009	Avg	2009	Avg	2009	Avg	2009	Avg	2009	Avg
2009:														
Apr 5											2	5		
Apr 12											4	12		
Apr 19											18	26		
Apr 26	2	3									31	47		
May 3	11	8									37	72		
May 10	22	22									57	88		
May 17	42	43					2	11			64	96		
May 24	59	66					16	28			94	99		
May 31	72	84					31	47			96	100		
Jun 7	83	92					55	63						
Jun 14	93	97					75	78						
Jun 21	97	100	6	11			87	88						
Jun 28			17	23			95	95						
Jul 5			30	38										
Jul 12			46	54										
Jul 19			60	70										
Jul 26			72	82										
Aug 2			81	89										
Aug 9			87	93										
Aug 16			93	96										
Aug 23			97	98										
Aug 30														
Sep 6														
Sep 13					3	2								
Sep 20					4	6							6	6
Sep 27					10	12							10	9
Oct 4					16	23			5	7			20	24
Oct 11					26	37			7	14			40	45
Oct 18					33	50			9	25			53	65
Oct 25					47	63			12	39			67	80
Nov 1					56	75			15	57			81	90
Nov 8					72	85			33	75			93	96
Nov 15					78	92			59	86			98	99
Nov 22					88	96			80	93				
Nov 29					92	98			90	96				
Dec 6					94				94	99				

[1] Planted the preceding fall.
NASS, Crops Branch, (202) 720–2127.

Appendix I

Telephone Contact List

Appreciation is expressed to the following agencies for their help in this publication. The information offices are listed to provide help to those users who require additional information about specific tables in this publication.

Agricultural Marketing Service:
USDA/AMS
Room 2632 South Bldg.
Washington, DC 20250
202–720–8998

Agricultural Research Service:
USDA/ARS
5601 Sunnyside Ave
Bldg 1, Rm 2251
Beltsville, MD 20705–5128
301–504–1638

Animal and Plant Health Inspection Service:
USDA/APHIS
4700 River Rd
Riverdale, MD 20737
301–734–7280

Center for Nutrition Policy and Promotion:
USDA/CNPP
3101 Park Center Drive
Alexandria, VA 22302
703–605–4266

Economic Research Service:
USDA/ERS
1800 M St, NW, Room 3051
Washington, DC 20036
202–694–5050

Farm Credit Administration:
FCA
1501 Farm Credit Dr.
McLean, VA 22102
703–883–4000

Farm Service Agency:
USDA/FSA
Room 3624 South Bldg.
Washington, DC 20250
202–720–7163

Food and Nutrition Service:
USDA/FNS
3101 Park Center Drive, Room 914
Alexandria, VA 22302
703–305–7600

Foreign Agricultural Service:
USDA/FAS
Room 5076 South Bldg.
Washington, DC 20250
202–720–7115

Forest Service:
USDA/FS
2nd Floor Central Wing, Yates Bldg.
Washington, DC 20250
202–205–8333

National Agricultural Statistics Service:
USDA/NASS
Room 5038 South Bldg.
Washington, DC 20250
202–720–3878

National Marine Fisheries Service:
USDC/NOAA/NMFS
1315 East/West Highway,
SSMC III - Room 12405
Silver Spring, MD 20910–3282
301–713–2328

Natural Resources Conservation Service:
USDA/NRCS
Room 6121 South Bldg.
Washington, DC 20250
202–720–2182

Rural Business-Cooperatives Service:
USDA/RECD/RBS
Room 4801 South Bldg.
Washington, DC 20250
202–720–1019

Rural Utilities Service:
USDA/RD/RUS
Room 5144 South Bldg.
Washington, DC 20250
202–720–1255

INDEX

Page

Agricultural commodity support prices IX–36–37
Agricultural conservation program
 See Conservation, Soil conservation, and water conservation.
Agricultural loans, *See* Loans.
Agricultural production, index numbers IX–16
Agricultural productivity IX–25
Agricultural products:
 Exports:
  Value XV–2–6, 10–12
 Foreign trade XV–2, 4–12
 Imports:
  Value XV–2, 7–8, 13–14
 See also Commodities, agricultural.
Agricultural Stabilization and Conservation Programs XII–3–17
Alaska crop statistics XV–33
 See also under specific commodities.
Alfalfa meal:
  Disappearance for feed I–41
  Average price per ton-bulk I–45
Alfalfa seed:
 Prices VI–9
Almonds:
 Area V–1, 38, IX–20
 Imports and exports V–38
 Prices, farm V–38
 Production V–38, IX–21
 Supply and utilization V–38
 Value V–38, IX–21
 Yield V–38, IX–20
 Shelled production in foreign countries V–38
Animal feeds, oils used:
 Per capita III–29
 Total III–29
Animal oils:
 Market prices III–30
 Animal proteins, disappearance for feed I–41
 Feed concentrates I–42
Animal units fed on farms I–42
Animal units fed:
 Grain consumption I–44
 Roughage consumption I–44
 Grain and roughage consumption I–44
Apples:
 Area V–1, IX–20
 Canned:
  Quantity V–4
 Cold-storage stocks V–58
 Consumption V–36
 Dried:
  Exports V–6
  Production V–4
  Quantity V–4
 Foreign trade V–5
 Frozen pack V–35
 Frozen, quantity V–35
 Imports V–5
 Juice V–4
 Prices:
  Farm V–3
 Production, US V–2–3, 5, IX–21
 Production, specified countries V–4
 Shipments V–35
 Utilization V–5, 34
 Value V–3, IX–21
 Yield IX–20
Apricots:
 Area V–1, IX–20
 Consumption V–36
 Cold storage V–56
 Canned:
  Exports V–7
  Quantity V–7, 35
 Dried:
  Exports V–7
  Production V–35
  Quantity sold V–7
 Fresh exports V–7
 Fresh imports V–7
 Frozen V–7, 35
 Prices, farm V–7
 Production V–2, 6–7, 34, IX–21
 Shipments V–34
 Trade, foreign V–6
 Utilization V–7
 Value V–6, IX–21
 Yield IX–20

Page

Area:
 Land in farms:
  Irrigated IX–7
  Total IX–2, 5
 Land utilization IX–6
 Principal crops, planted and harvested IX–17–18, 20, 22
 See also under specific crops.
Artichokes:
 Area IV–6, IX–22
 Consumption IV–34
 Frozen, commercial pack IV–36
 Price IV–6
 Production IV–6, IX–23
 Shipments IV–33
 Value IV–6, IX–23
 Yield IV–6, IX–22
Asparagus:
 Area IV–6–7, IX–22
 Cold storage IV–37
 Consumption IV–36
 Frozen pack IV–35
 Price IV–7
 Production IV–7, IX–23
 Shipments IV–33
 Value IV–7, IX–23
 Yield IV–7, IX–22
Acreage income to farm operator households IX–40
Avocados:
 Area V–1, IX–20
 Consumption V–36
 Imports V–8
 Prices, farm V–8
 Production V–2, 8, IX–21
 Shipments V–34
 Utilization V–34
 Value V–8, 35, IX–21
 Yield IX–20
Balance sheet, farming sector IX–12
Bananas:
 Area V–1, 9, IX–20
 Consumption V–36
 Prices V–9
 Production V–2, 9, 34, IX–21
 Value V–9 IX–21
 Yield V–9, IX–20
Banks, operating
 Amount of agricultural loans outstanding X–16
Barley:
 Area I–32–34, IX–18
 Consumption I–35
 Crop progress XV–38
 Crop ranking XV–34
 Disappearance I–32
 Exports I–32
 Feed concentrates I–42
 Imports I–32
 Prices:
  Farm I–32–33
  Market I–44
  Seed, average price paid VI–9
  Support operations I–35
 Production I–32–34, IX–19
 Stocks on and off farms I–32
 Supply I–32
 Trade, international I–34
 Value I–32–33, IX–19
 Yield I–32–34, IX–18
Barrows and gilts VII–21
Beans:
Butter beans:
Frozen commercial pack IV–36
 Dry edible:
  Area VI–10–11, IX–18
  Crop ranking XV–34
  Exports VI–12
  Prices:
   Farm VI–11
   Market VI–11
  Production VI–10–11, IX–19
  Value VI–11, IX–19
  Yield VI–10–11, IX–18
 Beans, green:
French cut:
 Frozen commercial pack IV–36
 Cold storage IV–38
Regular cut:
 Frozen commercial pack IV–36
 Cold storage IV–38
Wax:
 Frozen commercial pack IV–36
Italian:
 Frozen commercial pack IV–36

Page

Total:
 Cold storage..IV–37–38
Beans, whole:
Frozen commercial pack IV–37
 Lima, baby:
  Frozen commercial pack IV–37
  Cold storage..IV–37
 Lima, Fordhook:
  Frozen commercial pack IV–37
  Cold storage..IV–37
 Lima, fresh:
  Area...IX–22
  Consumption..IV–36
  Frozen pack...IV–37
  Production..IX–23
  Shipments IV–33
  Value...IX–23
  Yield..IX–22
 Lima, processing:
  Area...IV–7, IX–22
  Frozen pack...IV–37
  Price...IV–7
  Production..IV–7, IX–23
  Value...IV–7, IX–23
  Yield..IV–7, IX–22
 Snap, fresh:
  Area...IV–7–8, IX–22
  Canned and frozen IV–37–38
  Consumption,fresh....................................IV–36
  Prices...IV–7–8
  Production.....................................IV–7–8, IX–23
  Shipments IV–33
  Value..IV–7, IX–23
  Yield..IV–7, IX–22
 Snap, processing:
  Area...IV–8, IX–22
  Consumption, canned IV–36
  Prices, farm..IV–8
  Production..IV–8, IX–23
  Shipments IV–33
  Value...IX–23
  Yield..IV–8, IX–22
Beef:
 Cold storage holdings VII–57–58
 Consumption VII–49
 Feed consumed per head and unit...................I–43
 Red meat production....................................VII–40
 Production VII–41
 See also Beef and Veal, Meats and Veal.
Beef and veal:
 Exports................................VII–40–42, 45, 47
 Imports...VII–47
 Production, specified countries....................VII–41
 See also Beef, Meats, and Veal.
Beets, for canning:
 Canned IV–35
 Shipments IV–33
Berries:
 Acreage...V–1, IX–20
 Cold storage V–56–58
 Frozen pack..V–35
 Production...V–2, IX–21
 Value...IX–21
 Yield..IX–20
Blackberries:
 Cold Storage V–56–58
 Consumption...V–36
 Frozen pack..V–35
Blackstrap molasses:
 Average price per ton, bulk I–45
Boysenberries:
 Cold Storage V–56–58
 Frozen pack..V–35
Blueberries:
 Cold Storage V–56–58
 Consumption...V–36
 Frozen pack..V–35
 Shipments V–33
Bluegrass seed, Kentucky:
 Average price VI–9
Breakfast Program XIII–8
Brewers' dried grains:
 Average price per ton, bulk I–45
Broccoli:
 Area...IV–9, IX–22
 Cold Storage...IV–37–38
 Consumption...IV–35
 Frozen pack...IV–36
 Prices...IV–9
 Production..IV–9, IX–23
 Shipments IV–33
 Value...IV–9, IX–23
 Yield..IV–9, IX–22

Page

Broilers:
 Production...VIII–28
 Production, price, and income...................VIII–29
 Feed consumed per head and unit.................I–43
 Feed ratio...VIII–28
Brussels sprouts:
 Cold storage...IV–37–38
 Consumption...IV–36
 Frozen pack...IV–36
Buildings, farm, value of IX–8
Bulls:
 Number, Jan. 1......................................VII–1, 4
 Slaughter under Federal inspection VII–11–12
Butter:
 Consumption.........................III–29, VIII–15
 Cold-storage holdings...........................VIII–36–38
 Production in specified countries.............VIII–16
 Disappearance..VIII–15
 Exports VIII–21
 Imports VIII–19
 Prices, market.........................III–30, VIII–23
 Stocks on hand, Dec. 31 ....VIII–15
Cabbage:
 Shipments IV–33
 Consumption...IV–36
Cabbage, Chinese:
 Shipments IV–33
Cabbage, fresh market:
 Area...IV–10, IX–22
 Price...IV–10
 Production..IV–10, IX–23
 Value..IV–10, IX–23
 Yield..IV–10, IX–22
Cabbage, for kraut:
 Per capita utilization..................................IV–35
Cacao butter. See Cocoa butter.
Caloric sweeteners:
 Per capita consumption...............................II–19
Calves:
 Carcasses condemned VII–54
 Disposition VII–7, 10
 Heifers, number, Jan. 1 ....VII–1, 3
 Income from, cash and gross VII–7, 10
 Number born...VII–5
 Number, Jan. 1..VII–1–3
 Operations and inventory........................VII–11–13
 Prices, farm..VII–56
 Production VII–7, 10
 Receipts at livestock markets VII–8
 Slaughter..VII–54
 Under 500 pounds VII–1, 4
 Skins:
  Exports VII–51–52
  Imports VII–50
 Slaughter:
  Farm VII–8
  Total VII–8
  Under Federal inspection........................VII–8, 54
 Slaughtered, live weight:
  By States VII–9
Canned vegetable and commercial production:
 Vegetable consumption IV–35
Canola:
 Area...IX–18
 Production..IX–19
 Value...IX–19
 Yield..IX–18
Cantaloups:
 Area...IV–11, IX–22
 Consumption...IV–35
 Price...IV–11
 Production..IV–11, IX–23
 Shipments IV–33
 Value..IV–11, IX–23
 Yield..IV–11, IX–22
 See also Melons.
Carcasses, whole, number condemned under Federal
 inspection...VII–55
Carrots:
 Cold storage...IV–37–38
 Consumption IV–35, 37
 Frozen pack...IV–36
 Shipments IV–33
Carrots, fresh:
 Area...IV–12, IX–22
 Price...IV–12
 Production..IV–12, IX–23
 Value...IX–23
 Yield..IX–22
Carrots, processing:
 Area...IV–12, IX–22
 Price...IV–12
 Production..IV–12, IX–23
 Value...IX–23
 Yield..IX–22

Page

Casein:
Imports VIII–18
Cash rents:
Rents, cropland, and pasture IX–11
Catfish:
Farm-raised XV–24
No. operations...XV–25
Cattle:
Carcasses condemned VII–54
Hides:
Exports..VII–51
Imports...VII–50
Price per pound, Chicago VII–55
Income from cash and gross VII–7, 10
Number:
In specified countries (cattle and buffalo) VII–3–4
Jan. 1 ....VII–1–6, 54
On feed, by States...VII–5–6
Operations and inventory...............................VII–11–14
Prices:
Market VII–7
Production and disposition VII–7, 10
Receipts at public stockyards VII–8
Slaughter:
Farm VII–8, 11
Number...VII–8–9, 11
Under Federal inspection............................VII–8–9, 12
Slaughtered, dressed weight VII–12
Slaughtered, live weight VII–9
By States VII–9
Value:
Jan. 1 ....VII–1–2
Of production VII–7, 10, IX–25
Cauliflower:
Area IV–12–13, IX–22
Consumption...IV–35
Cold storage...IV–37
Frozen pack IV–36
Prices...IV–12–13
Production IV–12–13, IX–23
Shipments IV–33
Value IV–12–13, IX–23
Yield...IV–12, IX–22
Celery:
Area IV–13, IX–22
Consumption...IV–35
Frozen pack IV–36
Prices IV–13
Production IV–13, IX–23
Shipments IV–33
Value IV–13, IX–23
Yield...IV–13, IX–22
Cheese:
American, factory production...................................VIII–10
Cold-storage holdings...VIII–38
Consumption...VIII–15
Disappearance VIII–15
Exports VIII–22
Foreign production...VIII–16
Imports VIII–17–18
Prices, market VIII–23
Prices, support operations......................................VIII–23
Stocks on hand, Dec. 31 ....VIII–15
Cherries, total:
Area V–1, IX–20
Production...V–2, IX–21
Exports V–9
Imports V–9
Shipments V–34
Consumption...V–36
Utilization...V–34
Cherries, sweet:
Production V–10–11
Price V–10
Value V–10
Utilization...V–11
Frozen commercial pack V–35
Cold storage...V–57
Cherries, tart:
Production V–10–11
Price V–10
Value V–10
Utilization...V–11
Cold storage V–57

Page

Chickens:
Broiler production, price, and income VIII–28–29
Cold storage stocks...VIII–36–37
Consumption VIII–28
Exports VIII–28
Feed consumed per head and unit............................I–42
Layers 1 year old+...VIII–23
Meat, imports VIII–26
Meat, production VIII–26
Pullets, total VIII–24
Number lost VIII–27
Number, Dec. 1 ....VIII–24–26
Prices, live weight VIII–27
Sold, number...VIII–27
Supply and distribution VIII–28
Value:
Dec. 1 ....VIII–25, 27
Of production VIII–28–29 IX–25
Of sales VIII–27
See also Chicks and Poultry.
Chickpeas:
Exports...VI–13
Chicks:
Baby, prices VIII–29
Hatched by commercial hatcheries VIII–29
Value VIII–29
Child Feeding Programs...XIII–7–9
Cigarettes and cigars:
Consumption and total output II–26
Exports II–26
Manufactured II–26
Citrus fruits:
Area V–1, IX–20
Consumption...V–36
Crop insurance X–4
Exports V–15
Foreign production..V–14
Quantity processed..V–12
Prices V–12–13
Production V–2, 12–13, IX–21
Shipments V–34
Trade, foreign V–16
Value V–12, IX–21
Yield...IX–20
Citrus juices:
Concentrated, pack V–16
See also Grapefruits, Lemons, Limes, Oranges, and
Tangerines.
Clover seed (red and Ladino), prices VI–9
Coconut oil:
Prices...III–30
Used in manufacture of shortening..........................III–28
Coffee:
Area V–42, IX–18
Exports from principal producing countries V–42
Imports, origin V–42
Production V–42, IX–19
Price V–42
Value V–42, IX–19
Yield...V–42, IX–18
Cold-storage stocks:
Apples V–58
Dairy products VIII–38
Frozen fruits V–56
Frozen orange juice V–56
Meats VII–57–58
Nuts V–56
Poultry products..VIII–36
Vegetables...IV–37–38
Commercial feeds:
Disappearance from feed I–36
Commodities, agricultural:
Owned by Commodity Credit Corporation,
Dec. 31...XI–1
Purchases by Commodity Credit Corporation,
costs XI–4–5
Under Commodity Credit Corporation price-support
loan, Dec. 31 ....XI–1
See also Agricultural products.
Commodity Credit Corporation:
Cost value of export and commodity
disposition..XI–6–7
Inventory transactions XI–4–5
Loan programs..XI–2
Loan transactions...XI–3
Price support operations.......................................IX–36–37
Price support:
Commodities owned..XI–1
Commodities under loan.......................................XI–2
Operations, investments in XI–8
Collard greens:
Frozen commercial pack IV–36
Consumption..IV–35

Page

Commodity Credit Corporation—Continued
Concentrates:
  Fed per grain consuming animal unit I–42
  Consumed by type of feed I–42
Conservation:
  Conservation Reserve Program (CRP) XII–1–17
Consumer's price index, by groups and by years IX–1
Consumer prices, index numbers XIII–10
Consumption per capita:
  Apples V–36
  Barley I–35
  Beans:
    Snap, canned IV–37
  Beef VII–49
  Butter III–29, VIII–15
  Cheese VIII–15
  Chickens VIII–30
  Cigarettes II–24
  Cigars II–24
  Citrus fruits V–36
  Coffee and cocoa XIII–7
  Corn:
    Canned IV–38
    Cereal-hominy and grits I–35
    Meal-flour and meal I–35
    Sugar (dextrose) I–35
    Syrup I–35
  Cornstarch I–35
  Dairy products XIII–6
  Eggs XIII–6
  Fat products III–29
  Fats III–29
  Fish XIII–6
  Flour XIII–6
  Food XIII–6
  Fruit juices, canned V–36
  Fruits:
    Canned V–36
    Dried V–36
    Fresh V–34, XIII–6
    Frozen V–36
    Total V–36, XIII–6
  Grain products XIII–4–5
  Grains I–35
  Ice cream VIII–15
  Lamb and mutton VII–49
  Lard III–29, VII–49
  Margarine III–27
  Meats VII–49, XIII–6
  Melons IV–35
  Milk:
    Condensed VIII–15
    Dry whole VIII–15
    Evaporated VIII–15
    Fluid VIII–15
    Nonfat dry milk VIII–15
  Nutrients XIII–1, 4–5
  Oat products I–31
  Oil products III–29
  Oils III–27–30, XIII–6–7
  Paper XII–29
  Paperboard XII–29
  Peas IV–35
  Pork VII–49
  Potatoes IV–34
  Poultry XIII–4–5
  Rice I–35
  Rye I–13, 35
  Shortening III–29
  Smoking tobacco II–24–26
  Snuff II–24–26
  Sugar XIII–7
  Tobacco products II–24–26
  Tomatoes:
    Canned IV–35
    Fresh IV–34
  Turkeys VIII–37
  Veal VII–49
  Vegetables:
    Canned IV–35, XIII–6
    Fresh IV–34, XIII–6
    Frozen IV–36, XIII–6
    Total XIII–4–5
  Wheat:
    Flour I–35
Conversion factors, weights, and measures .... iv–ix
Cooperative organizations, farmers' X–11–13, 16

Page

Corn, for grain:
  Area I–21–22, 25, IX–18
  Consumption, total I–35
  Crop insurance X–4
  Crop progress XV–36
  Crop ranking XV–34
  Disappearance I–23
  Exports I–23, 26–27
  Feed concentrates I–40,–42–43
  Imports I–23, 26
  Loan program, Commodity Credit Corporation XI–2–8
  Oil:
    Prices, market III–30
    Used in margarine and shortening III–27–28
  Prices:
    Farm I–21, 25
    Market I–43
    Support operations I–27
  Production I–21–24, IX–19
  Products, consumption I–35
  Seed:
    Average price paid VI–9
  Silage I–21, 24
  Stocks on and off farms I–21
  Supply I–23
  Syrup:
    Trade, international I–26
  Value I–21, 25, IX–19
  Yield I–21–22, 26, IX–18
  Sweet corn:
    Area, production, and value IV–14, IX–22–23
    Cold storage IV–38
    Consumption IV–34
    Shipments IV–35
    Yield and value IV–14, IX–22–23
    Canned:
      Pack IV–35
      Frozen pack IV–35, 36
Corn-hog price ratios VII–18
Cornstarch, consumption, civilian, per capita I–35
Cotton:
  Area II–1, 3, III–1, IX–18
  American Pima, carryover and ginnings, by grade and
    staple length II–6
  Carryover II–5
  Consumption II–3
  Crop insurance X–3–9
  Crop progress XV–37
  Crop ranking XV–34
  Distribution II–3, 10
  Exports II–3, 7–8
  Foreign:
    Distribution II–10
    Production II–3
  Forward contracted percentages II–4
  Ginnings II–3
  Grade and staple length II–5
  Imports II–9–10
  Linters:
    Distribution II–10
    Exports II–7–8
    Imports II–10
    Percentage distribution of fiber strength II–4
    Production II–10
    Supply II–10
  Loan program, Commodity Credit
    Corporation XI–2–3
  Micronaire readings II–10
  Prices:
    Farm II–1–2
    Market II–9, 11
    Support operations II–2
  Production II–1, 3 IX–19
  Trade, international II–9
  Upland:
    Carryover, ginning, supply and disappearance:
      By grade II–5
      By staple length II–6
    Value II–1–2, IX–19
    Yield II–1, 3, IX–18
Cottonseed:
  Cake:
    Disappearance for feed I–41
    Exports by destination III–3
    Crushings III–2
    Exports III–3
  Meal:
    Exports III–3
    Prices III–2
    Production III–2

Page

Cottonseed—Continued
  Oil:
    Exports...III–3
    Prices..III–1, 30
    Production...III–1
    Used in margarine..III–27
    Used in shortening..III–28
    Prices, farm...III–1–2
    Production...III–1, 4, IX–19
  Products:
    Sales to mills..III–1
    Seed, prices paid by farmers...............................VI–9
    Value...III–1–2, IX–19
  World:
    Area...III–4
    Production..III–4
Cows and heifers that have calved:
  Beef cows..VII–1, 3
  Milk cows..VII–1, 3
Cowhides:
  Exports..VII–51
  Imports..VII–50
  Prices..VII–50
Cows, beef:
  Feed consumed..I–41
  Numbers, Jan. 1.....................................VII–1, 3, 5, 13
  Number of operations....................................VII–12–13
  Percent of inventory.......................................VII–12
  Prices, market..VII–7
  Replacements...VII–1, 3
  Slaughter under Federal inspection.........VII–8–9, 11–12
Cows, milk:
  Number:
    Average during year...............................VIII–1, 4–5
    Jan. 1...............................VII–1, 3, VIII–1–2
    Kept for milk-cow replacement.....................VIII–1–2
    That have calved...VIII–1–2
    Percent of inventory.......................................VIII–3
    Percent of production....................................VIII–4
    Number of operations.....................................VIII–1
  Prices, farm..VIII–11
  Slaughter, Federally Inspected.............................VII–11
  See also Cattle.
Cranberries:
  Area..V–1, 17–18, IX–20
  Consumption..V–36
  Shipments..V–34
  Prices, farm and quantity processed..................V–17–18
  Production............................V–2, 17–18, 34, IX–21
  Utilization..V–17
  Value...V–17, 34 IX–21
  Yield...V–17–18, IX–20
Crop insurance programs..X–3–9
Crop loan programs, Commodity Credit
  Corporation..XI–1–8
Crop losses..X–1–2
Cropland:
  Area..IX–17
  Cash rents..IX–11
  Land values..IX–10
Crop progress, 5-year average.............................XV–35–38
Crop ranking, major field crops, by production.......XV–34
Crops:
  Alaska..XV–33
  Area...IX–17–18, 20, 22
  Production.......................................IX–19, 21, 23
  Value of production...........................IX–19, 21, 23
  Yield...IX–18, 20, 22
Crops principal:
  Production, index numbers.................................IX–16
  See also specific crops.
Cucumbers:
  Consumption...IV–34
  Shipments...IV–33
Cucumbers, fresh market:
  Area...IV–15–16, IX–22
  Price..IV–15–16
  Production.......................................IV–15–16, IX–23
  Value...IV–16, IX–23
  Yield..IV–16, IX–22
Cucumbers, for pickles:
  Area...IV–15, IX–22
  Consumption...IV–35
  Price...IV–15
  Production.......................................IV–15, IX–23
  Stocks..IV–15
  Shipments...IV–33
  Value...IV–15, IX–23
  Yield..IV–15, IX–22
Currants:
  Exports by destination......................................V–21

Page

Dairy statistics:
  Cold storage holdings.....................................VIII–38
  Consumption..VIII–15
  Dairy product feed-price ratio............................VIII–2
  Disappearance..VIII–15
  Exports..VIII–20–22
  Herd improvement associations...........................VIII–3
  Imports..VIII–17–19
  Income from cash and gross............................VIII–7–8
  Manufactured...VIII–9
    Prices:
      Manufacturers..VIII–11
      Support operations....................................VIII–23
    Production, factory......................................VIII–10
    Average price per specified product................VIII–9
    Stocks, manufacturers..................................VIII–11
    Stocks on hand, Dec. 31...............................VIII–15
Dates:
  Area...V–1, 17, IX–20
  Consumption..V–36
  Dried, production...V–36
  Farm price...V–17
  Imports..V–17
  Production.......................................V–2,17, IX–21
  Utilization..V–34
  Value...V–17, IX–21
  Yield...V–17, IX–20
Distillers' dried grains:
  Average price per ton bulk.................................I–45
Ducks:
  Frozen..VIII–37
Economic trends..IX–1
Eggplant:
  Consumption...IV–35
  Shipments...IV–33
Eggs:
  Cold-storage stocks..VIII–36
  Consumption..VIII–33
  Egg-feed ratio...VIII–28
  Exports...VIII–31
  Imports...VIII–33
  Lay, rate..VIII–35
  Numbers used for hatching................................VIII–33
  Prices:
    Farm..VIII–33
    Per dozen..VIII–33
    Volume buyers, New York.............................VIII–33
  Production...VIII–34–35
  Products under Federal inspection.....................VIII–34
  Sales...VIII–35
  Value...VIII–35
Emergency conservation measures.........................XII–16
Employment, farm:
  Number of persons employed........................IX–13–15
  Total..IX–13
  See also Labor, farm and Workers, farm equipment,
    farm. See Farm equipment.
Escarole-Endive:
  Consumption...IV–35
  Shipments...IV–33
Expenses, farm production......................................IX–39
European Union:
  value of agricultural imports...............................XV–7
Ewes:
  Number, Jan. 1..VII–25
Family farm:
  Workers..IX–13–15
Farm:
  Balance sheet..IX–12
  Buildings, value..IX–8
  Cash receipts..IX–38
  Economic sales class...IX–2
  Income, farm operator......................................IX–39

  Prices, food..IX–24
  Employment..IX–13–15
  Equipment:
    Value..IX–12
  See also Machinery, farm.
    Family. See Family, farm.
    Income. See Income.
    Labor. See Labor, farm.
  Land:
    Area.......................................IX–2, 3, 5, 7
    Utilization......................................IX–6, 9
    Value..IX–8
  Mortgage. See Mortgages, farm.
  Operators tenure of:
  Output, index numbers......................................IX–16
  Population. See Population, farm.
  Production:
    Expenses..IX–39
    Index numbers..IX–16, 24

Page

Farm—Continued
 Products:
  Income:
   Cash..IX–1, 39
   Gross...IX–1, 39
  Nonmoney..IX–39
  Prices, farm..IX–27–29
 Property:
  Maintenance costs IX–39
  Taxes..IX–39
  Value..IX–8–10
 Real estate. See Real estate, farm.
 Loans:
  Outstanding..X–9–10
Farmers:
 Cooperative associations type, number and membership
  ....X–16
 Operator households...IX–40
 Marketing and purchasing associations membership and
  business...X–13
 See also Farm, operators.
Farmers Home Administration loans X–10
 Community projects..X–10
 Farming purposes...X–10
Farms:
 Number...IX–2, 4–5
 Tenure of operator...IX–3–4
Farm Service Agency:
 Payments to producers by program and commodity XI–8–9
 Loans made to individuals and associations for farming
  purposes...X–10
 Payment received XI–10
Fat products, consumption III–28–29
Fatty acids:
 Total and per capita...III–28–29
Fats:
 Baking and frying..III–29
 Consumption...III–28
 Prices:
  Market..III–30
  Wholesale...III–30
 Stocks, Jan. 1...III–28
 Supply and disposition III–28
 Use, food and industrial III–29
 Used in—
  Margarine...III–27
  Shortening..III–28
 See also Oils, shortening, and under specific kinds.
Feed:
 Bought, cost to farm operators IX–39
 Grains:
  Average price and selected markets...........................I–38, 42
  Disappearance..I–1
  Quantity consumed by livestock and poultry...................I–42
Feed concentrates:
 Fed to livestock and poultry.................................I–42
 Quantity fed per animal unit.................................I–42
 Total fed..I–42
Feed Grain Program, payments to producers......................XI–10
Field seeds:
 Aveage retail price VI–9
Feedstuffs:
 Commercial, disappearance I–41
 Prices, market...I–45
Fertilizer:
 Acres receiving applications.......XIV–1–2,4,6–8, 10–12, 14–15,
                                                           17–20
 Bought, cost to farm operators IX–32–33
Figs:
 Area harvested..IX–20
 Consumption...V–36
 Dried:
  Exports and imports..V–18
  Price...V–18
  Production V–2, 19, 38, IX–21
  Utilization...V–34
  Value...V–18, IX–21
  Yield IX–20
Fish:
 Consumption per capita..XIII–6
 Packaged, fresh and frozen production by
  species..XV–15
Fisheries, catch and disposition.............................XV–16–17
Fishermen and craft...XV–16

Page

Fishery products:
 Canned, production and value.............................XV–15–16
 Fresh..XV–15
 Exports..XV–20–21
 Frozen, production..XV–15
 Imports..XV–20–21, 24
 Landings..XV–14
 Production, processed XV–18
 Sales, inventory...XV–24
 Supply XV–17
Fishing trips XV–20
Fish caught by marine recreational fishermen XV–21, 24
 Value...XV–14–16, 18
Fish meal:
 Average price per ton bulk I–45
 Disappearance for feed I–41
Flaxseed:
 Area III–5, IX–18
 Crop insurance..X–5
 Crushed for linseed oil III–6–7
 Disappearance...III–5
 Exports III–5, 7, 26
 Imports III–5, 7, 27
 Prices:
  Farm...III–5–6
  Market..III–6
  Support operations III–6
 Production III–5, 7, IX–19
 Seed VI–9
 Stocks..III–5
 Supply III–5
 Used for seed...III–5
 Value...III–5–6, IX–19
 Yield III–5, IX–18
 See also Linseed.
Flour. See under specific kinds.
Flowers..V–45–55
Food:
 Consumption, pounds..XIII–6–7
 Nutrients, contributed by food groups......................XIII–4–5
 Plans, cost levels..XIII–7
 Stamp Program XIII–9
 Prices, farm-to-retail spread.................................IX–26
 Products, marketing spreads IX–26
 Food and Nutrition Service Programs XIII–8
 Distributions to States XIII–9–10
 Costs..XIII–8
 Persons participating...XIII–10
 Quantity of macronutrients...................................XIII–1
 Food grain prices, market.....................................I–26
Foreign trade. See Agricultural products, foreign trade;
  also under specific crops.
Forestry statistics..XII–16–27
 Land, total...XII–19–20
 Products, value..XII–25
French fries:
 Cold storage..IV–37–38
Frozen commercial pack:
 Fruits V–35
Frozen meat
 Cold storage holdings VII–58–59
Fruit:
 Juices, canned:
  Consumption...V–36
  Pack...V–16
 See under specific kinds.
Fruits:
 Area, bearing..V–1, IX–20
 Canned:
  Consumption...V–36
  Citrus...V–12–16
  Citrus products..V–12–16
  Cold-storage holdings...V–56–58
  Processed exports...V–25
  Prepared, misc...V–25
  Deciduous fruits...V–1
 Dried:
  Consumption...V–36
  Production V–35
 Frozen:
  Cold-storage holdings...V–56–58
  Consumption...V–36
  Pack...V–35
 Noncitrus:
  Production V–34
  Utilization...V–34
  Value...V–34
 Orders, marketing...XI–14
 Production..V–2
 Shipments..V–34
 Percent of acres receiving applications XIV–20
 See also under specific kinds.

Page

Garlic:
  Area...IV–16, IX–22
  Consumption...IV–35
  Price..IV–16
  Production...IV–16, IX–23
  Value..IV–16, IX–23
  Yield...IV–16, IX–22
Ginger root:
  Area...V–19, IX–18
  Price..V–19
  Production..V–19, IX–19
  Value...V–19, IX–19
  Yield...V–19, IX–18
Gluten:
  Average price per ton of feed, bulk.....................I–45
  Disappearance for feed...I–41
Goats:
  Average clip..VII–35
  Carcasses condemned..VII–54
  Numbers clipped...VII–35
  Operations, number
  Production, price and value...................................VII–35
  Slaughtered under Federal inspection...................VII–54
Government stocks:
  Barley..I–32
  Corn...I–22
  Sorghum...I–36
Grain products:
  Consumption...I–35
  Exports, destination..I–10, 19, 36
Grains:
  Consumption, civilian..I–35
  Course, international trade.......................................I–39
  Disappearance, total..I–1
  Supply..I–1
  Quantity for feeding...I–42
  Feed:
    Area..IX–17
    Prices, market...I–44
  Food, area...IX–17
  Food, price, selected markets and grades...............I–21
  Supply..I–1
  See also under specific kinds.
Grapefruit:
  Area...V–1, IX–20
  Consumption...V–36
  Canned:
  Exports..V–15–16
  Foreign production..V–15
  Imports..V–16
  Juice, canned:
  Juice, concentrated, pack..V–16
  Prices, farm..V–13
  Processed, quantity...V–12
  Production...................................V–2, 12–13, IX–21
  Shipments..V–34
  Value...V–12, IX–21
  Yield...IX–20
  See also Citrus fruits.
Grapes:
  Area...V–1, IX–20
  Cold storage..V–57
  Consumption...V–36
  Crushed for wine...V–20
  Dried, production (raisins).......................................V–20
  Exports..V–21
  Imports..V–21
  Prices...V–19
  Production...................................V–2, 19–20, IX–21
  Shipments..V–34
  Utilization...V–26
  Value...V–19, IX–21
  Yield...IX–20
Grazing:
  Fees..IX–40
  On national forests...XII–26
  Number of stock and receipts...............................XII–26
Grease:
  Supply and disappearance.............................III–27–28
Greens:
  Cold storage...IV–38
  Consumption...IV–35
  Shipments...IV–33
Guavas:
  Area...V–1, 21, IX–20
  Price..V–21
  Production....................................V–2, 21, IX–21
  Utilized production...V–34
  Value...V–21, IX–21
  Yield...V–21, IX–20
Hardwood, production...XII–28

Page

See also under specific commodities.
Hazelnuts:
  Acreage...V–1, 40, IX–20
  Exports..V–39
  Imports..V–39
  Price...V–39
  Production..V–39, IX–21
  Production in specified countries............................V–39
  Value...V–39, IX–21
  Yield...V–39, IX–20
Hay:
  Area...VI–1–4, IX–18
  Crop ranking...XV–34
  Forage, all- area harvested, yield, and production...............VI–5
  Forage, all alfalfa - area harvested, yield, and
    production...VI–6
  Haylage and greenchop, all- area harvested, yield, and
    production...VI–7
  Haylage and greenchop, all alfalfa- area harvested,
    yield, and production..VI–8
  Prices, farm..VI–2, 4
  Production..VI–1–4, IX–19
  Seeds..VI–9
  Stock on farms...VI–3
  Supply and disappearance, total and per animal unit
    ....VI–9
  Value..VI–2, 4, IX–19
  Yield..VI–1–3, IX–18
Heifers:
  Aveage dressed weight, Federally inspected...................VII–12
  Milk cows replacements.............................VII–1, 3, VIII–1, 2
  Number, Jan. 1...VII–1, 3
  Number that have calved......................................VIII–1, 2
  Number slaughter..VII–11
Hens:
  Feed consumed per head and unit.........................I–43
Hides and skins:
  Average price, Central....................................VII–55–56
  Exports..VII–51–52
  Imports..VII–50
  Mink pelts produced...VII–53
High protein feeds:
  Disappearance for feed...I–41
Hired farm workers:
  Number of workers...IX–17
  Median weekly earnings.......................................IX–17
Hogs:
  Carcasses condemned.....................................VII–54–55
  Disposition...VII–18–19
  Feed consumed per head and unit.........................I–43
  Income from, cash and gross..........................VII–18–19
  Marketings..VII–18–19
  Number:
    Dec. 1...VII–21
    For breeding..VII–15
    For market..VII–15
    In specified states..VII–20
    Operations and inventory.....................................VII–21
  Operations...VII–21
  Prices, farm...VII–18, 55–56
  Production..VII–18–19
  Receipts:
    At interior markets...VII–18
    At public stockyards...VII–18
  Shipments..VII–18–19
  Slaughter:
    Farm..VII–20
    Under Federal inspection.............................VII–16, 54
  Slaughtered live weight...................................VII–20–21
  Value, Dec. 1...VII–17, 20–21
  Value of production..VII–19
Hog-corn price ratio..VII–18
  See also Pig crop *and* Sows.
Hominy feed:
  Average price per ton, bulk....................................I–45
Honey:
  Exports and imports for consumption...................II–17
  Number of colonies, yield, production, and stocks...........II–18
  Price and value..II–18
Honeydews:
  Area...IV–17, IX–22
  Consumption...IV–35
  Price..IV–17
  Production...IV–17, IX–23
  Shipments...IV–33
  Value..IV–17, IX–23
  Yield..IV–17, IX–22

Page

Hops:
Area VI–15, IX–18
Exports VI–16
Imports VI–16
Prices, farm VI–15
Production VI–15, IX–19
Stocks on hand VI–15
Value VI–15, IX–19
Yield VI–15, IX–18
Horses:
Carcasses condemned VII–54
Slaughtered under Federal inspection VII–54
Horses and mules, feed consumed I–43
Ice cream:
Consumption VIII–15
Disappearance VIII–15
Exports VIII–21
Production VIII–10
Imports:
Agricultural products. See Agricultural products,
   imports.
See also under specific commodities.
Income:
Cash from—
   Beef VII–7, 10
   Calves VII–7, 10
   Cattle VII–7, 10
   Chickens IX–38
   Dairy products VIII–8
   Eggs IX–38
   Farming IX–39
   Hogs VII–18–19
   Lambs VII–26–27
   Mohair VII–35
   Sheep VII–26–27
   Veal VII–7, 10
   Personal IX–1
Government payments:
Gross from—
   Broilers VIII–29
   Calves VII–7, 10
   Cattle VII–7, 10
   Chickens VIII–28
   Farming IX–1
   Hogs VII–18–19
   Lambs VII–26–27
   Sheep VII–26–27
   Turkeys VIII–32
   Gross, farm IX–39
   National IX–1
   Net, farm IX–1
Index numbers:
Agricultural:
   Production IX–25
Farm:
   Employment IX–13–15
   Food consumption XIII–6–7
   Input IX–25
   Labor IX–14–15
   Production IX–25
   Real estate values per acre IX–9
   Wage rates IX–13–15
   Fats, prices III–30
Industrial production IX–1
Livestock:
   Production IX–25
   Products, production IX–25
Oils:
   Prices III–30
Oilseeds:
Parity ratio IX–30
Prices:
   Consumers' IX–1, XIII–10
   Paid by farmers IX–1, 31
   Producer IX–1, 30
   Received by farmers IX–1, 27–30
   Tobacco:
Industrial production IX–1
Insurance, crop:
Coverage X–3–9
International trade:
   Corn I–25
   Cotton II–9
   Rice I–19
   Soybeans III–19
   Wheat I–9
Irrigation:
Land irrigated IX–7
Kale:
   Frozen pack IV–36
   Utilization IV–35

Page

K–Early Citrus:
Area V–1, IX–20
Price V–13
Production V–2, 13–14, IX–21
Value V–12, IX–21
Quantity processed V–12
Skins:
Exports and imports VII–50–52
Kiwifruit:
Area V–1, 9, IX–20
Price V–9, 34
Production V–2, 9, 34, IX–21
Shipments V–34
Value V–9, IX–21
Yield V–9, IX–20
Labor, farm:
Number of workers IX–13–15
See also Employment, farm and Workers, farm.
Lamb:
Crop VII–25
Skin:
   Exports VII–51–52
   Imports VII–50
See also Lamb and mutton and Meat.
Lamb and mutton:
Cold-storage holdings VII–57–58
Consumption VII–49
Exports VII–40, 44
Imports VII–40
Lambs:
Cash receipts VII–26–27
Marketings VII–26–27
Number, Jan. 1, by classes and State VII–28–29
Prices:
   Market VII–27
   Shipments VII–26–27
Slaughter:
   Farm VII–28–29
   Under Federal inspection VII–29
See also Livestock and Sheep and lambs.
Land:
Utilization:
   Cropland IX–6, 9–11
   Economic class IX–3
   Forest land IX–6, 9
   In farms IX–7
   Pasture IX–6, 9
   Special uses IX–6, 9
Lard:
Consumption VII–49
Exports VII–21–22
Production VII–22, 49
Stocks VII–21
Supply VII–21
Trade VII–22
Used in:
   Food products III–29
   Margarine III–27
   Shortening III–28
Lemons:
Area V–1, IX–20
Consumption V–36
Exports V–14
Imports V–16
Prices V–12–13
Processed, quantity V–12
Production V–2, 12–13, IX–21
Production, foreign V–15–16
Shipments V–33–34
Value V–12, IX–21
Yield IX–20
See also Citrus fruits.
Lentils:
Exports VI–13
Lespedeza seed:
Average price paid VI–9
Lettuce:
Consumption IV–35
Shipments IV–33
Lettuce, head:
Area IV–17, IX–22
Price IV–17
Production IV–17, IX–23
Value IV–17, IX–23
Yield IV–17, IX–22
Lettuce, leaf:
Area IV–18, IX–22
Consumption IV–35
Price IV–18
Production IV–18, IX–23
Shipments IV–33
Value IV–18, IX–23
Yield IV–18, IX–22

Page

Lettuce, Romaine:
  Area IV–18, IX–22
  Consumption IV–35
  Price IV–18
  Production IV–18, IX–23
  Shipments IV–33
  Value IV–18, IX–23
  Yield IV–18, IX–22
Limes:
  Area V–1, IX–20
  Consumption V–36
  Exports V–16
  Imports V–16
  Prices V–12–13
  Processed quantity V–12
  Production V–2, 12–13, IX–21
  Shipments V–34
  Value V–12, IX–21
  Yield IX–20
  See also Citrus fruits.
Linseed:
  Cake and meal:
    Disappearance for feed I–36
    Exports III–7
    Imports III–7
    Quantity for feed I–41
    Average price per ton bulk I–45
    Production III–7
  Meal prices III–6
  Wholesale price III–30
  Oil:
    Exports III–7
    Prices, market III–6
    Production III–7
    Stocks, June 1 .... III–7
Linters. See Cotton, linters.
Livestock:
  Bought, cost to farm operators IX–39
  Grazed on national forests, and receipts XII–25
  Number on farms VII–54
  Prices VII–55–56
  Production IX–25
  Production, index numbers IX–25
  Products:
    Index numbers IX–25
    Production IX–25
    Value IX–25
  Slaughtered under Federal inspection VII–54
  Value:
    Of production IX–25
    Total and per head VII–55
  World. See specific kinds.
Loans and debt outstanding:
  Agricultural, Federal and other
    agencies X–9–10
  Commodity Credit Corporation XI–1–8, 11–13
  Economic opportunity X–10
  Emergency X–10
  Farm real estate, of all operating banks X–10
  Farmers' cooperative organizations X–16
  Insurance programs X–3–9
  Interest:
    Non-real estate X–9
    Rural Utilities Service X–14–16
  See also under specific type of commodity.
Loganberries:
  Frozen commercial pack V–35
Lumber:
  Production XII–28
Macadamia nuts:
  Area V–1, 39, IX–20
  Price V–39
  Production V–39, IX–21
  Value V–39, IX–21
  Yield V–39, IX–20
Machinery, farm:
  Number, Jan. 1 .... IX–12
  See also Farm, equipment.
Macronutrients:
  Quantity available XIII–1
Mangoes:
  Consumption V–36
Margarine:
  Consumption III–29
  Disappearance, domestic III–27
  Exports III–27
  Manufacture, materials used III–27
  Production III–27
  Supply III–27

Page

Marketing:
  Agreements and orders:
    Fruits, vegetables, and tree nuts XI–14
    Milk VIII–7–8
  Associations, membership and business X–11, 13
  Bill for farm food products IX–25
  Costs IX–24
  Marketings, farm, cash receipts IX–38, 40
Meals. See under specific kinds.
Measures:
  Equivalent weights .... v–vii
  Tables, explanation iv
Meat:
  Cold-storage holdings VII–57–58
  Trade, international VII–48
Meat and lard:
  Production and consumption VII–49
Meat meal:
  Average price per ton bulk I–45
Meats:
  Consumption VII–49
  Exports VII–51–52
  Imports VII–50
  United States VII–47, 50
Melons
  See Cantaloups, Honeydews and Watermelons
Milk:
  Fluid milk and cream VIII–10
  Total consumption VIII–10
  Disappearance for feed I–41
  Condensed:
    Consumption VIII–15
    Disappearance VIII–15
    Exports VIII–20
    Production, percent by size group VIII–1
    Stocks, manufacturers' VIII–11
    Stocks on hand, Dec. 31 .... VIII–15
  Dry:
    Nonfat:
      Consumption VIII–15
      Disappearance VIII–15
      Exports VIII–20
      Exports, destination VIII–20
      Prices VIII–9
      Production VIII–7
      Stocks, manufacturers' VIII–11
      Stocks on hand, Dec. 31 .... VIII–15
    Whole:
      Consumption VIII–15
      Disappearance VIII–15
      Exports VIII–20
      Prices VIII–11
      Stocks, manufacturers' VIII–11
      Stocks on hand, Dec. 31 .... VIII–15
  Evaporated:
    Consumption VIII–15
    Disappearance VIII–15
    Exports VIII–20
    Prices VIII–11
    Stocks, manufacturers' VIII–11
    Stocks on hand, Dec. 31 .... VIII–15
  Marketing orders, Federal VIII–12, 14
  Prices:
    Farm VIII–11–12
    Milk-feed price ration VIII–2
    Received by producers VIII–14
  Producer deliveries VIII–14
  Producers, number VIII–14
  Production VIII–10
  Production in specified countries VIII–16
  Sales VIII–11
  Supply and utilization VIII–13
  Utilization VIII–14
  Value VIII–7
Milkfat:
  Percentage in milk VIII–8
  Prices, farm VIII–8
  Production:
    Per cow VIII–4–5
    Sales from farms VIII–11
Millet:
  Area, yield and production I–40
Minerals:
  Consumption XIII–3
Mill products, disappearance I–41
Mink pelts, number produced VII–53
Mint oil III–26
Mixed grains:
  Area, yield and production I–40
Mohair:
  Price VII–35
  Price-support operations VII–35
  Production VII–35
  Value VII–35

Page

Mules. *See* Horses and mules.
Mushrooms:
  Area in production...V–44
  Frozen commercial pack......................................IV–36
  Prices...V–44
  Sales..V–44
  Specialty...V–44
  Value of production...V–44
Mustard greens:
  Frozen consumption pack....................................IV–35
  Utilization..IV–34
Mustardseed:
  Area..IX–18
  Production...IX–19
  Value...IX–19
  Yield...IX–18
Mutton. *See* Lamb and mutton *and* Meats.
National forests:
  Area by States...XII–24
  Payments to States and Puerto Rico...................XII–25
  Receipts..XII–25
  Stock grazed on...XII–25
  Timber cut..XII–21
National income..IX–1
National marketing bill for civilian purchases of food
  products..IX–25
Nectarines:
  Area and Acreage..V–1, IX–20
  Consumption...V–36
  Production.....................................V–2, 21, 35, IX–21
  Shipments...V–34
  Use, price, and value...V–21
  Value...V–21, 34, IX–21
  Yield...IX–20
Nutrients contributed by food groups.....................XIII–4–5
Nuts:
  Area, bearing.......................................V–1, IX–20
  Cold storage..V–58
  Production...IX–21
  Value...IX–21
  Yield...IX–20
  Tree:
    Commercial production in foreign countries.....V–37
    Supply and utilization..V–36
Oats:
  Area...I–28–31, IX–18
  Consumption, civilian..I–35
  Crop progress...XV–38
  Crop ranking..XV–34
  Exports...I–28, 31
  Feed concentrates..I–28
  Imports...I–28 ,31
  Prices:
    Farm..I–28–29
    Market...I–28–30
    Support operations..I–29
  Production.....................................I–28–31, IX–19
  Seed, prices paid by farmers...............................VI–9
  Stocks on and off farms......................................I–28
  Supply and disappearance...................................I–28
  Value...I–25, 30, IX–19
  Yield...I–28–31, IX–18
Oil products, consumption......................................III–29
Oils:
  Animal:
    Used in margarine...III–27
    Used in shortening..III–28
    Consumption...III–29
    Disappearance...III–28
  Prices:
    Wholesale..III–30
  Supply and disposition...III–28
  Use, food and nonfood:
    Margarine..III–27
    Shortening...III–28
  Vegetable:
    Used in margarine...III–27
    Used in shortening..III–28
  See also Fats, Shortening, *and under specific kinds.*
Oilseed cake and meal:
  Disappearance..I–41
  Quantity for feeding..I–42
  Feed concentrates..I–42
  See also under specific kinds.
Oilseeds:
  Disappearance for feed..I–41
Okra:
  Cold storage..IV–37
  Frozen pack...IV–37
  Shipments...IV–33
Oleomargarine. *See* Margarine.
Olive oil:
  Imports...V–22
  Prices, wholesale...III–30
  Production, world...III–27

Page

Olives:
  Area...V–1, IX–20
  Consumption...V–36
  Imports...V–22
  Prices...V–22
  Production...V–2, 22, IX–21
  Utilization...V–22, 34
  Value...V–22, 34, IX–21
  Yield...IX–20
Onions:
  Area...IV–19, IX–22
  Cold storage..IV–38
  Consumption...IV–35
  Exports...IV–20
  Imports...IV–20
  Loss..IV–19
  Prices...IV–19
  Production...IV–19, IX–23
  Shipments...IV–33
  Shrinkage...IV–19
  Value...IV–19, IX–23
  Yield...IV–19, IX–22
Onion rings:
  Cold storage..IV–38
Orange juice:
  Concentrated pack...V–16
  Frozen, cold-storage stocks................................V–56–57
Oranges:
  Area...V–1, IX–20
  Consumption...V–36
  Exports...V–16
  Imports...V–16
  Prices...V–12–13
  Processed, quantity...V–12
  Production...V–2, 12–13, IX–21
  Production, foreign..V–14
  Shipments...V–34
  Value...V–12, IX–21
  Yield...IX–20
  See also Citrus fruits.
Orchardgrass:
  Average price paid..VI–9
Palm oil:
  Prices, wholesale...III–30
  Used in manufacture of shortening....................III–28
Papayas:
  Area...V–1, 29, IX–20
  Consumption...V–36
  Prices...V–29
  Production...V–2, 29, 34, IX–21
  Shipments...V–34
  Utilization..V–29
  Value...V–29, IX–21
  Yield...IX–20
Paper:
  Consumption per capita.......................................XII–29
  Production and consumption...............................XII–29
Paperboard:
  Consumption per capita.......................................XII–29
  Production and consumption...............................XII–29
Pasture:
  Feed consumed by type of feed..........................I–43
  Land value...IX–10
  Cash rents..IX–11
Parity prices..IX–27–29
Parity ratio...IX–30
Parsley shipments...IV–33
Payments, Government..IX–39
Peaches:
  Area...V–1, IX–20
  Canned:
    Exports..V–23
    Quantity..V–23–24
  Dried:
    Exports..V–23
    Production..V–23
  Cold storage..V–57
  Consumption...V–36
  Exports...V–23–25
  Frozen..V–35
  Prices...V–24
  Production...V–2, 23–24, IX–21
  Shipments...V–34
  Utilization..V–23–24
  Value...V–23, IX–21
  Yield...IX–20
Peanut cake and meal:
  Production...III–9
  Quantity for feeding..I–41–42
  Stocks..III–9
Peanut oil:
  Exports...III–9
  Imports...III–9
  Production and stocks..III–9–10

Page

Peanuts:
Area..................................III–9, 11, 13, IX–18
Cold storage..V–58
Consumption..III–10
Crop progress..XV–55
Crop ranking...XV–34
Crushed..III–9–10
Disappearance..III–10–11
Disappearance for feed.................................I–41
Disposition..III–10
Exports..III–9–10
Foreign production.....................................III–13
Imports..III–9–10
Milled...III–9
Prices:
  Cleaned..III–10
  Farm...III–9, 12
  Shelled..III–11
  Support operations...................................III–12
Production...III–9, 11, 13, IX–19
Seed, prices paid by farmers...........................VI–9
Stocks on hand...III–9
Supply...III–9–10
Utilization, shelled...................................III–10–11
Value..III–9, 12, IX–19
Yield..III–9, 11, IX–18
Pears:
Area...V–1, IX–20
Canned:
  Exports..V–26
Cold storage...V–56
Consumption..V–36
Dried:
  Exports..V–26
  Production...V–26, 34
Exports..V–26
Imports..V–26
Prices:
  Farm...V–26
Production...V–2, 26, 29, IX–21
Production by country..................................V–29
Shipments..V–34
Utilization..V–26, 29, 34
Value..V–26, 34, IX–21
Yield..IX–20
Peas:
Blackeye, frozen pack..................................IV–36
Cold storage...IV–38
Dry:
  Exports..VI–14
  Shipments..IV–33
Green peas:
  Area...IV–20, IX–22
  Cold storage...IV–38
  Price..IV–20
  Production...IV–20, IX–23
  Shipments..IV–33
  Value..IV–20, IX–23
  Yield..IV–20, IX–22
Canned:
  Consumption..IV–35–36
  Frozen pack..IV–36
  Prices...IV–20
Pecans:
Cold storage...V–58
Exports..V–41
Imports..V–41
Prices...V–40
Production...V–40, IX–21
Value..V–40, IX–21
Peppermint:
Area...III–26, IX–18
Farm price...III–26
Production...III–26, IX–19
Value..III–26, IX–19
Yield..III–26, IX–18
Pepper, bell:
Area...IV–21–22, IX–22
Consumption..IV–35
Frozen pack..IV–36
Price..IV–21–22
Production...IV–21–22, IX–23
Shipments..IV–33
Value..IV–21–22, IX–23
Yield..IV–21–22, IX–22
Pepper, chili:
Area, yield, production, price and value...............IV–21
Consumption..IV–35
Persimmons:
Shipments..V–34
Pesticides:
Percent of acres receiving applications................XIV–1–20

Page

Pickles:
Canned pack..IV–35
Pig crop...VII–17
  See also Hogs and Sows.
Pineapples:
Acreage..V–1, 25, IX–20
Consumption..V–36
Price..V–25
Production...V–2, 25, IX–21
Shipments..V–34
Utilization..V–25, 34
Value..V–25, 34, IX–21
Pistachios:
Acreage..V–1, 41, IX–20
Price..V–41
Production...V–41, IX–21
Value..V–41, IX–21
Yield..V–41, IX–20
Plums:
Acreage..IX–20
Canned:
  Frozen, commercial pack..............................V–35
  Prices, farm...V–30
  Production...V–2, 30, IX–21
  Shipments..V–34
  Utilization..V–30
  Value..V–30, IX–21
  Yield..IX–20
Pomegrantes:
Shipments..V–34
Population:
Eating from civilian food supplies.....................XIII–1
Pork:
Cold-storage holdings..................................VII–57–58
Consumption..VII–49
Exports..VII–42–48
Imports..VII–48
Income from, cash and gross............................VII–18–19
Production:
  Specified countries..................................VII–41
  United States..VII–49
Potatoes:
Area...IV–22–23, IX–18
Cold storage...IV–38
Consumption..IV–35–36
Crop ranking...XV–34
Exports..IV–27
Farm disposition.......................................IV–22, 26
Frozen pack..IV–36
Imports..IV–27
Prices, farm...IV–22–23
Production...IV–22–23, 26, IX–19
Seed, prices paid by farmers...........................VI–9
Shipments..IV–33
Stocks...IV–22, 24
Trade, foreign...IV–27
Utilization..IV–25
Value..IV–22, IX–19
Yield..IV–22–23, IX–18
Poultry:
Cold-storage holdings..................................VIII–36–38
Consumption per capita.................................XIII–6
Exports..VIII–31
Feed consumed..I–42
Slaughtered under Federal inspection...................VIII–30
  See also Chickens, Chicks, and Turkeys.
Poultry-feed price ratios..............................VIII–28
Price index, consumers', by groups and years...........IX–16
Price support:
Commodities owned......................................XI–1
Commodities under loan.................................XI–2
Operations, investment in..............................XI–8
Prices:
Consumer, index numbers................................IX–25
Farm product...IX–27–29
Index numbers..IX–30
Paid by farmers..IX–1, 31, 32–35
Parity ratio...IX–30
Producer, index numbers................................IX–30
Received by farmers, index numbers.....................IX–1, 30–31
Production:
Credit associations, loans.............................XI–2
Expenses of farmers....................................IX–39
Operating loans..X–16
Index numbers..IX–16, 24
Proso millet:
Area...I–46, IX–18
Price..I–46
Production...I–46, IX–19
Value..I–46, IX–19
Yield..I–46, IX–18
Proteins (animal), disappearance.......................I–42

Page

Prunes:
Area..IX–20
Canned:
  Quantity...V–30
Dried:
  Exports...V–32
  Farm price..V–30
  Imports...V–32
  Production..V–30–31
Exports...V–32
Frozen:
  Pack...V–35
  Quantity...V–31
Imports...V–32
Prices, farm..V–30
Production..V–2, 30, IX–21
Shipments...V–34
Utilization...V–30
Value..V–30, IX–21
Yield...IX–20
Pullets, number, Dec. 1.................................VIII–24
Pulpwood consumption..............................XII–28
Pumpkin:
Area, yield, production, price and value........IV–28
Consumption..IV–36
Frozen pack...IV–36
Purees, noncitrus:
Frozen commercial pack................................V–35
Radishes:
Consumption..IV–34
Shipments...IV–33
Raisins:
Consumption...V–36
Exports..V–21
Production...V–20–21, 35
Rams:
Number, Jan 1...VII–23
Rapeseed:
Area...IX–18
Production..IX–19
Value..IX–19
Yield...IX–18
Raspberries:
Frozen cold pack...V–35
Cold storage...V–57
Real estate, farm:
Debt:
  Outstanding, by regions and total................X–9
Loans:
  Farmers Home Administration....................X–16
Value:
  By States..IX–9
  Index number...IX–25
Refrigeration, warehouse space.................XV–31
Apple and pear storage..............................XV–31
General storage..XV–32
Rhubarb, frozen pack...................................IV–36
Rice:
Area...I–14, 16, 20, IX–18
Consumption...I–35
Crop progress...XV–36
Crop ranking...XV–34
Millfeeds, disappearance.................................I–41
By length of grain:
  Area..I–14, 16
  Production...I–14, 16
  Stocks...I–15, 17
  Yield..I–14, 16
Disappearance...I–15
Exports:
  Destination..I–19
Imports...I–15, 20
Exports..I–15, 20
Prices:
  Farm...I–14, 17
  Market..I–21
  Support operations.......................................I–19
Production.............................I–14, 16, 20, IX–19
Seeds, average price paid..............................VI–9
Stocks..I–14–15, 17
Supply..I–15
Trade, international...I–20
Value...I–14, 19, IX–19
Yield..I–14, 16, 20, IX–18
Rural Utilities Service:
Borrowers, status of.......................................X–15
Electric borrowers...X–16
Expenses..X–16
Loans...X–16
Services..X–16

Page

Rye:
Area..I–11–13, IX–18
Consumption, civilian......................................I–35
Disappearance..I–11
Exports..I–11
Feed concentrates...I–42
Flour, consumption per capita.........................I–35
Imports...I–11
Prices:
  Farm...I–11–12
  Market..I–21
Production....................................I–11–13, IX–19
Supply...I–11
Trade, international...I–13
Value......................................I–11–12, IX–19
Yield...I–11–13, IX–18
Ryegrass seed:
Average price paid...VI–9
Safflower:
Area...IX–18
Production..IX–19
Value..IX–19
Yield...IX–18
Safflower oil:
Wholesale price..III–30
School lunch programs, quantity and costs.....XIII–8
Seeds:
Bought, cost to farm operators.....................IX–40
Field:
  Prices, paid by farmers...............................VI–9
See also under specific kinds.
Sheep:
Feed consumed per head and unit..................I–43
Marketings..VII–27
Number, Jan. 1...VII–29
Prices:
  Farm..VII–27
  Market...VII–27
  Shipments..VII–27
Slaughter:
  Farm..VII–28
  Under Federal inspection.....................VII–29, 54
See also Sheep and lambs and Livestock.
Sheep and lambs:
Breeding...VII–23–25, 30
Breeding inventory......................................VII–24
Carcasses condemned..................................VII–54
Disposition...VII–26–27
Income...VII–27
Number:
  Jan. 1, by classes and States...................VII–25
  Shorn for wool....................................VII–30, 33
Operations..VII–30
Prices, farm..VII–26
Production..VII–26–27
Receipts at public stockyards....................VII–26
Shipments..VII–26–27
Skins:
  Exports..VII–51–52
  Imports..VII–50
Slaughter:
  Farm..VII–29
  Under Federal inspection........................VII–29
Value:
  Jan. 1..VII–27
  Of production..VII–27
See also Lambs, Livestock, and Sheep.
Shortening:
Consumption:
  Disappearance...III–28
  Exports...III–28
  Manufacture, fats and oils used...............III–28
  Production..III–28
  Supply...III–28
See also Fats and Oils.
Skins:
Exports...VII–51–52
Imports...VII–50
Mink pelts produced..................................VII–53
Snuff:
Consumption...II–24–26
Manufactured..II–25–26
Soap:
Fats and oils used in..................................III–29
Per capita..III–29
Softwoods, production..............................XII–21
Soil, conservation:
Flood prevention operations......................XII–18
Watershed improvements...........................XII–18
Sorghum:
Area......................................I–36–37, 39, IX–18
Crop progress...XV–36
Crop ranking..XV–34
Feed concentrates..I–42

Page

Sorghum—Continued
Grain:
Disappearance ... I–32
Exports ... I–34
Prices, farm ... I–33, 36
Prices, market ... I–44
Stocks on and off farms ... I–36
Silage ... I–36–37
Supply ... I–36
Support operations ... I–38
Production ... I–36–37, IX–19
Seed, prices paid by farmers ... VI–9
Trade, international ... I–39
Value ... I–36, IX–19
Yield ... I–36, IX–18
Southern greens:
Cold storage ... IV–40
Sows:
Farrowing ... VII–15, 17
Slaughter ... VII–19–20
See also Hogs *and* Pigs.
Soybean cake and meal:
Cake and meal stocks, production, and exports ... III–16, 19
Disappearance for feed ... I–41
Meal, prices ... III–14
Trade, international ... III–21–22
Soybean oil:
Exports ... III–19, 21
Prices:
Market ... III–14
Wholesale ... III–30
Production ... III–16
Stocks ... III–16, 20
Trade, international ... III–21–22
Used in—
Margarine ... III–27
Shortening ... III–28
Soybeans:
Area ... III–14, 16, 18, IX–18
Crop insurance ... X–8
Crop progress ... XV–37
Crop ranking ... XV–34
Crushed ... III–16
Exports ... III–15–16, 18, 21
Imports ... III–21
Prices:
Farm and Market ... III–14, 17
Seed prices ... VI–9
Support operations ... III–15
Production ... III–14, 16, 18, IX–19
Stocks on and off farms ... III–14–15
Supply and disappearance ... III–15
Trade, international ... III–20
Value ... III–14, 17, IX–19
Yield ... III–14, 16, IX–18
Spearmint:
Area ... III–26, IX–18
Price ... III–26
Production ... III–26, IX–19
Value ... III–26, IX–19
Yield ... III–26, IX–18
Special Milk Program ... XIII–9–10
Spinach:
Cold storage ... IV–38
Consumption ... IV–35
Frozen pack ... IV–36
Shipments ... IV–33
Spinach, fresh:
Area ... IV–29, IX–22
Price ... IV–29
Production ... IV–29, IX–23
Value ... IV–29, IX–23
Yield ... IV–29, IX–22
Spinach, processing:
Area ... IV–29–30, IX–22
Price ... IV–29–30
Production ... IV–29–30, IX–23
Value ... IV–29, IX–23
Yield ... IV–29, IX–22
Squash:
Area, yield, production, price, and value ... IV–31
Cold storage ... IV–38
Consumption ... IV–34
Frozen pack ... IV–36
Shipments ... IV–33
Stags:
Slaughter under Federal inspection ... VII–11–12
Steers:
Aveage dressed weight ... VII–12
Beef, prices, market ... VII–7
Number, Jan. 1 ... VII–1, 4
Slaughter under Federal inspection ... VII–11–12

Page

Storage:
General ... XV–32
Strawberries:
Area ... V–1, 33, IX–20
Cold storage ... V–57
Consumption ... V–36
Frozen pack ... V–35
Prices, farm ... V–33
Production ... V–2, 32–33, IX–21
Shipments ... V–34
Value ... V–32–33, IX–21
Yield ... V–33, IX–20
Sudangrass seed:
Aveage price paid ... VI–9
Sugar:
Exports ... II–16
Imports ... II–16
Marketings, by source ... II–14
Prices:
Retail, United States ... II–15
Wholesale, at New York ... II–15
Production ... II–13
Stocks, receipts, meltings, and deliveries ... II–14
Trade, international ... II–16
Sugarbeets:
Area ... II–12, IX–18
Prices, farm ... II–12
Production ... II–12, IX–19
Exports ... II–16
Value ... II–12, IX–19
Yield ... II–12, IX–18
See also Beets.
Sugarcane:
Area ... II–13–14, IX–18
Area, production, and yield:
In Hawaii ... II–14
Prices, farm ... II–13
Production for sugar and seed ... II–13–14, IX–19
Stocks, receipts and meltings ... II–14
Value ... II–13, IX–19
Yield per acre ... II–13–14, IX–18
Sunflower:
Area ... III–23–24, IX–18
Cake and meal ... III–8
Disappearance for feed ... I–41
Meal, quantity for feeding ... I–42
Price ... III–24
Price for oil, wholesale ... III–30
Production ... III–24, IX–19
Seed, average price paid ... VI–9
Value ... III–24, IX–19
Yield ... III–24, IX–18
Sunflower:
Area and production in specified countries ... III–25
Exports ... III–7–8
Oil, production ... III–8
Support prices, commodity ... IX–36–37
Sweet potatoes:
Area ... IV–30, IX–18
Consumption ... IV–34
Frozen consumption pack ... IV–36
Prices ... IV–30
Production ... IV–30, IX–19
Shipments ... IV–33
Value ... IV–30, IX–19
Yield ... IV–30, IX–18
Syrup:
Corn:
Consumption, civilian, per capita ... I–35
Tall fescue seed:
Aveage price paid ... VI–9
Tallow:
Inedible:
Disposition ... III–28
Exports ... III–28
Factory consumption ... III–28
Production ... III–28
Stocks ... III–28
Supply ... III–28
Tallow oil, prices, wholesale ... III–30
Tangelos:
Area ... V–1, IX–20
Quantity processed ... V–12
Price ... V–12–13
Production ... V–2,12–13 IX–21
Shipments ... V–34
Value ... V–12, IX–21
Yield ... IX–20
Tangerine, juice, pack ... V–16, 35

Page

Tangerines:
  Area ... IX–20
  Concentrate, annual pack ... V–16
  Exports ... V–14
  Imports ... V–16
  Quantity processed ... V–12
  Prices ... V–12–13
  Production ... V–2, 12–13, IX–21
  Production, specified countries ... V–14
  Value ... V–12–13, IX–21
  Yield ... IX–20
Tankage and meat meal:
  Disappearance for feed ... I–41
Taro:
  Area ... IV–31, IX–18
  Production ... IV–31, IX–19
  Price ... IV–31
  Value ... IV–31, IX–19
  Yield ... IX–18
Tea:
  Imports, origin ... V–43
Temples:
    Area ... V–1
    Consumption ... V–36
    Quantity processed ... V–12
    Price ... V12–13
    Production ... V–2, 12–13, IX–21
    Shipments ... V–34
    Value ... V–2, 13
    Yield ... IX–20
Timber:
  Cut ... XII–25
  Prices ... XII–25, 27
  Products ... XII–25
  Removals and growth ... XII–22
  Volume ... XII–21
Timothy seed, prices ... VI–9
Tobacco:
  Area ... II–19–22, IX–18
  Chewing:
    Manufactured ... II–25
  Consumption ... II–24, 26
  Crop ranking ... XV–34
  Disappearance ... II–20–22
  Exports:
    Destination ... II–20–22, 26
  Imports:
    Origin ... II–26
  Prices:
    Farm ... II–19
    Support operations ... II–25
  Production ... II–19–22, IX–19
  Products:
    Consumption ... II–24
    Manufactured ... II–25–26
  Smoking:
    Consumption ... II–24
    Manufactured ... II–26
  Stocks ... II–20–24
  Supply ... II–20–22
  Value ... II–19, IX–19
  Yield ... II–19–22, IX–18
Trade, foreign:
  Value of total agricultural exports and imports ... XV–2
  Value of exports ... XV–2–6
  Value of imports ... XV–7–9
  See also Cigarettes, Cigars, *and* Snuff.
Tomato products:
  Catsup exports ... IV–32
  Juice:
    Exports ... IV–32
  Paste:
    Exports and imports ... IV–32
Tomatoes:
  Area ... IV–32, IX–22
  Canned:
    Consumption ... IV–35, 36
    Exports ... IV–32
    Imports ... IV–32
  Prices ... IV–32
  Production ... IV–32, IX–23
  Shipments ... IV–33
  Cherry and Plum tomatoes, shipments ... IV–33
  Value ... IV–33, IX–23
  Yield ... IV–32, IX–22
Trade, international:
  Corn ... I–26
  Cotton ... II–9
  Rice ... I–19
  Soybeans ... III–22
  Wheat and flour ... I–9
Tree planting:
  Acres seeded ... XII–19
  Acres of tree plantings ... XII–19

Page

Tree planting—Continued
  Trout
  Operation and sales ... XV–28
Truck crops. *See* Vegetable crops.
Tung oil:
  Prices, wholesale ... III–30
Turkeys:
  Cold storage ... VIII–37
  Consumption:
    Per capita and total ... VIII–31
  Exports ... VIII–31
  Feed price ratio ... VIII–28
  Number raised ... VIII–32
  Placed ... VIII–32
  Prices ... VIII–32
  Production ... VIII–32
  Raised, feed consumed per head and unit ... I–43
  Supply and distribution ... VIII–31
Turnip greens, frozen pack ... IV–36
  Utilization ... IV–34
Utilization of farm commodities:
  Apples ... V–4
  Apricots ... V–6
  Cherries ... V–10
  Grapes ... V–20
  Land ... IX–6
  Milk ... VIII–13
  Olives ... V–22
  Peaches ... V–23
  Pears ... V–29–30
  Plums and prunes ... V–30
  Total index ... IX–25
  See also individual items.
Value of agricultural commodities:
  Crops:
    Principal ... IX–18–19
    Specific. *See under name of specific crop.*
    Vegetable ... IV–2–5
  Exports ... XV–2–6, 10–12
  Imports ... XV–2, 7–9,13–14
  Livestock ... VII–54–56
Veal:
  Cold storage ... VII–58
  Consumption ... VII–49
  Production ... VII–41, 49
  See also Beef, Beef and Veal, *and* Meats.
Vegetable fats and oils:
  Fats. *See* Fats, vegetable.
  Oils. *See* Oils, vegetable.
Vegetables:
  Area ... IV–2–3, IX–22–23
  Canned:
    Consumption ... IV–34 –36
  Frozen:
    Cold-storage stocks ... IV–37
    Consumption ... IV–35
    Pack ... IV–36
  Per capita consumption ... IV–35
  Percent of acres receiving applications-fertilizer ... XIV–18
  Orders, marketing ... XI–14
  Production ... IV–2, 4, IX–23
  Shipments ... IV–33
  Value ... IV–2, 5, IX–23
  Yield ... IX–22
  See also under specific kinds.
Vitamins:
  Quantity available ... XIII–2
Wages, farm:
  Average earnings received ... IX–17
  Rates ... IX–13–15
Walnuts, English:
  Area ... V–1, 41, IX–21
  Exports ... V–41
  Imports ... V–41
  Prices ... V–41
  Production in foreign countries ... V–41
  Production ... V–41, IX–21
  Value ... V–41, IX–21
  Yield ... V–41, IX–20
Warehouse space, refrigerated ... XV–31–32
Water conservation. *See* Conservation.
Watermelons:
  Area ... IV–35, IX–22–23
  Consumption ... IV–34
  Price ... IV–35
  Production ... IV–35, IX–23
  Shipments ... IV–33
  Value ... IV–35, IX–23
  Yield ... IV–35, IX–22
Watershed improvements of Natural Resources
  Conservation Service ... XII–18
Weights:
  Equivalent measures ... v–vii
  Tables, explanation ... iv

Page

Wheat:
Area..I–1–2, 5–6, 8, IX–18
Cereal, consumption per capita I–35
Consumption...I–35
Crop insurance X–9
Crop progress...XV–35
Crop ranking XV–34
Disappearance...I–4–5
Exports I–5, 9–10
Feed concentrate...I–42
Flour:
    Consumption, civilian, per capita I–35
    Exports, destination.......................................I–10
Imports I–5, 8–10
Loan program, Commodity Credit
    Corporation..XI–1
Prices:
    Farm I–1–2, 7
    Market..I–21
    Support operations...I–7
Production I–1–2, 4–6, 8, IX–19
Seed, price paid by farmers VI–9

Page

Wheat—Continued
Stocks:
    On and off farms...I–3
Supply...I–4–5
Trade, international I–9
Value...I–1–2, 7, IX–19
Yield...I–1–2, 5–6, 8, IX–18
Wheat bran, average price per ton bulk.................I–45
Wheat, mill, average price per ton bulk..................I–45
Wheat, middlings, average price per ton bulk..........I–45
Wheat, millfeeds, disappearance for feed I–45
Whey:
    Dried exports VIII–22
Wood-pulp production XII–29
Wool:
    Consumption, total VII–31–32
Imports:
    By grades..VII–31
    Origin..VII–32
    Quantity for consumption VII–31–32
Prices:
    Delivered to U.S. mills VII–32
    Farm VII–34, 35
Price-support operations VII–30
Production, shorn......................................VII–30, 33
Value and weight per fleece...........................VII–33
Workers, farm:
    Average wage rate...................................IX–13–15
    Hired IX–13–15